T0137360

Lecture Notes in Computer Science 13355

More information about this series at https://link.springer.com/bookseries/558

Maria Mercedes Rodrigo ·
Noburu Matsuda · Alexandra I. Cristea ·
Vania Dimitrova (Eds.)

Artificial Intelligence in Education

23rd International Conference, AIED 2022
Durham, UK, July 27–31, 2022
Proceedings, Part I

Editors
Maria Mercedes Rodrigo
Ateneo De Manila University
Quezon, Philippines

Alexandra I. Cristea
Durham University
Durham, UK

Noburu Matsuda
Department of Computer Science
North Carolina State University
Raleigh, NC, USA

Vania Dimitrova
University of Leeds
Leeds, UK

ISSN 0302-9743 ISSN 1611-3349 (electronic)
Lecture Notes in Computer Science
ISBN 978-3-031-11643-8 ISBN 978-3-031-11644-5 (eBook)
https://doi.org/10.1007/978-3-031-11644-5

This Springer imprint is published by the registered company Springer Nature Switzerland AG
The registered company address is: Gewerbestrasse 11, 6330 Cham, Switzerland

Preface

The 23rd International Conference on Artificial Intelligence in Education (AIED 2022) was hosted by Durham University, UK. It was organized in a hybrid face-to-face and online format. This allowed participants to meet in person after two years of running AIED online only, which was a welcome change. However, as the world was only just emerging from the COVID-19 pandemic and travel for some attendees was still a challenge, online participation was also supported. AIED 2022 was the next in a longstanding series of annual international conferences for the presentation of high-quality research on intelligent systems and the cognitive sciences for the improvement and advancement of education. It was hosted by the prestigious International Artificial Intelligence in Education Society, a global association of researchers and academics who specialize in the many fields that comprise AIED, including computer science, learning sciences, educational data mining, game design, psychology, sociology, linguistics, and many others.

The theme for the AIED 2022 conference was "AI in Education: Bridging the gap between academia, business, and non-profit in preparing future-proof generations towards ubiquitous AI." The conference hoped to stimulate discussion on how AI shapes and can shape education for all sectors, how to advance the science and engineering of intelligent interactive learning systems, and how to promote broad adoption. Engaging with the various stakeholders – researchers, educational practitioners, businesses, policy makers, as well as teachers and students – the conference set a wider agenda on how novel research ideas can meet practical needs to build effective intelligent human-technology ecosystems that support learning.

AIED 2022 attracted broad participation. We received 243 submissions for the main program, of which 197 were submitted as full papers, 37 were submitted as short papers, and nine were submitted as extended abstracts. Of the full paper submissions, 40 were accepted as full papers and another 40 were accepted as short papers. The acceptance rate for both full papers and short papers was thus 20%.

Beyond paper presentations and keynotes, the conference also included a Doctoral Consortium Track, an Industry and Innovation Track, Interactive Events, Posters/Late-Breaking Results, and a Practitioner Track. The submissions for all these tracks underwent a rigorous peer-review process. Each submission was reviewed by at least two members of the AIED community, assigned by the corresponding track organizers who then took the final decision about acceptance. The conference also included keynotes, panels, and workshops and tutorials.

For making AIED 2022 possible, we thank the AIED 2022 Organizing Committee, the hundreds of Program Committee members, the Senior Program Committee members, the AIED Proceedings Chair Irene-Angelica Chounta, and our Program

Chair assistant Jonathan DL. Casano. They all gave of their time and expertise generously and helped with shaping a stimulating AIED 2022 conference. We are extremely grateful to everyone!

July 2022

Maria Mercedes (Didith) T. Rodrigo
Noboru Matsuda
Alexandra I. Cristea
Vania Dimitrova

Organization

General Chair

Vania Dimitrova — University of Leeds, UK

Program Co-chairs

Noboru Matsuda — North Carolina State University, USA
Maria Mercedes (Didith) T. Rodrigo — Ateneo de Manila University, Philippines

Doctoral Consortium Co-chairs

Olga C. Santos — UNED, Spain
Neil Heffernan — Worcester Polytechnic Institute, USA

Workshop and Tutorials Co-chairs

Ning Wang — University of Southern California, USA
Srećko Joksimović — University of South Australia, Australia

Interactive Events Co-chairs

Dorothy Monekosso — Durham University, UK
Genaro Rebolledo-Mendez — Institute for the Future of Education, Mexico
Ifeoma Adaji — University of British Columbia, Canada

Industry and Innovation Track Co-chairs

Zitao Liu — TAL Education Group, China
Diego Zapata-Rivera — Educational Testing Service, USA

Posters and Late-Breaking Results Co-chairs

Carrie Demmans Epp — University of Alberta, Canada
Sergey Sosnovsky — Utrecht University, The Netherlands

Practitioner Track Co-chairs

Jeanine A. DeFalco — Medidata Solutions, Dassault Systemes, USA
Diego Dermeval Medeiros da Cunha Matos — Federal University of the Alagoas, Brazil

Berit Blanc	German Research Centre for Artificial Intelligence (DFKI), Germany
Insa Reichow	German Research Centre for Artificial Intelligence (DFKI), Germany

Panel Chair

Wayne Holmes	University College London, UK

Local Organizing Chair

Alexandra I. Cristea	Durham University, UK

Proceedings Chair

Irene-Angelica Chounta	University of Duisburg-Essen, Germany

Web Chair

Lei Shi	Durham University, UK

Online Activities Chair

Guanliang Chen	Monash University, Australia

Publicity Co-chairs

Elle Wang	Arizona State University, USA
Elaine Harada Teixeira de Oliveira	Federal University of Amazonas, Brazil
Mizue Kayama	Sinshu University, Japan

Sponsorship Chairs

Craig Stewart	Durham University, UK
Ben du Boulay	University of Sussex, UK

Diversity and Inclusion Co-chairs

Eric Walker	Arizona State University, USA
Rod Roscoe	Arizona State University, USA
Seiji Isotani	University of São Paulo, Brazil

Senior Program Committee

Alireza Ahadi	University of Technology Sydney, Australia
Vincent Aleven	Carnegie Mellon University, USA
Laura Allen	University of New Hampshire, USA
Claudio Alvarez	Universidad de los Andes, Chile
Rafael D. Araújo	Universidade Federal de Uberlandia, Brazil
Tracy Arner	Arizona State University, USA
Luciana Assis	Universidade Federal dos Vales do Jequitinhonha e Mucuri, Brazil
Roger Azevedo	University of Central Florida, USA
Ryan Baker	University of Pennsylvania, USA
Tiffany Barnes	North Carolina State University, USA
Emmanuel Blanchard	IDÛ Interactive Inc., Canada
Nigel Bosch	University of Illinois Urbana-Champaign, USA
Steven Bradley	Durham University, UK
Christopher Brooks	University of Michigan, USA
Armelle Brun	Loria, Université de Lorraine, France
Maiga Chang	Athabasca University, Canada
Min Chi	BeiKaZhouLi, USA
Cesar A. Collazos	Universidad del Cauca, Colombia
Cristina Conati	University of British Columbia, Canada
Sidney D'Mello	University of Colorado Boulder, USA
Mihai Dascalu	Politehnica University of Bucharest, Romania
Jeanine A. DeFalco	Medidata Solutions, Dassault Systemes, USA
Mĭchel Desmarais	Ecole Polytechnique de Montreal, Canada
Vania Dimitrova	University of Leeds, UK
Tenzin Doleck	Atlas Lab, USA
Benedict du Boulay	University of Sussex, UK
Márcia Fernandes	Federal University of Uberlandia, Brazil
Rafael Ferreira Mello	Federal Rural University of Pernambuco, Brazil
Kobi Gal	Ben Gurion University, Israel
Dragan Gasevic	Monash University, Australia
Sébastien George	LIUM, Le Mans Université, France
Ashok Goel	Georgia Institute of Technology, USA
Art Graesser	University of Memphis, USA
Peter Hastings	DePaul University, USA
Yusuke Hayashi	Hiroshima University, Japan
Bastiaan Heeren	Open University, The Netherlands
Neil Heffernan	Worcester Polytechnic Institute, USA
Laurent Heiser	Université Côte d'Azur, Inspé de Nice, France
Ulrich Hoppe	University Duisburg-Essen, Germany
Sharon Hsiao	Santa Clara University, USA
Lingyun Huang	McGill University, Canada
Seiji Isotani	University of São Paulo, Brazil
Yang Jiang	Columbia University, USA

David Joyner	Georgia Institute of Technology, USA
Akihiro Kashihara	University of Electro-Communications, Japan
Judy Kay	University of Sydney, Australia
Mizue Kayama	Shinshu University, Japan
Min Kyu Kim	Georgia State University, USA
Stefan Küchemann	TU Kaiserslautern, Germany
Jakub Kužílek	Charles Technical University in Prague, Czechia
Amruth Kumar	Ramapo College of New Jersey, USA
Susanne Lajoie	McGill University, Canada
H. Chad Lane	University of Illinois at Urbana-Champaign, USA
Nguyen-Thinh Le	Humboldt-Universität zu Berlin, Germany
Francois Lecellier	Xlim Laboratory, France
James Lester	North Carolina State University, USA
Shan Li	McGill University, Canada
Tong Li	Arizona State University, USA
Carla Limongelli	Università Roma Tre, Italy
Sonsoles López-Pernas	Universidad Politécnica de Madrid, Spain
Vanda Luengo	LIP6, Sorbonne Université, France
Collin Lynch	North Carolina State University, USA
Wannisa Matcha	Prince of Sonkla University, Pattani, Thailand
Noboru Matsuda	North Carolina State University, USA
Gordon McCalla	University of Saskatchewan, Canada
Kathryn McCarthy	Georgia State University, USA
Bruce Mclaren	Carnegie Mellon University, USA
Agathe Merceron	Beuth University of Applied Sciences Berlin, Germany
Eva Millan	Universidad de Málaga, Spain
Caitlin Mills	University of New Hampshire, USA
Tanja Mitrovic	University of Canterbury, New Zealand
Kazuhisa Miwa	Nagoya University, Japan
Riichiro Mizoguchi	Japan Advanced Institute of Science and Technology, Japan
Negar Mohammadhassan	University of Canterbury, UK
Phaedra Mohammed	University of the West Indies, Jamaica
Bradford Mott	North Carolina State University, USA
Roger Nkambou	Université du Québec à Montréal, Canada
Jaclyn Ocumpaugh	University of Pennsylvania, USA
Amy Ogan	Carnegie Mellon University, USA
Elaine H. T. Oliveira	Universidade Federal do Amazonas, Brazil
Andrew Olney	University of Memphis, USA
Luc Paquette	University of Illinois at Urbana-Champaign, USA
Bernardo Pereira Nunes	Australian National University, Australia
Niels Pinkwart	Humboldt-Universität zu Berlin, Germany
Kaska Porayska-Pomsta	University College London, UK
Thomas Price	North Carolina State University, USA
Mladen Rakovic	Monash University, Australia
Ilana Ram	Technion – Israel Institute of Technology, Israel

Martina Rau	University of Wisconsin-Madison, USA
Traian Rebedea	Politehnica University of Bucharest, Romania
Genaro Rebolledo-Mendez	Tecnologico de Monterrey, Mexico
Steven Ritter	Carnegie Learning, Inc., USA
Maria Mercedes (Didith) T. Rodrigo	Ateneo de Manila University, Philippines
Ido Roll	Technion - Israel Institute of Technology, Israel
Rod Roscoe	Arizona State University, USA
Jonathan Rowe	North Carolina State University, USA
Olga C. Santos	UNED, Spain
Kazuhisa Seta	Osaka Prefecture University, Japan
Sergey Sosnovsky	Utrecht University, The Netherlands
Namrata Srivastava	University of Melbourne, Australia
Richard Tong	Yixue Education Inc., China
Stefan Trausan-Matu	Politehnica University of Bucharest, Romania
Maomi Ueno	University of Electro-Communications, Japan
Masaki Uto	University of Electro-Communications, Japan
Kurt Vanlehn	Arizona State University, USA
Alessandro Vivas	UFVJM, Brazil
Erin Walker	Arizona State University, USA
Diego Zapata-Rivera	Educational Testing Service, USA

Program Committee

Mark Abdelshiheed	North Carolina State University, USA
Ifeoma Adaji	University of British Columbia, Canada
Bunmi Adewoyin	University of Saskatchewan, Canada
Seth Adjei	Northern Kentucky University, USA
Jenilyn Agapito	Ateneo de Manila University, Philippines
Kamil Akhuseyinoglu	University of Pittsburgh, USA
Bita Akram	North Carolina State University, USA
Samah Alkhuzaey	University of Liverpool, UK
Isaac Alpizar Chacon	Utrecht University, The Netherlands
Nese Alyuz	Intel, USA
Sungeun An	Georgia Institute of Technology, USA
Antonio R. Anaya	Universidad Nacional de Educacion a Distancia, Spain
Roberto Araya	Universidad de Chile, Chile
Esma Aïmeur	University of Montreal, Canada
Michelle Banawan	Arizona State University, USA
Ayan Banerjee	Arizona State University, USA
Jordan Barria-Pineda	University of Pittsburgh, USA
Shay Ben-Elazar	Microsoft, USA
Ig Ibert Bittencourt	Federal University of Alagoas, Brazil
Geoffray Bonnin	Loria, Université de Lorraine, France
Anthony F. Botelho	University of Florida, USA
Jesus G. Boticario	UNED, Spain

François Bouchet	LIP6, Sorbonne Université, France
Bert Bredeweg	University of Amsterdam, The Netherlands
Julien Broisin	IRIT, Université Toulouse 3 Paul Sabatier, France
Okan Bulut	University of Alberta, Canada
James Bywater	James Madison University, USA
Daniela Caballero	McMaster University, Canada
Alberto Casas-Ortiz	UNED, Spain
Francis Castro	New York University, USA
Geiser Chalco Challco	ICMC/USP, Brazil
Pankaj Chavan	IIT Bombay, India
Guanliang Chen	Monash University, Australia
Jiahao Chen	TAL Education Group, China
Penghe Chen	Beijing Normal University, China
Heeryung Choi	University of Michigan, USA
Andrew Clayphan	University of Sydney, Australia
Keith Cochran	DePaul University, USA
Ricardo Conejo	Universidad de Malaga, Spain
Mark G. Core	University of Southern California, USA
Alexandra Cristea	Durham University, UK
Mutlu Cukurova	University College London, UK
Maria Cutumisu	University of Alberta, Canada
Anurag Deep	IIT Bombay, India
Carrie Demmans Epp	University of Alberta, Canada
Diego Dermeval	Federal University of the Alagoas, Brazil
M. Ali Akber Dewan	Athabasca University, Canada
Tejas Dhamecha	IBM, India
Barbara Di Eugenio	University of Illinois at Chicago, USA
Daniele Di Mitri	DIPF—Leibniz Institute for Research and Information in Education, Germany
Darina Dicheva	Winston-Salem State University, USA
Mohsen Dorodchi	University of North Carolina at Charlotte, USA
Fabiano Dorça	Universidade Federal de Uberlandia, Brazil
Alpana Dubey	Accenture, India
Yo Ehara	Tokyo Gakugei University, Japan
Ralph Ewerth	L3S Research Center, Leibniz Universität Hannover, Germany
Fahmid Morshed Fahid	North Carolina State University, USA
Stephen Fancsali	Carnegie Learning, Inc., USA
Arta Farahmand	Athabasca University, Canada
Effat Farhana	Vanderbilt University, USA
Mingyu Feng	WestEd, USA
Reza Feyzi Behnagh	University at Albany - SUNY, USA
Carol Forsyth	Educational Testing Service, USA
Reva Freedman	Northern Illinois University, USA
Maurizio Gabbrielli	University of Bologna, Italy
Cristiano Galafassi	Universidade Federal do Rio Grande do Sul, Brazil

Lucas Galhardi	State University of Londrina, Brazil
Yanjun Gao	University of Wisconsin-Madison, USA
Isabela Gasparini	UDESC, Brazil
Elena Gaudioso	UNED, Spain
Michael Glass	Valparaiso University, Chile
Benjamin Goldberg	United States Army DEVCOM Soldier Center, USA
Alex Sandro Gomes	Universidade Federal de Pernambuco, Brazil
Aldo Gordillo	Universidad Politécnica de Madrid, Spain
Monique Grandbastien	Loria, Université de Lorraine, France
Floriana Grasso	University of Liverpool, UK
André Greiner-Petter	University of Wuppertal, Germany
Nathalie Guin	LIRIS, Université de Lyon, France
Sandeep Gupta	Arizona State University, USA
Binod Gyawali	Educational Testing Service, USA
Hicham Hage	Notre Dame University-Louaize, Lebanon
Rawad Hammad	University of East London, UK
Yugo Hayashi	Ritsumeikan University, Japan
Martin Hlosta	The Open University, UK
Anett Hoppe	TIB – Leibniz Information Centre for Science and Technology and L3S Research Centre, Leibniz Universität Hannover, Germany
Tomoya Horiguchi	Kobe University, Japan
Daniel Hromada	Einstein Center Digital Future and Berlin University of the Arts, Germany
Stephen Hutt	University of Pennsylvania, USA
Chanyou Hwang	Riiid, South Korea
Tomoo Inoue	University of Tsukuba, Japan
Paul Salvador Inventado	California State University, Fullerton, USA
Mirjana Ivanovic	University of Novi Sad, Serbia
Johan Jeuring	Utrecht University, The Netherlands
Srećko Joksimović	University of South Australia, Australia
Yvonne Kammerer	Stuttgart Media University, Germany
Shamya Karumbaiah	University of Pennsylvania, USA
Hieke Keuning	Utrecht University, The Netherlands
Rashmi Khazanchi	Open University of the Netherlands and Mitchell County School System, The Netherlands
Hassan Khosravi	University of Queensland, Australia
Jung Hoon Kim	Riiid, South Korea
Simon Knight	University of Technology Sydney, Australia
Kazuaki Kojima	Teikyo University, Japan
Emmanuel Awuni Kolog	University of Ghana Business School, Ghana
Tanja Käser	EPFL, Switzerland
Sébastien Lallé	Sorbonne University, France
Andrew Lan	University of Massachusetts Amherst, USA
Jim Larimore	Riiid, South Korea
Hady Lauw	Singapore Management University, Singapore

Elise Lavoué	LIRIS, Université Jean Moulin Lyon 3, France
Seiyon Lee	University of Pennsylvania, USA
Marie Lefevre	LIRIS, Université Lyon 1, France
Blair Lehman	Educational Testing Service, USA
Sharona T. Levy	University of Haifa, Israel
Fuhua Lin	Athabasca University, Canada
Qiongqiong Liu	TAL Education Group, China
Zitao Liu	TAL Education Group, China
Nikki Lobczowski	University of Pittsburgh, USA
Yu Lu	Beijing Normal University, China
Aditi Mallavarapu	University of Illinois at Chicago, USA
Leonardo Brandão Marques	University of São Paulo, Brazil
Mirko Marras	University of Cagliari, Italy
Daniel Moritz Marutschke	Ritsumeikan University, Japan
Jeffrey Matayoshi	McGraw Hill ALEKS, USA
Wookhee Min	North Carolina State University, USA
Sein Minn	Inria, France
Tsegaye Misikir Tashu	Eötvös Loránd University, Hungary
Merav Mofaz	Microsoft, Israel
Abrar Mohammed	University of Leeds, UK
Dorothy Monekosso	Durham University, UK
Kasia Muldner	Carleton University, Canada
Anabil Munshi	Vanderbilt University, USA
Tricia Ngoon	Carnegie Mellon University, USA
Nasheen Nur	Florida Institute of Technology, USA
Negar Mohammadhassan	University of Canterbury, UK
Phaedra Mohammed	University of the West Indies, Jamaica
Bradford Mott	North Carolina State University, USA
Marek Ogiela	AGH University of Science and Technology, Poland
Urszula Ogiela	AGH University of Science and Technology, Poland
Christian Otto	TIB – Leibniz Information Centre for Science and Technology University Library, Germany
Ranilson Paiva	Universidade Federal de Alagoas, Brazil
Rebecca Passonneau	Pennsylvania State University, USA
Rumana Pathan	Indian Institute of Technology Bombay, India
Prajwal Paudyal	Arizona State University, USA
Terry Payne	University of Liverpool, UK
Radek Pelánek	Masaryk University, Czech Republic
Francesco Piccialli	University of Naples Federico II, Italy
Elvira Popescu	University of Craiova, Romania
Miguel Angel Portaz	UNED, Spain
Shi Pu	Education Testing Service, USA
Ramkumar Rajendran	IIT Bombay, India
Insa Reichow	Deutsches Forschungszentrum für Künstliche Intelligenz, Germany
Luiz Antonio Rodrigues	UniFil, Brazil

Qian Zhang	University of Technology Sydney, Australia
Guojing Zhou	University of Colorado Boulder, USA
Jianlong Zhou	University of Technology Sydney, Australia
Xiaofei Zhou	University of Rochester, USA
Stefano Pio Zingaro	Università di Bologna, Italy
Gustavo Zurita	Universidad de Chile, Chile

Additional Reviewers

Abdelshiheed, Mark
Afzal, Shazia
Anaya, Antonio R.
Andres-Bray, Juan Miguel
Arslan, Burcu
Barthakur, Abhinava
Bayer, Vaclav
Chung, Cheng-Yu
Cucuiat, Veronica
Demmans Epp, Carrie
Diaz, Claudio
DiCerbo, Kristen
Erickson, John
Finocchiaro, Jessica
Fossati, Davide
Frost, Stephanie
Gao, Ge
Garg, Anchal
Gauthier, Andrea
Gaweda, Adam
Green, Nick
Gupta, Itika
Gurung, Ashish
Gutiérrez Y. Restrepo, Emmanuelle
Haim, Aaron
Hao, Yang
Hastings, Peter
Heldman, Ori
Jensen, Emily
Jiang, Weijie
John, David
Johnson, Jillian
Jose, Jario
Karademir, Onur
Landes, Paul
Lefevre, Marie
Li, Zhaoxing
Liu, Tianqiao
Lytle, Nick
Marwan, Samiha
Mat Sanusi, Khaleel Asyraaf
Matsubayashi, Shota
McBroom, Jessica
Mohammadhassan, Negar
Monaikul, Natawut
Munshi, Anabil
Paredes, Yancy Vance
Pathan, Rumana
Prihar, Ethan
Rodriguez, Fernando
Segal, Avi
Serrano Mamolar, Ana
Shahriar, Tasmia
Shi, Yang
Shimmei, Machi
Singh, Daevesh
Stahl, Christopher
Swamy, Vinitra
Tenison, Caitlin
Tobarra, Llanos
Woodhead, Simon
Xhakaj, Franceska
Xu, Yiqiao
Yamakawa, Mayu
Yang, Xi
Yarbro, Jeffrey
Zamecnick, Andrew
Zhai, Xiao
Zhou, Guojing
Zhou, Yunzhan

International Artificial Intelligence in Education Society

Management Board

President

Vania Dimitrova	University of Leeds, UK

Secretary/Treasurer

Rose Luckin	University College London, UK

Journal Editors

Vincent Aleven	Carnegie Mellon University, USA
Judy Kay	University of Sydney, Australia

Finance Chair

Ben du Boulay	University of Sussex, UK

Membership Chair

Benjamin D. Nye	University of Southern California, USA

Publicity Chair

Manolis Mavrikis	University College London, UK

Tech and Outreach Officer

Yancy Vance Paredes	Arizona State University, USA

Executive Committee

Ryan Shaun Baker	University of Pennsylvania, USA
Min Chi	North Carolina State University, USA
Cristina Conati	University of British Columbia, Canada
Jeanine A. DeFalco	Medidata Solutions, Dassault Systemes, USA
Rawad Hammad	University of East London, UK
Neil Heffernan	Worcester Polytechnic Institute, USA
Christothea Herodotou	Open University, UK
Seiji Isotani	University of São Paulo, Brazil
Akihiro Kashihara	University of Electro-Communications, Japan
Amruth Kumar	Ramapo College of New Jersey, USA
Diane Litman	University of Pittsburgh, USA

Zitao Liu	TAL Education Group, China
Judith Masthoff	Utrecht University, The Netherlands
Manolis Mavrikis	University College London, UK
Bruce M. McLaren	Carnegie Mellon University, USA
Tanja Mitrovic	University of Canterbury, New Zealand
Kaska Porayska-Pomsta	University College London, UK
Maria Mercedes (Didith) T. Rodrigo	Ateneo de Manila University, Philippines
Olga Santos	UNED, Spain
Diego Zapata-Rivera	Educational Testing Service, USA
Erin Walker	University of Pittsburgh, USA
Ning Wang	University of Southern California, USA

Invited Keynotes

The Role of AI and Gamification in the Future of Healthcare

Lucia Pannese

imaginary, Milano, Italy
lucia.pannese@i-maginary.it

Abstract. In this keynote, Lucia Pannese shows how AI and Gamification support medical practices and the impact that has on several health and care interventions. Presenting a series of different examples about digital approaches to health and care, this talk looks at the future of healthcare and to how learning and development needs will be affected if AI and machine learning support decision making and behavioural change.

The talk will start with a focus on the pervasiveness of gamification in everyday life, something that usually people do not even recognize, given the narrow understanding currently attributed to this life skill. After sharing definitions, understanding and some examples of gamification, game-based approaches and enabling technologies for health and care, Lucia, who is a mathematician by profession, will point at a series of critical issues that are too often ignored in practical applications if machine learning and AI are applied in these contexts. She will provocatively introduce some concepts of usefulness, quantity of data, bias, measurement, clinical responsibility, clinical observation, instability of models to show how complex and risky it is to produce an AI based system.

This talk aims to trigger reflection and critical analysis at a time when everyone is talking about AI and the danger of this extremely complex concept just becoming a "buzzword" to attract attention without consideration of whether these solutions are genuinely innovative and useful.

Learning Engineering: Looking Back, Looking Forward

Kumar Garg

Schmidt Futures, New York, USA
kgarg@schmidtfutures.com

Abstract. In this fireside chat, Kumar Garg will discuss some of the biggest wins in learning engineering to date as well as discuss opportunities like addressing the lack of large n studies. He will also outline some of Schmidt Futures' recent efforts including a new Learning Engineering Virtual Institute.

Contents – Part I

Full Papers

Short Papers

Contents – Part II

Doctoral Consortium

Industry and Innovation Track

Workshops and Tutorials

Practitioner Track

Posters and Late-Breaking Results

Workshops Track Contributions

Full Papers

How to Give Imperfect Automated Guidance to Learners: A Case-Study in Workplace Learning

Jacob Whitehill(✉) and Amitai Erfanian

Worcester Polytechnic Institute, Worcester, MA, USA
{jrwhitehill,aberfanian}@wpi.edu

Abstract. In a workplace learning scenario in which workers in a simulated Material Recovery Facility learn to recognize and manipulate objects on conveyer belts, we studied how imperfect guidance from a machine learning (ML) assistant may impact learners' experience and behaviors. Specifically, in a randomized experiment ($n = 181$ participants from Amazon MTurk) we varied the assistant's False Positive (FP) and False Negative (FN) rates in detecting non-recyclable objects and assessed the impact on learners' performance, learning, and trust. We also explored a soft highlighting [8] condition, whereby the assistant provides fine-grained information about how confident it is. We found evidence that the FP/FN trade-off can impact learners' performance when working cooperatively with the assistant, and that the soft highlighting condition may generate less trust from learners compared to the other conditions. There was tentative evidence that workers' behaviors were impacted by the FP/FN trade-off of their assigned experimental condition even after the ML assistant was removed. Finally, in a follow-up study ($n = 27$) we found evidence that learners modulate their behaviors based on the fine-grained confidence values conveyed by the assistant.

Keywords: Perceptual learning · Simulation-based learning · AI-based feedback · Material recovery facility

1 Introduction

As the field of machine learning (ML) continues to grow and proliferate into daily life, the number and diversity of subjects in which AI systems for education (AIEd) can provide learners with automated help will expand as well. Whereas classical AI methods such as expert systems were instrumental in creating intelligent tutoring systems (ITS) in highly structured domains such as computer programming and high school mathematics in the 1980s–90s [10], ML has opened new possibilities to provide scaffolding, guidance, and feedback in more flexible and open-ended domains such as medicine [14], sign language [4], teacher training [1,15,17], waste management [11], and many more. It also has the potential to augment standard ITS with new sensors that can better estimate students' emotions and thereby provide a more tailored learning experience [7].

M. M. Rodrigo et al. (Eds.): AIED 2022, LNCS 13355, pp. 3–14, 2022.
https://doi.org/10.1007/978-3-031-11644-5_1

In AIEd systems based on classical AI techniques, the automated feedback strategies are either manually programmed or inferred systematically from the rules of the subject matter (e.g., laws of algebra). In contrast, ML provides more flexibility because it can infer the correctness of a student's answer, and suggest helpful learning strategies, by harnessing swaths of both real and simulated data from previous learners – all without the need for manually crafted heuristics. ML, as the backbone of modern AIEd systems, has yielded powerful new feedback mechanisms in intelligent learning platforms, e.g., automated testing methods for novice programmers building interactive computer programs [13], and classroom observation systems that can give teachers feedback about the quality of their teaching based on audio [15] or video [1] of classroom interactions. The reach of AIEd can thereby expand from traditional classroom-oriented subject matter into more diverse fields, including the space of *workplace learning* where human workers may acquire not just cognitive but also perceptual and motor skills [16] to perform their jobs. This can manifest, for example, in an ITS or perhaps a collaborative "assistant" that provides visual cues about which types of objects in an airport screener are dangerous [2], or by suggesting a Python implementation to a function whose specification was entered by the user [3].

ML in AIEd – Useful but Imperfect: Along with the great potential of ML to expand the impact of AIEd come new challenges. One of the most severe is that, since the system's behavior is learned statistically so as to generalize to new scenarios (e.g., to new students with answers similar, but not identical, to those in the training set), its feedback is no longer guaranteed to be correct – sometimes the AIEd system may make mistakes. Concrete instances of these mistakes include telling a student that their solution to a computer programming task is incorrect when in fact it is correct; suggesting to them a solution path to a math problem that is either wrong or unnecessarily complicated; failing to recognize the hand gesture of a student learning sign language; etc. Mistakes on the part of the AIEd system may directly inhibit students' learning by misguiding and confusing them. Due to bidirectional influence of *learning* and *trust* between a student and their teacher [12], such mistakes can also indirectly and negatively impact learning if the student loses trust in the AI's ability to help them. Given an ML-based AIEd system's fallibility, it is important to consider how to set the learners' expectations of and structure their interactions with the system so as to optimize their learning.

How to Present Feedback on Binary Decisions: When developing ML-based AIEd systems, the question arises of (1) whether, on some binary decision, the machine should tend to err with more false positives (FP) or more false negatives (FN). Binary decisions are ubiquitous in intelligent learning platforms, e.g., judging whether a student's hand-written solution in a mathematics ITS is correct/correct, showing novice teachers a moment in a classroom observation video when a teacher seems to speak angrily (or not) to their student, or flagging an object as dangerous/non-dangerous in an augmented reality display to train airport security officers. False negatives (misses) can result in missed learning opportunities and overly optimistic self-assessments of learning, whereas false positives can confuse the student and damage trust in the AI. Related to the optimum FP/FN trade-off is the question of (2) whether and how to provide to the learner information about how *confident* the machine is in its own judgment; might this information help the learner to interpret the feedback more judiciously and preserve trust?

Case Study on Material Recovery Facilities: In this paper, we investigate these two questions within a workplace learning case-study on material recovery facilities (MRFs), i.e., recycling plants. MRFs sort objects by their materials so that they can be bailed and reprocessed, and they form an important part of waste management. MRF workers assist in this process by manually picking out items that were incorrectly sorted by machines, as well as removing objects that are dangerous. MRF work is arduous, with long shifts (often 10+ hours) and physically demanding conditions, and can be dangerous (e.g., due to syringes and other sharp objects on the belts). Because of high employee turnover, new MRF workers must frequently be trained. Due to continually new kinds of manufactured products that arrive at MRFs, workers must learn to recognize and physically manipulate objects with different materials and appearance. Helping new MRF workers to correctly recognize objects could thus boost the efficiency of the MRF, improve waste management, and improve safety for workers (Fig. 1).

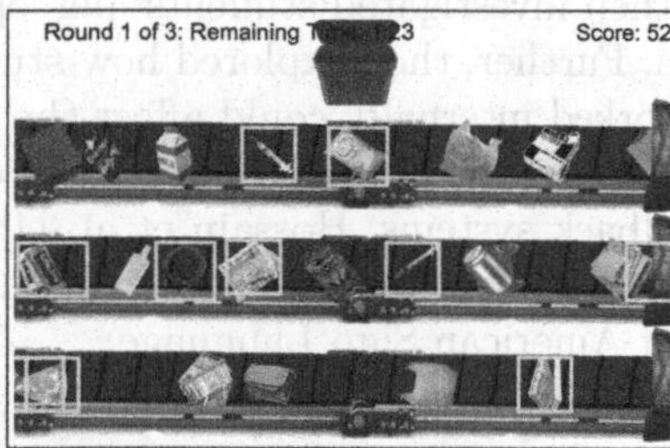

Fig. 1. Left: A material recovery facility (MRF) in which workers sort through items on conveyer belts. **Center**: The simulated MRF [11] used in our experiment on worker training with automated guidance. **Right**: Examples of soft highlighting [8] by the machine learning assistant to express its confidence to the learner about whether each highlighted object should be removed from the conveyer belts. (Color figure online)

Research Contributions: The central research question that we examine is: **How does the automated guidance provided by a machine learning assistant for an object detection task (recyclable vs. non-recyclable objects on a MRF conveyer belt) affect learners' performance, learning, trust, and behavior?** Aligned with our goal of examining *workplace learning*, we recruit our research participants ($n = 181$) from Amazon Mechanical Turk, a marketplace for online work. By varying the system's detection threshold (higher FP/lower FN; lower FP/higher FN; or a threshold-free "soft highlighting" [8] approach) while keeping its overall discriminability constant, we can explore how the learners in this task may respond differently to the information they receive in terms of their willingness to *complete* the task, their ability to *perform* the task when guidance is available, and the degree to which they *learn* to do the task independently. Further, we explore how much the learners' *trust* the system. In a follow-up study ($n = 27$), we also explored, for just the soft highlighting condition, whether participants use the confidence information conveyed by the highlight intensity to modulate their decisions about which objects to move. Our paper is, to our knowledge, one of the first to explore how imperfect guidance from an AIEd system can affect learners' experiences.

2 Related Work

Imperfect AI and Trust: The past few years have seen growing interest in the ML, human-computer interaction, as well as the AIEd communities, in how humans trust AI and how they can work together cooperatively. Kocielnik et al. [9] investigated how the FP/FN trade-off affected users' perceptions of the machine's accuracy as well as their acceptance of a ML-based tool that detects scheduling requests from free-text emails. They found that users had more favorable impressions of the system when the detection threshold was adjusted to give a lower FN rate in exchange for a higher FP rate. Hsu et al. [5] trained an automatic auto-grader of students' short-answer responses about computer science topics; they then investigated students' perceptions of the accuracy and fairness of the system. Further, they explored how students' (mis-)understanding of how the system worked internally could affect the answers that students constructed and submitted. Finally, in a research study toward increasing learners' trust in AI-based feedback systems, Hossain et al. [4] studied how experts perceive the feedback generated from an automatic explainable hand-gesture feedback system for learners of American Sign Language.

Soft Highlighting to Convey the Machine's Confidence: In research within the intersection of data visualization, ML, and human-computer interaction, Kneusel et al. [8] explored whether human workers benefit more from a ML-based object detector on visual perception tasks, such as examining satellite imagery for specific objects, when they have access to the *confidence* of the machine's predictions. They devised a "soft highlighting" mechanism, whereby the machine's detections were colored with an intensity proportional to the confidence of the predictions, and found that human workers performed better on the task with "soft" highlights than with binary (hard) object predictions.

MRF Training System: Our work builds on a prior workplace learning study by Kyriacou et al. [11], who created a MRF simulator to compare different training strategies for human workers who are learning to sort different objects on the conveyer belts. Similarly to our work, they examined how different accuracy characteristics of a machine learning assistant can affect learners' performance. In contrast to their work, we hold the overall discriminability of the assistant constant and manipulate the FP/FN trade-off in isolation. Moreover, we introduce a new soft-highlighting condition and explore how the learners may benefit from and mimic the behaviors exhibited by the assistant.

3 Experiment I

We conducted a randomized experiment in which participants learned to recognize and manipulate garbage objects of different types in a simulated MRF. The simulator we used was based on that of [11] but extended to support a soft highlighting condition (described above). During the training rounds, the participants received automated guidance from a ML assistant about whether each object is recyclable. This can potentially help them to learn to recognize the object types more quickly and thereby perform better on the task.

3.1 Participants

The participants ($n = 181$) in our study were adults ($\geq$18 years) on Amazon Mechanical Turk who had earned a "Masters" qualification.

3.2 Experimental Conditions

At the outset of the experiment, each participant was randomly assigned to one of three different conditions: (1) FP = 0/FN = 0.11, i.e., the machine learning assistant will correctly highlight 89% of the non-recyclable objects and miss 11%; it will never falsely flag a recyclable object as non-recyclable; (2) FP = 0.11/FN = 0; the assistant never misses a non-recyclable object but occasionally falsely flags recyclable object as non-recyclable; and (3) Soft highlighting: every single object on the conveyer belt is highlighted, but the intensity of the highlight corresponds to the confidence of the machine's prediction. Importantly, the overall *discriminability* of the object detector used for automated guidance was held constant across all three conditions. The particular FP/FN values were chosen based on previous work [11] with this simulator, which suggested that the discriminability of the ML assistant needs to very high in order to be useful.

3.3 MRF Simulator

The simulator contains three conveyer belts, each of which has a different speed and moves a never-ending stream of objects from left to right. In the task, the goal is to remove only and all the non-recyclable items (syringes, broken glass, coat-hangers, batteries, etc.) from the conveyer belts into a trash-can (a green bucket shown at the top of the screen).

Machine Learning Assistant: During the training rounds (see Sect. 3.4), the learner is aided by a simulated machine learning assistant that detects (with imperfect accuracy) whether each object is recyclable or not. Across all three experimental conditions, the discriminability of the assistant is held constant at 0.945. Specifically, we quantified *discriminability* as the widely used Area Under the Receiver Operating Characteristics Curve (AUROC) metric. The AUROC is equivalent to the probability, in a 2-alternative forced-choice task, that the detector can correctly recognize the non-recyclable object from a random pair of objects (one recyclable, one not). In the two experimental conditions corresponding to FP = 0/FN = 0.11 and FP = 0.11/FN = 0, a random number generator determines whether each object is highlighted, depending on the FP and FN rates and on whether the object is recyclable. In the third condition corresponding to soft highlighting, no threshold is used, and instead the intensity of the red color is proportional to a real-valued confidence score C for that object in the interval [0, 1]. In particular, if the object is recyclable, then C is drawn from a Beta distribution, i.e., $C \sim \text{Beta}(2, 4.675)$; if it is non-recyclable, $C \sim \text{Beta}(4.675, 2)$. Under this generative process, non-recyclable objects tend to have higher C values and thus are highlighted with a brighter red box. Moreover, the parameters for the probability distributions were chosen so that the discriminability of the assistant is exactly 0.945 (i.e., $P(C_a > C_b) = 0.945$ where C_a and C_b are the confidence values of a random non-recyclable and recyclable object, respectively).

Scoring: In the pre-test, post-test, and each training round, the user's score starts at 100 and decreases by 1 whenever (a) the user misses a non-recyclable object and it moves off the screen to the right; or (b) the user incorrectly moves a recyclable item from the belt into the trash-can. Hence, the maximum score in each round is 100, and it can decrease to 0 (or even below) if the participant makes many mistakes. The training and testing rounds were essentially the same, except that (a) the pre-test was shorter (1 min) than the other rounds (3 min), and (b) the ML assistant was available only during the training rounds.

Good performance in the task requires quick and accurate visual recognition of the objects and manipulation of the objects (using the mouse) from the belt to the trash-can. There is also some strategy that can be beneficial to task performance, e.g., prioritizing one belt over another due to the different speeds, moving objects slightly so as to reduce occlusion, etc. Participants received $1.50 for completing the task; to incentivize good performance, they could also earn a reward that increased linearly with their scores up to a maximum of $2.50.

3.4 Procedures

The experiment consisted of (1) a study overview; (2) task instructions; (3) pre-test (1 min), during which no help from the ML assistant was provided; (4) two rounds of training (3 min each), during which the machine learning assistant provided automated guidance; (4) post-test (3 min), during which no help from the machine learning assistant was provided; and (5) questionnaire about trust in the system. All in all, the study takes about 12–15 min to complete.

3.5 Task Instructions

The instructions to the learners describe and show examples of the objects that they should move from the belts into the green trash bucket. In addition, the instructions explain that the machine learning assistant automatically flags certain objects as likely non-recyclable using a red rectangle. To set the users' expectations and promote effective usage of the assistant, the instructions explain:

"The object detector is imperfect. You may find it useful to you as you play the game as a way of focusing your attention on the correct objects, but you should not rely on it completely. In particular, sometimes the detector will miss non-recyclable objects and therefore not flag it. It may also make 'false alarms' and highlight a recyclable object."

3.6 Trust Questionnaire

To assess participants' trust in the AI system after completing the task, we asked them to complete a validated questionnaire on trust in technology [6], which consists of 12 Likert items (e.g., strongly disagree to strongly agree with: "the system is deceptive") about the users' impressions of the following attributes: *deceptive*, *underhanded*, *suspicious*, *wary*, *harm*, *confident*, *security*, *integrity*, *dependable*, *reliable*, *trust*, and *familiar* (some of these items are negatively scored). The maximum score (most trusting) is 84 and minimum score is 12.

3.7 Measures

The simulator automatically records the participant's score after every pre-test, post-test, and training round. The scores on the pre-test tend to be higher than in the other phases of the experiment since the pre-test is shorter (only 1 min), and hence the participants have less time to make mistakes. The simulator also records the responses to the trust questionnaire.

3.8 Results

A total of $n = 181$ total participants completed the task; in the FP = 0/FN = 0.11, FP = 0.11/FN = 0, and soft-highlighting conditions, their numbers were 56, 60, and 65, respectively. These numbers are not stat. sig. different ($\chi^2(2) = 0.67403$, $p = 0.7139$) from a uniform distribution, and hence we do not conclude that any condition caused participants to drop out more often than another.

Figure 2 (**left**) shows the mean score (along with its standard error) across the three rounds separately for each experimental condition. The difference in mean pre-test scores across condition was not statistically significant ($F(2) = 0.906$, $p = 0.406$). As a general trend, participants in the FP = 0/FN = 0.11 condition tended to receive higher scores, followed by those in the soft highlighting condition, and then the FP = 0.11/FN = 0 condition.

Figure 2 (**right**) shows the mean trust score (and s.e.) for each condition.

Task Performance with Automated Guidance We used a linear mixed-effect model (repeated-measures design with subject id as the random effect and pre-test score as a covariate) to analyze the scores during the training rounds 1 & 2 when workers had access to the assistant. We found that: (1) scores were stat. sig. higher when participants received guidance from the assistant compared to on the post-test, when they did not receive guidance

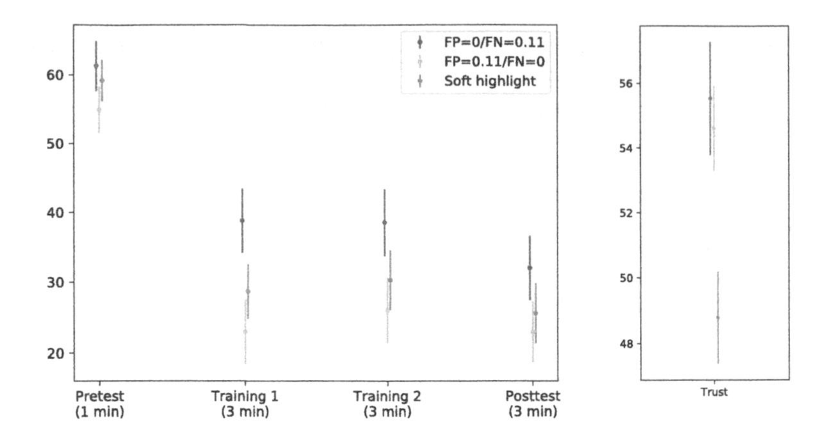

Fig. 2. Left: Mean scores (and error bars for standard errors, shifted slightly for readability) for the different phases of the study, split by condition. Note that, as the pre-test is only 1 min and participants have less time to make mistakes, the scores tend to be higher. **Right**: Mean (and s.e.) trust questionnaire results, split by condition.

($t(360) = 3.325$, $p < 0.001$); and (2) participants in the FP $= 0.11$/FN $= 0$ condition received stat. sig. lower scores compared to those in the other conditions ($t(178) = -1.992$, $p = 0.048$). The soft-highlighting condition was also associated with a lower score, but the effect was not stat. sig. ($t(178) = -1.341$, $p = 0.1816$).

In terms of participants' actions, we found a stat. sig. difference in the number of non-recyclable objects moved into the trash-can ($F(2) = 6.782$, $p = 0.001$), but not in the number of recyclable objects ($F(2) = 1.301$, $p = 0.275$), that depends on the experimental condition. In particular, participants in the FP $=$ 0.11/FN $= 0$ condition moved more non-recyclable objects (an average of 67.3 combined over rounds 2 & 3) compared to those in the FP $= 0$/FN $= 0.11$ (average of 59.9) or soft highlighting (58.1) conditions.

Learning to Perform the Task Without Guidance. We used ANOVA (with pre-test score as a covariate) to assess whether participants' scores on the post-test differed with their experimental conditions. The difference was not stat. sig. ($F(2) = 0.62$, $p = 0.539$). In terms of participants' behaviors, we found a borderline effect in the number of trashed non-recyclable objects ($F(2) = 2.734$, $p = 0.0677$): the average number of such objects was higher for participants in the FP $= 0.11$/FN $= 0$ condition (32.4 objects) and soft highlighting (31.7) conditions compared to those in the FP $= 0$/FN $= 0.11$ (27.6) condition.

Trust. With an ANOVA, we found that the trust score calculated from the questionnaire responses was stat. sig. related to the experimental condition $F(2) = 4.173$, $p = 0.0177$). In particular, the soft-highlighting condition was stat. sig. negatively associated with trust ($t = -2.778$, $p = 0.006$).

3.9 Discussion

Workers benefited from the guidance of the ML assistant, as evidenced by their higher scores during training rounds compared to the post-test. The fact that, during the training rounds, the mean scores in the FP $= 0/\text{FN} = 0.11$ condition were less than 82 (which, based on how the simulator was constructed, is the expected value if the participants perfectly followed the assistant's guidance) could indicate that participants were reluctant to accept all the assistant's advice, or that doing so was too difficult (possibly due to the fast pace of the simulation). Also, the fact that mean scores in the FP $= 0.11/\text{FN} = 0$ condition during training rounds were greater than 0 (which, due to the higher number of recyclable compared to non-recyclable objects, was the expected value for perfectly following the assistant in this condition), but were lower during the post-test than during training rounds, suggests that participants in this condition did use the MLA, but they did not follow its guidance blindly. In our experiment, a higher False Positive rate of the assistant was associated with lower performance. This contrasts with the result of [9], thus suggesting that the optimal FP/FN trade-off is likely task-dependent based on the cost of each kind of mistake.

There was no stat. sig. impact of condition on *learning*. However, there was a borderline effect of condition on the number of non-recyclable objects that were trashed during the post-test. This result is encouraging, especially given the short study duration, and suggests that the training effect of the MLA may linger in users' behaviors even after the MLA has been removed.

The soft highlighting condition was advantageous neither in terms of task performance during training rounds nor in terms of learning. This contrasts with [8], who found that participants could perceive more effectively when provided with soft highlights rather than hard detections. Learners indicated that they trusted the soft highlighting condition the least among the three assistant types.

4 Experiment II

Soft highlighting gives the learner information about the machine's confidence in its guidance. Does the learner's likelihood of picking an object and moving it to the trash-can increase when the object is highlighted with higher confidence? To explore this, we conducted a follow-up experiment ($n = 27$ participants on Mechanical Turk) consisting of just the soft highlighting condition. We assessed whether the probability distribution $P(\text{pick} \mid c, \text{isRecyclable})$ is equal to $P(\text{pick} \mid \text{isRecyclable})$, where pick indicates whether or not the participant picked the object; C is the confidence score of the object; and isRecyclable is either recyclable or non-recyclable. If the distributions are equal (for both values of isRecyclable), then by definition, the participants' decisions to pick or not pick the objects is independent of the confidence scores. From Bayes' rule, we obtain:

$$P(\text{pick} \mid c, \text{isRecyclable}) = \frac{P(c \mid \text{pick}, \text{isRecyclable})P(\text{pick} \mid \text{isRecyclable})}{P(c \mid \text{isRecyclable})}$$

Hence, for either value of isRecyclable, we can evaluate the hypothesis that the decision to pick/not pick is independent of c by testing whether $\frac{P(c \mid \text{pick,isRecyclable})}{P(c \mid \text{isRecyclable})}$ equals 1 for all values of C. The numerator can be estimated from the log data of participants' actions in the MRF simulator about the confidence scores of the objects they picked and whether they were recyclable or not; the denominator is given by the Beta distributions as described in Sect. 3.3.

4.1 Results

To estimate $P(c \mid \text{pick, isRecyclable})$ for isRecyclable = false and pick = true, we computed a histogram over the confidence scores of the picked objects using a bin width of 0.1 over the interval [0, 1]. We were primarily interested in whether there was a population-level association between picking behaviors and the confidence scores, and hence we computed a histogram over the events of all $n = 27$ participants, summed over both training rounds, separately for the recyclable and non-recyclable objects. We then computed the corresponding histogram based on the generative process of the confidence scores themselves using the appropriate Beta distribution (see Sect. 3.3). A χ^2-test showed a stat. sig. difference between the distributions, both for isRecyclable = true ($\chi^2(9) = 28.30$, $p < 0.001$) and for isRecyclable = false ($\chi^2(9) = 38.33$, $p < 0.001$). Also, Fig. 3 shows the probability ratio $\frac{P(c \mid \text{pick,isRecyclable})}{P(c \mid \text{isRecyclable})}$ both for the recyclable (**left**) and non-recyclable (**right**) cases. Despite some outliers that are partly due to small numbers of observations in the outer-most histogram bins, participants' likelihood of picking an object tends to increase with larger C.

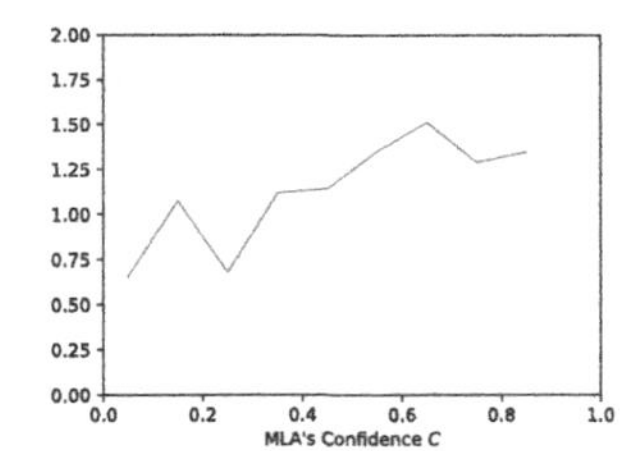

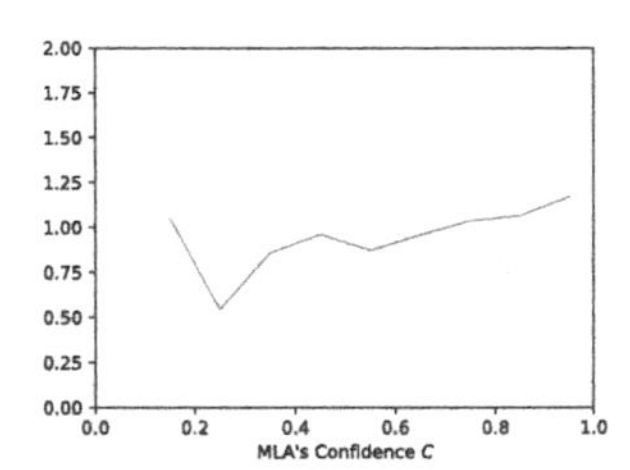

Fig. 3. Left: Relative probability increase of picking up and disposing of a *recyclable* object (**left**) or a *non-recyclable* object (**right**), given the assistant's confidence C.

4.2 Discussion

We found stat. sig. support for the hypothesis that learners do take into account the assistant's fine-grained confidence, as conveyed by the soft highlights, when deciding whether or not to pick or an object. This effect was observed for both the recyclable and the non-recyclable objects, indicating that the influence of the confidence scores goes beyond simple distinctions in the object's recyclability.

From the perspective of the designer of an AIEd system to provide better guidance and feedback to the user, this result is encouraging as it shows an example of when the confidence information is perceived by and acted on by the learner.

5 Conclusions

We conducted two experiments ($n = 181$, $n = 27$) on a workplace learning task in which participants learned to recognize and manipulate objects in a simulated MRF. We varied the FP/FN trade-off of the ML assistant that automatically, but imperfectly, highlighted objects that were deemed to be non-recyclable and thus should be moved to the trash-can. Moreover, we explored whether and how soft highlighting, rather than binary thresholding, of the guidance may impact participants' performance, learning, and trust. Our results indicate that (1) the FP/FN trade-off can affect learners' cooperative task performance with the assistant; (2) there was tentative evidence that the assigned experimental condition affected learners' behaviors – in terms of how many non-recyclable objects were trashed – even after the assistant was removed; and (3) soft highlighting impacted learners' trust as well as their behaviors in the task.

Limitations: The experiments we conducted were short (about 12–15 min). Moreover, the learning task we examined was about perception and motor control in a fast-paced environment. It is possible that different trends would be found for different learning tasks (e.g., cognitive rather than perceptual).

5.1 Future Research

Better Framing of the AI Guidance/Feedback: Our results suggest that learners trust the extra confidence information provided by the soft highlighting condition less than a simpler binary guidance mechanism. In order to be both more trustworthy and effective, the learners might need more thorough framing, and possibly even some form of "pre-training", about how the system works, what its intentions are, and how they ought to make use of it.

Explaining the Optimal FP/FN Trade-Off: Given the contrasting results of our study compared to [9], it would be interesting to change the scoring system so that different mistakes (FP versus FN) resulted in different penalties, and to investigate whether this affected the optimal FP/FN trade-off.

Acknowledgements. This research was supported by the NSF National AI Institute for Student-AI Teaming (iSAT) under grant DRL #2019805. The opinions expressed are those of the authors and do not represent views of the NSF. This research was also supported by NSF award #1928506 (FW-HTF-RL).

References

1. Ahuja, K., et al.: EduSense: practical classroom sensing at scale. In: Proceedings of the ACM on Interactive, Mobile, Wearable and Ubiquitous Technologies, vol. 3, no. 3, pp. 1–26 (2019)
2. Akcay, S., Breckon, T.: Towards automatic threat detection: a survey of advances of deep learning within X-ray security imaging. Pattern Recogn. **122**, 108245 (2022)
3. Chen, M., et al.: Evaluating large language models trained on code. arXiv preprint arXiv:2107.03374 (2021)
4. Hossain, S., Kamzin, A., Amperayani, V.N.S.A., Paudyal, P., Banerjee, A., Gupta, S.K.S.: Engendering trust in automated feedback: a two step comparison of feedbacks in gesture based learning. In: Roll, I., McNamara, D., Sosnovsky, S., Luckin, R., Dimitrova, V. (eds.) AIED 2021. LNCS (LNAI), vol. 12748, pp. 190–202. Springer, Cham (2021). https://doi.org/10.1007/978-3-030-78292-4_16
5. Hsu, S., Li, T.W., Zhang, Z., Fowler, M., Zilles, C., Karahalios, K.: Attitudes surrounding an imperfect AI autograder. In: CHI Conference on Human Factors in Computing Systems (2021)
6. Jian, J.Y., Bisantz, A.M., Drury, C.G.: Foundations for an empirically determined scale of trust in automated systems. Int. J. Cogn. Ergon. **4**, 53–71 (2000)
7. Karumbaiah, S., Lizarralde, R., Allessio, D., Woolf, B., Arroyo, I., Wixon, N.: Addressing student behavior and affect with empathy and growth mindset. International Educational Data Mining Society (2017)
8. Kneusel, R.T., Mozer, M.C.: Improving human-machine cooperative visual search with soft highlighting. ACM Trans. Appl. Percept. (TAP) **15**, 1–21 (2017)
9. Kocielnik, R., Amershi, S., Bennett, P.N.: Will you accept an imperfect AI? Exploring designs for adjusting end-user expectations of AI systems. In: CHI Conference on Human Factors in Computing Systems (2019)
10. Kulik, J.A., Fletcher, J.: Effectiveness of intelligent tutoring systems: a meta-analytic review. Rev. Educ. Res. **86**(1), 42–78 (2016)
11. Kyriacou, H., Ramakrishnan, A., Whitehill, J.: Learning to work in a materials recovery facility: can humans and machines learn from each other? In: Learning Analytics and Knowledge (LAK) Conference, pp. 456–461 (2021)
12. Landrum, A.R., Eaves, B.S., Jr., Shafto, P.: Learning to trust and trusting to learn: a theoretical framework. Trends Cogn. Sci. **19**(3), 109–111 (2015)
13. Nie, A., Brunskill, E., Piech, C.: Play to grade: testing coding games as classifying Markov decision process. In: Neural Information Processing Systems, vol. 34 (2021)
14. Randhawa, G.K., Jackson, M.: The role of artificial intelligence in learning and professional development for healthcare professionals. In: Healthcare Management Forum, vol. 33, pp. 19–24. SAGE Publications, Los Angeles (2020)
15. Schlotterbeck, D., Uribe, P., Jiménez, A., Araya, R., van der Molen Moris, J., Caballero, D.: TARTA: teacher activity recognizer from transcriptions and audio. In: Roll, I., McNamara, D., Sosnovsky, S., Luckin, R., Dimitrova, V. (eds.) AIED 2021. LNCS (LNAI), vol. 12748, pp. 369–380. Springer, Cham (2021). https://doi.org/10.1007/978-3-030-78292-4_30
16. Schmidt, R.A., Wrisberg, C.A.: Motor Learning and Performance: A Situation-Based Learning Approach. Human Kinetics (2008)
17. Zylich, B., Whitehill, J.: Noise-robust key-phrase detectors for automated classroom feedback. In: International Conference on Acoustics, Speech and Signal Processing (2020)

A Causal Inference Study on the Effects of First Year Workload on the Dropout Rate of Undergraduates

Marzieh Karimi-Haghighi[1(✉)], Carlos Castillo[1,2], and Davinia Hernández-Leo[1]

[1] Universitat Pompeu Fabra, Roc Boronat 138, Barcelona, Spain
{marzieh.karimihaghighi,carlos.castillo,davinia.hernandez-leo}@upf.edu
[2] ICREA, Pg. Lluís Companys 23, Barcelona, Spain

Abstract. In this work, we evaluate the risk of early dropout in undergraduate studies using causal inference methods, and focusing on groups of students who have a relatively higher dropout risk. We use a large dataset consisting of undergraduates admitted to multiple study programs at eight faculties/schools of our university. Using data available at enrollment time, we develop Machine Learning (ML) methods to predict university dropout and underperformance, which show an AUC of 0.70 and 0.74 for each risk respectively. Among important drivers of dropout over which the first-year students have some control, we find that first year workload (i.e., the number of credits taken) is a key one, and we mainly focus on it. We determine the effect of taking a relatively lighter workload in the first year on dropout risk using causal inference methods: Propensity Score Matching (PSM), Inverse Propensity score Weighting (IPW), Augmented Inverse Propensity Weighted (AIPW), and Doubly Robust Orthogonal Random Forest (DROrthoForest). Our results show that a reduction in workload reduces dropout risk.

Keywords: University dropout · Machine learning · Causal inference · Average treatment effect

1 Introduction

Research on actionable indicators that can lead to interventions to reduce dropout has received increased attention in the last decade, especially in the Learning Analytics (LA) field [18,29,31,32,34]. These indicators can help provide effective prevention strategies and personalized intervention actions [17,27]. Machine Learning (ML) methods, which identify patterns and associations between input variables and the predicted target [25], have been shown to be effective at this predictive task in many LA studies [1,4,10,15,23,26].

Dropout is a serious problem especially in higher education, leading to social and financial losses impacting students, institutions, and society [7]. In particular, the early identification of vulnerable students who are prone to fail or drop

M. M. Rodrigo et al. (Eds.): AIED 2022, LNCS 13355, pp. 15–27, 2022.
https://doi.org/10.1007/978-3-031-11644-5_2

their courses is necessary to improve learning and prevent them from quitting and failing their studies [20].

We remark that among students who discontinue their studies, some subgroups are over-represented, something that needs to be considered when designing dropout-reduction interventions. For example, in the UK, older students at point of entry (over 21 years) are more likely to drop out after the first year compared to younger students who enter university directly from high school [17], something that we also observe in our data. In the US, graduation rates among ethnic minority university students are lower than among White students [30]. Disparities in dropout risk have been studied in previous work [11,13,14,16]. Recent studies [8,21,22,24] look at the influence on student's performance and dropout of factors such as having a scholarship or being employed. In our work, we consider the increased dropout risk of older students and of students who do not enter university immediately after high school, and we study the effects of some features such as age and workload (i.e., number of credits taken on the first year).

Research Contribution. In this work, we use causal inference methods to study the effects of several features on the risk of early dropout in undergraduates students. We consider students enrolled between 2009 and 2018 in eight centers at our university. The average dropout rate we observe among these students is 15.3%, which is lower than the European average (36%) [35]. The originality of our contribution relies on its focus on students who have higher risk, the combination of features, the use of causal inference methods, and the size and scope of our dataset.

Specifically, we predict the risks of early dropout (i.e., not enrolling on the second year) and underperformance (failing to pass two or more subjects in the first year in the regular exams[1]) using Machine Learning (ML) methods. ML models are created using features available at the time of enrolment and the predictive performance of the models is evaluated in terms of AUC-ROC (Area Under ROC Curve). For the sake of space, we focus our exposition on dropout.

Among features available at the time of enrolment, we obtain the most important features for predicting dropout in our setting, which are the workload (number of credits taken) in the first year, admission grade, age, and study access type. Focusing on the workload, which is the most important feature and one over which first-year students have some level of control (only a minimum number of credits is established), we compute its effect on dropout risk in different age and study access type groups. We use causal inference methods to test the effects of combinations of theses features, and calculate the average treatment effect on dropout; the methods we use are the most used in the literature [2,3] including Propensity Score Matching (PSM) [28], Inverse-Propensity score Weighting (IPW) [6], Augmented Inverse-Propensity Weighted (AIPW) [12], and Doubly Robust Orthogonal Forest Estimation (DROrthoForest) [5] methods.

[1] These students have an opportunity of taking a resit exam which may finally result in passing or failing the subject, but given that passing the regular exam at the end of the course is expected, we consider failing the regular exam as underperforming.

The rest of this paper is organized as follows. After outlining related work on Sect. 2, the dataset used in this study is described and analysed in Sect. 3. The methodology is presented in Sect. 4. Results are given in Sect. 5, and finally, the results are discussed and the paper is concluded in Sect. 6.

2 Related Work

Machine Learning (ML) methods have been used to predict dropout and detect students at risk in higher education and play essential roles in improving the students' performance [1]. In a reference [4], the impact of ML on undergraduate student retention is investigated by predicting students dropout. Using students' demographics and academic transcripts, different ML models result in AUCs between 0.66 and 0.73. Another reference [7] develops a model to predict real-time dropout risk for each student during an online course using a combination of variables from the Student Information Systems and Course Management System. Evaluating the predictive accuracy and performance of various data mining techniques, the study results show that the boosted C5.0 decision tree model achieves 90.97% overall predictive accuracy in predicting student dropout in online courses. In a study [23], early university dropout is predicted based on available data at the time of enrollment using several ML models with AUCs from 0.62 to 0.81. Similarly, in a recent study [10], several ML methods are used to predict the dropout of first-year undergraduate students before the student starts the course or during the first year.

Some studies look at the features driving dropout. A reference [9] identifies factors contributing to dropout and estimates the risk of dropout for a group of students. By presenting the computed risk and explaining the reasons behind it to academic stakeholders, they help identify more accurately students that may need further support. In a research [33], the potential relationship between some features (academic background, students' performance and students' effort dimensions) and dropout is investigated over time by performing a correlation analysis on a longitudinal data collected spanning over 11 years. The results show that the importance of features related to the academic background of students and the effort students make may change over time. On the contrary, performance measures are stable predictors of dropout over time. Influential factors on student success are identified in a reference [19] using subgroup discovery; this uncovers important combinations of features known before students start their degree program, such as age, sex, regional origin or previous activities.

Recent work uses sophisticated statistical methods including causal inference. In a very recent paper [21], using propensity score matching (PSM) it is investigated whether university dropout in the first year is affected by participation in Facebook groups created by students. The estimated effect indicates that participation in social media groups reduces dropout rate. Another recent paper [24], implements an uplift modeling framework to maximize the effectiveness of retention efforts in higher education institutions, i.e., improvement of academic performance by offering tutorials. Uplift modeling is an approach for estimating the incremental effect of an action or treatment on an outcome of interest at the individual level (individual treatment effect). They show promising results in tailoring retention efforts in higher education over conventional predictive modeling approaches. In a study, the effect of grants on university dropout rates is studied [22]. The average treatment effect is estimated using blocking on the propensity score with regression adjustment. According to their results, grants have a relevant impact on the probability of completing college education.

In our paper, we carefully measure the effect of the most important features (the number of credits in the first year, age, and study access type) on the early risk of dropout in undergraduate studies. This effect is obtained for combinations of these features. The Average Treatment Effect (ATE) is measured using multiple causal inference methods [2,3] as discussed in the introduction. It is noteworthy that according to a recent survey, the methods we use in this paper have not been applied in related studies so far [1].

3 Dataset

The anonymized dataset used in this study has been provided by Universitat Pompeu Fabra and consists of 24,253 undergraduate students who enrolled between 2009 to 2018 to 21 different study programs offered by eight academic centers. From this population, about 5% of cases were discarded for various reasons: 54 had an external interruption in their education between the first and second study year, 469 students did not have grade records (dropped out before starting), 560 students were admitted but did not enroll for the first trimester, and 74 cases did not have a study access type. Finally, 23,096 cases remained.

Students were admitted to university through four access types: type I students took a standard admission test (81%), type II students moved from incomplete studies in another university or were older than 25 (10%), type III students completed vocational training before (7%), and type IV students completed a different university degree before (2%). First year courses add up to a total of 60 credits across all study programs, this is also the median number of credits taken by first year students. However, students are also free to take additional credits out of different educational offers at the university such as languages, sports, and solidarity action.

The main studied outcome is dropout and consists of students who enroll in the first year but not in the second year. We also studied underperformance, which we defined as failing two or more subjects of the first year in the regular exams. Out of 23,096 cases, 3,531 students drop out (15.3%) and 6,652 students underperform (28.8%). Per-center dropout, underperformance, and other features are shown in Table 1. There are various differences among centers.[2] The students in the School of Engineering and Faculty of Humanities have the highest dropout and underperformance rates and the Faculty of Communication has

[2] ENG: Engineering, HUM: Humanities, TRA: Translation and Language Sciences, POL: Political and Social Sciences, HEA: Health and Life Sciences, ECO: Economics and Business, COM: Communication.

Table 1. Per-center statistics: number of students, drop-out rate, underperformance rate, percentage of national students, percentage of men, average age, average first year credits, average grade on the first year, and percentage of students in access type I.

Center	N	Dropout rate	Underperf. rate	National %	Male %	Avg. age	Avg. credits	Avg. grade	Access type I
ENG	2,444	41%	56%	89%	79%	19.4	63.4	4.6	65%
HUM	1,749	22%	33%	90%	32%	20.3	63.1	5.9	76%
TRA	2,292	16%	28%	88%	18%	19.3	62.9	6.3	83%
POL	1,683	14%	27%	94%	55%	18.8	63.1	6.2	87%
HEA	1,206	14%	16%	93%	25%	19.0	60.2	7.2	82%
LAW	5,479	12%	32%	92%	33%	19.3	62.5	6.0	79%
ECO	5,707	9%	26%	93%	47%	18.5	62.9	6.3	88%
COM	2,536	7%	7%	96%	27%	18.8	61.7	7.5	84%
All	**23,096**	**15%**	**29%**	**92%**	**40%**	**19.1**	**62.6**	**6.2**	**81%**

the lowest dropout rate and the best performance. In the Faculty of Communication, which has the lowest dropout and underperformance rates, there are more national students compared to other schools. In the School of Engineering, with the highest dropout and underperformance rates, males are in the majority. The average age in the two centers with the highest dropout and underperformance rates (School of Engineering and Faculty of Humanities) is higher compared to other faculties. In these two centers, the percentage of students admitted through a standard test (study access type I) is lower than other centers, and we can observe higher average number of credits and lower average grades in their first year compared to others. In the Faculty of Humanities, 22% of the students drop out (that includes 38% of those who underperform), while in the Faculty of Law, with almost the same underperformance rate, only 12% of the students drop out (including 18% of those who underperform). This might be partially explained because in the Faculty of Law, students are one year younger (19.3 vs 20.3 years old on average) and are also slightly more likely to come directly from high school (study access type I: 79% vs 76%).

4 Methodology

Our study focuses on modeling dropout and underperformance risks using data available at the time students enrol. The feature set for our two models consists of demographics (gender, age, and nationality), study access type, study program, number of first year credits, and average admission grade. Different ML algorithms: logistic regression (LR), multi-layer perceptron (MLP), and decision trees are used to predict the risks. Both ML models are trained using students enrolled between 2009 to 2015 (16,273 cases), and tested on students enrolled in 2016, 2017, and 2018 (6,823 cases). Due to space consideration and because of the severity of dropout, we mainly focus on this risk. Using a feature selection

Table 2. Dropout rate (%) across groups defined by age, workload (number of credits), and access type. Differences of ten percentage points or more appear in **boldface**.

Center	ENG	HUM	TRA	POL	HEA	LAW	ECO	COM	All
Age > Avg. age	45	28	24	**21**	13	**21**	16	12	**26**
Age ≤ Avg. age	39	21	15	**12**	15	**10**	8	6	**13**
Access types III/IV	44	28	**27**	23	16	18	**21**	11	**24**
Access types I/II	40	22	**16**	14	14	11	**9**	6	**14**
Credits > 60	47	29	22	22	21	19	11	7	18
Credits ≤ 60	39	20	15	13	13	10	9	6	14
Age > Avg. age & credits > 60	**53**	29	**33**	**29**	13	**33**	18	13	**32**
Others	**39**	22	**16**	**13**	14	**11**	9	6	**14**
Acc. types III/IV & credits > 60	**51**	**36**	**61**	**27**	15	**33**	**23**	10	**30**
Others	**40**	**22**	**16**	**14**	14	**11**	**9**	7	**15**

method based on decision trees (CART), we find that among the features available at the time of enrolment, the most important features in predicting dropout risk are the number of credits in the first year (workload), admission grade, age, and study access type.

In Table 2, we compare the dropout rate of different student groups in terms of these features and some of their combinations (due to space constraints, we omit some combinations). This comparison shows the following results. Students older than the average age have higher rate of dropout than younger students, across all centers except the Faculty of Health and Life Sciences (HEA). Students admitted through study access types III and IV have a higher dropout rate compared to the cases admitted through access types I and II; and students taking more credits than the median also have higher dropout rate. Considering combinations of these features, we can see that mostly older students with a number of credits larger than the average, as well as students admitted through access types III and IV who take a larger number of credits than the average have higher dropout rates. Results for underperformance (omitted for brevity) are similar, except in two senses: they do not hold for Engineering (ENG) and Humanities (HUM), possibly in part due to the overall lower grades in these centers compared to all others (Table 1), and they do not hold for credits alone, but for credits in combination with other features.

We aim to determine the causal effects on dropout of the features we studied by the following intervention: taking a workload in the first year of less credits than the median. The number of credits taken is a feature over which students have some degree of control at the enrolment time. Since higher dropout rates are observed among older students and students with access types III and IV, we are interested in the following scenarios:

- Scenario 1: in this scenario, the study group is limited to the first-year students who are older than the mean. Among these, those with less workload (credits < median) are considered as treated and those with more workload (credits ≥ median) are regarded as a control group.

- Scenario 2: in this scenario, the study group are all students. Older students taking less workload (credits < median) plus all younger students are considered as treated, and older students with more workload (credits $\geq$ median) are regarded as a control group.
- Scenario 3: in this scenario, the study group is limited to students from access types III and IV. Among these, students with less workload (credits < median) are considered as treated and students with more workload (credits $\geq$ median) are regarded as a control group.

The propensity of treatment is estimated in each scenario using Machine Learning (ML) models and input features including demographics (gender and nationality), study programs, and average admission grade. In scenarios 1 and 2, study access type is also added as a feature, and in scenario 3, age is added as a feature. We compute the Average Treatment Effect (ATE) of each treatment on the dropout probability using various causal inference methods:

- The propensity score matching method [28], in which data is sorted by propensity score and then stratified into buckets (five in our case). In our work, we obtain ATE by subtracting the mean dropout of non-treated (control) cases from treated ones in each bucket.
- Inverse-Propensity score Weighting (IPW) [6]: The basic idea of this method is weighting the outcome measures by the inverse of the probability of the individual with a given set of features being assigned to the treatment so that similar baseline characteristics are obtained. In this method, the treatment effect for individual i is obtained using the following equation:

$$TE_i = \frac{W_i Y_i}{p_i} - \frac{(1 - W_i) Y_i}{1 - p_i} \tag{1}$$

  W_i shows treatment (1 for treated and 0 for control cases), p_i represents probability of receiving treatment (propensity score of treatment), and Y_i shows dropout (1 if drop out and 0 if not drop out) for individual i.
- Augmented Inverse-Propensity Weighted (AIPW) [12]: This method combines both the properties of the regression-based estimator and the IPW estimator. It has an augmentation part $(W_i - p_i)\widehat{Y}_i$ to the IPW method, in which $\widehat{Y}_i$ is the estimated probability of dropout using all features applied to the propensity score model plus the treatment variable. So, this estimator can lead to doubly robust estimation which requires only either the propensity or outcome model to be correctly specified but not both. We can compute the treatment effect on individual i as:

$$TE_i = \frac{W_i Y_i - (W_i - p_i)\widehat{Y}_i}{p_i} - \frac{(1 - W_i) Y_i - (W_i - p_i)\widehat{Y}_i}{1 - p_i} \tag{2}$$

- Causal forests from EconML package [5]: This method uses Doubly Robust Orthogonal Forests (DROrthoForest) which are a combination of causal forests and double machine learning to non-parametrically estimate the treatment effect for each individual.

Table 3. AUC-ROC of the prediction of dropout and underperformance across centers. Centers are sorted left-to-right by decreasing dropout rate.

Center	All	ENG	HUM	TRA	POL	HEA	LAW	ECO	COM
Dropout	**0.70**	0.72	0.72	0.68	0.67	0.57	0.64	0.67	0.68
Underperformance	**0.74**	0.82	0.80	0.73	0.69	0.53	0.64	0.69	0.76

Table 4. AUC-ROC of propensity score prediction.

	Scenario 1	Scenario 2	Scenario 3
N	3,866	23,096	1,963
Model	MLP	MLP	LR
AUC	0.75	0.91	0.75

In IPW, AIPW, and DROrthoForest, we obtain the individual treatment effect TE_i, which is the difference between the outcomes if the person is treated (treatment) and not treated (control). In other words, this effect is the difference of dropout probability when the student is treated and not treated; a negative value shows a reduced dropout risk and a positive value indicates an increased dropout risk. The resulting ATE is the average over individual treatment effects.

5 Results

The ML-based models of dropout and underperformance obtained using an MLP (Multi-Layer Perceptron) with 100 hidden neurons show the best predictive performance, with AUC-ROC of 0.70 and 0.74 for each risk respectively. Table 3 shows the AUC-ROC per center, and we observe that the AUC-ROC is in general higher for centers with higher dropout and underperformance rates. We also observe that dropout and underperformance predictions are not reliable for some centers, particularly Health and Life Sciences (HEA), and Law, where the AUC is less than 0.65.

For the three scenarios introduced in Sect. 4, the best predictive performance results obtained for the propensity score of the related treatment are shown on Table 4 in terms of AUC-ROC. Propensity is better predicted for scenarios 1 and 2 with the Multi-Layer Perceptron (MLP) and for scenario 3 with the Logistic Regression (LR). In each scenario, we removed study programs with relatively low predictive performance. According to the AUC values, ML models show accurate results in all of the scenarios, especially in scenario 2. In all scenarios, there is an overlap in the distribution of the propensity scores of treatment and control groups to find adequate matches (figure omitted for brevity). This is a necessary condition to be able to apply some of our methods.

Our goal is to determine whether these "treatments," which have a common feature of involving less workload, reduce dropout rate. The Average Treatment

Table 5. ATE obtained using Propensity Score Matching with five buckets.

Propensity	1. Low	2. Med-low	3. Med	4. Med-high	5. High
Scenario 1	0.18	0.04	−0.05	0.02	−0.08
Scenario 2	−0.04	0.03	0.00	0.02	−0.42
Scenario 3	0.04	−0.08	−0.17	0.30	−0.22

Table 6. IPW, AIPW, and DROrthoForest results estimating the Average Treatment Effect (ATE) and its 95% confidence interval [lower-ci, upper-ci] in three scenarios.

Scenario	IPW			AIPW			DROrthoForest		
	lower-ci	ATE	upper-ci	lower-ci	ATE	upper-ci	lower-ci	ATE	upper-ci
Scenario 1	−0.06	0.02	0.11	−0.01	0.07	0.15	−0.07	−0.06	−0.05
Scenario 2	−0.03	0.03	0.09	−0.06	0.01	0.08	−0.04	−0.04	−0.03
Scenario 3	−0.12	−0.01	0.10	−0.10	0.01	0.12	−0.07	−0.05	−0.03

Effect (ATE) obtained using propensity score matching is shown on Table 5. Across all three scenarios we can see mixed results, as in some propensity buckets the treatment increases the risk of dropout (scenario 1, bucket "1. Low"; scenario 3, bucket "4. Med-high") while in other cases the results are neutral or large reduction. In general, the results suggest that in high propensity to treatment conditions (bucket "5. High" i.e., students who are already likely to take less workload) there is a substantial reduction of the probability of dropout, particularly in scenarios 2 and 3.

The ATE values obtained from IPW, AIPW, and DROrthoForest methods are shown in Table 6 for all scenarios. In the case of IPW and AIPW, we can see that the 95% confidence intervals (from "lower-ci" to "upper-ci" in the table) contain the value zero. This means that the uncertainty in these methods is large and we cannot establish with them whether there is a change in the dropout risk due to the treatment. However, the results with the DROrthoForest method, which is a combination method of causal forests and doubly robust learner, are all negative with confidence intervals that do not contain the zero; indeed, they show a reduction of the probability of dropout of about 5% points in all three scenarios because of the treatment.

6 Discussion, Conclusions, and Future Work

In this study, we first created ML models to predict dropout (students who enroll in the first year but do not show up in the second year) and underperformance (failing two or more subjects in the regular exams of the first year), using only information available at the time of enrollment. The obtained AUC-ROC of our models were 0.70 and 0.74 for dropout and underperformance risks respectively, which shows a relatively reliable prediction of students at risk. This is particularly true for centers having large risk of dropout or underperformance, while the performance of the same models for centers having lower risk is lower. This is to some extent expected and in those cases we are modeling a phenomenon that is more rare.

Next, we focused in dropout risk prediction and found that workload (first year credits) was an important feature. We also compared dropout risk across various groups of students. The comparison showed that to a large extent there is higher probability of dropout in older students (age > average-age), in students taking a higher workload (more first year credits than the established minimum and the median), and in students admitted through access types III and IV.

We considered three scenarios using a combination of these features. In these scenarios, interventions were designed having the common characteristic of a reduced workload for students. In each scenario, the propensity score of the treatment was obtained with AUC-ROC of 0.75–0.91 using ML-based models. Then, for each scenario, the Average Treatment Effect (ATE) on dropout was computed using causal inference methods. The results suggest a negative effect, i.e., a reduction of risk of dropout, following a lower number of credits taken on the first year. An actionable recommendation that these results suggest is to ask students at risk (in this study, older students and students admitted through access types III and IV) to consider taking a reduced workload (e.g., the minimum established), or to ask educational policy makers to consider revising the regulations that establish the minimum number of credits (e.g., to reduce the current minimum).

In addition to creating ML models for early prediction of dropout and underperformance risks that exhibit high predictive performance, the originality of this contribution is focusing on the vulnerable groups of students prone to dropout, studying combinations of different features such as workload, age, and study access type, and using different causal inference models to calculate the effects of these features on dropout in terms of ATE. Causal inference methods such as the ones we used provide a path towards effectively supporting the students. They also allow to perform observational studies, as education is a domain in which some types of direct experimentation might be unethical or harmful. We also used a large dataset and our results hold across substantially diverse study programs. We stress that the methodology we described is broadly applicable. Our findings are likely to be specific to this particular dataset, but show the general effectiveness of the methodology in this setting.

More scenarios can be defined in terms of other combinations of the relevant features, to determine their effects on dropout or underperformance. Additionally, the causal inference methods used in this study can also be applied to other risks faced by higher education students.

Acknowledgements. This work has been partially supported by: the HUMAINT programme (Human Behaviour and Machine Intelligence), Joint Research Centre, European Commission; "la Caixa" Foundation (ID 100010434), under the agreement LCF/PR/PR16/51110009; and the EU-funded "SoBigData++" project, under Grant Agreement 871042. In addition, D. Hernández-Leo acknowledges the support by ICREA under the ICREA Academia programme, and the National Research Agency of the Spanish Ministry (PID2020-112584RB-C33/MICIN/AEI/10.13039/501100011033).

Ethics and Data Protection. We remark that the Data Protection Authority of the studied university performed an ethics and legal review of our research.

References

1. Albreiki, B., Zaki, N., Alashwal, H.: A systematic literature review of student' performance prediction using machine learning techniques. Educ. Sci. **11**(9), 552 (2021)
2. Athey, S.: Machine learning and causal inference for policy evaluation. In: Proceedings of the 21th ACM SIGKDD International Conference on Knowledge Discovery and Data Mining, pp. 5–6 (2015)
3. Athey, S., Wager, S.: Estimating treatment effects with causal forests: an application. Observational Stud. **5**(2), 37–51 (2019)
4. Aulck, L., Velagapudi, N., Blumenstock, J., West, J.: Predicting student dropout in higher education. arXiv preprint arXiv:1606.06364 (2016)
5. Battocchi, K., et al.: EconML: A Python Package for ML-Based Heterogeneous Treatment Effects Estimation, version 0.x (2019). https://github.com/microsoft/EconML
6. Bray, B.C., Dziak, J.J., Patrick, M.E., Lanza, S.T.: Inverse propensity score weighting with a latent class exposure: estimating the causal effect of reported reasons for alcohol use on problem alcohol use 16 years later. Prev. Sci. **20**(3), 394–406 (2019). https://doi.org/10.1007/s11121-018-0883-8
7. Bukralia, R., Deokar, A.V., Sarnikar, S.: Using academic analytics to predict dropout risk in E-learning courses. In: Iyer, L.S., Power, D.J. (eds.) Reshaping Society through Analytics, Collaboration, and Decision Support. AIS, vol. 18, pp. 67–93. Springer, Cham (2015). https://doi.org/10.1007/978-3-319-11575-7_6
8. Choi, Y.: Student employment and persistence: evidence of effect heterogeneity of student employment on college dropout. Res. High. Educ. **59**(1), 88–107 (2018). https://doi.org/10.1007/s11162-017-9458-y
9. Chounta, I.A., Uiboleht, K., Roosimäe, K., Pedaste, M., Valk, A.: From data to intervention: predicting students at-risk in a higher education institution. In: Companion Proceedings 10th International Conference on Learning Analytics & Knowledge (LAK20) (2020)
10. Del Bonifro, F., Gabbrielli, M., Lisanti, G., Zingaro, S.P.: Student dropout prediction. In: Bittencourt, I.I., Cukurova, M., Muldner, K., Luckin, R., Millán, E. (eds.) AIED 2020. LNCS (LNAI), vol. 12163, pp. 129–140. Springer, Cham (2020). https://doi.org/10.1007/978-3-030-52237-7_11
11. Gardner, J., Brooks, C., Baker, R.: Evaluating the fairness of predictive student models through slicing analysis. In: Proceedings of the 9th International Conference on Learning Analytics & Knowledge, pp. 225–234 (2019)
12. Glynn, A.N., Quinn, K.M.: An introduction to the augmented inverse propensity weighted estimator. Polit. Anal. **18**(1), 36–56 (2010)
13. Hutt, S., Gardner, M., Duckworth, A.L., D'Mello, S.K.: Evaluating fairness and generalizability in models predicting on-time graduation from college applications. International Educational Data Mining Society (2019)

14. Karimi-Haghighi, M., Castillo, C., Hernandez-Leo, D., Oliver, V.M.: Predicting early dropout: calibration and algorithmic fairness considerations. In: ADORE Workshop at the International Conference on Learning Analytics & Knowledge (LAK) (2021)
15. Kemper, L., Vorhoff, G., Wigger, B.U.: Predicting student dropout: a machine learning approach. Eur. J. High. Educ. **10**(1), 28–47 (2020)
16. Kizilcec, R.F., Lee, H.: Algorithmic fairness in education. arXiv preprint arXiv:2007.05443 (2020)
17. Larrabee Sønderlund, A., Hughes, E., Smith, J.: The efficacy of learning analytics interventions in higher education: a systematic review. Br. J. Edu. Technol. **50**(5), 2594–2618 (2019)
18. Leitner, P., Khalil, M., Ebner, M.: Learning analytics in higher education—a literature review. In: Peña-Ayala, A. (ed.) Learning Analytics: Fundaments, Applications, and Trends. SSDC, vol. 94, pp. 1–23. Springer, Cham (2017). https://doi.org/10.1007/978-3-319-52977-6_1
19. Lemmerich, F., Ifl, M., Puppe, F.: Identifying influence factors on students success by subgroup discovery. In: Educational Data Mining 2011 (2010)
20. Márquez-Vera, C., Cano, A., Romero, C., Noaman, A.Y.M., Mousa Fardoun, H., Ventura, S.: Early dropout prediction using data mining: a case study with high school students. Expert. Syst. **33**(1), 107–124 (2016)
21. Masserini, L., Bini, M.: Does joining social media groups help to reduce students' dropout within the first university year? Socioecon. Plann. Sci. **73**, 100865 (2021)
22. Modena, F., Rettore, E., Tanzi, G.M.: The effect of grants on university dropout rates: evidence from the Italian case. J. Hum. Cap. **14**(3), 343–370 (2020)
23. Nagy, M., Molontay, R.: Predicting dropout in higher education based on secondary school performance. In: 2018 IEEE 22nd International Conference on Intelligent Engineering Systems (INES), pp. 000389–000394. IEEE (2018)
24. Olaya, D., Vásquez, J., Maldonado, S., Miranda, J., Verbeke, W.: Uplift modeling for preventing student dropout in higher education. Decis. Support Syst. **134**, 113320 (2020)
25. Pal, S.: Mining educational data to reduce dropout rates of engineering students. Int. J. Inf. Eng. Electron. Bus. **4**(2), 1 (2012)
26. Plagge, M.: Using artificial neural networks to predict first-year traditional students second year retention rates. In: Proceedings of the 51st ACM Southeast Conference, pp. 1–5 (2013)
27. Romero, C., Ventura, S.: Guest editorial: special issue on early prediction and supporting of learning performance. IEEE Trans. Learn. Technol. **12**(2), 145–147 (2019)
28. Rosenbaum, P.R., Rubin, D.B.: The central role of the propensity score in observational studies for causal effects. Biometrika **70**(1), 41–55 (1983)
29. Sclater, N., Peasgood, A., Mullan, J.: Learning analytics in higher education. Jisc, London, p. 176 (2016). Accessed 8 Feb 2017
30. Shapiro, D., et al.: Completing college: a national view of student completion rates-fall 2011 cohort (2017)
31. Siemens, G.: Learning analytics: the emergence of a discipline. Am. Behav. Sci. **57**(10), 1380–1400 (2013)
32. Syed, M., Anggara, T., Lanski, A., Duan, X., Ambrose, G.A., Chawla, N.V.: Integrated closed-loop learning analytics scheme in a first year experience course. In: Proceedings of the 9th International Conference on Learning Analytics & Knowledge, pp. 521–530 (2019)

33. Tanvir, H., Chounta, I.-A.: Exploring the importance of factors contributing to dropouts in higher education over time. Int. Educ. Data Min. Soc. (2021). ERIC
34. Viberg, O., Hatakka, M., Bälter, O., Mavroudi, A.: The current landscape of learning analytics in higher education. Comput. Hum. Behav. **89**, 98–110 (2018)
35. Vossensteyn, J.J., et al.: Dropout and completion in higher education in Europe: main report (2015)

Adaptive Scaffolding in Block-Based Programming via Synthesizing New Tasks as Pop Quizzes

Ahana Ghosh[1(✉)], Sebastian Tschiatschek[2], Sam Devlin[3], and Adish Singla[1]

[1] Max Planck Institute for Software Systems, Saarbrücken, Germany
{gahana,adishs}@mpi-sws.org
[2] University of Vienna, Vienna, Austria
sebastian.tschiatschek@univie.ac.at
[3] Microsoft Research, Cambridge, UK
sam.devlin@microsoft.com

Abstract. Block-based programming environments are increasingly used to introduce computing concepts to beginners. However, novice students often struggle in these environments, given the conceptual and open-ended nature of programming tasks. To effectively support a student struggling to solve a given task, it is important to provide adaptive scaffolding that guides the student towards a solution. We introduce a scaffolding framework based on pop quizzes presented as multi-choice programming tasks. To automatically generate these pop quizzes, we propose a novel algorithm, PQuizSyn. More formally, given a reference task with a solution code and the student's current attempt, PQuizSyn synthesizes new tasks for pop quizzes with the following features: (a) *Adaptive* (i.e., individualized to the student's current attempt), (b) *Comprehensible* (i.e., easy to comprehend and solve), and (c) *Concealing* (i.e., do not reveal the solution code). Our algorithm synthesizes these tasks using techniques based on symbolic reasoning and graph-based code representations. We show that our algorithm can generate hundreds of pop quizzes for different student attempts on reference tasks from *Hour of Code: Maze Challenge* [11] and *Karel* [9]. We assess the quality of these pop quizzes through expert ratings using an evaluation rubric. Further, we have built an online platform for practicing block-based programming tasks empowered via pop quiz based feedback, and report results from an initial user study.

Keywords: Block-based visual programming · Scaffolding · Task synthesis

1 Introduction

The emergence of block-based visual programming platforms has made coding more interactive and appealing for novice students. Block-based programming uses "code blocks" that reduce the burden of syntax and focuses on key programming concepts. Led by the success of languages like *Scratch* [33], initiatives like *Hour of Code* by Code.org [12], and online courses like *Intro to Programming with Karel* by CodeHS.com [9,25], block-based programming has become integral to introductory CS education.

M. M. Rodrigo et al. (Eds.): AIED 2022, LNCS 13355, pp. 28–40, 2022.
https://doi.org/10.1007/978-3-031-11644-5_3

Programming tasks on these platforms are conceptual and open-ended, requiring multi-step deductive reasoning to solve, thereby making them challenging for students. To effectively support a struggling student to solve a particular task, it is important to provide feedback on their attempts. However, on platforms that have millions of students, it is infeasible for human tutors to provide feedback. Hence, there is a critical need for automated feedback generation systems to provide personalized support to students [13,22]. Existing work in the domain has explored various methods of personalized feedback generation within a task, such as providing next-step hints in the form of next code blocks to use in a student attempt [26,31,35,36,44], providing adaptive worked examples [28,32,43], and providing data-driven analysis of a student's misconceptions [5,6,16,24,39–41].

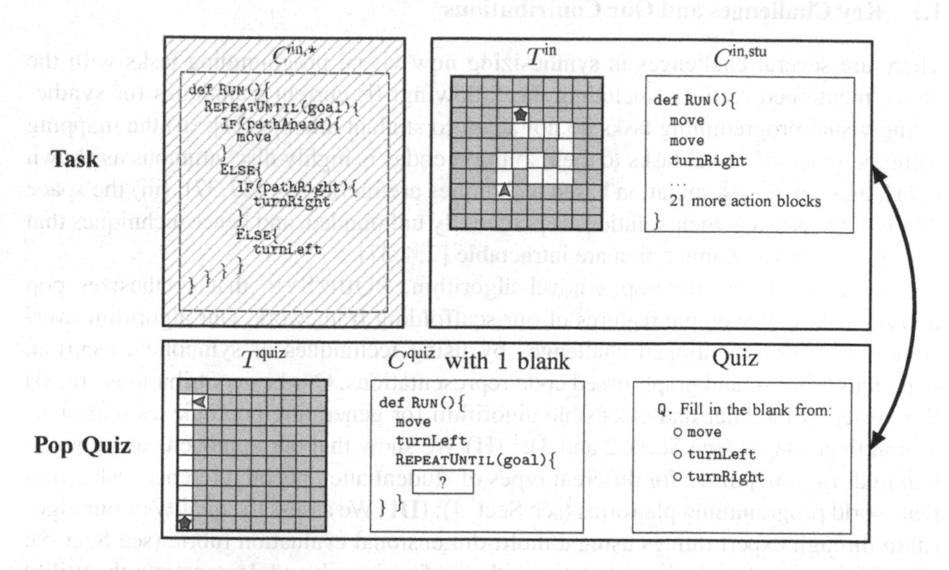

Fig. 1. Illustration of our pop quiz based framework. The "Task" panel shows an input task T^{in} from HOC [11], the student's current attempt $C^{in,stu}$, and the solution code $C^{in,\star}$ (not revealed to the student). The student is currently unsuccessful in solving the task: the current attempt $C^{in,stu}$ does not solve the visual puzzle within the maximal number of permitted blocks (7 blocks) and does not use any of the required constructs (REPEATUNTIL and IFELSE constructs). The "Pop Quiz" panel shows a pop quiz generated by our algorithm in the form of task-code pair (T^{quiz}, C^{quiz}) along with a multiple choice question, introducing the REPEATUNTIL construct. After the student solves the pop quiz, they resume working on the input task. The framework would be invoked when a student needs help; importantly, the pop quizzes presented to the student are adaptive w.r.t. the student's current attempt $C^{in,stu}$. Moreover, our algorithm generates pop quizzes that are easy to comprehend and solve, and C^{quiz} sufficiently conceals $C^{in,\star}$.

In this paper, we investigate an alternate method of personalized feedback generation that guides a student towards a task's solution while involving inquiry-driven and problem-solving aspects [14]. In particular, we introduce a scaffolding framework based on pop quizzes that contain new programming tasks presented as multi-choice questions.[1] Our framework is inspired by prior studies that showed the efficacy of multi-choice questions in helping novice students learn to code [17,19,30,38,42]. The framework is designed to be invoked as follows: Given a task and a student's current unsuccessful attempt, the framework can help the student by presenting a pop quiz intended to resolve their misconception. For the scaffolding to be effective, we center the design of the new programming task for a pop quiz around three features: *Adaptive*, *Comprehensible*, and *Concealing*; see details in Fig. 1 and Sect. 2.1. However, hand-crafting these new quizzes is time-consuming and potentially error-prone when required for a large number of tasks and different student attempts. To this end, we seek to *automatically* generate these pop quizzes by synthesizing new programming tasks.

1.1 Key Challenges and Our Contributions

There are several challenges in synthesizing new visual programming tasks with the above mentioned features, including the following: (i) current techniques for synthesizing visual programming tasks do not adapt to student attempts [1]; (ii) the mapping from the space of visual tasks to their solution codes is highly discontinuous as shown in [1], and hence task mutation based techniques are ineffective [27,37]; (iii) the space of possible tasks and their solutions is potentially unbounded, and hence techniques that rely on exhaustive enumeration are intractable [2,4,37].

In this work, we develop a novel algorithm, PQUIZSYN, that synthesizes pop quizzes with the desirable features of our scaffolding framework. Our algorithm overcomes the above-mentioned challenges by using techniques of symbolic execution, search algorithms, and graph-based code representations. Our key contributions are: **(I)** We present a modular and extensible algorithm for generating pop quizzes that operates in three stages (see Sects. 2 and 3);[2] **(II)** We show that our approach can generate hundreds of pop quizzes for different types of student attempts on reference tasks from real-world programming platforms (see Sect. 4); **(III)** We assess the quality of our algorithm through expert ratings using a multi-dimensional evaluation rubric (see Sect. 5); **(IV)** We have built an online platform with our framework and demonstrate the utility of pop quiz based feedback through an initial user study (see Sect. 6).[3]

1.2 Additional Related Work

Feedback via Modelling Programming Concepts. Apart from the above-mentioned methods such as next-step hints, there has been extensive work on feedback generation via modelling programming concepts. Here, several techniques have been proposed, including: (a) detecting challenging concepts by analyzing student attempts [6,39,40]; (b) discovering student misconceptions using task-specific rubrics and neural program embeddings [41]; (c) defining concepts through knowledge components [3,15,34].

[1] We refer to these multi-choice questions as "pop quizzes" as the framework could present these quizzes whenever a student needs help [7].

[2] Implementation of the algorithm is publicly available at https://github.com/machine-teaching-group/aied2022_pquizsyn_code.

[3] https://www.teaching-blocks-hints.cc/.

Evaluation of Feedback Methods. An important aspect to consider when developing feedback generation methods is their evaluation criteria. Most next-step feedback generation methods are evaluated based on expert annotations or automated procedures [24,26,29]. In contrast, example-driven feedback techniques are typically evaluated using a multi-dimensional rubric [32,43]. In our work, we evaluate the scaffolding framework through expert ratings using a rubric, as well as an initial user study.

2 Problem Setup and Definitions

In this section, we formalize our objective and introduce important technical definitions.

2.1 Problem Setup

Task Space. We define the space of tasks as $\mathbb{T}$. A task $T \in \mathbb{T}$ consists of a visual puzzle and a set of available types of code blocks (e.g., `move`, RepeatUntil) allowed in the solution code. Additionally, the solution code must be within a certain size threshold in terms of the number of code blocks. We denote the current task that a student is solving as $T^{\text{in}} \in \mathbb{T}$; see T^{in} in Fig. 1. In this work, we use tasks from *Hour of Code: Maze Challenge* [11] by Code.org [10] and *Intro to Programming with Karel* [9] by CodeHS.com [8]; henceforth, we refer to them as HOC and Karel tasks, respectively.

Code Space. We define the space of all possible codes as $\mathbb{C}$ and represent them using a *Domain Specific Language* (DSL) [20]. In particular, for codes relevant for HOC and Karel tasks, we use a DSL based on [1]. A code $C \in \mathbb{C}$ has the following attributes: C_{blocks} is the set of types of code blocks used in C, C_{size} is the number of blocks used, and C_{depth} is the depth of the *Abstract Syntax Tree* of C. We denote a distance metric in this space as $D_{\mathbb{C}}$. For a given $C \in \mathbb{C}$ and a positive integer l, we define a neighborhood function as $\mathcal{N}_{\mathbb{C}}(C, l) = \{C' \mid D_{\mathbb{C}}(C', C) \leq l\}$. The solution code $C^{\text{in},\star} \in \mathbb{C}$ for the task T^{in} solves the visual puzzle using the allowed types of code blocks within the specified size threshold. A student attempt for T^{in} is denoted as $C^{\text{in,stu}} \in \mathbb{C}$.

Objective. For an input task T^{in} with solution code $C^{\text{in},\star}$ and given the current student attempt $C^{\text{in,stu}}$, our objective is to generate a pop quiz in form of a new task-code pair $(T^{\text{quiz}}, C^{\text{quiz}})$ designed on the basis of the following features: (i) *Adaptive*, i.e., C^{quiz} accounts for $C^{\text{in},\star}$ and $C^{\text{in,stu}}$, ensuring that C^{quiz} is individualized to the student's current attempt; (ii) *Comprehensible*, i.e., C^{quiz} solves T^{quiz} correctly and the pop quiz is easy to comprehend/solve without confusing the student; (iii) *Concealing*, i.e., $D_{\mathbb{C}}(C^{\text{quiz}}, C^{\text{in},\star})$ is high, ensuring that C^{quiz} sufficiently conceals the solution code $C^{\text{in},\star}$ and does not directly reveal it in order to encourage problem-solving aspects.

2.2 Technical Definitions

Sketch Space. We capture the key conceptual elements of a code using a higher level abstraction called a *sketch* [2,37]. The sketch of a code preserves its important programming constructs. Similar to the code DSL, we define the sketch space $\mathbb{S}$ using a sketch DSL based on [1]. Similar to the *Abstract Syntax Tree* representation of a code, we represent a sketch as a tree having the programming constructs as its nodes. The mapping from the code space to the sketch space is captured by the many-to-one map, $\Psi: \mathbb{C} \rightarrow \mathbb{S}$, i.e., the representation of a code C in $\mathbb{S}$ is given by $\Psi(C)$. As $\mathbb{S}$ is an abstraction of $\mathbb{C}$, multiple elements of $\mathbb{C}$ can correspond to a single element in $\mathbb{S}$. Similar to $D_{\mathbb{C}}$ and $\mathcal{N}_{\mathbb{C}}$, we denote a distance metric in the sketch space as $D_{\mathbb{S}}$ and a neighborhood function as $\mathcal{N}_{\mathbb{S}}(S, l) = \{S' \mid D_{\mathbb{S}}(S', S) \leq l\}$ for a given $S \in \mathbb{S}$ and a positive integer l.

Sketch Substructures. For a sketch S, we define a substructure as a sub-tree containing the nodes of S up to a particular depth and sharing the same root node; note that a substructure of a sketch is also a sketch. We denote the set of all substructures of S as $\textsc{SubStructs}(S) \subseteq \mathbb{S}$; the size of the set $\textsc{SubStructs}(S)$ is typically small. For example, the sketch shown in Fig. 2b has the following 4 substructures: (i) {RUN}, (ii) {RUN {REPEATUNTIL(goal)}}, (iii) {RUN {REPEATUNTIL(goal){IFELSE (B)}}}, and (iv) {RUN {REPEATUNTIL(goal){IFELSE (B){{}; {IFELSE (B)}}}}}.

Code Reductions. For a code $C \in \mathbb{C}$ with sketch $S := \Psi(C)$, consider one of the sketches $S_{\text{sub}} \in \textsc{SubStructs}(S)$. We define the set of code reductions of C w.r.t. sketch S_{sub} as all codes obtained by removing one or more nodes of C while preserving the sketch S_{sub}; note that the reduction of a code is also a code. We denote the set of all reductions as $\textsc{RedCodes}(C \mid S_{\text{sub}}) \subseteq \mathbb{C}$. For example, for $C^{\text{in},\star}$ in Fig. 1 and $S_{\text{sub}} =$ {RUN{REPEATUNTIL(*goal*)}}, the set $\textsc{RedCodes}(C^{\text{in},\star} \mid S_{\text{sub}})$ has the following 3 codes: (i) {RUN {REPEATUNTIL(goal){move}}}, (ii) {RUN{REPEATUNTIL(goal){turnRight}}}, and (iii) {RUN{REPEATUNTIL(goal){turnLeft}}}.

3 Our Algorithm PQuizSyn

In this section, we present our algorithm that generates pop quizzes via synthesizing new tasks. One might be tempted to synthesize tasks by first generating a new visual puzzle and then obtaining its solution code. As discussed in Sect. 1 and shown in [1], the mapping from the space of visual tasks to their solution codes is highly discontinuous and reasoning about desirable tasks directly in the task space is ineffective. However, the task synthesis algorithm from [1] is not applicable to our work as we seek to generate tasks that also account for the student's current attempt. To this end, we develop a novel algorithm PQuizSyn (*Programming Pop Quizzes via Synthesis*) that generates tasks adaptive to the student's current attempt. Our algorithm operates in three stages: (i) Stage 1 generates a sketch based on the task's solution code and the student's current attempt; (ii) Stage 2 instantiates this sketch in the form of a new task-code pair; (iii) Stage 3 generates the pop quiz from the new task-code pair. Figure 2a illustrates these stages, and details are provided below.

3.1 Stage 1: Generating the Pop Quiz Sketch S^{quiz}

We begin by describing Stage 1 of our algorithm as illustrated in Fig. 2a. In this stage, `GetSketch()` routine returns a suitable sketch S^{quiz} that is instantiated in the later stages. The input to the routine is the student sketch $S^{\text{in,stu}} := \Psi(C^{\text{in,stu}})$ and solution sketch $S^{\text{in},\star} := \Psi(C^{\text{in},\star})$. By operating on the sketch space first, we can generate meaningful and adaptive codes in the later stages. To generate pop quizzes based on the features mentioned in Sect. 2.1, we require the sketch of the pop quiz S^{quiz} to have the following attributes: (i) S^{quiz} should direct the student towards the solution sketch $S^{\text{in},\star}$, i.e., $D_{\mathbb{S}}(S^{\text{quiz}}, S^{\text{in},\star})$ should be low; (ii) S^{quiz} should be adaptive w.r.t. the student's sketch $S^{\text{in,stu}}$, i.e., $S^{\text{quiz}} \in \mathcal{N}_{\mathbb{S}}(S^{\text{in,stu}}, l)$ for a low value of l. While these conditions ensure that S^{quiz} directs the student towards the solution sketch and is adaptive, it could potentially lead to a sketch that does not belong to the set of substructures of the solution sketch, i.e., $S^{\text{quiz}} \notin \text{SUBSTRUCTS}(S^{\text{in},\star})$—in that case, there is no valid code reduction of $C^{\text{in},\star}$ w.r.t. S^{quiz} (see Sect. 2.2) and this makes it challenging to instantiate sketches into desirable codes C^{quiz} (see algorithm variant PQS-ONEHOP in Sect. 5 and Footnote 4). Hence, `GetSketch()` generates S^{quiz} as follows (see Fig. 3):

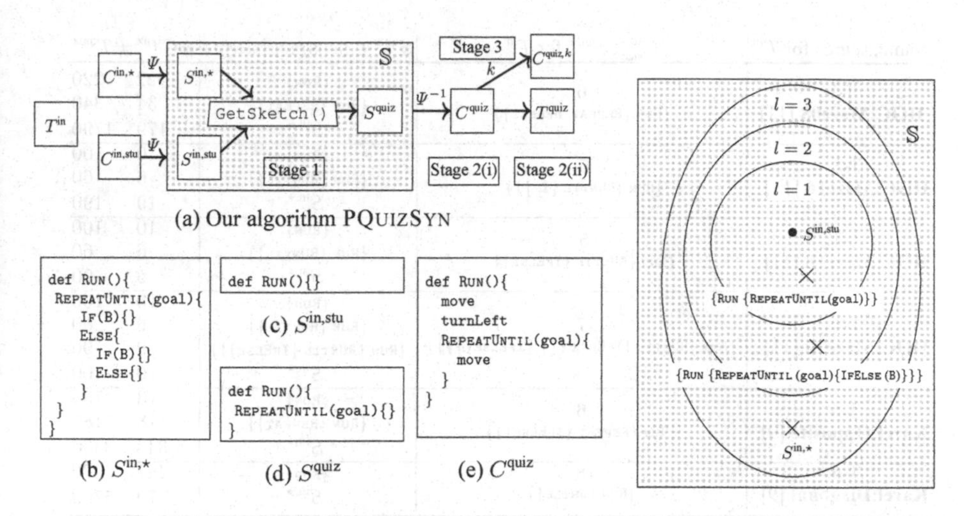

Fig. 2. (a) illustrates PQUIZSYN. In particular, we can instantiate the presented algorithm using input task T^{in}, its solution code $C^{\text{in},\star}$, and the current student attempt $C^{\text{in,stu}}$ from Fig. 1. The sketch of $C^{\text{in},\star}$ is shown in (b), sketch of $C^{\text{in,stu}}$ is shown in (c), sketch of C^{quiz} is shown in (d), and the code of the pop quiz C^{quiz} is shown in (e).

Fig. 3. PQUIZSYN Stage 1 for the scenario shown in Fig. 1. X shows substructures of $S^{\text{in},\star}$ in l-hop neighborhoods of $S^{\text{in,stu}}$ for $l \in \{1, 2, 3\}$. Details are provided in Sect. 3.1.

(i) Pick $\hat{l}$ as $\min l \in \{1, 2, \ldots\}$ s.t. $\mathcal{N}_{\mathbb{S}}(S^{\text{in,stu}}, l) \cap \text{SUBSTRUCTS}(S^{\text{in},\star})$ is non-empty.
(ii) Generate $S^{\text{quiz}} \in \operatorname{argmin}_{S \in \mathcal{N}_{\mathbb{S}}(S^{\text{in,stu}},\, \hat{l}) \,\cap\, \text{SUBSTRUCTS}(S^{\text{in},\star})} D_{\mathbb{S}}(S, S^{\text{in},\star})$.

3.2 Stage 2: Synthesizing $(T^{\text{quiz}}, C^{\text{quiz}})$ from S^{quiz}

Next, we describe Stage 2 of our algorithm. We first generate C^{quiz} from S^{quiz}, as illustrated in Stage 2(i) of Fig. 2a. Specifically, for a sketch S^{quiz} generated in Stage 1, we employ the *code mutation* methodology proposed in [1] to obtain a code C^{quiz}. However, this methodology requires a meaningful starting code C^{seed}. Since $S^{\text{quiz}} \in \text{SUBSTRUCTS}(S^{\text{in},\star})$ by the design of Stage 1, we begin by picking C^{seed} from the set

REDCODES($C^{\text{in},\star} \mid S^{\text{quiz}}$).[4] The methodology of [1] provides us multiple code mutations of C^{seed}. The extent to which these code mutations differ from C^{seed} and $C^{\text{in},\star}$ is controlled by the constraints imposed based on the values of the boolean variables, conditionals, and action blocks (`move`, `turnLeft`, `turnRight`, `pickMarker`, `putMarker`) of C^{seed}, as well as constraints on the size of the obtained code. Specifically, these mutations allow us to control the extent to which $D_{\mathbb{C}}(C^{\text{quiz}}, C^{\text{in},\star})$ varies, which is a desired feature as stated in Sect. 2.1.

Next, we generate a new task T^{quiz} from a code C^{quiz} as illustrated in Stage 2(ii) of Fig. 2a. Specifically, we generate T^{quiz} such that its solution code is C^{quiz}. We achieve this using techniques of symbolic execution and best-first search, building on the task synthesis methodology presented in [1].

Name, source for T^{in}	$C^{\text{in},\star}_{\text{size}}$, $S^{\text{in},\star}$ for T^{in}	$S^{\text{quiz}} \in$ SUBSTRUCTS($S^{\text{in},\star}$)	$\#C^{\text{quiz}}$	$\#T^{\text{quiz}}$
T-1 **HOC:Maze08** [11]	6 {RUN {REPEAT; REPEAT}}	{RUN}	22	220
		{RUN {REPEAT}}	34	340
		$S^{\text{in},\star}$	179	1790
T-2 **HOC:Maze16** [11]	5 {RUN {RUNTIL {IF}}}	{RUN}	10	100
		{RUN {RUNTIL}}	6	60
		$S^{\text{in},\star}$	19	190
T-3 **HOC:Maze18** [11]	5 {RUN {RUNTIL {IFELSE}}}	{RUN}	10	100
		{RUN {RUNTIL}}	6	60
		$S^{\text{in},\star}$	9	90
T-4 **HOC:Maze20** [11]	7 {RUN {RUNTIL {IFELSE {{};{IFELSE}}}}}	{RUN}	10	100
		{RUN {RUNTIL}}	6	60
		{RUN {RUNTIL {IFELSE}}}	9	90
		$S^{\text{in},\star}$	10	100
T-5 **Karel:Opposite** [9]	6 {RUN {REPEAT {IFELSE}}}	{RUN}	73	730
		{RUN {REPEAT}}	118	1180
		$S^{\text{in},\star}$	343	3430
T-6 **Karel:Diagonal** [9]	8 {RUN {WHILE}}	{RUN}	447	4470
		$S^{\text{in},\star}$	579	5790

Fig. 4. PQUIZSYN applied to six HOC and Karel reference tasks; see Sect. 4 for details. For brevity, sketches have been abbreviated, e.g., REPEATUNTIL(`goal`) as RUNTIL.

3.3 Stage 3: Generating Multi-choice Question from ($T^{\text{quiz}}, C^{\text{quiz}}$)

In this stage, we generate a pop quiz with a fixed set of answer choices; see Figs. 1 and 5. We pick a task-code pair ($T^{\text{quiz}}, C^{\text{quiz}}$), and expose only a part of C^{quiz} determined by an exposure parameter k, i.e., C^{quiz} contains k blanks. These blanks must be filled out by the student from the set of answer choices in a manner that would solve T^{quiz}. Specifically, we generate the pop quiz with $k = 1$ blanks. To obtain the blank for the quiz, we do an in-order traversal of C^{quiz} and leave out the last leaf node as blank.

[4] When $S^{\text{quiz}} \notin$ SUBSTRUCTS($S^{\text{in},\star}$), we set C^{seed} as a random instantiation of S^{quiz} – see algorithm variant PQS-ONEHOP in Sect. 5..

4 PQUIZSYN on Real-World Tasks

In this section, we present the performance of PQUIZSYN on six reference tasks taken from real-world block-based programming platforms: HOC [11] and Karel [9]. The set of these tasks along with their sources are mentioned in Fig. 4. These tasks differ in complexity, measured in terms of the programming constructs of their solution code as illustrated by the diversity of their respective solution sketches $S^{\text{in},\star}$. For the exhaustive set of substructures of $S^{\text{in},\star}$, Fig. 4 lists the total number of pop quizzes, in the form of unique task-code pairs $(T^{\text{quiz}}, C^{\text{quiz}})$, generated by our algorithm. As can be seen in the figure, our algorithm generates 50 to 1000s of pop quizzes for each substructure. For any potential student attempt on these tasks, Stage 1 of PQUIZSYN would generate one of these task-specific substructures by design – hence, for every attempt we can present several unique yet adaptive pop quizzes to the student. Note that, our algorithm generates higher number of tasks than codes for each substructure. This is because the task synthesis methodology used in Stage 2(ii) can generate more than one task for a single code in Stage 2(ii) of Fig. 2a. In particular, for each new code, we obtain 10 diverse tasks. For instance, Fig. 1 and Fig. 5 illustrate pop quizzes generated by PQUIZSYN for the specific student attempts on tasks T-4 and T-5, respectively.

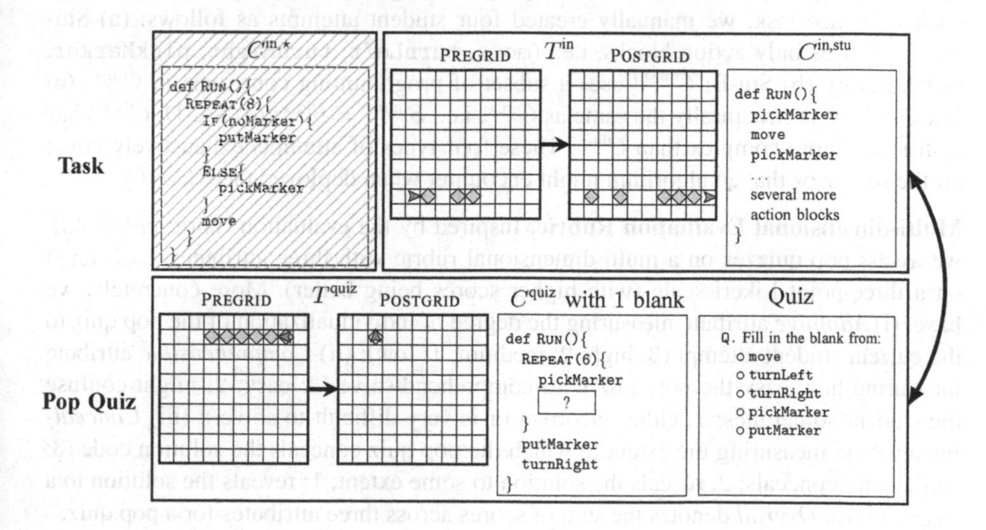

Fig. 5. Analogous to Fig. 1, here we illustrate our framework on a Karel task, T-5 (see Fig. 4). Karel tasks [25] comprise of a pair of visual grids, (PREGRID, POSTGRID), and the objective is to write code that, when executed, transforms PREGRID to POSTGRID.

5 Expert Study via Multi-dimensional Rubric

In this section, we evaluate PQUIZSYN w.r.t. the desired features specified in the objective, i.e., *Adaptive*, *Comprehensible*, and *Concealing* (see Sect. 2.1). In particular, we seek to compare PQUIZSYN with its variants resulting from different design choices in Sect. 3. To this end, we conduct an expert study via a multi-dimensional rubric.

Variants of PQUIZSYN Algorithm. We compare the performance of PQUIZSYN with the following variants: PQS-FULLHOP, PQS-ONEHOP, and PQS-REDCODE. PQS-FULLHOP and PQS-ONEHOP differ from PQUIZSYN only in the `GetSketch()` routine used in Stage 1 of Fig. 2a when generating S^{quiz}. In particular, Stage 1 of PQS-FULLHOP always returns the sketch of the solution code, i.e., $S^{\text{quiz}} := S^{\text{in},\star}$; Stage 1 of PQS-ONEHOP returns a sketch directly from the 1-hop neighborhood of $S^{\text{in,stu}}$, i.e., $S^{\text{quiz}} \in \mathcal{N}_{\mathbb{S}}(S^{\text{in,stu}}, 1)$. The third baseline, PQS-REDCODE, differs from PQUIZSYN only in Stage 2(i) of Fig. 2a when generating C^{quiz} from S^{quiz}. In particular, Stage 2(i) of PQS-REDCODE generates C^{quiz} as a direct reduction of the solution code w.r.t. the sketch obtained in Stage 1, i.e., $C^{\text{quiz}} \in \text{REDCODES}(C^{\text{in},\star} \mid S^{\text{quiz}})$.

Simulated Student Attempts. For this expert evaluation, we simulated unsuccessful student attempts as seen in block-based programming domains [26]. In particular, for each reference task, we manually created four student attempts as follows: (a) Stu-A: $C^{\text{in,stu}}$ uses only action blocks, i.e., (`move`, `turnLeft`, `turnRight`, `pickMarker`, `putMarker`); (b) Stu-B: $C^{\text{in,stu}}$ uses a subset of programming constructs in $C^{\text{in},\star}$; (c) Stu-C: $C^{\text{in,stu}}$ is structurally the same as $C^{\text{in},\star}$, i.e., $S^{\text{in,stu}} = S^{\text{in},\star}$; (d) Stu-D: $C^{\text{in,stu}}$ has a structure more complex than $C^{\text{in},\star}$. These four types of attempts exhaustively cover all the scenarios that an algorithm might encounter when deployed (see Sect. 6).

Multi-dimensional Evaluation Rubric. Inspired by the evaluation rubric in [32,43], we assess pop quizzes on a multi-dimensional rubric with three attributes, each rated on a three-point Likert scale (with higher scores being better). More concretely, we have: (i) *Adaptive* attribute measuring the degree of individualization of the pop quiz to the current student attempt (3: high; 2: medium; 1: low); (ii) *Comprehensible* attribute measuring how easy the pop quiz is to comprehend/solve (3: easy; 2: might confuse the student sometimes; 1: either incorrect or is very difficult to solve.); (iii) *Concealing* attribute measuring the extent to which the pop quiz conceals the solution code (3: sufficiently conceals; 2: reveals the solution to some extent; 1: reveals the solution to a large extent). *Overall* denotes the sum of scores across three attributes for a pop quiz.

Expert Study Setup. We picked three tasks spanning different types of constructs and complexity: T-1, T-4, and T-5 from Fig. 4. Thus, in total we evaluated 48 scenarios: 4 algorithm variants $\times$ 4 student types $\times$ 3 tasks (see Figs. 1 and 5 as example scenarios). Two researchers, with experience in block-based programming, evaluated each of the 48 scenarios independently. The evaluation was done through a web survey where a scenario was introduced at random, and assessed based on the rubric.

Expert Study Results. First, we validate the expert ratings using the quadratic-weighted Cohen's kappa inter-agreement reliability value [32] for each attribute: 0.62 (*Adaptive*), 0.69 (*Comprehensible*), 0.79 (*Concealing*), and 0.7 (*Overall*). The values indicate *substantial agreement* between the raters. The average ratings are presented in Fig. 6 and PQUIZSYN has the highest *Overall* score. We analyze these ratings per attribute based on the Kruskal-Wallis significance test [21]; the results discussed next are statistically significant with $p < 0.01$. On the *Adaptive* attribute, PQS-FULLHOP performs significantly worse because it does not account for the student attempt (see Sect. 3.1). On the *Comprehensible* attribute, PQS-ONEHOP performs significantly worse because there are instances where no valid code reduction of $C^{\text{in},\star}$ w.r.t. S^{quiz} is found (see Footnote 4, Sect. 3.2). Finally, on the *Concealing* attribute, PQS-REDCODE performs significantly worse because it obtains C^{quiz} via a direct reduction of $C^{\text{in},\star}$ without any mutation (see Sect. 3.2).

Algorithm	*Adaptive*	*Comprehensible*	*Concealing*	*Overall*
PQS-FULLHOP	2.0(0.7)	2.8(0.1)	3.0(0.0)	7.8(0.8)
PQS-ONEHOP	2.8(0.1)	2.5(0.6)	3.0(0.0)	8.3(0.7)
PQS-REDCODE	2.7(0.3)	3.0(0.0)	1.5(0.4)	7.2(0.7)
PQUIZSYN	2.7(0.2)	3.0(0.0)	2.9(0.1)	8.6(0.3)

Fig. 6. Mean (variance) attribute ratings for different algorithms. Higher scores are better. PQUIZSYN performs well across all three attributes and has the highest *Overall* score; see Sect. 5 for details.

6 User Study via Online Platform

We have built an online platform with our PQUIZSYN framework using the Blockly Games library [18]. The online platform is publicly accessible – see Footnote 3, Sect. 1.1. The platform provides an interface for a participant to practice block-based programming tasks, and receive pop quiz based feedback when stuck. In this section, we report results from an initial user study to assess the efficacy of our scaffolding framework in comparison to other feedback methods.

Participation Session and Feedback Methods. A single session on our platform comprises of three steps. In STEP-A, the participant is presented with a task and has 10 execution tries to solve it. If a participant fails to solve the task at STEP-A, they proceed to STEP-B with a randomly assigned feedback method (NOHINT, NEXTSTEP, and PQUIZSYN as discussed below). After STEP-B, the participant resumes their attempt on the task in STEP-C with 10 additional execution tries. Note that the feedback method is invoked only once in a single session. Next, we describe different feedback methods at STEP-B. NOHINT represents a baseline where the participant is directed to STEP-C without any feedback. NEXTSTEP corresponds to next-step hints as feedback where the participant's code is updated to bring it closer to a solution code [26,31,35,36,44]; we prioritized next-step edits involving programming constructs (e.g., `REPEATUNTIL`) over basic actions (e.g., `move`). PQUIZSYN is our pop quiz based feedback.

User Study Results. We conducted an initial user study with participants recruited from Amazon Mechanical Turk; an IRB approval was received before the study. The participants were US-based adults, without expertise in block-based visual programming. Due to the costs involved (over 3 USD per participant), we selected two tasks for the study: T-3 and T-5 from Fig. 4. We present the detailed results in Fig. 7. In total, we had 575 unique participants; out of these, 0.774 fraction failed to solve the task at STEP-A and proceeded to STEP-B. PQUIZSYN was assigned to 148 participants in STEP-B (0.60 fraction successfully solved the presented pop quiz). Subsequently, 0.128 fraction of these participants solved the task in STEP-C. Here, 0.128 measures the success rate of participants assigned to PQUIZSYN; in comparison, it is 0.082 for NEXTSTEP and 0.046 for NOHINT – see Fig. 7. Overall, the performance of PQUIZSYN is better than NEXTSTEP (the gap is not significant w.r.t. χ^2-test, $p = 0.19$) and NOHINT (the gap is significant w.r.t. χ^2-test, $p = 0.01$) [23]. These initial results demonstrate the utility of providing pop quiz based feedback.

Feedback	Total (**STEP-B**)			Fraction solved (**STEP-C**)		
	Both	T-3	T-5	Both	T-3	T-5
NOHINT	151	63	88	0.046	0.079	0.023
NEXTSTEP	146	63	83	0.082	0.127	0.048
PQUIZSYN	148	62	86	0.128	0.177	0.093

Fig. 7. Results for tasks T-3 and T-5 ("Both" represents aggregated results). In **STEP-A**, we had a total of 575 (293 for T-3, 282 for T-5) participants; about 0.774 (0.642 for T-3, 0.911 for T-5) fraction failed to solve the task at **STEP-A** and proceeded to **STEP-B/STEP-C** with a randomly assigned feedback method.

7 Conclusions and Outlook

We proposed a novel scaffolding framework for block-based programming based on pop quizzes that involve inquiry-driven and problem-solving aspects. We developed a modular synthesis algorithm, PQUIZSYN, that generates these pop quizzes. After conducting an expert assessment using a multi-dimensional rubric, we developed an online platform empowered by our scaffolding framework. While initial user study results with our platform demonstrate the utility of our pop quiz based framework, there are several interesting directions to continue this study, including: (i) extending our platform to provide multiple rounds of feedback within a single participation session and measuring the efficacy of different methods; (ii) comparing our synthesized pop quizzes with those generated by experts; (iii) conducting longitudinal studies with novice students to measure long-term improvements in problem solving skills; (iv) extending our framework to more complex block-based programming domains.

Acknowledgments. We would like to thank the reviewers for their feedback. Ahana Ghosh was supported by Microsoft Research through its PhD Scholarship Programme. Adish Singla acknowledges support by the European Research Council (ERC) under the Horizon Europe programme (ERC StG, grant agreement No. 101039090).

References

1. Ahmed, U.Z., et al.: Synthesizing tasks for block-based programming. In: NeurIPS (2020)
2. Ahmed, U.Z., Gulwani, S., Karkare, A.: Automatically generating problems and solutions for natural deduction. In: IJCAI (2013)
3. Akram, B., et al.: Automated assessment of computer science competencies from student programs with gaussian process regression. In: EDM (2020)
4. Alvin, C., Gulwani, S., Majumdar, R., Mukhopadhyay, S.: Synthesis of geometry proof problems. In: AAAI (2014)
5. Bunel, R., Hausknecht, M.J., Devlin, J., Singh, R., Kohli, P.: Leveraging grammar and reinforcement learning for neural program synthesis. In: ICLR (2018)
6. Cherenkova, Y., Zingaro, D., Petersen, A.: Identifying challenging CS1 concepts in a large problem dataset. In: SIGCSE (2014)
7. Cicirello, V.A.: On the role and effectiveness of pop quizzes in CS1. In: SIGCSE (2009)
8. CodeHS.com: CodeHS - Teaching Coding and CS. https://codehs.com/
9. CodeHS.com: Intro to Programming with Karel the Dog. https://codehs.com/info/curriculum/introkarel
10. Code.org: Code.org - Learn Computer Science. https://code.org/
11. Code.org: Hour of Code - Classic Maze Challenge. https://studio.code.org/s/hourofcode
12. Code.org: Hour of Code Initiative. https://hourofcode.com/
13. Cody, C., Maniktala, M., Lytle, N., Chi, M., Barnes, T.: The Impact of looking further ahead: a comparison of two data-driven unsolicited hint types on performance in an intelligent data-driven logic tutor. IJAIED (2021)
14. Cordova, L., Carver, J.C., Gershmel, N., Walia, G.: A comparison of inquiry-based conceptual feedback vs. traditional detailed feedback mechanisms in software testing education: an empirical investigation. In: SIGCSE (2021)
15. Crichton, W., Sampaio, G.G., Hanrahan, P.: Automating program structure classification. In: SIGCSE (2021)
16. Efremov, A., Ghosh, A., Singla, A.: Zero-shot learning of hint policy via reinforcement learning and program synthesis. In: EDM (2020)
17. Ene, A., Stirbu, C.: Automatic generation of quizzes for Java programming language. In: ECAI (2019)
18. Games, B.: Games for Tomorrow's Programmers. https://blockly.games/
19. Grover, S.: Toward a framework for formative assessment of conceptual learning in K-12 computer science classrooms. In: SIGCSE (2021)
20. Gulwani, S., Polozov, O., Singh, R.: Program synthesis. Found. Trends® Program. Lang. **4**(1–2), 1–119 (2017)
21. MacFarland, T.W., Yates, J.M.: Kruskal–Wallis H-test for oneway analysis of variance (ANOVA) by ranks. In: Introduction to Nonparametric Statistics for the Biological Sciences Using R, pp. 177–211. Springer, Cham (2016). https://doi.org/10.1007/978-3-319-30634-6_6
22. Marwan, S., Gao, G., Fisk, S., Price, T., Barnes, T.: Adaptive immediate feedback can improve novice programming engagement and intention to persist in computer science. In: ICER (2020)
23. McHugh, M.L.: The Chi-square test of independence. Biochemia Medica **23**, 143–149 (2013)
24. Paaßen, B., Hammer, B., Price, T.W., Barnes, T., Gross, S., Pinkwart, N.: The continuous hint factory - providing hints in continuous and infinite spaces. JEDM **10**, 1–35 (2018)
25. Pattis, R.E.: Karel the Robot: A Gentle Introduction to the Art of Programming. Wiley (1981)

26. Piech, C., Sahami, M., Huang, J., Guibas, L.J.: Autonomously generating hints by inferring problem solving policies. In: L@S (2015)
27. Polozov, O., O'Rourke, E., Smith, A.M., Zettlemoyer, L., Gulwani, S., Popovic, Z.: Personalized mathematical word problem generation. In: IJCAI (2015)
28. Price, T.W., Dong, Y., Lipovac, D.: iSnap: towards intelligent tutoring in novice programming environments. In: SIGCSE (2017)
29. Price, T.W., et al.: A comparison of the quality of data-driven programming hint generation algorithms. IJAIED **29**, 368–395 (2019)
30. Price, T.W., Williams, J.J., Solyst, J., Marwan, S.: Engaging students with instructor solutions in online programming homework. In: CHI (2020)
31. Price, T.W., Zhi, R., Barnes, T.: Evaluation of a data-driven feedback algorithm for open-ended programming. In: EDM (2017)
32. Price, T.W., Zhi, R., Barnes, T.: Hint generation under uncertainty: the effect of hint quality on help-seeking behavior. In: AIED (2017)
33. Resnick, M., et al.: Scratch: programming for all. Commun. ACM **52**, 60–67 (2009)
34. Rivers, K., Harpstead, E., Koedinger, K.R.: Learning curve analysis for programming: which concepts do students struggle with? In: ICER (2016)
35. Rivers, K., Koedinger, K.R.: Automating hint generation with solution space path construction. In: Trausan-Matu, S., Boyer, K.E., Crosby, M., Panourgia, K. (eds.) ITS 2014. LNCS, vol. 8474, pp. 329–339. Springer, Cham (2014). https://doi.org/10.1007/978-3-319-07221-0_41
36. Rivers, K., Koedinger, K.R.: Data-driven hint generation in vast solution spaces: a self-improving Python programming tutor. IJAIED **27**, 37–64 (2017)
37. Singh, R., Gulwani, S., Rajamani, S.: Automatically generating algebra problems. In: AAAI (2012)
38. Soltanpoor, R., Thevathayan, C., D'Souza, D.J.: Adaptive remediation for novice programmers through personalized prescriptive quizzes. In: ITiCSE (2018)
39. Wiese, E.S., Rafferty, A.N., Fox, A.: Linking code readability, structure, and comprehension among novices: it's complicated. In: ICSE (2019)
40. Wiese, E.S., Rafferty, A.N., Kopta, D.M., Anderson, J.M.: Replicating novices' struggles with coding style. In: ICPC (2019)
41. Wu, M., Mosse, M., Goodman, N.D., Piech, C.: Zero shot learning for code education: rubric sampling with deep learning inference. In: AAAI (2019)
42. Zhang, L., Li, B., Zhang, Q., Hsiao, I.: Does a distributed practice strategy for multiple choice questions help novices learn programming. iJET **15**, 234 (2020)
43. Zhi, R., Marwan, S., Dong, Y., Lytle, N., Price, T.W., Barnes, T.: Toward data-driven example feedback for novice programming. In: EDM (2019)
44. Zimmerman, K., Rupakheti, C.R.: An automated framework for recommending program elements to novices (N). In: ASE (2015)

Learning Profiles to Assess Educational Prediction Systems

Amal Ben Soussia(✉), Célina Treuillier(✉), Azim Roussanaly(✉), and Anne Boyer(✉)

Université de Lorraine, CNRS, LORIA, Campus scientifique, 54506 Vandoeuvre-lès-Nancy, France
{amal.ben-soussia,celina.treuillier,azim.roussanaly,anne.boyer}@loria.fr

Abstract. Distance learning institutions record a high failure and dropout rate every year. This phenomenon is due to several reasons such as the total autonomy of learners and the lack of regular monitoring. Therefore, education stakeholders need a system which enables them the prediction of at-risk learners. This solution is commonly adopted in the state of the art. However, its evaluation is not generic and does not take into account the diversity of learners. In this paper, we propose a complete methodology which objective is a more detailed evaluation of a proposed educational prediction system. This process aims to ensure good performances of the system, regardless of the learning profiles. The proposed methodology combines both the identification of personas existing in a learning context and the evaluation of a prediction system according to it. To meet this challenge, we used a real dataset of k-12 learners enrolled in a french distance education institution.

Keywords: Learning analytics · Assessment methodology · Risk prediction · Learning profiles · K-12 learners

1 Introduction

Nowadays, schools and universities are moving towards online learning due to the generalization of digital infrastructures and learning platforms which allow to better meet the needs of learners. However, this learning modality is facing many challenges, and the most widespread is the high failure rate among learners. This phenomenon is due to many reasons such as the large diversity of student profiles expressing different needs and requiring personalized support [23].

Virtual learning environments (VLE) store learner's online activity. The corresponding data, called learning traces, is very diversified and is used by Learning Analytics (LA) [19]. One aim among others is to provide educational stakeholders with intelligent technology-based solutions to help them in identifying at-risk of failure learners as early as possible. These solutions need to take into consideration all learners behaviors. Therefore, a major issue is: **does a system perform equally with all learners profiles?**

To answer this research question, we propose a methodology which is based on the identification of personas, defined as learners profiles representations [8,22]

M. M. Rodrigo et al. (Eds.): AIED 2022, LNCS 13355, pp. 41–52, 2022.
https://doi.org/10.1007/978-3-031-11644-5_4

and on the evaluation of model's performances for each persona. We illustrate the methodology on a case study (evaluation of a prediction model). To resume, our main contribution relies on a more precise evaluation of educational systems, taking into account the different learning profiles and based on a broad range of metrics. We proceeded according to the following steps:

- Given the disparity of available learning traces, we defined several learning indicators characterizing a learner's behavior. Then, we identified homogeneous groups of learners sharing similar behaviors according to these indicators. These learners groups are finally characterized into personas.
- We reviewed the existing assessment indicators and identified new ones to complete the evaluation.
- We conducted a precise evaluation on a specific use case relying on a weekly prediction approach.

We carried out our experimentation using real data of k-12 learners, enrolled in a French distance learning center (CNED). This institution is characterized by the multi-modality of learning and the total autonomy of its learners.

This paper is organized as follows. The Sect. 2 presents the general methodology and the used dataset. Section 3 and 4 present the first and second steps of the methodology respectively. The results of the evaluation are detailed in Sect. 5. A general conclusion and several perspectives are given in Sect. 6.

2 Evaluation Methodology

In this section, we start by describing the proposed the methodology and its different steps. Then, we present our case study for the experimental part.

2.1 Methodology Description

In order to achieve our assessment objective, the methodology is organized around three main steps (See Fig. 1):

1. **Identification of learner profiles** from learning. The profiles are characterized by personas, containing key information about learners' behaviors.
2. **Run of the prediction system** on the data and measurement of a complete set of metrics, containing both precision metrics, as accuracy, and new time-dependent ones (earliness, stability).
3. **Deeper evaluation of the system** according to the identified learners profiles and the various performance metrics.

2.2 Case Study

The case study concerns the k-12 learners enrolled in the physics-chemistry module during the 2017–2018 school year within the French center for distance education (CNED) [2]. It offers a large variety of fully distance courses to numerous

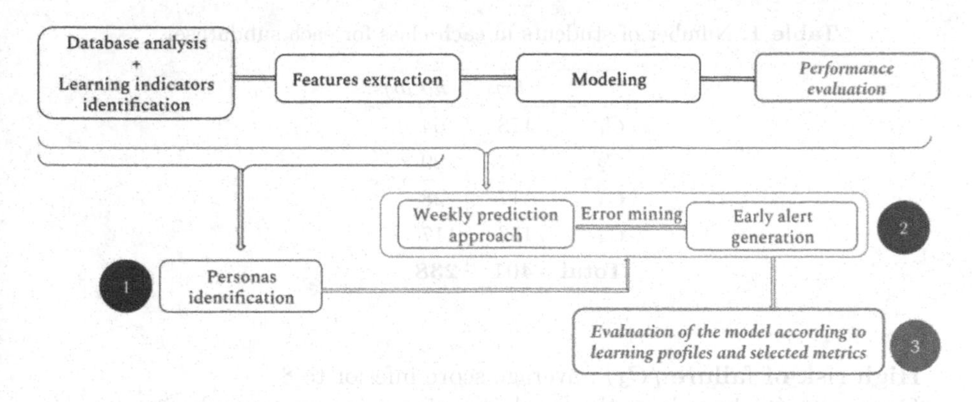

Fig. 1. The in-depth methodology phases.

physically dispersed learners. The courses contents are both available online and in printed papers which gives the learner the freedom to choose the learning mode which suits him/her the best. Given the large number of learners and the specificity of learning, it is highly time consuming for teachers to provide their students with an effective and personalized feedback.

2.3 Data Description

The learning traces are collected from two data sources. The first one is the Moodle platform, which generates the *logs* and the interaction traces between the learner and the learning content. The second platform is GAEL, which is a management system where all performance data, including grades, are stored. In CNED, learners don't start the school year at the same time t_0 [2]. We select learners with t_0 between Sept. 1st and Oct. 31st, as they share similar learning paths and characteristics. According to this information, our database gathers learning traces of 639 learners. The learning period of the physics-chemistry course is 300 d, during which 6 exams could be submitted. On average, learners only submit around 4.51 assignments. The average mark on the submitted exams is 13.73. However, if we consider setting the grade of 0 to the unsubmitted assignments, this average is lowered to 10.21. The bi-modality (digital or paper-based) of the learning makes the study of the dataset difficult. Indeed, learners who use the course exclusively in paper format may not produce any logs and it is therefore not relevant to compare them with active learners on the VLE. In our dataset, we noticed that 37.25% of the population never logged in. To handle this particularity, the dataset was divided into two subdatasets: one containing data about learners who made at least one log on the VLE, and the other one about those who have never logged in. Finally, the learners were classified into 4 classes according to their mean performance:

- **Success** (C_1) : average score superior to 12.
- **Medium risk of failure** (C_2) : average score between 8 and 12.

Table 1. Number of students in each class for each subdataset.

	logs	*no_logs*
C_1	178	64
C_2	53	29
C_3	17	28
C_4	153	117
Total	**401**	**238**

- **High risk of failure** (C_3) : average score inferior to 8.
- **Drop out** (C_4) : at least the two last assignments are not submitted.

The Table 1 summarizes the number of learners from *logs* and *no_logs* subdatasets belonging to each class. The process of identifying learners profiles within these classes is described in the following section.

3 Methodology Step 1: Identification of Learner Profiles

3.1 State of the Art

Learners' behaviors are observed through their online learning traces. In LA, multiple studies exploit this data to compare learners based on various indicators, such as engagement [11], performance [4] or regularity [7]. In our context, learning behaviors need to be described according to a set of indicators, allowing a more detailed characterization of learners [22,24]. For this reason, we define learners personas corresponding to typical learners identified through Machine Learning classification processes [8]. In one hand, the identification of such personas enables a more precise description of the corpus, especially in terms of learners profiles representation. In another hand, these personas meet the need for an ethical learning analytics implementation [20], and ensure fair support between learners and provide useful tools to the field stakeholders who need to help their learners with equal support [10]. However, the diversity of the available data makes the task tricky: the variety of recorded data does not allow for the same indicators to be computed all the time. The indicators we calculated for the case study are described in the following subsection.

3.2 Study and Selection of Learning Indicators

Learning traces available in the CNED dataset are diverse and contain both logs and performance data. This data was first used to define five absolute indicators (e.g. calculated for each learner):

- **Engagement** reflects the learner's activity on the VLE (logs) [11].
- **Regularity** translates the learner's constancy of connection between the beginning and the end of the course (frequency of connection) [2].

- **Curiosity** expresses the intrinsic motivation of the learner to consult various educational resources (variety of accessed content) [17].
- **Performance** corresponds to learner's scores in the exams.
- **Reactivity** provides information about learner's responsiveness during course-related events (timeliness of the assessments)[7].

To go further, we completed these absolute indicators with a set of **relative indicators**. The average of each indicator is computed for all learners, and the associated relative indicator gives information about the behavior of a specific learner profile comparing to his/her peers (negative or positive difference in relation to the rest of the group). Both types of indicators were used as a basis for the identification of learners profiles, described in the following subsection.

Obviously, engagement, curiosity and regularity (on the VLE) indicators, based on the logs, were not computed for the *no_logs* subdataset as associated learners have never logged in.

3.3 Identification of Learners Profiles

For each subdataset, the study of learning indicators enables the identification of learners profiles. These profiles correspond to homogeneous subsets of learners, sharing similar behaviors, and are identified through different steps:

- **Data-preprocessing:**
  - *Data normalization:* use of the RobustScaler[1] method (ScikitLearn [15]) to improve the model's performance.
  - *Outliers identification:* Use of the IsolationForest[2] algorithm [14] to set apart the atypical data and increase model's performance. This step is crucial because outliers' atypicity does not allow them to be associated with other students.
- **Identification of homogeneous groups of learners:** k-means Algorithm [13] is used to identify homogeneous groups of learners. Results are evaluated with Silhouette analysis [18] and Davies-Bouldin Criterion [9]. We run the algorithm with values from 2 to 15 and selected the one giving the best performance.
- **Description of learners profiles:** each of the identified clusters are then characterized by a size (number of associated students), its proportion in dataset to which it belongs and a set of learning indicators. Outliers are not discarded but are studied individually.

Applying this methodology, we identified 12 outliers among the 639 learners. Each of the remaining learners is associated to one of the 21 identified learners profiles. Some statistics are given in Table 2.

---

[1] https://scikit-learn.org/stable/modules/generated/sklearn.preprocessing.RobustScaler.html.

[2] https://scikit-learn.org/stable/modules/generated/sklearn.ensemble.IsolationForest.html.

Table 2. Clustering results by subdatasets and classes.

	Logs				No_logs			
	C1	C2	C3	C4	C1	C2	C3	C4
Number of inliers	176	52	16	151	63	28	27	115
Number of outliers	2	1	2	2	1	1	1	2
Optimal value of k	2	2	2	2	3	3	3	4
Silhouette index	0,44	0,30	0,85	0,28	0,43	0,36	0,36	0,34
Davies-bouldin index	1,00	1,45	0,07	1,32	0,85	0,96	0,92	1,04

3.4 Personas: Examples

Each persona contains a large variety of information: narrative description of the learning behavior, its proportion in the dataset it belongs to, visual indicators of the risk of failure, and learning modality (See Fig. 2).

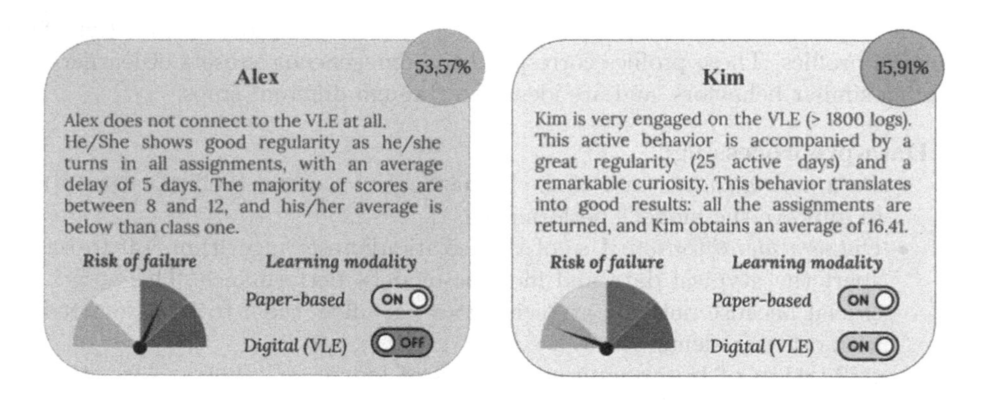

Fig. 2. Example of two personas.

The utility of such personas is threefold. In addition to providing valuable information about learners behaviors, they contribute to the improvement of the results interpretation of a LA system. Finally, they are particularly interesting for our study because they can be used to refine the evaluation of an educational system. The results presented in the Sect. 5 confirm this last point.

4 Methodology Step 2: Earliness and Stability Measurements

In addition to the usual performance measures, this section defines new metrics for a deeper evaluation of an educational prediction system. These metrics consider the importance of the temporal evolution of the prediction.

4.1 State of the Art

The main objective of the majority of educational prediction systems is the early identification of at-risk of failure or dropout learners. Static and precision Machine Learning (ML) metrics such as accuracy are mainly used to evaluate the performance of educational prediction systems. For example, [5] studied the accuracy of early warning system (EWS) on identifying at-risk students in a real educational setting. The study of [12] aimed to improve the performance of a dropout EWS by evaluating the trained classifiers with both receiver operating characteristic (ROC) curves and precision-recall (PR) curves. [3] compared the performance of a developped EWS on two different subjects based on the accuracy, the true negative rate (TNR) and the true positive rate (TPR) measures. [1] compares the performance of different ML model in analyzing the problems faced by at-risk learners enrolled in online university. This performance assessment is based on accuracy, precision, recall, support and f-score results. The majority of education prediction systems uses static and precision ML metrics for performance evaluation. However, both learning and prediction are time-evolving. Consequently, we need to consider the temporal dimension in the performance measures and illustrate the evolution of the whole process over the learning period. For this aim, we propose new metrics to evaluate the prediction and which the definition is based on the regular tracking of the prediction results.

4.2 Metrics Description

Prediction Earliness: Researchers work on providing stakeholders with the most accurate prediction results. A common theoretical definition of the early prediction is the right time to identify at risk learners. The earliness of the right prediction depends always on the studied context. We propose to measure the earliest time to predict as accurate as possible the classes of learners. We define the earliness of prediction as the mean time from which we start to correctly predict the learners classes [6]. While defining this measure, we focus on at-risk learners to best respond to the objectives of our study.

Prediction Stability: Stability is usually related to small changes in system output when changing the training set [16]. In our context, we are interested in temporal stability referring to the capacity of a classifier to give the same output over time when training the same dataset [21]. We measure temporal stability as the average of the longest sequences of successive right predictions [6].

5 Methodology Step 3: In-Depth Evaluation of a Prediction System

This section presents the whole methodology from the prediction system description to the modeling and assessments steps. It ends up by a comparative study based on the obtained results.

5.1 Short Description of the Prediction Model

Our system is based on a weekly prediction model of at-risk of failure or dropout learners. As explained, learners of the cohort are classified into four classes. First, we went through both processes of features extraction and selection. Going through these processes is important to select the activity features most correlated to the learner's final result as well as to minimize noise in the model. Thus, each week w_i, a learner is represented by a vector X composed of features going from f_1 to f_n and the class y to which he belongs to. Each learner belongs to one and only class over the year.

$$X = < f_1, f_2, ..., f_n, y >$$

Each feature f_1 to f_n represents one learning activity till the prediction time w_i. For each prediction time w_i, the value of one feature is added to that of prediction time w_{i-1}: we proceed to an accumulation of values. Based on the accuracy results of [2], we use the Random Forest (RF) as a ML model for our system.

5.2 Results

In the first evaluation phase, we divided the test dataset population into two groups *(logs, no_logs)* as explained in the Sect. 2.3. In this experimental part, we report on the results of 3 metrics: *accuracy*, *earliness* and *stability.*

Accuracy Analysis. The curves of the Fig. 3 show a difference in the accuracy between the test dataset of the total population, *logs* and *no_logs* subdatasets. Indeed, we notice that until almost the week 15, classes of learners who belong to the *no_log* group are the best predicted. In fact, the dropout class is the most predictable one and is highly represented in the *no_log* subdataset (cf. Table 1). However, the further we advance in the school year, the more the prediction results of *logs* and *no_logs* converge towards almost the same values.

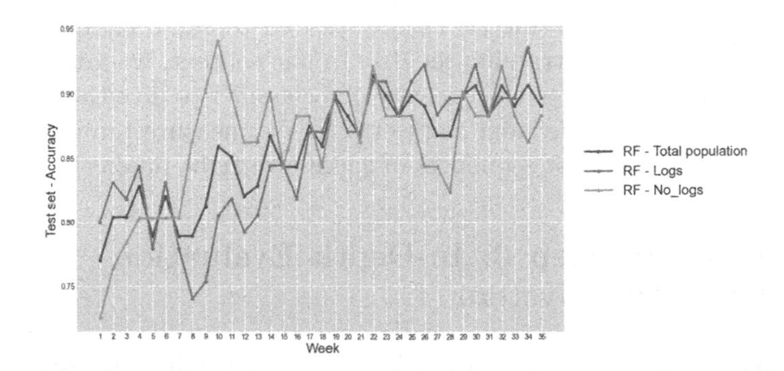

Fig. 3. Accuracy evaluation with total population and the two subdatasets (*logs*, *no_logs*).

The Fig. 4 shows the curves of the evolution of the accuracy of the personas identified in *logs* (A) and *no_logs* (B) subdatasets. To ensure the figure lisibility, we only present the results of one persona per class and by subdataset: personas 1, 4, 5 and 8 were selected for the *logs* subdataset, and personas 10, 13, 16 and 20 were selected for the *no_logs* subdataset. From the different curves, we can clearly notice that personas belonging to the same profile group do not have the same prediction accuracy. Differently from the results shown in Fig. 3, even at the end of the learning period, the accuracy curves do not converge towards a same value for all the personas.

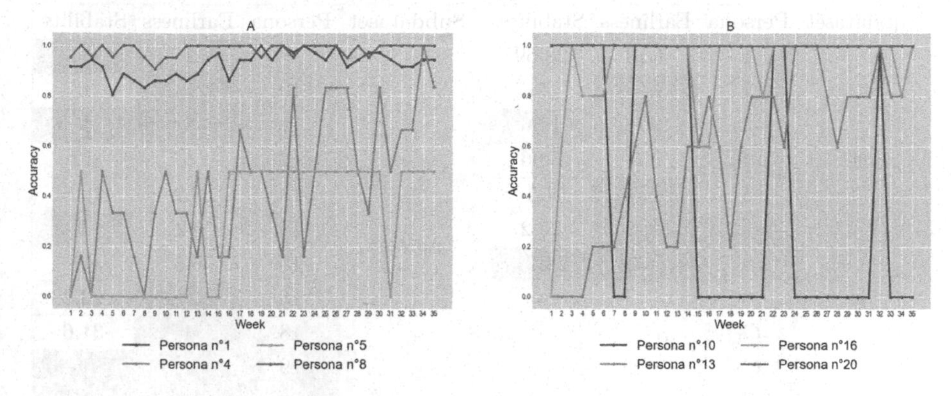

Fig. 4. Accuracy evolution of personas of *logs* (A) and *no_logs* (B) subdatasets.

Earliness and Stability Analysis The Table 3 shows the results of earliness and stability metrics of the test dataset, *logs* and *no_logs* subdatasets. We can notice that both *logs* and *no_logs* have different values for the earliness. Furthermore, we can see from this table that the stability performances of the system are different from one profile group to another. In addition, whatever the subdataset is, the algorithm has the same stability and earliness performance for each class. Thus, the dropout class has always the best metrics values, whereas the medium risk class has the worst results.

Table 3. Earliness and Stability measurement of each class of a profile group.

	Total		*logs*		*no_logs*	
	Earliness	Stability	Earliness	Stability	Earliness	Stability
Dropout	1.01	31.44	1.03	31	1	31.88
High Risk	3	16.55	1	7.5	3.57	19.14
Medium Risk	8.06	6.62	6.77	4.33	8.28	9.57
Success	1.1	28.38	1.12	28.69	1	28.6
Total	2.06	25.35	1.75	26.10	2.35	26.43

The Table 4 shows the results of the earliness and stability metrics of each persona belonging to *logs* or *no_logs* subdatasets. We notice that the measures are different from one persona to another. In addition, the difference is even more tangible when it comes to the personas of medium *(C_2, in pink in Table 4)* and high risk learners *(C_3, in yellow in Table* 4). Due to the lack of pages, we cannot report all the results: we only present a selection in order to illustrate the kind of results that we can provide with the presented methodology.

Table 4. Earliness and stability for each persona.

Subdataset	Persona	Earliness	Stability
logs	1	1.13	28.69
	2	1	26
	3	12.66	1.66
	4	5.5	5.66
	5	1	7.5
	7	1	25.2
	8	1.04	32.31
no_logs	9	1	35
	10	1	6
	11	1	35
	12	5	12
	13	8.2	10.8
	14	12	1
	15	6.5	8.5
	16	24	23.2
	18	1	31.6
	19	1	35
	20	1	35
	21	1	33.16

Legend
C_1
C_2
C_3
C_4

5.3 Discussion

The previous tables and figures showed that the prediction algorithm out performs globally (up to 93% of accuracy). However, the prediction algorithm doesn't exhibit the same performance with each learner profile. For example, the successful learners and those who dropout are much better predicted than those who are at-risk of failure. In addition, learners who belong to the *log* group are also more accurately predicted. Earliness and stability results show that the algorithm performance is dependent on the learners profiles. In order to provide education stakeholders with accurate and reliable results over time, the prediction system has to take into consideration the different learning profiles existing within a cohort.

6 Conclusion and Perspectives

The identified learners profiles, characterized by personas, within our dataset were diversified and confirmed that learners adopt different behaviors and must receive an adapted support. In addition, the prediction model evaluation reveals that the algorithm's performances were not the same for all personas and classes.

The obtained results answer our research question and confirm the interest of personas in LA tools assessment. Furthermore, indicators such as earliness and stability, which have been introduced, give information about the confidence that a user can have in the system. Indeed, the usual accuracy metrics are insufficient to evaluate the weekly results of an educational prediction system. It's a reason why, we plan to investigate several research directions relying either on personas or on new refinements in LA assessment. First, we believe that the personas identified in year N could also be used as a basis for evaluating classes in year N+1, assuming that the behaviors observed from one year to the next are similar. This research context deserves attention because it would help to provide quick feedbacks for teachers about their learners' situations. This early information could help them to promptly develop solutions for students considered at risk. Secondly, we wonder how much the separation of the initial dataset according to learning modalities (*logs*, *no_logs*) and classes (C_1, C_2, C_3, C_4) influences the performance of the learning systems, and particularly of the prediction system in our case study. Therefore, it seems interesting to compare the results with different partitions of the dataset. In one hand, it could allow to highlight the key features which are essential to the good functioning of the model. In another hand, this would further improve the explainability by allowing teachers and academics to select the appropriate partition according to their pedagogical objectives. Finally, both indicators presented (earliness and stability) provided additional information about systems' behavior. In that way, a further work on these indicators and especially on their generalization seems to be necessary, so that they can be used in a wider range of areas.

References

1. Adnan, M., et al.: Predicting at-risk students at different percentages of course length for early intervention using machine learning models. IEEE Access **9**, 7519–7539 (2021)
2. Ben Soussia, A., Roussanaly, A., Boyer, A.: An in-depth methodology to predict at-risk learners. In: De Laet, T., Klemke, R., Alario-Hoyos, C., Hilliger, I., Ortega-Arranz, A. (eds.) EC-TEL 2021. LNCS, vol. 12884, pp. 193–206. Springer, Cham (2021). https://doi.org/10.1007/978-3-030-86436-1_15
3. Anjeela Jokhan, B.S., Singh, S.: Early warning system as a predictor for student performance in higher education blended courses. Stud. High. Educ. **44**(11), 1900–1911 (2018)
4. Arnold, K.E., Pistilli, M.D.: Course signals at purdue: using learning analytics to increase student success. In: Proceedings of the 2nd International Conference on Learning Analytics And Knowledge. pp. 267–270 (2012)
5. Bañeres, D., Rodríguez, M.E., Guerrero-Roldán, A.E., Karadeniz, A.: An early warning system to detect at-risk students in online higher education. Appl. Sci. **10**(13), 4427 (2020)
6. Ben soussia, A., Labba, C., Roussanaly, A., Boyer, A.: Assess performance prediction systems: Beyond precision indicators. In: CSEDU (2022)

7. Boroujeni, M.S., Sharma, K., Kidziński, Ł, Lucignano, L., Dillenbourg, P.: How to quantify student's regularity? In: Verbert, K., Sharples, M., Klobučar, T. (eds.) EC-TEL 2016. LNCS, vol. 9891, pp. 277–291. Springer, Cham (2016). https://doi.org/10.1007/978-3-319-45153-4_21
8. Brooks, C., Greer, J.: Explaining predictive models to learning specialists using personas. In: Proceedins of the Fourth International Conference on Learning Analytics And Knowledge - LAK 2014, pp. 26–30. ACM Press (2014). https://doi.org/10.1145/2567574.2567612,http://dl.acm.org/citation.cfm?doid=2567574.2567612
9. Davies, D.L., Bouldin, D.W.: A cluster separation measure. IEEE Trans. Pattern Anal. Mach. Intell. PAMI **1**(2), 224–227 (1979)
10. Holmes, W., et al.: Ethics of AI in education: Towards a community-wide framework. Int. J. Artifi. Intell. Educ. 1–23 (2021)
11. Hussain, M., Zhu, W., Zhang, W., Abidi, S.M.R.: Student engagement predictions in an e-learning system and their impact on student course assessment scores. In: Computational Intelligence and Neuroscience (2018)
12. Lee, S., Chung, J.Y.: The machine learning-based dropout early warning system for improving the performance of dropout prediction. Appl. Sci. **9**(15), 3093 (2019)
13. Likas, A., Vlassis, N., Verbeek, J.: The global k-means clustering algorithm. Pattern Recogn. **36**, 451–461 (2003). https://doi.org/10.1016/S0031-3203(02)00060-2
14. Liu, F.T., Ting, K.M., Zhou, Z.H.: Isolation forest. In: 2008 Eighth IEEE International Conference on Data Mining, pp. 413–422 (2008). https://doi.org/10.1109/ICDM.2008.17, ISSN: 2374-8486
15. Pedregosa, F., et al.: Scikit-learn: machine learning in Python. J. Mach. Learn. Res. **12**, 2825–2830 (2011)
16. Philipp, M., Rusch, T., Hornik, K., Strobl, C.: Measuring the stability of results from supervised statistical learning. J. Comput. Graph. Stat. **27**(4), 685–700 (2018)
17. Pluck, G., Johnson, H.L.: Stimulating curiosity to enhance learning. GESJ: Educ. Sci. Psychol. **2**(19) (2011). ISSN 1512-1801
18. Rousseeuw, P.J.: Silhouettes: a graphical aid to the interpretation and validation of cluster analysis. J. Comput. Appli. Mat. **20**, 53–65 (1987). https://doi.org/10.1016/0377-0427(87)90125-7
19. Siemens, G., Long, P.: Penetrating the fog: analytics in learning and education. EDUCAUSE Rev. **46**(5), 30 (2011)
20. Slade, S., Prinsloo, P.: Learning analytics: ethical issues and dilemmas. Am. Behav. Sci. **57**(10), 1510–1529 (2013). https://doi.org/10.1177/0002764213479366
21. Teinemaa, I., Dumas, M., Leontjeva, A., Maggi, F.M.: Temporal stability in predictive process monitoring. Data Min. Knowl. Disc. **32**(5), 1306–1338 (2018). https://doi.org/10.1007/s10618-018-0575-9
22. Treuillier, C., Boyer, A.: Identification of class-representative learner personas. In: Learning Analytics for Smart Learning Environments (LA4LSE) Workshop, at EC-TEL 2021 (2021)
23. Xu, D., Jaggars, S.S.: Performance gaps between online and face-to-face courses: differences across types of students and academic subject areas. J. High. Educ. **85**(5), 633–659 (2014). https://doi.org/10.1080/00221546.2014.11777343
24. Zhang, N., Biswas, G., Dong, Y.: Characterizing students' learning behaviors using unsupervised learning methods. In: André, E., Baker, R., Hu, X., Rodrigo, M.M.T., du Boulay, B. (eds.) AIED 2017. LNCS (LNAI), vol. 10331, pp. 430–441. Springer, Cham (2017). https://doi.org/10.1007/978-3-319-61425-0_36

Identifying Student Struggle by Analyzing Facial Movement During Asynchronous Video Lecture Viewing: Towards an Automated Tool to Support Instructors

Adam Linson[1(✉)], Yucheng Xu[2], Andrea R. English[3], and Robert B. Fisher[2]

[1] School of Computing and Communications, Open University, Edinburgh EH3 7QJ, UK
adam.linson@open.ac.uk

[2] School of Informatics, University of Edinburgh, Edinburgh EH8 9LE, UK
{yucheng.xu,r.b.fisher}@ed.ac.uk

[3] Moray House School of Education and Sport, University of Edinburgh, Edinburgh EH8 8AQ, UK
andrea.english@ed.ac.uk

Abstract. The widespread shift in higher education (HE) from in-person instruction to pre-recorded video lectures means that many instructors have lost access to real-time student feedback for the duration of any given lecture (a 'sea of faces' that express struggle, comprehension, etc.). We hypothesized that this feedback could be partially restored by analyzing student facial movement data gathered during recorded lecture viewing and visualizing it on a common lecture timeline. Our approach builds on computer vision research on engagement and affect in facial expression, and education research on student struggle. Here, we focus on individual student struggle (the effortful attempt to grasp new concepts and ideas) and its group-level visualization as student feedback to support human instructors. Research suggests that instructor supported student struggle can help students develop conceptual understanding, while unsupported struggle can lead to disengagement. Studies of online learning in higher education found that when students struggle with recorded video lecture content, questions and confusion often remain unreported and thus unsupported by instructors. In a pilot study, we sought to identify group-level student struggle by analyzing individual student facial movement during asynchronous video lecture viewing and mapping cohort data to annotated lecture segments (e.g. when a new concept is introduced). We gathered real-time webcam data of 10 student participants and their self-paced intermittent click feedback on personal struggle state, along with retrospective self-reports. We analyzed participant video with computer vision techniques to identify facial movement and correlated the data with independent human observer inferences about struggle-related states. We plotted all participants' data (computer vision analysis, self-report, observer annotation) along the lecture timeline. The visualization exposed group-level struggle patterns in relation to lecture content, which could help instructors identify content areas where students need additional support, e.g. through student-centered interventions or lecture revisions.

Keywords: Video analysis · Data visualization · Facial expression · Student struggle · Reflective teaching · Human-centered computing

M. M. Rodrigo et al. (Eds.): AIED 2022, LNCS 13355, pp. 53–65, 2022.
https://doi.org/10.1007/978-3-031-11644-5_5

1 Introduction

1.1 Project Description

When instructors in higher education (HE) deliver in-person lectures, they can access a form of real-time student feedback simply by 'reading the room'. That is, by scanning their students' nonverbal cues, instructors can make inferences on the assumption that the 'sea of faces' may express various cognitive and affective states relevant to group instruction, such as struggle and comprehension. However, the widespread shift in HE from in-person instruction to pre-recorded video lectures means that many instructors have lost access to this form of feedback. We sought to test the hypothesis that this feedback could be partially restored by analyzing student facial movement data gathered during recorded lecture viewing and visualizing it on a common lecture timeline. To this end, we developed a software prototype, PUZZLED, inspired by current research in education, computer vision, and data visualization. The system was designed to identify when and in what respect students are struggling, and to analyze and visualize results to provide insights to human instructors.

In this report, we describe the prototype design and piloting in a small-scale exploratory and feasibility study (N = 10), funded by the University of Edinburgh Regional Skills program. As a key contribution of this paper, the study showed that visual evidence can be extracted from video of student facial movement (e.g. eye gaze aversion) that aligns temporally with aspects of a corresponding viewed lecture video. That is, moments in the lecture video containing conceptually challenging content, omitted background information, or other difficulties posed to student viewers (e.g. blurry text) led to measurable student facial movements (e.g. expression changes). These contrasted with student facial movement data captured during introductory or otherwise straightforward segments of the lecture video.

A second key contribution of this paper relates to the visualization of the data. Current data visualization interfaces for time-based models are primarily anchored in either absolute time (e.g. audience feedback during an in-person lecture, which uses global timestamps), or abstract task time (e.g. time solving a problem, which is averaged across individuals). Here, we integrate both by using global timestamps indexed against a common reference timeline (feedback from asynchronous viewings of a video lecture). This timeline is used to visualize group-level feedback from a student cohort at a granular level. In this application, the visualization can inform instructors about how student struggle relates to specific segments of the lecture.

1.2 Research Context in Education

Research on student learning, including in HE, suggests that struggle – the effortful attempt to grasp new concepts and ideas – is important to the learning process [1–5]. More precisely, there is differentiation between 'unproductive' struggle, such as unresolved confusion that leads to task disengagement, and 'productive' struggle, as when a challenging task is cognitively engaging. Productive struggle is important for the development of students' critical thinking and deep understanding [6–9, see also 10]. It can also indicate an appropriate level of challenge that maintains learner engagement, a key factor in HE student retention [11, 12]. In addition, research has found that equity-oriented teaching both successfully challenges all students to engage in struggle, and also supports all students through struggle [3, 5].

In online HE instruction, however, research indicates that when students struggle with recorded video lecture content (when a question or confusion arises), it often remains unreported, and thus unsupported by instructors, leading to lower student engagement and greater attrition, relative to in-person courses [13, 14]. A growing number of students are negatively impacted by this problem, given that recorded lectures are now a "mainstream" part of online HE provision [15]. 94% of UK universities make recorded lectures available to students year-round, in part "as a catalyst for inclusivity" [16]. In addition, European HE reform has promoted the expansion of e-Learning provision, including online recorded lectures (75% of universities), as a means of "widening access" to HE, while acknowledging that quality "teacher support" is needed to maintain learning with "critical thinking" and "deep understanding" in online contexts [17, 18].

In principle, instructors can only effectively support student struggle in online courses of students who self-report. This poses a serious problem, as the move to online HE places more demands on students to make their struggle known to instructors. Yet students who do not have knowledge of the subject area, and/or are not skilled in self-regulated learning, are less likely to self-assess and self-report their difficulties [19]. Additional research suggests that self-reporting may privilege the subset of students who are vocal about their struggles [20]. Thus, many students' learning needs remain unsupported in relation to recorded lecture content online [13]. Overall, this situation highlights the tension between the benefits and drawbacks of online learning and its constitutive technologies [21].

1.3 Research Context in Informatics

A range of literature has used estimations of student engagement or affect, primarily using well-developed standards for assessing facial action units, along with self- and/or observer-coded higher-level states correlated with quantitative face and postural data [10, 22–25]. We take inspiration from these methods and bellwethers of feasibility. To our knowledge, other studies using computer vision related analysis of student faces involve participants interacting with virtual tutors or games, but not pre-recorded lecture videos with human instructors, as we have done.[1]

The choice of educational material 'delivery mode' in our study (i.e. lecture videos with a human instructor) can be understood in relation to our primary aim of supporting human instructors, as compared to the aims of similar studies that seek to investigate learning itself or to enhance virtual tutors. For our purposes, rather than seeking to determine whether a student is 'really' struggling, we instead focus on providing inferences about the student cohort that could motivate an instructor to re-evaluate aspects of their lecture (e.g. content, delivery, pre-requisites, etc.) or provide corresponding non-lecture-based student support (e.g. forums, further readings, etc.).

[1] See our source video lectures (1a) and (1b), which played sequentially without interruption: (1a) https://media.ed.ac.uk/media/1_yd7krol3 (1b) https://media.ed.ac.uk/media/1_ktjie97s.

In educational technology research that does not involve face data, there are related applications of data visualization that aim to provide insights to instructors. We draw from an example that uses a horizontal time axis and vertical axis of individual students to surface salient patterns within a vast and complex student cohort (e.g. in a MOOC) [13]. We also build directly on a concept for annotating synchronously viewed content. The latter implementation uses crowd-sourced structured tagging (live audience feedback) that is visualized on a horizontal time axis, aligned with a video replay interface [21, 26].

Outside of education, computer vision research has sought to measure affective indicators in participant video under 'real world' capture conditions. In an approach related to ours, independent algorithms for classifying events in multiple low-level data streams (face, posture, etc.) are followed by fusion and high-level affect classification [27]. Our approach also relates to research on stress using multimodal biosignal information extracted from videos [28], in that we aspired to improve the accuracy of student struggle estimation by correlating low-level objective measures (e.g. gaze direction) with real-time and retrospective self-report. (Similar classification techniques have been used in education research with sophisticated instruments in controlled environments [10], in contrast to our use of webcams in everyday locations).

1.4 Research Context on Facial Expressions and Eye Gaze

Overwhelmingly, research on eye gaze centers on visual content fixation, to understand how people look at text, mathematical formulas, videos, interfaces, etc. In contrast, a smaller body of research considers eye gaze as an indicator of affective or cognitive state, primary in terms of whether or not gaze is averted. In computer vision affect detection, gaze aversion metrics have been used to improve automated classification of emotional state [29]. In developmental psychology, research has confirmed that young children judge faces to be engaged in thinking when viewing photos with subjects averting their gaze [30]. (The study uses a similar approach to ground truth as the present one, in that cognitive-affective states are inferred from images by independent raters). Educational psychology research has also identified gaze aversion in young children as indicative of thinking; this is suggested as a cue to instructors who must gauge how soon to expect a verbal response following a question [31]. Research on inferring mental states from gaze aversion supports the idea that disengaging from perceptual demands (such as looking at an instructor) facilitates thinking [32].

2 Methodology

2.1 Participants and Study Design

Following ethics approval, we recruited university students (N = 10) from a UK postgraduate degree program in informatics. We offered a £10 voucher in exchange for roughly 45 min of participation. 7 female and 3 male volunteers responded with informed consent (demographic data was self-reported in free text). 90% of ages fell in the range of 22–29, plus one 37-year-old participant, and ethnicity was entered as South Korean (1), Chinese (2), Indian (2), or white/Caucasian (5). 70% of participants had subject matter experience ranging from 1–4 years; a further two had 0.25 years, and one had 8 years. 80% selected a multiple-choice answer to report having a "little" experience learning through video lectures (20% chose a "lot", and none chose "it's new to me", the sole remaining option).

We invited the participants via an emailed link (with an anonymous unique identifier) to start and complete their session in a single sitting. The web interface provided instructions for positioning the webcam using a live feed from their own device ("ensure that your face is in the middle of the image"). No webcam video was displayed during recorded lecture viewing. A short 3-min practice session was offered with a separate lecture not used for the study, to provide experience viewing and using the click feedback interface, which offered the following button options and instructions:

- "Feels easy" - *click when you think the content is easy for you to understand*
- "Feels challenging" - *click when you are able to follow the content, but it is not too easy for you*
- "I'm lost" - *click when you cannot follow at all what the lecturer is talking about*

The click feedback interface could be repositioned from left-to-right, while remaining in a fixed row beneath the lecture video (Fig. 1). A further instruction stated "*When viewing the lecture, as your feeling changes, continue to click these buttons, to show how you feel*". Following the practice lecture video, the participants were shown two different lecture videos (ca. 3 min + ca. 6 min = total ca. 10 min) from an existing MSc computer vision course taught by the last author. The lecture videos played back-to-back automatically, while participant webcam and click feedback was captured in real-time during viewing. Participants were not informed that the lectures were not originally intended for consecutive viewing, as the second lecture summarized a previous lecture not shown to participants. This provided an experimental control indicator of 'challenged' reactions not related to challenging subject matter. Upon completion, an 'exit survey' was provided for retrospective self-report on the lecture material and feedback on the study interface design.

There were some problems with both the interface code (cross-browser compatibility) and comprehensibility of the instructions to participants. The data collection succeeded as intended for participants 1–5 (50%). Participant 6 successfully provided webcam data without click feedback, and we did not request a repeat session. For participants 7–10, we requested a repeat session; 7–9 succeeded in their second session and participant 10 succeeded in a third session. Thus, data on 40% of participants did not relate to their first viewing of the lecture materials. However, for an exploratory pilot and feasibility study, this was not a serious obstacle to completing its objectives.

Our hypothesis was that granular student cohort feedback on video lecture materials could be provided to instructors by students who viewed videos asynchronously. We sought to analyze student facial movement data gathered during recorded lecture viewing, and to visualize the cohort feedback on a common lecture timeline. Our expectation was that computer vision techniques could identify positional changes in head orientation and eye gaze direction, along with facial expression changes indicating struggle, and that these could be correlated with independent observations inferring struggle-related states, to provide a pathway to increasingly automated recognition. With all participants' data (computer vision analysis, self-report, observer annotation) plotted along the lecture timeline, evidence for our hypothesis would be found if group-level patterns emerged in relation to segments of lecture content. A toolchain of data processing and visualization steps would then provide an overview of the student feedback in relation to the lecture material. This could help instructors identify content that corresponded to a potential requirement for further student support.

2.2 Data Analysis Techniques

Facial feature analysis was performed with OpenVINO, using models adapted from Open Model Zoo.[2] Pre-existing trained models for facial landmark localization, face detection, gaze estimation, head pose estimation, and eyes open-or-closed state were integrated with PUZZLED, to analyze participant video frames for eye gaze direction and head orientation. While further analysis is needed to uncover potentially relevant patterns in head orientation, we found a notable correlation between eye gaze direction (aversion) and possible student struggle (see context in Sect. 1.4 and results in Sect. 3).

We normalized and smoothed out eye gaze direction to indicate a baseline bandwidth, above which was classified as upward gaze and below as downward. We then filtered out gaze direction data within the baseline band, treating it as direct video viewing (including left-right patterns of reading on-screen text). We also filtered out downward gaze data, since our self-report click feedback interface was below the video (Fig. 1), and participants appeared to be saccading to the interface when contemplating or performing click feedback. The remaining upward gaze data was plotted against real-time and retrospective self-report and observer annotations. Notably, we did not treat it as a universally valid measure of struggle (see below).

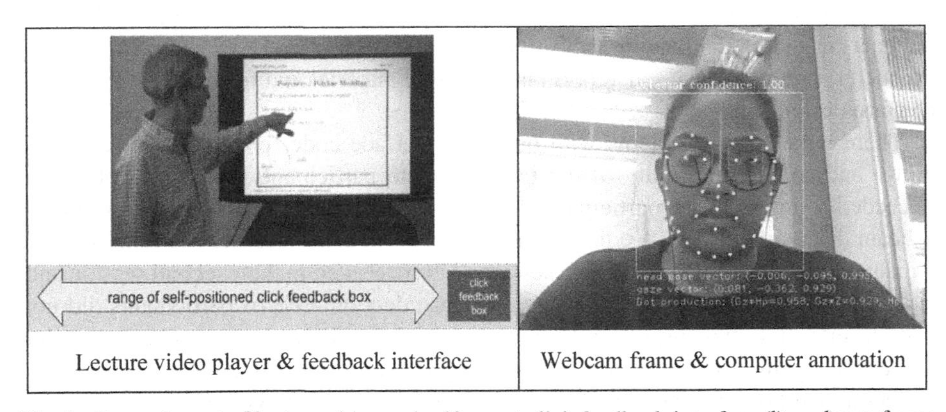

Fig. 1. Frame layout of lecture video and self-report click feedback interface (l); webcam frame of face with features detected and gaze direction (r). A planned production version would not remotely transmit facial images, only non-visual data from webcam analyses performed locally.

[2] https://docs.openvino.ai/latest/omz_demos.html

Individual Classifications

Our approach is inclusive, in that it does not depend on generalizations across ethnicity, culture, gender, etc., and can even remain robust with respect to individual difference. For example, for participant '052', upward gaze likely indicates struggle, whereas for '409', it does not. This classification for '052' is in part established through correlations with the other within-subject data points (e.g. retrospective self-report that the first lecture was challenging and the second was not, and real-time self-report of challenge coinciding with independent observer annotations of challenge).

Cohort Classifications

Similar to related work [33], which uses a percentage of frames with a target classification ('stress') in a time window to generalize the classification to the window (stress segment), we use a relative threshold of events with a target classification ('struggle') within and across students to identify a relevant lecture segment in which students struggled. In our case, we allow for sparse target classification events in different data streams (Fig. 2): individual students (grey horizontal bars) self-reports of being challenged (blue stars), computer-detected upward gaze aversion events (cyan dots), and independent annotations of inferred challenge (red circles and dots), aggregated across students for each bounded lecture segment (within green vertical lines).

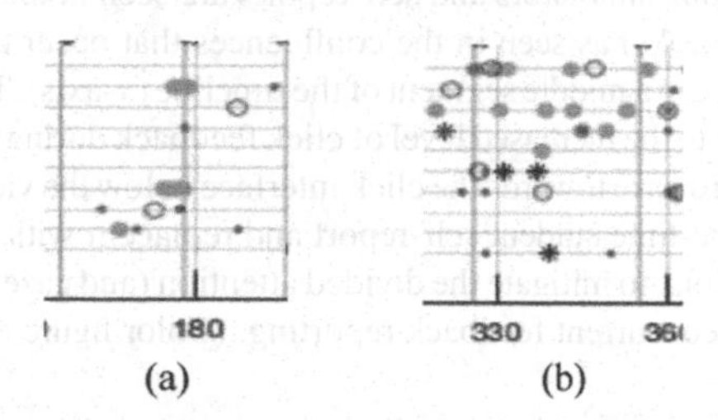

(a) (b)

Fig. 2. Example of two plotted segments, a (left), b (right). (Color figure online)

In the above illustration, relatively sparse struggle events do not meet the threshold within the ca. 40 s window depicted (Fig. 2a, two lecture segments, 155 s–195 s), while for a similar time window (Fig. 2b, one lecture segment, 320 s–360 s), relatively dense struggle events generalize to a segment classification of cohort struggle.

Independent observations were done by three members of the research team (AE, RF, AL), by viewing all the participant videos (without the lecture video), and using the same click feedback interface as the participants. A further annotation option of "bored" was added to the interface to allow for distinguishing between comprehension and disengagement, and thereby to increase robustness in correlating data points for the 'feels easy' option. A manual segmentation and annotation of the lecture was performed based solely on its content, independent of any participant or observer data (e.g. '3:00–3:25, introduction of new term'; '3:26–4:10, description of applying a technique'). Finally, data from annotations, computer vision, real-time and retrospective self-reports, and independent observations was plotted on the lecture timeline.

3 Results

Given the structure of our study, there are two relevant sets of results. The first set of results relates to the correspondence between computer vision data (eye gaze direction) and manual annotation data (including student-self report and observer annotations). These results indicate that the computer vision algorithm shows promise, and could be developed in future work to increase automated analysis and tagging.

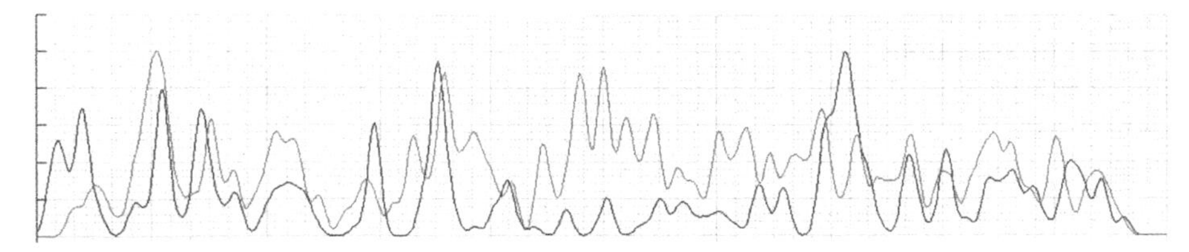

Fig. 3. The horizontal axis represents the lecture timeline from start time to end time. The black line is a normalized and smoothed plot of all students' vertical gaze direction, where the lowest points are nearest to the median gaze bandwidth (i.e. looking towards any point left-to-right on the horizon), and the highest points indicate peak vertical upward gaze. The red line is a normalized and smoothed density plot of all struggle annotations (student self-report and three independent observers). Agreement among annotators and self-reports are seen in the red peaks, and typically track upward gaze (black peaks) as seen in the confluences that occur throughout, apart from a notable red/black divergence in a middle segment of the timeline (x-axis). This period of divergence hypothetically corresponds to the increased level of click feedback during that time window, which appears to cause a downward gaze towards the click interface below the video (see Fig. 1). In future work, we will eliminate real-time student self-report and replace it with retrospective self-report during a second video viewing, to mitigate the divided attention (and gaze patterns) between initial lecture video viewing and concurrent feedback reporting. (Color figure online).

With respect to eye gaze direction, our results suggest initial evidence for a hypothesis that intermittent periods of upward gaze aversion could be related to a 'struggle state', perhaps related to increased mental effort (see Sect. 1.4), based on the correspondence depicted in Fig. 3. Data from each individual is used to establish their own baseline (median) vertical gaze direction, such that upwards gaze is a relative measure. Our aim is to use multiple indicators of struggle that may vary across individuals, but that occur consistently for a single individual. For example, an individual who does not exhibit upwards gaze when they struggle might exhibit a different indicator. Our approach should therefore be receptive to individual differences, whether cultural or idiosyncratic. If heterogenous indicators have a greater density for a given lecture segment, this becomes noteworthy for the instructor. (At present, head position data was too noisy to identify any reliable correlations).

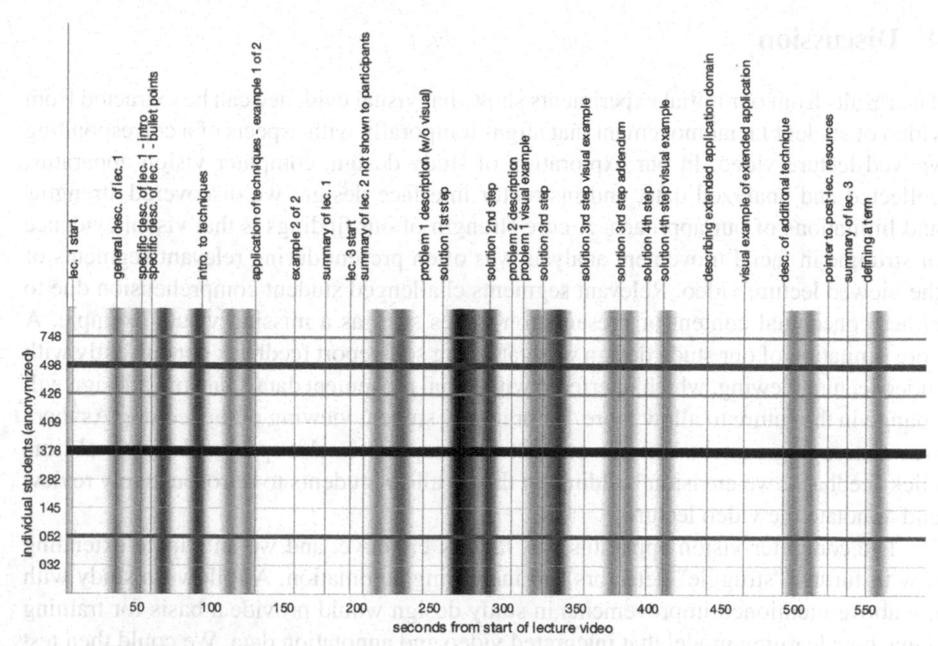

Fig. 4. Prototype of instructor data visualization interface (darker/wider bars = more self-reported student struggle). Vertical bars are density plots of the cohort. For example, at around 260 s, in the lecture segment labelled 'solution 1st step', a greater proportion of students reported struggle than elsewhere. This would suggest to the instructor that the lecture video segment could be re-examined, to understand if the struggle was part of the pedagogical design (e.g. providing 'food for thought'), or alternatively, if students need further support, and how to support them. Support could include student-centered interventions (e.g. adding a forum or group study session, linking to additional resources), or revising the lecture content as needed, by clarifying a term, adding a visual example, etc. Horizontal bars indicate frequency of report. For example, student '378' reported 'feels challenging' more frequently than all other students. The range of horizontal bars gives the instructor an overview of the student cohort, which in this figure, suggests a large proportion of the students are well-matched to the lecture content. Empirical survey data of the students' retrospective self-reports bear out this reading.

Our second set of results indicates how a data visualization of student struggle could benefit instructors (Fig. 4). At present, for clarity, we illustrate a minimalist version of an instructor interface prototype using only student self-report data. The figure caption describes the visualization in detail.

A full-featured interface will include embedded lecture video, for the instructor to 'seek' to relevant video positions for lecture review. It could also include anonymized background information for individual students who elect to disclose it (e.g. disabilities, non-native speakers relative to lecture language, experience level in the subject matter). These and other features would help instructors understand the overview.

4 Discussion

The results from our initial experiments show that visual evidence can be extracted from video of student facial movement that aligns temporally with aspects of a corresponding viewed lecture video. In our exploration of study design, computer vision apparatus, collected and analyzed data, and instructor interface design, we discovered strengths and limitations of our approach. A core strength of our findings is that visual evidence of struggle in facial movement analysis was often present during relevant segments of the viewed lecture video. Relevant segments challenged student comprehension due to either conceptual content or presentation issues such as a missing visual example. A core limitation of our study design was gathering self-report feedback concurrently with video lecture viewing, which interfered with facial movement data. Our study design will change in the future to allow more 'naturalistic' student viewing of the lecture. As there was a strong agreement between students' retrospective self-report and their real-time click feedback, we envision building on this to allow students to retrospectively review and annotate the video lecture.

The computer vision apparatus was largely effective, and we anticipate extending it with further "struggle" detectors and increasing automation. A follow-on study with the above-mentioned improvements in study design would provide a basis for training a machine learning model that integrated video and annotation data. We could then test how well it generalized to new student video.

We were often surprised by data we collected and analyzed, in terms of what it contained and the patterns it revealed. For example, we did not expect students to exhibit such pronounced facial movements when viewing a video alone (e.g. nodding their head). It was also interesting to see apparent visual manifestations of struggle correspond so closely to aspects of the video lecture, ranging from the use of unexpected, unusual, or new terms, to unclear lecture video imagery (e.g. blurry text), to future-oriented references such as abstract descriptions, concretized with visual examples in a following slide (a transition also reflected in the student data analysis).

Finally, taking the patterns we found in the data and visualizing them in an instructor interface was not trivial. At times, when a relevant pattern in the data was strong conceptually, it was opaque when the data was visualized. Other times, patterns in the data were easy to overlook until they were visualized. While both of these issues are typical of data visualization in the sciences, we were not able to fully anticipate how they would arise in an interface for instructors to gain insights about a student cohort.

As the last author was also the instructor who wrote and delivered the video lectures, it is of interest to report his takeaway from the study, irrespective of bias. His report is suggestive of the potential benefits to instructors that we plan to explore systematically in future work. He notes that in reflecting on his past in-person teaching, he indeed made inferences about 'face-to-face' student cohorts by observing behaviors, e.g. different forms of nodding in seeming comprehension, or less positive indicators such as paper rustling, mobile phone usage, or staring at the desk.

For both in-person and recorded video instruction, he received positive student feedback, collected following his lectures. Having done this study, he now sees how student feedback evolved from one recorded lecture segment to another, rather than being a gestalt post-lecture impression. He also sees at a glance how many of the times that students indicated challenge corresponded to a lecture segment with presentation issues as opposed to those with genuine challenging content (see Fig. 4).

5 Conclusion

This implementation of the PUZZLED prototype realized the aims of its design. It provided an instructor with insights about how segments of their lectures related to the students' experience of them. Conceptually challenging lecture segments corresponded to pronounced student struggle patterns initially, which then transitioned back to a baseline. In context, for the lecturer who structured the content, this indicated that students were at least coping with and potentially learning advanced techniques in the subject matter. A few students who struggled more often than others may have needed further support to the get the most from the lesson. In still other instances, the content or order of slides could be modified to provide (e.g.) visual examples at key moments.

We imagine that PUZZLED could also indicate where instructors might increase the challenge level of under-challenging content, to encourage productive struggle. In addition, we believe it can help contribute to more inclusive online HE instruction, by facilitating instructors' ability to receive non-verbal feedback from all students using recorded video lectures. A second prototype will be tested with more instructors.

References

1. English, A.R.: Discontinuity in Learning. Cambridge University Press, Cambridge (2013)
2. Boaler, J.: Mathematical Mindsets: Unleashing Students' Potential Through Creative Math, Inspiring Messages and Innovative Teaching. Wiley, Hoboken (2015)
3. Hiebert, J., Grouws, D.A.: The effects of classroom mathematics teaching on students' learning. In: Second Handbook of Research on Mathematics Teaching and Learning, vol. 1, pp. 371–404 (2007)
4. Shulman, L.S.: Those who understand: knowledge growth in teaching. J. Educ. **193**, 1–11 (2013)
5. Warshauer, H.K.: Productive struggle in middle school mathematics classrooms. J. Math. Teacher Educ. **18**(4), 375–400 (2014). https://doi.org/10.1007/s10857-014-9286-3
6. Alexander, R.: Towards Dialogic Teaching: Rethinking Classroom Talk. Dorchester Publishing Company, Incorporated (2008)
7. Murdoch, D., English, A.R., Hintz, A., Tyson, K.: Feeling heard: inclusive education, transformative learning, and productive struggle. Educ. Theory **70**, 653–679 (2020)
8. Lipman, M.: Thinking in Education. Cambridge University Press, Cambridge (2003)
9. Oser, F., Spychinger, M.: Lernen ist schmerzhaft. Zur Theorie der Fehlerkultur und zur Praxis des Negativen Wissens, Beltz (2005)
10. D'Mello, S., Graesser, A.: Dynamics of affective states during complex learning. Learn. Instr. **22**, 145–157 (2012)
11. Nelson Laird, T.F., Chen, D., Kuh, G.D.: Classroom practices at institutions with higher-than-expected persistence rates: what student engagement data tell us. New Dir. Teach. Learn. **115**, 85–99 (2008)
12. Trowler, P., Trowler, V.: Student engagement evidence summary. The Higher Education Academy (2010)

13. Chen, Y., Chen, Q., Zhao, M., Boyer, S., Veeramachaneni, K., Qu, H.: DropoutSeer: visualizing learning patterns in massive open online courses for dropout reasoning and prediction. In: 2016 IEEE Conference on Visual Analytics Science and Technology (VAST), pp. 111–120 (2016)
14. Lee, Y., Choi, J.: A review of online course dropout research: implications for practice and future research. Educ. Technol. Res. Dev. **59**(5), 593–618 (2011). https://doi.org/10.1007/s11423-010-9177-y
15. Rios-Amaya, J., Secker, J., Morrison, C.: Lecture recording in higher education: risky business or evolving open practice. LSE/University of Kent (2016)
16. Newland, B.: Lecture Capture in UK HE 2017. HeLF UK (2017)
17. Gaebel, M., Kupriyanova, V., Morais, R., Colucci, E.: E-learning in European higher education institutions: results of a mapping survey conducted in October–December 2013. European University Association (2014)
18. High Level Group on the Modernisation of Higher Education: Report to the European Commission on improving the quality of teaching and learning in Europe's higher education institutions. Publications Office of the European Union (2013)
19. Meltzer, D.E., Manivannan, K.: Transforming the lecture-hall environment: the fully interactive physics lecture. Am. J. Phys. **70**, 639–654 (2002)
20. Phuong, A.E., Nguyen, J., Marie, D.: Evaluating an adaptive equity-oriented pedagogy. J. Effective Teach. **17**, 5–44 (2017)
21. Bourgatte, M., Fournout, O., Puig, V.: Les technologies du numérique au service de l'enseignement: Vers un apprentissage instrumenté et visuel. In: Châteauvert, J., Delavaud, G. (eds.) D'un écran à l'autre, pp. 437–456. l'Harmattan (2016)
22. Grafsgaard, J.F., Wiggins, J.B., Boyer, K.E., Wiebe, E.N., Lester, J.C.: Automatically recognizing facial indicators of frustration: a learning-centric analysis. In: 2013 Humaine Association Conference on Affective Computing and Intelligent Interaction, pp. 159–165 (2013)
23. Whitehill, J., Serpell, Z., Lin, Y., Foster, A., Movellan, J.R.: The faces of engagement: automatic recognition of student engagementfrom facial expressions. IEEE Trans. Affect. Comput. **5**, 86–98 (2014)
24. Nezami, O.M., Dras, M., Hamey, L., Richards, D., Wan, S., Paris, C.: Automatic recognition of student engagement using deep learning and facial expression. In: Brefeld, U., Fromont, E., Hotho, A., Knobbe, A., Maathuis, M., Robardet, C. (eds.) ECML PKDD 2019. LNCS (LNAI), vol. 11908, pp. 273–289. Springer, Cham (2020). https://doi.org/10.1007/978-3-030-46133-1_17
25. Bosch, N., D'Mello, S.K., Ocumpaugh, J., Baker, R.S., Shute, V.: Using video to automatically detect learner affect in computer-enabled classrooms. ACM Trans. Interact. Intell. Syst. **6**, 1–26 (2016)
26. Huron, S., Isenberg, P., Fekete, J.D.: PolemicTweet: video annotation and analysis through tagged tweets. In: Kotzé, P., Marsden, G., Lindgaard, G., Wesson, J., Winckler, M. (eds.) Human-Computer Interaction – INTERACT 2013. LNCS, vol. 8118, pp. 135–152. Springer, Heidelberg (2013). https://doi.org/10.1007/978-3-642-40480-1_9
27. Schuller, B., et al.: Being bored? Recognising natural interest by extensive audiovisual integration for real-life application. Image Vis. Comput. **27**, 1760–1774 (2009). https://doi.org/10.1016/j.imavis.2009.02.013
28. Nagasawa, T., Takahashi, R., Koopipat, C., Tsumura, N.: Stress estimation using multimodal biosignal information from RGB facial video. In: IEEE/CVF Conference on Computer Vision and Pattern Recognition Workshops, pp. 292–293 (2020)
29. Zhao, Y., Wang, X., Petriu, E.M.: Facial expression analysis using eye gaze information. In: 2011 IEEE International Conference on Computational Intelligence for Measurement Systems and Applications (CIMSA) Proceedings, pp. 1–4 (2011)

30. Baron-Cohen, S., Cross, P.: Reading the eyes: evidence for the role of perception in the development of a theory of mind. Mind Lang. **7**, 172–186 (1992)
31. Doherty-Sneddon, G., Phelps, F.G.: Teachers' responses to children's eye gaze. Educ. Psychol. **27**, 93–109 (2007)
32. Glenberg, A.M., Schroeder, J.L., Robertson, D.A.: Averting the gaze disengages the environment and facilitates remembering. Mem. Cogn. **26**, 651–658 (1998). https://doi.org/10.3758/BF03211385
33. Gao, H., Yüce, A., Thiran, J.-P.: Detecting emotional stress from facial expressions for driving safety. In: 2014 IEEE International Conference on Image Processing (ICIP), pp. 5961–5965 (2014)

Preparing Future Learning with Novel Visuals by Supporting Representational Competencies

Jihyun Rho(✉), Martina A. Rau, and Barry D. Van Veen

University of Wisconsin-Madison, 1025 W Johnson Street, Madison, WI 53706, USA
{jrho6,marau,bvanveen}@wisc.edu

Abstract. Many STEM problems involve visuals. To benefit from these problems, students need representational competencies: the ability to understand and appropriately use visuals. Support for representational competencies enhances students' learning outcomes. However, it is infeasible to design representational-competency supports for entire curricula. This raises the question of whether these supports enhance future learning from novel problems. We addressed this question with an experiment with 120 undergraduates in an engineering class. All students worked with an intelligent tutoring system (ITS) that provided problems with interactive visual representations. The experiment varied which types of representational-competency supports the problems provided. We assessed future learning from a subsequent set of novel problems that involved a novel visual representation. Results show that representational-competency support can enhance future learning from the novel problems. We discuss implications for the integration of these supports in educational technologies.

Keywords: Visualizations · Representational competencies · Future learning

1 Introduction

Instruction in STEM domains heavily relies on visual representations because much of the content knowledge in such domains is visuospatial [1]. As a result, students encounter multiple visual representations to learn about foundational concepts [1, 2]. For instance, when learning about sinusoids, engineering students typically encounter the time-domain visual and phase-domain visual shown in Fig. 1.

Unfortunately, students often do not benefit from these visual representations. Students' difficulties in understanding visual representations are a major obstacle to their success in STEM domains [1, 3], including engineering [4]. Such difficulties result from a lack of *representational competencies*, that is, knowledge about how visuals reveal information relevant to scientific concepts and practice [5, 6].

Further, challenges that are caused by lack of representational competencies are particularly severe for students with low spatial skills [7]. For example, when translating between visuals in Fig. 1, students need to mentally rotate a phasor and project a sinusoid's amplitude to the magnitude of a phasor [8, 9].

M. M. Rodrigo et al. (Eds.): AIED 2022, LNCS 13355, pp. 66–77, 2022.
https://doi.org/10.1007/978-3-031-11644-5_6

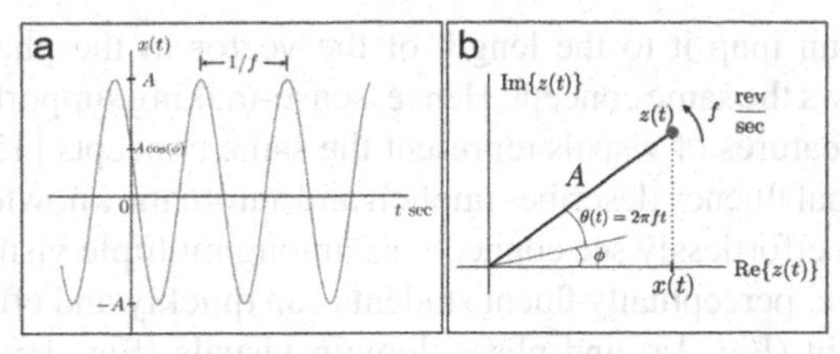

Fig. 1. Visual representations: (a) time-domain visual showing a sinusoid as a function of time; (b) phase-domain visual showing sinusoid as rotating vector.

More crucially, lack of representational competencies could subsequently impede students' future learning because the concepts they learn today are the basis for their later learning from novel problems. For example, students who fail to understand time-domain and phase-domain visuals (Fig. 1) will likely struggle to learn about more advanced concepts building on an understanding of these visuals, such as phasor addition.

Educational technologies offer a solution to this problem. They can provide adaptive support for representational competencies while students interact with visuals [10]. Prior research has established effective technology-based supports for students' representational competencies [10]. However, experimental evidence shows that designing adequate supports requires substantial time and effort [10]. Consequently, it is infeasible to design representational-competency supports for entire curricula. This raises the question of whether the effectiveness of representational-competency supports generalizes by enhancing students' future learning of novel concepts with novel visuals. Addressing this question will yield novel insights into the practicality of integrating supports for representational competencies in technology-based curricula.

Given that issues due to lack of representational competencies are particularly severe for students with low spatial skills, it is important to explore how spatial skills moderate the effects of representational-competency supports on students' future learning. Addressing this question will yield novel insights into how representational-competency supports relate to equity issues in STEM fields because students with low spatial skills are disproportionally women [11] or have low socioeconomic status [12].

2 Literature Review

2.1 Supporting Representational Competencies

Previous research identified two broad types of representational competencies that play an important role in learning with visuals in STEM [6]: sense-making competencies and perceptual fluency. Since these competencies derive from different learning processes, they should be supported by different types of instructional activities [13].

First, sense-making competencies describe explicit, analytical knowledge that allows students to explain how visual features of representations map to domain concepts [14]. Sense-making competencies also involve the ability to connect multiple visuals based on conceptual features [1, 6]. For example, students with sense-making competencies understand that the y-maximum in the time-domain visual (Fig. 1a) shows the amplitude of a sinusoid and can map it to the length of the vector in the phase-domain visual (Fig. 1b), which shows the same concept. Hence, sense-making supports prompt students to explain how the features of visuals represent the same concepts [15, 16].

Second, perceptual fluency describes implicit and automatic knowledge allowing students to quickly and effortlessly see connections among multiple visual representations [17, 18]. For example, perceptually fluent students can quickly and effortlessly translate between time-domain (Fig. 1a) and phase-domain visuals (Fig. 1b). Such perceptual fluency frees cognitive resources that students can invest for higher-order thinking, creative problem solving, or learning advanced concepts [18]. Perceptual-fluency supports expose students to a large number of simple recognition or classification problems that involve various types of visual representations. Through repeated practice, students learn to induce which visual features carry meaningful information [18].

Thus far, research has only examined whether these representational-competency supports enhance learning from the problems that provide these supports [10]. Hence, it remains unknown whether representational-competency supports are effective beyond the duration of the support. This question relates to transfer research that has examined how to prepare students for future learning experiences.

2.2 Transfer and Preparation for Future Learning

Current transfer research focuses on how instruction can prepare students to optimally benefit from future learning experiences [19]. This research developed in response to traditional transfer research, which defined transfer as the direct application of prior knowledge or skills to novel problems [20]. However, students rarely demonstrated this type of transfer, which led to criticisms of the traditional transfer definition [21]. The critiques argued that traditional transfer studies accept only specific evidence as the "right" form of transfer by prioritizing models of expert performance [22]. Instead, students often adapt their prior knowledge in a way that helps them learn about new concepts [20]. In line with this, "preparation for future learning" (PFL) research examines how instruction can support students' knowledge in a way that enhances their future learning from novel problems [19].

However, little research has investigated transfer of representational competencies. The few studies that have investigated this question rely on the traditional transfer framework [23]. For example, Cromley [23] tested whether representational-competency support enhances students' understanding of visuals they did not encounter during instruction, as assessed by a transfer posttest. Results showed advantages of representational-competency supports on the transfer posttest. However, this research leaves open whether representational-competency supports enhance students' learning from novel problems in subsequent instruction. Research on expert problem solving suggests that representational competencies contribute to experts' adaptive thinking about novel problems [17]. First, sense-making competencies enable experts to analyze the deep structure of a problem [24], allowing them to use representations to generate creative solutions [25]. Second, perceptual fluency has been linked to adaptive thinking because the ability to quickly process information from given representations frees cognitive resources to flexibly apply prior knowledge when solving new problems [18]. Thus, supporting students' representational competencies may equip them with knowledge that enhances their subsequent learning.

3 Research Questions

Our review of prior research shows that there is a gap between research on representational-competency supports and research on transfer, especially from a PFL perspective on transfer. Consequently, the following research question (RQ) remains open:

RQ1: Do problems that support sense-making competencies and perceptual fluency enhance students' learning from novel problems?

Further, given that issues due to a lack of representational competencies are particularly severe for students with low spatial skills, we explore:

RQ2: Do spatial skills moderate the effect of representational-competency supports?

4 Methods

4.1 Participants

The experiment was conducted as part of an introductory engineering course on signal processing at a university in the Midwestern U.S. All 120 undergraduate students enrolled in the course participated. The course involved two 75-min class meetings per week. The intervention took place in the first 3 weeks that covered sinusoids.

4.2 Signals Tutor: An ITS for Undergraduate Electrical Engineering

We conducted an experiment in the context of five units of Signals Tutor, an ITS for undergraduate electrical engineering. Signals Tutor provides problems in which students learn about sinusoids by manipulating time-domain and phase-domain visuals (Fig. 1). Both visuals play an important role in learning advanced engineering concepts such as Fourier analysis, circuit analysis, and single-frequency analysis of system. Signals Tutor involves three types of problems.

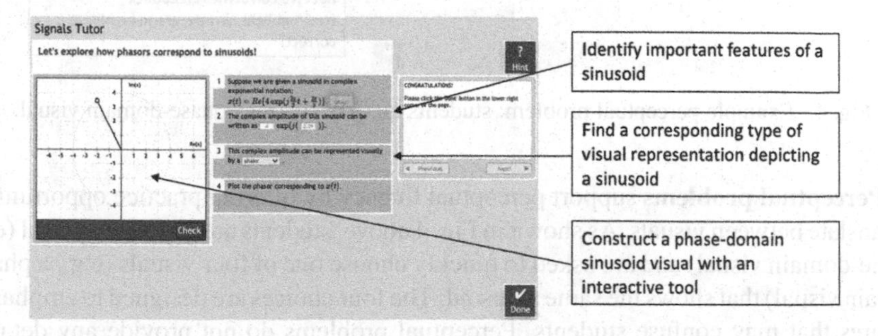

Fig. 2. Example individual problem: students construct a phase-domain visual.

Individual problems provide one visual representation per problem. While these problems do not specifically support representational competencies, they familiarize students with one visual at a time while asking students to relate the visuals to corresponding equations. As shown in Fig. 2. Above, individual problems ask students to answer questions about sinusoids and to construct a visual representation based on an equation by using an interactive visualization tool. Students receive error-specific feedback and on-demand hints on all problem-solving steps, including the visuals they construct.

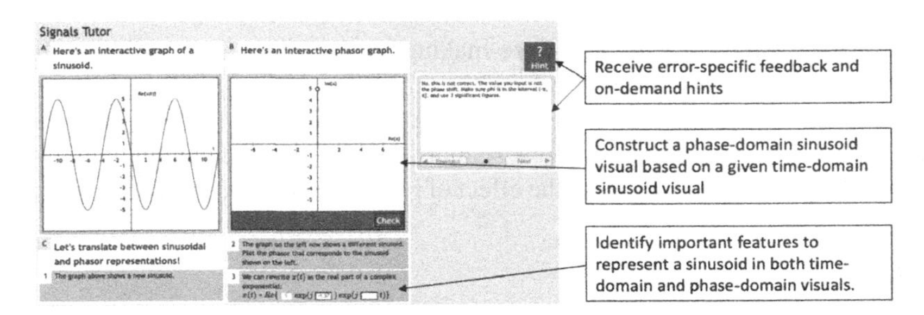

Fig. 3. Example sense-problem: students reflect on time-domain and phase-domain visuals.

Sense problems support sense-making competencies. As shown in Fig. 3 above, sense problems have two parts. First, students are given one visual (e.g., a time-domain visual) and are asked to construct a second visual (e.g., a phase-domain visual) of the same sinusoid. Second, students are prompted to reflect on how the two visuals represent corresponding and complementary concepts related to sinusoids. Similar to individual problems, students receive error-specific feedback and on-demand hints.

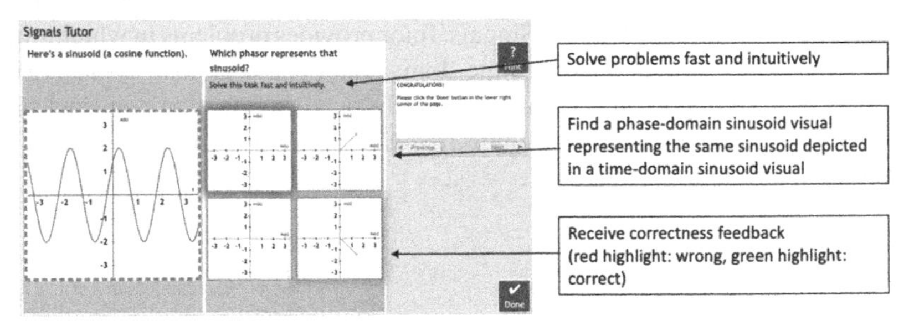

Fig. 4. Example perceptual problem: students quickly choose a phase-domain visual.

Perceptual problems support perceptual fluency by offering practice opportunities to translate between visuals. As shown in Fig. 4 above, students are given one visual (e.g., a time-domain visual) and are asked to quickly choose one of four visuals (e.g., a phase-domain visual) that shows the same sinusoid. The four choices are designed to emphasize features that may confuse students. Perceptual problems do not provide any detailed feedback or hints. Students only receive correctness feedback because explanations could disrupt perceptual processing [26]. Students receive many of these short problems with numerous examples.

4.3 Experimental Design and Procedure

To investigate the effect of representational-competency support on students' future learning, we used a 2 (sense problems: yes/no) × 2 (perceptual problems: yes/no) design. This yielded four conditions: (1) The *control condition* received only individual problems without representational-competency supports. (2) The *sense condition* received individual and sense problems. (3) The *perceptual condition* received individual and perceptual problems. (4) The *sense-perceptual condition* received individual, sense, and perceptual problems. Across conditions, we adjusted the number of steps in each problem so that all conditions received the same number of problem-solving steps.

The sequence of problems was organized as follows across the five Signals Tutor units. As detailed in Table 1, Units 1–4 provided time-domain and phase-domain visuals. Unit 1 was an introductory unit that familiarized students with basic sinusoids and with time-domain and phase-domain visuals. Unit 1 was identical across conditions.

Unit 2 provided only time-domain visuals. Because individual problems ask students to translate between equations and visuals, there were no sense problems for Unit 2. Yet, Unit 2 offered perceptual problems that asked students to quickly translate between equations and time-domain visuals. Students in the control and sense conditions received only individual problems. By contrast, students in the perceptual and sense-perceptual conditions received individual problems followed by perceptual problems.

Units 3 and 4 provided both types of visuals. For each of these units, students in the control condition received only individual problems. Students in the sense condition received individual problems followed by sense problems. Students in the perceptual condition received individual problems followed by perceptual problems. Students in the sense-perceptual condition received individual, then sense, then perceptual problems. Across Units 3–4, we implemented sense problems before perceptual problems following prior research suggesting that this sequence is more effective [27].

Finally, Unit 5 provided instructional problems on phasor addition, a novel, more complicated concept that builds on the content covered in Units 2–4. Students used a vector graph, a novel type of visual. Unit 5 served to assess students' preparation for future learning and was identical across conditions.

In the first course meeting, students were greeted by the research team and informed about the study. Then, they worked on one Signals Tutor unit per meeting for the first five meetings of the course. For Units 2–5, students received a pretest prior to the Signals Tutor problems and a posttest immediately after. As Unit 1 was an introductory unit administered in the first course meeting, it did not include a pretest or posttest. The spatial skills test was given prior to the Unit 2 pretest.

Table 1. Overview of Signals Tutor units

Unit	Content	Sinusoid visuals	Experimental factors
1	Sinusoids, sinusoid visuals	Time/phase domain	None
2	Sinusoids as function of time	Time domain	Perceptual (y/n)
3	Multiple sinusoid visuals	Time/phase domain	Sense (y/n); perceptual (y/n)
4	Complex numbers	Time/phase domain	Sense (y/n); perceptual (y/n)
5	Sum of sinusoids	Vector graph	None

4.4 Measures

We assessed students' learning gains with pretests and posttests for each unit (except for the introductory Unit 1). Isomorphic test versions were counterbalanced across test times (i.e., the versions had structurally identical items but used different examples). Each test had ten multiple-choice items assessing students' ability to internally visualize and manipulate sinusoids. Some items provided a visual of a sinusoid and asked students to mentally modify it to answer questions about the sinusoid. Other items provided an equation and asked students to mentally visualize the corresponding sinusoid to answer questions. Students were not allowed to draw or use calculators. We computed accuracy scores as the percentage of correctly answered items on each test. We computed efficiency scores to take response time into account following [28]:

$$\text{efficiency score} = \frac{Z(\text{average correct responses}) - Z(\text{average response time per test item})}{\sqrt{2}} \tag{1}$$

Finally, we assessed spatial skills with the Vandenberg & Kuse mental rotation test [29], which is a common measure in engineering education research [30].

5 Results

We excluded students from analysis who were absent from any test, whose test performance was a statistical outlier (i.e., 2 standard deviations above or below the median), or who dropped the course. As a result, a total of $N = 117$ students were included in the data set (control: $n = 28$, sense: $n = 28$, perceptual: $n = 32$, sense-perceptual: $n = 29$). We report partial η^2 (p. η^2) for effect sizes, with .01 corresponding to a small, .06 to a medium, and .14 to a large effect [31]. Table 2 shows efficiency scores by unit.

5.1 Prior Checks

First, we checked for differences between conditions on the pretests for Units 2–5. A multivariate ANOVA showed no significant effects of condition ($ps > .10$). However, each unit's pretest significantly correlated with the posttest (ranging from $r = .274$ to $r = .726$; $ps < .01$). Thus, we included pretest as a covariate in the analyses for each unit.

Second, we checked whether students showed learning gains after working with Signals Tutor. We used a repeated measure ANOVA with test-time (pretest, posttest) as the repeated, within-subject factor and average test scores across units as the dependent measure. Results showed significant gains, $F(1{,}116) = 87.871, p < .001$, p. $\eta^2 = 431$. Separate repeated measure ANOVAs for Units 2–5 showed significant gains for all units (*ps* < .01) with effect sizes ranging from p. $\eta^2 = .09$ to p. $\eta^2 = .24$.

Third, we checked whether representational-competency supports enhanced students' learning from Units 2–4; that is, on the units where these supports were present. We conducted separate ANCOVAs for Unit 3 and 4, with pretest as covariate, the sense (y/n) and perceptual (y/n) factors as independent variables, and posttest as dependent measure. For Unit 2, we conducted similar ANCOVA but used only the perceptual factor (y/n) as an independent variable. For accuracy, results revealed a significant interaction between the sense and perceptual factors in Unit 4, $F(1{,}116) = 4.499$, $p = .036$ p. $\eta^2 = .039$. Predefined contrasts showed that students in the sense condition showed marginally higher accurate posttest performance than students in the sense-perceptual condition ($p = .09$). No other effects were significant (*ps* > .10). For efficiency, we found no significant effects (*ps* > .10).

Table 2. Each unit's means and standard deviations (in parentheses) of efficiency scores

Unit	Test	Control	Sense	Perceptual	Sense-perceptual
2	Pre	−0.199 (0.691)	−0.127 (0.882)	−0.681 (1.046)	0.108 (1.023)
	Post	0.302 (0.927)	0.032 (0.778)	0.097 (.782)	0.528 (1.037)
3	Pre	−0.338 (0.891)	−0.495 (1.017)	−0.345 (1.172)	−0.341 (0.923)
	Post	0.216 (0.958)	0.312 (1.135)	0.359 (1.011)	0.621 (0.823)
4	Pre	−0.464 (1.190)	−0.380 (1.014)	−0.526 (.987)	0.100 (1.085)
	Post	0.064 (1.057)	0.380 (0.881)	0.273 (1.010)	0.564 (0.886)
5	Pre	−0.267 (1.160)	−0.575 (1.013)	−0.529 (.927)	−0.401 (0.880)
	Post	0.608 (1.036)	0.223 (1.235)	0.210 (1.167)	0.763 (1.118)

5.2 Effects on Future Learning

To test whether representational-competency supports enhance students' learning from novel problems (RQ1), we used an ANCOVA with Unit 5 pretest as covariate, sense and perceptual factors as independent variables, and Unit 5 posttest as dependent measure. On the accuracy measure, results showed no significant effects (*ps* > .10). On the efficiency measure, students who had received sense problems in Units 3–4 (i.e., students in sense and sense-perceptual conditions) had significantly higher posttest efficiency than students who had not received sense problems (i.e., students in control, perceptual conditions), $F(1, 116) = 7.366$, $p = .008$, p. $\eta 2 = .063$. Further, the sense and perceptual factors interacted, $F(1, 116) = 5.386$, $p = .022$, p. $\eta^2 = .047$. As shown in Fig. 5a, students who had received both sense and perceptual problems in Units 3–4 had the highest posttest efficiency in Unit 5.

Next, we tested whether students' spatial skills moderate the effect of representational-competency supports (RQ2). To this end, we included spatial skills as a covariate to the ANCOVA and an aptitude-treatment interaction of spatial skills with the sense factor and the perceptual factor. This tests whether the continuous spatial skills variable moderates the effect of sense problems and perceptual problems. For efficiency, there was a significant interaction between spatial skills and the sense factor, $F(1,116) = 8.989$, $p = .003$, p. $\eta^2 = .076$ (Fig. 5b). To understand this effect, we computed effect slices that estimate the effect of the sense factor for specific levels of spatial skills. Students with high spatial skills ($\geq 80^{th}$ percentile of the sample, $p = .026$) showed a significant benefit from receiving sense problems (i.e., sense and sense-perceptual conditions). By contrast, there was no significant benefit of sense problems for students with low spatial skills ($\leq 20^{th}$ percentile of the sample, $p = .207$).

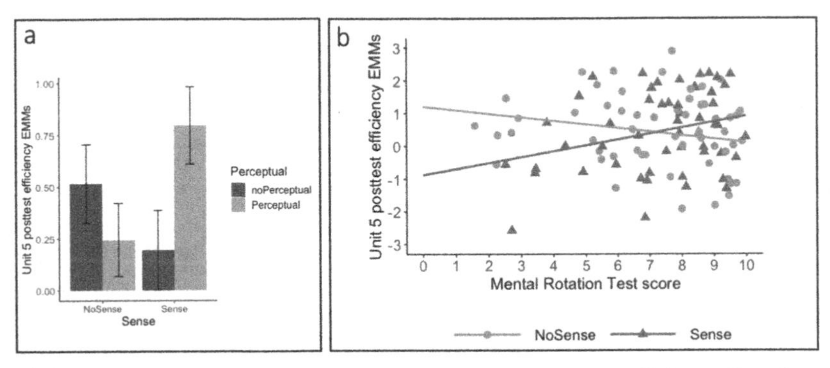

Fig. 5. (a) Interaction between sense and perceptual factors on posttest efficiency. Error bars show standard errors of the Estimated Marginal Means (EMMs); (b) effect of sense factor for levels of spatial skills. EMMs were computed controlling for covariates.

6 Discussion

The goal of this paper was to investigate whether representational-competency supports enhance students' future learning from novel problems with novel visuals (RQ1). We examined the effects of two types of representational-competency supports that were provided in the form of sense and perceptual problems. Our results show that students who received a combination of both problems showed more efficient posttest performance after learning from novel visuals, compared to students who received problems with no or with only one type of support. We interpret these findings in terms of the preparation for future learning (PFL) transfer framework [19]. Students learned how to make sense of representations through sense problems and how to quickly see meaning in the visuals through perceptual problems. Students appeared to be able to adapt these representational competencies when learning about sums of sinusoids using an unfamiliar vector graph. The finding that the combination of sense-making and perceptual-fluency supports was most effective suggests that both types of representational competencies are relevant to future learning experiences. Based on expertise research [18, 24, 25], we conjecture that sense-making competencies allow students to analyze a novel problem to generate a solution, whereas perceptual fluency frees cognitive resources for them to adapt prior knowledge to novel problems.

Further, we investigated whether spatial skills moderate the effect of representational-competency supports (RQ2). We found that students with high spatial skills benefited from sense problems, whereas students with low spatial skills did not. This suggests that the sense problems disadvantaged students with low spatial skills; that is, students who are already at a disadvantage in STEM domains such as engineering. What might explain this unfortunate effect? Sense problems support students in constructing mental models of multiple visuals [32]. Students with high spatial skills might have the necessary cognitive resources to spatially integrate multiple visuals in their mental models. This may have allowed them to efficiently incorporate a new visual in their mental model when learning from Unit 5. In contrast, students with low spatial skills may find it more cognitively demanding to integrate new visuals into their working memory. This finding suggests that research needs to focus on students with low spatial skills. It is possible that our sense problems did not offer optimal support for these students. For example, sense problems could visually highlight correspondences between visuals after students make mistakes in connecting the visuals. This may help low-spatial-skills students to understand spatially distributed correspondences. Future research should examine whether redesigned sense problems are effective for low-spatial-skills students. In the absence of redesigned sense problems, low-spatial-skills students may need continued sense-making support when they encounter novel visuals.

Finally, the results on the PFL assessment (Unit 5) differ from the results on the manipulation checks (Units 2–4), where we only found an advantage of sense problems on posttest accuracy (Unit 4). It is possible that the effectiveness of the sense problems only appeared after students had sufficient practice in reflecting on how the two visuals show sinusoid concepts (i.e., after Unit 4). However, the effectiveness of perceptual problems was not apparent immediately in Units 2–4, but only when students encountered novel problems with a novel visual in Unit 5. Thus, it seems that the ability to process familiar visuals quickly and effortlessly did not pay off when the visuals were familiar. However, it enabled students to solve novel problems more efficiently.

In sum, our study highlights the importance of assessing future learning. An intervention that seems effective for all may lack long-term benefits for some students (e.g., low-spatial-skills students). An intervention that seems ineffective (e.g., perceptual problems) may have long-term benefits, including for students with low spatial skills. These findings also have important implications for the design of adaptive educational technologies. Designing supports in a way that ensures long-term benefits may resolve the impracticality of providing representational-competency supports for entire curricula, which is infeasible because of the significant development costs.

Our study has several limitations. First, if focused on individual learning, whereas STEM instruction often involves collaboration. Future research should test effects of collaborative representational-competency supports on future learning. Second, our study only assessed students' improvement of content knowledge. Future research should additionally assess students' learning sense-making competencies and perceptual fluency. Finally, our study revealed the risk of disadvantaging students with low spatial skills. Future research should examine how representational-competency supports can prepare these students for future learning experiences.

To conclude, our findings suggest that integrating sense-making supports and perceptual-fluency supports in educational technologies enhances students' learning with novel visuals in novel tasks. This study is the first to show that representational-competency supports have the potential to enhance future learning. However, our study cautions that sense-making supports need to be designed in a way that better serves low-spatial-skills students. Without research that examines long-term effects of representational-competency supports, we may widen rather than close the achievement gap in STEM domains.

Acknowledgements. This work was supported by NSF DUE 1933078. We also thank Bernie Lesieutre and his teaching assistants for their help with our study.

References

1. Ainsworth, S.: The educational value of multiple-representations when learning complex scientific concepts. In: Gilbert, J.K., Reiner, M., Nakhleh, M. (eds.) Visualization: Theory and Practice in Science Education. MMSE, vol. 3, pp. 191–208. Springer, Dordrecht (2008). https://doi.org/10.1007/978-1-4020-5267-5_9
2. Kozma, R.: The material features of multiple representations and their cognitive and social affordances for science understanding. Learn. Instr. **13**, 205–226 (2003)
3. Gilbert, J.K.: Visualization: a metacognitive skill in science and science education. In: Gilbert, J.K. (ed.) Visualization in Science Education. MMSE, vol. 1, pp. 9–27. Springer, Dordrecht (2005). https://doi.org/10.1007/1-4020-3613-2_2
4. McCracken, W.M., Newstetter, W.C.: Text to diagram to symbol: representational transformations in problem-solving. In: 31st Annual Frontiers in Education Conference. Impact on Engineering and Science Education. Conference Proceedings (Cat. No. 01CH37193), p. F2G-13 (2001)
5. diSessa, A.A.: Metarepresentation: native competence and targets for instruction. Cogn. Instr. **22**, 293–331 (2004). https://doi.org/10.1207/s1532690xci2203_2
6. Rau, M.A.: Conditions for the effectiveness of multiple visual representations in enhancing STEM learning. Educ. Psychol. Rev. **29**(4), 717–761 (2017). https://doi.org/10.1007/s10648-016-9365-3
7. Kozhevnikov, M., Motes, M.A., Hegarty, M.: Spatial visualization in physics problem solving. Cogn. Sci. **31**, 549–579 (2007)
8. Hegarty, M., Waller, D.A.: Individual Differences in Spatial Abilities. Cambridge University Press, Cambridge (2005)
9. Stieff, M.: Mental rotation and diagrammatic reasoning in science. Learn. Instr. **17**, 219–234 (2007)
10. Rau, M.A.: A framework for educational technologies that support representational competencies. IEEE Trans. Learn. Technol. **10**, 290–305 (2017)
11. Steiner, S., Wagaman, M.A., Lal, P.: Thinking spatially: teaching an undervalued practice skill. J. Teach. Soc. Work **34**, 427–442 (2014)
12. Levine, S.C., Vasilyeva, M., Lourenco, S.F., Newcombe, N.S., Huttenlocher, J.: Socioeconomic status modifies the sex difference in spatial skill. Psychol. Sci. **16**, 841–845 (2005)

13. Koedinger, K.R., Corbett, A.T., Charles, P.: The knowledge-learning-instruction framework: bridging the science-practice chasm to enhance robust student learning. Cogn. Sci. **36**, 757–798 (2012)
14. Bodemer, D., Faust, U.: External and mental referencing of multiple representations. Comput. Hum. Behav. **22**, 27–42 (2006)
15. Ainsworth, S.: DeFT: a conceptual framework for considering learning with multiple representations. Learn. Instr. **16**, 183–198 (2006)
16. Berthold, K., Eysink, T.H.S., Renkl, A.: Assisting self-explanation prompts are more effective than open prompts when learning with multiple representations. Instr. Sci. **37**, 345–363 (2009). https://doi.org/10.1007/s11251-008-9051-z
17. Goldstone, R.L., Landy, D.H., Son, J.Y.: The education of perception. Top. Cogn. Sci. **2**, 265–284 (2010)
18. Kellman, P.J., Massey, C.M.: Perceptual learning, cognition, and expertise. In: Ross, B.H. (ed.) Psychology of Learning and Motivation, pp. 117–165. Academic Press (2013)
19. Schwartz, D.L., Martin, T.: Inventing to prepare for future learning: the hidden efficiency of encouraging original student production in statistics instruction. Cogn. Instr. **22**, 129–184 (2004)
20. Hohensee, C.: Transfer of Learning: Progressive Perspectives for Mathematics Education and Related Fields. Springer, Cham (2013). https://doi.org/10.1007/978-3-030-65632-4
21. Bransford, J., Schwartz, D.: Rethinking transfer: a simple proposal with multiple implications, vol. 24. American Educational Research Association, Washington DC (1999)
22. Lobato, J.: How design experiments can inform a rethinking of transfer andviceversa. Educ. Res. **32**, 17–20 (2003)
23. Cromley, J.G., Perez, T.C., Fitzhugh, S.L., Newcombe, N.S., Wills, T.W., Tanaka, J.C.: Improving students' diagram comprehension with classroom instruction. J. Exp. Educ. **81**, 511–537 (2013)
24. Chi, M.T.H., Feltovich, P.J., Glaser, R.: Categorization and representation of physics problems by experts and novices. Cogn. Sci. **5**, 121–152 (1981)
25. Arcavi, A.: The role of visual representations in the learning of mathematics. Educ. Stud. Math. **52**, 215–241 (2003). https://doi.org/10.1023/A:1024312321077
26. Chin, J.M., Schooler, J.W.: Why do words hurt? Content, process, and criterion shift accounts of verbal overshadowing. Eur. J. Cogn. Psychol. **20**, 396–413 (2008)
27. Rau, M.A.: Sequencing support for sense making and perceptual induction of connections among multiple visual representations. J. Educ. Psychol. **110**, 811 (2018)
28. Van Gog, T., Paas, F.: Instructional efficiency: Revisiting the original construct in educational research. Educ. Psychol. **43**, 16–26 (2008)
29. Peters, M., Laeng, B., Latham, K., Jackson, M., Zaiyouna, R., Richardson, C.: A redrawn Vandenberg and Kuse mental rotations test - different versions and factors that affect performance. Brain Cogn. **28**, 39–58 (1995)
30. Sorby, S.A., Baartmans, B.J.: The development and assessment of a course for enhancing the 3-D spatial visualization skills of first year engineering students. J. Eng. Educ. **89**, 301–307 (2000)
31. Cohen, J.: Statistical Power Analysis for the Behavioral Sciences. Academic Press (2013)
32. Schnotz, W.: An integrated model of text and picture comprehension. In: The Cambridge Handbook of Multimedia Learning, vol. 49, no. 69 (2005)

Leveraging Student Goal Setting for Real-Time Plan Recognition in Game-Based Learning

Alex Goslen[1(✉)], Dan Carpenter[1], Jonathan P. Rowe[1], Nathan Henderson[1], Roger Azevedo[2], and James Lester[1]

[1] North Carolina State University, Raleigh, NC, USA
{amgoslen,dcarpen2,jprowe,lester}@ncsu.edu
[2] University of Central Florida, Orlando, FL, USA
roger.azevedo@ucf.edu

Abstract. Goal setting and planning are integral components of self-regulated learning. Many students struggle to set meaningful goals and build relevant plans. Adaptive learning environments show significant potential for scaffolding students' goal setting and planning processes. An important requirement for such scaffolding is the ability to perform student plan recognition, which involves recognizing students' goals and plans based upon the observations of their problem-solving actions. We introduce a novel plan recognition framework that leverages trace log data from student interactions within a game-based learning environment called CRYSTAL ISLAND, in which students use a drag-and-drop planning support tool that enables them to externalize their science problem-solving goals and plans prior to enacting them in the learning environment. We formalize student plan recognition in terms of two complementary tasks: (1) classifying students' selected problem-solving goals, and (2) classifying the sequences of actions that students indicate will achieve their goals. Utilizing trace log data from 144 middle school students' interactions with CRYSTAL ISLAND, we evaluate a range of machine learning models for student goal and plan recognition. All machine learning-based techniques outperform the majority baseline, with LSTMs outperforming other models for goal recognition and naive Bayes performing best for plan recognition. Results show the potential for automatically recognizing students' problem-solving goals and plans in game-based learning environments, which has implications for providing adaptive support for student self-regulated learning.

Keywords: Plan recognition · Game-based Learning · Self-regulated learning

1 Introduction

Self-regulated learning (SRL) describes learning that is guided by metacognition, strategic action, and motivated behavior [17, 20]. A key attribute of SRL is its focus on goal-driven learning. Self-regulated learners formulate goals and develop plans for achieving them, which are monitored and adapted based upon learners' self-evaluated progress [22]. Goal setting and planning is particularly important in scientific inquiry where

M. M. Rodrigo et al. (Eds.): AIED 2022, LNCS 13355, pp. 78–89, 2022.
https://doi.org/10.1007/978-3-031-11644-5_7

learning is guided by students' curiosity and motivation for acquiring knowledge, and where students need well-defined plans to carry out productive investigations [9]. Self-regulated learners set goals and sub-goals to complete a learning task [23]. To achieve their goals, students build plans that outline approaches, such as strategies or sequences of actions they intend to enact [22].

Learning environments that support goal setting and planning foster positive emotions and can create opportunities for student success [4]. Adaptive learning environments provide a way to scaffold student goal setting and planning in a manner that is individualized to each student. An important component of adaptive scaffolding is recognizing student goals and plans while the learner solves problems within the learning environment [1]. The task of plan recognition is focused upon predicting an individual's high-level goal, and the plan for achieving it, based on lower-level observations of the individual's strategies and actions. Goal recognition is considered a special case of plan recognition where the prediction task is focused only on recognizing high-level goals [3]. While there has been considerable work on modeling student knowledge in adaptive learning environments, limited research has been done on student plan recognition.

This paper presents a novel student plan recognition framework that uses machine learning to build goal and plan recognition models to predict students' problem-solving goals and the series of actions students intend to achieve them. The framework is evaluated with CRYSTAL ISLAND, a game-based learning environment for middle school microbiology, in which students utilize a novel planning support tool that encourages them to externalize their goal setting and planning processes during science problem solving. We utilize trace log data from students' interactions with the planning support tool, as well as their other problem-solving actions in the game, to train multi-label classification models to predict students' goals and plans. Specifically, we predict labels derived from student goals and a cluster-based representation of planned actions for the goal recognition and plan recognition tasks, respectively. We present results from a comparison of six machine learning-based classification techniques (support vector machines, random forest, naive Bayes, logistic regression, multilayer perceptron, long short-term memory networks) for modeling student goals and plans in CRYSTAL ISLAND. Our findings indicate that long short-term memory (LSTM) networks show promise in both goal and plan recognition tasks, which have potential to inform real-time scaffolding to support student goal setting and planning.

2 Related Work

Goals and plans are critical in SRL. Winne and Hadwin's Information Processing Theory of SRL posits that, throughout goal setting, planning, and enactment, students are continually monitoring and controlling how their learning is unfolding so that they are in control of their learning processes, and they are monitoring how effective these processes are in contributing to learning, information processing, and task completion [21, 22]. This implies that students know to set subgoals, use the appropriate and effective cognitive and metacognitive SRL strategies, and adapt the use of these strategies. However, how middle school students set goals and plans during science problem solving is not well understood, leaving key questions regarding how to effectively support student goal setting and planning in science learning environments [20].

Despite the importance of student goal setting and planning in SRL, there has been relatively little work on devising computational models of student plan recognition in adaptive learning environments. An important exception is work on Andes, an intelligent tutoring system for physics, which utilized Bayesian networks to model student plans and make predictions about student actions during problem solving [6]. This work exemplifies a successful application of plan recognition that informs adaptive support to provide students with specialized help through hints.

Prior work has also investigated a restricted form of student plan recognition, i.e., student goal recognition, using trace log data from student interactions with a game-based learning environment. A set of eleven goals were inferred from player activity. Authors explored a variety of event representations, models, and different evaluation metrics for accuracy and efficiency [10, 13, 15]. The most recent work found using one-hot encoding vectors to represent in-game events as input for LSTMs achieved the best performance predicting these game activity-derived goals [14]. Additionally, prior work has highlighted similarities between natural language processing and plan recognition, demonstrating the effectiveness of applying various natural language processing techniques (NLP) to plan recognition tasks [2, 7].

In this work, we extend these findings by devising a novel student plan recognition framework that uses students' in-game actions and planning support tool usage as observed input and leverages neural embedding-based representations of student action sequences from students' externalized plans to produce target labels. This framework utilizes two multi-label classifiers to compare six machine learning-based classification techniques for modeling student goals and plans in CRYSTAL ISLAND. Our aim is to demonstrate that a machine learning-based framework for student plan recognition can accurately model student goals and plans during science problem solving in a game-based learning environment.

3 Goal Setting and Planning in CRYSTAL ISLAND

3.1 Planning Support Tool in CRYSTAL ISLAND

To investigate predictive models of student goal setting and planning during science problem-solving, we utilize a game-based learning environment for middle school microbiology. CRYSTAL ISLAND features an interactive science mystery that engages students in a process of scientific inquiry as they investigate the source of a mysterious disease outbreak on a remote island research station. Students assume the role of an infectious disease investigator who is tasked with diagnosing the outbreak and recommending a treatment and prevention plan.

In order to support student goal setting and planning in CRYSTAL ISLAND, we have developed a planning support tool that incorporates design concepts from visual programming languages [19] and AI planning [8]. Specifically, students utilize a block-based visual interface to assemble hierarchical (i.e., two-layer) plans consisting of high-level goals and low-level sequences of actions that can be enacted in CRYSTAL ISLAND (Fig. 1). Students choose from a palette of pre-defined goal and action blocks in the tool. The goal blocks represent possible subgoals that students may wish to achieve on their way to solving the mystery, which are the overarching goal of the problem-solving scenario. Example goals include "Learn about outbreak" and "Report evidence-based diagnosis". Each action block lists specific steps that students can take to achieve a goal. Example actions include "Read about how diseases spread" and "Use scanner to test objects". Goal and action blocks are connected to form plans. For example, if a student sets a goal to "Explore Island", they can place movement actions such as "Go to Infirmary" under the goal block to indicate a necessary step needed to complete the specified goal.

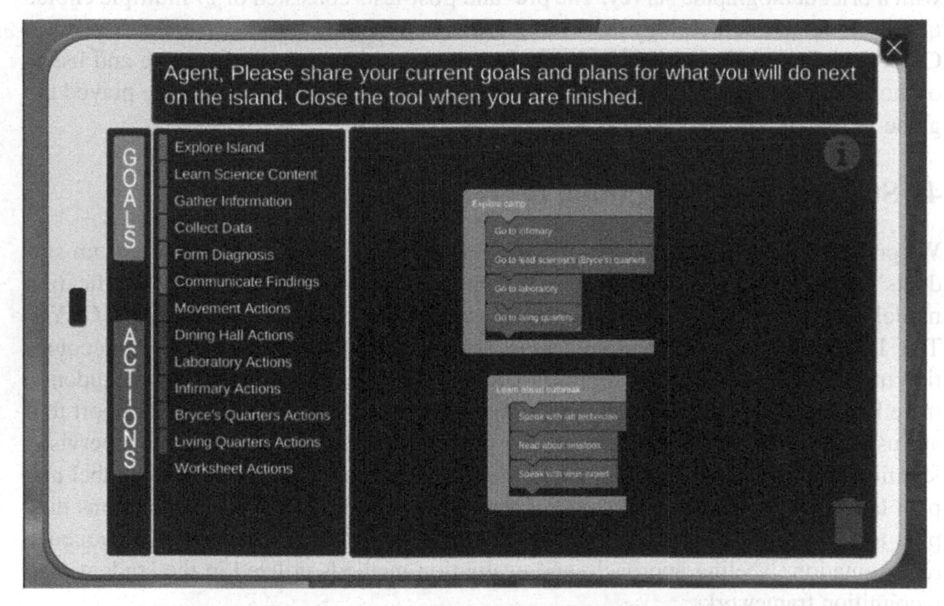

Fig. 1. Planning support tool in the CRYSTAL ISLAND learning environment.

Prior to engaging with CRYSTAL ISLAND, students watch a short, narrated video that introduces the planning support tool and demonstrates how to use the tool to build a plan. Once students begin using the game, they are prompted early on to set their own goal(s) and build plans using the tool. Students use the tool by dragging and dropping goal and action blocks onto a virtual canvas that serves as the planning area. After they have formulated a plan, they can close the tool and choose to enact their plan (or not) within the CRYSTAL ISLAND virtual environment. If students complete a goal or want to remove a goal that they previously chose, they can drag the block to a trash icon in the planning support tool. Upon deleting a goal block, students are prompted to indicate whether they reached the discarded goal or not. Students are presented with mandatory prompts to use the tool at major milestones in the science mystery, as well as every thirty minutes during gameplay, and may also voluntarily access the planning support tool at any time.

3.2 Goal Setting and Planning Dataset

A study was conducted with 144 middle school students in the United States. Of these students, 60% were female and the average age was 13.2 years. Students played CRYSTAL ISLAND remotely during asynchronous science class time due to a transition to remote learning during the COVID-19 pandemic. Students were instructed to access the game over a two-day span and were not given a time limit to complete the game. Students also completed pre- and post-tests to assess science content knowledge, along with a brief demographic survey. The pre- and post-tests consisted of 17 multiple choice questions about microbiology that could be answered based on the curricular content in CRYSTAL ISLAND. Interaction logs of students' actions within the game and usage of the planning support tool were logged automatically. Students on average played the game for 94.7 min (SD = 47.7).

4 Student Plan Recognition in Game-Based Learning

We present a student plan recognition framework that utilizes trace log data from students' planning support tool usage and gameplay to induce multi-label classification models to predict student goals and plans during science problem solving in the CRYSTAL ISLAND game-based learning environment. The input to the student plan recognition models is a feature vector representation of student actions distilled from students' trace log data from the game. Students' goals and plans from the planning support tool are used to devise labels for training the plan recognition models using a supervised learning approach. Specifically, each student action is annotated with a goal label and plan label that signify the goal students are attempting next and the set of actions they plan to take to achieve that goal, respectively. Below we describe the event sequence representation, labeling approach, and evaluation methods utilized in the student plan recognition framework.

4.1 Event Sequence Representation

Student interactions with CRYSTAL ISLAND generate trace log data that consists of timestamped sequences of actions taken by students while playing the game. We refer to these as event sequences. Based on prior work, each student action in an event sequence is represented by three types of features: action types, action arguments, and locations [14].

- **Action type.** Action type refers to categories of in-game activities undertaken by the student within the learning environment. These actions ranged from viewing posters and reading articles about viruses and bacteria to scanning items and talking to characters. For example, "Movement" signifies moving to a particular location or "Conversation" means a student had a conversation with a non-playable character in the game. There were 9 total action types.
- **Action argument.** Action arguments provide more details about the action type. For example, if the action type is "BooksAndArticles", the title of the book or article the student read is included as the action argument. There were 108 unique action arguments.

- **Location.** Location represents the region of the virtual island where the action took place. If the action type is "Movement", the location is the place where the student moved to. There were 24 unique locations in the game.

To prepare the dataset for student plan recognition, event sequences were segmented according to student usage of the planning support tool. The intuition for this approach is that students externalize their goals and plans using the planning support tool. Afterward, they enact their plans by performing actions in the game. An event sequence concludes when the student next reopens the planning support tool and changes their goals or plans, thereby initiating a new event sequence. In other words, an event sequence begins with the first student action after the planning support tool is closed. The event sequence concludes with the last student action before next opening the planning support tool. In total, there were 400 event sequences across all students. The length of event sequences ranged from 1 to 454, with a median of 30. The event sequences were constructed cumulatively to allow for action-level prediction, with the maximum length of a sequence being 30. For example, events one through 30 between planning support tool uses would translate to 30 rows of data, the first row only containing the first event, the second containing the first and second event, and so on up to 30. Because LSTMs require fixed-length input sizes, sequences of less than length 30 were zero-padded. Once the event sequences were created, we used one-hot encoding to convert student actions into a vector representation. One-hot encoding vectors have been shown to work effectively in prior work on student goal recognition in game-based learning environments [14].

Each plan that students constructed in the planning support tool consisted of a goal and a set of actions. We utilized student goals from the planning support tool to devise labels for the goal recognition task, and we used sets of actions from the planning support tool to devise labels for the plan recognition task. Event sequences were assigned labels based upon students' plans from their prior use of the planning sup-port tool. To illustrate, consider the following example. A student opens the planning support tool and creates a plan consisting of a goal and a set of actions (i.e., Plan 1). The event sequence that follows this planning support tool interaction is assigned a goal and plan label based upon the goals and set of actions that are included in Plan 1.

4.2 Goal Recognition Labels

The planning support tool allows students to select from 20 possible goals and was designed so that each goal falls into one of 5 categories: (1) Collect Data, (2) Communicate Findings, (3) Form Diagnosis, (4) Learn Science Content, and (5) Gather Information. For our analysis, these five categories serve as goal labels, rather than using all 20 lower-level goals. Since students can create multiple plans at a time, we formalized goal recognition as a multi-label classification task, assigning each event sequence a binary label vector in which each element of the vector corresponds to a possible goal category. The dataset had the following distribution of goal categories: (1) Collect Data: 22%, (2) Communicate Findings: 4%, (3) Form Diagnosis: 13%, (4) Learn Science Content: 22%, and (5) Gather Information: 40%.

4.3 Plan Recognition Labels

The planning support tool allows students to select from 55 possible actions to build plans for achieving their intended goals. Similar to goals, the palette of actions in the planning support tool was divided across six action categories. We utilized these higher-level categories to represent the actions in students' plans. Students' plans typically contained more than one action associated with a goal, with an average of 2.58 (SD = 1.96) actions per goal. To convert the action sets into labels for student plan recognition, the following procedure was applied. First, all actions in a plan were concatenated using the same order that students specified in the planning support tool. Next, SpaCy word embeddings were applied to each categorical action set [18]. The resulting embeddings were averaged for each set of actions in a plan. Next, k-means clustering was applied to the word embeddings to separate the plans into clusters. The number of clusters was determined visually using the Elbow method, resulting in 4 distinct groups of action sets [5]. The resulting clusters were used to derive 4 possible class labels for plan recognition.

When reviewing patterns of action categories within the clustering, it seemed that the most used action category in each plan aligned within the clusters. Cluster 0 (9%) represents plans that mostly contain "Read Science Content". Cluster 1 (30%) represents primarily "Explore" action category usage. Cluster 2 (33%) represents plans that contain mostly "Gather and Scan Items", and Cluster 3 (28%) represents plans that contain mostly "Speak with Characters".

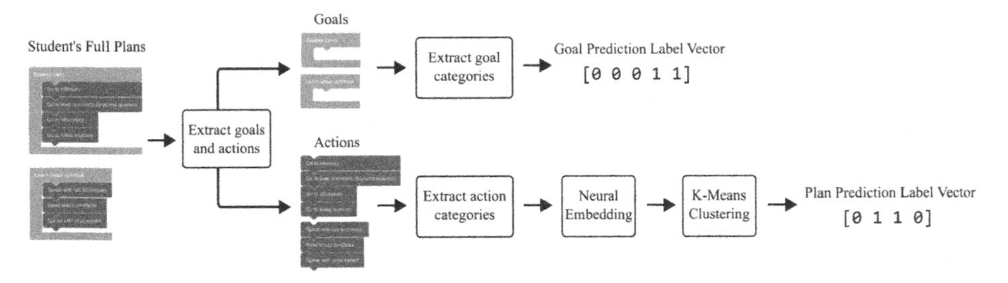

Fig. 2. Procedure for translating student plans into multi-label vectors for student goal recognition (top) and student plan recognition (bottom).

These labels were assigned to event sequences in a multi-label fashion, similar to the goal recognition task. Figure 2 illustrates the process for translating students' plans into label vectors for goal recognition and plan recognition, respectively.

4.4 Model Selection and Evaluation

We examined six different supervised learning techniques to induce multi-label classifiers for student goal recognition and plan recognition: support vector machines (SVM), random forest (RF), naive Bayes (NB), logistic regression (LR), multi-layer perceptron (MLP), and long short-term memory (LSTM) networks. These models were selected to establish a general baseline of results. Since this task has not been completed previously, we chose mostly non-sequential models to analyze patterns of overall performance. We performed nested 5-fold cross validation using an iterative grid search for hyperparameter tuning of all six models. Due to the limited representation of some of the labels, we could not choose a k any greater than 5 without having one of the classes no longer represented in the training or test set. We used a stratified student-level split within the nested cross validation to maintain a similar class distribution across the training and test sets and to prevent data leakage between folds. For the non-LSTM models, we took the sum of the one-hot encoding vector across events to handle different lengths of sequences and created a single vector representing the number of times each type of action occurs in a sequence. The LSTM received the entire one-hot encoding vector as input.

We utilized the macro-average F-measure to evaluate the models. F-measure has been shown to be a good indicator of model performance in multi-label classification tasks because it highlights incorrectly classified labels by basing the calculations on false positives and false negatives [11, 12]. Since false positives and false negatives are instances that can create user frustration, they are important indicators of performance in an adaptive learning environment. In addition, we have an uneven distribution of classes for both the goal and plan recognition tasks. Macro-average F-measure works well on imbalanced datasets because it computes the average for each class label separately and then aggregates them together [16]. Therefore, this metric is well suited for evaluating models intended for use in adaptive learning environments.

5 Results

To investigate the effectiveness of the machine learning-based goal recognition and plan recognition models, we compared all models against a baseline model that always predicts the majority class.

5.1 Goal Recognition Results

Goal recognition results for all six models are shown in Table 1. All models except random forest improved on the baseline in four out of five goal categories. Random forest appeared to overfit to the majority class, and it performed similarly to the baseline model. In some cases, an imbalance of the class labels causes classifiers to ignore the less-represented classes, which could cause a model to overfit to the majority class. Because random forest makes decisions based on information gain, it makes sense that it would often favor choosing the majority class. The LSTM was among the top two highest-performing models for four out of five classes, including one of the least represented goal categories (i.e., Form Diagnosis). SVM, NB, LR and MLP all improved on the baseline with respect to the macro-average F-measure. The LSTM showed the greatest improvement on the baseline with a 42% relative improvement in the F-measure.

5.2 Plan Recognition Results

Table 2 shows the plan recognition results for all six machine learning models, as well as the baseline. For the plan classes 0, 1 and 2, all machine learning-based models improved on the baseline. Naive Bayes showed the highest macro-average F-measure for plan

Table 1. Average F-measure for each classification model and goal category in the student goal recognition task. Distributions in results represent the test set and are averaged across 5 folds of cross validation.

	Collect data	Comm. findings	Form diagnosis	Learn science content	Gather info.	Overall
N dis.t	21%	3%	3%	24%	49%	
	F	F	F	F	F	Macro F
Maj.	0.00	0.00	0.00	0.00	0.74	0.15
SVM	0.20	0.07	0.22	0.20	0.71	0.28
RF	0.00	0.00	0.00	0.00	**0.74**	0.15
NB	**0.42**	0.12	0.23	**0.43**	0.58	0.35
LR	0.24	0.16	0.40	0.27	0.67	0.35
MLP	0.29	0.19	0.31	0.16	0.64	0.32
LSTM	0.32	**0.35**	**0.47**	0.35	0.62	**0.42**

classes 0 and 1. This could be due to the model attributing most input actions to all four plan classes, causing the results to be improved. The multi-layer perceptron outperformed the baseline model on the majority plan class, which indicates it more precisely predicted the majority plan class than any other approach. The LSTM performed best again for the least represented plan class. All models improved on the macro-average F-measure compared to the majority baseline.

Table 2. Average F-measure for each classification model and plan class in the student plan recognition task. Distributions in results represent the test set and are averaged across 5 folds of cross validation.

Plan class	0	1	2	3	Overall
N dist.	8%	27%	28%	36%	
	F	F	F	F	Macro F
Maj.	0.00	0.00	0.00	0.55	0.14
SVM	0.36	0.35	0.20	0.29	0.30
RF	0.31	0.41	0.00	0.18	0.22
NB	**0.53**	**0.54**	0.17	0.48	**0.43**
LR	0.46	0.50	0.21	0.43	0.40
MLP	0.29	0.19	**0.31**	**0.64**	0.32
LSTM	0.48	0.47	**0.31**	0.38	0.40

6 Discussion

Overall, the machine learning-based models show clear improvement with respect to macro-averaged F-measure over a naive baseline on the student goal and plan recognition tasks. Prior work on student goal recognition found LSTMs to be the best performing model on a multiclass goal recognition task [14]. Our work extends these findings by showing that LSTMs also perform effectively for goal recognition in a multi-label context. Student plan recognition proved to be a more difficult task than student goal recognition. Unlike goal recognition, there was not a single model that performed best across all plan classes. For example, naive Bayes showed the highest macro-average F-measure, but its predictions were consistently every plan class for a given set of input actions. This type of prediction is not ideal to inform run-time scaffolding because it does not provide a precise indication of what students are planning.

The imbalanced labels in the dataset presented challenges in training and evaluating the models for student goal recognition and plan recognition. However, it is representative of the types of plans generated by students through their use of the planning support tool in CRYSTAL ISLAND. Notably, we saw planning support tool usage decrease over time, with students trending toward using the tool frequently in the first half of the game, but less so as time went on. There were also different levels of granularity associated with the different goal categories and plan classes. For example, goals related to gathering information typically occurred early in the game, and they encompassed a relatively broad set of possible actions. In comparison, goals in the Communicate Findings category ideally occurred after a student formed a hypothesized diagnosis, which typically occurs later in the game. The steps involved to communicate findings are directly outlined in the game, and as a result, one would expect plans related to this goal to occur less frequently. Encouragingly, the results show the promise of using machine learning-based multi-label classification techniques for student goal and plan recognition despite the inherent challenges of imbalanced data.

The wide variety of student plans also presented distinctive challenges for plan recognition. Some students frequently used the planning support tool and updated plans without being prompted, while other students opened and closed the planning support tool only when required. This limits our framework because if students do not update their plans, our framework interprets all input actions as being towards the same goal and plan. Similarly, if students use the planning support tool sparingly, then the goal and action labels might not be fully representative of the event sequences enacted in between planning support tool uses Further enhancements to the framework could be added by identifying when a plan has been completed through gameplay or a goal, so it is not singularly relying on students to update their goals and plans. Additionally, more work could be done to predict goal abandonment based on how long a goal or plan persists in the planning support tool interactions. Such improvements could alter the distribution in goal and plan labels and potentially help with recognition performance. Additionally, more work could be done to predict goal abandonment based on how long a goal or plan persists in the planning support tool interactions. Such improvements could alter the distribution in goal and plan labels and potentially help with recognition performance.

7 Conclusion

Goal setting and planning are key components of self-regulated learning. Adaptive learning environments show significant promise for adaptively scaffolding students' goal setting and planning processes, but they require computational models of student plan recognition to do so. This work presents a student plan recognition framework that leverages student goals and plans captured during interactions with a novel planning support tool in a game-based learning environment for middle school microbiology. Students' goals and plans were used to derive labels to formalize goal and plan recognition as multi-label classification tasks. Several machine learning techniques were evaluated to predict students' goal and plan labels based upon observations of their problem-solving actions in the game. In both tasks, we saw significant improvement on the majority baseline with most machine learning models. LSTMs showed particular promise in both the goal recognition and plan recognition tasks with respect to their ability to perform well across all classes.

The results indicate the potential of integrating student plan recognition models into real-time adaptive learning environments. Plan recognition models could be used to drive adaptive scaffolding in the form of open learner models of student goal setting and planning processes, or they could drive adaptive hints and prompts related to student SRL. Additionally, future work could investigate additional nuances of student goal setting and planning, which will contribute to more robust models because students can work towards multiple goals and plans at a time or abandon goals and plans without updating their planning support tool. Lastly, exploring additional sequential models and a multi-task learning approach to student goal recognition and plan recognition is a promising direction for future work.

Acknowledgements. This research was supported by funding from the National Science Foundation under grant DUE-1761178. Any opinions, findings, and conclusions expressed in this material are those of the authors and do not necessarily reflect the views of the NSF.

References

1. Azevedo, R., Wiedbusch, M.: Theories of metacognition and pedagogy applied in AIED systems. In: du Boulay (ed.) Handbook of Artificial Intelligence in Education. Springer, The Netherlands (in Press)
2. Baikadi, A., et al.: Towards a computational model of narrative visualization. In: Workshops at the Seventh Artificial Intelligence and Interactive Digital Entertainment Conference, pp. 2–9, October 2011
3. Blaylock, N., Allen, J.: Corpus-based, statistical goal recognition. In: International Joint Conference on Artificial Intelligence, vol. 3, pp. 1303–1308, August 2003
4. Boekaerts, M., Pekrun, R.: Emotions and emotion regulation in academic settings. Handb. Educ. Psychol. **3**, 76–90 (2015)
5. Bholowalia, P., Kumar, A.: EBK-means: a clustering technique based on elbow method and k-means in WSN. Int. J. Comput. Appl. **105**(9) (2014)
6. Conati, C., Gertner, A., Vanlehn, K.: Using Bayesian networks to manage uncertainty in student modeling. User Model. User-Adap. Inter. **12**(4), 371–417 (2002)

7. Geib, C., Steedman, M.: On natural language processing and plan recognition. In: International Joint Conference on Artificial Intelligence, vol. 2007, pp. 1612–1617, January 2007
8. Ghallab, M., Nau, D., Traverso, P.: Automated Planning and Acting. Cambridge University Press (2016)
9. Graesser, A., McNamara, D.: Self-regulated learning in learning environments with pedagogical agents that interact in natural language. Educ. Psychol. **45**(4), 234–244 (2010)
10. Ha, E.Y., Rowe, J.P., Mott, B.W., Lester, J.C.: Goal recognition with Markov logic networks for player-adaptive games. In: Seventh Artificial Intelligence and Interactive Digital Entertainment Conference, pp. 32–39, October 2011
11. Liu, S.M., Chen, J.H.: A multi-label classification based approach for sentiment classification. Expert Syst. Appl. **42**(3), 1083–1093 (2015)
12. Madjarov, G., Kocev, D., Gjorgjevikj, D., Džeroski, S.: An extensive experimental comparison of methods for multi-label learning. Pattern Recogn. **45**(9), 3084–3104 (2012)
13. Min, W., Ha, E.Y., Rowe, J., Mott, B., Lester, J.: Deep learning-based goal recognition in open-ended digital games. In: Tenth Artificial Intelligence and Interactive Digital Entertainment Conference, pp. 37–43, September 2014
14. Min, W., et al. Multimodal goal recognition in open-world digital games. In: Thirteenth Artificial Intelligence and Interactive Digital Entertainment Conference, pp. 80–86, September 2017
15. Min, W., Mott, B.W., Rowe, J.P., Liu, B., Lester, J.C.: Player goal recognition in open-world digital games with long short-term memory networks. In: Twenty-Fifth International Joint Conference on Artificial Intelligence, pp. 2590–2596, July 2016
16. Pereira, R.B., Plastino, A., Zadrozny, B., Merschmann, L.H.: Correlation analysis of performance measures for multi-label classification. Inf. Process. Manage. **54**(3), 359–369 (2018)
17. Schunk, D.H., Greene, J.A.: Historical, Contemporary, and Future Perspectives on Self-Regulated Learning and Performance, pp. 1–15. Routledge, New York (2017)
18. Srinivasa-Desikan, B.: Natural Language Processing and Computational Linguistics: A Practical Guide to Text Analysis with Python, Gensim, spaCy, and Keras. Packt Publishing Ltd.. (2018)
19. Weintrop, D., Wilensky, U.: Transitioning from introductory block-based and text-based environments to professional programming languages in high school computer science classrooms. Comput. Educ. **142**, 103646 (2019)
20. McComas, W.F.: Metacognition. In: McComas, W.F. (ed.) The language of science education, pp. 63–63. SensePublishers, Rotterdam (2014). https://doi.org/10.1007/978-94-6209-497-0_55
21. Winne, P.H., Hadwin, A.E.: Studying as Self-Regulated Learning, pp. 291–318. Routledge, New York (2018)
22. Winne, P.H., Hadwin, A.F.: The weave of motivation and self-regulated learning. In: Schunk, D.H., Zimmerman, E.J. (eds.) Motivation and Self-Regulated Learning: Theory, Research, and Applications, pp. 297–314. Lawrence Erlbaum Associates Publishers (2008)
23. Winne, P.H.: Theorizing and researching levels of processing in self-regulated learning. Br. J. Educ. Psychol. **88**(1), 9–20 (2018)

Identifying and Comparing Multi-dimensional Student Profiles Across Flipped Classrooms

Paola Mejia-Domenzain[1](✉), Mirko Marras[2], Christian Giang[1,3], and Tanja Käser[1]

[1] EPFL, Lausanne, Switzerland
{paola.mejia,christian.giang,tanja.kaeser}@epfl.ch
[2] University of Cagliari, Cagliari, Italy
mirko.marras@acm.org
[3] University of Applied Sciences and Arts of Southern Switzerland, Locarno, Switzerland

Abstract. Flipped classroom (FC) courses, where students complete pre-class activities before attending interactive face-to-face sessions, are becoming increasingly popular. However, many students lack the skills, resources, or motivation to effectively engage in pre-class activities. Profiling students based on their pre-class behavior is therefore fundamental for teaching staff to make better-informed decisions on the course design and provide personalized feedback. Existing student profiling techniques have mainly focused on one specific aspect of learning behavior and have limited their analysis to one FC course. In this paper, we propose a multi-step clustering approach to model student profiles based on pre-class behavior in FC in a multi-dimensional manner, focusing on student effort, consistency, regularity, proactivity, control, and assessment. We first cluster students separately for each behavioral dimension. Then, we perform another level of clustering to obtain multi-dimensional profiles. Experiments on three different FC courses show that our approach can identify educationally-relevant profiles regardless of the course topic and structure. Moreover, we observe significant academic performance differences between the profiles.

Keywords: Clustering · Time series · Self-regulated learning

1 Introduction

Flipped Classrooms (FC) courses are a form of blended learning where students complete pre-class activities before attending interactive face-to-face sessions. These courses allow students to conveniently access learning resources and independently manage their studying time, which requires a high degree of self-regulation. While pre-class activities are essential for course success [2,16], students often do not engage with such activities due to a lack of motivation, time,

M. M. Rodrigo et al. (Eds.): AIED 2022, LNCS 13355, pp. 90–102, 2022.
https://doi.org/10.1007/978-3-031-11644-5_8

or necessary skills [22]. Understanding student behavior in pre-class activities can hence help the teaching staff identify these reasons and timely intervene.

Nevertheless, a large number of prior studies on FC have mainly focused on the effectiveness and implementation of the approach rather than on students' learning strategies during the course [22]. Moreover, the studies on learning strategies have formerly used self-evaluating questionnaires [9,26], which can be biased and do not acknowledge the dynamic nature of learning. Fewer works have used log data from pre-class activities to predict student success. For instance, [3] showed that the video usage frequency is correlated to student success, [1] predicted homework grades by modeling student strategies as clickstream event n-grams, and [17,28] identified at-risk students based on clickstream features.

Regarding clustering approaches to profile student learning behavior in FC, [12] identified student learning strategies by examining the distribution of learning actions in students' pre-class online sessions. In subsequent work, the same authors examined student regularity of pre-class activities and its association with course grades [13]. Other works used clustering techniques to analyze student time management skills [6], study the evolution of video usage indicators [23], and analyze consistency in student learning [25].

However, most of the aforementioned studies have investigated one specific FC course only (e.g., [12,23]) and/or focused on one specific aspect of student learning behavior (e.g., consistency [25], time management [13]). In other digital learning environments, such as massive open online courses, [18] identified rule-based clusters and explored the movement of students across clusters over time. However, no groups of students with similar changing behavior were studied. In contrast, [5] analyzed how students changed their studying strategies during the course, but did not incorporate multiple student behavioral aspects like [18].

In this paper, we investigate the integration of multiple dimensions of student behavior, including self-regulated learning (SRL), in data-driven student profiles. To this end, we propose a multi-step clustering pipeline based on previous findings on SRL in online education. In the first step, we model students' log data as time series and cluster student behavior individually in terms of effort, consistency, regularity, proactivity, control, and assessment. Through a second level of clustering, we integrate the obtained behavioral patterns into interpretable multi-dimensional profiles. With our approach, we aim to combine multiple behavioral dimensions to obtain interpretable student profiles in FC and study how these profiles compare across FC courses (**RQ1**); as well as analyze the relationship between the detected profiles and academic performance (**RQ2**). Our analysis on three FC courses shows that profiles integrating multiple dimensions can be identified and interpreted using clusters' prototypes and that sometimes similar profiles emerge regardless of the course topic and structure. We also find a significant variance in academic performance across profiles. The obtained profiles hence contribute to teachers' understanding of student behavior, enabling better-informed course decisions and student interventions.

2 Method

To investigate student behavior in FC, we propose the multi-step clustering pipeline depicted in Fig. 1. We first extract features from pre-class log data to explain relevant dimensions of learning behavior (Sect. 2.1). Instead of clustering the multiple features in a single step, we propose a multi-step approach that allows a better interpretation and understanding of the cluster composition and characteristics. Thus, we perform a first clustering step separately for each dimension (Sect. 2.2); and a second clustering step, in which we integrate the obtained behavioral patterns into multi-dimensional profiles (Sect. 2.3). Source code accompanying this paper: https://github.com/epfl-ml4ed/fc-clustering.

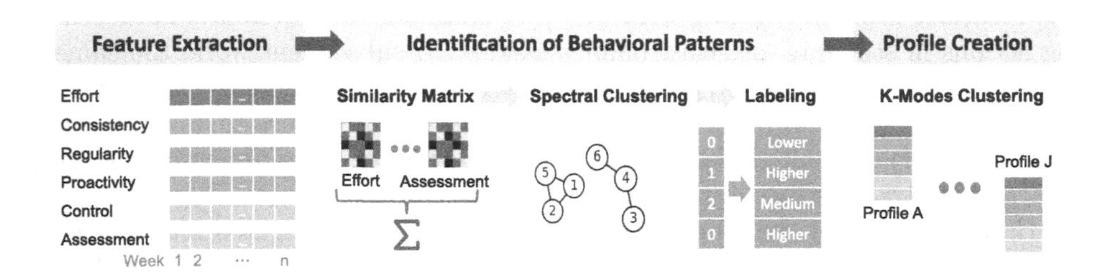

Fig. 1. Overview of the clustering pipeline.

2.1 Feature Extraction

Different aspects of SRL have been researched extensively (e.g., [9,25]). In a meta-analysis on online education, [7] found significant associations with academic achievement for five sub-scales of SRL: effort regulation (persistence in learning), time management (ability to plan study time), metacognition (awareness and control of thoughts), critical thinking (ability to carefully examine material), and help-seeking (obtaining assistance if needed). Based on these findings, we use the following dimensions to represent student behavior: effort regulation (*Effort*), time management (*Consistency*, *Regularity*, *Proactivity*), and metacognition (*Control*). The nature of our log data does not allow us to represent critical thinking and help-seeking. Assuming that there will be a significant association between performance in pre-class activities and course grades (e.g., [16,17,28]), we add a sixth dimension (*Assessment*) to our representation of student behavior. We measure these dimensions using features that proved to be relevant in prior work analyzing learning strategies in online or blended learning (e.g., [6,8,15,17,20]). Table 1 shows the dimensions and their respective features.

The first dimension, ***Effort***, aims to monitor the intensity of student engagement in the course, which is fundamental for learning success [9]. In contrast, ***Consistency*** is concerned with the relative shape of student events, measuring how student effort varies over time. Specifically, it estimates the intra-course time management skills of the students, an important SRL aspect [7,25]. The ***Regularity*** dimension is also associated with time management; it estimates the

Table 1. Features are grouped into six different dimensions. Each feature stems from a relevant prior study and is accompanied by a short description.

Dimension[a]	Feature	Description
Effort	Total time online [8]	Sum of session durations
	Total video clicks [8]	Video events (play, pause, stop, seek)
Consistency	Mean session duration [8]	Time measured in minutes
	Relative time online	Unit vector of total time online
	Relative video clicks	Unit vector of total video clicks
Regularity	Periodicity of week day [6]	Studying on certain day(s) of the week
	Periodicity of week hour [6]	Studying at certain hours of the day
	Periodicity of day hour [6]	Studying on certain day(s) & hours of the week
Proactivity	Content anticipation [17]	Fraction of videos (from subsequent weeks) watched before the scheduled due date
	Delay in lecture view [6]	Time interval between the first views and the due date of videos of prior weeks
Control	Fraction spent [20]	Real time spent watching the video divided by its duration, averaged across videos
	Pause action frequency [15]	Mean number of pauses divided by the time spent watching a video per video
	Average change rate [20]	Mean playback speed used to watch videos
Assessment	Competency strength [17]	Highest grade achieved by the student on a quiz divided by the number of attempts
	Student shape [17]	Student's tendency of obtaining the maximum grade in a quiz in the first attempt

[a]Features names taken from original papers and implementation from [17].

intra-week and intra-day time management patterns (i.e., capturing whether a student is regularly engaged on specific weekdays or day times), which have been proved to be predictive of student success in MOOCs [6] and FC [13]. Another dimension of time management, ***Proactivity***, attempts to measure the extent to which students are on time or ahead of the schedule [11]. Engagement in pre-class activities has shown to be associated with exam performance [2,16]. The ***Control*** dimension models the in-video behavior as a proxy of student ability to control the cognitive load of video lectures (metacognition). The flow of video information can result in cognitive overload and thus regular pauses can improve learning outcomes [4]. In the platform, students are provided with functionalities (e.g., pause button) to control video flows [4]. Finally, the ***Assessment*** dimension assumes that there is a relation between student performance in voluntary non-graded online quizzes and the final course grade (e.g., [16,17,28]). Given that learning is dynamic in nature [27], we model features as week-wise time series (length equal to the number of course weeks). The only exceptions are the *Regularity* features, whose computation requires evidence from all course weeks and thus are computed for the whole course as a scalar.

2.2 Identification of Behavioral Patterns

The first clustering step is done separately per dimension: we compute a pairwise similarity matrix, feed it into a spectral clustering, and interpret its labels.

Similarity Matrix. First, we compute a pairwise distance matrix between students separately for each feature. To compute distances between student time series, we use the Dynamic Time Warping (DTW) distance [25]. DTW can identify similar patterns (e.g., peaks) regardless of small variations (shifts) in time. In contrast, we use the Euclidean distance for the *Regularity* features, since they are scalars and not time series. Second, we apply a Gaussian kernel to transform the distance matrix into a similarity matrix. The standard deviation (σ) of the kernel controls the blurring degree, which is useful to reduce the impact of students with extreme behavior. We then add the similarity matrices of the features of each dimension to get the dimension similarity matrix. We optimize the DTW window size (w) and the width of the Gaussian kernel (σ) per dimension via a grid search maximizing the clusters' *Silhouette* score (see next paragraph).

Spectral Clustering. We apply *Spectral Clustering* [21] to cluster the similarity matrix of each dimension separately. This clustering algorithm treats points as nodes in a graph and then solves the graph partitioning problem. Unlike *K-Means*, it is not limited to convex clusters. The algorithm outputs a vector containing the cluster identifiers for each student. In total, there are as many vectors as behavioral dimensions, and each vector length is equal to the number of students. We perform a grid search separately for each dimension using $k = 2, ..., 10$ clusters. We use the *Silhouette* score [24] to determine the optimal number of clusters as this heuristic is easy to interpret (higher scores indicate high separability between clusters).

Labeling. We label the obtained clusters for each dimension according to the intensity, shape (including peaks), and relation to key aspects of the course (e.g., exams), by thoroughly inspecting the time series of the students in each cluster. When the patterns differ in more than one attribute (e.g., intensity and shape), we choose the attribute that better explains each dimension. Labels are created relative to other clusters and not in absolute terms. For instance, labeling a cluster as *Higher Effort* does not mean effort exceeds a given threshold, but that students in this cluster work more intensively than those in the other clusters.

2.3 Profile Creation

The second clustering step integrates all dimensions into a single learner profile, enabling us to describe student behavior across dimensions (e.g., a cluster with *Higher Effort*, *Lower Assessment*, *Higher Control*, etc.). We are hence able to gain insights into the dependencies across dimensions. We take as input the five/six annotated labels (one per dimension) from Sect. 2.2 and cluster them using *K-Modes* (selecting K as in Sect. 2.2). *K-Modes* extends K-Means to use the mode (most frequent element) instead of the mean to compute cluster centroids from categorical data. These centroids provide insights into the cluster composition (e.g., [25]) and will be analyzed in the next section.

3 Experimental Evaluation

We evaluated our approach on three different FC courses. We first analyzed and compared the obtained profiles across courses (**RQ1**) and then investigated their relation to academic performance (**RQ2**). The study was approved by the institutional review board (HREC No. 058-2020/10.09.2020).

Table 2. Characteristics of the FC courses.

Course	Year	Semester	Students	Female	Event type	No. events	Fail
LA	2018/19	1	292	29%	Video + Quiz	1,033,962	41%
FP	2018	3	216	20%	Video	464,115	2%
PC	2019	4	147	14%	Video	156,375	11%

Data Set. Our analysis is based on the log data collected from an EPFL online institutional platform (custom Open edX installation) that tracked student pre-class activities (watching video lectures and solving quizzes) in three FC courses. The log entries are tuples reporting the user, the activity, and the timestamp (e.g., user: 10, activity: play video 32, timestamp: 05-03-2018 12:06:01). The three considered FC courses (Table 2) are compulsory courses for the Computer Science and Communication Systems Bachelor degrees in EPFL. The first data set was collected from two consecutive FC editions of the *Linear Algebra* (LA) course, taught by the same lecturer and with a flipped duration of 10 weeks. Among the three courses, this is the only one including online quizzes. The second data set was collected from the FC edition of a *Functional Programming* (FP) course with a flipped duration of 11 weeks. The third data set stems from a FC course in *Parallelism and Concurrency* (PC) lasted 15 weeks. It is important to note that this course was taught in a traditional way between weeks 4–7.

3.1 Behavioral Patterns and Multi-dimensional Profiles

We first examined the profiles obtained for LA and then compared the profiles and behavioral patterns across courses (LA, FP, PC). Table 3 shows the characteristics of the identified profiles for all courses, i.e., the centroids from the *K-Modes* clustering. The centroid is the mode (majority label) of each learning dimension. For instance, for profile *A*, the majority of students were labeled *Lower Effort*.

Profiling for LA. We identified five profiles (A, B, C, D and E) for LA. To visualize their patterns, we inspected the barycenters (centroid) of each cluster. To compute the barycenter, we used the DTW Barycenter Averaging method that averages time series considering the DTW alignment and window constraint.

Figure 2 shows the barycenters as lines and the Euclidean mean of each week as bars for *Effort*, *Assessment*, and *Control*. Concerning the *Effort* dimension, the students with *Lower Effort* were less active (in terms of online time and number of video clicks) than the students with *Higher Effort*. One profile (C) exhibits

Table 3. Percentage of students per profile for each course and profile description.

Profile	%			Dimension					
	LA	FP	PC	*Effort*	*Consistency*	*Regularity*	*Proactivity*	*Control*	*Assessment*
A	24			Lower	Uniform	Lower Peaks	Delayed	Lower	Lower
B	18	28	35	Lower	Uniform	Lower Peaks	Delayed	Higher	Higher
C	19		18	Higher	Uniform	Higher Peaks	Anticipated	Higher	Higher
D	21			Lower	Uniform	Higher Peaks	Delayed	Higher	Higher
E	18			Lower	Uniform	Higher Peaks	Anticipated	Higher	Higher
F		15	27	Higher	Midterm	Higher Peaks	Delayed	Higher	
G		25		Higher	Midterm	Lower Peaks	Anticipated	Higher	
H		14		Lower	Midterm	Lower Peaks	Delayed	Lower	
I		18		Higher	Midterm	Higher Peaks	Anticipated	Higher	
J			20	Lower	Midterm	Lower Peaks	Anticipated	Lower	

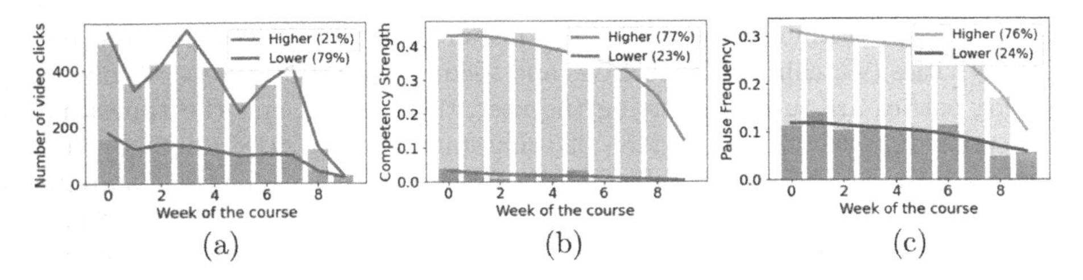

Fig. 2. Patterns for *Effort* (a), *Assessment* (b), and *Control* (c) for LA.

patterns of higher effort compared to the other four profiles. For *Assessment*, the difference between the detected patterns again lies mainly in the intensity (Fig. 2b). We observed two clusters, with one cluster (denoted as *Higher Assessment*) exhibiting a higher pattern than the other cluster (labeled as *Lower Assessment*). Different from *Effort*, most profiles showed *Higher Assessment.* The difference in competency strength between the two clusters is very large, with the *Lower Assessment* cluster having very low values. This observation could suggest that *Assessment* is reflecting the differences in students' willingness to solve the quizzes rather than measuring their actual quiz performance. For the *Control* dimension (Fig. 2c), we observed two groups: the *Higher* cluster (76%) had a greater ratio indicating that it pauses the video more often than the *Lower* cluster (24%). Higher pause frequency and longer pauses can be a result of students taking time to reflect on unclear or interesting parts of a video [15]. It is worth noting that *Control* and *Assessment* are the only dimensions that are paired. It is not surprising that the students that have *Higher Control* and manipulate the video content more are also the students with *Higher Assessment* that are engaged with the optional quizzes.

While the aforementioned dimensions mostly capture the differences in the intensity of student activity, *Consistency* captures differences in the relative intensity (in terms of online session time and video clicks) over the whole course. We obtained two distinct patterns shown in Fig. 3a. The majority of students (84%) worked consistently over time with little or no peaks (*Uniform*), while only a few students (16%) worked considerably more in the last week of the semester (*Final Exam*). Interestingly, all the LA profiles are labeled with *Uniform Consistency* (see Table 3). This means the *Uniform Consistency* students outnumber the *Final Exam Consistency* students in all profiles, indicating that the differences in other dimensions were more significant or separable.

Regarding the *Regularity* patterns, Fig. 3b shows an example of the relative frequency of events per day of the week for two example students. The (*Higher Peaks*) student worked only on Sundays, Mondays, and Tuesdays. In contrast, the student with (*Lower Peaks*) worked some weeks on Saturdays and other weeks on the other days of the week without a clear pattern (Fig. 3b). The in-person part of the course was taught on Tuesdays; this can explain the relative peak in activity for (*Higher Peaks*) on Monday. Students in profiles *A*, and *B* exhibit less regular working patterns than students in profiles *C*, *D*, and *E*.

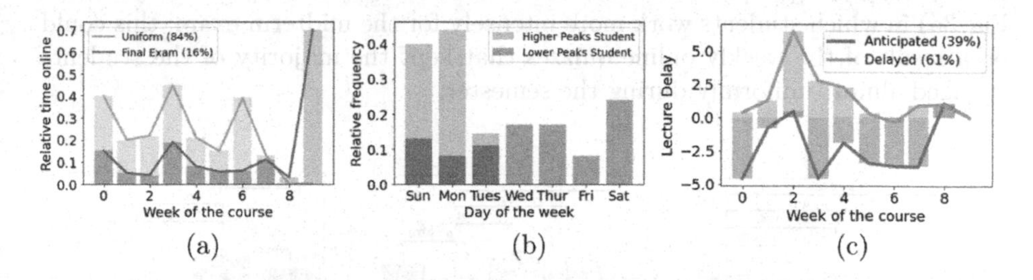

Fig. 3. Patterns for *Consistency* (a), *Regularity* (b) and *Proactivity* (c) for LA.

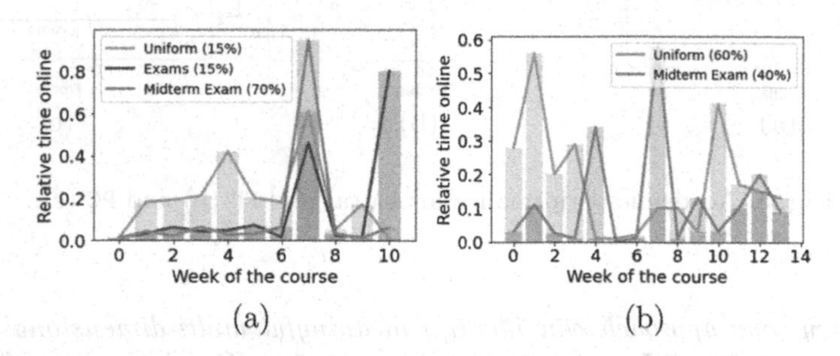

Fig. 4. *Consistency* patterns for FP (a) and PC (b). In FP, most of the students show increased activity for exams. In PC, the majority works consistently.

Unlike other dimensions, *Proactivity* includes two features with contrasting behavior: content anticipation and delay in lecture view. Figure 3c shows that the *Anticipated* cluster (39%) has negative values in delay in lecture view, while the *Delayed* cluster (61%) has positive values with a peak in the beginning.

Comparison Across Courses. In a second analysis, we compared the profiles from LA with the ones identified for the other two courses (FC and the PC). We obtained a total of 10 profiles, listed in Table 3. Profile *B* was found in all three courses and profile *C* and *F* were found in two out of three courses.

From Table 3, it seems that *Consistency* has peculiar behavioral patterns between courses. Figure 4 presents the relative time online for FP and PC. In FP, three different patterns were identified (Fig. 4a). The students that worked strongly for the midterm (*Midterm*), those that had more activity before both the midterm and final exam (*Exams*), and those that had a normal-shaped activity with a visible peak one week before the midterm (*Uniform*). For PC, we observed two distinct behaviors (Fig. 4b). A group of students worked more during the weeks before the midterm (*Midterm*), whereas another group worked more consistently over the semester (*Uniform*). Note that there were no videos from weeks 4 to week 7 in this course, which explains the drop in activity during these weeks for the *Uniform* group. In contrast to FP and PC, there is no pattern in LA (see Fig. 3a) in which students work more intensely for the midterm exam; this could be a result of the weekly online quizzes that kept the majority of the students engaged almost uniformly during the semester.

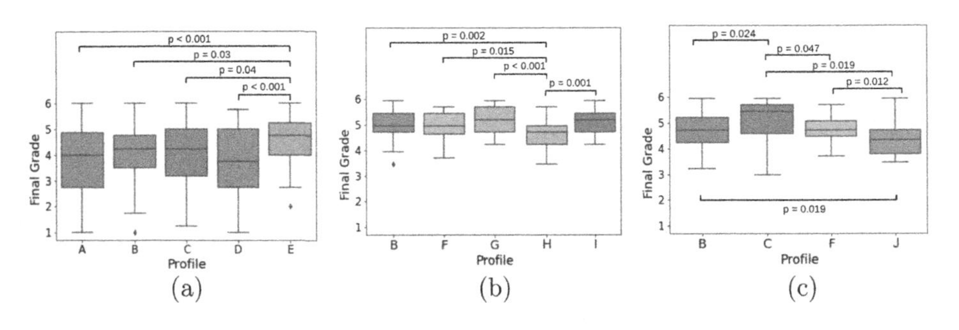

Fig. 5. Academic performance for LA (a) and FP (b) and PC (c).

*In summary, our approach can identify meaningful multi-dimensional profiles across courses with different topics and structures. We observed profiles with varying compositions and no completely aligned dimensions. The multiple combinations of dimensions reflect the complexity of learning behavior **(RQ1)**.*

3.2 Relation to Academic Performance

We finally explored the relationship between the profiles and academic performance, measured using the students' final course grade[1]. For LA, a significant Shapiro-Wilk normality test ($W = 0.96$, p = 2.6e−07) indicated that the grades were not normally distributed. We hence used the non-parametric Kruskal-Wallis test to identify significant differences between profiles ($\chi^2(4) = 12.7$, p = 5.3e−03). We then performed pairwise comparisons between profiles using the Wilcoxon Rank Sum test[2]. Subsequently, we replicated the analysis using the grades from PC and FP[3]. Figure 5 shows the distribution of grades per profile.

In LA, students in profile E have significantly higher grades than the students in profiles A, B, C, and D. These differences in performance also emerge in the failure rate for each profile. Students in profile E have a lower chance of failing the course (failure rate: 19%) compared to students in profiles D (57%), A (48%), C (38%), and B profiles (36%). When we compare the four other profiles to profile E, we observe that *Proactivity* is the only difference between profiles E (the best performing profile) and D (the profile with the highest failure rate). Therefore, it seems that delaying lecture material and not being proactive results in worse academic performance. Likewise, profile E and C only differ in the *Effort* dimension, but surprisingly, profile E with *Lower Effort* outperforms profile C with *Higher Effort*. We hypothesize that in this case, *Effort* is an indicator of students struggling rather than a measure of commitment as expected [9].

In PC, profile C outperforms the other three profiles B, F, and J; while students in J perform poorly compared to profiles B, C, and F. In FP, students in profile H perform significantly worse than the students in B, I, F, and G. As shown in Table 3, these poor performing profiles (profiles H and J) are quite similar. The results are as expected since both have *Lower Effort*, *Lower Peaks in Regularity*, *Lower Control*, and increased activity before the midterm exam. For PC, it is hard to identify the dimension responsible for the worse academic performance of profile J, as the other profiles differ in several dimensions. For example, it would be inaccurate to say that profile F outperforms profile J despite having *Delayed Proactivity* because it is not the only dimension that varies. For PC and FP, the combination of dimensions explains the differences in performance rather than an isolated dimension.

In summary, we found significant differences in academic performance in all three courses. Although the level of significance varies across courses, we found coherent results between the shared profiles ***(RQ2)***.

---

[1] Grades range from 1 to 6, with 6 being the best and 4 being the passing grade.
[2] Correcting for multiple comparisons via a Benjamini-Hochberg (BH) procedure.
[3] Shapiro-Wilk test for FP: $W = 0.97$, $p = 5.6\text{e}{-05}$; and PC: $W = 0.97$, $p = 3.1\text{e}{-03}$. Kruskal-Wallis Test for FP: $\chi^2(4) = 21.8$, $p = 2.2\text{e}{-04}$; PC: $\chi^2(3) = 13.4$, $p = 3.8\text{e}{-03}$.

4 Discussion and Implications

In this work, we combined multiple behavioral dimensions to obtain interpretable student profiles in FC and analyzed how these profiles and their behavioral patterns compare across courses (**RQ1**). We then showed the relation between those profiles and academic performance (**RQ2**). Unlike prior work mostly focusing on a single FC course [12,23], we applied our pipeline to three courses (LA, FP and PC) with different topics, instructors, FC period length, and study level.

Our results showed that our pipeline can identify interpretable student profiles in FC, with some of them showing similar behavior across different courses and others associated with a behavior unique to a specific course (**RQ1**). In addition, our results emphasize the importance of taking into account the dependencies between learning dimensions and analyzing them in combination rather than focusing on an isolated dimension. It is noteworthy that despite *Effort*, *Consistency*, *Regularity* and *Proactivity* are SRL skills, they do not always go hand-to-hand in the profiles description. For example, *Effort* appears to be constant in several profiles, and the profiles with the same effort magnitude differ based on other dimensions (e.g., *Consistency*). This is in line with [19], where three groups showed the same effort but a different consistency. Interestingly, a profile with a *Lower*, *Decreasing*, and *Delayed* patterns in all dimensions was also found among university students with high dropout rates [14] and profiles A and H resemble the minimalist behavior identified by [23].

Our analyses also confirmed that there were some significant differences in academic performance between the profiles (**RQ2**). From a pedagogical perspective, these results are mostly coherent with findings from prior work (e.g., [4,6,10,25]) showing that achievement is significantly higher for students with high SRL skills (focusing on a single dimension). In LA, surprisingly, counter to the work of [9], keeping all the other dimensions equal, the *Lower Effort* profile (E) outperformed the *Higher Effort* profile (C). In contrast, in PC, profile C (with *Higher Effort*) was the best performing profile. These differences exemplify how the proposed pipeline expresses the profiles relative to the classmates of each course. Likewise, the results from LA showed that *Proactivity* appeared to be the most indicative behavioral dimension for academic performance: watching the lecture videos ahead of schedule (like profile E) was associated with good academic performance, while delaying lecture material (like profile D) was related to inferior academic performance, in line with [17]. Nevertheless, in PC, profile F with *Delayed Proactivity* outperformed profile J with *Anticipated Proactivity.* This does not rule out the importance of *Proactivity* but rather the limitations of analyzing dimensions separately. Profiles F and J also differ in *Effort, Regularity* and *Control*, thus, the differences in academic performance in PC and FP can be better explained with multi-dimensional profiles. Instructors should acknowledge this to foster learning profiles beneficial to their course (e.g., profile E), and prevent counterproductive behaviors (e.g., profile H).

In this work, we used three different data sets to provide a diverse evaluation of our approach. From a research perspective, we proposed a method to help both researchers and practitioners improve the understanding of student learning in FC. From a teacher's perspective, this study enables data-driven course modifications (e.g., weekly quizzes) and better-informed student interventions. In addition, students could receive automatic personalized feedback and recommendations depending on their profile. Overall, our work contributes to the ongoing research of reusable analytics and to the generality of theories and patterns of SRL. Nevertheless, further work is needed to assess the generalizability of our results in other educational contexts.

References

1. Akpinar, N., Ramdas, A., Acar, U.: Analyzing student strategies in blended courses using clickstream data. In: Proceedings of the EDM, pp. 6–17 (2020)
2. Bassett, K., Olbricht, G.R., Shannon, K.B.: Student preclass preparation by both reading the textbook and watching videos online improves exam performance in a partially flipped course. CBE Life Sci. **19**(3), ar32 (2020)
3. Beatty, B.J., Merchant, Z., Albert, M.: Analysis of student use of video in a flipped classroom. TechTrends **63**(4), 376–385 (2019)
4. Biard, N., Cojean, S., Jamet, E.: Effects of segmentation and pacing on procedural learning by video. Comput. Hum. Behav. **89**, 411–417 (2018)
5. Boroujeni, M.S., Dillenbourg, P.: Discovery and temporal analysis of latent study patterns in MOOC interaction sequences. In: Proceedings of the LAK, pp. 206–215 (2018)
6. Boroujeni, M.S., Sharma, K., Kidziński, Ł., Lucignano, L., Dillenbourg, P.: How to quantify student's regularity? In: Proceedings of the EC-TEL, pp. 277–291 (2016)
7. Broadbent, J., Poon, W.L.: Self-regulated learning strategies & academic achievement in online higher education learning environments: a systematic review. Internet High. Educ. **27**, 1–13 (2015)
8. Chen, F., Cui, Y.: Utilizing student time series behaviour in learning management systems for early prediction of course performance. J. Learn. Anal. **7**(2), 1–17 (2020)
9. Cho, M.H., Shen, D.: Self-regulation in online learning. Distance Educ. **34**(3), 290–301 (2013)
10. Corrin, L., de Barba, P.G., Bakharia, A.: Using learning analytics to explore help-seeking learner profiles in MOOCs. In: Proceedings of the LAK, pp. 424–428 (2017)
11. Geertshuis, S., Jung, M., Cooper-Thomas, H.: Preparing students for higher education: the role of proactivity. Int. J. Teach. Learn. High. Educ. **26**(2), 157–169 (2014)
12. Jovanovic, J., Gasevic, D., Dawson, S., Pardo, A., Mirriahi, N.: Learning analytics to unveil learning strategies in a flipped classroom. Internet High. Educ. **33**, 74–85 (2017)
13. Jovanovic, J., Mirriahi, N., Gašević, D., Dawson, S., Pardo, A.: Predictive power of regularity of pre-class activities in a flipped classroom. Comput. Educ. **134**, 156–168 (2019)
14. Khalil, M., Ebner, M.: Clustering patterns of engagement in Massive Open Online Courses (MOOCs): the use of learning analytics to reveal student categories. J. Comput. High. Educ. **29**(1), 114–132 (2016). https://doi.org/10.1007/s12528-016-9126-9

15. Lallé, S., Conati, C.: A data-driven student model to provide adaptive support during video watching across MOOCs. In: Proceedings of the AIED, pp. 282–295 (2020)
16. Lee, J., Choi, H.: Rethinking the flipped learning pre-class: its influence on the success of flipped learning and related factors. Br. J. Educ. Technol. **50**, 934–945 (2019)
17. Marras, M., Vignoud, J.T.T., Käser, T.: Can feature predictive power generalize? Benchmarking early predictors of student success across flipped and online courses. In: Proceedings of the EDM, pp. 150–160 (2021)
18. McBroom, J., Yacef, K., Koprinska, I.: DETECT: a hierarchical clustering algorithm for behavioural trends in temporal educational data. In: Proceedings of the AIED, pp. 374–385 (2020)
19. Mojarad, S., Essa, A., Mojarad, S., Baker, R.S.: Data-driven learner profiling based on clustering student behaviors: learning consistency, pace and effort. In: Nkambou, R., Azevedo, R., Vassileva, J. (eds.) ITS 2018. LNCS, vol. 10858, pp. 130–139. Springer, Cham (2018). https://doi.org/10.1007/978-3-319-91464-0_13
20. Mubarak, A.A., Cao, H., Ahmed, S.A.: Predictive learning analytics using deep learning model in MOOCs' courses videos. Educ. Inf. Technol. **26**(1), 371–392 (2021)
21. Ng, A.Y., Jordan, M.I., Weiss, Y., et al.: On spectral clustering: analysis and an algorithm. Adv. Neural. Inf. Process. Syst. **2**, 849–856 (2002)
22. O'Flaherty, J., Phillips, C.: The use of flipped classrooms in higher education: a scoping review. Internet High. Educ. **25**, 85–95 (2015)
23. Pardo, A., Gašević, D., Jovanovic, J., Dawson, S., Mirriahi, N.: Exploring student interactions with preparation activities in a flipped classroom experience. IEEE Trans. Learn. Technol. **12**(3), 333–346 (2018)
24. Rousseeuw, P.J.: Silhouettes: a graphical aid to the interpretation and validation of cluster analysis. J. Comput. Appl. Math. **20**, 53–65 (1987)
25. Sher, V., Hatala, M., Gašević, D.: Analyzing the consistency in within-activity learning patterns in blended learning. In: Proceedings of the LAK, pp. 1–10 (2020)
26. Sletten, S.R.: Investigating flipped learning: student self-regulated learning, perceptions, and achievement in an introductory biology course. J. Sci. Educ. Technol. **26**(3), 347–358 (2017)
27. Vermunt, J.D., Donche, V.: A learning patterns perspective on student learning in higher education: state of the art and moving forward. Educ. Psychol. Rev. **29**(2), 269–299 (2017)
28. Wan, H., Liu, K., Yu, Q., Gao, X.: Pedagogical intervention practices: improving learning engagement based on early prediction. IEEE Trans. Learn. Technol. **12**(2), 278–289 (2019)

Embodied Learning with Physical and Virtual Manipulatives in an Intelligent Tutor for Chemistry

Joel P. Beier(✉) and Martina A. Rau

Department of Educational Psychology, University of Wisconsin, Madison, USA
jpbeier@wisc.edu

Abstract. Blended educational technologies can leverage complementary benefits of physical and virtual manipulatives. However, it is not clear how best to combine these manipulatives. Prior research has focused on combining physical and virtual manipulatives by offering them sequentially based on whether they make a specific concept salient. This research has mostly ignored embodied learning mechanisms that can ground students' conceptual understanding in bodily actions. To address this issue, we conducted a lab experiment on chemistry learning with 80 undergraduate students. We compared different ways of sequencing virtual and physical manipulatives in ways that first engaged students in embodied experiences or made the target concepts salient. Results suggest that providing embodied experiences early in the learning sequence enhances conceptual learning. These findings extend extant theory on blending physical and virtual manipulatives and provide practical advice for developers of blended interactive educational technologies.

Keywords: Blended technologies · Physical/virtual manipulatives · Embodiment

1 Introduction

Blended educational technologies that combine physical and virtual experiences are becoming increasingly popular [1, 2]. This has revived a century-old debate about when physical manipulatives enhance learning [3]. For example, chemistry students may interact with physical or virtual manipulatives while learning about atoms (Fig. 1). Physical manipulatives are tangible objects that students construct with their hands (Fig. 1a). Virtual manipulatives are displayed on a screen and are manipulated by mouse, keyboard, or touchscreen (Fig. 1b). The goal of blended technologies is to combine these manipulatives in a way that leverage their complementary benefits [1, 2, 4].

A prevalent way of blending physical and virtual manipulatives is to provide them sequentially [5–7]. However, prior studies yield conflicting results as to how physical or virtual manipulatives should be sequenced (e.g., [5, 7]). To resolve these conflicts, a dominant blending framework [1, 4] proposes that students should work with the manipulative that makes task-relevant concepts salient by drawing students' attention to

M. M. Rodrigo et al. (Eds.): AIED 2022, LNCS 13355, pp. 103–114, 2022.
https://doi.org/10.1007/978-3-031-11644-5_9

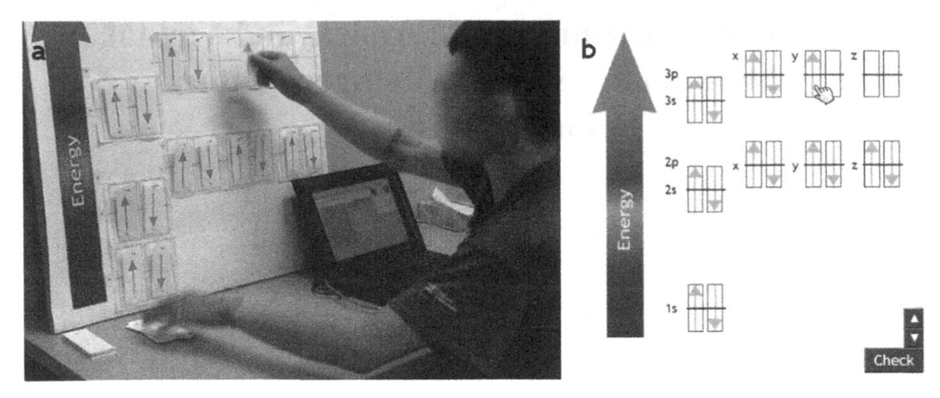

Fig. 1. Physical (a) and virtual (b) manipulatives showing an atomic orbital energy diagram.

the concepts. When students switch to a different task, they may switch to a different manipulative that better aligns with the concepts relevant to the new task.

A limitation of this blending framework is that it solely focuses on conceptual learning processes. Yet, embodied processes also affect students' learning with physical and virtual manipulatives [8]. Most prior research on blended educational technologies has focused on conceptual processes (e.g., [9–11]) while disregarding embodied processes [8]. The lack of research that integrates both processes is problematic. First, focusing on only a subset of relevant processes may lead to confounded experiments, which may contribute to conflicting results from prior studies. Second, research needs to compare the relative strength of these processes to determine which process accounts for the observed sequence effects. Without such knowledge, we cannot make recommendations for when students should receive a physical or virtual manipulative. Further, such knowledge will determine which process adaptive blended technologies should trace to assign physical or virtual manipulatives based on an individual's learning progress.

To achieve these goals, we present an experiment that systematically varied design features of manipulatives that affect conceptual and embodied processes. We tested sequences of physical and virtual manipulatives within an intelligent tutoring system.

2 Theoretical Background

2.1 Learning Processes Affected by Physical and Virtual Manipulatives

A recent review [8] showed that prior studies mostly focus on how physical and virtual manipulatives make concepts salient while fewer studies focus on embodied processes.

Conceptual salience describes the capacity of a visual representation to draw students' attention to visual features that depict conceptually relevant information [8]. Concepts may become salient because visual design features of the manipulative draw students' attention to them [12] or because students' interactions with the manipulative draw attention to a specific feature that depicts conceptual information [13].

According to this perspective, whether a physical or virtual manipulative is more effective depends on which makes a concept more salient [1]. For instance, physical manipulatives are more effective if they allow students to experience spatial concepts [14, 15] or offer concrete experiences relevant to the target concept [16]. As mentioned, the dominant blending framework [1] recommends to match physical and virtual manipulatives to learning tasks based on whether they make the target concept conceptually salient. Indeed, this way of blending physical and virtual manipulatives leads to higher learning gains than working with only physical or only virtual manipulatives [4].

In sum, the dominant view is that the type of manipulative that makes the target concept salient should be most effective.

Embodied theory assumes that cognition evolved for humans to mentally simulate effects of their actions [17, 18]. Hence, abstract thinking builds on mental simulations of body actions. For example, understanding growth functions builds on experiences of growth and increase in the real world. We distinguish two tenets of this theory [8, 13].

Explicit embodiment emphasizes the importance of explaining relationships between kinesthetic experiences and concepts [8]. Physical manipulatives may allow students to experience a target concept through the sense of touch and motion [19]. Students can explicitly connect these embodied experiences to the concept. For example, suppose manipulating a physical manipulative involves lifting an object. Prompting students to explain how the physical effort associated with this action relates to concepts of kinetic and potential energy can help students understand these concepts. Explicit embodied experiences can perceptually ground students' understanding of abstract concepts [20]; that is, students' gradual understanding of abstract concepts based on concrete experiences becomes increasingly stylized [21, 22]. Indeed, perceptual grounding enhances learning outcomes [23]. Thus, explicit embodiment suggests that physical manipulatives are advantageous if they allow students to explain connections between the target concepts and experiences of touch and movement.

Implicit embodiment emphasizes the importance of body movements without requiring that students are aware of the connections between the movement and the concept [8]. Building on the idea that thought is a mental simulation of action [24, 25], even abstract concepts (e.g., justice) are based on real-world experiences (e.g., balance), often without our awareness of this connection [26]. This implies that instruction should invoke embodied schemas relevant to the target concept [27]. Embodied schemas can be invoked by metaphors, body movements, or gesture [28, 29]. Students' learning of a concept is enhanced if they receive instruction on the concept while moving their body in ways that are synergistic to the associated embodied schema, even if they are not aware that their movement related to the concept [20, 30]. For example, moving one's hand upwards may activate an embodied schema related to increase, which can help students learn concepts related to growth.

Implicit embodiment is not only afforded by physical but also by virtual manipulatives. When virtual manipulatives are manipulated in ways that invoke synergistic embodied metaphors, students learn the target concept better than when manipulating the same manipulative with less synergistic movements [31]. Because physical and virtual manipulatives often engage students in different movements, implicit embodiment has implications for which type of manipulative is most effective. For example, to manipulate a physical manipulative, a student may move their hand vertically, which implicitly invokes embodied schemas related to growth and increase. In contrast, a virtual manipulative may require a sideways movement that invokes embodied schemas of balance and equality. Depending on which embodied schema matches the target concept, one or the other type of manipulative may be more effective [8].

In sum, implicit embodiment suggests that manipulatives are more effective if they invoke embodied schemas that match the target concept without requiring awareness of the match.

2.2 Blending Physical Versus Virtual Manipulatives

There is no empirical basis for the superiority of physical or virtual manipulatives [8]. Many studies showed that physical and virtual manipulatives complement each other by making different concepts salient [1, 6, 9]. Hence, research investigated how to blend these manipulatives by sequencing them in a way that best leverages their strengths [7–13]. This yielded the dominant blending framework [1, 4], which suggests that manipulatives should be chosen based on their ability to make concepts salient.

Yet, the dominant blending framework is limited because it is based on studies that focused only on conceptual salience of the target concepts and thus conflated ways that the manipulatives affected embodied processes [8]. Our prior work [32] started addressing this limitation. We systematically varied whether physical and virtual manipulatives implicitly induced embodied schemas that were synergistic to the target concepts, offered explicit embodied experiences of the concepts, and provided visual cues that made the concepts salient. We found that implicit embodiment yielded higher learning gains on a reproduction test. However, physical manipulatives that offered explicit embodied experiences yielded higher gains on a transfer test. If explicit embodiment was not available for a given concept, manipulatives (physical or virtual) that made the concept salient yielded higher transfer gains. We interpreted the findings based on the complexity of the learning outcome [13, 33]: Implicit embodiment enhanced simple learning outcomes (i.e., reproduction). In contrast, explicit embodiment and conceptual salience (both explicit processes) enhanced complex outcomes (i.e., transfer). We consider explicit embodiment more complex than conceptual salience because it allows students to make more connections between the manipulative and the concept (embodied plus visual experience vs visual experience only). This explains the benefit of explicit embodiment compared to the effects of conceptual salience.

3 Research Questions and Hypotheses

Our prior study suggests that effects of physical and virtual manipulatives affect learning outcomes not only via conceptual processes but also via embodied processes. Further, the different processes affect different learning outcomes. This raises the question of how manipulatives should be sequenced to best leverage implicit and explicit embodiment as well as conceptual salience. Our prior study suggests two hypotheses:

On the one hand, instruction often progresses from simple to complex. This yields the *simple-first (SF) hypothesis:* Students should first work with manipulatives that engage simple learning processes by implicitly inducing embodied schemas relevant to the concept. Then, they should work with manipulatives that engage complex processes by offering explicit embodied experiences of the concept. If explicit embodiment is unavailable, the manipulative should make the concept salient. This should enhance students' ability to construct correct manipulatives ($H_{SF\text{-}1}$) and learning gains ($H_{SF\text{-}2}$).

On the other hand, students may need to acquire deep understanding of a complex concept before they should practice simple recall. This yields the *complex-first (CF) hypothesis:* Students should first work with manipulatives that engage complex processes by offering explicit embodied experiences of the target concept (or, if not available, make the concept salient). Then, they should work with manipulatives that engage simple processes by implicitly inducing embodied schemas. This should enhance students' ability to construct correct manipulatives ($H_{CF\text{-}1}$) and learning gains ($H_{CF\text{-}2}$).

The goal of the present study is to systematically test these hypotheses. To this end, we conducted an experiment on students' use of manipulatives in a chemistry lesson.

4 Methods

4.1 Participants

Eighty undergraduate students were recruited from our institution via flyers and emails. Screening questions ensured they were naïve to the content and the manipulatives.

4.2 Experimental Design

In line with our prior study [32], we created four types of energy diagram manipulatives. For two concepts (A and B), they offered either conceptual or implicit-embodied experiences: two physical manipulatives (physical$_{\text{conceptual}}$, P_C; and physical$_{\text{implicit-embodied}}$ P_{IE}), and two virtual manipulatives (V_C; V_{IE}). As detailed below and shown in Table 1, P_C and P_{IE} offered explicit-embodied experiences for concept A but not for concept B.

Table 1. Overview of physical (P_C/P_{IE}) vs virtual (V_C/V_{IE}) manipulatives and target concepts.

Process		**Concept A**				**Concept B**			
Complexity	*Type of experience*	P_C	P_{IE}	V_C	V_{IE}	P_C	P_{IE}	V_C	V_{IE}
Complex	Explicit-embodied	✓	✓	-	-	-	-	-	-
↕	Conceptual	✓	-	✓	-	✓	-	✓	-
Simple	Implicit-embodied	-	✓	-	✓	-	✓	-	✓

Concept A: Electrons Randomly Fill Equal-Energy Orbitals.
Atomic properties are determined by the location of their electrons in subatomic regions called orbitals. Energy diagrams show the location of electrons and the relative energies of orbitals (Fig. 1). Electrons fill lower energy orbitals before higher energy orbitals.

Because equal-energy orbitals are equally likely to be filled, many atoms have alternative energy diagrams. A common misconception is that electrons fill equal-energy orbitals from left to right, rather than randomly. Target concept A was for students to learn that electrons randomly fill equal-energy orbitals.

To construct $\mathbf{P_C}$, students moved cards that showed electrons from the bottom up to put them in orbitals (Fig. 1A). P_C makes the *concept salient* because planning the motor action involved in the vertical action requires attention to the height of the orbital when students put a card in an orbital. However, this vertical action implicitly induces a *conflicting embodied schema* because it aligns with a metaphor of increase [26] that conflicts with the concept of equality.

To construct $\mathbf{P_{IE}}$, students held the cards next to the orbitals and moved their hands horizontally to put them in orbitals. The horizontal action makes the *concept less salient* because it does not require paying attention to the height of the orbitals. However, this horizontal action implicitly induces *beneficial embodied schemas* for the concept because horizontal actions induce a metaphor of equality [26].

Both $\mathbf{P_C}$ and $\mathbf{P_{IE}}$ offer *explicit embodied experiences* of concept A because students can physically experience the height of the orbital.

To construct $\mathbf{V_C}$, students had to click a button at the bottom of the interface each time before moving the mouse up to put arrows in orbitals. This vertical action makes the *concept more salient* but implicitly induces a *conflicting embodied schema.*

To construct $\mathbf{V_{IE}}$, students had to move the mouse horizontally to click in equal-energy orbitals (Fig. 1B). V_{IE} makes the *concept less salient* but implicitly induces a *beneficial embodied schema.*

$\mathbf{V_C}$ and $\mathbf{V_{IE}}$ offer *no explicit embodied experience* of concept A.

Concept B: Up and Down Spins Have Equal Energy. Electrons in the same orbital have opposite spins (shown by up and down arrows; Fig. 1). Up and down spins are equally likely. A common misconception is that the first electron in an orbital always has an up spin. Hence, target concept B was for students to learn that both spins are equally likely.

For $\mathbf{P_C}$, the card stack was sorted so that all cards had an up arrow. This makes the concept *more salient* because students had to purposefully flip the arrows to show that the spins are equally likely, which requires explicit attention. Yet, this implicitly induces a *conflicting embodied schema* because it takes two actions to show a down spin (i.e., more effort) and only one action to show an up spin (i.e., less effort).

For $\mathbf{P_{IE}}$, the card stack was not sorted, so that up and down arrows were random. This makes the *concept less salient* because the spin is already random and does not require attention to a related action. Yet, this implicitly induces a *beneficial embodied schema* because it takes the same number of actions and hence the same amount of effort to show up or down spin.

In $\mathbf{V_C}$, students clicked to add arrows. The first click added an up arrow, the second click flipped it to a down arrow. V_C makes the *concept more salient* because students had to purposefully flip the arrows. Yet, this implicitly induces a *suboptimal embodied schema* because it took two clicks to show a down spin (more effort) but only one click (less effort) to show an up spin. V_C offers *no explicit embodied experience* of spin.

In V_{IE}, the first click created an arrow with random spin and the second click flipped it. This makes the *concept less salient* but induces a *beneficial embodied schema.* V_{IE} offers *no explicit embodied experience* of spin.

P_C, P_{IE}, V_C, and V_{IE} offer *no explicit embodied experience* of spin.

Experimental Design: Sequences of Manipulatives. The experiment involved two sessions. Session 1 covered concept A and session 2 covered concept B. The experiment varied the sequence of mode (P-V vs V-P) and of design (IE-C vs C-IE).

Specifically, students were randomly assigned to one of four conditions for session 1: P_C-V_{IE}, P_{IE}-V_C, V_C-P_{IE}, or V_{IE}-P_C. For session 2, students were assigned to a condition that offered manipulatives they had not encountered in session 1. For example, if students received P_C-V_{IE} in session 1, they received either P_{IE}-V_C or V_C-P_{IE} in session 2. This ensured that all students received each manipulative. Further, this design allowed us to test the two competing hypotheses in the following manner. The *simple-first hypothesis* predicts an advantage for V_{IE}-P_C and P_{IE}-V_C over P_C-V_{IE} and V_C-P_{IE} because these sequences engage students in simple learning processes first (i.e., V_{IE} and P_{IE}), and then engage students in complex learning processes (i.e., P_C and V_C). For concept A, this advantage should be particularly pronounced for V_{IE}-P_C because P_C offers explicit embodied experiences in addition to making the concept salient.

By contrast, the *complex-first hypothesis* predicts an advantage for P_C-V_{IE} and V_C-P_{IE} over V_{IE}-P_C and P_{IE}-V_C because these sequences engage students in complex learning processes first (i.e., P_C and V_C), and then in simple learning processes (i.e., V_{IE} and P_{IE}). For concept A, this advantage should be particularly pronounced for P_C-V_{IE} because P_C offers explicit embodied experiences in addition to making the concept salient.

4.3 Materials

Intelligent Tutoring System: Chem Tutor. All students worked with Chem Tutor, an intelligent tutoring system for undergraduate chemistry [34, 35]. Chem Tutor engages students in iterative representation-reflection practices by asking them to construct manipulatives and reflect on how the manipulative shows the target concepts.

Students worked through a sequence of eight problems focused on concept A and five problems focused on concept B. Each problem asked students to construct an energy diagram. Physical manipulatives (P_C/P_{IE}) were placed next to the computer (Fig. 1a). Virtual manipulatives (V_C/V_{IE}) were embedded in Chem Tutor. Chem Tutor provided feedback and on-demand hints on all problem-solving steps, including the manipulatives. For physical manipulatives, the experimenter provided scripted feedback and hints that matched those provided by Chem Tutor.

Measures. We assessed students' *conceptual knowledge* with a pretest, immediate posttest, and delayed posttest for each concept. For each concept, the tests included a reproduction scale (i.e., assessing recall of information about the concept) and a transfer scale (i.e., assessing the ability to apply the information to novel problems).

Further, as instruction was self-paced, we measured *time on task* for each concept.

Finally, we computed *errors* as the proportion of mistakes per step in manipulating V_C and V_{IE} with log data and in manipulating P_C and P_{IE} based on video recordings.

4.4 Procedure

The experiment involved three sessions in a research lab. In session 1, students completed the concept A pretest, Chem Tutor problems on concept A, and took the concept A posttest. In session 2 (2–5 days later), students completed the concept A delayed posttest, the concept B pretest, Chem Tutor problems on concept B, and the concept B posttest. In session 3 (2–5 days later), students took the concept B delayed posttest.

5 Results

5.1 Prior Checks

One student was excluded for scoring 2 standard deviations above the median. Repeated measures ANOVAs with pretest, immediate, and delayed posttest as dependent measures showed learning gains for all concepts and scales ($ps < .01$) with effect sizes ranging from p. $\eta^2 = .568$ to p. $\eta^2 = .876$. For concept A, we found no significant condition effects on pretest measures and time on task ($ps > .10$). For concept B, there were no significant differences on the pretests ($ps > .10$), but a significant effect on time on task ($p = .01$). Post-hoc comparisons showed that students in the P_C-V_{IE} condition took significantly longer than students in the V_C-P_{IE} condition ($p = .008$). Time on task correlated with posttests ($r = -.244$ to $-.558$). Thus, we use it as covariate in our analyses.

5.2 Effects on Error Rates During Interactions with Manipulatives

We used a repeated ANCOVA with mode-sequence (P-V vs V-P) and design-sequence (IE-C vs C-IE) as independent factors, mode-type (P vs V) as repeated measures, pretest and time on task as covariates, and errors as dependent measure. For *concept A*, the effect of mode-sequence was significant, $F(1, 72) = 5.309$, $p = .024$, p. $\eta^2 = .069$. Students who received physical manipulatives first made fewer errors, which partially supports $H_{CF\text{-}1}$. For *concept B*, the effect of design-sequence was significant, $F(1, 72) = 6.664$, $p = .012$, p. $\eta^2 = .085$. Students who received implicit-embodied manipulatives first made fewer errors. This finding supports $H_{SF\text{-}1}$. Figure 2a-b illustrate these results.

5.3 Effects on Learning Outcomes

We used a repeated ANCOVA with mode-sequence (P-V vs V-P) and design-sequence (IE-C vs C-IE) as independent factors, test-time (immediate, delayed posttest) and scale (reproduction, transfer) as repeated factors, pretest and time on task as covariates, and test scores as dependent measures. For *concept A*, there were no main effects of mode-sequence and design-sequence ($ps > .10$), but mode-sequence interacted with test-scale, $F(1, 72) = 9.644$, $p = .003$, p. $\eta^2 = .045$. Pairwise comparisons showed that the P-V sequence yielded better transfer, $F(1, 72) = 6.568, p = .012$, p. $\eta^2 = .084$, but did not affect reproduction ($F < 1$). This effect held for P_C-V_{IE} and P_{IE}-V_C. This finding partially supports $H_{CF\text{-}2}$. For *concept B*, there were no effects of mode-sequence or design-sequence (Fs < 1), thus supporting neither $H_{SF\text{-}2}$ nor $H_{CF\text{-}2}$. Figure 2c illustrates the results.

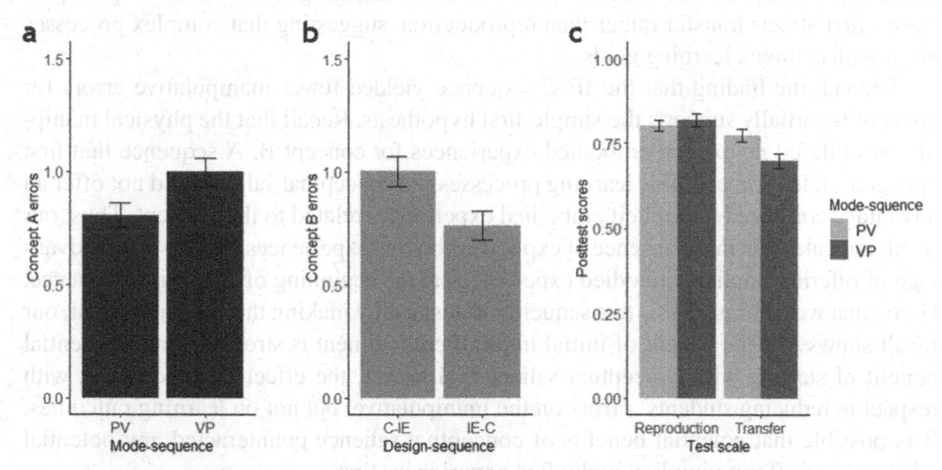

Fig. 2. (a) Effect of mode-sequence on concept A errors; (b) effect of design-sequence on concept B errors; (c) effects of mode-sequence on reproduction and transfer posttests. All bars show estimated marginal means. Error bars show standard errors of the mean.

6 Discussion and Conclusion

Prior research recommends blending physical and virtual manipulatives by sequencing them in a way that makes the target concepts salient. However, a mostly separate line of research shows that explicit and implicit types of embodied processes also affect learning with manipulatives. A severe limitation of prior research is that it had not investigated all three types of processes together. Our prior research had contrasted effects of conceptual, explicit-embodied, and implicit-embodied processes on learning with manipulatives. Results had indicated that these processes differently affect learning outcomes of varying complexity. This gave rise to two competing hypotheses about sequencing physical and virtual manipulatives either so that they engage students in simple learning processes first (i.e., via implicit-embodiment) or so that they engage students in complex learning processes first (i.e., preferably via explicit-embodiment or else via conceptual salience). While the results of the present experiment seem to be complex, two relatively simple patterns emerge. First, explicit embodiment has a strong effect on both errors and learning gains. Second, whether in the form of explicit or implicit embodiment, some type of embodied experience at the beginning of the learning sequence is advantageous. In the following, we discuss each pattern in turn.

First, the finding that the P-V sequence yielded fewer manipulative errors and higher transfer gains for concept A than the V-P sequence partially supports the complex-first hypothesis. Recall that this hypothesis had also predicted an advantage of sequences that start with conceptual salience (i.e., in addition to a main effect of P-V > V-P, an advantage of P_C-V_{IE} > P_{IE}-V_C). Yet, our results suggest that starting with physical manipulatives that offer explicit embodied experiences of the target concept is sufficient. Engaging students in additional complex processes with the concept early in the sequence is not necessary. Further, in line with our prior research, engaging students in complex processes first affects transfer rather than reproduction, suggesting that complex processes align with complex learning goals.

Second, the finding that the IE-C sequence yielded fewer manipulative errors for concept B partially supports the simple-first hypothesis. Recall that the physical manipulatives offered no explicit embodied experiences for concept B. A sequence that first engaged students in complex learning processes via conceptual salience did not offer an advantage compared to implicit embodied experiences related to the concept. Thus, our result indicates that in the absence of explicit embodied experiences, there is some advantage of offering implicit embodied experiences at the beginning of a learning sequence. Given that we contrasted this to a sequence that starts by making the concept salient, our result shows that the benefit of initial implicit embodiment is stronger than a potential benefit of starting with conceptual salience. However, the effect only bears out with respect to reducing students' errors on the manipulative, but not on learning outcomes. It is possible that potential benefits of conceptual salience counteracted any potential advantage of offering implicit embodied experiences first.

Our findings expand research on blending physical and virtual manipulatives in at least two ways. First, our research is the first to consider conceptual salience as well as explicit and implicit embodied experiences, yielding a systematic comparison of sequences. Moreover, no prior research has compared explicit and implicit embodied processes, even though they appear to yield dramatically different outcomes. Second, our findings suggest that blending should not be done purely based on conceptual salience. Wherever possible, manipulatives should first offer explicit embodied experiences of target concepts. Otherwise, implicit embodied experiences can offer some advantages. Consequently, adaptive blended learning technologies should not only trace students' conceptual learning but should also trace their embodied engagements by assessing movement and touch.

Our findings should be interpreted in light of several limitations. First, we focused on one combination of concepts and manipulatives. Other manipulatives lend themselves to studying different combinations of conceptual and embodied designs. For example, we did not include a manipulative that offered implicit embodied experiences while also making the target concept salient. Future research should examine whether it is possible to combine benefits of implicit embodiment and conceptual salience, especially when explicit embodiment is not available. Second, our experiment was conducted in a research lab and should be replicated in a realistic educational context. Third, while long for a lab experiment, our intervention was relatively short for realistic instruction. Future research should examine sequence effects over longer periods.

In conclusion, blended educational technologies offer novel opportunities for combining physical and virtual experiences. The dominant framework that guides extant integrations of physical and virtual manipulatives focuses on conceptual salience while disregarding emerging findings about the importance of embodied engagement. Our research systematically juxtaposed conceptual salience with two types of embodied engagements. Our findings show that explicit embodied engagements early in a learning sequence can significantly enhance students' learning with manipulatives.

Acknowledgements. This work was supported by NSF IIS 1651781.

References

1. Olympiou, G., Zacharia, Z.C.: Blending physical and virtual manipulatives: an effort to improve students' conceptual understanding through science laboratory experimentation. Sci. Educ. **96**, 21–47 (2012)
2. de Jong, T., Linn, M.C., Zacharia, Z.C.: Physical and virtual laboratories in science and engineering education. Science **340**, 305–308 (2013)
3. Huxley, T.H.: Scientific education: notes of an after-dinner speech. In: Huxley, T.H. (ed.) Collected Essays: Science and education, vol. 3, pp. 111–133. Appleton, New York (1897)
4. Zacharia, Z.C., Michael, M.: Using physical and virtual manipulatives to improve primary school students' understanding of concepts of electric circuits. In: Riopel, M., Smyrnaiou, Z. (eds.) New Developments in Science and Technology Education. ISET, vol. 23, pp. 125–140. Springer, Cham (2016). https://doi.org/10.1007/978-3-319-22933-1_12
5. Jaakkola, T., Nurmi, S.: Fostering elementary school students' understanding of simple electricity by combining simulation and laboratory activities. J. Comput. Assist. Learn. **24**, 271–283 (2008)
6. Gire, E., et al.: The effects of physical and virtual manipulatives on students' conceptual learning about pulleys. In: Gomez, K., Lyons, L., Radinsky, J. (eds.) 9th International Conference of the Learning Sciences, vol. 1, pp. 937–943. International Society of the Learning Sciences (2010)
7. Winn, W., Stahr, F., Sarason, C., Fruland, R., Oppenheimer, P., Lee, Y.L.: Learning oceanography from a computer simulation compared with direct experience at sea. J. Res. Sci. Teach. **43**, 25–42 (2006)
8. Rau, M.A.: Comparing multiple theories about learning with physical and virtual representations: conflicting or complementary effects? Educ. Psychol. Rev. **32**(2), 297–325 (2020). https://doi.org/10.1007/s10648-020-09517-1
9. Chini, J., Madsen, A., Gire, E., Rebello, N., Puntambekar, S.: Exploration of factors that affect the comparative effectiveness of physical and virtual manipulatives in an undergraduate laboratory. Phys. Educ. Res. **8**, 010113 (2012)
10. Renken, M.D., Nunez, N.: Computer simulations and clear observations do not guarantee conceptual understanding. Learn. Instr. **23**, 10–23 (2013)
11. Yuan, Y., Lee, C., Wang, C.: A comparison study of polyominoes explorations in a physical and virtual manipulative environment. Compu. Assist. Learn. **26**, 307–316 (2010)
12. Mautone, P.D., Mayer, R.E.: Cognitive aids for guiding graph comprehension. J. Educ. Psychol. **99**, 640–652 (2007)
13. Rau, M.A., Herder, T.: Under which conditions are physical versus virtual representations effective? Contrasting conceptual and embodied mechanisms of learning. J. Educ. Psychol. **113**, 1565–1586 (2021)
14. Schneider, B., Sharma, K., Cuendet, S., Zufferey, G., Dillenbourg, P., Pea, R.: Using mobile eye-trackers to unpack the perceptual benefits of a tangible user interface for collaborative learning. In: ACM Transactions on Computer-Human Interaction (TOCHI), vol. 23, p. 39 (2016)

15. Stull, A.T., Hegarty, M., Dixon, B., Stieff, M.: Representational translation with concrete models in organic chemistry. Cogn. Instr. **30**, 404–434 (2012)
16. Stusak, S., Schwarz, J., Butz, A.: Evaluating the memorability of physical visualizations. In: Begole, B., Kim, J., Inkpen, K., Woo, W. (eds.) Proceedings of the 33rd Annual ACM Conference on Human Factors in Computing Systems, pp. 3247–3250. ACM (2015)
17. Glenberg, A.M., Witt, J.K., Metcalfe, J.: From the revolution to embodiment 25 years of cognitive psychology. Perspect. Psychol. Sci. **8**, 573–585 (2013)
18. Wilson, M.: Six views of embodied cognition. Psychon. Bull. **9**, 625–636 (2002)
19. Zaman, B., Vanden Abeele, V., Markopoulos, P., Marshall, P.: Editorial: the evolving field of tangible interaction for children: the challenge of empirical validation. Pers. Ubiquit. Comput. **16**, 367–378 (2012)
20. Hayes, J.C., Kraemer, D.J.M.: Grounded understanding of abstract concepts: the case of STEM learning. Cogn. Res. Princip. Implicat. **2**(1), 1–15 (2017). https://doi.org/10.1186/s41235-016-0046-z
21. Goldstone, R.L., Schyns, P.G., Medin, D.L.: Learning to bridge between perception and cognition. Psychol. Learn. Motiv. **36**, 1–14 (1997)
22. Harnad, S.: The symbol grounding problem. Phys. D **42**, 335–346 (1990)
23. Han, I.: Embodiment: a new perspective for evaluating physicality in learning. J. Educ. Comput. Res. **49**, 41–59 (2013)
24. Abrahamson, D., Lindgren, R.: Embodiment and embodied design. In: Sawyer, R.K. (ed.) The Cambridge handbook of the Learning Sciences, pp. 358–376. Cambridge University Press, Cambridge (2014)
25. Clark, A.: Whatever next? Predictive brains, situated agents, and the future of cognitive science. Behav. Brain Sci. **36**, 181–204 (2013)
26. Lakoff, G., Johnson, M.: Metaphors We Live by. University of Chicago Press (1980)
27. Johnson-Glenberg, M., Birchfield, D., Tolentino, L., Koziupa, T.: Collaborative embodied learning in mixed reality motion-capture environments: two science studies. J. Educ. Psychol. **106**, 86–104 (2014)
28. Black, J.B., Segal, A., Vitale, J., Fadjo, C.L.: Embodied cognition and learning environment design. In: Jonassen, D.H., Land, S.M. (eds.) Theoretical Foundations of Learning Environments, pp. 198–223. Routledge Taylor & Francis Group, New York (2012)
29. Nathan, M.J., Walkington, C., Boncoddo, R., Pier, E.L., Williams, C.C., Alibali, M.W.: Actions speak louder with words. Learn. Instr. **33**, 182–193 (2014)
30. Nathan, M.J., Walkington, C.: Grounded and embodied mathematical cognition. Cogn. Res. Princip. Implicat. **2**, 1–20 (2017)
31. Segal, A., Tversky, B., Black, J.: Conceptually congruent actions can promote thought. J. Appl. Res. Mem. Cogn. **3**, 124–130 (2014)
32. Rau, M.A., Schmidt, T.A.: Disentangling conceptual and embodied mechanisms for learning with virtual and physical representations. In: Isotani, S., Millán, E., Ogan, A., Hastings, P., McLaren, B., Luckin, R. (eds.) AIED 2019. LNCS (LNAI), vol. 11625, pp. 419–431. Springer, Cham (2019). https://doi.org/10.1007/978-3-030-23204-7_35
33. Koedinger, K.R., Corbett, A.T., Perfetti, C.: The knowledge-learning-instruction Framework. Cogn. Sci. **36**, 757–798 (2012)
34. Rau, M.A., Michaelis, J.E., Fay, N.: Connection making between multiple graphical representations. Comput. Educ. **82**, 460–485 (2015)
35. Rau, M.A.: A framework for discipline-specific grounding of educational technologies with multiple visual representations. IEEE Trans. Learn. Technol. **10**, 290–305 (2017)

How do A/B Testing and Secondary Data Analysis on AIED Systems Influence Future Research?

Nidhi Nasiar(✉), Ryan S. Baker, Jillian Li, and Weiyi Gong

Graduate School of Education, University of Pennsylvania, Philadelphia, USA
nasiar@upenn.edu

Abstract. Recent years have seen a surge in research conducted on intelligent online learning platforms, with a particular expansion of research conducting A/B testing to decide which design to use, and research using secondary platform data in analyses. This scientometric study aims to investigate how scholarship builds on these two different types of research. We collected papers for both categories - A/B testing, and educational data mining (EDM) on log data- in the context of the same learning platform. We then collected a randomized stratified sample of papers citing those A/B and EDM papers, and coded the reason for each citation. On comparing the frequency of citation categories between the two types of papers, we found that A/B test papers were cited more often to provide background and context for a study, whereas the EDM papers were cited to use past specific core ideas, theories, and findings in the field. This paper establishes a method to compare the contribution of different types of research on AIED systems such as interactive learning platforms.

Keywords: Scientometrics · A/B testing · Online learning · AIED systems

1 Introduction

1.1 Research on Interactive Learning Platforms

Large-scale platforms for interactive online learning have become a core part of educational practice, a trend that has accelerated due to the pandemic-related shutdowns of educational institutions. There are several benefits of interactive learning platforms for learners. They make learning significantly more accessible [29] for learners unable to travel, learners whose work constraints make class attendance infeasible, and learners at home in quarantine. They are often also beneficial even when learners can attend class in-person, enabling classroom instruction to be enhanced by using data from online activities given as homework or in-class [34, 37]. Research-based platforms such as intelligent tutoring systems tend to lead to substantial learning benefits, an average of 0.76 standard deviations better than traditional curricula [33].

M. M. Rodrigo et al. (Eds.): AIED 2022, LNCS 13355, pp. 115–126, 2022.
https://doi.org/10.1007/978-3-031-11644-5_10

Even beyond these benefits, AIED learning platforms provide opportunities for enhancing learning through research [32] and can support it by iterative refinement through A/B tests and secondary data analysis. A large number of automated experiments have been conducted on these online learning platforms. Initially, it was common for single research groups to use their own platforms for research [2, 25]. In the early 2000s, the Pittsburgh Science of Learning Center (PSLC) built an infrastructure enabling hundreds of studies to be conducted in classrooms [20], albeit in a relatively resource-intensive fashion where researchers visited individual classrooms. In recent years, the ASSISTments learning platform has developed a research platform that allows automatic deployment of studies across the web. This platform has been used by dozens of external researchers to carry out their studies in thousands of math classrooms [27]. Increased support for A/B studies has also been incorporated into MOOC platforms [30], leading to large-scale studies such as [18], which tested an intervention in over 200 courses with millions of enrolled learners.

There has been an even larger expansion in the use of AIED learning platform data in secondary analyses by educational data mining (EDM) researchers. Initial research within the educational data mining conference was heavily based on data sets from the PSLC [19], with 14% of total analyses using DataShop data [1]. Over time, a range of learning platforms have moved towards sharing their data publicly, increasing the number of research questions that can be investigated by researchers without direct access to a large-scale platform. Specific data sets have become standards for comparing algorithms across papers – for instance, many papers have used a specific public data set from ASSISTments, to study student knowledge modeling [17, 38, 39], and Cognitive Tutor data has been used to compare ways to automatically refine knowledge structures [14, 22].

Both A/B testing infrastructure and secondary data analyses have facilitated and expedited research in the learning sciences, but the full details of how these trends have impacted the field are not fully known. We know there are more papers, but how do these papers influence the field? And do these two innovations influence future research in similar ways or do they have different types of influence? In this paper, we investigate the question of how the research afforded by these learning platforms impacts scientists and projects even beyond the specific papers that are produced. In other words, what is the scientific impact of each type of research, and is there a different impact on the science of learning from A/B tests versus EDM analyses?

1.2 Scientometrics in Secondary Data Analysis

In answering this question, we draw upon methods and past work in scientometrics, the field of scientific study which investigates the properties of scientific publications in order to better understand science more broadly. One of the core and long-standing questions and contributions of scientometrics has been in terms of comparing papers in terms of citation counts [15, 31] and comparing the relative contribution of different scientists [6]. This has been a prominent area of analysis in the learning analytics community. For example, research studies have looked at what learning analytics and EDM papers are most cited [1, 8, 36], and have analyzed the quantity of research output and collaboration in order to rank universities and scholars [11, 36]. This work has been highly useful to researchers in understanding the state and scope of the field of learning analytics. However, it does not answer our current research question around how this field makes progress scientifically.

A second category of scientometric research in EDM has focused on which topics are being studied, and how these EDM topics have shifted over time [8, 9], building on similar long-standing trends in scientometrics more broadly [4]. Furthermore, researchers have looked at the differences between the topics studied in learning analytics and educational data mining [5, 10], which sub-community's papers are cited more often [5, 10], and the relationships between published topics [36].

A third category of existing scientometric research in learning analytics has investigated equity in the field's practices. Concurrently with an increase in interest within scientometrics more broadly in whether gender, race, and ethnicity influence publication and citation patterns [16], learning analytics researchers have investigated the diversity in the field [5, 24, 36]. Recent work has also studied the degree to which diversity in samples is considered in secondary data analytics research (or even reported) [28]. The results of [28] indicated that most papers in the field do not even mention the background of learners, much less check for algorithmic biases, which makes it challenging to gauge the generalizability and transferability of our findings.

However, despite the considerable interest in scientometrics within communities closely aligned with the AIED community, there has not yet been research on analyzing citations to understand how researchers in these communities build on each others' research or on why papers are cited. In other words, there has been research on who is conducting research in these communities, and what they are researching, but not how they are building upon each others' research. Fortunately, there is considerable work in the scientometrics community that we can build on in analyzing this question for EDM and A/B testing research. Starting with [12], scientometricians have attempted to identify lists of reasons for why a scholar might choose to cite a specific paper. [3] expanded upon a list by Garfield [12] in an extensive review, which [21] then distilled into a manageable coding scheme. In this literature, one of the key steps towards understanding why a citation occurs was developing methods for the qualitative analysis of a citation's context [3, 7]. This literature found that researchers choose to cite a paper for a wide variety of reasons, including both scientific reasons (crediting key past contributions, refuting previously published ideas) and political reasons (citing an important member of the field, citing papers from the venue being submitted to). Political citations can be quite common – for example, a review of citations in computer science education found that few citations actually involved building on the contributions in previous papers [23].

In this paper, we built on this past work to investigate our research question of why researchers cite EDM and A/B testing papers, and what the differences are between the citations to each type of paper. We do so by collecting a corpus of citations of work to each type (citations all to work occurring in the same learning platform, to reduce confounds), qualitatively coding the reasons for each citation, and then statistically comparing the proportion of each reason for citation.

2 Methods

2.1 Research Context

In this paper, we analyze the citations received by papers presenting research conducted in the context of the ASSISTments platform [28]. ASSISTments is an online learning system with 500K students and 20K teachers currently, primarily used for mathematics. ASSISTments has users in over 20 countries, but the majority of learners are in the United States of America. Randomized controlled studies have demonstrated positive learning gains for students using the platform on an ongoing basis [26]. Learners using ASSISTments complete mathematics problems, and can receive multi-step hints or scaffolding on demand or after making errors. ASSISTments provides support for mastery learning, where learners continue working on a skill until they demonstrate they can answer correctly three times in a row, and offers spiraling practice/review functionality as well.

Among AIED learning systems, ASSISTments offers substantial support for external researchers. Learning analytics and educational data mining researchers are able to download (as of this writing) fourteen publicly available data sets named Open Released Datasets,[1] which offer extensive interaction log data, combined in some cases with additional data such as field observations of student affect or longitudinal student outcomes. Dozens of external researchers have used data from the ASSISTments system in further analyses.

ASSISTments also offers substantial support for A/B testing research, enabling a researcher to conduct randomized experiments on learners across the United States, using E-Trials, the Ed-Tech Research Infrastructure to Advance Learning Science [41]. A substantial number of external educational psychology and learning sciences researchers have used the ASSISTments system to conduct A/B tests on a wide range of research questions. The large scale of ASSISTments' use in both learning analytics and A/B testing research makes it an appropriate context to compare the scientific impact of these two types of research.

2.2 Articles Studied

In this study, we compared the types of scientific impact achieved by two categories of papers, referred from here onwards as the "target" papers. We selected all the papers published up until March 2021 (when we pulled our data set for analysis) that leverage the ASSISTments platform for conducting the two different kinds of research. We filtered out the papers which did not fall into either category.

The first type of papers (referred to as A/B papers) compare the impact of two versions of a learning activity within the ASSISTments system. For the A/B papers, students are experimentally assigned to one condition or the other, to evaluate the impact of intervention on student learning or other outcomes.

[1] The open released data sets are publicly available at https://www.etrialstestbed.org/resources/featured-studies/dataset papers.

The second type of papers (henceforth referred to as secondary data analysis or EDM papers), use interaction log data from the ASSISTments system to investigate a range of research questions, including the impact of different behaviors on student outcomes, the accuracy of different knowledge modeling algorithms, and the linguistic attributes of ASSISTments math problems.

All the target papers for both categories were obtained from the publicly available ASSISTments website, which provides a list of papers that use their Open Released Datasets, as well as a repository of all the published randomized controlled experiments using ASSISTments. This yielded a total of 27 target A/B papers, and 32 target EDM papers. In March 2021, we used Google Scholar to obtain a list of papers citing each of these target articles. An article was considered if the full text could be obtained either openly over the internet, through the University of Pennsylvania library, or through interlibrary loan. Both peer-reviewed and non-peer-reviewed (i.e. dissertations, xArxiv, white papers) documents were included. Only articles in English were considered for the review process. Duplicates were filtered out if a single paper was citing one target paper more than once, however, if a single paper was citing different target papers multiple times, then each citation was considered separately. This gave a total of 2418 citations across all of the target papers (756 total citations for A/B papers, or 28 per paper; 1662 total citations for EDM papers, or 51.9 per paper).

We conducted statistical power analysis in order to determine how many citing papers to sample from this large number of articles for qualitative coding. An initial analysis of the citations of two highly-cited papers was used to choose parameters for the statistical power analysis. Statistical power was calculated using G*Power 3.1.9.4, assuming an effect size where papers in one category were cited 50% more often for one reason than the other paper category, with a baseline of 40% citation for the less common reason (i.e. 40% versus 60%; risk ratio = 1.5), with the allocation ratio set to one (i.e. we will sample approximately the same number of papers of each type), and α set to 0.05, using the Z test of the significance of the difference between two independent proportions (this test is mathematically equivalent to $\chi 2$ with one degree of freedom – they provide the exact same p values). For this test, statistical power of 0.8 would be achieved with samples of 97 and 97. Given this goal number of papers, we conducted stratified random sampling (stratified to equalize the number of citing papers per target paper as much as possible). This resulted in a data set of 174 papers citing A/B papers and 167 papers citing EDM papers for coding, moderately larger than the goal sample size.

2.3 Coding Scheme

We identified all the citations of any target papers within each article that cited one or more of the target papers. In many cases a citing article cited multiple target papers, in most cases all from the same type of paper (A/B or EDM) and in exactly one case from both.

Next, we developed a coding scheme to identify the reasons why a citation might cite an article. Our first step towards developing this coding scheme was to take an extensive list of reasons why people cite published articles [21], which had been distilled from a review of 30 studies on citing behavior [3]. We then eliminated reasons not found in our citing articles or that would not be explicitly stated in the text surrounding a citation. For instance, [21] notes that authors may choose which paper to cite based on the availability of full text for that paper, a reason that would be difficult to identify from how a paper is cited within the text. We then removed or merged categories that were not clearly differentiated from each other, and categories that did not seem to occur in our papers. This yielded our final coding scheme for citations. As will be noted below, not all of the categories we chose to code were ultimately found in our sample of citing papers. The final coding scheme was:

Publication-Dependent Reasons

Citation due to some attribute of the publication being cited (in the target article).

P1: The target paper was the original publication in which an idea or concept was discussed – a "classic" article.

P2: Using/giving credit to ideas, concepts, theories, methodology, and empirical findings by others.

P3: Earlier work on which current work builds.

P4: Providing background, to give "completeness" to an introduction or discussion.

P5: Empirical findings that justified the author's own statements or assumptions.

P6: Refuting or criticizing the work or ideas of others.

P7: Mentions of other work ("see also", "see for example", "cf", "e.g.", "i.e.") without further discussion.

P8: Used target paper's dataset for secondary analysis.

Author-Dependent Reasons

Citation due to some attribute of the author being cited (in the target article).

A1: Paying homage to a pioneer in the research area/giving general credit for related work.

A2: Ceremonial citation, the author of the cited publication is regarded as "authoritative".

A3: Self-citation: one of the authors was also an author on the target article.

Note that this coding scheme is not exhaustive; some citations may not be coded as representing any of these categories (for instance, articles cited as a part of the systematic review of studies) for both types of paper.

Initially, a subset of citations for each target paper was coded[2] in terms of this coding scheme by two coders (the first and third authors), to establish inter-rater reliability, and then the first author coded all the papers. If a coder judged that a paper was cited for multiple reasons – for instance, if it was cited in different parts of the paper – multiple codes were given. However, if a citing paper cited the same target paper multiple times for the same reason, it was counted a single instance – i.e., if the citing paper cited the target paper for reason P2 in four different places, it was treated as a single citation because of reason P2.

The proportion of each citation category found across citing papers was compared using the chi-squared test, between the two types of target papers (i.e. A/B versus EDM). Both Bonferroni and Benjamini and Hochberg corrections were applied (separately).

[2] The data set created is publicly available at https://osf.io/rmswe/?view_only=d496417aef1e4046907d2271b8a86cbb.

Inter-rater reliability (Cohen's Kappa) was calculated for each coding category, treating each category as independent (i.e. a set of binary codes) since coding was non-exclusive. The average Kappa across categories was 0.77 for A/B and 0.72 for EDM, 0.75 overall. Kappa was above 0.6 for every category. Categories P1, A1, and A2 were never coded for any citation by either of the two coders. For A1 and A2, this might be due to difficulty in identifying an author-dependent reason for citation from the text of the paper; much of the research on author-dependent reasons for citation has involved self-report rather than content analysis ([36, see review in [3]). The lack of application of P1 may similarly be due to the difficulty of identifying it from the paper text. Although the original reason for citing a paper may be its classic status, the practice of academic writing may result in a paper being discussed in terms of a different reason.

3 Analysis and Results

After inter-rater reliability was established, the first coder coded every citation in every paper. We next analyzed the prevalence of each citation category for each type of paper, and whether the prevalence of any citation category was statistically significantly different between the two types of papers. As mentioned above, within analysis we considered each citing paper/reason combination only once for each target paper, even if a target paper was cited for the same reason more than once in the same citing paper.

Table 1 shows that the most common citation category, for both papers, was P2, using/giving credit to specific ideas, concepts, theories, methodology, and empirical findings by others. It was seen in around more than half of the citations (averaged at the level of citing papers) for target EDM papers, and 35.6% for A/B papers. P4 appeared in a substantial 32.2% of citations for A/B papers, and about half of that in EDM papers. Two categories were seen between 15% and 25% of the time for both types of papers: P3, Earlier work on which current work builds, and A3, Self-citations. The remaining three categories were seen less than 10% of the time for both papers.

Statistically significant differences between the two paper types are given in boldface.

We then compared the prevalence of each citation category between paper A/B and paper EDM using a chi-squared test. This test assumes that paper A/B and paper EDM are cited by different sets of papers. In practice, only 1 paper in our sample cited both of these two categories of papers (out of a total of 341 papers), so this seemed like a safe assumption rather than a situation where a significantly more complex method tailored to partial overlap of data sets would be warranted. The statistically significant categories are P2 and P4. Category P2 stands for using/giving credit to specific ideas, concepts, theories, methodology, and empirical findings by others, which was cited 35.6% of the time by A/B papers, and 58.1% by the EDM papers, $\chi 2$ (df = 1, N = 341) = 17.26, p = 0.00003. Category P4 represents providing background, to give "completeness" to an introduction or discussion, and it was about twice as commonly cited in A/B papers (32.2%) than in the EDM papers (16.2%), $\chi 2$ (df = 1, N = 341) = 11.87, p = 0.0005. The full pattern of statistical evidence is given in Table 1.

Table 1. The prevalence of different Citation Categories for each of the two paper types

Reason for citation	Average prevalence (paper AB)	Average prevalence (paper EDM)	p value
P2: Using/giving credit to specific ideas, concepts, theories, methodology, and empirical findings by others	**35.6%**	**58.1%**	**0.00003**
P3: Earlier work on which current work builds	18.4%	15.6%	0.49
P4: Providing background, to give "completeness" to an introduction or discussion	**32.2%**	**16.2%**	**0.00057**
P5: Empirical findings that justified the author's own statements or assumptions	9.8%	6.0%	0.20
P6: Refuting or criticizing the work or ideas of others	1.2%	3.6%	0.14
P7: Mentions of other work ("see also", "see for example", "cf", "e.g.", "i.e.") without further discussion	8.0%	9.0%	0.76
P8: Used target paper's dataset for secondary analysis	4.0%	1.8%	0.22
A3: Self-citation	19.5%	24.6%	0.35

There is an inflated risk of Type I error since we ran eight statistical tests. To address this risk, we applied Benjamini and Hochberg and Bonferroni post-hoc controls. No significant tests became non-significant after the post-hoc test. Categories P2 & P4 were found to have $p < 0.001$, so they remain significant after post-hoc control. All other tests were non-significant, even without a post-hoc correction.

4 Conclusions and Discussions

In this study, we have investigated the reasons behind why scientists cited two types of papers using AIED systems for research. One category of papers used the platform to conduct automated A/B tests, the other category of papers used the platform's data to do secondary learning analytics (EDM) research.

We distilled a list of eleven reasons on why a paper is cited from prior literature on scientometrics, and then applied this list of reasons (as citation categories) to a sample of papers that cited one of the two types of target papers, within the same learning platform, with two coders who established inter-rater reliability for each code. Within this learning platform, the EDM papers were cited almost twice as much as the A/B papers, which may reflect several factors, including the relative contribution of each type of work, the ease in building on work of each type, or the size of the large and flourishing learning analytics research community.

In our findings, both types of papers were cited primarily for publication-based reasons rather than author-based reasons (except for self-citation). However, this may simply be due to the difficulty in identifying author-based reasons for citation. For example, a paper may have been cited because of its author's political power, but that citation may then be justified within the paper in terms of some scientific aspect of the paper, such as category P7 (citations to a paper as an example of some more general category, without further discussion). As such, determining if a citation is author-based probably depends on other forms of data collection such as anonymous surveys [32].

In comparing the two types of articles, two statistically significant differences were found: the EDM type of papers were cited for reason P2 (Using/giving credit to specific ideas, concepts, theories, methodology, and empirical findings by others) over 50% of the time, which was 1.6 times more than A/B papers cited for that reason. This finding suggests that EDM papers are more prevalent in generating ideas, concepts, and empirical findings that other researchers in the field find useful. This type of research directly contributes to the field moving forward.

On the other hand, category P4 (Providing background, to give "completeness" to an introduction or discussion) was cited as a reason twice as many times by the A/B papers than the EDM papers. These citations were primarily found in the 'Introduction' or the 'Literature Review' section of the papers. The findings might indicate that A/B papers are being cited for related work, and to cover the breadth of the research related to that topic, instead of directly building on previous work.

Overall, these findings seem to highlight the different types of contributions the two types of papers make – EDM type of papers seem to have a larger impact on subsequent research than A/B papers. A/B research studies seem to be carried out more independently from prior work. One possible explanation for this pattern can be because the range of potential design modifications is large and varies based on the original design of the system being studied, whereas EDM algorithms tend to either compare algorithms (directly using previous work) or develop an analysis across papers (like work on defining wheel-spinning and studying it). It is also possible that as the community of learning platform A/B researchers develops, they will converge to a smaller set of designs and begin to use P2 citations more often.

A limitation of this study remains that it investigated the citation reasons for two types of research on a single learning platform. It is possible that some aspect of the design of ASSISTments facilitated conducting work that would receive citations for specific ideas more in EDM research than A/B research (although ASSISTments is one of the learning platforms currently most committed to supporting external A/B researchers). It is also possible that the learning domain (of mathematics) influenced the contributions made by the work, or that the community of researchers drawing upon mathematics education research influenced this paper's results. To draw more substantial conclusions, this work must be replicated within a wide variety of learning platforms (also varying by subject matter). However, there are currently only a small number of learning platforms used at scale both for A/B testing and learning analytics research, though this number is increasing. In future work, we recommend that researchers focus analysis on single platforms, as in this paper. Comparing between different platforms raises confounds not present in single-platform analysis.

Other factors in the field may of course also impact how studies are cited. For example, differences in the expectations of reviewers in venues that see more A/B studies versus EDM studies may impact how authors cite papers when submitting to these venues. The time it takes to conduct A/B studies may also explain the lower total quantity of citation for A/B studies, although not why the type of citation differed.

Another limitation to the study was a possible lack of statistical power. Although a power analysis was conducted prior to research, some rare categories had seeming differences that were not statistically significant (i.e. 1.2% versus 3.6% for category P6). Unfortunately, this limitation was unavoidable for the overall data set, even if we had coded every example in the data (an arduous task). Power of 0.8 would only have been achieved by category P6 if we had been able to code 878 examples of both A/B and EDM, larger than the total current population for A/B, even if we had skipped the necessary step of conducting a post-hoc correction. P5, the next closest category to significance, would have required 1088 examples of each category. Thus, investigating differences in categories this rare would require a substantially larger data set. It is possible that this paper's work can eventually contribute to such a goal, by developing a categorization scheme and building a corpus of codes that can be used as a training set for an eventual NLP approach that can automatically detect why one paper cites another [13]. Ultimately, the work presented here suggests that EDM papers and A/B testing papers are cited for different reasons. More comprehensively investigating this topic – and investigating subcategories within these broader categories of research – may help us to understand how scientific progress occurs, in our field and more broadly.

References

1. Baker, R.S., Yacef, K.: The state of educational data mining in 2009: a review and future visions. J. Educ. Data Mining **1**(1), 3–17 (2009)
2. Beck, J.E., Arroyo, I., Woolf, B.P., Beal, C." An ablative evaluation. In: Proceedings of the 9th International Conference on Artificial Intelligence in Education, pp. 611–613 (1999)
3. Bornmann, L., Daniel, H.: What do citation counts measure? A review of studies on citing behavior. J Document. **64**, 45–80 (2009)
4. Cambrosio, A., Limoges, C., Courtial, J., Laville, F.: Historical scientometrics? Mapping over 70 years of biological safety research with coword analysis. Scientometrics **27**(2), 119–143 (1993)
5. Chen, G., Rolim, V., Mello, R.F., Gašević, G.: Let's shine together! a comparative study between learning analytics and educational data mining. In: Proceedings of the Tenth International Conference on Learning Analytics & Knowledge, pp. 544–553 (2020)
6. Cole, J.R., Cole, S.: The Ortega hypothesis: Citation analysis suggests that only a few scientists contribute to scientific progress. Science **178**(4059), 368–375 (1972)

7. Cronin, B.: The Citation Process: The Role and Significance of Citations in Scientific Communication, p. 103. Taylor Graham, London (1984)
8. Romero, C., Ventura, S.: Educational data mining and learning analytics: an updated survey. Wiley Interdiscipl. Rev. Data Mining Knowl. Discov. **10**(3) (2020)
9. Peña-Ayala, A.: Educational data mining: a survey and a data mining-based analysis of recent works. Expert Syst. Appl. **41**(4), 1432–1462 (2014)
10. Dormezil, S., Khoshgoftaar, T., Robinson-Bryant, F.: Differentiating between educational data mining and learning analytics: a bibliometric approach. In: Proceedings of the Workshops of the International Conference on Educational Data Mining, pp. 17–22 (2019)
11. Fazeli, S., Drachsler, H., Sloep, P.: Socio-semantic networks of research publications in the learning analytics community. In: Proceedings of the LAK Data Challenge (2019)
12. Garfield, E.: Can citation indexing be automated. In: Symposium Proceedings of the Statistical Association Methods for Mechanized Documentation, pp. 189–192 (1965)
13. Garzone, M., Mercer, R.E.: Towards an automated citation classifier. In: Conference of the Canadian Society for Computational Studies of Intelligence, pp. 337–346 (2000)
14. Goel, G., Lallé, S., Luengo, V.: Fuzzy logic representation for student modelling. In: Proceedings of the International Conference on Intelligent Tutoring Systems, pp. 428–433 (2012)
15. Gross, P.L.K., Gross, E.M.K.: College libraries and chemical education. Science **66**(1713), 385–389 (1927)
16. Hopkins, A.L., Jawitz, J.W., McCarty, C., Goldman, A., Basu, N.: Disparities in publication patterns by gender, race and ethnicity based on a survey of a random sample of authors. Scientometrics **96**(2), 515–534 (2013)
17. Khajah, M., Lindsey, R.V., Mozer, M.C.: How deep is knowledge tracing? In: Proceedings of the International Conference on Educational Data Mining (2016)
18. Kizilcec, R., et al.: Scaling up behavioral science interventions in online education. Proc. Natl Acad. Sci. **117**(26), 14900–14905 (2020)
19. Koedinger, K.R., Baker, R.S., Cunningham, K., Skogsholm, A., Leber, B., Stamper, J.: A data repository for the EDM community: the PSLC DataShop. In: Romero, C., Ventura, S., Pechenizkiy, M., Baker, Ryan S.J.d. (eds.) Handbook of Educational Data Mining, pp. 43–56. CRC Press, Boca Raton (2010)
20. Koedinger, K.R., Corbett, A.T., Perfetti, C.: The Knowledge-Learning-Instruction framework: bridging the science-practice chasm to enhance robust student learning. Cogn. Sci. **36**(5), 757–798 (2012)
21. Lindgren, L.: If Robert Merton said it, it must be true: A citation analysis in the field of performance measurement. Evaluation **17**(1), 7–19 (2011)
22. Liu, R., Koedinger, K.R.: Closing the Loop: Automated Data-Driven Cognitive Model Discoveries Lead to Improved Instruction and Learning Gains. Journal of Educational Data Mining **9**(1), 25–41 (2017)
23. Malmi, L., Sheard, J., Kinnunen, P., Sinclair, S., Sinclair, J.: Theories and models of emotions, attitudes, and self-efficacy in the context of programming education. In: Proceedings of the 2020 ACM Conference on International Computing Education Research, pp. 36–47 (2020)
24. Maturana, R.A., Alvarado, M.E., López-Sola, S., Ibáñez, M.J., Elósegui, L.R.: Linked data based applications for learning analytics research: Faceted searches, enriched contexts, graph browsing and dynamic graphic visualisation of data. In: Proceedings of the LAK Data Challenge (2013)
25. Mostow, J., Beck, J.E., Valeri, J.: Can automated emotional scaffolding affect student persistence? A baseline experiment. In: Proceedings of the Workshop on "Assessing and Adapting to User Attitudes and Affect: Why, When and How?" at the 9th International Conference on User Modeling (UM'03), pp. 61–64 (2003)

26. Murphy, R., Roschelle, J., Feng, M., Mason, C.A.: Investigating efficacy, moderators and mediators for an online mathematics homework intervention. J. Res. Educ. Effect. **13**(2), 235–270 (2020)
27. Ostrow, K., Heffernan, N., Williams, J.J.: Tomorrow's edtech today: establishing a learning platform as a collaborative research tool for sound science. Teach. Coll. Rec. **119**(3), 300–306 (2017)
28. Paquette, L., Ocumpaugh, J., Li, Z., Andres, A., Baker, R.S.: Who's Learning? Using Demographics in EDM Research. J. Educ. Data Min. **12**(3), 1–30 (2020)
29. Park, J., Choi, H.J.: Factors influencing adult learners' decision to drop out or persist in online learning. J. Educ. Technol. Soc. **12**(4), 207–217 (2009)
30. Reich, J.: Rebooting MOOC research. Science **347**(6217), 34–35 (2015)
31. Shockley, W.: On the statistics of individual variations of productivity in research laboratories. Proc. IRE **45**(3), 279–290 (1957)
32. Stamper, J.C., et al.: The rise of the super experiment. In: Proceedings of the International Conference on Educational Data Mining Society (2012)
33. VanLehn, K.: The relative effectiveness of human tutoring, intelligent tutoring systems, and other tutoring systems. Educ. Psychol. **46**(4), 197–221 (2011)
34. Verbert, K., Duval, E., Klerkx, J., Govaerts, S., Santos, J.L.: Learning analytics dashboard applications. Am. Behav. Sci. **57**(10), 1500–1509 (2013)
35. Vinkler, P.: A quasi-quantitative citation model. Scientometrics **12**(1–2), 47–72 (1987)
36. Waheed, H., Hassan, S., Aljohani, N.R., Wasif, M.: A bibliometric perspective of learning analytics research landscape. Behav. Inf. Technol. **37,** 10–11 (2018)
37. Wise, A.F., Jung, Y.: Teaching with analytics: Towards a situated model of instructional decision-making. J. Learn. Anal. **6**(2), 53–69 (2019)
38. Yeung, C., Yeung, D.: Addressing two problems in deep knowledge tracing via prediction-consistent regularization. In: Proceedings of ACM Conference on Learning at Scale, pp. 1–10 (2018)
39. Zhang, J., Shi, X., King, I., Yeung, D.: Dynamic key-value memory networks for knowledge tracing. In: Proceedings of the 26th International Conference on World Wide Web, pp. 765–774 (2017)
40. Zouaq, A., Joksimovic, S., Gasevic, D.: Ontology learning to analyze research trends in learning analytics publications. In: Proceedings of the LAK Data Challenge (2013)
41. Krichevsky, N., Spinelli, K., Heffernan, N., Ostrow, K., Emberling, M.R.: E-TRIALS, Doctoral dissertation, Worcester Polytechnic Institute (2020)

Investigating the Effectiveness of Visual Learning Analytics in Active Video Watching

Negar Mohammadhassan(✉) and Antonija Mitrovic(✉)

University of Canterbury, Christchurch, New Zealand
Negar.mohammadhassan@pg.canterbury.ac.nz,
Tanja.mitrovic@caterbury.ac.nz

Abstract. Video-based Learning (VBL) is a popular form of online learning, which may lead to passive video watching and low learning outcomes. Besides potential low engagement, VBL often provides very limited feedback on student's progress. As a way to overcome these challenges, we present student-facing visual learning analytics (VLA) designed for the AVW-Space VBL platform. Using a quasi-experimental design, we compared data collected in the same first-year university course in 2020 (control group, 294 participants using the original version of AVW-Space) to the 2021 data when 351 participants used the enhanced version of AVW-Space (experimental group). We analysed various measures of engagement (number of watched videos, comments, etc.) and learning (pre/post-study knowledge scores). The findings show that VLA encourage constructive behaviour and increase learning. This research contributes to using student-facing VLA in VBL platforms to boost engagement and learning.

Keywords: Video-based learning · Visual learning analytics · Student model

1 Introduction

Learning by watching videos is increasingly popular due to its flexibility in time and place. Many studies show that video-based learning (VBL) increases motivation, engagement and learning [1, 2]. However, the lack of interaction with videos and humans, as well as the lack of feedback and personalisation can turn VBL into a passive form of learning, with learners simply watching the videos and not engaging deeply [2, 3]. Several approaches have been used in VBL to overcome the engagement challenge, such as integrating annotation tools [3] and quizzes [4]. Although these approaches address the lack of human interaction and interactivity with videos, they do not provide personalised feedback to learners. One way to address these shortcomings is integrating visual learning analytics (VLA) into VBL to boost engagement by providing feedback [5].

VLA can provide insights on learning resources [3, 6] and the student's learning progress. The former is the same for all learners, while the latter provides more personalisation [7]. Visualisation of the learner model provides up-to-date information to the learner, such as progress in learning activities, knowledge and affective states [8, 9].

M. M. Rodrigo et al. (Eds.): AIED 2022, LNCS 13355, pp. 127–139, 2022.
https://doi.org/10.1007/978-3-031-11644-5_11

Visualising this information helps learners assess their learning to make informed decisions about what to do next to achieve their learning goals [9], and increases engagement and learning [7, 10]. Although the effectiveness of student-facing VLA has been studied in various educational platforms [11, 12], only a few studies have investigated the use of VLA in video-based learning.

AVW-Space [13] is a VBL platform that supports engagement via note-taking, peer-reviewing and personalised prompts [14, 15]. In AVW-Space, the learner can watch and comment on videos, and rate peers' comments. An early AVW-Space study [14] found that students who commented on videos learnt more than those who watched videos passively. Thus, a histogram and timeline of the class's comments were added to AVW-Space to encourage commenting and help students recognise the highly-attended video parts [16]. However, these visualisations do not convey information about the learner's progress to help them regulate their learning activities.

This paper investigates the effectiveness of new VLA integrated into AVW-Space. We present the effects of visualisations on students' engagement and learning as well as their subjective opinions on visualisations. This research contributes to the utilisation of VLA to tackle engagement challenges in VBL. We defined three research questions:

RQ1. Do VLA increase engagement and foster constructive behaviour?

RQ2. What is the effect of the visual learning analytics on learning?

RQ3. What is the students' opinion on different visualisations?

2 Related Work

The visualisation of the learning process offers evocative insight and allows students to monitor and control learning [17]. Various types of information can be presented in VLA [17, 18]. Competency tracking and displaying learning difficulties are examples of cognitive visual analytics [19]. Behavioural visual analytics includes the progression in learning tasks (e.g. watched videos) [20]. Some VLA go beyond the domain knowledge and present the learner's metacognitive state, such as study tactics and planning [21]. Other visualisations indicate students' emotional status to increase emotional awareness [22, 23]. Some visualisations provide analytics of social models such as comparisons to the class [24]. However, the effectiveness of visualisations depends on their explainability. Some studies suggest that learners find it hard to interpret the VLA to inform their learning strategies [25]. Visualisations may even harm students' motivation; some research found that VLA caused social anxiety when students were presented with their peers' performance compared to themselves [26, 27].

Visual analytics has been applied to the learners' interaction with video lectures, attitudes, and learning performance to find the most difficult parts of the video [28]. However, these VLA were not displayed to the students. CourseMapper [3] uses students' interaction with video to provide a heatmap on the video scrub bar to help students identify the most viewed parts. CourseMapper also uses annotations timeframe and counts to display an annotation map, which illustrates portions of videos which received more annotations and likely contain interesting information. However, these visualisations are the same for all students and do not provide any personalised information. A VBL platform used in a flipped classroom [29] provides a simple visualisation of quiz scores and video completion rates to support students' self-assessment. An experiment with this visualisation showed that the learners who had access to the visualisation showed higher engagement levels pre-class (e.g. watching videos and answering quizzes) and in-class sessions (e.g., team discussion) without the instructor's reminders. However, these visualisations provided limited information on students' performance. Thus, we propose more detailed VLA to support engagement in VBL.

AVW-Space is a VBL platform designed for teaching transferable skills [13]. To create a learning space in AVW-Space, the teacher first selects videos from YouTube. AVW-Space supports engagement in two phases: 1) Personal space (Fig. 1), where students watch videos and make comments, and 2) Social space (Fig. 2), where students review and rate their peers' comments. Personal Space is always available for students, while Social Space becomes available after the teacher selects comments for review.

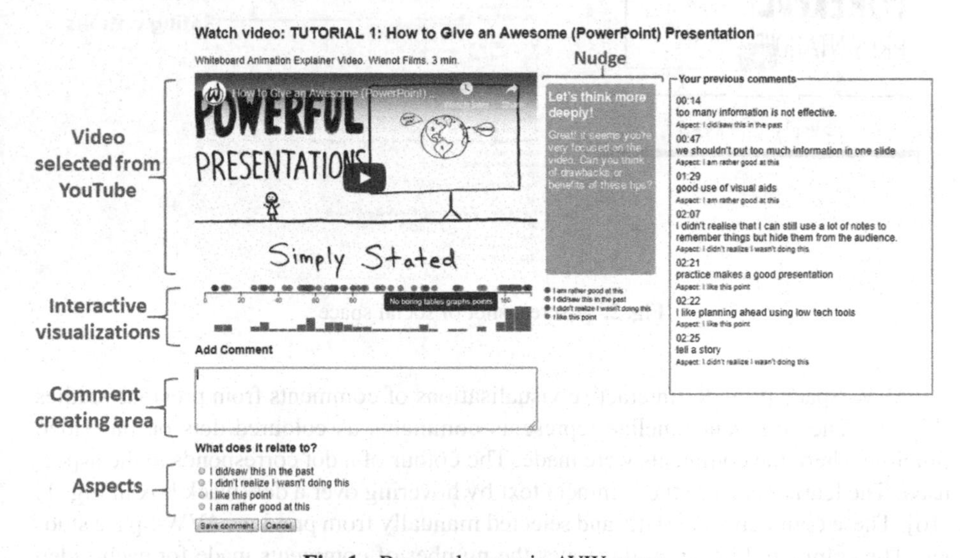

Fig. 1. Personal space in AVW-space

In Personal Space, students tag each comment by an aspect that the teacher defined. The aspects are micro-scaffolds for directing students to reflect on key points of the video or their experience. A previous study on AVW-Space showed commenting and using aspects had a positive effect on engagement and learning [14]. The analysis of comments from previous studies differentiated low-quality comments, which merely repeat the video content, from high-quality comments, in which students elaborated critically on the video and reflected on their experience or planned for future improvement [30, 31]. Students who made high-quality comments learnt more [31]. We added nudges to AVW-Space, to foster good commenting behaviours (i.e. writing high-quality comments using various aspects) [15, 16]. For example, if a student is passively watching a video and has made no comments, the student will receive a nudge stating that commenting is beneficial for learning. If the student used only one aspect when commenting, a nudge will draw the student's attention to other aspects. AVW-Space analyses comment quality as students write them, using machine learning classifiers we developed to predict the comment quality [15, 30]. When students write comments that merely repeat video content, they will receive a nudge suggesting to think more critically about the video (as in Fig. 1). A student who is watching the last part of the video and has made no self-reflective comments, will receive a nudge to reflect on their previous experience. Previous studies showed nudges significantly increased the number and quality of comments [15, 16]. However, the student might overlook nudges and not benefit from them, since they are visible for a few seconds. Allowing students to review nudges they received could help them understand the expectations for commenting.

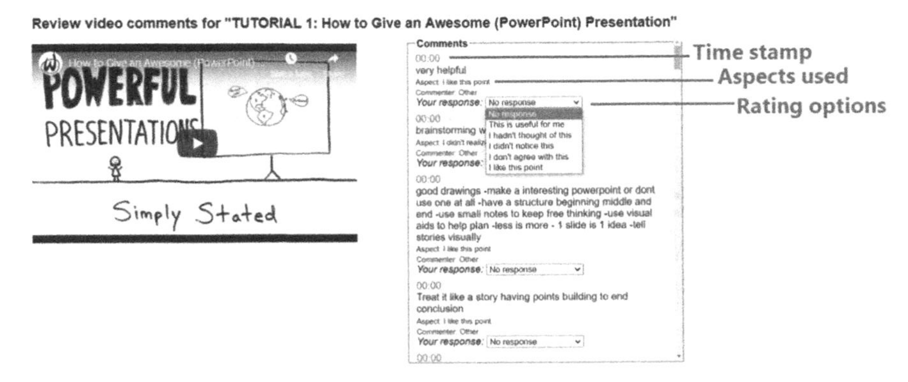

Fig. 2. A screenshot of social space

AVW-Space provides interactive visualisations of comments from previous studies (Fig. 1). The comment timeline represents comments as coloured dots on the video timeline where the comments were made. The colour of a dot corresponds to the aspect used. The learner can see the comment text by hovering over a dot (black box in Fig. 1) [16]. These comments are static and selected manually from previous AVW-Space studies. The comment histogram illustrates the number of comments made for each video segment. This way, AVW-Space supports social learning by allowing the student to see what other students wrote about the same video. However, it is possible for a student to use the comment timeline only for learning others' opinions and still make no comment. Thus, the timeline visualisation needs improvements to clarify its purpose.

In Social Space, students can rate comments by choosing the rating options defined by the teacher. Students can see their own comments but cannot rate them or see their received ratings. An early study showed that rating brings an additional benefit to learning on top of commenting [14]. Hence, visualising the student's progress in commenting and rating could encourage more engagement. Moreover, visualising received ratings could motivate students to write high-quality comments.

AVW-Space is based on the ICAP framework [32], which categorises learners' overt behaviour into Interactive, Constructive, Active and Passive. The more engaged students are, the more they learn (i.e. Passive < Active < Constructive < Interactive). Passive learners receive information via merely watching videos. Active students perform additional actions like note-taking, but their annotations repeat the received information with no elaboration. Constructive learners add new information that was not explicitly taught, by reflecting on their knowledge and making connections. The last category, Interactive, is not relevant for our research as AVW-Space does not support direct interaction between students. Previous AVW-Space studies showed that nudges increased constructive engagement [15, 16], but no research has investigated the effectiveness of VLA on constructive behaviour. One of our goals is to address this gap.

3 Enhancing Visualisations in AVW-Space

To identify what VLA to integrate into AVW-Space, we analysed the students' feedback from previous studies. Most students requested a progress visualisation to monitor videos they have watched and reviewed. Some students complained that nudges disappear before they read them thoroughly, so they wanted to revisit them. Moreover, students wanted to see ratings they received from their peers in the Social Space. We also decided to provide a personal timeline visualisation to allow students to compare their comment timeline to the others' comments timeline. We conducted rapid-prototyping and evaluated prototypes by brainstorming and interviewing five domain experts. The visualisations went through three iterations: a paper-based mock-up, a digital mock-up and functional visualisation developed using D3.js and JavaScript.

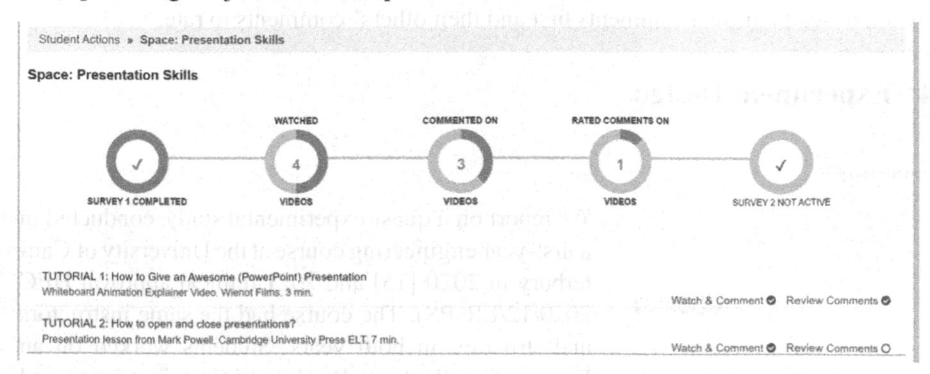

Fig. 3. Video page with progress visualisation

Figure 3 shows the new progress visualisation at the top of the page, the list of videos (unchanged) and the new green indicators of visited Personal and Social Space. Each student can only see their own progress report, showing the number of watched videos, commented videos, and videos on which comments are rated as well as whether the surveys have been completed. The tasks in the progress report are presented in the preferred order: watch a video, comment, and rate peers' comments; since a previous study showed some students rated peers' comments before making their own comments [33].

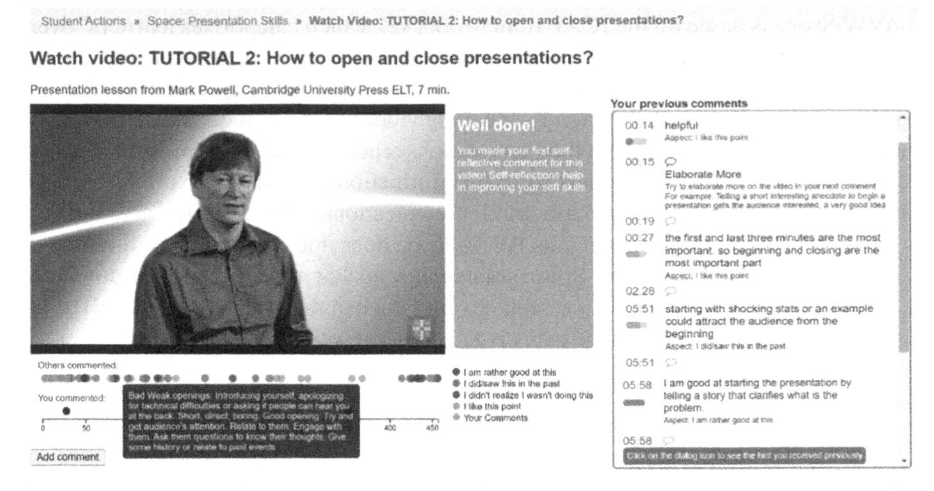

Fig. 4. The enhanced personal space interface with the new visualisation

We replaced the comment histogram with a visualisation showing the student's own comments (Fig. 4). The comment list now shows the quality of each comment and the nudges received. Students can read the nudge message by clicking on the dialogue icon. The quality indicators are in three colours: red (off-topic), yellow (reflecting on the video) and green (self-reflective or self-regulating). We also modified the Social Space interface (Fig. 5). The student can now see a pie chart for their own comments, showing ratings received from others. The number of ratings received for a particular option is shown by hovering over the rating option on the pie chart. Students can also use a toggle switch to see their own comments first and then others' comments to rate.

4 Experiment Design

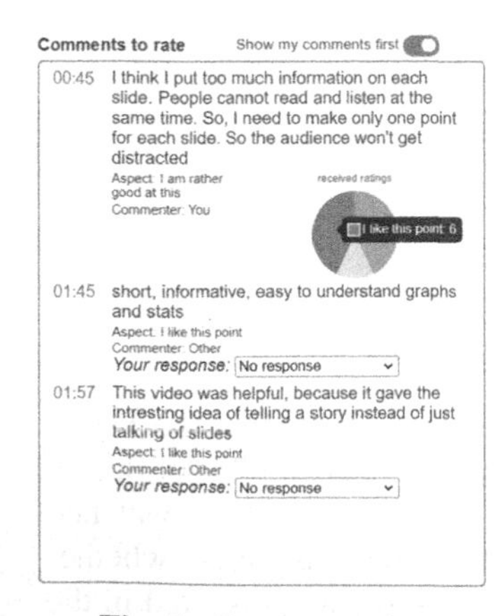

Fig. 5. Comment rating

We report on a quasi-experimental study, conducted in a first-year engineering course at the University of Canterbury in 2020 [15] and 2021 (ethical approval HEC 2020/12/LR-PS). The course had the same instructors and structure in both years. Students worked on an Engineering Without Borders project in teams, and needed to give a group presentation in the last week of the course. In both years, the students were notified about the online training for presentation Skills. The students who watched at least one video on AVW-space received 1% of the final course grade. The control group consisted of the 2020 participants, who used the original version of AVW-space presented in Sect. 2. The experimental group consisted of the 2021 participants who could see the new VLA. The learning materials and procedure were identical in both years. There were four tutorial videos on how to give presentations, and four example videos of real presentations [13]. Participants provided informed consent and completed Survey 1, containing questions about the participant's demographic, knowledge, experience and training in giving presentations. Participants were instructed to 1) watch and comment on the tutorial videos, 2) critique the example videos, and 3) rate peers' comments. Finally, participants completed Survey 2, containing the same questions about giving presentations. In the surveys, the participants had one minute to list concepts about presentation skills. The students' answers were marked automatically, using the ontology developed in previous work [34]. The marks for conceptual knowledge questions are used as the pre-/post-tests scores (CK1 and CK2).

Regarding the first research question (RQ1), we expected that the VLA will result in a significant increase in the number of videos watched and commented/rated on in 2021 (Hypothesis H1). Secondly, because the quality of comments written is visualised in the enhanced version of AVW-Space, we expected to see a significantly higher proportion of constructive students in 2021 (Hypothesis H2). Our second research question (RQ2) focuses on the impact of VLA on learning. A previous study [15] found that students who engaged more deeply with AVW-Space and wrote high-quality comments learnt more; we expected to see the same effect in 2021 (Hypothesis H3). We also expected that more interactions with VLA will lead to more videos watched/commented on, and more high-quality comments (Hypothesis H4). The last research question (RQ3) focuses on the students' feedback on VLA collected in Survey 2.

5 Results

351 students from the 2021 course and 294 students from the 2020 course completed Survey 1 and watched at least one video. However, only 277 students in 2021 and 147 students in 2020 completed both surveys. The increase in survey completion in 2021 could be due to the survey status in the progress visualisation. There were no significant differences between the groups in students' demographics, CK1, training and experience scores.

Effects of VLA on Engagement (RQ1): Table 1 presents the summary of interactions with AVW-Space. The 2021 students watched more videos and wrote more comments than the 2020 students, but there was no significant difference on the number of ratings made. Students in 2021 made significantly more comments and received significantly more nudges, which could be attributed to the visualisations in the Personal Space. In addition, 2021 students watched and made ratings on significantly more videos, indicating that the progress visualisation may have motivated students to complete commenting and rating. The number of days spent on AVW-Space increased significantly in 2021, and Hypothesis H1 is confirmed.

We categorised the students post-hoc into three categories, using the ICAP framework (Table 2). Students who watched videos without making any comments were classified as Passive. To distinguish Constructive from Active students, we used the median number of high-quality comments made on tutorial videos, which was 2 in both years. We defined Constructive students as those who wrote three or more high-quality comments, and

Active students as those who wrote up to two high-quality comments. A chi-square test of homogeneity between years and ICAP categories revealed a significant difference (Chi-square = 45.24, $p < .001$) with effect size (Phi) of .26 ($p < .001$). A post hoc analysis showed a significant increase of Constructive students and a significant decrease of Passive students in 2021, confirming Hypothesis H2.

Table 1. Activities (mean and standard deviation)

	2020 (294)	2021 (351)	t-test
Unique videos	5.26 (2.74)	6.98 (2.24)	$t = 8.54$, $p < .001$
Comments	10.29 (14.78)	14.04 (11.43)	$t = 3.34$, $p < .001$
Nudges	19.76 (16.06)	23.26 (12.62)	$t = 3.02$, $p < .01$
Ratings	21.74 (73.39)	23.55 (52.26)	$t = .36$, $p = .72$
Videos commented	3.78 (3.29)	6.60 (2.74)	$t = 11.67$, $p < .001$
Videos rated	1.16 (2.31)	6.44 (3.06)	$t = 24.91$, $p < .001$
Days on AVW-Space	3.08 (1.93)	4.29 (2.93)	$t = 5.73$, $p < .001$

Table 2. The distribution of ICAP categories in 2020/2021

ICAP categories	2020 (294)	2021 (351)	Significance
Passive	75 (25.5%)	25 (7.1%)	$p < .001$
Active	114 (38.8%)	141 (40.2%)	$p = .68$
Constructive	105 (35.7%)	185 (52.7%)	$p < .001$

Table 3 reports how each ICAP category interacted with VLA (hovering for longer than 5 s or clicking). Constructive students interacted with all visualisations significantly more than Active students ($p < .05$), except nudge visualisations. There was no significant difference on interactions with progress visualisation between the Active and Passive groups. However, Active students interacted significantly more with the others' comments timeline visualisation than Passive students ($p < .001$). The Passive group neither used the personal timeline nor the rating visualisations since they made no comments.

Effects of VLA on Learning (RQ2): We developed a model (Fig. 6) for the 2021 class, based on the hypotheses H3 and H4. The nodes represent the number of interactions with visualisations (progress, personal space, rating or nudges), the number of videos watched, the number of high-quality comments, and the conceptual knowledge score at the end of the study (CK2). The circles represent latent variables, curved bidirectional arrows for correlations and straight arrows link a predicting to a predicted variable.

The model was evaluated in IBM SPSS AMOS using the data from 277 students who completed both surveys. The mean of CK1/CK2 for these students was 14.18

Table 3. VLA interactions performed by different ICAP categories in 2021

Visualisation	Passive (25)	Active (141)	Constructive (185)	ANOVA
Progress	6.00 (7.56)	7.24 (6.84)	9.73 (8.76)	$F = 5.17, p < .01$
Others' timeline	4.72 (5.69)	18.99 (13.67)	33.06 (22.47)	$F = 39.69, p < .001$
Personal timeline	0	.58 (1.11)	1.73 (2.32)	$F = 20.76, p < .001$
Previous nudges	1.8 (3.09)	.93 (1.85)	1.00 (1.64)	$F = 2.33, p = .01$
Received ratings	0	.63 (2.08)	2.25 (4.78)	$F = 9.58, p < .001$

$\pm$ 6.05/13.53 $\pm$ 6.47, respectively. The chi-square test (14.01) for this model (df = 9, 19 estimated parameters) shows that the model's predictions were not statistically significantly different from the data ($p = .12$). The Comparative Fit Index (CFI) was .99, and the Root Mean Square Error of Approximation (RMSEA) was .04. Hence, the model is acceptable: CFI is greater than 0.9, and RMSEA is less than .06 [35]. The model indicates that a higher number of high-quality comments is associated with a higher CK2 score ($p < .01$). The number of interactions with rating visualisations positively affects CK2 ($p < .001$). Other links are all significant at $p < .001$ except *Progress visualisation → Video* ($p < .05$) and *Progress visualisation → Personal timeline visualisation* ($p < .05$). The covariances with *e8* show that a student who interacts with one visualisation is likely to interact with other ones. The model shows that the number of videos watched, received nudges, and interactions with visualisations affect the number of high-quality comments and consequently CK2, confirming hypotheses H3 and H4.

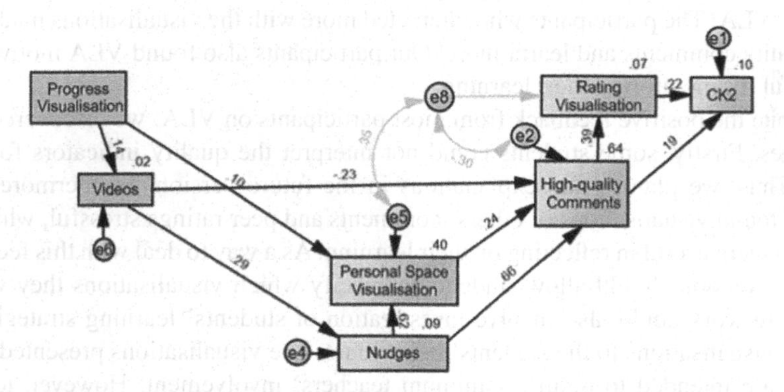

Fig. 6. The path diagram for investigating how VLA affects learning

Subjective Opinions on VLA (RQ3): We collected the students' feedback on visualisations in Survey 2. The progress visualisation received the most positive feedback (90.97%) among all visualisations. Students reported that the progress visualisations increased their motivation and facilitated learning organisation. 75% of feedback on the timeline visualisations was positive, stating that timelines helped them compare their progress with the others and inspired them to comment (e.g. "*It initially helped me to grasp an idea of what kind of comments were being looked for. It also encouraged me to try pick up on points in areas of the video which had less comments made.*"). However, some students found the timeline visualisations cluttered, and suggested adding filtering functionality. Some students disliked seeing others' comments, since they wanted to form their own ideas. 70% of feedback on nudges visualisations and comment quality indicators was positive, and students noted using them as feedback to improve commenting and stay on track (e.g. "*This tool can help you notice a pattern in the nature of hints you are getting, to identify whether there is something you consistently fail to recognise, or something you always comment on.*"). Students who criticised quality indicators suggested more explanation on their criteria. Lastly, 69% of feedback on rating visualisations were positive since they helped students see if their comments were useful to their peers (e.g. "*To motivate people to write good comments and also so they feel good about the work they've done*"). Some students found the rating visualisation unhelpful since they disliked being judged.

6 Discussion and Conclusions

We proposed visual learning analytics for VBL in order to increase engagement and learning. The visualisations are intended to assist students in monitoring and managing their learning. We conducted a study to investigate the effectiveness of the proposed VLA, which confirmed our hypotheses. The newly introduced VLA enhanced engagement: the experimental group participants watched more videos, wrote more comments, and spent more days on AVW-Space than participants who did not receive VLA. Furthermore, the proportion of students who wrote high-quality comments was higher when students had access to VLA. The participants who interacted more with the visualisations made more high-quality comments and learnt more. Our participants also found VLA motivational and useful for monitoring their learning.

Despite the positive feedback from most participants on VLA, we discovered some challenges. Firstly, some students could not interpret the quality indicators for comments. Thus, we plan to add explanations in the future version. Furthermore, some students found visualisations of others' comments and peer ratings stressful, while others found them useful in reflecting on their learning. As a way to deal with this feedback, the future version should allow students to specify which visualisations they want to see. Future work could also involve investigation of students' learning strategies and adapting visualisations to the students' behaviours. The visualisations presented in this research are intended to require minimum teachers' involvement. However, teachers must manually select comments shown in others' comments timeline. Hence, potential solutions for automating this task should be explored in future research.

The main challenge in studying the effectiveness of visualisations is in measuring the interactions with them. Since most interaction types are in the form of hovering, it is difficult to identify which interactions were intentional. Analysing eye gaze is one way to investigate interactions more precisely, but it is time-consuming and impractical for large classes. A limitation of our study was a single domain (presentation skills). Future work will investigate the effectiveness of the visualisations in other domains.

As discussed earlier, the literature on visualising the student model in VBL is limited. Our research addresses this gap and contributes to using student-facing VLA in video-based educational platforms to boost engagement and learning. The source of requirement elicitation for designing VLA in this research is mainly from students' feedback. The visualisations suggested in this research are applicable to any VBL platform that supports commenting and peer-reviewing.

References

1. Scagnoli, N.I., Choo, J., Tian, J.: Students' insights on the use of video lectures in online classes. Br. J. Edu. Technol. **50**, 399–414 (2019)
2. Yousef, A.M.F., Chatti, M.A., Schroeder, U.: The state of video-based learning: a review and future perspectives. Adv. Life Sci. **6**, 122–135 (2014)
3. Chatti, M.A., et al.: Video annotation and analytics in CourseMapper. Smart Learn. Environ. **3**, 10 (2016)
4. Cummins, S., Beresford, A.R., Rice, A.: Investigating engagement with in-video quiz questions in a programming course. IEEE Trans. Learn. Technol. **9**, 57–66 (2016)
5. Giannakos, M.N., Sampson, D.G., Kidziński, Ł.: Introduction to smart learning analytics: foundations and developments in video-based learning. Smart Learn. Environ. **3**(1), 1–9 (2016). https://doi.org/10.1186/s40561-016-0034-2
6. Wang, M., Peng, J., Cheng, B., Zhou, H., Liu, J.: Knowledge visualization for self-regulated learning. J. Educ. Technol. Soc. **14**, 28–42 (2011)
7. Hooshyar, D., Pedaste, M., Saks, K., Leijen, Ä., Bardone, E., Wang, M.: Open learner models in supporting self-regulated learning in higher education: a systematic literature review. Comput. Educ. **154**, 103878 (2020)
8. Bodily, R., et al.: Open learner models and learning analytics dashboards: a systematic review. In: Proceedings of 8th International Conference on Learning Analytics and Knowledge, pp. 41–50 (2018)
9. Bull, S., Kay, J.: Open learner models. In: Advances in Intelligent Tutoring Systems, pp. 301–322 (2010). https://doi.org/10.1007/978-3-642-14363-2_15
10. Aguilar, S., Karabenick, S.A., Teasley, S.D., Baek, C.: Associations between learning analytics dashboard exposure and motivation and self-regulated learning. Comput. Educ. **162**, 104085 (2021). https://doi.org/10.1016/j.compedu.2020.104085
11. Aguilar, S., Lonn, S., Teasley, S.D.: Perceptions and use of an early warning system during a higher education transition program. In: Proceedings of Learning Analytics and Knowledge, pp. 113–117 (2014)
12. Ruiz, J.S., Díaz, H.J.P., Ruipérez-Valiente, J.A., Muñoz-Merino, P.J., Kloos, C.D.: Towards the development of a learning analytics extension in open EdX. In: Proceedings of 2nd International Conference on Technological Ecosystems for Enhancing Multiculturality, pp. 299–306 (2014)
13. Mitrovic, A., Dimitrova, V., Weerasinghe, A., Lau, L.: Reflective experiential learning: using active video watching for soft skills training. In: International Conference on Computers in Education, pp. 192–201 (2016)
14. Mitrovic, A., Dimitrova, V., Lau, L., Weerasinghe, A., Mathews, M.: Supporting constructive video-based learning: requirements elicitation from exploratory studies. In: André, E., Baker, R., Hu, X., Rodrigo, M.M.T., du Boulay, B. (eds.) AIED 2017. LNCS (LNAI), vol. 10331, pp. 224–237. Springer, Cham (2017). https://doi.org/10.1007/978-3-319-61425-0_19
15. Mohammadhassan, N., Mitrovic, A., Neshatian, K.: Investigating the effect of nudges for improving comment quality in active video watching. Comput. Educ. **176**, 104340 (2022)

16. Mitrovic, A., Gordon, M., Piotrkowicz, A., Dimitrova, V.: Investigating the effect of adding nudges to increase engagement in active video watching. In: Isotani, S., Millán, E., Ogan, A., Hastings, P., McLaren, B., Luckin, R. (eds.) AIED 2019. LNCS (LNAI), vol. 11625, pp. 320–332. Springer, Cham (2019). https://doi.org/10.1007/978-3-030-23204-7_27
17. Matcha, W., Uzir, N.A., Gašević, D., Pardo, A.: A systematic review of empirical studies on learning analytics dashboards: a self-regulated learning perspective. IEEE Trans. Learn. Technol. **13**, 226–245 (2020)
18. Sedrakyan, G., Malmberg, J., Verbert, K., Järvelä, S., Kirschner, P.A.: Linking learning behavior analytics and learning science concepts: designing a learning analytics dashboard for feedback to support learning regulation. Comput. Hum. Behav. **107**, 105512 (2020)
19. Chou, C.-Y., et al.: Open student models of core competencies at the curriculum level: using learning analytics for student reflection. IEEE Trans. Emerg. Top. Comput. **5**, 32–44 (2017)
20. Majumdar, R., Akçapınar, A., Akçapınar, G., Flanagan, B., Ogata, H.: LAView: learning analytics dashboard towards evidence-based education. In: Companion Proceedings of 9th International Conference on Learning Analytics & Knowledge, pp. 68–73 (2019)
21. Broos, T., Peeters, L., Verbert, K., Van Soom, C., Langie, G., De Laet, T.: Dashboard for actionable feedback on learning skills: scalability and usefulness. In: Zaphiris, P., Ioannou, A. (eds.) LCT 2017. LNCS, vol. 10296, pp. 229–241. Springer, Cham (2017). https://doi.org/10.1007/978-3-319-58515-4_18
22. Ez-zaouia, M., Tabard, A., Lavoué, E.: EMODASH: a dashboard supporting retrospective awareness of emotions in online learning. Human-Comput. Stud. **139**, 102411 (2020)
23. Ruiz, S., Charleer, S., Urretavizcaya, M., Klerkx, J., Fernández-Castro, I., Duval, E.: Supporting learning by considering emotions: tracking and visualization a case study. In: International Conference on Learning Analytics & Knowledge, pp. 254–263 (2016)
24. Guerra, J., Hosseini, R., Somyurek, S., Brusilovsky, P.: An intelligent interface for learning content: combining an open learner model and social comparison to support self-regulated learning and engagement. In: Intelligent User Interfaces, pp. 152–163 (2016)
25. Corrin, L., de Barba, P.: Exploring students' interpretation of feedback delivered through learning analytics dashboards. In: Rhetoric and Reality: Critical Perspectives on Educational Technology. Proceedings ASCILITE, pp. 629–633 (2014)
26. Lim, L., Dawson, S., Joksimovic, S., Gašević, D.: Exploring students' sensemaking of learning analytics dashboards: does frame of reference make a difference? In: International Conference on Learning Analytics & Knowledge, pp. 250–259 (2019)
27. Lonn, S., Aguilar, S., Teasley, S.D.: Investigating student motivation in the context of a learning analytics intervention during a summer bridge program. Comput. Hum. Behav. **47**, 90–97 (2015)
28. Srivastava, N., Velloso, E., Lodge, J.M., Erfani, S., Bailey, J.: Continuous evaluation of video lectures from real-time difficulty self-report. In: Proceedings of Human Factors in Computing Systems, pp. 1–12 (2019)
29. Yoon, M., Hill, J., Kim, D.: Designing supports for promoting self-regulated learning in the flipped classroom. J. Comput. High. Educ. **33**(2), 398–418 (2021). https://doi.org/10.1007/s12528-021-09269-z
30. Mohammadhassan, N., Mitrovic, A., Neshatian, K., Dunn, J.: Automatic assessment of comment quality in active video watching. In: International Conference on Computers in Education, pp. 1–10 (2020)
31. Taskin, Y., Hecking, T., Hoppe, H.U., Dimitrova, V., Mitrovic, A.: Characterizing comment types and levels of engagement in video-based learning as a basis for adaptive nudging. In: Scheffel, M., Broisin, J., Pammer-Schindler, V., Ioannou, A., Schneider, J. (eds.) EC-TEL 2019. LNCS, vol. 11722, pp. 362–376. Springer, Cham (2019). https://doi.org/10.1007/978-3-030-29736-7_27

32. Chi, M.T.H., Wylie, R.: The ICAP framework: linking cognitive engagement to active learning outcomes. Educ. Psychol. **49**, 219–243 (2014)
33. Mohammadhassan, N., Mitrovic, A.: Discovering differences in learning behaviours during active video watching using epistemic network analysis. In: Wasson, B., Zörgő, S. (eds.) Advances in Quantitative Ethnography. ICQE 2021. Communications in Computer and Information Science, vol. 1522. Springer, Cham (2022). https://doi.org/10.1007/978-3-030-93859-8_24
34. Dimitrova, V., Mitrovic, A., Piotrkowicz, A., Lau, L., Weerasinghe, A.: Using learning analytics to devise interactive personalised nudges for active video watching. In: User Modeling, Adaptation and Personalization, pp. 22–31 (2019)
35. Hu, L., Bentler, P.M.: Cutoff criteria for fit indexes in covariance structure analysis: conventional criteria versus new alternatives. Struct. Equ. Model. **6**, 1–55 (1999)

Debiasing Politically Motivated Reasoning with Value-Adaptive Instruction

Nicholas Diana[1(✉)], John Stamper[2], Ken Koedinger[2], and Jessica Hammer[2]

[1] Colgate University, Hamilton, USA
ndiana@colgate.edu
[2] Carnegie Mellon University, Pittsburgh, USA
http://nickdiana.com/

Abstract. While there is a substantial appetite in the United States for improving media consumption skills, little work has focused on the biases that can make inaccurate or misleading claims feel true. This skill is particularly difficult to teach, as effective instruction requires the instructor to adapt course content to the specific beliefs of individual students, a process that is unscalable in most classrooms. Here we examine the impact of a novel method of user-centered personalized instruction that uses value-adaptivity to highlight and address user bias in the context of a civics education game. This intervention uses estimates of player and content values to predict when players may be most susceptible to biased reasoning and then intervene in those instances. We found that the intervention successfully reduced bias among high bias-regulators with practice. These results suggest that value-adaptive systems may be able to support debiasing instruction in an effective, scalable way.

Keywords: Myside bias · Confirmation bias · Personalization · Educational games · Civic technology · Civics education

1 Introduction

The co-opting of social media platforms in large-scale disinformation campaigns has spurred the development of novel tools and methods for more responsible media consumption. Much of this work focuses on the media content itself, with researchers developing sophisticated machine learning models for classifying patently false information [15,20]. Other work focuses on the opposing side of the media equation: the user. This work examines methods for improving media literacy (i.e., their ability to evaluate the credibility of the information they are consuming) [10].

Less work, however, has focused on the dynamic relationship between the media consumer and the content they are consuming, and in particular: how that relationship can be a powerful source of bias and how best to mitigate

M. M. Rodrigo et al. (Eds.): AIED 2022, LNCS 13355, pp. 140–152, 2022.
https://doi.org/10.1007/978-3-031-11644-5_12

those biases. Recognizing and reducing bias (sometimes called "debiasing") [14] is a critical component of any comprehensive 21st century civics education. Civics teachers (along with those in the English Department) are often tasked with equipping students with the media literacy skills they need to navigate an increasingly fraught media landscape. One common approach to debiasing in civics class is to ask a student to defend a political position that they themselves do not hold (or actively oppose). This act of *perspective taking* can be powerful [11], but efficient perspective taking requires that the teacher: 1) knows the positions of each student with respect to each topic, and 2) takes the time to match students to positions individually. In a class of 30 students that might discuss a topic a week, a systematic adherence to this approach is likely unscalable.

In the current experiment, we used Moral Foundations Theory [9] in conjunction with natural language processing methods to model student and content values. We used the relationship between those two sets of values to power a value-adaptive debiasing intervention. This debiasing intervention was integrated into an educational game designed to help students practice engaging in productive civil discourse. We tested the efficacy of this novel approach to debiasing by examining students' bias regulation, or their ability to ignore an intuitively correct option (biased response) in favor of the actual correct option. Specifically, we hypothesize that the bias regulation of students who saw the debiasing intervention will improve over time relative to their peers who did not see the intervention.

The primary contribution of this work is the demonstration of a scalable approach to debiasing instruction in civics education that is powered by a novel method of personalized instruction: value-adaptive instruction.

2 Related Work

The fallibility of human rationality has long been established as an important and consequential area of study [1,4]. In many cases, human cognition fails in regular and predictable ways. As such, it is not unreasonable to attempt to identify the circumstances under which we may be most susceptible to these cognitive biases and to develop training programs designed to mitigate the impact of the most common or most critical biases. Despite the large body of work pertaining to the identification and measurement of cognitive biases, the body of work pertaining to the development and testing of so-called *debiasing* training programs is relatively small [14]. This may be due to the fact that many cognitive biases are quite robust, persisting even in the face of explicit debiasing instruction [5].

With respect to debiasing instruction, tasks that require participants to consider the opposing viewpoint may mitigate the impact of biased reasoning. This strategy shares many features with a skill called *perspective taking*, a common instructional goal in civics curricula that can be complicated by the personal nature of political beliefs. For example, a civics instructor might ask a student who is anti-immigration to defend a pro-immigration stance. The serious consideration of opposing perspectives may reduce the impact of bias. This kind of individualized debiasing instruction is an example of what we call *value-adaptive instruction* (i.e., instruction that is adapted to the specific values of the learner).

Unfortunately, the traditional approach to value-adaptive instruction described above is simply unscalable in classrooms of 20–30 students and in courses that might cover 40 topics over the span of a school year. One potential solution is to use technology to support the estimation of student and content values. These estimations will never be as accurate as those from an expert human instructor, but reasonably accurate estimations may allow us to provide numerous individualized practice opportunities to students in a scalable way. Moreover, it would allow us to use educational technologies to:

1. Estimate the impact of bias on informal reasoning tasks,
2. Predict when students may be most susceptible to biased reasoning, and
3. Provide targeted debiasing interventions precisely in those moments of vulnerability.

Myside Bias in Civics Education. In this study, we explore a particular type of bias, termed *Myside Bias*, that is often found in civil discourse. In brief, *Myside Bias* refers to one's tendency to evaluate claims or evidence more favorably if the claim or evidence supports one's own beliefs or worldview [19]. Myside bias has been characterized as both a more accurate term for *Confirmation Bias* [16] and a subclass of *Confirmation Bias* [19] in various works. In the context of civil discourse, myside bias can manifest as one's inability to speak across ideological lines to the values that motivate the beliefs of those they disagree with. It is our tendency to reach for the argument that seems strongest to us, rather than the argument that would appeal most strongly to whomever we are trying to persuade.

Effectively choosing arguments that will be most persuasive to those with a differing ideology requires two skills:

1. The ability to identify the values that underpin the beliefs of your interlocutor
2. The ability to choose an argument that best aligns with those values

Inherent in this second skill is the challenge of overcoming myside bias (in this context, our tendency to choose an argument that aligns with our own values instead of more persuasive options). In the current study, we examine a value-adaptive intervention designed to mitigate the impact of myside bias when choosing effective arguments in civil discourse. This intervention was integrated into an educational game designed to give students opportunities to practice these key discourse skills (Fig. 1).

3 Methods

A total of 87 students from high schools located in the Northeastern Region of the United States participated in the study. Note that all demographics questions were optional, and a small number of students chose not to answer some questions. The students where evenly split with respect to sex (41 females, 43

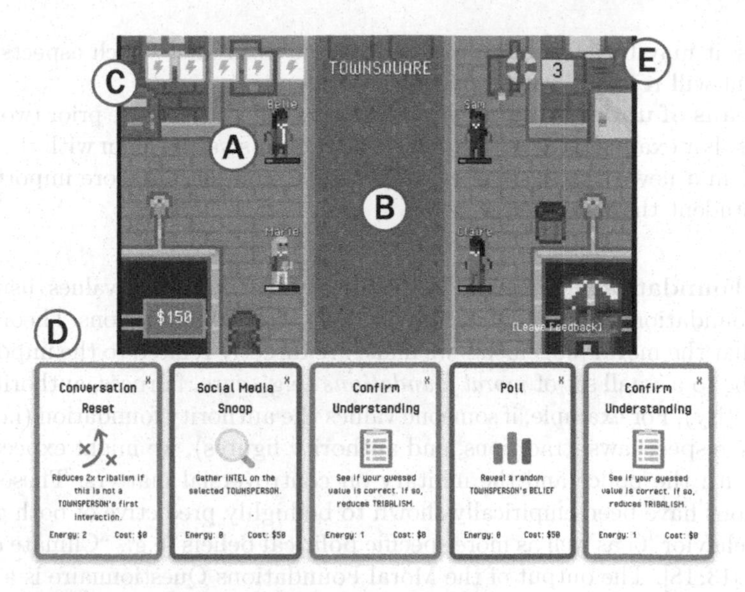

Fig. 1. An annotated screenshot of a scenario. In each scenario, players must persuade NPCs like Belle (A) to move into the TOWNSQUARE (B). To do this players must identify which argument from the opposing side appeals to what Belle values. Persuading NPCs costs Energy (C). The bar positioned below each NPC represents their political tribalism. Players must reduce an NPC's tribalism before attempting to persuade them. They can do this by playing Discourse Cards (D) like *Conversation Reset*. Finally, the action menu (E) allows the player to request a hint, end the day/turn, or reference information in their notebook.

males, 1 other), and reflective of the racial demographics of the area (9 black or African American students, 72 white or Caucasian students, and 4 students identifying as other/more than one race). Students from six classes (three English classes and two Social Studies classes) participated in the study.

3.1 Leveraging NLP Alongside Theories of Moral Judgments

Any given student will, by definition, only exhibit myside bias when presented with information that aligns with their beliefs. As such, crafting instructional events that give students the opportunity to wrestle with their biases requires three critical pieces of information:

1. **An estimate of a particular student's values**. That is, if we were to cover a new topic in class, can we be relatively confident that we could use our understanding of their values to predict the kind of belief this student will espouse?
2. **An estimate of the political values latent in the content** we are presenting to this student. In the case of commonly debated topics, these latent political values may be obvious. However, in the case of uncommon or novel

topics, it may be substantially more difficult to predict which aspects of the content will resonate with a particular student.
3. **A means of understanding the *relationship*** between the prior two sets of values. For example, to what extent do a student's values align with the values latent in a news article? Will one aspect of a problem be more important to this student than another aspect?

Moral Foundations Theory. We estimate the student's values using the Moral Foundations Theory Questionnaire [9]. Moral Foundations Theory [8,9] argues that the moral judgements we make are directly related to the importance we ascribe to a small set of *moral foundations* (e.g., care, fairness, authority, loyalty, sanctity). For example, if someone values the authority foundation (i.e., they generally respect laws, traditions, and authority figures), we might expect them to side with the police and the military on controversial matters. These moral foundations have been empirically shown to be highly predictive of both general voting behavior [6] as well as more specific political beliefs (e.g., "Climate change is real") [13,18]. The output of the Moral Foundations Questionnaire is a vector of five scores, representing the degree to which the student values each of the five foundations when making moral judgments.

It is worth reiterating that these are *estimates* of user values. The reasons humans hold beliefs are numerous and personal. As such, we can say with relative certainty that the model of human beliefs (based on values) employed in this study is incomplete and flawed. What remains to be seen is whether the model provides a good enough estimate of user beliefs to be useful in the context of debiasing instruction.

Distributed Dictionary Analysis. We estimate the values latent in text content using natural language processing, specifically distributed dictionary representations (DDR) [7]. DDR builds off of Word2Vec [17], which involves modeling a large corpus of text data in a low-dimensional space, where each word can be represented as a point in that semantic space. DDR was created to model psychological constructs (such as the foundations in Moral Foundations Theory) using this semantic space. That is, the foundation referred to as *Fairness* actually encompasses more than just the concept of fairness; it includes equality, injustice, rights, and fraud. To find the point in the semantic space that matches this more nuanced concept that we label *Fairness*, we first generate a concept dictionary (i.e., a list of terms that approximate the meaning of the concept). Because each word in the concept dictionary can be represented as a vector, we can simply average across all word vectors in the dictionary to find the vector that corresponds to our operational definition of the concept *Fairness*.

In this work, we follow the original procedure outlined in [7] to generate a vector for each of the five moral foundations. Next, to estimate the values latent in a piece of text, we compute the cosine distance between the representations of each of the five moral foundations and the average representation of all words in the text. Like the Moral Foundations Questionnaire, the output of this process is a vector of five scores, representing the degree to which the text was semantically similar to each of the five foundations (see [2] for a more complete discussion of this process). In our analysis, we used the pre-trained Google News corpus (approximately 100 billion words) Word2Vec model[1], and a Python implementation of Word2Vec [17] called *gensim*.

Computing Alignment. Because the outputs of the Moral Foundations Questionnaire and the DDR analysis are two vectors of equal length, measuring the relationship between the two vectors can be done simply by computing the cosine similarity between them. The extent to which the student's values are similar to the values latent in the content is termed *Alignment*. Previous work has shown that *Alignment* is predictive of bias in argument evaluation tasks [3]. In the current study, we use *Alignment* to predict where students might be most susceptible to biased reasoning during gameplay and to adapt the debiasing intervention accordingly. It is worth clarifying that *Alignment* is essentially measuring the presence of foundational concepts and their relationship to the user's estimated values. As such, we expect it to fail in the face of sufficiently nuanced language. What the current experiment aims to test is whether or not the resulting model of bias is good enough to be useful.

3.2 A Value-Adaptive Debiasing Intervention

The debiasing intervention was incorporated into *Persuasion Invasion*, an educational game called designed to help students practice productive civil discourse skills. The goal of each level in *Persuasion Invasion* is to persuade ideologically entrenched townspeople to engage with those they disagree with. Successfully persuading a townsperson requires that the student 1) identify which of the five moral foundations the townsperson values most, and 2) identify which of three arguments from the opposing side appeals most to someone who values that foundation. We expect that players may be biased to choose the argument that aligns most to their own values rather than the argument that aligns with the values of the townsperson they are attempting to persuade.

To mitigate the impact of bias, we integrated a value-adaptive debiasing intervention into this *Persuasion* interaction. All students were randomly assigned to one of two conditions: an adaptive condition or a control condition. The two conditions were identical in every respect with one exception: When players in the adaptive condition were asked to choose which of the three listed arguments would be most persuasive to a townsperson, they saw one of the options presented in orange-colored text with an additional piece of instruction that read:

> **Caution:** Orange-colored **options might seem more persuasive to you (based on your values). Remember to choose the best response for [NPC NAME].**

[1] The pre-trained Google News model can be found here: https://code.google.com/p/word2vec/.

The orange-colored option corresponds to the option with the highest *Alignment* score (i.e., the option that, based on this methodology, aligns most closely to this specific player's values). The color orange was chosen because it is attention-grabbing, but isn't traditionally associated with correctness (as colors like red and green are in the United States). Importantly, the orange-colored option was not any more or less likely to be the correct answer; this intervention is simply designed to elicit a more critical analysis of the options.

3.3 A Composite Measure of Bias Regulation

In previous work, *Alignment*-based estimates of potential bias have represented the extent to which the user's values align with the values of the correct response. This made a direct comparison between *Alignment* and other baseline measures possible. However, this *Alignment*-based estimate of bias is limited in that it fails to account for the alignment between the user and the other potential options. Imagine, for example, a case in which the correct option happens to also be the option with the highest alignment. We would expect that, in this case, the choice is easy, as there is no conflict between the intuitive choice and the correct choice. This case also tells us nothing about the user's ability to regulate their own bias. Contrast this with a scenario in which the correct option happens to be the option with the lowest alignment (i.e., least congruent with the user's values). Choosing the correct option in this case, may require the user to overcome their own bias.

We used the alignment scores of all options presented to the user to generate a more nuanced estimate of the amount of potential bias a student may be overcoming at each opportunity. This novel composite metric, which we call the *Bias Regulation Index* (BRI), is computed as follows:

$$BRI = (A_{highest} - A_{chosen}) + (A_{correct} - A_{chosen}) \tag{1}$$

Here $A_{highest}$ represents the alignment score of the option with the highest alignment score (i.e., the option we would expect a completely biased player to pick). Similarly, A_{chosen} represents the alignment score of the option the player chose, and $A_{correct}$ represents the alignment score of the correct option. The first set of parentheses in this equation essentially gives the player credit for choosing an option that isn't the option with the highest alignment, and gives them more credit the farther away their choice's score is from that highest score. This first set of parentheses cannot penalize players, as they cannot chose an option with a score higher than the highest score.

The second set of parentheses penalizes the player if they chose an option with a higher alignment score than the correct option. If $A_{correct} < A_{chosen}$, then the result of this second set of parenthesis is set equal to 0 to keep the metric from crediting the players for choosing an incorrect option with lower alignment than the correct option. Importantly, players are neither penalized nor credited in this metric for choosing the correct option. The resulting sum of these two sets of parentheses represents a student's ability to overcome bias to

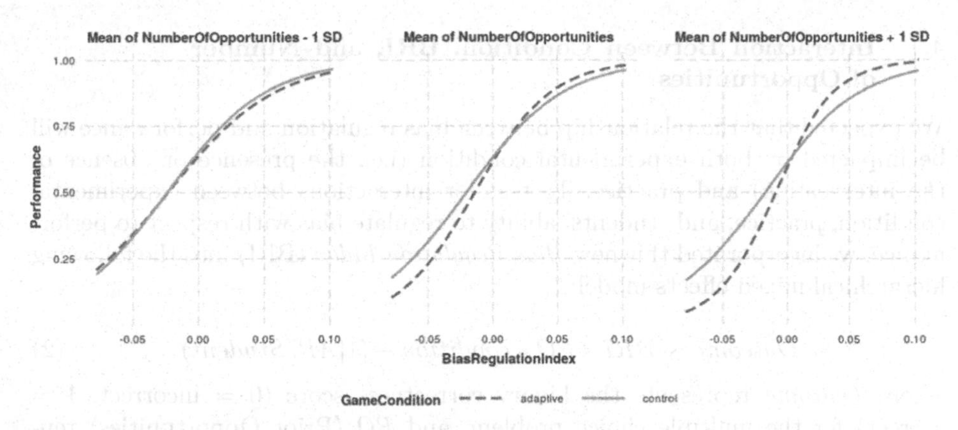

Fig. 2. The three-way interaction between condition, opportunities and *Bias Regulation Index* (BRI). We see that, while the relationship between BRI and performance remains relatively constant with additional practice in the control condition, the relationship seems to change in the adaptive condition. As the number of practice opportunities increase, students in the adaptive condition with low bias-regulation appear to do worse than their peers in the control condition, whereas students in the adaptive condition with high bias-regulation appear to benefit from the intervention compared to their peers in the control condition. (Color figure online)

choose the correct answer. Positive scores on this metric capture those instances in which a student chooses the low-aligned correct score over the high-aligned incorrect one. Negative scores capture those instances in which the player chooses a high-aligned incorrect option over a lower-aligned correct one.

This metric is more nuanced than simply including the *Alignment* score of the correct option, as it mitigates the impact of the option's correctness on choice. That is, did the player choose this because it is the correct option, or because it aligned with their values. When the correct option is also the option with the highest alignment score, the choice is easy and uninteresting. In contrast, this metric focuses on instances in which the choice is difficult. We expect that bias regulation will improve over time for students in the adaptive intervention condition.

4 Results

We examined the impact of an intervention (designed to reduce bias) on in-game performance. Recall that students in the adaptive condition had an in-game experience identical to those in the control condition with one exception: during *Persuade* actions, students in the adaptive condition saw an additional piece of instruction that highlighted the option that most aligned with their values (i.e., had the highest computed *Alignment*) alongside a message warning the player that they may be biased to select the highlighted option.

4.1 Interaction Between Condition, BRI, and Number of Opportunities

We expected that the relationship between bias regulation and performance will be impacted by both experimental condition (i.e., the presence or absence of the intervention) and practice. To test for interactions between experimental condition, practice, and students' ability to regulate bias with respect to performance, we incorporated this new *Bias Regulation Index* (BRI) into the following hierarchical mixed effects model:

$$Outcome \sim BRI * PO * condition + (1|AP/Student) \quad (2)$$

where *Outcome* represents the binary correctness score (0 = incorrect, 1 = correct) for the multiple choice problem, and *PO* (Prior Opportunities) represents the number of times, prior to the current opportunity, that the player has attempted a *Persuade* action. The model also includes the nested random effects of *AP* status[2] and the *Student* identifier. Table 1 shows the model results.

Table 1. Results from the hierarchical mixed effects model. There was a significant three way interaction between Prior Opportunities, Bias Regulation and Condition.

	Estimate	SE	P-val	Sig
PO	−0.015	0.009	0.070	.
BRI	23.339	4.954	0.000	***
Condition (control)	0.067	0.145	0.646	
PO:BRI	1.558	0.611	0.011	*
PO:Conditioncontrol	0.000	0.013	0.973	
BRI:Condition (control)	3.163	6.603	0.632	
PO:BRI:Condition (control)	−1.583	0.806	0.049	*

As expected, we found a significant three-way interaction between *Bias Regulation Index*, the number of prior practice opportunities, and experimental condition ($\beta = -1.583, p < .05$). We used the R library *interactions* to explore and visualize this interaction. Figure 2 shows the relationship between BRI and Performance at three different opportunity counts. We see that, while the relationship between performance and BRI remains relatively stable across practice opportunities in the control condition, the relationship between these variables changes with practice in the adaptive condition. Recall that BRI scores below zero indicate opportunities in which the student chose a high-aligned incorrect option over a low-aligned correct one, and positive scores indicate opportunities in which the student chose a low-aligned correct option over a high-aligned incorrect one. This graph suggests that the intervention may have caused students with low bias-regulation to perform worse (potentially choosing the visually

[2] AP Status was shown to be predictive of performance in previous work.

salient orange-colored option more). However, students with high bias-regulation seemed to benefit from seeing the intervention, outperforming their peers in the control condition.

5 Discussion

We found that, by comparing estimates of student values to estimates of the values latent in text content, we could provide an adaptive intervention that appears to have reduced the impact of bias on task performance (for high bias-regulators). In the context the educational game, this effect appears to be gradual, increasing with additional practice opportunities. This suggests that regulating bias in this context is likely a skill that can be learned, but that it may also require many practice opportunities to hone.

To measure the intervention's impact, we developed a novel metric, the *Bias Regulation Index* (BRI). BRI more accurately captures the difficulty of bias-prone tasks, allowing us to measure a student's capacity to overcome (or regulate) their own biases. In the future, value-adaptive systems may use BRI to provide additional practice opportunities or individualized feedback to students exhibiting low bias-regulation.

Why the adaptive debiasing intervention had a differential impact on low and high bias-regulators remains an open question. While it may be tempting write this off as another example of the "rich get richer" effect that can occur in educational technology work, this would not explain the discrepancy in the performance of low bias-regulators across conditions. That is, the intervention seems to not only have made the rich richer, but the poor poorer as well. One potential explanation for this effect is a simple misunderstanding about the nature of the intervention. Low bias-regulators may have incorrectly interpreted the orange color as an indicator of an option's correctness (e.g., assumed it was a hint), when in fact, there was no such relationship between correctness and color. This may explain why low bias-regulators in the adaptive condition displayed worse performance than their peers in the control condition.

Such confusion may have been avoidable with additional instruction about the nature of the intervention. However, because students in the same class were randomly assigned to either the control or adaptive condition, drawing attention to the debiasing intervention (seen by those in the adaptive condition) may have tainted the independence of the control condition.

5.1 Limitations

Perhaps the largest limitation of this debiasing intervention is the absence of powerful social influences. It was unfortunately necessary to test the intervention at the individual level, separate from peer-influence simply because half the students within a classroom were assigned to the control group (no intervention). Thus, the instruction pertaining to bias was given to each student in the adaptive condition individually (via the interface). Future experiments might instead provide the bias instruction to the class as a whole, which might add social pressure to make unbiased decisions.

Other limitations pertain to our participant population and experiment structure. While we believe our sample was representative of late high school-aged students, there are known interactions between bias and age [12] that leave us unable to confidently generalize these results to a population that includes older players. Similarly, an important part of debiasing research is the longevity of the effects [14]. As part of this work, we had originally planned to return to our participants' classrooms both one week and three weeks later to examine potential effects of the game on real-world classroom discussions. However, the onset of the COVID-19 pandemic cut our original data collection plan short. Both of these limitations are important areas of future work.

5.2 Potential Applications

This work has several potential applications. First and foremost, we believe that educational technologies that implement value-adaptive debiasing interventions allow instructors to provide students with opportunities to recognize and overcome their biases. These technology-based interventions will never be as nuanced as an intervention from an expert human instructor, but unlike traditional instruction, technology-based interventions like the one described in the current study are scalable to any number of students. Because of these tradeoffs, we see value-adaptive debiasing systems ultimately as a tool for supporting the critical, real-world classroom discussions.

Beyond the classroom, value-adaptive debiasing systems might be embedded into our interactions with media content. Here, such systems could make the content consumer aware of the degree to which the content they are consuming aligns with their own values. Alternatively, the system could alert the user to engage their critical thinking faculties when the alignment between the content's values and their own is above a certain threshold. What remains to be seen is how users will react to these kinds of interventions absent the affordances of game environments.

6 Conclusion

In this study, we examined the impact of a value-adaptive debiasing intervention on myside bias in the context of an educational game designed to teach productive civil discourse skills. We found a significant three-way interaction between the number of prior practice opportunities, our measure of bias (BRI), and condition (adaptive vs. control). Further investigation revealed that students in the adaptive condition (i.e., who saw the adaptive intervention) got better at mitigating the impact of bias with practice relative to their peers in the control condition. However, this was only true for high bias-regulators. While further improvements are necessary to ensure that the impact of debiasing interventions is equitable, this encouraging result demonstrates that value-adaptivity, this novel method of personalized learning, may be a useful tool for *scalable* debiasing instruction. Value-adaptivity allows us to craft instruction that recognizes and reacts to the dynamic relationship between the media content and the media consumer. With it, we can provide the rich, user-centered practice necessary for any comprehensive media literacy education.

References

1. Baron, J.: Thinking and Deciding. Cambridge University Press (2000)
2. Diana, N., Stamper, J., Koedinger, K.: Towards value-adaptive instruction: a data-driven method for addressing bias in argument evaluation tasks. In: Proceedings of the 2020 CHI Conference on Human Factors in Computing Systems, pp. 1–11 (2020)
3. Diana, N., Stamper, J.C., Koedinger, K.: Predicting bias in the evaluation of unlabeled political arguments. In: CogSci, pp. 1640–1646 (2019)
4. Evans, J.S.B.T., Barston, J.L., Pollard, P.: On the conflict between logic and belief in syllogistic reasoning. Mem. Cogn. **11**(3), 295–306 (1983). https://doi.org/10.3758/BF03196976
5. Evans, J.S.B., Newstead, S., Allen, J., Pollard, P.: Debiasing by instruction: the case of belief bias. Eur. J. Cogn. Psychol. **6**(3), 263–285 (1994)
6. Franks, A.S., Scherr, K.C.: Using moral foundations to predict voting behavior: regression models from the 2012 US presidential election. Anal. Soc. Issues Public Policy **15**(1), 213–232 (2015)
7. Garten, J., Hoover, J., Johnson, K.M., Boghrati, R., Iskiwitch, C., Dehghani, M.: Dictionaries and distributions: combining expert knowledge and large scale textual data content analysis. Behav. Res. Meth. **50**(1), 344–361 (2017). https://doi.org/10.3758/s13428-017-0875-9
8. Graham, J., et al.: Moral foundations theory: the pragmatic validity of moral pluralism. In: Advances in Experimental Social Psychology, vol. 47, pp. 55–130. Elsevier (2013)
9. Haidt, J.: The emotional dog and its rational tail: a social intuitionist approach to moral judgment. Psychol. Rev. **108**(4), 814 (2001)
10. Hone, B., Rice, J., Brown, C., Farley, M.: Factitious (2018). factitious.augamestudio.com
11. Johnson, D.W., Johnson, R.T., Tjosvold, D.: Constructive controversy: the value of intellectual opposition (2000)
12. Klaczynski, P.A., Robinson, B.: Personal theories, intellectual ability, and epistemological beliefs: adult age differences in everyday reasoning biases. Psychol. Aging **15**(3), 400 (2000)
13. Koleva, S.P., Graham, J., Iyer, R., Ditto, P.H., Haidt, J.: Tracing the threads: how five moral concerns (especially purity) help explain culture war attitudes. J. Res. Pers. **46**(2), 184–194 (2012)
14. Lilienfeld, S.O., Ammirati, R., Landfield, K.: Giving debiasing away: can psychological research on correcting cognitive errors promote human welfare? Perspect. Psychol. Sci. **4**(4), 390–398 (2009)
15. McGrew, S., Ortega, T., Breakstone, J., Wineburg, S.: The challenge that's bigger than fake news: civic reasoning in a social media environment. Am. Educ. **41**(3), 4 (2017)
16. Mercier, H.: Confirmation bias-myside bias (2017)

17. Mikolov, T., Chen, K., Corrado, G., Dean, J.: Efficient estimation of word representations in vector space. arXiv preprint arXiv:1301.3781 (2013)
18. Rottman, J., Kelemen, D., Young, L.: Tainting the soul: purity concerns predict moral judgments of suicide. Cognition **130**(2), 217–226 (2014)
19. Stanovich, K.E., West, R.F., Toplak, M.E.: Myside bias, rational thinking, and intelligence. Curr. Dir. Psychol. Sci. **22**(4), 259–264 (2013)
20. Wang, Y., et al.: EANN: event adversarial neural networks for multi-modal fake news detection. In: Proceedings of the 24th ACM SIGKDD International Conference on Knowledge Discovery & Data Mining, pp. 849–857. ACM (2018)

Towards Human-Like Educational Question Generation with Large Language Models

Zichao Wang[1(✉)], Jakob Valdez[1], Debshila Basu Mallick[2], and Richard G. Baraniuk[1,2]

[1] Rice University, Houston, USA
{jzwang,jpv3,richb}@rice.edu
[2] OpenStax, Houston, USA
debshila@rice.edu

Abstract. We investigate the utility of large pretrained language models (PLMs) for automatic educational assessment question generation. While PLMs have shown increasing promise in a wide range of natural language applications, including question generation, they can generate unreliable and undesirable content. For high-stakes applications such as educational assessments, it is not only critical to ensure that the generated content is of high quality but also relates to the specific content being assessed. In this paper, we investigate the impact of various PLM prompting strategies on the quality of generated questions. We design a series of generation scenarios to evaluate various generation strategies and evaluate generated questions via automatic metrics and manual examination. With empirical evaluation, we identify the prompting strategy that is most likely to lead to high-quality generated questions. Finally, we demonstrate the promising educational utility of generated questions using our concluded best generation strategy by presenting generated questions together with human-authored questions to a subject matter expert, who despite their expertise, could not effectively distinguish between generated and human-authored questions.

1 Introduction

Practice questions and quizzes have been vital instruments for the assessment of learning [1,20,27]. Engaging in retrieval practice by answering expert-designed questions has shown to be more effective at improving learning outcomes [9,10], by providing opportunities for recall of knowledge, applying knowledge to novel scenarios, and critical thinking and writing skills. The learning benefits are greater than other means of pedagogy such as passively re-reading course materials or studying notes [4,8–10,12,13] or watching instructional videos [21]. However, these questions are also known to be challenging to create: they usually take subject matter experts (SMEs) a significant amount of time, which is both costly and

Z. Wang and J. Valdez—Contributed equally.

M. M. Rodrigo et al. (Eds.): AIED 2022, LNCS 13355, pp. 153–166, 2022.
https://doi.org/10.1007/978-3-031-11644-5_13

labor-intensive [20]. Therefore, this question generation process does not easily generalize and scale to the continually expanding repositories of educational content that need large banks of assessments to be effective sources of instruction.

To create a scalable question generation process, several recent works leveraged artificial intelligence (AI) methods for *automatically* generating questions. For example, some prior works [5,25,26] focused on generating factual questions using recurrent neural network (RNN) architectures. [28] designed a method to select highly interesting phrases which a generated question is supposed to ask about. The implications of these works are far-reaching. In addition to reducing the labor and cost for producing assessment questions, automatic question generation methods have the potential to create a more engaging learning experience by generating (i) personalized questions that adapt to each student's learning trajectory [7] and (ii) real-time pop-up quizzes while the student is reading a textbook or watching instructional videos. Once trained, these methods have been shown to perform well on question generation tasks. However, they require custom model design and (sometimes significant) computational resources for training, making them a less appealing option for practitioners who desire a "plug-and-play" AI-assisted question generation process that allows them to easily interact with an AI system without the need for model training.

Recently, a new paradigm in text generation using large pretrained language models (PLMs), such as GPT-3 [3], is now making such "plug-and-play" question generation a possibility. These PLMs have been pretrained on web-scale data which equip the model with abundant knowledge of the language compared to their earlier counterparts. Furthermore, they can be easily and effectively adapted to various generation tasks via the "prompting" technique, where the user simply specifies the generation task that they would like to perform as a prompt. A *prompt* usually contains, in addition to a "query" from which the PLM will generate the outcome, a series of examples in an input-output structure that "teach" the model how to generate the output given the input specific to a particular task. Figure 1 gives an example of using prompting to adapt a PLM for machine translation and arithmetic question answering. Prompting provides an easy interface and high controllability for users to interact with PLMs and customize it for different generation tasks. Because of its simplicity and practicality, prompting techniques to adapt PLMs for downstream generation tasks have attracted increasing attention in the past few years [11,16,18,19]. Figure 1 shows an example of prompting for machine translation, question answering.

Unfortunately, using prompts to adapt PLMs for question generation is challenging due to the open-ended nature of the process, i.e., it does not have a clearly defined input-output structure. This poses certain challenges such as, what content should the questions be generated from, how should we deal with the fact that multiple different questions can be asked about the same concept, etc. This open-ended nature makes question generation unique in contrast with other generation tasks commonly studied in existing literature (e.g., in machine translation, input and output are simply texts in the source and target languages, respectively). As a result, unlike other generation tasks where adapting PLMs via prompting is straightforward (e.g., see Fig. 1 for an illustration), it is

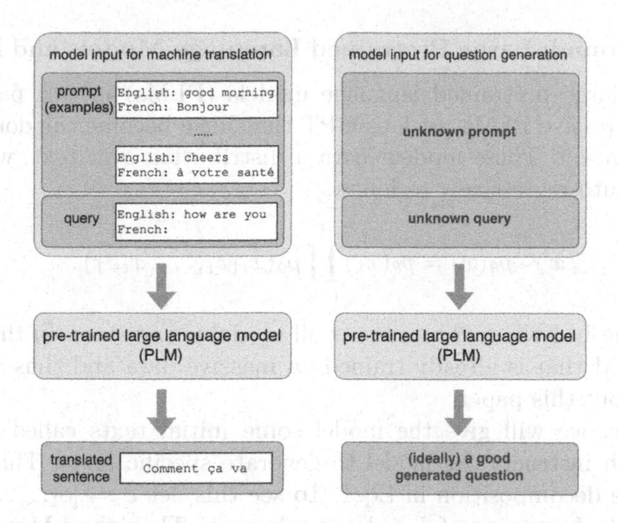

Fig. 1. Illustrations of adapting PLMs for machine translation and the challenges in designing prompts to adapt PLMs for educational question generation.

unclear how to design effective prompts for PLMs in order for question generation. To the best of our knowledge, to date no existing literature has investigated the modification of prompting strategy for question generation. To harness the power of AI for educational question generation, prompt design for question generation by PLMs is an exciting open problem.

1.1 Contributions

In this paper, we investigate the problem of effectively prompting a PLM to generate desirable, high-quality, educational practice questions. An effective prompt strategy will enable us to leverage the power of PLMs with minimal effort and without having to conduct model training with large volumes of domain-focused content. We start with the core question: how do we design prompts such that a PLM can generate the most desirable and effective practice questions? We answer this question by proposing 5 different generation settings with a specific prompting strategy for each. We conduct a series of manual examinations of the generated questions as well as automatic evaluations, which lead to the empirical conclusion of the best combinations of our prompting strategy. This strategy serves as an empirical guideline for practitioners to set up PLMs to generate the best practice questions for educational purposes. Furthermore, we evaluated the educational value of PLM-generated questions by presenting them alongside human-authored questions for SMEs to discern the human-authored from machine-authored questions. Evaluation by the respective SMEs (biology, psychology, and history) demonstrated that the generated questions achieved similar educational value relative to the human-authored ones, setting a strong case for their practical utility. In essence, we emulate how real practitioners and educators might be able to use these models to generate questions that meet their need in a practical setting.

1.2 Background: Large Pretrained Language Models and Prompting

We focus on large pretrained language models (PLMs) in this paper, specifically, auto-regressive PLMs, such as GPT that have become the dominant tools for text generation. These models learn a distribution over text, which can be decomposed auto-regressively as follows:

$$\boldsymbol{x} \sim p_\theta(\boldsymbol{x}) = p_\theta(x_1) \prod_{t=2}^{T} p_\theta(x_t | x_1, \ldots, x_{t-1}) \,. \tag{1}$$

where p_θ is the LM where θ represents all model parameters. In this paper, we focus on an LM that is already trained on massive data and thus assume p_θ is fixed throughout this paper.

In practice, we will give the model some initial texts called a "prompt" as input which instructs the model to generate specific texts. This is possible because of the decomposition in Eq. 1. To see this, let $\boldsymbol{c} := [c_1, \ldots, c_L]$ denotes the prompt which consists of L ordered tokens c_l. Then the LM models a conditional distribution as follows:

$$p_\theta(\boldsymbol{x}|\boldsymbol{c}) = p_\theta(x_1|\boldsymbol{c}) \prod_{t=2}^{T} p_\theta(x_t | x_1, \ldots, x_{t-1}, \boldsymbol{c}) \,. \tag{2}$$

Equation 2 makes it possible to adapt an LM for a wide range of generation tasks: depending on the interpretation of **c**, we can adapt a pretrained LM for a wide range of tasks. [3] shows that, without further fine-tuning p_θ, simply changing $\boldsymbol{c}$ for different tasks perform on par with fine-tuning p_θ. This makes it very easy to use the LM because we only need to change the input to the model to adapt it for a variety of tasks. See Fig. 1 for an illustration. The question now is how to design such a prompt for question generation.

2 Exploring Prompting Strategies in Question Generation

Table 1. Summary of the four factors in our prompting strategy and the choices under consideration for each factor.

Example structure for question generation	Data source in the examples	Number of examples	Lengths of context and question in each example
CAQ: context (C) and an answer (A) and the output contains a question (Q)	Content agnostic (SQuAD)	One-Shot	Small (avg. 15 words)
CTQA: (C) and a target (T) and the output contains a question (Q) and an answer (A)	Content specific	Few-Shot	Medium (avg. 25 words)
		Five-Shot	Large (40 and above)
		Seven-Shot	

In the remainder of the paper, we set out to answer the question: how do we design effective prompts for educational question generation? Answers to this question will provide practitioners with clear guidance on how to better control off-the-shelf PLMs for high-quality question generation. We take an empirical approach and design a series of experiments to systematically investigate various factors that impact the effectiveness of prompting strategies for question generation with PLMs. We propose four factors that are crucial considerations to prompt design for question generation. Below, we detail these factors and the possible choices that we study for each factor (see Table 1 for a high-level summary). In contrast to automated prompting methods as in existing literature, our prompting design is interpretable and flexible, enabling practitioners to explicitly control and iteratively refine the generation process as needed.

2.1 Example Structure for Question Generation

The first factor we investigate is the question generation formulation, i.e., the input-output structure in each example that we will use to instruct and adapt the PLM for question generation. Different formulations will likely impact the generated questions' quality. In this work, we focus on contextualized question generation, in which a question is asked and the answer to it can be found within a given paragraph. We compare two different generation setups. In the first setup, labeled as CAQ, the input contains a context (C) and an answer (A) and the output contains a question (Q). The context can be a short excerpt from a textbook and the answer should correctly answer the generated question. This setup has been considered in a wide range of question generation tasks [5,26,28]. In the second setup, referred to as CTQA, the input contains a context (C) and a target (T) and the output contains a question (Q) and an answer (A). The target does not need to be the answer to the generated question but guides the model to generate a question to ask *about* the particular part in the context specified by the target. The model also generated an answer in addition to the question. The intuition behind this setup is that the model may generate more on-topic and relevant questions because it is forced to also generate the answer. This setup is reminiscent of prior work that leverages question answering modeling for question generation [6,17].

2.2 Data Source in the Examples

The second factor we investigate is the data source in each example, i.e., where do the context, question, answer (target) come from? This question arises when a user wants to generate questions for different subjects; depending on the subject, the examples in the prompt may need to change so that PLM is given the appropriate domain knowledge. We are most interested in whether we can use the same set of examples that come from a generic source for question generation across different subjects/content. We thus compare a *content-agnostic and a* content-specific selection of examples. In the content-agnostic setup, we choose examples from SQuAD [24], a generic, widely used question answering dataset that can also be used for question generation. In the content-specific setup, we choose examples in the same subject as the one in which the PLM will generate questions.

2.3 Number of Examples

The third factor we investigate is the number of examples to include in the prompt. Usually, PLMs' performance improves with more examples. Nevertheless, because of the open-ended nature of question generation, it is unclear to what point increasing the number of examples will help. We thus consider four setups including One-shot, Few-shot, Five-shot, Seven-shot where "shot" refers to the number of examples.

2.4 Lengths of Context and Question in Each Example

The last factor we investigate is the length of context and question in each example. A context or question that is too short may limit the diversity and complexity of the generated questions. A context or question that is too long may contain irrelevant information which may confuse the PLMs, potentially leading to generated questions that are irrelevant or off-topic. We thus compare three different setups including small, medium, large contexts and questions depending on the length of texts they contain. Small corresponded to questions about 15 words in length, medium questions were around 25 words long, and large questions were about 40 words long on average. Small contexts consist of around 2 sentences, medium contexts around 4–5 sentences of information, and large contexts usually a full paragraph or multiple paragraphs.

3 Experiments

We recommend the best prompt setting for each generation strategy that yielded the best-generated questions. Code scripts, additional clarifications, and additional results such as examples of generated questions are publicly available.[1]

Experiment Setup. We choose biology as the subject to generate questions and use the Openstax Biology 2e (Bio 2e) Textbook as the source for most of our example content. In this paper, we focus on generating open-ended questions of Bloom's level below three because higher-order Bloom's questions typically involve making connections across larger content [2,14]. Generating diverse types of potentially more challenging questions is left for future work. We also limit

[1] https://github.com/openstax/research-question-generation-gpt3.

our investigation to textual content and remove images, tables, links, and references from the textbook. During generation, we first pre-select a fixed number of examples from the textbook (and SQuAD, for the data source experiment; see Sect. 2.2). During generation for all setups under each factor, we randomly pick a fixed number of examples to serve as the prompt and another two queries, i.e., with only the context (possibly also the target; see Sect. 2.1) from which the PLM is asked to generate questions. Unless otherwise noted, for each query in each setup under each factor, the PLM generates 75 questions for evaluation. When generating questions for a factor, all the other factors are set to the same value to ensure fairness in comparison. Throughout our experiments, we use the GPT-3 Davinci API from OpenAI with temperature = 0.9 and top_p = 1.

Evaluation Protocol. We primarily evaluate the quality and diversity of the generated questions. For quality, we report **perplexity** and **grammatical error**. Perplexity is inversely related to the coherence of the generated text; the lower the perplexity score, the higher the coherence. To make the process computationally efficient, we computed perplexity using a GPT-2 language model for all generations. We computed grammatical error using the Python Language Tool [22] which counts the number of grammatical errors averaged over all generated questions in each setup under each factor. For diversity, we report the **Distinct-3** score [15], which counts the average number of distinct 3-grams in the generated questions. Furthermore, we believe that ensuring the generated questions are safe, i.e., without profanity or inappropriate language is critical for high-stakes educational applications. Therefore, we report the **toxicity** of the generated questions, using the Perspective API [23], which is often missing from the evaluation in existing question generation literature. Last but not least, we perform a preliminary human evaluation to mark **percentage of acceptable questions** for each setup under each factor. A question is considered acceptable if it is coherent, on-topic, answerable, grammatically correct, and appropriate. We conduct a more comprehensive human evaluation in Sect. 3.3.

3.1 Empirical Observations

Table 2. Results for the example structure comparisons, which show that the CTQA structure is distinctly better than the CAQ structure.

Gen. format	Diversity ↑	Perplexity ↓	Toxicity ↓	Gramm. error ↓	% acceptable ↑
CAQ	0.895	64.683	0.153	**0.053**	26.7%
CTQA	**0.898**	**29.900**	0.153	0.080	**54.7%**

Structure of Examples in the Prompt. Recall that this experiment compared CAQ and CTQA structures of the examples in the prompt (Sect. 2.1). The results, presented in Table 2, show that, although the CTQA structure produces questions of comparable diversity, quality, and toxicity, it generates about twice as many acceptable questions as the CAQ structure. This comparison suggests that CTQA is a superior example structure and confirms our earlier hypothesis that asking PLMs to generate the answer in addition to only the question is beneficial for improving the quality of generated questions. Additionally, the generated answers can be potentially useful for evaluating a student's performance on the generated question. Ensuring that the generated answer correctly answers the generated question is important ongoing work.

Table 3. Results for the example data source comparisons. Using content specific examples gives superior generation performance compared to content agnostic example.

Gen. format	Diversity ↑	Perplexity ↓	Toxicity ↓	Gramm. error ↓	% acceptable ↑
SQuAD	0.884	102.840	0.201	0.093	18.0%
OpenStax	**0.895**	**64.683**	**0.153**	**0.053**	**26.7%**

Data Source in Examples. Recall that this experiment compared whether the examples come from the same subject (Bio 2e) as the query or a generic dataset (SQuAD) (Sect. 2.2). The results in Table 3 showed that when a prompt consists of examples from the same subject, the PLM can generate questions about twice as effective as when using SQuAD examples across all metrics. These results suggest that a generic set of examples may not adapt to question generation for various domains and that appropriately choosing examples from desired subjects is a better setup for question generation.

Table 4. Results for the number of examples comparisons. Five- and seven-example settings yield better questions compared to one- and three-example settings.

# Examples	Diversity ↑	Perplexity ↓	Toxicity ↓	Gramm. error ↓	% acceptable ↑
1 example	0.897	37.954	0.384	0.182	24.9%
3 examples	0.924	36.586	0.232	0.151	37.8%
5 examples	**0.938**	35.990	0.208	0.119	**51.6%**
7 examples	0.918	**30.731**	**0.176**	**0.076**	44.9%

Number of Examples. Table 4 shows the results comparing one-, three-, five-, and seven-shots, i.e., the number of examples in the prompt. The results show that one- and three-shots are ineffective; we observe that they produce a majority of unacceptable questions. The five-shot condition results were optimal followed closely by the seven-shot, with the one-shot being most inefficient. We prefer using the five-shot condition because here, the PLM generated more varied questions that are also of high quality. For example, although the model was only given free-response questions, it could produce a small number of multiple-choice or true-or-false questions.

Table 5. Results for the context and question length comparisons. We see that, in general, short context and question lengths in the examples improve generation quality.

Context length	Diversity ↑	Perplexity ↓	Toxicity ↓	Gramm. error ↓	% acceptable ↑
Short	0.861	33.452	0.329	**0.380**	22.0%
Medium	**0.878**	30.692	**0.214**	0.410	**24.0%**
Long	0.877	**30.385**	0.331	0.420	**24.0%**
Question length	Diversity ↑	Perplexity ↓	Toxicity ↓	Gramm. error ↓	% acceptable ↑
Short	**0.906**	34.275	**0.246**	**0.377**	**30.0%**
Medium	0.893	33.704	0.318	0.487	23.7%
Long	0.885	**30.38**	0.295	0.610	14.7%

Lengths of Context and Question in Each Example. Table 5 shows the results comparing different lengths of the question and context in each example, respectively. In terms of question lengths, results suggested that a smaller question length generally yields the best performance. In terms of context lengths, results are mixed. This is likely because longer contexts contain information that is not directly useful for generating questions and because longer texts lead to longer prompts, which makes it more difficult to instruct the model to adapt to the question generation task.

3.2 Discussions

From the above quantitative results, we obtain a good understanding of how the different choices, while constructing the prompt for each generation strategy, will impact the quality of the generated questions. It is clear that when preparing examples to instruct and adapt PLMs for question generation, the PLM is likely to generate higher quality questions given the prompt design: if prompt contains five to seven examples that are in CTQA format, are chosen from the desired subject, rather than generic content, and contain relatively short contexts and questions. This recommendation has the potential to serve as a guideline for practitioners when adapting off-the-shelf PLMs for their unique question generation needs.

3.3 Human Expert Evaluation for Multiple Subjects

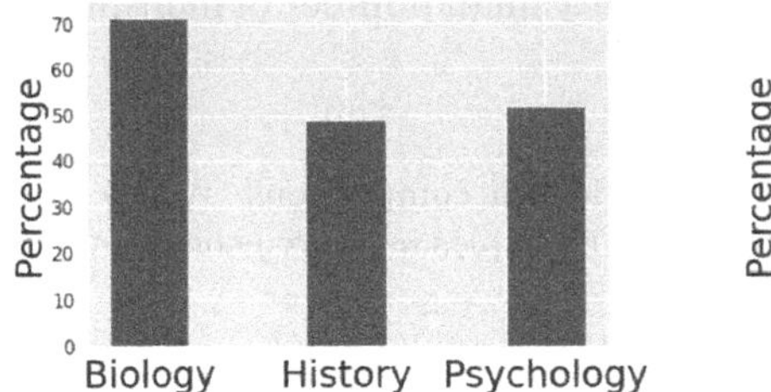

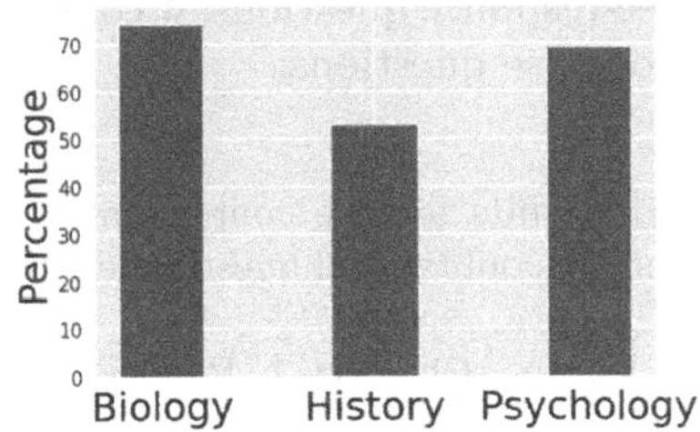

Fig. 2. Human evaluation results. **Left**: the percentage of PLM-generated questions that are recognized as human-authored by SMEs. **Right**: the percentage of PLM-generated questions that SMEs considered as ready-to-use in their classes.

To validate the utility of the generated questions as well as to investigate whether our best prompt strategy would result in good question generations across domains (e.g., natural sciences, social sciences, and humanities), we engaged biology, psychology, and history subject matter experts (SMEs) respectively to evaluate the quality of questions from these domains generated using the best prompting strategy.

For each domain, we chose 5 examples as the prompt and another 5 examples with only the question and target as the query. For each query, we generated and selected the 10 best questions. We then mixed these 50 questions together with 5 reference human-authored questions (one for each query) and presented them to SMEs. SMEs were asked to (1) distinguish whether each question is machine or human authored and (2) evaluate whether they would use the question in the assessment they create for their class.

The results, shown in Fig. 2, suggest that the majority of PLM-generated questions look "human-like" for SMEs and can be readily used in their respective classes. This result is encouraging because it suggests that our prompting strategy can effectively adapt a publicly available, general-purpose text generation model for high-quality educational question generation. Table 6 demonstrates a few successful generated questions.

Table 6. Examples of three generated questions using our recommended prompting strategy along with the context, target, and reference human-authored questions.

Context (Biology): At 0.1 to $5.0\,\mu\text{m}$ in diameter, prokaryotic cells are significantly smaller than eukaryotic cells, which have diameters ranging from 10 to $100\,\mu\text{m}$ (Fig. 4.6). The prokaryotes' small size allows ions and organic molecules that enter them to quickly diffuse to other parts of the cell. Similarly, any wastes produced within a prokaryotic cell can quickly diffuse. This is not the case in eukaryotic cells, which have developed different structural adaptations to enhance intracellular transport
Target: size
Reference human-authored question: Some unusually large cells do exist. How might such cells solve the problem of diffusion?
Machine-generated question: Compare the size of prokaryotic and eukaryotic cells to describe why diffusion is more rapid through prokaryotic cells
Context (Psychology): Sleep debt and sleep deprivation have significant negative psychological and physiological consequences. As mentioned earlier, lack of sleep can result in decreased mental alertness and cognitive function. In addition, sleep deprivation often results in depression-like symptoms. These effects can occur as a function of accumulated sleep debt or in response to more acute periods of sleep deprivation. It may surprise you to know that sleep deprivation is associated with obesity, increased blood pressure, increased levels of stress hormones, and reduced immune functioning
Target: sleep deprivation
Reference human-authored question: Which animal would be most likely to die from a lack of sleep?
Machine-generated question: Sleep debt and sleep deprivation are associated with significant negative psychological and physiological consequences. What are some of the potential consequences of a person having less than optimal sleep?
Context (U.S. History): Opponents made clear their resistance to Garrison and others of his ilk; Garrison nearly lost his life in 1835, when a Boston anti-abolitionist mob dragged him through the city streets. Anti-abolitionists tried to pass federal laws that made the distribution of abolitionist literature a criminal offense, fearing that such literature, with its engravings and simple language, could spark rebellious Black people to action. Their sympathizers in Congress passed a "gag rule" that forbade the consideration of the many hundreds of petitions sent to Washington by abolitionists. A mob in Illinois killed an abolitionist named Elijah Lovejoy in 1837, and the following year, ten thousand protestors destroyed the abolitionists' newly built Pennsylvania Hall in Philadelphia, burning it to the ground
Target: anti-abolitionists
Reference human-authored question: How did anti-abolitionists respond to the abolition movement?
Machine-generated question: What were some of the responses to the abolitionist movement?

3.4 Limitations and Future Work

Question generation is complex and is an active ongoing area of research. In this work, we proposed four factors as the prompt design framework and conducted a controllable scientific study on those factors. However, our investigation was by no means exhaustive and several other factors have the potential to impact question generation such as question difficulty, type (i.e., multiple-choice, fill in the blank, etc.), and modality (i.e., potentially including tables and equations). Investigations of more factors and complex interactions among them are left for future work. In addition, our human evaluation was a small-scale experiment because we were only able to engage the SMEs for a short time. The next step is to conduct a large-scale evaluation that involves both instructors and students

Table 7. Examples of failed cases and the failing reasons. Our prompting strategy can still generate questions that contain grammatical errors and other types of errors.

(Biology): What is the correct statement is about centrosomes? (Multiple-choice question with no options and bad grammar)
(Psychology): Sleep deprevation can lead to serious changes in the body. Which one of these changes characterized by sleep deprivation? (grammatical and spelling errors)
(History): During the Gold Rush, the Forty-Niners did not find wealth so easy to come by, most did not. (not a question)

in a safe environment to obtain a better understanding of the educational utility of machine-generated questions. Lastly, our prompting strategy generated questions with grammatical errors and other problems at times; we show some failed examples in Table 7. A promising future direction is to develop automated filters capable of removing undesirable generated questions and only select the highest quality ones, preferably also personalized to each student and instructor.

4 Conclusion

In this work, we investigate the best practices to prompt a PLM for educational question generation. We develop and empirically study a prompting strategy consisting of four different factors. Based on a series of quantitative experiments, we recommended the choices for each factor under our prompting strategy that led to high-quality generated questions. Human evaluations by subject experts in three different educational domains suggest that most of the questions generated by a PLM with our recommended prompting strategy are human-like and ready-to-use in real-world classroom settings. Our results indicate that properly prompting existing off-the-shelf PLMs is a promising direction for high-quality educational question generation with many exciting future research directions.

Acknowledgements. This work is supported by NSF grants 1842378, 1917713, 2118706, ONR grant N0014-20-1-2534, AFOSR grant FA9550-18-1-0478, and a Vannevar Bush Faculty Fellowship, ONR grant N00014-18-1-2047. We thank Prof. Sandra Adams (Excelsior College), Prof. Tyler Rust (California State University), Prof. Julie Dinh (Baruch College, CUNY) for contributing their subject matter and instructional expertise. Thanks to the anonymous reviewers for thoughtful feedback on the manuscript.

References

1. Adesope, O.O., et al.: Rethinking the use of tests: a meta-analysis of practice testing. Rev. Educ. Res. **87**(3), 659–701 (2017)
2. Bloom, B.S., Engelhart, M.D., Furst, E., Hill, W.H., Krathwohl, D.R.: Handbook I: Cognitive Domain. David McKay, New York (1956)

3. Brown, T., et al.: Language models are few-shot learners. In: Larochelle, H., Ranzato, M., Hadsell, R., Balcan, M.F., Lin, H. (eds.) Advances in Neural Information Processing Systems, vol. 33, pp. 1877–1901 (2020)
4. Connor-Greene, P.A.: Assessing and promoting student learning: blurring the line between teaching and testing. Teach. Psychol. **27**(2), 84–88 (2000)
5. Du, X., Shao, J., Cardie, C.: Learning to ask: neural question generation for reading comprehension. In: Proceedings of the ACL, pp. 1342–1352 (July 2017)
6. Duan, N., Tang, D., Chen, P., Zhou, M.: Question generation for question answering. In: Proceedings of the Conference on EMNLP, pp. 866–874 (September 2017)
7. Huang, Y.T., Chen, M.C., Sun, Y.S.: Bringing personalized learning into computer-aided question generation (2018)
8. Karpicke, J.D.: Retrieval-based learning: active retrieval promotes meaningful learning. Curr. Dir. Psychol. Sci. **21**(3), 157–163 (2012)
9. Karpicke, J.D., Blunt, J.R.: Retrieval practice produces more learning than elaborative studying with concept mapping. Science **331**(6018), 772–775 (2011)
10. Karpicke, J.D., Roediger, H.L., III.: The critical importance of retrieval for learning. Science **319**(5865), 966–968 (2008)
11. Keskar, N.S., McCann, B., Varshney, L.R., Xiong, C., Socher, R.: CTRL: a conditional transformer language model for controllable generation (2019)
12. Koedinger, K.R., Kim, J., Jia, J.Z., McLaughlin, E.A., Bier, N.L.: Learning is not a spectator sport: Doing is better than watching for learning from a MOOC. In: Proceedings of the Conference on Learning at Scale, pp. 111–120 (2015)
13. Kovacs, G.: Effects of in-video quizzes on MOOC lecture viewing. In: Proceedings of the Conference on Learning at Scale, pp. 31–40 (2016)
14. Krathwohl, D.R.: A revision of bloom's taxonomy: a overview. Theor. Pract. **41**(4), 212–218 (2002)
15. Li, J., Galley, M., Brockett, C., Gao, J., Dolan, B.: A diversity-promoting objective function for neural conversation models. In: Proceedings of the 2016 Conference of the North American Chapter of the Association for Computational Linguistics: Human Language Technologies, pp. 110–119 (Jun 2016)
16. Li, X.L., Liang, P.: Prefix-tuning: optimizing continuous prompts for generation. In: Proceedings of the ACL. pp. 4582–4597 (August 2021)
17. Li, Y., Duan, N., Zhou, B., Chu, X., Ouyang, W., Wang, X.: Visual question generation as dual task of visual question answering. arXiv e-prints (2017)
18. Liu, P., et al.: Pre-train, prompt, and predict: a systematic survey of prompting methods in natural language processing (2021)
19. Liu, X., Ji, K., Fu, Y., Du, Z., Yang, Z., Tang, J.: P-tuning v2: prompt tuning can be comparable to fine-tuning universally across scales and tasks (2021)
20. Lu, O.H., Huang, A.Y., Tsai, D.C., Yang, S.J.: Expert-authored and machine-generated short-answer questions for assessing students learning performance. Educ. Technol. Soc. **24**(3), 159–173 (2021)
21. Martin, L., Mills, C., D'Mello, S.K., Risko, E.F.: Re-watching lectures as a study strategy and its effect on mind wandering. Exp. Psychol. **65**(5), 297–305 (2018)
22. Morris, J.: Python language tool (2021). https://github.com/jxmorris12/language_tool_python
23. Perspective: Using machine learning to reduce toxicity online (2021). https://www.perspectiveapi.com/
24. Rajpurkar, P., et al.: SQuAD: 100,000+ questions for machine comprehension of text. In: Proceedings of the Conference on EMNLP, pp. 2383–2392 (November 2016)

25. Serban, I.V., et al.: Generating factoid questions with recurrent neural networks: the 30M factoid question-answer corpus. In: Proceedings of the ACL, pp. 588–598 (August 2016)
26. Wang, Z., Lan, A.S., Nie, W., Waters, A.E., Grimaldi, P.J., Baraniuk, R.G.: QG-Net: a data-driven question generation model for educational content. In: Proceedings of the Conference on Learning at Scale (2018)
27. Wiklund-Hörnqvist, C., Jonsson, B., Nyberg, L.: Strengthening concept learning by repeated testing. Scand. J. Psychol. **55**(1), 10–16 (2014)
28. Willis, A., et al.: Key phrase extraction for generating educational question-answer pairs. In: Proceedings of the Conference on Learning at Scale (2019)

Towards the Automated Evaluation of Legal Casenote Essays

Mladen Raković[1](✉), Lele Sha[1], Gerry Nagtzaam[2], Nick Young[2], Patrick Stratmann[2], Dragan Gašević[1,3,4], and Guanliang Chen[1](✉)

[1] Centre for Learning Analytics, Faculty of Information Technology, Monash University, Clayton, Victoria, Australia
{mladen.rakovic,lele.sha1,dragan.gasevic,guanliang.chen}@monash.edu

[2] Faculty of Law, Monash University, Clayton, Victoria, Australia
{gerry.nagtzaam,nick.young,pstr0001}@monash.edu

[3] School of Informatics, University of Edinburgh, Edinburgh, UK

[4] Faculty of Computing and Information Technology, King Abdulaziz University, Jeddah, Saudi Arabia

Abstract. A legal casenote essay is a commonly assigned writing task to first-year law students aiming to promote their understanding of legal reasoning and help them acquire writing skills in a legal domain. To ensure law students master the legal casenote writing, it is critical that instructors monitor and evaluate students' progress, and provide a timely and specific feedback. This is, however, a challenging task to many instructors as they often need to dedicate excessive time and effort to evaluate writing of and provide formative feedback to each individual student. We posit a computational tool that can afford at-scale evaluation of legal casenote writing may help remedy this challenge. Although quite some automatic writing evaluation (AWE) tools have been applied in the domain of education, the AWE tool that can analyse rhetoric of a legal casenote essay (i.e., specific rhetorical elements required by this task) is yet to be developed. We made the first step towards developing such a tool. We manually annotated each sentence in a corpus of 1,020 authentic casenote essays written over 6 offerings of the first-year legal writing course and developed one traditional machine learning classifier (Random Forest) and two deep learning classifiers (based on vanilla BERT and Legal BERT pre-trained language models). We found that the deep learning classifier based on Legal BERT could correctly identify more than 86% of rhetorical moves in a casenote. Our findings may be of a particular interest for educational researchers and practitioners who seek to use the methods of artificial intelligence to support legal writing education.

Keywords: Legal casenote writing · Automated writing evaluation · Rhetorical moves · Machine learning · Deep learning

1 Introduction

In contemporary society, lawyers are often required to analyse different legal texts and clearly articulate this analysis in writing [28]. For this reason, writing legal

M. M. Rodrigo et al. (Eds.): AIED 2022, LNCS 13355, pp. 167–179, 2022.
https://doi.org/10.1007/978-3-031-11644-5_14

documents of different genres – e.g., casenotes, memos, appeals and judgements – is deemed to require a critical set of skills for aspiring lawyers. To provide law students with the opportunity to acquire and hone their legal writing skills, and to learn how to develop different legal documents, law educators administer many writing assignments.

A legal casenote essay is usually the first writing assignment assigned to law students in their freshman year. In this task, students analyse multiple documents that describe an authentic legal case decided at court, and write a summary (i.e., casenote) that provides essential information about the case including the material facts, procedural history, court reasoning, and court decision. The presence of these elements, commonly referred to as rhetorical moves [44], will determine the casenote assignment mark. Importantly, this writing task is not only considered potent to help first-year law students understand and communicate legal reasoning [8,31], but also to acclimate them to the culture of legal writing, e.g., use of specialized vocabulary and formal syntax [36].

To ensure law students master the legal casenote writing, it is critical that instructors monitor and evaluate students' progress on this task, and provide a timely and specific feedback [23,31] on how the casenote essay should be improved. For example, an instructor may notice that the *court decision* move was not provided in the casenote draft, and advise a student on how to include this move in the next draft. This is, however, a challenging task to many instructors as they often need to dedicate excessive time and effort to evaluate writing and provide formative feedback to each individual student [18,33]. This challenge is further amplified in large-enrollment courses, such as the first-year legal writing course at many law schools, where instructors typically need to evaluate a few hundred casenote submissions in a limited time frame.

We posit a computational tool that provides at-scale evaluation of legal casenote writing may help remedy this challenge and benefit law students and instructors. For instance, as the presence of rhetorical moves determines the quality of a casenote, such computational tool may be applied to detect rhetorical moves in a casenote draft, and, on that basis, tailor writing analytics to help students identify areas for improvement (e.g., missing required rhetorical moves), and thus lessen marking burden for instructors. Following recent advances in computational technologies for text analysis, especially those grounded in machine learning (ML), quite some automatic writing evaluation (AWE) tools have been developed and applied in the domain of education (e.g., [4,15,25,37]). Whereas researchers have documented empirical benefits of these tools in supporting student writing in different genres and subjects, to our knowledge, the existing tools appear to be limited in supporting casenote writing (for details, see Sect. 2.2.). More research is hence needed to develop an AWE tool that can analyse the specific rhetoric of a legal casenote genre and provide at-scale support to law instructors and students working on this task.

In the present study, we took the first step towards developing such a tool. Specifically, we investigated the affordances of state-of-the-art ML and deep learning (DL) algorithms to automatically evaluate students' responses to a legal casenote assignment. To this end, we collected a corpus of authentic casenote essays written by first-year law students over 6 offerings of the legal writing course. In these essays, we manually labeled each sentence with a corresponding rhetorical move. Using this data set, we developed and evaluated performance of several classifiers that can identify the presence of each rhetorical move in a casenote essay with a high accuracy.

2 Related Work

2.1 Rhetorical Moves

A rhetorical move is theorised as a segment of text that fulfills a particular communicative purpose within a genre [16]. In other words, rhetorical moves are building blocks of a genre and they jointly contribute to a coherent understanding of a text [16,44]. Students are often required to provide a set of genre-specific rhetorical moves in their essays. For instance, an argumentative essay may contain the *claim*, *evidence* and *rebuttal* moves [1], whereas a literature review essay may contain the *summary of prior research*, *gaps in prior research* and *justification for a new study* moves [46]. The presence of rhetorical moves within the essay often determines the communicative quality of that essay and a subsequent grade the essay will receive. For this reason, educational researchers have utilised rhetorical moves to theoretically guide the analysis of student writing across a range of genres and disciplines [35]. Following this approach, in the present study, we analysed the legal casenote writing from the perspective of rhetorical moves.

2.2 Automated Analysis of Rhetorical Moves in Student Writing

The AWE tools developed to date have mainly examined students' writing on the micro-level, e.g., lexical and syntactical errors, and only a few have examined students' writing on the macro-level, e.g., text structure and rhetorical moves required by a task [2,43]. For example, the Academic Writing Analytics (AWA) and AcaWriter [25] tools use a rule-based parser [38] to detect rhetorical moves in students' essays of different genres and tailor context-sensitive feedback to learners to help them improve their writing. The Mover [4] and the Research Writing Tutor (RWT; [15]), on the other hand, harness supervised ML algorithms to analyse rhetorical structure of research essays. These tools can identify rhetorical moves that occur in a draft (e.g., "Establishing a Territory", "Identifying a Niche" and "Addressing the Niche") and present these moves to a learner (e.g., as a color-coded text). In this way, a learner can gain a deeper insight into the rhetorical structure of their draft, engage in metacognitive evaluation of the draft (e.g., by appraising the extent to what the draft aligns to assessment criteria), and improve their text accordingly (e.g., "It seems like I still have to address the niche."). For a more detailed overview of the aforementioned tools, see [24].

Whereas these tools have been found promising to support learners to revise and improve their drafts of different genres, including argumentative law essays [26,40], business reports [40], sections of research articles [15] and reflective essays [29], the tool that can analyse rhetorical characteristics specific to legal casenote writing and support learners in these assignments is yet to be developed. For instance, as legal casenotes are brief summaries of decided court cases, learners are expected to concisely communicate essential information about the court judgement in these documents [7,45]. To this end, many learners rarely use in-text linguistic cues (e.g., "As a consequence...") to signal rhetorical moves to readers. For this reason, the existing rule-based systems that rely upon linguistic cues to determine rhetorical moves do not appear to be fully applicable to the analysis of legal casenote responses. Instead, supervised ML may be considered a more viable approach towards creating a robust tool for automated evaluation of rhetorical moves in casenote writing. Even though a small group of researchers has already utilised supervised ML to classify rhetorical moves in student writing, and the systems developed (e.g., [4,15]) achieved good performance relative to standards in discourse analysis (80% overlap with human evaluators recommended in [5]), the classification tasks these systems attempted to address were not explicitly related to legal writing. Moreover, the sets of texts researchers used to train these systems did not appear to include legal documents. Adding to this line of research, we collected and manually annotated a corpus of student responses to casenote assignments, and used this dataset to create the classifier that classifies rhetorical moves in a casenote. For this purpose, we explored both traditional ML and DL approaches. Specifically, we attempted to address the following **research question**: *To what extent can a machine learning/deep learning classifier accurately identify rhetorical moves in a legal casenote essay?*

3 Method

3.1 Learning Task and Code Book

In a casenote assignment, students were required to analyse multiple documents that describe an authentic legal case with court decision, identify major points of the case and write up to a 500-word casenote that includes the following rhetorical moves: (1) *material facts* that gave the rise to the original cause of action (2) *procedural history* representing the proceedings that arose as a result of the cause of action, e.g., arguments made by counsel (3) *court reasoning*, e.g., the reasoning of the Justices of the High Court, and (4) *court decision and relief*. The presence of the rhetorical moves (1)–(4) determined the casenote assignment mark. In addition, the students were required to use *footnotes* in their essays to explicitly refer to relevant documents from the case. The use of footnotes counted towards the assignment mark, too. Informed by these rhetorical and marking requirements, we developed a codebook to define and categorise rhetorical moves in a legal casenote. We included two additional categories that did not count towards the mark, *Title & Introduction* that contained the essay title, introductory comments and/or signposting in the first paragraph, and *Other Information* that contained any information that could not be categorised into any other category. The code book with examples is provided in Table 1.

Table 1. Rhetorical moves in legal casenotes

Label	Definition	Example
Title & introduction	Title, introductory comments to the essay or signposting	LAW CASE REPORT The case of McHale Watson involved an unfortunate event that brought to light the issues of a minor's capability to be held to the same standard of care under the tort of negligence
Material facts	The material facts giving rise to the original cause of action	During the school holidays on January 21st 1957, the respondent, Barry Watson, aged twelve years old was playing with the appellant, Susan McHale, and another girl both of whom aged nine years old
Procedural history	The proceedings and arguments that arose as a result of the cause of action	This was an action commenced in the original jurisdiction of the High Court of Australia before Windeyer J in McHale v Watson (1964)
Court reasoning	Court reasoning that includes obiter and ratio	The appeal was argued on two grounds, firstly that Windeyer J had errored in finding a variance between an adult and child standard of care, and, secondly that His Honour should have found negligence, regardless of the measured standard
Court decision	The actual decision reached and the relief granted to the parties as a result	Justice Windeyer's judgement was held by a majority-of-three-to-one by McTiernan CJA, Kitto J and Owen J, with only Menzies J dissenting
Footnote	References to the case documents cited	Wrongs Act 1958 (Vic) s 26

3.2 Dataset

We collected 5,800 responses to casenote assignments written by the first-year law students at a large Australian university over 6 offerings of a legal writing course between 2017 and 2020. Of these, we randomly selected 1,020 responses (i.e., 170 responses from each offering) for manual labeling and ML/DL classifier modeling. The average length of a casenote essay was 290.93 words (SD = 72.69). This study was implemented with the approval from the Human Research Ethics Committee (Project ID: *29451*).

Two human raters, experts in this writing task, randomly selected 113 casenotes, i.e., approximately 10% of the dataset, trained together on how to apply the code book for annotation (Table 1), and then separately annotated these essays. The raters used sentences as the unit of analysis, because, compared to paragraphs, sentences can afford more fine-grained analysis of a text [10]. Specifically, the raters labeled each sentence in the essay with a corresponding category from the code book. The inter-rater reliability between the two raters was nearly 1 in all the categories, measured by Cohen's *kappa*. The reason for this almost perfect overlap may be because the casenotes were generally short (approximately 26 sentences, on average) and some students introduced optional headings to their essays to explicitly signal rhetorical moves. This, in turn, could have made the training and annotation process more straightforward. One of the raters proceeded with the manual annotation and annotated the remaining casenote essays. As the sentences belonging to the *Other Information* category were not identified in the annotated casenotes, this category has been removed from the final dataset. The descriptive statistics of the dataset is provided in Table 2.

Table 2. The descriptive statistics of the casenote dataset. **Total:** Total Number; **Avg:** Average Number; **Title**: Title & Introduction; **Material**: Material Facts; **Procedural**: Procedural History; **CourtR**: Court Reasoning; **CourtD**: Court Decision. Standard deviation is provided in parentheses.

	Total	Title	Material	Procedural	CourtR	CourtD	Footnote
Total casenote	1,020	609	848	859	856	722	836
Avg. words per casenote	290.93 (72.69)	18.27 (21.60)	47.67 (22.35)	73.82 (42.65)	126.29(57.06)	14.92 (17.92)	35.94(33.78)
Avg. unique words per casenote	156.63 (29.47)	15.02 (15.66)	38.76 (15.73)	54.27 (27.05)	86.53 (33.16)	13.16 (12.81)	14.90 (11.56)
Avg. sentences per casenote	26.23 (10.70)	1.44 (1.09)	3.86 (2.05)	5.32 (3.17)	8.98 (4.76)	1.43 (1.33)	7.06 (7.49)
Total sentences	23,347	966	3,873	5,460	9,167	1,142	2,739
Avg. words per sentence	13.28 (8.09)	13.73 (9.67)	12.51 (6.65)	13.97 (7.81)	14.02 (7.26)	11.36 (5.91)	11.16 (11.75)
Avg. unique words per sentence	12.21 (6.32)	12.12 (7.21)	11.89 (5.79)	13.07 (6.61)	13.31 (6.27)	10.68 (5.23)	7.89 (4.65)

3.3 Model Implementation

We utilised the manually annotated casenote dataset to train three classifiers to identify rhetorical moves. In particular, one classifier was based on the traditional ML algorithm Random Forest [11] and the other two were based on the deep learning (DL) algorithm BERT SCL (Bidirectional Encoder Representations from Transformers with Single Classification Layer; [17]). We describe the implementation procedure below.

Random Forest. As Random Forest is deemed the one of the most widely used traditional ML algorithms for classification tasks [42], we elected to implement Random Forest in this study and compare its classification performance with the state-of-the-art Deep Learning algorithms. We implemented the Random Forest model using the `sklearn`[1] Python package. We tuned the model hyper parameters using `GridSearchCV`[2] Python package. To obtain input features for traditional ML models, researchers typically need to perform an extensive, manual feature extraction from a raw text. We surveyed prior literature [3,12,20,30,49] and identified two groups of features commonly used in the development of traditional ML models: (1) `LIWC`[3] and (2) `N-GRAM` features. `LIWC` [34] is a predefined dictionary of 84 features reflecting a frequency of different linguistic choices a student made, e.g., function words, summary, affect, relativity and time orientation. N-GRAM features contain the 1000 most frequent unigrams and bigrams extracted from a case note. We included both groups of features as input to our Random Forest classifier.

BERT. Although DL classification models have demonstrated a considerable performance in classifying legal texts (e.g., [32]), these models typically rely upon a large amount of training data. In recent years, the development of pre-trained language models, e.g., BERT [17], provided researchers with the opportunity to

[1] https://scikit-learn.org/.
[2] https://scikit-learn.org/stable/modules/grid_search.html.
[3] https://liwc.wpengine.com/.

develop high-performing DL classifiers without necessarily using large datasets for training. For this reason, we utilised the pre-trained BERT language models in the current study. Specifically, we created two DL models based on BERT: (1) BERT Base that uses a version of the BERT pre-trained on general texts (e.g., Wikipedia articles and books) and (2) BERT Legal[4], a version of the BERT pre-trained on the legal documents (e.g., legislation, court cases, contracts). To fine-tune legal BERT, we followed the broader hyper-parameter search space procedure proposed in [13]. Given that our dataset was relatively small, we set batch size to 8 without setting a fixed maximum number of epochs to avoid under-fitting. Then, an early stopping mechanism was applied based on validation loss. We applied a low learning rate of 1e−5 and a high drop-out rate of 0.2 which have been shown to improve regularization in [13].

Data Pre-Processing. In data pre-processing stage, we first split each casenote essay into sentences using the `sentence-tokenize`[5] routine of NLTK Python package [9]. Each sentence was mapped to a corresponding label, i.e., manually annotated rhetorical move: *Title and Introduction*, *Material Facts*, *Procedural History*, *Court Reasoning*, *Court Decision* and *Footnotes*. Following recommendations in [22], we randomly split the dataset using the 80%:20% train-test ratio, i.e., 80% of data were used for model training and the remaining 20% were used for model testing.

Model Training and Testing. We trained a single multi-label classification model to classify the six rhetorical moves within a casenote. We detected an unequal representation of outcome classes in the training sample, e.g., the number of sentences labeled as *procedural history* was nearly five times higher than the number of sentences labelled as *title & introduction*. To mitigate the class imbalance issue, we applied the *SMOTE* algorithm [14], motivated by prior research that successfully applied this algorithm to balance outcome classes in educational classification tasks (e.g., [6,27]). Using SMOTE, we over-sampled minority classes and reached a parity in sample sizes between minority and majority classes. In this way, we ensured that the model was evenly trained on both minority and majority classes. We tested and compared the models' performance using the testing sample. To this end, we computed the four classification performance metrics: Accuracy, Cohen's κ (denoted as Kappa), AUC, and F1 score. To answer our research question, we evaluated the overall model performance in classifying multiple labels at once, but also the model performance in classifying each individual label, e.g., material facts vs. other rhetorical moves.

4 Results

The evaluation results show that BERT Legal outperformed the Base BERT and Random Forest models, as indicated by the all four metrics (Table 3). In

[4] https://huggingface.co/nlpaueb/legal-bert-base-uncased.
[5] https://www.nltk.org/.

particular, the AUC of 0.956 achieved by BERT Legal indicates an outstanding classification performance on the test dataset. Moreover, a better performance of BERT Legal compared to BERT Base may have been expected given that BERT Legal is pre-trained on legal texts.

We further evaluated how the three models classified each individual label. The evaluation results are presented in Table 4. We thus found that DL models outperformed Random Forest in correctly identifying all the labels. Base and Legal BERT performed similarly well in detecting all the rhetorical moves, except in detecting court decision (CourtD) where Legal BERT outperformed Base BERT (0.971 vs 0.788, measured by classification accuracy). Overall, our results indicate that Legal BERT classifier[6] can correctly classify at least 86% of all the sentences into one of the rhetorical moves, and, as such, may be a preferable DL model for automatic detection of rhetorical moves in legal casenotes. We show the confusion matrix of rhetorical moves predicted by Legal BERT in Fig. 1.

Table 3. The classification performance of Random Forest, BERT Base and BERT Legal measured by Accuracy, Kappa, AUC and F1 score

Model	Label	Accuracy	Kappa	AUC	F1
Random forest	MULTI-6	0.743	0.691	0.945	0.734
BERT base	MULTI-6	0.816	0.751	0.951	0.820
BERT legal	MULTI-6	**0.835**	**0.777**	**0.956**	**0.836**

Note. **MULTI-6:** Classifying each sentence into one of the 6 rhetorical moves. The highest classification scores for each metric are in bold.

Table 4. Accuracy and AUC scores the models achieved when classifying each rhetorical move individually

Rhetorical move	Random forest		BERT base		BERT legal	
	Accuracy	AUC	Accuracy	AUC	Accuracy	AUC
Material	0.834	0.840	0.958	0.946	0.957	0.940
Procedural	0.728	0.758	0.874	0.829	0.868	0.818
CourtR	0.792	0.793	0.889	0.882	0.892	0.886
Title	0.834	0.866	0.984	0.945	0.984	0.938
CourtD	0.770	0.832	0.788	0.858	0.971	0.869
Footnotes	0.966	0.944	0.986	0.986	0.996	0.997

[6] Source files of the casenote classifier developed in this study are publicly available at https://bit.ly/3roDWTC.

5 Discussion

5.1 Interpretation of the Results

Our results indicate it is possible to develop a highly accurate classifier of rhetorical moves in legal casenotes written by first-year law students, a major contribution of our study. More specifically, we found that deep learning classifiers based on the state-of-the-art pre-trained language models outperformed Random Forest, a traditional machine learning classifier. This finding conforms to prior research suggesting that pre-trained language models can boost text classification performance compared to more traditional approaches [39], even when using a relatively small corpus for model training [21], like in our study. We also demonstrated that the deep learning classifier based on Legal BERT [13], a language model pre-trained on a few hundred thousand diverse legal texts, can be very accurate in identifying rhetorical moves in students' casenotes. We note that, while Legal BERT has been successfully applied in several analytical tasks that involve legal texts (e.g., identifying topics in legal documents [41]), resolving domain name disputes [47] and labeling factual information in legal cases [48], to our knowledge, our study was the first to successfully apply a Legal BERT model to detect rhetorical moves specific to legal casenotes.

Our results corroborate prior evidence (e.g., [48]) that Legal BERT generally performs better on legal classification tasks compared to Base BERT, a language model pre-trained on generic texts. In particular, we found that Legal BERT outperformed the Base BERT over nearly 19% in detecting the *court decision* move. This may indicate that the Legal BERT was more adept to specific vocabulary and syntax of court decisions, compared to the base model. We, however, note that Legal BERT tended to confuse mainly between *procedural history* and *court reasoning* rhetorical moves (Fig. 1). We speculate this might be due to the similar vocabulary that legal writers may use to describe the court procedure and court reasoning, e.g., it may be possible that the arguments provided in procedural history are reiterated in court reasoning to justify the final court decision. This speculation should be tested in future work.

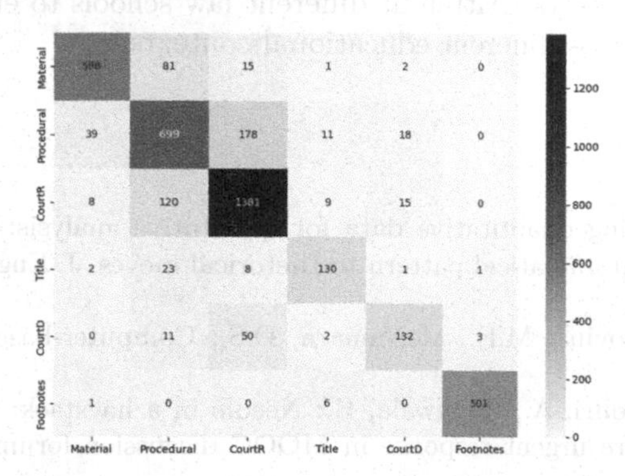

Fig. 1. The Confusion Matrix of predicted rhetorical moves based on the LEGAL BERT model on the testing set

5.2 Implications for Research and Practice

Our findings may be of a particular interest for educational researchers and practitioners who seek to use the methods of artificial intelligence to develop at-scale writing support to law students. For instance, the theoretical framing based on rhetorical moves, and dataset creation and model development approaches reported in this study can be applied in the context of legal writing tasks other than the casenote, e.g., legal memos or argumentative essays, as these genres also involve rhetorical moves. Equally important, the publicly available Legal BERT classifier developed in this study can be used as a part of a future writing analytics tool that can analyse a casenote draft. For example, the tool may color-code the rhetorical moves in a draft, making it easier for instructor or a student to identify the moves that still need to be included, a potential benefit towards more efficient marking and formative feedback.

6 Limitations and Future Work

We identified a few major limitations to our study that may be addressed in future research. First, as per our data labelling approach, we categorised each individual sentence into only one rhetorical move. However, we acknowledge it is possible that one sentence sometimes can be categorised into multiple rhetorical moves, e.g., a compound sentence elaborating on *court reasoning* and *court decision*. To address this challenge in future studies, researchers may need to label discourse units that are more fine-grained than a sentence, e.g., idea unit [19]. Next, even though we recorded in our analysis whether students provided citations in their responses or not, the quality of citation use was not measured. For example, our analysis could not distinguish whether a student merely copied or more deeply analysed and transformed information from documents cited in their casenotes. To remedy this issue, researchers may compute semantic similarity between a casenote draft and each document cited in the draft, and use this value as an additional feature to describe rhetorical moves in a casenote. Last, we acknowledge that the classifier developed in this study should be validated on casenote essays written at different law schools to ensure it performs comparably well across different educational contexts.

References

1. Ädel, A.: Selecting quantitative data for qualitative analysis: a case study connecting a lexicogrammatical pattern to rhetorical moves. J. Engl. Acad. Purp. **16**, 68–80 (2014)
2. Allen, L.K., Jacovina, M.E., McNamara, D.S.: Computer-based writing instruction(2016)
3. Almatrafi, O., Johri, A., Rangwala, H.: Needle in a haystack: identifying learner posts that require urgent response in MOOC discussion forums. Comput. Educ. **118**, 1–9 (2018)

4. Anthony, L., Lashkia, G.V.: Mover: a machine learning tool to assist in the reading and writing of technical papers. IEEE TPC **46**(3), 185–193 (2003)
5. Artstein, R., Poesio, M.: Inter-coder agreement for computational linguistics. Comput. Linguist. **34**, 555–596 (2008)
6. Barbosa, G., et al.: Towards automatic cross-language classification of cognitive presence in online discussions. In: LAK, pp. 605–614 (2020)
7. Bhatia, V.K.: Simplification v. easification-the case of legal texts1. Appl. Linguis. **4**(1), 42–54 (1983)
8. Bhatia, V.K.: Analysing genre: Language use in professional settings. Routledge (2014)
9. Bird, S., Loper, E.: Nltk: the natural language toolkit. ACM (2004)
10. Bransford, J.D., Barclay, J.R., Franks, J.J.: Sentence memory: a constructive versus interpretive approach. Cogn. Psychol. **3**(2), 193–209 (1972)
11. Breiman, L.: Random forests. Mach. Learn. **45**(1), 5–32 (2001)
12. Caines, A., Pastrana, S., Hutchings, A., Buttery, P.J.: Automatically identifying the function and intent of posts in underground forums. Crime Sci. **7**(1), 1–14 (2018). https://doi.org/10.1186/s40163-018-0094-4
13. Chalkidis, I., Fergadiotis, M., Malakasiotis, P., Aletras, N., Androutsopoulos, I.: Legal-bert: The muppets straight out of law school (2020). arXiv preprint arXiv:2010.02559
14. Chawla, N.V., Bowyer, K.W., Hall, L.O., Kegelmeyer, W.P.: Smote: synthetic minority over-sampling technique. J. Artifi. Intell. Res. **16**, 321–357 (2002)
15. Cotos, E., Huffman, S., Link, S.: Understanding graduate writers' interaction with and impact of the research writing tutor during revision. J. Writing Res. **12**(1), 187–232 (2020)
16. Crossley, S.: A chronotopic approach to genre analysis: an exploratory study. Engl. Specif. Purp. **26**(1), 4–24 (2007)
17. Devlin, J., Chang, M.W., Lee, K., Toutanova, K.: Bert: Pre-training of deep bidirectional transformers for language understanding (2018). arXiv preprint arXiv:1810.04805
18. Driessen, E., Van Der Vleuten, C.: Matching student assessment to problem-based learning: lessons from experience in a law faculty. Stud. Contin. Educ. **22**(2), 235–248 (2000)
19. Dunlosky, J., Hartwig, M.K., Rawson, K.A., Lipko, A.R.: Improving college students' evaluation of text learning using idea-unit standards. Quart. J. Exp. Psychol. **64**(3), 467–484 (2011)
20. Ferreira, M., Rolim, V., Mello, R.F., Lins, R.D., Chen, G., Gašević, D.: Towards automatic content analysis of social presence in transcripts of online discussions. In: LAK, pp. 141–150 (2020)
21. Hao, Y., Dong, L., Wei, F., Xu, K.: Visualizing and understanding the effectiveness of bert (2019). arXiv preprint arXiv:1908.05620
22. Haykin, S., Lippmann, R.: Neural networks, a comprehensive foundation. Int. J. Neural Syst. **5**(4), 363–364 (1994)
23. Hendry, G.D., Bromberger, N., Armstrong, S.: Constructive guidance and feedback for learning: the usefulness of exemplars, marking sheets and different types of feedback in a first year law subject. Ass. Evalu. High. Educ. **36**(1), 1–11 (2011)
24. Knight, S., Abel, S., Shibani, A., Goh, Y.K., Conijn, R., Gibson, A., Vajjala, S., Cotos, E., Sándor, Á., Shum, S.B.: Are you being rhetorical? a description of rhetorical move annotation tools and open corpus of sample machine-annotated rhetorical moves. J. Learn. Analy. **7**(3), 138–154 (2020)

25. Knight, S., Shibani, A., Abel, S., Gibson, A., Ryan, P., Sutton, N., Wight, R., Lucas, C., Sandor, A., Kitto, K., et al.: Acawriter: a learning analytics tool for formative feedback on academic writing. J. Writing Res. **12**(1), 141–186 (2020)
26. Knight, S., Shum, S.B., Ryan, P., Sándor, Á., Wang, X.: Designing academic writing analytics for civil law student self-assessment. Int. J. AIED **28**(1), 1–28 (2018)
27. Kovanović, V., et al.: Towards automated content analysis of discussion transcripts: A cognitive presence case. In: LAK, pp. 15–24 (2016)
28. Levine, J.M.: Legal writing as a discipline: Past, present, and future. ABA Legal Writing Sourcebook, 3rd ed., Duquesne University School of Law Research Paper (2020)
29. Lucas, C., Shum, S.B., Liu, M., Bebawy, M.: Implementing acawriter as a novel strategy to support pharmacy students' reflective practice in scientific research. In: JPE (2021)
30. Neto, V., et al.: Automated analysis of cognitive presence in online discussions written in portuguese. In: Pammer-Schindler, V., Pérez-Sanagustín, M., Drachsler, H., Elferink, R., Scheffel, M. (eds.) EC-TEL 2018. LNCS, vol. 11082, pp. 245–261. Springer, Cham (2018). https://doi.org/10.1007/978-3-319-98572-5_19
31. Neumann Jr., R.K., Margolis, E., Stanchi, K.M.: Legal reasoning and legal writing. Aspen Publishers (2021)
32. Nguyen, T.-S., Nguyen, L.-M., Tojo, S., Satoh, K., Shimazu, A.: Recurrent neural network-based models for recognizing requisite and effectuation parts in legal texts. Arti. Intell. Law **26**(2), 169–199 (2018). https://doi.org/10.1007/s10506-018-9225-1
33. Pardo, A., Jovanovic, J., Dawson, S., Gašević, D., Mirriahi, N.: Using learning analytics to scale the provision of personalised feedback. Br. J. Edu. Technol. **50**(1), 128–138 (2019)
34. Pennebaker, J.W., Francis, M.E., Booth, R.J.: Linguistic inquiry and word count: Liwc 2001, vol. 71. Lawrence Erlbaum Associates, Mahway (2001)
35. Ren, H., Li, Y.: A comparison study on the rhetorical moves of abstracts in published research articles and master's foreign-language theses. Engl. Lang. Teach. **4**(1), 162–166 (2011)
36. Robson, R.: Law students as legal scholars: an essay/review of scholarly writing for law students and academic legal writing. NY City L. Rev. **7**, 195 (2004)
37. Roscoe, R.D., McNamara, D.S.: Writing pal: Feasibility of an intelligent writing strategy tutor in the high school classroom. J. Educ. Psychol. **105**(4), 1010 (2013)
38. Sándor, Á.: Modeling metadiscourse conveying the author's rhetorical strategy in biomedical research abstracts. Revue française de linguistique appliquée **12**(2), 97–108 (2007)
39. Sha, L., et al.: Which hammer should i use? a systematic evaluation of approaches for classifying educational forum posts. In: EDM (2021)
40. Shibani, A., Knight, S., Shum, S.B.: Contextualizable learning analytics design: a generic model and writing analytics evaluations. In: LAK, pp. 210–219 (2019)
41. Silveira, R., Fernandes, C., Neto, J.A.M., Furtado, V., Pimentel Filho, J.E.: Topic modelling of legal documents via legal-bert. In: Proceedings CEUR 1613 (2021)
42. Speiser, J.L., Miller, M.E., Tooze, J., Ip, E.: A comparison of random forest variable selection methods for classification prediction modeling. Expert Syst. Appl. **134**, 93–101 (2019)
43. Strobl, C., Ailhaud, E., Benetos, K., Devitt, A., Kruse, O., Proske, A., Rapp, C.: Digital support for academic writing: a review of technologies and pedagogies. Comput. Educ. **131**, 33–48 (2019)

44. Swales, J.M.: Aspects of article introductions. No. 1. University of Michigan Press (2011)
45. Tessuto, G.: Investigating English legal genres in academic and professional contexts. Cambridge Scholars Publishing (2013)
46. Tessuto, G.: Generic structure and rhetorical moves in English-language empirical law research articles: sites of interdisciplinary and interdiscursive cross-over. Engl. Specif. Purp. **37**, 13–26 (2015)
47. Vihikan, W.O., Mistica, M., Levy, I., Christie, A., Baldwin, T.: Automatic resolution of domain name disputes. In: NLP Workshop 2021, pp. 228–238 (2021)
48. Wenestam, A.: Labelling factual information in legal cases using fine-tuned bert models (2021)
49. Xing, W., Tang, H., Pei, B.: Beyond positive and negative emotions: looking into the role of achievement emotions in discussion forums of MOOCS. Internet High. Educ. **43**, 100690 (2019)

Fully Automated Short Answer Scoring of the Trial Tests for Common Entrance Examinations for Japanese University

Haruki Oka[1(✉)], Hung Tuan Nguyen[2], Cuong Tuan Nguyen[2], Masaki Nakagawa[2], and Tsunenori Ishioka[3]

[1] Recruit Co., Ltd., Tokyo, Japan
haruki_oka@r.recruit.co.jp
[2] Tokyo University of Agriculture and Technology, Tokyo, Japan
nakagawa@cc.tuat.ac.jp
[3] The National Center for University Entrance Examinations, Tokyo, Japan
tunenori@rd.dnc.ac.jp

Abstract. Studies on automated short-answer scoring (SAS) have been conducted to apply natural language processing to education. Short-answer scoring is a task to grade the responses from linguistic information. Most answer sheets for short-answer questions are handwritten in an actual educational setting, which is a barrier to SAS. Therefore, we have developed a system that uses handwritten character recognition and natural language processing for fully automated scoring of handwritten responses to short-answer questions. This is the most extensive scoring data for responses to short-answer questions, and it may be the largest in the world. Applying the Cohen's kappa coefficient to the graded evaluations, the results show 0.86 in the worst case, and approximately 0.95 is recorded for the remaining five question answers. We observe that the fully automated scoring system proposed in our study can also score with a high degree of accuracy comparable to that of human scoring.

Keywords: Short answer scoring · Natural language processing · Handwritten character recognition

1 Introduction

Considering the current educational field, descriptive questions are often introduced to properly evaluate the abilities developed in linguistics. Moreover, to improve the scoring process's efficiency and stability, the effective use of computers and artificial intelligence has recently been increasing. There are approximately two types of descriptive questions: "essays without a correct answer" and "short-answer questions with correct answers." Many systems have been developed and have been practicalized for essays, especially in the United States.

H. Oka—Work done while at The University of Tokyo.

M. M. Rodrigo et al. (Eds.): AIED 2022, LNCS 13355, pp. 180–192, 2022.
https://doi.org/10.1007/978-3-031-11644-5_15

Some of the systems include the e-rater [2], IntelliMetric [19], intelligent essay assessors (IEA) [7], and CRASE [13]. Although the importance of short-answer questions has been recognized, various technical issues remain unsolved, such as semantic incomprehension.

On the other hand, short-answer questions are often used in several cases. Because short-answer questions are widely regarded as more orthodox, authentic, and reliable than the traditional multiple-choice tests [6], they have the potential to be used if the technical challenges for scoring are overcome. Automated short-answer scoring (SAS) techniques for English language have undergone technical improvements. Since the proposal of SAS that uses deep learning, its (SAS) performance has improved [1,5,17]. Particularly, SAS was devised using a massive transformer-based language model [8,12,21,22]. The demand of SAS is immeasurable and is not limited to new tests. Therefore, recent studies on SAS for practical purposes in Japan use data from actual mock examinations [8,15].

However, these studies have two unresolved problems. First, SAS requires additional manual work. It takes time and effort to convert handwritten data into electronic media because most of the descriptive answers in the educational domain are handwritten. The conventional SAS method aims to reduce the effort involved in scoring and requires extra effort. Furthermore, annotations were added as a guide for scoring to ensure accuracy. Considering its practical use in education, SAS requires improvements to eliminate these efforts. We have produced a fully automated scoring system that reliably eliminates data processing (such as annotations) and converts handwritten responses into text data. Second, the data handled in actual educational settings were too few to be verified on a large scale. When considering the privacy viewpoint, the amount of data was limited, and the verification was limited to a small scale. We conducted an experiment using data from a nationwide test and clarified that we could guarantee high prediction accuracy, even with large-scale data from actual educational settings.

The contributions of our research are as follows:

- We have developed a fully automated scoring system for handwritten responses, making it possible to grade many handwritten responses with high accuracy cost effectively.
- Large-scale data collected from two trial tests of entrance examinations nationwide were used to verify the practicality of the method in education.

Section 2 describes the large dataset used for the trial test of Japanese common entrance examinations. Section 3 explains the handwriting recognition technology and the scoring model used. The recognition evaluation criteria were also added. Section 4 presents the evaluation results, Sect. 5 describes the ablation studies, and Sect. 6 concludes the paper.

2 Trial Test Dataset for University Common Entrance Examinations

2.1 Overview

We used the written answers in Japanese in the trial test for the university common entrance examination conducted in 2017 and 2018. These exams are for national and private Japanese universities and are jointly conducted by the National Center for University Entrance Examinations, an independent administrative organization in Japan.

Approximately 500,000 examinees nationwide take the exams annually. These exams are considered essential for admission to national universities. Moreover, many leading private universities base their admission on these exams. Japanese exam questions comprise only first-appearing questions and are conducted once annually. While, SAT and ACT use test items repeatedly, and carry out many times a year.

We used trial test data for university common entrance exams conducted in 2017 and 2018. The test questions were prepared in a manner similar to the production, and the quality of the test questions was rigorously examined. Regarding the trial test, items on the national language (i.e., Japanese), mathematics, geography, history, civics, science, and foreign languages (only in 2018) were included. Descriptive questions were used only in Japanese and mathematics. Approximately 38% of high schools in Japan participated in this trial test; nonetheless, candidates did not have to take all the subjects. We analyzed the national language, which was taken by approximately 60,000 people. This is an unprecedented number of short-answer data for analysis.

2.2 Short-Answer Questions

The national language test in the trial test consisted of five test sets, known as the item bundles. One of these questions was a short-answer question. The test set consisted of three test questions. In 2017, these three test questions needed to be answered within 50, 25, and 120 characters, respectively. In 2018, the answers were to be of 30, 40, and 120 characters. Two Japanese characters are roughly equivalent to one English word. Figure 1 demonstrates a short-answer question administered in 2018.

3 Method

3.1 Task Settings

We input the answers to a short-answer question converted into text data using the automated handwriting recognition, and we output the corresponding predicted score. Subsequently, we demonstrate that our scoring model can predict the scores correctly by comparing the manual scores based on the rubric or scoring criteria. Regarding all the questions, we applied a single scoring model.

Sentence

"...Consider a situation where you travel to a country
where you do not speak their language at all, and
you have to ask for something. Regarding this case,
pointing will work like magic..."

Question

What does the underlined phrase,
"pointing will work like magic" indicate?

Student answer

It indicates that you can communicate your intentions,
even if you do not speak the language.

Score: 3/3

Fig. 1. Example of a short-answer question conducted in 2018. It is originally written in Japanese and has been translated into English for reader's understanding.

Figure 2 shows the task flow. We evaluate the performance using the score, without modifying the character answer data and without adding any annotation to the answer. The part that should be correctly identified as "“や”" was identified as "“に”". The quality of the written letters was sometimes insufficient. This is because of stains that remained in the paper.

3.2 Handwriting Recognition

We employ the extracting, transforming, and loading (ETL) database, which has offline Japanese handwritten single characters. This database consists of nine datasets collected under different conditions [18]. Because the collected samples are written in separate boxes similar to the answer sheet of university entrance exams, the ETL database is appropriate for building an offline Japanese handwriting recognizer. This database covers the most common Japanese characters belonging to 2965 kanji (Japanese Industrial Standards : JIS Level 1) and 94 kana categories. Although there are more kanji categories, the characters of JIS Level 1 are mostly used daily and in examinations, whereas other kanji characters are rarely used.

Based on the success of the ensemble convolutional neural networks (CNNs) for Japanese historical character recognition [16], we also used an ensemble of multiple well-known CNN models. Our recognizer consists of a visual geometric group (VGG), MobileNet, residual network (ResNet), and ResNext networks with 16, 24, 34, and 50 layers, respectively [9,10,20,23].

To train these CNNs, we applied multiple transformations such as rotating, shearing, scaling, blurring, contrasting, and noise addition to avoid over-fitting problems because the database had only approximately one million samples in total. After training these CNNs using the ETL database, we fine-tuned them using 100 manually labeled samples from our collected Japanese handwritten answer database.

A trained neural network provides a prediction output as a k-dimensional vector of probabilities, where k is the number of categories for each character sample. These prediction outputs are averaged together with an equal weight of 1.0 to form an ensemble prediction output. Thus, the top-most prediction is the category with the highest probability in the ensemble prediction output. Figure 3 shows the procedure in which the CNN using 16, 24, and 50 layers is judged as "指 ."Here, the CNN using 34 layers is judged as "提 ," and finally it is judged correctly as "指 ."

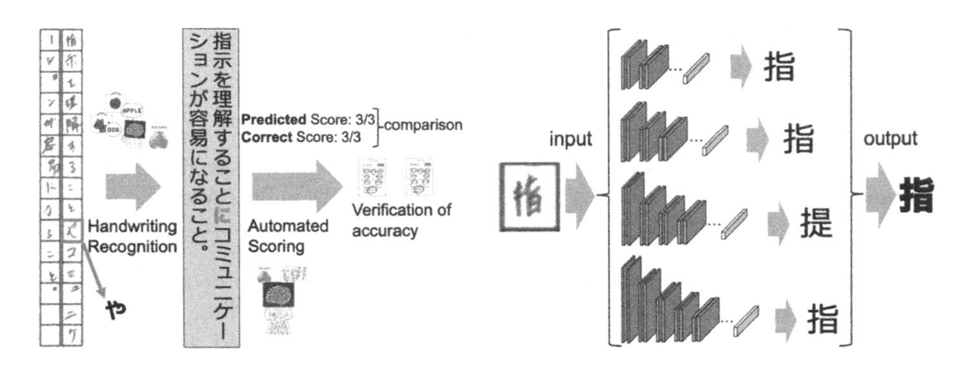

Fig. 2. Task flow

Fig. 3. Ensemble CNN handwriting recognition

Owing to the ambiguities of some characters, we also use an N-gram language model to correct misrecognized characters using the linguistic context. Considering every character of a text line, we computed the combined score based on the recognition and language scores of each character. First, the recognition score is the probability product of the previously recognized characters produced by the ensemble CNN recognizer. Second, the language score is the probability product of previous characters based on a five-gram Japanese language model that has been pre-trained by the Japanese Wikipedia corpus. Although N-grams are simple, they are sufficiently effective. Third, the combined score is a linear combination of the recognition and language scores with a weight of $\alpha \in [0, 1]$. Based on the combined score, we employ the beam search algorithm along the text line with a beam width of ten to export the top-ten candidates with the highest combined scores. However, only the highest combined score candidate was used for scoring in this experiment.

3.3 Scoring Procedure

The methods by [8] and [15], which perform the same type of SAS in Japanese, use an attention mechanism added to the bidirectional long-short term memory (Bi-LSTM). Their method outputs a predicted score based on each scoring criterion or rubric. However, our method does not accumulate scores for each scoring criterion. It predicts the overall score. We explicitly utilize a multi-label classification model by fine-tuning it with Bidirectional Encoder Representations from Transformers (BERT) [4], which is pre-trained on Japanese Wikipedia.[1] If we consider the operation in large-scale tests, the scoring model should be implemented more efficiently. Nevertheless, we must utilize a better language model that is as accurate as possible.

The procedure is as follows (Fig. 4):

1. $\boldsymbol{x} = \{x_1, x_2, \ldots, x_n\}$ is input as the written answer converted to text data using handwriting recognition, and the predicted score $s \in C = \{0, \ldots, N\}$ for the answer is provided as the output of the label.
2. The sentence $\boldsymbol{x}$ of the written answer is decomposed for each token, and a special token known as [CLS] is provided at the beginning of the sentence.
3. These token IDs are provided, and they are entered into the pre-trained BERT using the Japanese Wikipedia. Thereafter, we converted them into series of 768-dimensional vectors.
4. Whereas BERT is composed of all 12 layers, we concatenate the vectors of the [CLS] tokens of the last four layers of the hidden layer. Considering [4], combining them improved the document classification accuracy, compared to using only the [CLS] token vector in the final layer. Adam was used to optimize the model. The batch size was 16, and the number of epochs was five.
5. The vector of the combined classification [CLS] tokens is input into the classifier, and the predicted score s is output.

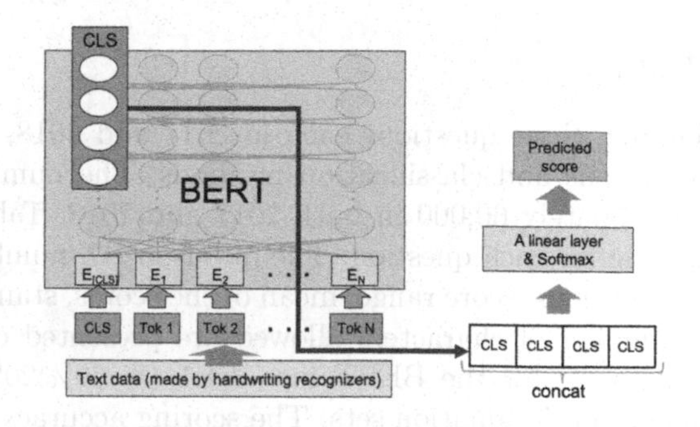

Fig. 4. Short-answer scoring model

3.4 Evaluation

The quadratic weighted kappa (QWK) [3] is often used as an evaluation index in SAS, and we used it in this study. The QWK is used for multilabel classification when an order relationship exists between labels. This index shows how well

[1] https://github.com/huggingface/transformers.

the correct and prediction labels match. The higher the value is, the better the prediction.

The QWK is calculated as:

$$\kappa = 1 - \frac{\sum_{i,j} W_{i,j} O_{i,j}}{\sum_{i,j} W_{i,j} E_{i,j}}, \tag{1}$$

where i and j represent the correct and predicted labels, respectively. $\boldsymbol{O}$ represents the ratio of each cell to the labels in the confusion matrix composed of the correct and predicted labels. $\boldsymbol{E}$ represents the expected value of the label belonging to each cell of the confusion matrix, assuming that the predicted and correct labels are independent.

$\boldsymbol{W}$ represents the penalty when the prediction is incorrect, and it is expressed as follows:

$$W_{i,j} = \frac{(i-j)^2}{(N-1)^2}, \tag{2}$$

where N represents the number of label classifications. $\boldsymbol{W}$ increases because the difference between the correct and predicted labels increases.

The QWK score is a ratio that can consider a value between −1 and 1. A negative QWK score indicates that the model is "worse than random." A random model should provide a score close to zero. Finally, the perfect predictions yielded a score of one. According to [14], Cohen suggested a kappa result of 0.81–1.00, which is interpreted as an approximately perfect agreement.

4 Experiments

4.1 Question Data

Six questions, including three questions each in 2017 and 2018, were classified based on these conditions and classification methods. The number of answers processed was approximately 60,000 in both 2017 and 2018. Table 1 shows the statistics of the scoring for each question. The question ID, number of answers, number of scoring conditions, score range, mean of the scores, standard deviation of the scores, and number of characters allowed are presented chronologically. We divided the data used for the BERT into 3:1:1 (= 60%:20%:20%) as the training, development, and evaluation sets. The scoring accuracy was evaluated using the QWK.

Table 1. Descriptive statistics on scoring for each question

Questions	# of answers	# of scoring conditions	Score range	Mean	# of characters allowed
2017 #Q1	62,222	4	0–6	4.46 ± 1.67	<50
2017 #Q2	61,777	3	0–2	1.51 ± 0.86	<25
2017 #Q3	59,791	4	0–5	0.43 ± 1.10	80–120
2018 #Q1	67,332	3	0–3	2.51 ± 0.88	<30
2018 #Q2	66,246	3	0–3	1.87 ± 1.14	<40
2018 #Q3	58,159	5	0–3	0.76 ± 1.07	80–120

4.2 Evaluation Results

Considering this experiment, the number of answer characters required at the university entrance level is relatively large, and the content is not plain. Regarding such cases, it is essential to know how large a sample is needed to guarantee the accuracy of the estimation.

Therefore, the sample size was changed to 50,000, 10,000, 5,000, 1,000, and 500, and the change in the QWK was observed. Table 2 shows the results, including the full-size data of approximately 60,000. The bold text indicates the best values.

Table 2. QWK for scoring each question

Questions	Sample size					
	Full size	50,000	10,000	5,000	1,000	500
2017 #Q1	0.978	**0.979**	0.967	0.946	0.883	0.679
2017 #Q2	**0.963**	0.949	0.934	0.922	0.818	0.884
2017 #Q3	**0.866**	0.836	0.705	0.680	0.473	0.276
2018 #Q1	**0.976**	0.968	0.974	0.914	0.863	0.820
2018 #Q2	**0.954**	0.945	0.923	0.903	0.796	0.724
2018 #Q3	**0.944**	0.929	0.916	0.894	0.783	0.753

The following can be obtained from the steps above.

1. We observe that the accuracy is kept high by the method for all six questions, regardless of the type of question. Even in the worst case of Q3 in 2017, the QWK is 0.86; otherwise, it is 0.94 or higher.
2. Essentially, the larger the sample size is, the better the accuracy. This indicates that the accuracy does not converge, which is an unexpected result. The sample size of 60,000 seems large enough in a typical test. Nevertheless, it shows that a more significant number is needed to improve the accuracy of the prediction.

This indicates in a sentence of a certain length, the variation in expressions is highly diverse. Because the number of characters increased, the number of variations increased exponentially, even if we have sufficient answer patterns that would not be significant. Therefore, the learning never converges.

3. The easier the question is, the higher the scoring rate, and the better the estimation accuracy. In both 2017 and 2018, Q1 was the easiest, and Q3 was the most difficult. The accuracy of Q1 was higher than that of Q3. This tendency did not depend on the number of scores.

5 Ablation Study

We observed the effect on scoring accuracy in our model from two perspectives. First, we considered the accuracy of handwriting recognition. We examined how the recognition rate affected the overall scoring accuracy. Second, we considered the position of the layer in the language-processing model. We changed the information position extracted from the 12 layers of the BERT model and verified how the change affected the overall scoring accuracy.

5.1 Effect of the Handwriting Recognition Models Used

To investigate the effect of the handwritten character recognition part on the scoring accuracy, we compare the original ensemble model of four methods with other methods. The compared methods are as follows:

1. No language model: This is a character recognition model without correction of misrecognized characters by the N-gram language models.
2. VGG only: This is a single character recognition model without ensemble learning.
3. DenseNet only: This is also a single character recognition model without ensemble learning.
4. Ensemble 5: This is a character recognition model with ensemble learning of five character recognition models.

Table 3 compared the QWK using each of the output results.

Table 3. Comparison of QWK by five methods

Questions	The handwriting recognition models				
	Original	No language model	VGG	DenseNet	Ensemble5
2017#Q1	0.978	0.975	0.977	0.974	**0.980**
2017#Q2	**0.963**	0.957	0.957	0.952	0.959
2017#Q3	**0.866**	0.847	0.844	0.820	0.830
2018#Q1	**0.976**	0.973	0.972	0.970	0.970
2018#Q2	**0.954**	0.950	0.952	0.953	0.953
2018#Q3	**0.944**	0.937	0.933	0.935	0.941

This shows that the model with ensemble learning with multiple character recognition models has a higher overall accuracy than the model with a single character recognition model. In addition, the results show that the accuracy of the models with the modification in the language model is higher than that of the models without modification. Moreover, increasing the number of ensemble learning models did not change significantly, considering the accuracy. Considering these results, we observed that the overall accuracy was affected by both the language model changes and character recognition model quality. Moreover, we found that the overall accuracy was limited by improving the quality of the character recognition model.

5.2 Effect of the Information Retrieved from the BERT Model

We investigated the effect of the different linguistic information retrieved from BERT on the scoring accuracy. The BERT used in our study consists of 12 layers, and each layer is known to contain different information [11]. Specifically, the layers close to the input, middle, and output parts possess morphological information, syntactic information, and information that focuses on the semantic information, respectively. We divided the BERT model into three parts: a layer near the input, a middle part, and a layer near the output. Thereafter, we examined the differences in the scoring accuracy between the three parts. Layers 1–4, 5–8, and 9–12 were extracted from the input section. The output from each layer was input into the linear layer, and the score was predicted. Table 4 lists the results for each accuracy. We observed that the scoring accuracy was the highest when the information of layers 9–12 was extracted for each problem.

Table 4. Comparison of QWK by the different extraction layers

Questions	The part of layers		
	1–4	5–8	9–12
2017#Q1	0.977	0.977	**0.978**
2017#Q2	0.952	0.955	**0.963**
2017#Q3	0.830	0.832	**0.866**
2018#Q1	0.969	0.972	**0.976**
2018#Q2	0.951	0.950	**0.954**
2018#Q3	0.936	0.939	**0.944**

This indicates that the system is paying particular attention to the semantic information when performing automatic scoring tasks. Particularly, QWK in 2017#Q3 was different by 3.0 or more among all the questions, and the difference was outstanding.

6 Summary and Conclusions

We have investigated a fully automated scoring method for short-answers using handwriting recognition data and have evaluated the system's performance using a large-scale national test. "Fully" indicates that there is no need to annotate the scoring data or convert the handwritten text manually. We used very large data conducted in two trial tests for university common entrance examinations and used a pre-trained BERT model for scoring. We made the following observations.

1. When the data is sufficiently large, our method increases the scoring accuracy without annotation and converts the handwritten text manually.
2. When we consider 25 to 120 character answers, learning often does not converge, even with a data size of 50,000.
3. Even if some errors are caused by handwriting recognition, the accuracy of scoring is guaranteed to some extent using the current technology.

This study reports the actual accuracy at the current technical level in a procedure without human intervention. Despite the variety in the types of questions we considered, such as the number of characters in the answer and the difficulty level, we could predict the scores with high accuracy in all cases. This suggests that our procedure is effective for all short-answer questions, and SAS is suitable for large-scale testing using the current technology. In addition, our study demonstrates the usefulness of the method for utilizing handwritten character recognition models in SAS. We can serve as an opportunity to develop a new learning method for educational application settings, where students often use handwriting.

Acknowledgements. This work was supported by JSPS KAKENHI Grant Number JP20H04300 and JST A-STEP Grant Number JPMJTM20ML.

References

1. Alikaniotis, D., Yannakoudakis, H., Rei, M.: Automatic text scoring using neural networks. In: Proceedings of the 54th Annual Meeting of the Association for Computational Linguistics (Volume 1: Long Papers), pp. 715–725. Association for Computational Linguistics, Berlin (2016). https://doi.org/10.18653/v1/P16-1068
2. Burstein, J., Tetreault, J., Madnani, N.: The e-rater automated essay scoring system. In: Shermis, M.D., Burstein, J. (eds.) Handbook of Automated Essay Evaluation, Chap. 4, pp. 55–67. Edwards Brothers Inc., New York (2013)
3. Cohen, J.: Weighted kappa: nominal scale agreement with provision for scaled disagreement or partial credit. Psychol. Bull. **7**(4), 213–220 (1968)
4. Devlin, J., Chang, M.W., Lee, K., Toutanova, K.: BERT: pre-training of deep bidirectional transformers for language understanding. In: Proceedings of the 2019 Conference of the North American Chapter of the Association for Computational Linguistics: Human Language Technologies, Volume 1 (Long and Short Papers), pp. 4171–4186. Association for Computational Linguistics, Minneapolis (2019). https://doi.org/10.18653/v1/N19-1423

5. Dong, F., Zhang, Y.: Automatic features for essay scoring - an empirical study. In: Proceedings of the 2016 Conference on Empirical Methods in Natural Language Processing, pp. 1072–1077. Association for Computational Linguistics, Austin (2016). https://doi.org/10.18653/v1/D16-1115
6. Drid, T.: The fundamentals of assessing EFL writing. Psychol. Educ. Stud. **11**(1), 292–305 (2018). https://doi.org/10.35156/1192-011-001-017 https://doi.org/10.35156/1192-011-001-017 https://doi.org/10.35156/1192-011-001-017 https://doi.org/10.35156/1192-011-001-017
7. Foltz, P.W., Streeter, L.A., Lochbaum, K.E., Landauer, T.K.: Implementation and applications of the intelligent essay assessor. In: Shermis, M.D., Burstein, J. (eds.) Handbook of Automated Essay Evaluation, Chap. 5, pp. 55–67. Edwards Brothers Inc., New York (2013)
8. Funayama, H., et al.: Preventing critical scoring errors in short answer scoring with confidence estimation. In: Proceedings of the 58th Annual Meeting of the Association for Computational Linguistics: Student Research Workshop, pp. 237–243. Association for Computational Linguistics (2020). https://doi.org/10.18653/v1/2020.acl-srw.32
9. He, K., Zhang, X., Ren, S., Sun, J.: Deep residual learning for image recognition. In: Proceedings of the IEEE Conference on Computer Vision and Pattern Recognition, pp. 770–778 (2016)
10. Howard, A.G., et al.: Mobilenets: efficient convolutional neural networks for mobile vision applications. arXiv preprint arXiv:1704.04861 (2017)
11. Jawahar, G., Sagot, B., Seddah, D.: What does BERT learn about the structure of language? In: Proceedings of the 57th Annual Meeting of the Association for Computational Linguistics, pp. 3651–3657. Association for Computational Linguistics, Florence (2019). https://doi.org/10.18653/v1/P19-1356
12. Li, Z., Tomar, Y., Passonneau, R.J.: A semantic feature-wise transformation relation network for automatic short answer grading. In: Proceedings of the 2021 Conference on Empirical Methods in Natural Language Processing, pp. 6030–6040. Association for Computational Linguistics, Online and Punta Cana, Dominican Republic (2021)
13. Lottridge, S., Wood, S., Shaw, D.: The effectiveness of machine score-ability ratings in predicting automated scoring performance. Appl. Measur. Educ. **31**(3), 215–232 (2018)
14. McHugh, M.L.: Interrater reliability: the kappa statistic. Biochem. Med. (Zagreb) **22**(3), 276–282 (2012)
15. Mizumoto, T., et al.: Analytic score prediction and justification identification in automated short answer scoring. In: Proceedings of the Fourteenth Workshop on Innovative Use of NLP for Building Educational Applications, pp. 316–325 (2019)
16. Nguyen, H.T., Ly, N.T., Nguyen, K.C., Nguyen, C.T., Nakagawa, M.: Attempts to recognize anomalously deformed kana in Japanese historical documents. In: Proceedings of the 4th International Workshop on Historical Document Imaging and Processing, pp. 31–36. HIP2017, Association for Computing Machinery, New York (2017). https://doi.org/10.1145/3151509.3151514
17. Riordan, B., Horbach, A., Cahill, A., Zesch, T., Lee, C.M.: Investigating neural architectures for short answer scoring. In: Proceedings of the 12th Workshop on Innovative Use of NLP for Building Educational Applications, pp. 159–168. Association for Computational Linguistics, Copenhagen (2017). https://doi.org/10.18653/v1/W17-5017

18. Saito, T., Yamada, H., Yamamoto, K.: On the database ETL9 of handprinted characters in JIS Chinese characters and its analysis. Trans. IECE Jpn. **J68-D**(4), 757–764 (1985)
19. Schultz, M.T.: The intellimetric automated essay scoring engine - a review and an application to chinese essay scoring. In: Shermis, M.D., Burstein, J. (eds.) Handbook of Automated Essay Evaluation, Chap. 6, pp. 55–67. Edwards Brothers Inc, New York (2013)
20. Simonyan, K., Zisserman, A.: Very deep convolutional networks for large-scale image recognition. arXiv preprint arXiv:1409.1556 (2014)
21. Sung, C., Dhamecha, T.I., Mukhi, N.: Improving short answer grading using transformer-based pre-training. In: Isotani, S., Millán, E., Ogan, A., Hastings, P., McLaren, B., Luckin, R. (eds.) AIED 2019. LNCS (LNAI), vol. 11625, pp. 469–481. Springer, Cham (2019). https://doi.org/10.1007/978-3-030-23204-7_39
22. Uto, M., Okano, M.: Robust neural automated essay scoring using item response theory. In: Bittencourt, I.I., Cukurova, M., Muldner, K., Luckin, R., Millán, E. (eds.) AIED 2020. LNCS (LNAI), vol. 12163, pp. 549–561. Springer, Cham (2020). https://doi.org/10.1007/978-3-030-52237-7_44
23. Xie, S., Girshick, R., Dollár, P., Tu, Z., He, K.: Aggregated residual transformations for deep neural networks. In: Proceedings of the IEEE Conference on Computer Vision and Pattern Recognition, pp. 1492–1500 (2017)

Machine Learning Techniques to Evaluate Lesson Objectives

Pei Hua Cher(✉), Jason Wen Yau Lee, and Fernando Bello

Duke-NUS Medical School, Singapore, Singapore
{peihua.cher,jason.lee,f.bello}@duke-nus.edu.sg

Abstract. The advancement of knowledge in medicine presents an important challenge when identifying gaps and deciding what content to include in a medical school curriculum and how to establish learning outcomes. Monitoring alignment between lesson objectives, the curriculum and achievement of intended outcomes can be difficult. A system that can automatically evaluate lesson objectives would be highly beneficial. We aim to assess the efficacy of using machine learning techniques to evaluate individual lesson objectives to a graduate entry allopathic medical school curriculum. The school's curriculum objectives consist of 11 categories and 356 curriculum objectives sentences. We considered the first year courses with a total of 1888 lesson objectives. Using various word embeddings (TF-IDF, word2vec, fastText, BioBERT), we then use cosine similarity to map each lesson objective to the curriculum objectives. Cognitive levels of lesson objectives were compared against the school's curriculum using Bloom's Taxonomy verbs. After implementation, 319 lesson objectives from each approach were randomly sampled (sample size, 95% CL, 5% CI) to examine match with curriculum objectives and curriculum categories. BioBERT performed best with 46.71% and 80.56% match between lesson objectives and curriculum objectives, and lesson objectives and categories, respectively. Further validation by a domain expert shows 80% match (without order). Visualisation of the Bloom's Taxonomy cognitive levels of lesson objectives and school's curriculum objectives showed a good match. Machine learning can be used to evaluate lesson and curriculum and automatically mapping lesson objectives to the medical school curriculum and analysing cognitive levels of lesson objectives.

Keywords: Natural language processing · Evaluate curriculum mapping · Evaluate cognitive learning outcome

1 Introduction

A school's curriculum acts like a blueprint for all the lessons. With the ever-increasing amount of content and frequency of updates in the medical school, keeping track of what was taught and checking if it is in-line with the school's curriculum has become an immense and immediate challenge. Over the years, lessons delivered might diverge from the school's main curriculum. Therefore,

M. M. Rodrigo et al. (Eds.): AIED 2022, LNCS 13355, pp. 193–205, 2022.
https://doi.org/10.1007/978-3-031-11644-5_16

there is a need for the administrators to be aware of outdated or latest updates in the lessons, to quickly and more frequently process all lesson objectives and ensure it is in-line with the school's curriculum. The aim of this paper is to assess the efficacy of using machine learning techniques to automatically evaluate and map lesson objectives with the school's curriculum objectives. This is so that we can quickly identify gaps, misalignment and bring to the users' attention for further action.

According to the constructive alignment framework [1], every lesson starts with clearly written lesson outcomes that communicates what is expected from the students by the lesson. Using a feedback loop, the lesson outcomes, teaching approaches and assessments are constantly evaluated so that they are aligned to the outcomes. In this paper, we will refer to learning outcomes as learning objectives as it is the terminology that is used in our institution. We acknowledge that there is a difference and debate within the field and would focus the discussion on the use of machine learning techniques for curriculum mapping.

Machine learning techniques in natural language processing (NLP) are automation techniques used to extract, represent and process semantics of natural texts. They are used in sentiment analysis, question and answering machine, etc. Although to human it is easy to understand a sentence, it is a very difficult task for the machine to understand that a word "stop" can have different meaning in "bus stop" and "please stop". NLP has improved tremendously into developing more natural language understanding (NLU) techniques. NLU adopts transformers that provides general-purpose architectures with pre-train models to help in representing words in multidimensional sentence or word embeddings. With a good sentence or word representation we can potentially capture the semantic context or meaning of each sentence or word then generate a mapping between the school's curriculum objectives and lesson objectives.

Another way to evaluate the lesson against the curriculum is to compare the cognitive levels based on well-established knowledge framework such as the Bloom's Taxonomy [8]. Bloom's Taxonomy is a hierarchical model that classifies learning into six domains of different complexity and specificity. Each learning objective begins with an action verb that would classify the different level of learning expectation by the student. If we can automate the classification of the objective sentences, we can visualize and compare the differences.

In this paper, we evaluate the lesson objectives against the curriculum objectives by exploring different pre-trained word representation techniques and mapping them, as well as comparing their cognitive levels. In the next section, we will expound on the related work in this area. Section 3 will explain in detail our methodology applying different word representation techniques and a cognitive classification model using a knowledge framework. The results will be presented in Sect. 4. We will then discuss, provide recommendations and state the limitations of this project in Sect. 5 and conclude in Sect. 6.

The main contributions in this paper are:

- automated extraction of information from both the school's curriculum objectives and the first-year lesson objectives
- mapping between the first-year lesson objectives and the school's curriculum objectives
- categorization and comparison of the Bloom's taxonomy cognitive domain categories used in the school's curriculum objectives and the first-year lesson objectives
- consolidation and gap analysis between what was planned at the school level and what was taught at the lesson level

2 Related Works

In this section we will provide a background on the related literatures in curriculum mapping, word representation techniques and knowledge framework using Bloom's taxonomy. With this background, we will highlight the gaps that we are attempting to bridge in the literature.

2.1 Curriculum Mapping

To align lesson objectives with school curriculum objectives, we looked at curriculum mapping literature. According to some authors [2,7], there are limited studies in the mapping of a medical school curriculum. Due to the lack of digitization and understanding of the bigger picture in the school's curriculum, Chan [2] mentioned it is difficult to introduce new content or review current taught content. This had previously led Komenda [7] to create a framework to enhance curriculum innovations and mapping using a network graph approach, but it lacked in reporting the performance of their curriculum mapping algorithm. In this work, we attempt to create curriculum mapping models that maps the content, as well as comparing cognitive levels using state-of-the-art natural language processing techniques, and evaluating the performance of the model.

2.2 Sentence Representation

Before mapping the objective statements, it is necessary to accurately represent the semantics of the sentence in the computer. For a computer to analyse a sentence, words in the sentence need to be represented as a number or vector. Komenda [7] proposed both a framework and a curriculum mapping model using term-frequency inverse document frequency (TF-IDF) to convert sentences into features to train their model. By using TF-IDF, the features do not take into account the context of a word in the sentence, but simply count the occurrence of each word without considering the order or the meaning of the words in a sentence. We aim to explore and compare other methods to represent sentences like Word2vec [10,12,13], fastText [5,11] and Bidirectional Encoder Representations from Transformers (BERT) [4].

Mikolov [10,12,13] started a major breakthrough in word representation using Word2vec. This algorithm uses a neural network model to learn the n-dimensional vector of each word in a large corpus of text using word associations. The team by Joulin and Mikolov [5,11] continued to improve the word representation and created fastText.

BERT [4] word embedding is the state-of-the-art pre-trained language representation model. The model is trained using a general-purpose "language understanding" model on a large text corpus (like Wikipedia). BERT captures the context of each word in a sentence as it is trained bidirectionally (uses both left and right of the sentence). Each word embedding depends on all other words in the sentence. There are several other versions of BERT that are trained on different domains. BioBERT [9] is trained on biomedical literature in PubMed and will be more relevant for representing our medical school texts.

2.3 Knowledge Framework - Bloom's Taxonomy Cognitive Domain Levels

In addition to the automated extraction and mapping, we also wanted to explore classifying lesson objectives and school's curriculum objectives into different cognitive levels. Bloom's taxonomy is often used to measure cognitive levels in lesson objectives [8]. Several papers have created classification models to automatically classify lesson objectives into levels [3,14]. None of these papers have looked at comparing curriculum objectives with lesson objectives.

Recognising the importance of aligning lesson objectives to curriculum objectives [1], and the challenges due to the vast amount of text to manually read and analyse, we explored ways to automate this process. Currently, there is a lack of literature in the area of medical curriculum mapping with empirical approach. In order to map curricula, we will look at various state-of-the-art pre-trained word representation models to quantify the curriculum mapping. Lastly, to the best of our knowledge, no work has automatically classified and compared cognitive levels of medical lesson objectives against the school's curriculum objectives.

3 Methods

Figure 1 provides a summary of our approach. In this paper, the constructive alignment framework's "Intended Learning Objectives" and "Teaching & Learning Activities" will be viewed as the school's curriculum objectives and lesson objectives respectively. We evaluate the lesson objectives and school's curriculum objectives in two different ways. Firstly, we use an automation to map the lesson objectives to the school's curriculum using an unsupervised transfer learning model to generate sentence embeddings. We then derive the similarity between the sentences using cosine similarity of the embeddings. Secondly, we use Bloom's taxonomy cognitive domain to categorize all the lesson objectives and all the school's curriculum objectives and compare their cognitive levels.

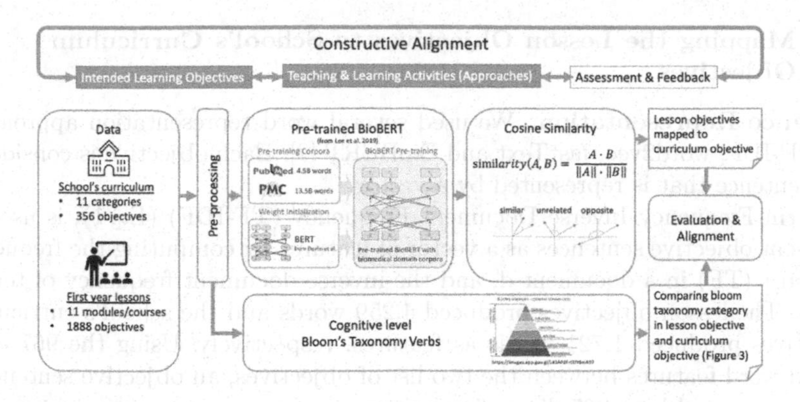

Fig. 1. Summary of curriculum mapping approach. Lesson objectives and the school's curriculum are pre-processed prior to being mapped and classified using BioBERT and Cosine Similarity. They are also mapped to Bloom's Taxonomy levels and evaluated.

3.1 Data

School's Curriculum Objectives. We obtained the school's curriculum objectives and first-year lesson objectives from the school's administrator. The school's curriculum objectives consist of 11 categories as shown in Table 1a. Categories are a collection of related lesson objectives. The total number of objective sentences are 356.

Lesson Objectives. The first-year lessons consist of 11 courses (Table 1b). There are total of 1888 lesson objectives with 1687 unique objective sentences.

Each objective sentence is pre-processed by removing punctuations, numbers, Unicode characters, null entry, convert to lowercase, remove unnecessary space, carriage return, tab and stop-words (e.g. common words like "a", "the", etc., that does not add information to the text).

Table 1. Listing of (a) categories in school's objectives and (b) first-year courses.

(a) Categories in school's objectives

Category	Category name
a	SCIENTIFIC FOUNDATIONS
b	BASIC CLINICAL SKILLS
c	PREVENTION, (HEALTH PROMOTION AND OCCUPATIONAL HEALTH)
d	DIAGNOSIS
e	TREATMENT - ACUTE AND CHRONIC
f	PATIENT SAFETY AND QUALITY IMPROVEMENT
g	INFORMATION MANAGEMENT
h	ETHICS, HUMANITIES, AND THE LAW
i	PROFESSIONALISM
j	LEADERSHIP
k	SCHOLARSHIP AND LIFELONG LEARNING

(b) First-year courses

ID	Courses
1	CARE
2	Transition 1
3	Fundamentals of Clinical Practice
4	Molecules, Cells, and Tissues
5	Human Structure and Function
6	Brain and Behaviour
7	Body and Disease
8	Fundamentals of Research and Scholarship
9	Innovation and Design Thinking
10	Scholarly Development Programme
11	Research Methods and Analysis

3.2 Mapping the Lesson Objectives to School's Curriculum Objectives

Sentence Representation. We used several word representation approaches like TF-IDF, word2vec, fastText and BioBERT [9]. Each objective is considered as a sentence that is represented by a vector.

Term Frequency Inverse Document Frequency (TF-IDF) (Eq. 1), is used to represent objective sentences as a vector of features by computing the frequency of term t (TF) in a document d, and the inverse document frequency of term t (IDF). The lesson objectives produced 4,259 words and the school's curriculum objectives produced 1,721 words as features, respectively. Using the 967 overlapped word features between the two list of objectives, an objective sentence is thus represented by a 967 dimension vector.

$$TDIDF(t,d) = TF(t,d) * IDF(t) \tag{1}$$

Word2vec [10,12,13] is a pretrained word representation that was trained on Google News (about 100 billion words). The model contains vectors for 3 million words and phrases, each represented by a 300 dimension vector. Sentence vectors are obtained by summing up the vector for each word in a sentence.

fastText [5,11] is another pretrained word representation model that was trained on Wikipedia 2017, University of Maryland, Baltimore County (UMBC) webbase corpus, and statmt.org news dataset (16 billion tokens). It contains one million word vectors with each word represented by a 300 dimension vector. Similarly, each objective statement is represented by a sentence vector that is a summation of the vector of each word in a sentence.

BERT [4] sentence embedding is a state-of-the-art pre-trained language representation model. BERT is trained on non-medical related words. As our objective sentences contain medical terms not found in common literature, we chose the BioBERT-Large v1.1 (+ PubMed 1M) [9], which is pre-trained on biomedical literature in PubMed. Each objective is treated like a sentence and represented by a 768-dimensional sentence vector generated by BioBERT using the **sentence_vector** function of the **pytorch_pretrained_bert** python package.

Distance Measures - Cosine Similarity. Once vector representations are obtained for both lesson objectives and school's curriculum objectives, it is possible to compare them through suitable distance measures, such as cosine similarity. Cosine similarity was chosen as it considered the normalised vector orientation and magnitude when computing two vectors. Its value will be closer to 1 for the most similar sentences. Equation 2 shows the calculation of similarity score between two vectors A and B.

$$similarity(A,B) = (A * B)/(\| A \| * \| B \|) \tag{2}$$

From the above vector representations, we mapped each first-year lesson objective to the most similar school curriculum objectives sentences using cosine similarity. We evaluate the mapping by randomly sample 319 lesson objectives (sample size, 95% confidence level, 5% confidence interval) in 1687 unique sentences in a total number of 1888 lesson objectives. We labeled 319 unique lesson objectives that mapped to curriculum objectives.

We further evaluated the results by enlisting a domain expert, who is an experienced faculty member in the MD Program to provide expert input on the mapping. The domain expert was shown 20 lesson objectives and each lesson objectives' top 10 mapped curriculum objectives. The top 10 mapped curriculum objectives are shuffled to reduce bias.

3.3 Bloom's Taxonomy Cognitive Domain

We used a keyword match rule-based classification model to classify the sentences into Bloom's taxonomy cognitive levels (Algorithm 1.). The action verbs (Table 2) are based on the six cognitive domain levels in the revised version of Bloom's Taxonomy [8]. The pre-processed objective sentences are joined into one corpus and action verbs are keyword matched and counted for each level. Action verbs found in stop-words are removed from stop-words.

Algorithm 1. Classifying Bloom's taxonomy cognitive levels using action verbs

1. For each lesson objective
2. For each Blooms taxonomy category
3. For each action verb
4. If verb found in lesson objective
5. Lesson objective flagged in Blooms category

Table 2. Action verbs.

Remembering	Understanding	Applying	Analysing	Evaluating	Creating
Define	Ask	Administer	Analyse	Appraise	Adapt
Describe	Associate	Apply	Break	Argue	Arrange
Enumerate	Classify	Calculate	Down	Assess	Assemble
Identify	Compare	Chart	Classify	Choose	Combine
Label	Contrast	Choose	Compare	Compare	Compile
List	Convert	Collect	Conclude	Consider	Compose
Match	Describe	Compute	Connect	Convince	Construct
Name	Differentiate	Construct	Contrast	Critique	Create
Observe	Discuss	Determine	Correlate	Debate	Design
Read	Distinguish	Discover	Criticize	Defend	Develop
Recall	Estimate	Employ	Deduce	Distinguish	Devise
Recite	Explain	Examine	Devise	Evaluate	Formulate
Recognize	Identify	Explain	Differentiate	Judge	Generalize
Select	Illustrate	Illustrate	Discriminate	Justify	Hypothesize
State	Indicate	Interpret	Distinguish	Persuade	Integrate
Tell	Infer	Interview	Experiment	Rate	Invent
	Interpret	Manipulate	Explain	Recommend	Justify
	Paraphrase	Modify	Infer	Select	Produce
	Relate	Relate	Plan		Propose
	Summarize	Report	Question		Reorganize
	Transform	Simulate	Select		Organize
	Translate	Solve	Survey		Rearrange
		Teach			Report
		Use			Simulate
					Solve
					Speculate

4 Results

4.1 Mapping Lesson Objectives to School's Curriculum Objectives

Through the above methods, we mapped each first-year lesson objective to a list of top 10 similar school curriculum objectives sentences, and a list of most similar school curriculum objective sentences.

The results of the word representation algorithms, namely TF-IDF, word2vec, fastText and BioBERT, are shown in Table 3. A comparison between the ground truth resulting from the manual labelling of human labelers, with the automated mapping using cosine similarity is shown in Table 3.

Table 3. Evaluation of lesson objectives mapped to most similar curriculum objective.

	Lesson objectives (n = 319, 95% CL, 5% CI)			
	BioBERT	TF-IDF	Word2vec	fastText
Curriculum categories (%)	**80.56**	35.42	69.28	52.04
Curriculum objectives (%)	**46.71**	25.39	45.45	20.69

The evaluation by the expert shows 80% of the sentences matched without considering the order, whereas 45% matched with the most similar objectives and 20% were found to have no match between lesson objectives and school curriculum objectives. Table 4 shows an example of expert mapping of a lesson objective to top 10 mapped curriculum objectives. The corresponding cosine similarity scores are shown in Table 4 but these were not made known to the expert.

4.2 Comparing Bloom's Taxonomy Distribution

Using the action verbs (Table 2) in each objective, we plotted the frequency of each Bloom's Taxonomy cognitive levels (Fig. 2). The school's curriculum objective and lesson objectives consists of mostly understanding verbs (200, 1000 respectively) which is followed by remember verbs (100, 800 respectively). Both objective statements have least creating verbs (less than 50).

5 Discussions

In this paper, we have demonstrated the use of advanced machine learning technique to map lesson objective sentences with school's curriculum objective sentences. Best percentage match for the approach was 80.56% and 46.71% for curriculum categories and curriculum objectives, respectively. We have also obtained and compared the cognitive levels of both lesson and curriculum objectives using Bloom's Taxonomy's cognitive domain framework.

Table 4. An example of expert mapping of a lesson objective to top 10 mapped curriculum objectives. The top 10 mapped curriculum objectives are randomly ranked. The expert was not shown the cosine similarity scores. Top 10 are based on cosine similarity score.

Lesson objective	School categories	School objectives	1 - best match blank - otherwise	Cosine similarity score
Be able to define the basic steps of the transcription cycle	SCIENTIFIC FOUNDATIONS	Describe how errors in cell division (meiosis, mitosis) can result in human disease		0.8989
	SCIENTIFIC FOUNDATIONS	Describe protein folding and its relation to protein structure and function		0.8954
	SCIENTIFIC FOUNDATIONS	Describe the biochemical and genetic basis of heritable diseases, recognizing normal process/structure and contrasting with the disease state		0.8943
	SCIENTIFIC FOUNDATIONS	Describe the structure and function of membrane ion channels, along with their roles in specific diseases		0.9045
	SCIENTIFIC FOUNDATIONS	Describe the structure and function of membrane receptors, along with pathways of signal transduction	1	0.9180
	SCIENTIFIC FOUNDATIONS	Describe the structure of cytoplasmic filaments, tubules, and motor proteins, along with their roles in cellular function and response to drug interventions		0.8951
	SCIENTIFIC FOUNDATIONS	Describe the structure of membrane and membranous organelles, along with their role in cellular function and response to drug interventions		0.9026
	SCIENTIFIC FOUNDATIONS	Describe the structure, synthesis, control, and importance of key molecules in basic cellular processes		0.9118
	SCIENTIFIC FOUNDATIONS	Discuss the processes of transcription and translation and their relation to human disease		0.9088
	SCIENTIFIC FOUNDATIONS	Discuss the role of chromosome instability and oncogenes in the development of cancer		0.8935

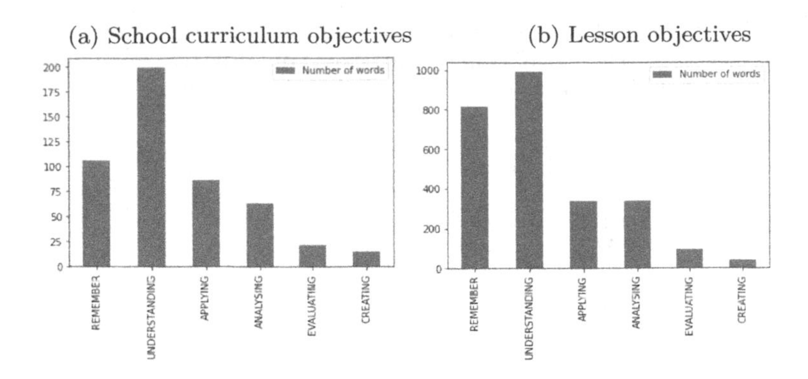

Fig. 2. Comparing frequency of using Bloom's taxonomy category in (a) school curriculum objectives and (b) lesson objectives.

5.1 Evaluating Through Curriculum Mapping

Using the current school's curriculum objectives and lesson objectives, we represent each sentence applying different word representation approaches (TF-IDF, word2vec, fastText, BioBERT) and find the closest matching sentence by calculating their cosine similarity. In Table 3, BioBERT is the best performing model followed by Word2vec, fastText and TF-IDF. One possible reason why BioBERT performed best is that it was pre-trained with medical words found in PMC and PubMed. Word2vec is the next best performing pre-trained word representation with mapped curriculum categories and objectives scoring 69.28% and 45.45%, respectively. Word2vec is trained on Google News with 3 billion words. However, words in the news might not contain medical terms, which may explain why the model did not perform as well. Likewise, fastText (52.04% category mapped, 20.69% objectives mapped) is trained on Wikipedia, UMBC webbase corpus and statmt.org news dataset, which recognises one million words, but there could be limited medical terms in this dataset. The reason why TF-IDF did not perform as well (35.42% category mapped, 25.39% objectives mapped) could be due to the limited number of words (967 words), since only words that appear in both corpus are included.

The map between the lesson and curriculum categories are derived from the mapping between lesson and curriculum objectives with BioBERT showing the best performing mapping algorithm at 80.56% (Table 3). This shows that, when we map at the sentence level, we should get a map at a higher level such as the categories.

Using the current school's curriculum objectives and lesson objectives, we represent each sentence with BioBERT [9] and use cosine similarity to find the closest matching sentence. We have automated the mapping of the two objectives and the percentage match shows that the algorithm matches well with curriculum categories. After further evaluation of the objectives mapping by a domain expert, 80% match without order was achieved. This shows that we can effectively use this algorithm to return the top 10 curriculum objectives for the users to choose. This reduces the cognitive load of individuals who are trying to map their lesson objectives to the curriculum objectives. Over time, this solution may potentially assist in the identification of gaps in the curriculum or lesson objectives. The effectiveness of the approach can be seen by removing one keyword from the school curriculum objective, with the cosine similarity score dropping from 0.79 to 0.69 (Fig. 3).

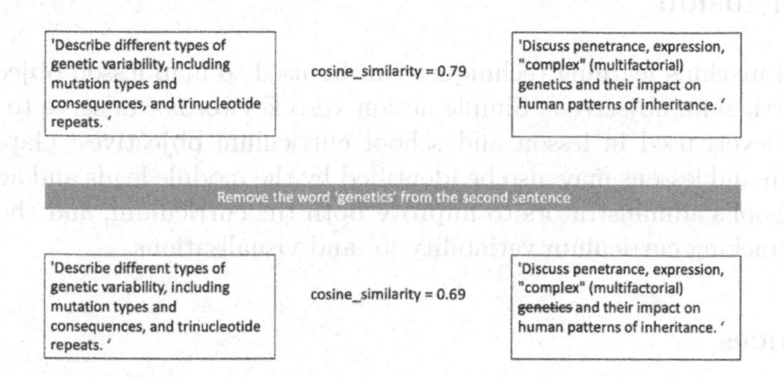

Fig. 3. The change in cosine similarity between two sentences when an important word is removed.

5.2 Evaluating Through Cognitive Levels in Bloom's Cognitive Domain

Comparing the distribution between the two objectives (Fig. 2), we can observe that they are similar, and that both use a lower level of the Bloom's taxonomy's cognitive domain in remembering and understanding, then applying. For first-year lesson objectives, it is expected that students will undergo basic scientific training. However, there may be a need to review the school curriculum objectives to raise the level of cognitive domains used in later years. It would be interesting to examine the actual execution of the lessons from student feedback to see if the content delivered matches the distribution of both objectives.

5.3 Recommendations and Limitations

There are several limitations to the keyword matching of action verbs as it might be double counted. Especially if there is a higher level of action verbs used in an objective, the higher level should be considered, as it might be implicit that the lower level is already fulfilled.

We assumed that lesson and curriculum objectives are well-written and that the lesson objectives are delivered according to what was written, which may not be the case. Further work needs to be done to investigate what was taught. This could be done through student feedback, student assessment and lesson objectives to complete the loop in the constructive alignment model.

For future experiments, word representation performance may be improved by including the school's category and goals in the word representation to enhance the mapping. Furthermore, in our pre-processed words, we have removed punctuations, stop-words and converted to lower case all sentences. We could rerun the experiments to explore if retaining these features affects the performance of the mapping.

6 Conclusion

Advanced machine learning techniques can be used to map lesson objectives to school curriculum objectives. Simple action verb keywords can serve to identify cognitive levels used in lesson and school curriculum objectives. Gaps in the curriculum and lessons may also be identified by the module leads and addressed by the school's administrators to improve both the curriculum, and the lessons through tracking curriculum variability [6] and visualisations.

References

1. Biggs, J.: Constructive alignment in university teaching. HERDSA Rev. High. Educ. **1**, 5–22 (2014)
2. Chan, K.S., Zary, N.: Applications and challenges of implementing artificial intelligence in medical education: integrative review. JMIR Med. Educ. **5**(1), e13930 (2019)
3. Das, S., Das Mandal, S.K., Basu, A.: Cognitive complexity analysis of learning-related texts: a case study on school textbooks. In: Vittorini, P., Di Mascio, T., Tarantino, L., Temperini, M., Gennari, R., De la Prieta, F. (eds.) MIS4TEL 2020. AISC, vol. 1241, pp. 74–84. Springer, Cham (2020). https://doi.org/10.1007/978-3-030-52538-5_9
4. Devlin, J., Chang, M.W., Lee, K., Google, K.T., Language, A.I.: BERT: pre-training of deep bidirectional transformers for language understanding. In: Proceedings of NAACL-HLT 2019. pp. 4171–4186 (2019)
5. Joulin, A., Grave, É., Bojanowski, P., Mikolov, T.: Bag of Tricks for Efficient Text Classification (2017)
6. Khan, R.A., Spruijt, A., Mahboob, U., van Merrienboer, J.J.: Determining 'curriculum viability' through standards and inhibitors of curriculum quality: a scoping review. BMC Med. Educ. **19**(1), 336 (2019)
7. Komenda, M., et al.: Curriculum mapping with academic analytics in medical and healthcare education. PLoS ONE **10**(12) (2015). https://doi.org/10.1371/journal.pone.0143748
8. Krathwohl, D.R.: A revision of bloom's taxonomy: an overview. Theory into Pract. **41**(4), 212–218 (2002)
9. Lee, J., et al.: BioBERT: a pre-trained biomedical language representation model for biomedical text mining. Bioinformatics **36**(4), 1234–1240 (2019)
10. Mikolov, T., Chen, K., Corrado, G., Dean, J.: Efficient estimation of word representations in vector space. In: Proceedings of Workshop at ICLR (2013)
11. Mikolov, T., Grave, É., Bojanowski, P., Puhrsch, C., Joulin, A.: Advances in Pre-Training Distributed Word Representations (2018)

12. Mikolov, T., Sutskever, I., Chen, K., Corrado, G.S., Dean, J.: Distributed Representations of Words and Phrases and their Compositionality. Adv. Neural Inf. Process. Syst. **26** (2013)
13. Mikolov, T., Yih, W.t., Zweig, G.: Linguistic Regularities in Continuous Space Word Representations (2013)
14. Omar, N., et al.: Automated analysis of exam questions according to Bloom's taxonomy. Proc. Soc. Behav. Sci. **59**, 297–303 (2012)

Towards Generating Counterfactual Examples as Automatic Short Answer Feedback

Anna Filighera(✉), Joel Tschesche, Tim Steuer, Thomas Tregel, and Lisa Wernet

Multimedia Communications Lab, Technical University of Darmstadt, Darmstadt, Germany
anna.filighera@kom.tu-darmstadt.de
https://www.kom.tu-darmstadt.de

Abstract. Receiving response-specific, individual improvement suggestions is one of the most helpful forms of feedback for students, especially for short answer questions. However, it is also expensive to construct manually. For this reason, we investigate to which extent counterfactual explanation methods can be used to generate feedback from short answer grading models automatically. Given an incorrect student response, counterfactual models suggest small modifications that would have led the response to being graded as correct. Successful modifications can then be displayed to the learner as improvement suggestions formulated in their own words. As not every response can be corrected with only minor modifications, we investigate the percentage of correctable answers in the automatic short answer grading datasets SciEntsBank, Beetle and SAF. In total, we compare three counterfactual explanation models and a paraphrasing approach. On all datasets, roughly a quarter of incorrect responses can be modified to be classified as correct by an automatic grading model without straying too far from the initial response. However, an expert reevaluation of the modified responses shows that nearly all of them remain incorrect, only fooling the grading model into thinking them correct. While one of the counterfactual generation approaches improved student responses at least partially, the results highlight the general weakness of neural networks to adversarial examples. Thus, we recommend further research with more reliable grading models, for example, by including external knowledge sources or training adversarially.

Keywords: Explainable AI · Short answer grading · Feedback

1 Introduction

Feedback is essential for learning as it helps uncover misconceptions, knowledge gaps and avenues for improvement [25]. However, providing feedback is

This research is funded by the Bundesministerium für Bildung und Forschung in the project: Software Campus 2.0 (ZN 01 S17050), Microproject: DA-VBB.

M. M. Rodrigo et al. (Eds.): AIED 2022, LNCS 13355, pp. 206–217, 2022.
https://doi.org/10.1007/978-3-031-11644-5_17

Table 1. Example student answer, common feedback and generated counterfactual.

Question:	What happens to the volume of the sound if you pluck a rubber band harder?
Reference:	The volume increases. The sound is louder
Response:	It vibrates more and it gets lower. → *Incorrect*
Counterfactual:	It vibrates more and it makes louder sound. → *Correct*

expensive for constructed response questions where each unique answer has to be considered carefully. Nevertheless, since constructed response questions are better suited to measuring complex skills compared to multiple-choice questions [16], the compromise is often to provide only verification feedback and a reference solution. Generally, verifying responses is much faster than formulating individual improvement suggestions. An example of verification feedback including a reference solution can be found in Table 1. It stems from the SciEntsBank [4] short answer grading dataset.

However, it can be hard to deduce one's mistakes from comparing with a reference solution. Depending on the reference's level of detail and exhaustiveness, a learner's response may not be covered by the solution or key differences may be drowned out by too many details. As there are often multiple correct solutions to short answer questions, learners may also have difficulties comprehending the particular solution provided by the teacher, especially when the teacher uses different terminology [31]. Thus, improvement suggestions in each learner's own words would likely be more helpful for learners [25].

Thus, this work proposes automatically generating counterfactual explanations as response-specific improvement suggestions. Inspired by human counterfactual reasoning [1], counterfactual explanation techniques essentially answer the question "What if the model's input would have looked like this instead?". The goal is to find small changes to the input features that would have changed the initially predicted output to the desired outcome [30]. For instance, given a learner response classified as incorrect by an automatic short answer grading (ASAG) model, what small changes to the learner's response would have led to the answer being predicted as correct? An example can be seen in Table 1.

However, not every learner's response lends itself to counterfactual feedback. Some answers may be so far from correct, such as "I don't know", that only massive changes would flip the predicted label to "correct". Other responses may be close to unreliable decision boundaries and lead to adversarial examples [6] that are only predicted as correct but do not actually improve the response. For this reason, this work addresses the following research question:

***RQ**: To which extent can we generate automatic feedback with counterfactual explanations?*

To this end, we make the following contributions:

- We show that counterfactual generation methods can modify student answers to be automatically graded as correct by comparing three counterfactual

generation models and a paraphrasing model on benchmark automatic short answer grading datasets (Sect. 3.4).
- Having an expert reevaluate a subset of the modified responses shows that almost all generated counterfactuals are adversarial examples instead of genuine improvements (Sect. 3.5).

2 Generating Counterfactual Feedback

The main idea of this work is to generate counterfactuals of incorrect student responses and explore their use as feedback. We develop and apply four approaches for this purpose. First are two approaches we developed based on Minimal Contrastive Editing (MiCE) [23]. They aim to iteratively replace the most impactful tokens in a response until it is graded as correct. Next is Polyjuice [32], a framework trained to perform pre-specified modifications, such as negating or shuffling entities in a sentence. Lastly, we develop an approach based on paraphrasing that generates novel responses instead of replacing parts of the original answer.

2.1 Contrastive Infilling

The main idea behind contrastive infilling approaches is finding the input parts detrimental to predicting the target class based on the model's gradients and replacing them with an editor model. MiCE [23] does this in two steps. In the first step, an editor model is trained to reproduce original data inputs. For this purpose, the most impactful tokens for the predicting model are masked so that the editor can learn to fill in critical sections of a response. The editor also receives the input's label to learn to produce responses of a specific class. In the second step, the editor iteratively fills in masked responses to find minimal modifications that cause the predictor to output a target label [23].

Inspired by MiCE, we implement two infilling models, one utilizing target labels and one without labels. The main idea behind cutting the labels used in MiCE is to simplify the task by only correcting wrong responses. Adding the target label does not carry any information in that case; it will always be the class "*correct*". However, one loses the ability to produce partially correct counterfactuals. Cutting the label requires a modification to the editor training proposed. While the label model is trained to reproduce all student answers by infilling masked parts of the student answers similar to MiCE, the other model is only trained to reproduce correct student responses. For both models, we randomly mask 20–55% of the student answer, and both models receive the reference answer in addition to the masked response. The label model is additionally conditioned on the target label in the following format: "label: *target label.* input: *masked student answer* </s> *reference answer*". Since the other model is only trained on correct responses and does not need a label, instead the input is formatted as follows: "input: *masked student answer* </s> *reference answer*".

In the second step, we use the previously fine-tuned models to modify incorrect student answers and perform up to four modification rounds. First, consecutive spans of tokens in the original student answer are masked based on importance scores provided by the gradient attribution method Integrated Gradients [27]. We create four masked versions in each round with 15, 30, 45, and 60% of the tokens masked. We generate seven candidates for each masked student answer using a combination of top-k = 30 and top-p = 0.95 sampling. At the end of each modification round, the candidates are graded using an ASAG model, keeping only the candidate with the highest target class probability. The modification process is terminated when the candidate's target class probability exceeds the classification threshold or the maximum number of rounds is reached.

2.2 Polyjuice

In contrast to the previous approach, Polyjuice [32] aims to control the modification process through control codes. Instead of masking the tokens with the highest impact and generating arbitrary replacements, Polyjuice uses a predefined set of possible modifications, such as *negating* the meaning of the input or *shuffling* key phrases or entities around. The type of modification also controls where modifications can be made in the input so that the generated counterfactual should be fluent. Since the modification process is more constrained and, thus, may not be applicable to all student answers, we expect this method to yield less counterfactuals overall compared to the other approaches. However, any counterfactuals found should be more natural. We utilize Wu et al.'s [32] implementation[1] of Polyjuice to generate counterfactuals for incorrect student responses allowing all predefined modification codes: *negation, quantifier, shuffle, lexical, resemantic, insert, delete* and *restructure.*

2.3 Paraphrasing

Finally, we trained a T5 [22] model to paraphrase correct responses. In contrast to the counterfactual methods described above, this model does not fill in masked parts of the student response but generates a novel response instead. The main idea behind this approach was to explore whether a model trained to generate various correct responses to a question could also correct incorrect student answers. For this purpose, we treat correct student responses and reference answers as paraphrases of each other. While this is likely not accurate in practice as reference answers tend to be more comprehensive than student answers, the idea is for the model to learn the characteristics of correct answers. During training, it receives either the student or reference answer and generates the respective counterpart. After training, it gets incorrect student responses as input instead.

3 Experiments

The goal of our experiments is to determine to which extent feedback can be generated with counterfactual explanation methods. For this purpose, we first

[1] https://github.com/tongshuangwu/polyjuice.

introduce the datasets and the ASAG model whose grading predictions will be explained by the counterfactual approaches. Then we introduce the metrics used to compare the approaches automatically, followed by insights gained from having a domain expert manually reevaluate the generated counterfactuals.

3.1 Datasets

We select three diverse ASAG datasets for our experiments: SciEntsBank [4], Beetle [4] and the English half of the Short Answer Feedback dataset (SAF) [7]. All three datasets offer a 3-way classification task with *correct* and *incorrect* responses. While the third class for SAF is *partially_correct*, the other datasets include *contradictory* as final class. The datasets offer multiple test sets, aimed at different grading scenarios. The unseen answer test split measures how well models perform on new answers to questions they were trained for, while the unseen questions split contains completely novel questions. Since SciEntsBank, in contrast to the others, contains multiple science domains, it also includes an unseen domain test split. Beetle, on the other hand, only contains basic electrical engineering questions and SAF is a computer science dataset in the *communication network* field. While SAF and Beetle consist of undergraduate responses, SciEntsBank's responses stem from American students in the grades 3 to 6. In contrast to the other datasets, Beetle includes multiple reference answers per question. In our experiments, we consider all reference answers.

3.2 Automatic Short Answer Grading Models

For each dataset, we train a BERT model that receives a student and reference answer as input and predicts the response's correctness. These three models form the predictors for the counterfactual search and, thus, should be as reliable as possible. For this reason, we follow the fine-tuning procedure used by Sung et al. [28] and achieve the predictive performance depicted in Table 2.

Table 2. Accuracy (Acc), macro-averaged F1 (M-F1) and weighted F1 (W-F1) of the automatic short answer grading models in percent.

Dataset	Unseen answers			Unseen questions			Unseen domains		
	Acc	M-F1	W-F1	Acc	M-F1	W-F1	Acc	M-F1	W-F1
SAF	77.1	75.5	77.1	52.9	57.5	52.9	–	–	–
Beetle	71.4	69.7	71.4	54.8	54.8	56.6	–	–	–
SciEnts.	72.9	70.9	72.9	59.7	50.9	59.7	61.5	54.6	61.5

3.3 Evaluation Measures and Experimental Setup

This paper focuses on two dimensions of counterfactuals that influence feedback quality: **validity** and **proximity**. Counterfactuals are often considered valid when they lead to the desired prediction [30]. Thus, validity is usually measured by calculating the percentage of counterfactuals that flipped the predicted label to the desired outcome irrespective of the class predicted priorly. While that works well for tasks where the predictors achieve nearly perfect accuracy, it would overestimate the generators' performance in our case, as the ASAG model already misclassifies some of the incorrect student responses as *correct* without any modification. For this reason, we exclude all answers already predicted as *correct* from the evaluation. Furthermore, we hypothesize that counterfactual feedback will work better for student responses that are closer to being correct in the first place, such as *partially_correct* responses in contrast to *incorrect* ones. Therefore, we calculate the **flip rate** for each class separately.

Additionally, generated responses should be as close to the original student answer as possible to ensure that only required changes are made and the response follows the learner's wording beyond that. Following related work [23], we also use the word-level Levenshtein **distance** to measure the counterfactual's proximity to the original answer. It provides the minimum number of deletions, insertions and substitutions needed to equalize two strings. The count is then divided by the number of words in the original response to normalize it. As long as the generated response is not longer than the original response, it can be seen as the percentage of words modified.

All models introduced in Sect. 2 are trained on two Nvidia GX 2080 Ti cards with 11 GB of RAM using gradient accumulation and mixed-precision floating-point numbers. The exact hyperparameters used for each approach can be found in our implementation.[2]

3.4 Comparison Results

Table 3 compares the counterfactuals generated by the Polyjuice, paraphrasing and contrastive infilling approaches introduced in Sect. 2 on the SAF dataset. It can be seen that the paraphrasing model succeeds in flipping the most labels to *correct* with flip rates between 50% and 100%. However, it also generates counterfactuals that differ vastly from the original student answer with an average distance of 2.22 across test splits and classes. Polyjuice is the opposite, generating counterfactuals that are very close to the original with an average distance of 0.02 but seldom flip the label to *correct*. The contrastive infilling methods seem to be more balanced, with an average flip rate of 24.2% without labels (infill) and 21.9% with labels (label infill) and average distances of 0.15 and 0.13, respectively. They also show the expected behaviour of flipping more partially correct responses than completely incorrect ones. While the paraphrasing model actually generates more flips on incorrect student answers compared to partially correct ones, they seem to be even more distant from the original responses.

Table 4 and Table 5 show the same comparison on the BEETLE and SCIENTSBANK datasets. The infilling approaches perform slightly better on SCIENTSBANK compared to SAF, flipping on average 28.2% of the predictions without

[2] https://github.com/joeltsch/CASAF-AIED2022.

Table 3. Flip rate (FR) and average distance (Dist) for counterfactuals generated on SAF's partially correct (Partial) and incorrect responses. Sample sizes are in brackets.

Approach	Unseen answers				Unseen questions			
	Partial (52)		Incorrect (9)		Partial (31)		Incorrect (8)	
	FR	Dist	FR	Dist	FR	Dist	FR	Dist
Paraphrase	**50.0**	1.72	**77.8**	3.89	**96.8**	1.60	**100**	1.66
Infill	25.0	0.19	11.1	0.11	35.5	0.12	25.0	0.19
Label infill	19.2	0.14	11.1	0.10	32.3	0.12	25.0	0.16
Polyjuice	0.0	**0.01**	11.1	**0.01**	3.2	**0.03**	0.0	**0.01**

using labels and 28.8% utilizing labels, with a comparable average distance of 0.15 for both approaches. On BEETLE, the infilling approaches flip considerably more predictions on average - at the cost of the much higher average edit distances. The labelless approach has an average flip rate of 55.9% and an average distance of 2.37. The approach with labels flips 41.0% of the predictions on average with an edit distance of 0.37. The paraphrasing model shows a similar behaviour of high flip rates and high edit distances on all datasets, with distances between 8 and 14 on BEETLE. Additionally, Polyjuice produces few counterfactuals on all datasets but has higher average edit distances on BEETLE with 0.14 and SCIENTSBANK with 0.12 compared to SAF.

Table 4. Flip rate (FR) and average distance (Dist) for counterfactuals generated on BEETLE's contradictory (Contra) and incorrect responses. Sample sizes are in brackets.

Approach	Unseen answers				Unseen questions			
	Contra (453)		Incorrect (480)		Contra (740)		Incorrect (830)	
	FR	Dist	FR	Dist	FR	Dist	FR	Dist
Paraphrase	**74.2**	8.56	**78.3**	10.63	**76.9**	8.04	**74.7**	13.27
Infill	60.9	2.77	63.3	2.18	46.5	2.38	52.8	2.14
Label infill	44.8	0.42	41.9	0.39	39.1	0.34	38.0	0.33
Polyjuice	1.8	**0.11**	2.1	**0.14**	1.8	**0.12**	3.3	**0.17**

3.5 Expert Regrading

While the flip rate indicates how many modifications lead to successful counterfactuals, it only considers the predictor's judgement and not whether the predictor was fooled into an incorrect prediction. For this reason, we asked one of the communication network experts involved in the original data annotation to reevaluate the generated counterfactuals for the SAF dataset. We selected SAF because it is the only dataset that includes elaborated feedback explaining why the response was graded as incorrect. This dramatically simplifies the

Table 5. Flip rate (FR) and average distance (Dist) for counterfactuals generated on SciEntsBank's contradictory (Contra) and incorrect responses. Sample sizes are in brackets.

Approach	Unseen answers				Unseen questions			
	Contra (48)		Incorrect (202)		Contra (35)		Incorrect (238)	
	FR	Dist	FR	Dist	FR	Dist	FR	Dist
Paraphrase	**72.9**	1.54	**74.8**	1.82	**65.7**	1.69	**68.5**	1.65
Infill	31.2	0.16	33.2	0.17	17.1	0.12	31.1	0.15
Label infill	29.2	0.15	31.7	0.18	20.0	**0.11**	34.5	0.17
Polyjuice	2.1	**0.14**	1.0	**0.12**	5.7	0.12	2.5	**0.11**

reevaluation since the expert only has to determine whether the modification corrects the mistake instead of regrading the responses from scratch.

The expert evaluated all counterfactuals the ASAG model predicted accurately prior to modification and as *correct* after modification. There were 59 examples for the paraphrasing model, 1 for Polyjuice, 21 for the label infilling approach and 25 for infilling without labels. In total, 106 examples were regraded.

Nearly all generated samples (N = 103) were adversarial examples and not genuine corrections of the response. Of the 3 correct examples, 2 stem from the paraphrasing model simply generating the reference answer to the question instead of modifying the student answer. In general, the paraphrases were often vastly different from the student responses, which matches the observations from Sect. 3.4. Sometimes the paraphrasing model would also mix reference solutions to multiple questions, which may be one of the reasons why it is so successful at fooling the predictor. Humorously, some of the content added to the response by the paraphrasing model was utterly absurd, such as "*... 56.648 * 64 bit/s = 128 bit processing tables = 276 bit data transfer tables + 3 * 1.31 s to reach the destination system ...*".

The infilling models also mostly produced adversarial examples with senseless modifications. For example, "*... the issue with this* ***case*** *is ...*" was replaced with "*... the issue with this* ***narcotic*** *is ...*" which does not make any sense in the communication network domain. Sometimes the model would also replace words with special tokens, such as "<extra_id_34>". However, not all modifications made by the infilling models were adversarial. Some modifications truly improved student responses partially, even if they were still incorrect overall. For example, "*extension headers are* ***the way to put additional information in the packet****...*" was correctly replaced with "*extension headers are* ***used to extend the fixed ipv6 header with additional options****...*".

4 Related Work

In recent years, the need for understandable machine learning models has given rise to diverse approaches aiming to explain the inner workings of neural networks. Such explanations can be used to increase the transparency and trustworthiness of automatic predictions [24]. The branch of explainable AI most relevant to our work is based on counterfactual reasoning, revolving around how an input's features would have to differ as to change a model's prediction. As there are countless counterfactual explanation techniques and we already describe the most relevant ones in Sect. 2, we recommend one of the excellent surveys summarizing the state-of-the-art [2,12,26,30] for further reading and focus on related work on generating elaborated feedback.

4.1 Elaborated Feedback Generation

Especially in the intelligent tutoring community, generating elaborated feedback has been a hot topic for many years [3,8,14,20,29]. Older approaches mainly focused on hand-crafting domain models and manually tailoring feedback systems to specific tasks [5,10,17]. More recently, research is exploring more flexible feedback systems for structured answer formats, such as programming exercises [13], proofs [18], or multiple-choice questions [15,33]. Here the structure of the response is exploited to automatically identify the kind of mistakes made, for example, by using a compiler. The most similar to our work here is an approach proposed by Olney [21]. They automatically generate elaborated feedback for cloze-style questions by first generating a question about the relationship between the correct cloze solution and the incorrect term provided by the student. The answer to the synthetic question provided by an automatic question answering system is then included as elaborated feedback.

For unstructured question formats, like essays and short answer questions, flexible feedback systems mainly focus on a response's language and style [11], identifying justifications [19] or discovering which topics are covered in an essay [9]. Only recently, a deep learning system to automatically generate elaborated feedback for short answer questions was introduced [7]. However, it relies on feedback data which is still unavailable for most domains.

5 Conclusion and Future Work

In summary, this work compared four approaches to providing counterfactual feedback to short answer questions. Three out of the four methods successfully generated counterfactuals for at least a fifth of the incorrect responses in three diverse short answer grading datasets. Around a quarter of incorrect responses could be modified until the automatic grading model judged them correct without diverging too far from the initial student response. However, a domain expert still deemed nearly all modified responses incorrect. This result illustrates the need for human evaluation of generated counterfactuals. In related work, counterfactuals are mainly evaluated using flip rates and automatic proximity measures [12]. However, considering the high rate of adversarial examples observed in this study, automatic metrics are not sufficient to capture the true usefulness of generated counterfactuals. Thus, future work should include human judgements.

Regarding the research question posed in this work, we conclude that counterfactual explanations are unsuitable as feedback at the current state. However, they can be even more helpful for teachers aiming to employ automatic grading models in practice. The generated counterfactuals could be used to identify critical weaknesses of the grading model. For example, the humorous example from Sect. 3.5 may indicate an inability to evaluate mathematical expressions correctly. Generated counterfactuals could also be added to the training data to facilitate adversarial training of more robust grading models. More robust grading models may, in turn, produce better counterfactuals. Since we observed genuine partial improvements in student responses in our experiments, incentivizing the counterfactual model to search beyond adversarial modifications seems like a promising avenue of future research.

Finally, the counterfactual generation methods themselves could be improved. We showed that counterfactual generators vary greatly in the number of label flips they entice and how dissimilar the modifications are to the original. Thus, other approaches may yield more or better counterfactuals. Especially approaches utilizing external knowledge sources and other neuro-symbolic methods may be beneficial for the short answer feedback task. The additional knowledge could inform the search for sensible modifications or help identify which parts of a student's response are incorrect and, thus, should be replaced.

References

1. Buchsbaum, D., Bridgers, S., Skolnick Weisberg, D., Gopnik, A.: The power of possibility: causal learning, counterfactual reasoning, and pretend play. Philos. Trans. R. Soc. B Biol. Sci. **367**(1599), 2202–2212 (2012). https://doi.org/10.1098/rstb.2012.0122
2. Chou, Y.L., Moreira, C., Bruza, P., Ouyang, C., Jorge, J.: Counterfactuals and causability in explainable artificial intelligence: theory, algorithms, and applications. Inf. Fus. **81**, 59–83 (2022)
3. Deeva, G., Bogdanova, D., Serral, E., Snoeck, M., De Weerdt, J.: A review of automated feedback systems for learners: classification framework, challenges and opportunities. Comput. Educ. **162**, 104094 (2021)
4. Dzikovska, M., et al.: SemEval-2013 task 7: the joint student response analysis and 8th recognizing textual entailment challenge. In: 2nd Joint Conference on Lexical and Computational Semantics (*SEM), Volume 2: Proceedings of the 7th International Workshop on Semantic Evaluation, SemEval 2013, Atlanta, Georgia, USA, pp. 263–274. Association for Computational Linguistics (June 2013). https://aclanthology.org/S13-2045
5. Dzikovska, M., Steinhauser, N., Farrow, E., Moore, J., Campbell, G.: BEETLE II: deep natural language understanding and automatic feedback generation for intelligent tutoring in basic electricity and electronics. Int. J. Artif. Intell. Educ. **24**(3), 284–332 (2014). https://doi.org/10.1007/s40593-014-0017-9
6. Filighera, A., Ochs, S., Steuer, T., Tregel, T.: Cheating automatic short answer grading: on the adversarial usage of adjectives and adverbs (2022). https://doi.org/10.48550/ARXIV.2201.08318

7. Filighera, A., Parihar, S., Steuer, T., Meuser, T., Ochs, S.: Your answer is incorrect... would you like to know why? Introducing a bilingual short answer feedback dataset. In: Proceedings of the 60th Annual Meeting of the Association for Computational Linguistics (Volume 1: Long Papers), Dublin, Ireland, pp. 8577–8591. Association for Computational Linguistics (May 2022)
8. Hasan, M.A., Noor, N.F.M., Rahman, S.S.B.A., Rahman, M.M.: The transition from intelligent to affective tutoring system: a review and open issues. IEEE Access **8**, 204612–204638 (2020). https://doi.org/10.1109/ACCESS.2020.3036990
9. Hellman, S., et al.: Multiple instance learning for content feedback localization without annotation. In: Proceedings of the 15th Workshop on Innovative Use of NLP for Building Educational Applications, Seattle, WA, USA, pp. 30–40. Association for Computational Linguistics (July 2020)
10. Jordan, S., Mitchell, T.: e-assessment for learning? The potential of short-answer free-text questions with tailored feedback. Br. J. Edu. Technol. **40**(2), 371–385 (2009)
11. Ke, Z., Ng, V.: Automated essay scoring: a survey of the state of the art. In: Proceedings of the 28th International Joint Conference on Artificial Intelligence, IJCAI-19, pp. 6300–6308. International Joint Conferences on Artificial Intelligence Organization (July 2019). https://doi.org/10.24963/ijcai.2019/879
12. Keane, M.T., Kenny, E.M., Delaney, E., Smyth, B.: If only we had better counterfactual explanations: five key deficits to rectify in the evaluation of counterfactual XAI techniques. In: Zhou, Z.H. (ed.) Proceedings of the 30th International Joint Conference on Artificial Intelligence, IJCAI-21, pp. 4466–4474. International Joint Conferences on Artificial Intelligence Organization (August 2021)
13. Keuning, H., Jeuring, J., Heeren, B.: A systematic literature review of automated feedback generation for programming exercises. ACM Trans. Comput. Educ. (TOCE) **19**(1), 1–43 (2018). https://doi.org/10.1145/3231711
14. Kulik, J.A., Fletcher, J.: Effectiveness of intelligent tutoring systems: a meta-analytic review. Rev. Educ. Res. **86**(1), 42–78 (2016). https://doi.org/10.3102/0034654315581420
15. Ling, W., Yogatama, D., Dyer, C., Blunsom, P.: Program induction by rationale generation: learning to solve and explain algebraic word problems. In: Proceedings of the 55th Annual Meeting of the Association for Computational Linguistics (Volume 1: Long Papers), Vancouver, Canada, pp. 158–167. Association for Computational Linguistics (July 2017). https://doi.org/10.18653/v1/P17-1015
16. Livingston, S.A.: Constructed-response test questions: why we use them; how we score them. R&D Connections, vol. 11 (September 2009)
17. Lu, X., Di Eugenio, B., Ohlsson, S., Fossati, D.: Simple but effective feedback generation to tutor abstract problem solving. In: Proceedings of the 5th International Natural Language Generation Conference, Salt Fork, Ohio, USA, pp. 104–112. Association for Computational Linguistics (June 2008)
18. Makatchev, M., Jordan, P.W., VanLehn, K.: Abductive theorem proving for analyzing student explanations to guide feedback in intelligent tutoring systems. J. Autom. Reason. **32**(3), 187–226 (2004)
19. Mizumoto, T., et al.: Analytic score prediction and justification identification in automated short answer scoring. In: Proceedings of the 14th Workshop on Innovative Use of NLP for Building Educational Applications, Florence, Italy, pp. 316–325. Association for Computational Linguistics (August 2019). https://doi.org/10.18653/v1/W19-4433

20. Mousavinasab, E., Zarifsanaiey, N., Kalhori, S.R.N., Rakhshan, M., Keikha, L., Saeedi, M.G.: Intelligent tutoring systems: a systematic review of characteristics, applications, and evaluation methods. Interact. Learn. Environ. **29**(1), 142–163 (2021). https://doi.org/10.1080/10494820.2018.1558257
21. Olney, A.M.: Generating response-specific elaborated feedback using long-form neural question answering. In: Proceedings of the 8th ACM Conference on Learning @ Scale, L@S 2021, New York, NY, USA, pp. 27–36. Association for Computing Machinery (2021). https://doi.org/10.1145/3430895.3460131
22. Raffel, C., et al.: Exploring the limits of transfer learning with a unified text-to-text transformer. J. Mach. Learn. Res. **21**(140), 1–67 (2020)
23. Ross, A., Marasović, A., Peters, M.: Explaining NLP models via minimal contrastive editing (MiCE). In: Findings of the Association for Computational Linguistics, ACL-IJCNLP 2021, pp. 3840–3852. Association for Computational Linguistics (August 2021). https://doi.org/10.18653/v1/2021.findings-acl.336
24. Shin, D.: The effects of explainability and causability on perception, trust, and acceptance: implications for explainable AI. Int. J. Hum Comput Stud. **146**, 102551 (2021). https://doi.org/10.1016/j.ijhcs.2020.102551
25. Shute, V.J.: Focus on formative feedback. Rev. Educ. Res. **78**(1), 153–189 (2008). https://doi.org/10.3102/0034654307313795
26. Stepin, I., Alonso, J.M., Catala, A., Pereira-Fariña, M.: A survey of contrastive and counterfactual explanation generation methods for explainable artificial intelligence. IEEE Access **9**, 11974–12001 (2021)
27. Sundararajan, M., Taly, A., Yan, Q.: Axiomatic attribution for deep networks. In: International Conference on Machine Learning, pp. 3319–3328. PMLR (2017)
28. Sung, C., Dhamecha, T.I., Mukhi, N.: Improving short answer grading using transformer-based pre-training. In: Isotani, S., Millán, E., Ogan, A., Hastings, P., McLaren, B., Luckin, R. (eds.) AIED 2019. LNCS (LNAI), vol. 11625, pp. 469–481. Springer, Cham (2019). https://doi.org/10.1007/978-3-030-23204-7_39
29. VanLehn, K.: The relative effectiveness of human tutoring, intelligent tutoring systems, and other tutoring systems. Educ. Psychol. **46**(4), 197–221 (2011)
30. Verma, S., Dickerson, J., Hines, K.: Counterfactual explanations for machine learning: a review. arXiv preprint arXiv:2010.10596 (2020)
31. Winstone, N.E., Nash, R.A., Parker, M., Rowntree, J.: Supporting learners' agentic engagement with feedback: a systematic review and a taxonomy of recipience processes. Educ. Psychol. **52**(1), 17–37 (2017)
32. Wu, T., Ribeiro, M.T., Heer, J., Weld, D.: Polyjuice: generating counterfactuals for explaining, evaluating, and improving models. In: Proceedings of the 59th Annual Meeting of the Association for Computational Linguistics and the 11th International Joint Conference on Natural Language Processing (Volume 1: Long Papers), pp. 6707–6723. Association for Computational Linguistics (August 2021). https://doi.org/10.18653/v1/2021.acl-long.523
33. Xie, Z., Thiem, S., Martin, J., Wainwright, E., Marmorstein, S., Jansen, P.: WorldTree V2: a corpus of science-domain structured explanations and inference patterns supporting multi-hop inference. In: Proceedings of the 12th Language Resources and Evaluation Conference, Marseille, France, pp. 5456–5473. European Language Resources Association (May 2020)

Combining Artificial Intelligence and Edge Computing to Reshape Distance Education (Case Study: K-12 Learners)

Chahrazed Labba[1](✉), Rabie Ben Atitallah[2], and Anne Boyer[1]

[1] University of Lorraine, CNRS, LORIA, Campus scientifique, 54506 Vandoeuvre-lès Nancy, France
{chahrazed.labba,anne.boyer}@loria.fr

[2] GSU - GalataSaray University, Istanbul, Turkey
rabie.ben-atitallah@expertisefrance.fr

Abstract. Nowadays, the use of distance learning is increasing, especially with the recent Covid-19 pandemic. To improve e-learning and maximise its effectiveness, artificial intelligence (AI) is used to analyse learning data stored in central repositories (e.g. in cloud). However, this approach provides time-lagged feedback and can lead to a violation of user privacy. To overcome these challenges, a new distributed computing paradigm is emerging, known as *Edge Computing* (EC), which brings computing and data storage closer to where they are required. Combined with AI capabilities, it can reshape the online education by providing real-time assessments of learners to improve their performance while preserving their privacy. Such approach is leading to the convergence of *EC* and *AI* and promoting *AI at the Edge*. However, the main challenge is to maintain the quality of data analysis on devices with limited memory capacity, while preserving user data locally. In this paper, we propose an Edge-AI based approach for distance education that provides a generic operating architecture for an AI unit at the edge and a federated machine learning model to predict at real-time student failure. A real-world scenario of K-12 learners adopting 100% online education is presented to support the proposed approach.

Keywords: Real-time feedback · Privacy · Federated learning

1 Introduction

In recent years, there has been a massive use of online courses, particularly with the current Covid-19 pandemic. Although the distance education helped in maintaining a certain continuity of the learning process, however it comes with several challenges related to the infrastructure cost and preserving learners privacy and security [6,9–11]. Conventionally, data are collected and stored in

M. M. Rodrigo et al. (Eds.): AIED 2022, LNCS 13355, pp. 218–230, 2022.
https://doi.org/10.1007/978-3-031-11644-5_18

centralised repositories to be later analyzed to fulfill various learning analytics objectives. However, with the excessive use of Internet of Things (IoT) technologies, there is an increasing amount of multi-source and heterogeneous data collected and analysed by educational institutions. This increase in data creates a risk of bandwidth saturation, which increases latency and leads to overuse of computing resources. Recently, an alternative computing paradigm, Edge Computing (EC) is being proposed to solve the aforementioned issues. It consists in bringing computing and data storage closer to where they are required, which increase performance while reducing operating costs [13]. Combined with the use of AI, the Edge AI has multiple advantages such as bringing more security and confidentiality by allowing the filtering and the aggregation of data before sharing it at the network. The Edge AI can reshape the world of education today by offering the potential to preserve privacy and to improve student performance, confidence and mental well-being by delivering real-time feedback. The main idea is to use analytical models at the edge. These models as well as the way of their training on a distributed data and heterogeneous systems need to be redesigned. To meet this challenge, Federated Learning (FL) [7] is emerging as a promising technology. FL is a Machine Learning (ML) technology that enables collaborative learning of a common model by a number of entities (users, organizations) holding data locally. Unlike the traditional centralized ML approaches, FL does not require data to be uploaded to a central repository. This feature addresses our need for data privacy in online education.

In the frame of this work, we present how the Edge AI combined with FL can be used to reshape the distance education and ensure more data privacy while minimising the infrastructure usage. The proposed approach tackles at first the architecture of the AI unit to be used at the edge. An AI unit represents the device a learner can use in distance education. Secondly, we present a new scenario of using Edge AI with FL to predict k12-learners' failure. Indeed, the real case study consists of learning data collected and stored in a centralized repository within the National Center for Distance Learning (Cned[1]). To adapt the data to a federated use, we used the TensorFlow Federated (TFF)[2] and the Artificial Neural Network (ANN) model to anticipate student failure as early as possible. The federated ANN was evaluated under different client sample selection strategies. The experimental results show that with proper selection of training samples in a federated setting, the federated model can be as good as the centralized model in anticipating students failure.

The rest of the paper is organized as follows: The Sect. 2 presents the related works. Section 3 introduces our Edge AI based approach to reshape the distance education. In Sect. 4, we present the case study of K-12 learners enrolled within the Cned as well as the results of using a federated ANN model to predict student failure. Section 5 presents the conclusions, the threats to validity and the future works.

[1] CNED: Centre national d'enseignement à distance.
[2] https://www.tensorflow.org/federated?hl=fr.

2 Related Works

Techniques such as machine learning and data mining have been widely applied in the context of e-learning [5,8,14]. Despite the diversity of the AI techniques, the used methodology to apply them is common. It consists in collecting and cleaning the data, then extracting the features and applying the AI algorithms. Usually, the data is stored in central repositories (e.g. Cloud), which may result in a breach of students privacy. According to [11], six distinct ethical concerns are identified within the context of big data and personalized learning, which are as follows: information privacy, anonymity, surveillance, autonomy, non-discrimination, and ownership of information. These concerns have been confirmed and discussed in numerous works [6,9,10]. Under the principle of data protection, in many fields such as healthcare and industry, data are not shared but stored and explored locally. Thus, in this case, we lose all the benefits brought by the use of big data technology. To address this problem, FL [7] is gaining momentum, especially with the emergence of the Edge AI paradigm. The principle of FL is that many entities collaboratively form a common model using their local learning data and communicate the updated model weights to a central server. No data is shared or exchanged between the different entities, thus reducing the risks related to privacy. An entity in FL can be a user (e.g. IoT device) or an organization. Depending on the level of granulation of the FL application, we distinguish two types of research work. On one hand the works that focus on the inter-organisational FL such as in [2], The authors highlight the confidentiality issues that hinder data sharing between different industrial organizations. To address this challenge, they present how FL can be used to predict production line failures in different organizations. In [3], the authors proposed a FL-based education data analysis framework that can be used to build data analysis federations between many institutions. In [12,15], FL based approaches have been proposed to address privacy issues and fully exploit the potential of AI in healthcare domain. On the other hand, other research works focus on the inter-devices (users) FL such as in [4], the authors used FL to predict the next word prediction in a virtual keyboard for smartphones. In [16,17], FL is used to provide personalized recommendations to users.

In this work, we focus on the use of FL on Edge computing-based system for the distributed analytics in order to support real-time students' assessment. To our knowledge, we are the first to consider the application of FL at a fine-grained level to reshape online education. According to the literature review, FL in education [3] has only been addressed at the inter-institutional level.

3 Distributed Analytics and Edge Intelligence

In our work, we consider a generic distributed EC-based system composed of N AI units at the edge EUs and one Coordinator Server CS as illustrated in Fig. 1. In such system, data analytics is distributed over all the nodes and conventionally only aggregated data or model parameters should be exchanged. However, if a

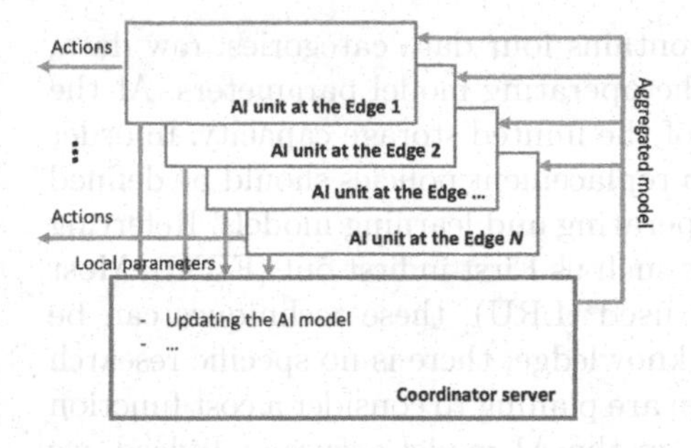

Fig. 1. Generic distributed EC-based system

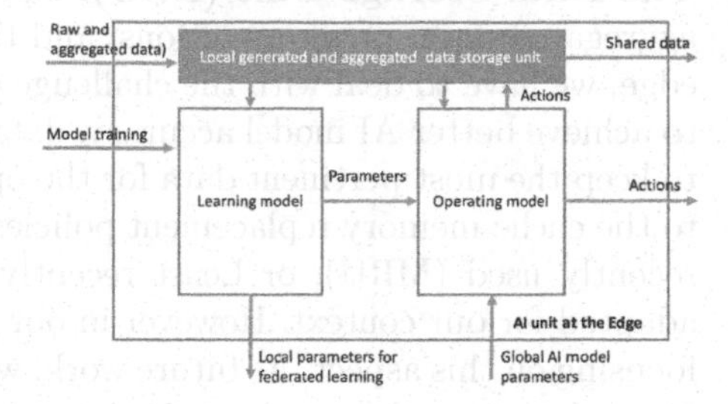

Fig. 2. AI-Edge unit architecture

raw data related to a given user is generated elsewhere, for example at the CS level, it can be transferred to the corresponding edge unit with a specific message. The EUs can be homogeneous or heterogeneous (PC, mobile terminal, industrial computer, IoT device, etc.) connected to the CS with a communication infrastructure (Ethernet, WIFI, LTE, etc.).

3.1 AI-Edge Unit Architecture

The key architectural feature of the distributed system is the EU. Let us consider $\mathcal{U}$ the set of EUs. Each unit EU_i is characterized with a memory storage capacity M_i, a processing power P_i, and a communication bandwidth B_i with the CS. It incorporates three main functions: the operating model, the data storage unit, and the learning model as shown in Fig. 2.

The Operating Model (OM): Depending on the application objective, this model can perform different type of actions recommendation, alerting, prediction, decision making, etc. In the literature, a multitude of techniques and algorithms are developed for each type of action regarding the size of the available data (SAD) as well as respecting some functional and non-functional constraints (e.g. real-time and energy consumption). In the Edge computing paradigm, the limited hardware resources lead rule-setting for the appropriate choice taking into consideration the computation cost of the algorithm and the SAD. These two parameters should fit respectively to the P_i and M_i of the corresponding EU_i. The execution of the operating model can be synchronised with a clock frequency f, the arrival of a new data, or the user request.

The operating model can be performed using the global AI model parameters collaboratively extracted on the CS or using the local parameters extracted from the learning model function. This last case is quite pertinent while considering specific user profile (e.g. disabled person).

After running the operating model, the generated actions will be communicated to the user as well as saved on the data storage unit for the next model training process (see Fig. 2).

The Data Storage Unit (DSU): It contains four data categories: raw data, aggregated data, previous actions, and the operating model parameters. At the edge, we have to deal with the challenge of the limited storage capacity. In order to achieve better AI model accuracy, data replacement policies should be defined to keep the most pertinent data for the operating and learning models. Referring to the cache memory replacement policies such us First in first out (FIFO), Most recently used (MRU), or Least recently used (LRU), these techniques can be adapted for our context. However in our knowledge, there is no specific research focusing on this aspect. As future work, we are planing to consider a cost function for each data and to evaluate its impact on the AI model accuracy. Indeed, we think that the data replacement at the Edge is an application-oriented challenge and it should be resolved according to the use case scenario.

The Learning Model (LM): It covers a large spectrum of learning techniques: symbolic (e.g. production rules) and machine learning (e.g. ANN, random forest, SVM). As same for the OM, the LM algorithm should 1) be adequate to the execution support characteristics (operating frequency, P_i and M_i), 2) satisfy non-functional constraints (e.g. energy consumption and thermal dissipation) while considering embedded devices (tracking static and dynamic obstacles for autonomous car), and 3) take into consideration the available small data (e.g. no sufficient data for running a deep learning algorithm).

As our main objective is to keep data closer to where they are generated, we propose a federated learning model well-traced on the distributed analytics EC-based system in order to respect the privacy of data and to train the AI model collaboratively with a subset of EUs $A \subset \mathcal{U} \longrightarrow A \in \mathcal{P}(\mathcal{U})$. The M EUs samples of A can be selected randomly among the N EUs of $\mathcal{U}$ or according to a guided strategy of sampling ($M << N$).

3.2 Communication Protocol and High Level System Reconfiguration

In order to ensure data transmission between the CS and the EUs, we propose a generic and open communication protocol dedicated to the distributed Edge computing-based system. It is conceived at a high level of abstraction to be transport technologies-independent, thus allowing easy integration of emerging technologies. The encapsulation of the protocol in software components will guarantee the evolutionary as well as the scalability of the system according to the number of EUs. Communication between nodes is provided by high-level frames consisting of a sequence of fields that indicate:

- **Header:** The beginning of a new frame.
- **Type:** There are different types of frame: 1) *RD* (Raw Data), 2) *AD* for sharing Aggregated Data with EUs as it respects users' privacy, 3) *TLM* is a command for a collaborative training of the learning model on a subset of the EUs, 4) *LMP* Local Model Parameters from an EU, 5) *MR* (Model

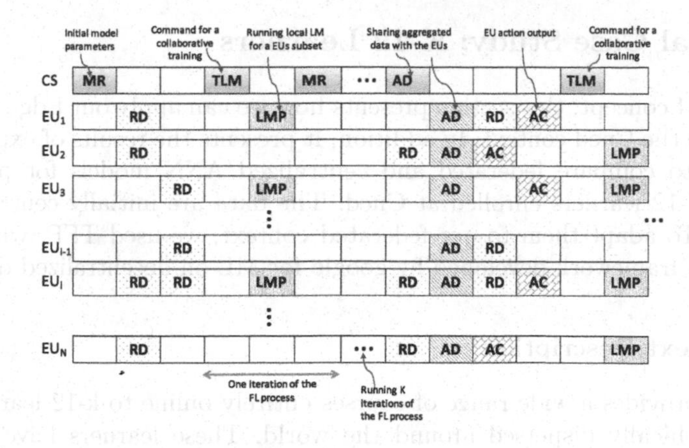

Fig. 3. Operating scenario with frame settings

Reconfiguration) is a broadcast of the global learning model to the EUs, and 6) *AC* (Action) corresponds to EU output.

- **Sender:** can be an EU or the CS.
- **Receiver:** that can be: 1) a specific EU (e.g. related raw data generated on the CS), 2) a subset of EUs (for the collaborative training process), 3) all EUs (updating the parameters of OM), or 4) the CS (receiving the local model parameters from an EU). We highlight that when a generated data should be stored locally the Sender and Receiver fields should have the same value.
- **Data:** According to the Type field, the transmitted information can be a raw data, aggregated data, actions (OM output), the global OM parameters, or the local extracted parameters from an EU.
- **Time:** All the generated and transmitted data are performed with time annotation for functional verification and the overall system synchronisation. For that reason, a Data Distribution Service (DDS) can be used in the distributed analytics system for better real-time performance.
- **Security:** allows encrypted data for an enhanced security.
- **Footer:** The end of the frame.

Figure 3 offers an operating scenario with frame settings covering the initialisation, the federated learning, and the operating model phases. At the beginning, the CS initiates a frame of Type *MR* that allows the EUs to download the initial OM parameters. After that, all the nodes start generating raw data through the users activities. Having enough data distributed on the system will trigger the frame *TLM* allowing several iterations of the FL process. In each iteration, the EUs share their local parameters (frame *LMP*) and receive the global extracted model (frame *MR*) from the CS. Over the time, the accuracy of the LM will be improved with the incremental process of learning and with sharing aggregated data (frame *AD*).

4 A Real Case Study: K-12 Learners

As a proof of concept, this section presents how we can apply our Edge-AI based approach to the Cned context. In addition, it presents the results of experiments conducted to compare federated and centralized ANN models for predicting failure of K-12 learners enrolled at Cned. The data are initially centralized on the cloud. To adapt them to our federated context, we used TFF, which is an open source framework developed by google for ML on decentralized data.

4.1 Context Description

The Cned provides a wide range of courses entirely online to k-12 learners who are geographically dispersed around the world. These learners have different demographic profiles and are unable to attend regular schools for many reasons. The Cned offers the courses through a Learning Management System (LMS) and provides with it a set of applications such as the education management system GAEL that allows administrative tracking of the students. All data are stored in a central repository and then analysed using ML techniques. For example, one of the main concerns of the Cned is to reduce the high failure rate among K-12 learners [1]. The Cned online teaching system has many limitations. First, as learners are physically dispersed around 173 countries, they do not make profit from the same quality of internet connection. Second, given the number of learners and the range of levels offered by the Cned, there is a huge amount of data generated on daily basis which has to be sent to a central repository without being filtered. For such a process, a powerful infrastructure in terms of bandwidth and storage resources is required. Third, sending user data containing sensitive information to a central repository may be exposed to security issues that could result in a breach of user privacy. Given all of these challenges, we believe that using our Edge-AI based approach is appropriate for reshaping distance learning at Cned.

The proposed approach (Sect. 3) can be easily transposed to the Cned context. Indeed, students connect to the LMS and other applications via their terminals (e.g. smartphone, tablet, personal computer), which represent the EUs. Whereas the Cned storage infrastructure is considered as the CS. The number of EUs is equal to the number of students. An open question is whether the use of FL on EC-based systems allows building a reliable alert system that predicts student performance on a weekly-basis. Data is collected from two sources: the LMS platform and the education management system GAEL. In the frame of this work, our case study consists of K-12 learners enrolled in the physics-chemistry course during the 2017–2018 school year. In total, there are 46 weeks in the school year and 671 enrolled students that represents the number of EUs in our system.

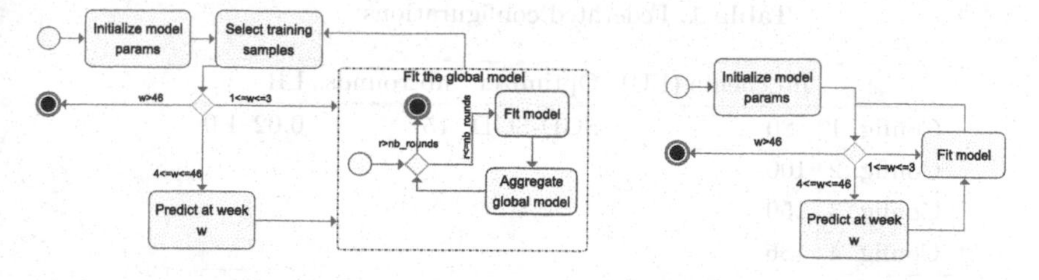

Fig. 4. Federated approach

Fig. 5. Centralized approach

4.2 Our Approach to Predict Students Performance per Week

To predict students performances on weekly-basis, the problem is formalized as a n-classification problem. We adopted the same classification as well as features introduced in [1]. The classification consists of three classes: high failure risk, medium failure risk and success. On each week w_i, a student is defined by a tuple $X = (f_1, .., f_m, y)$ where $f_1, .., f_m$ are the features and y is the class to predict. The student class may vary from one week to another. The selected features are extracted from the two data sources including the LMS and GAEL (the selected features are out of the scope of this work). For the reader information, the same features have been used to train and test both centralized and federated ANN models. Our main goal is to build an incremental ML model to predict student performance over time. To mimic a real situation, two approaches have been proposed: i) the centralized approach: as shown in Fig. 5, we first initialize the model parameters, then we fit the model to the first three weeks of data. The choice of this number is not arbitrary and was set based on experimentation. Indeed, from week 4 on, the model starts to make good predictions. Each week, the model is tested on the current week's dataset and then trained on it. At the end of each week, the model is updated and used to make predictions for the following week. ii) the federated approach: as shown in Fig. 4, we adopt the same principle of weekly validation and training. The main differences with the centralised approach are: the use of federated data to train and validate the model as well as the way the model is built through the communication of model updates between the clients (EUs) and the server (CS). The process starts with the initialization of the model parameters (CS) and then selects the set of clients (EUs) that will participate in the model training phase. During the first three weeks, the model is simply adjusted to the data. From week 4 on, the model is tested to make predictions on the current week's data, and then it is trained on the local data of the selected EUs. Each EU fits the model on its own data and then sends the model updates to the CS. The CS aggregates all the updates and sends the model back to the clients participating in the training phase. The training phase can last several rounds that are set on the basis of experimentation. The number of selected EUs for the training phase may vary from round to round as well as from week to week.

Table 1. Federated configurations

	nb_clients (EU)	Optimizer	nb_rounds	LR
Config. 1	80	SGD-SGD	15	0.02–1.0
Config. 2	100			
Config. 3	150			
Config. 4	186			

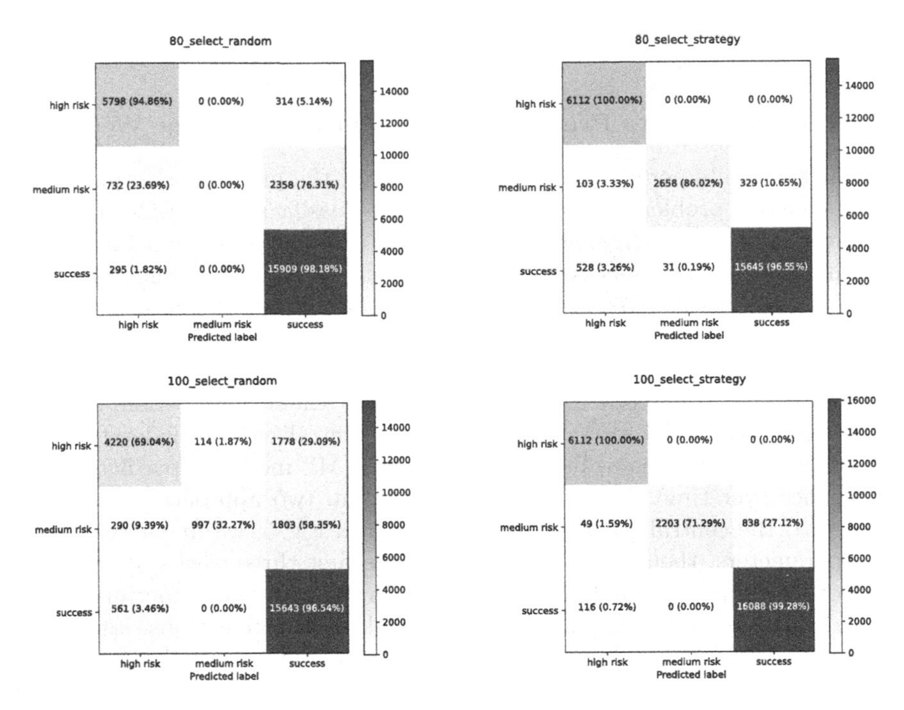

Fig. 6. Confusion matrix

4.3 Experiment Results

Different configurations have been used to build and evaluate the federated incremental ANN model. A set of experiments have been conducted to find the best parameters for our model, including defining the optimizer, the learning rate and the number of rounds. Due to space limitations, the suitable parameters are directly presented in the Table 1 and will not be discussed. The configurations differ mainly in terms of the number of selected EUs (80, 100, 150, 186) to be used during the training phase. In a federated context, this number is dynamic and may vary during the training sessions, since EUs can connect/disconnect at any time.

Impact of the Client Selection Strategy on the Quality of the Predictions: The used data present imbalances with respect to the “medium risk failure” class. Consequently, this class is not well detected, especially when EUs participating in the training phase are randomly selected. First, we compared the accuracy of the federated model using different numbers of EUs (80 and 100) with two selection strategies. The first strategy consists in randomly taking a set of client samples, for the training, without checking the proportions of the 3 classes taken in it. The second strategy consists in selecting a set of clients with 30% of the samples belonging to the high-risk of failure class, 30% to the medium-risk of failure class and the rest to the success class. The Fig. 7 represents the federated models accuracy on the test data. We note that models with a guided strategy for selecting client (EU) samples for training perform better than this with random selection for both fixed numbers of samples (80 and 100). Since the accuracy measure does not distinguish between the number of correct labels from different classes, we present, in Fig. 6, the confusion matrix for the four experiments. The matrices present the cumulative measures over all weeks. The main objective is to detect students at high and medium risk of failure in order to alert teachers about them and take the right action. We find that with a random selection strategy, there is a problem in classifying medium risk students and most of the time they are classified as successful. This classification is due to the fact that in the random sample, the medium-risk class is underrepresented compared to the rest of the classes. Therefore, we tried to guide the model during the learning phase by selecting samples that are balanced in terms of class representation. As shown in Fig. 6, the medium risk class is better predicted under the guided selection strategy. In addition, we obtained 100% correct predictions for the high-failure risk.

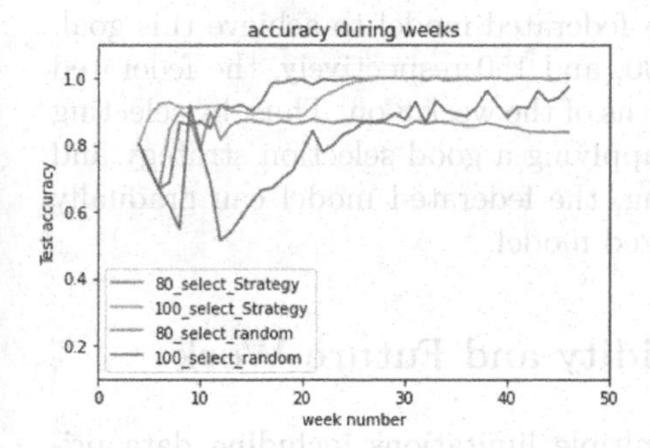

Fig. 7. Impact of clients selection strategy on accuracy

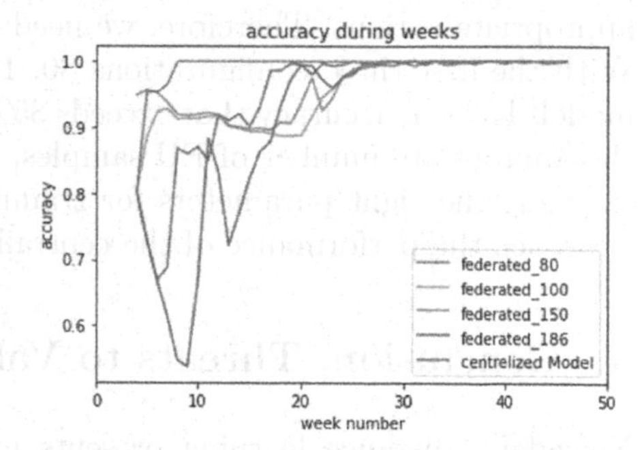

Fig. 8. Accuracy comparison: centralized model vs federated ones

Impact of the Number of Client Samples on the Quality of the Predictions: As shown in the Table 1, we used four configurations (80, 100, 150, 186). The guided selection strategy, presented above, is used for this experiment. As shown in Fig. 8, by varying the number of selected samples, the test accuracy over the weeks also changes. Indeed, we find that selecting more samples does not necessarily improve the accuracy of the model. For the performed experiments, training the model with 80 samples performs better, in terms of test accuracy, than training it with larger numbers of samples. Indeed, this may be a consequence of the used selection strategy. During the first weeks, some classes are poor in terms of the number of students that belong to them. Therefore, we cannot always have the total number of samples defined by the fixed rate (e.g. 30% of the number of samples as medium-risk). However, with a smaller number of samples, during the learning phase, we can reach the full proportions of the different classes more quickly than by using a larger number of samples. The rapidity is addressed in terms of the number of the week at which we begin to have a complete representation of all classes of students in the selected samples with respect to the predefined rate for each class. We believe it is important to determine the appropriate threshold that should be used as the number of EU samples to train the federated model.

Accuracy Comparison: Centralized Model vs Federated Model
As shown in the Fig. 8 and as expected, the centralized model performs better than the federated models in terms of test accuracy. Each week, the centralized model is trained on all available data, while the federated models are trained on a subset of the data. However, the first thing to notice is that all the federated models eventually converge to reach an accuracy very close or even equal to that of the centralized model. In our context, the goal is to predict as early as possible when students are at high or medium risk of failure in order to take appropriate actions. Therefore, we need a federated model to achieve this goal. With the first three configurations 80, 100, and 150 respectively, the federated models have an accuracy that exceeds 85% as of the week 8 on. Thus, by selecting the appropriate number of EU samples, applying a good selection strategy and choosing the right parameters for training, the federated model can gradually approach the performance of the centralized model.

5 Conclusion, Threats to Validity and Future Work

Nowadays, distance learning presents multiple limitations including data privacy risks and high infrastructure costs in terms of bandwidth and computing resources to store big data. To overcome these challenges, we proposed to use an AI-based approach at the edge to perform distributed analytics while keeping the data stored where it is generated. Further, we presented a new scenario of using Edge AI with FL to predict k12-learners' failure. The experimental results are promising and show that with appropriate parameter settings in FL, we can still obtain good performance as the centralized approach.

The current work presents some limitations that we tried to mitigate when possible: i) As a proof of concept, one ML model (ANN) has been used. In the short term, we intend to apply other ML models such as decision trees. ii) The results presented are for the physics-chemistry course for the academic year 2017–2018. We plan to expand our work and consider students performance in different courses as well as across modules. iii) The results of the federated models are determined through the simulation environment TFF. The models need to be adapted for future use on IoT devices. Thus, a complexity study and encapsulation is required for future embedded implementation.

References

1. Ben Soussia, A., Roussanaly, A., Boyer, A.: An in-depth methodology to predict at-risk learners. In: De Laet, T., Klemke, R., Alario-Hoyos, C., Hilliger, I., Ortega-Arranz, A. (eds.) EC-TEL 2021. LNCS, vol. 12884, pp. 193–206. Springer, Cham (2021). https://doi.org/10.1007/978-3-030-86436-1_15
2. Ge, N., Li, G., Zhang, L., Liu, Y.: Failure prediction in production line based on federated learning: an empirical study. J. Intell. Manuf., 1–18 (2021). https://doi.org/10.1007/s10845-021-01775-2
3. Guo, S., Zeng, D., Dong, S.: Pedagogical data analysis via federated learning toward education 4.0. Am. J. Educ. Inf. Technol. **4**(2), 56 (2020)
4. Hard, A., et al.: Federated learning for mobile keyboard prediction. arXiv:1811.03604 (2018)
5. Jiang, W., Pardos, Z.A.: Time slice imputation for personalized goal-based recommendation in higher education. In: Proceedings of the 13th ACM Conference on Recommender Systems, RecSys 2019, pp. 506–510. Association for Computing Machinery, New York (2019). https://doi.org/10.1145/3298689.3347030
6. Joksimović, S., Marshall, R., Rakotoarivelo, T., Ladjal, D., Zhan, C., Pardo, A.: Privacy-driven learning analytics. In: McKay, E. (ed.) Manage Your Own Learning Analytics. SIST, vol. 261, pp. 1–22. Springer, Cham (2022). https://doi.org/10.1007/978-3-030-86316-6_1
7. McMahan, B., Moore, E., Ramage, D., Hampson, S., Arcas, B.A.: Communication-efficient learning of deep networks from decentralized data. In: Artificial Intelligence and Statistics, pp. 1273–1282. PMLR (2017)
8. Mrhar, K., Abik, M.: Toward a deep recommender system for MOOCs platforms. In: Proceedings of the 2019 3rd International Conference on Advances in Artificial Intelligence, pp. 173–177 (2019)
9. Pardo, A., Siemens, G.: Ethical and privacy principles for learning analytics. Br. J. Educ. Technol. **45**(3), 438–450 (2014). https://doi.org/10.1111/bjet.12152
10. Priedigkeit, M., Weich, A., Schiering, I.: Learning analytics and privacy—respecting privacy in digital learning scenarios. In: Friedewald, M., Schiffner, S., Krenn, S. (eds.) Privacy and Identity 2020. IAICT, vol. 619, pp. 134–150. Springer, Cham (2021). https://doi.org/10.1007/978-3-030-72465-8_8
11. Regan, P.M., Jesse, J.: Ethical challenges of edtech, big data and personalized learning: twenty-first century student sorting and tracking. Ethics Inf. Technol. **21**(3), 167–179 (2018). https://doi.org/10.1007/s10676-018-9492-2
12. Rieke, N., et al.: The future of digital health with federated learning. NPJ Digit. Med. **3**(1), 1–7 (2020)

13. Shi, W., Dustdar, S.: The promise of edge computing. Computer **49**(5), 78–81 (2016). https://doi.org/10.1109/MC.2016.145
14. Wang, J., Xie, H., Au, O.T.S., Zou, D., Wang, F.L.: Attention-based CNN for personalized course recommendations for MOOC learners. In: 2020 International Symposium on Educational Technology (ISET), pp. 180–184. IEEE (2020)
15. Wu, Q., Chen, X., Zhou, Z., Zhang, J.: FedHome: cloud-edge based personalized federated learning for in-home health monitoring. IEEE Trans. Mob. Comput. **21**, 2818–2832 (2022)
16. Zhao, S., Bharati, R., Borcea, C., Chen, Y.: Privacy-aware federated learning for page recommendation. In: 2020 IEEE International Conference on Big Data (Big Data), pp. 1071–1080 (2020). https://doi.org/10.1109/BigData50022.2020.9377942
17. Zhou, P., Wang, K., Guo, L., Gong, S., Zheng, B.: A privacy-preserving distributed contextual federated online learning framework with big data support in social recommender systems. IEEE Trans. Knowl. Data Eng. **33**(3), 824–838 (2021). https://doi.org/10.1109/TKDE.2019.2936565

Plausibility and Faithfulness of Feature Attribution-Based Explanations in Automated Short Answer Scoring

Tasuku Sato[1,2](✉), Hiroaki Funayama[1,2], Kazuaki Hanawa[2], and Kentaro Inui[1,2]

[1] Tohoku University, Sendai, Japan
{tasuku.sato.p6,h.funa}@dc.tohoku.ac.jp, inui@tohoku.ac.jp
[2] RIKEN, Tokyo, Japan
kazuaki.hanawa@riken.jp

Abstract. Automated short answer scoring (SAS) is the task of automatically assigning a score as output to a given input answer. In this work, we tackle the challenging task of outputting the basis (i.e., *justification cues*) as an explanation for scoring given a defined rubric for assigning a score. In previous studies, researchers explored the performance of scoring via *justification identification* and constructed their own datasets, but their studies focused solely on limited experiments. On the basis of previous studies, we consider justification identification as an *explanation by feature attribution methods* for *explainable artificial intelligence* research. We conduct a comprehensive experiment consisting of multiple explanation generation methods, supervised learning of explanations, and evaluating on two axes, *plausibility* and *faithfulness*, which are important for automatic SAS. Our results indicate that we can improve the plausibility of gradient-based methods by supervised learning. However, methods with high plausibility and high faithfulness are still different methods, so it is crucial to select an appropriate method depending on which perspective of explainability is essential.

Keywords: Short answer scoring · Explainability · Feature attribution

1 Introduction

Automated short answer scoring (SAS) is the task of automatically assigning a score as output to a given input answer. In domains such as education, automated SAS has attracted considerable owing to its ability to reduce the workload for teachers and ensure fair grading without human error [6,13,16].

In this work, we tackle the challenging task of outputting the basis as an explanation for scoring given a predefined rubric for assigning a score. Figure 1 shows an example of a prompt from the ASAP-SAS dataset [2], which is a commonly used SAS dataset. Here, the rubric has some key elements, and the task is to score the answer by the number of key elements contained within

M. M. Rodrigo et al. (Eds.): AIED 2022, LNCS 13355, pp. 231–242, 2022.
https://doi.org/10.1007/978-3-031-11644-5_19

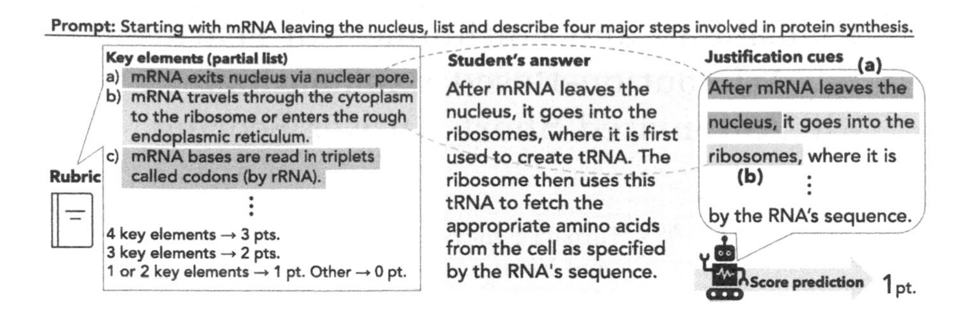

Fig. 1. Score prediction and justification identification in SAS. Each key element and its corresponding justification cue are highlighted in the same color.

the answer. In such a case, one can consider explaining the score by showing which parts of the student answer match the key elements. In the figure, "After mRNA ..." and "it goes into ...", highlighted by two different colors, are such basis parts because they include the contents of the key elements described in the rubric.

For outputting a basis of explanation for scoring in SAS, ideally a model would identify and present a basis of parts for a given student's answer in addition to predicted scores. Such an identification task is referred to as *justification identification* in the literature [13], where basis parts of a student's answer are referred to as *justification cues*. By identifying the various cues for scoring, end users (students and teachers) would feel more confident with the scoring results and could use them as hints for review. For example, in Fig. 1, a SAS system would ideally be able to automatically calculate the number of points for the student's answer and inform them the specific reasoning for their overall score (e.g., *You were awarded 1 point because your answer included important elements* (a) *and* (b)). In addition, system administrators could utilize justification cues in order to debug the system. In previous SAS research, however, how the current technologies can be augmented with the ability to explain the scoring results has rarely been explored, and the datasets commonly used for SAS research, including ASAP-SAS [2], do not contain annotations of justification cues.

In this study, we first reformulated the task of justification identification with the notion of *feature attribution* developed in the context of *Explainable AI* (XAI). We then rigorously conducted comprehensive experiments to improve SAS systems consisting of the following: i) several explanation generation methods, ii) supervised learning of explanations, and iii) evaluation of two axes, *plausibility* and *faithfulness*. Our results indicate that we can improve the plausibility of gradient-based methods by supervised learning. However, methods with high plausibility and high faithfulness are still different methods, so it is crucial to select an appropriate method depending on which perspective of explainability is essential. The code for our experiments is publicly available at https://github.com/cl-tohoku/Explainability_of_SAS.git.

2 Related Work

Presenting the decision process of deep learning models in a form that humans can interpret has recently attracted attention in various fields, including law, medicine, and education [22], which is called *explainability* or *interpretability*. Models with high explainability are actively explored in the context of XAI [22].

Explainability is often discussed in terms of two axes. One axis focuses on *how convincing the explanation for the prediction is to humans*, which includes human interpretability [11], persuasiveness [8], plausibility [9], and others. The other axis focuses on *to what extent do the explanations for the predictions reflect the predictive process of the model*, which includes fidelity [7], faithfulness [9], and others. In this paper, we follow the definition of Jacovi et al. [9] and refer to the former axis strictly as *plausibility* and the latter as *faithfulness*.

As in other subfields of natural language processing (NLP), SAS has substantially advanced with recent deep-learning-based technologies, e.g., [6,16,23, etc.]. However, these previous studies mainly focused on scoring performance, and explainability for SAS has rarely been addressed so far despite its potential significance in educational contexts. To the best of our knowledge, one exception aimed at interpreting the process of predicting model scores in SAS is the work of Mizumoto et al. [13]. They defined the task of *justification identification* (Sect. 3) and constructed a dataset for evaluating the performance of that task. However, their study also focused mainly on scoring performance, and only limited experiments were conducted on justification identification only from the plausibility perspective. In this study, we (i) recast justification identification as a special case of *feature attribution*, (ii) newly explore a range of recent feature attribution methods for this under-explored task, and (iii) empirically evaluate the quality of the methods by faithfulness as well as plausibility.

Feature attribution is an approach for XAI, which assesses the importance of each feature of input against the output and uses that value to explain the prediction to help humans understand the model's internal behavior. Feature attribution has been studied mainly in the field of computer vision, and various methods [18–20] have been proposed. In NLP, it is also common to use attention weights to analyze which words in an input text are effective in prediction such as text classification [24], which can also be seen as a method for feature attribution. In this paper, we present the first study in which the behavior of these feature attribution methods on SAS, is comprehensively investigated.

3 Justification Identification: Task Setting

SAS is to predict a score $s \in \{0, 1, ..., S\}$ for a given answer $\boldsymbol{x} = (x_1, x_2, ..., x_T)$ to a prompt, where x_t is the t-th word, and S is an allocation of score. We use quadratic weighted kappa (QWK) [4] to evaluate the scoring performance.

Justification identification is the task of providing the basis (i.e., justification cues) as an explanation for the score prediction. In Fig. 2, the model predicts a score of 2 points for the answer and presents "After mRNA leaves" as justification cues. Formally, the task is to generate the justification cue label

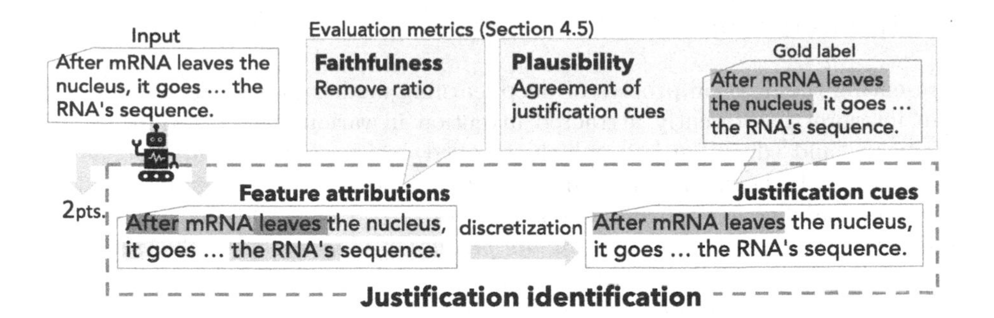

Fig. 2. Task setting of justification identification. The feature attributions are continuous values, and the color depth represents the magnitude of the value. We generate the justification cues shown in blue by discretizing these values at a certain threshold.

$\boldsymbol{r} = (r_1, r_2, ..., r_T)$, $r_t \in \{0, 1\}$ for the score prediction in the input answer $\boldsymbol{x}$. Here, when $r_t = 1$, the word x_t is the justification cue.

We evaluate justification cues in terms of *plausibility* and *faithfulness* [9]. In SAS, plausibility can be considered the dimension of evaluating whether the generated justification cues are consistent with the rubric. We thus evaluate the plausibility of generated justification cues on the basis of whether they match the gold labels.

Faithfulness, on the other hand, refers to whether the generated justification cues are consistent with the model's behavior. There is no guarantee that the model is actually predicting the score on the basis of the predicted justification cues. Therefore, it is crucial to evaluate faithfulness when discussing explainability. The details of these two metrics are discussed in Sect. 4.5.

As illustrated in Fig. 2, we realize justification identification by (i) conducting feature attribution and (ii) discretizing the results of the feature attribution with a threshold determined by maximizing the agreement of justification cues (Sect. 4.5) in the development data. We normalize the feature attribution generated from the model arbitrary and represent it as $\boldsymbol{f} = (f_1, f_2, ..., f_T)(0 < f_t < 1, \sum_{t=1}^{T} f_t = 1)$. With $f_{\max}$ as its maximum value and h as its threshold value, we set r_t to 1 if $f_{\max} - f_t < h$, and 0 otherwise.

The task setting of justification identification in the context of SAS allows us to assume special situations that have not been considered in previous XAI research. In SAS, one can consider cases where the gold labels of explanations (i.e., justification cues) are available for training and evaluating the model. One reason is that, in the context of using SAS for education, the quality of justification identification is as important as scoring accuracy, and the generation of gold label justification data to improve justification can be well worth the cost. Furthermore, human graders must be doing the work of identifying which parts of a given student answer match the rubric when they grade the answer, and the overhead for labeling the identified parts as justification cues in the process of generating training data may well be not very significant. Therefore, in

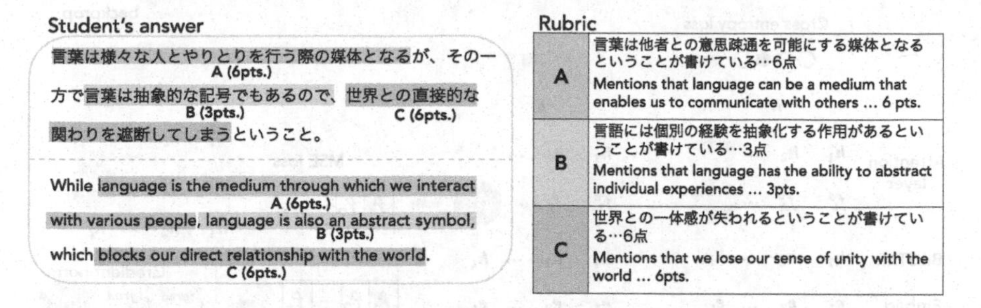

Fig. 3. Overview of RIKEN Dataset for Short Answer Assessment. The highlighted parts in the student's answer are justification cues and indicate parts of information included in the rubric. The rubric consists of several items such as A, B, and C.

this study, we also explore the effectiveness of using gold justification cue labels for training the feature attribution component jointly with the score prediction model (see Sect. 4.4). Note that such supervised settings for feature attribution have been rarely considered in the XAI literature.

4 Experimental Settings

In this study, we comprehensively experiment with several methods for generating justification cues and supervised learning methods for models, and we investigate whether there is a method that is both plausible and faithful in SAS. All experimental setting is publicly available at the GitHub site.

4.1 Dataset

Most SAS datasets [2,3,15] do not have justification cue labels. To the best of our knowledge, RIKEN-SAA [1][1] is the only dataset that contains such labels. We use this dataset in our experiments. In Fig. 3, we show an example of a student's answer taken from RIKEN-SAA along with its corresponding rubric. This dataset consists of the following: the answer text for each prompt, the score assigned by the grader, and the justification cues for assigning the score. Each prompt has multiple items such as "A, B, C, ...", and each answer is manually assessed for each item. Since each item is associated with an independent rubric and is annotated independently, for our experiments, we train and evaluate models independently for each item. Henceforth, we refer to the combination of prompts and items as *questions*.

RIKEN-SAA includes 36 questions with about 2000 student answers each and 21 questions with 500 answers. Out of this original dataset, we selected 28 questions with 500 answers for our experiment considering the following two criteria.

[1] https://www.nii.ac.jp/dsc/idr/rdata/RIKEN-SAA/. This study is the first to focus on the task of justification identification with this dataset, while it has been used in prior studies whose main interest is in SAS accuracy [6,13,23].

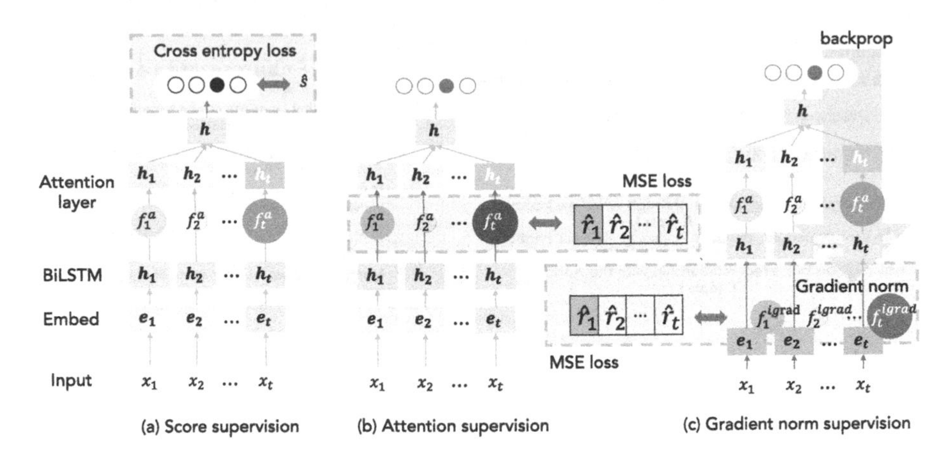

Fig. 4. Diagram of learning methods. (a) computes the cross-entropy loss between the model's final output and the gold score label s_g. (b) computes the mean squared error between the attention weights and the gold justification cues. (c) calculates the mean squared error between the gradient norms and the gold justification cues. We use Integrated Gradients for gradient-based feature attribution.

One was the quality of human annotations. The dataset provides annotations by two annotators. We calculated the human-to-human agreement of each question with the QWK for the given score labels and the f1 score for the justification cue labels (see Sect. 4.5 for the definition of the f1 score), and we filtered out questions with QWK lower than 0.7 or f1 score lower than 0.85. The second criterion was the ratio of non-zero scored answers. In this dataset, answers with a score of zero are deemed to contain no evidence to give the score, and no annotation of the evidence is given to them. Therefore, in this experiment, which focuses on generating justification cues, we excluded questions with more than half of the answers scored zero. The details of this data selection and sampling process are publicly available at the GitHub experiment report site. For each question, we divided the data into 200 instances for training, 100 for development, and 200 for evaluation. One may suspect that the training and development sets are relatively small. However, note that SAS in a supervised setting is essentially a low-resource problem because those who prepare the training data are usually the educators themselves who use the model. It is crucially important to ensure the model's performance with such a small scale of training signals.

4.2 SAS Model

We used a standard classification model with word embeddings, one-layer biLSTM, and one attention layer as shown in Fig. 4(a), following Riordan et al. [16]. The word embeddings were obtained by running word2vec [12] on Japanese Wikipedia data following Mizumoto et al. [13]. The dimensionality of the word embeddings was 100, and the hidden layer dimensionality was 300. We supervised the model's final output with gold score labels using the loss function:

$\mathcal{L}_{\text{score}} = \frac{1}{N}\sum_{n=1}^{N} \text{CE}(s_p^{(n)}, \hat{s}^{(n)})$, where CE is a cross-entropy error, N is the number of data, s_p is the predicted probability of scoring, and $\hat{s}$ is the gold score. We used RMSProp as the optimizer with a learning rate of 1.0×10^{-3}, a batch size of 32, and trained on 50 epochs. To perform early stopping, we selected the checkpoints with the highest accuracy on the development data.

4.3 Feature Attribution

We used several common methods for feature attribution as described below:

Attention Weight. Considered in the attention layer as the importance of each word to the prediction and use it as the feature attribution

Saliency Map. [19] Generated from the gradient of each input to the loss

Input X Gradient. [18] Multiplication of the Saliency Map by an input. It is possible to generate a feature attribution that reflects the features of the input

Integrated Gradients [20]. Generated by integrating gradients from the origin to the input direction, which satisfies the axioms of *Implementation Invariance* and *Sensitivity*, and is said to be faithful

Random (baseline). Feature attribution generated from a uniform distribution

Among these, the justification identification method used by Mizumoto et al. [13] can be categorized as Attention Weight. In this study, in contrast, we also explore gradient-based methods: the Saliency Map, Input X Gradient, and Integrated Gradients. The gradients computed in these methods represent *in which direction and by how much the embedding of each word changes, the prediction probability changes significantly.* We used the norm of this gradient as a value of importance to the corresponding input word.

4.4 Supervision of Justification Cues

Feature attribution is typically used to analyze which features are used by a model to make a prediction, so for that purpose, it is inherently unsupervised. In justification identification, on the other hand, the objective is to ensure the quality of the justification cues, and therefore, one can naturally consider the setting where the model is supervised with human-labeled justification cues. Fortunately, the RIKEN-SAA dataset includes human-labeled gold justification cues, so we tried using it.

There are at least two possible ways to train the model with gold justification cues, as illustrated in Figs. 4 (b) and (c).

Attention Supervision. When considering using the justification cue labels as supervisory signals, the first thing to consider is to learn the weights of the attention layer. Since weights of the attention layer are normalized so that the summation becomes 1, we normalize the justification cue labels by the number of words in the justification cues, and we calculate their mean squared error as the loss. If $\hat{r}$ is the gold justification cues, T is the sentence length, and f^a is the attention weight, the formula is as follows.

$$\mathcal{L}_{\text{attn}} = \frac{1}{N}\sum_{n=1}^{N}\sum_{t=1}^{T}\left({f^a}_t^{(n)} - \frac{\hat{r}_t^{(n)}}{\sum \hat{\boldsymbol{r}}^{(n)}}\right)^2,$$

Gradient Norm Supervision. We also consider using justification cue labels to supervise the gradient norm. A previous study [21] showed that gradient supervision can improve the performance of sentiment analysis, natural language inference, and image task, but there has been no research on gradient norm supervision with explanation labels. It is thus intriguing whether one can effectively supervise gradient norms for the sake of justification identification analogously to the attention layer. In the experiment, we determine the performances of the models obtained by minimizing the errors between the feature attributions with Integrated Gradients and gold justification cues. The newly incorporated loss was:

$$\mathcal{L}_{igrad} = \frac{1}{N}\sum_{n=1}^{N}\sum_{t=1}^{T}\left({f^{igrad}}_t^{(n)} - \frac{\hat{r}_t^{(n)}}{\sum \hat{\boldsymbol{r}}^{(n)}}\right)^2,$$

where f^{igrad} is the feature attribution by Integrated Gradients.

In this experiment, we compare four learning methods: 1, $\mathcal{L}_{\text{score}}$ (unsup); 2, $\mathcal{L}_{\text{score}}+\mathcal{L}_{\text{attn}}$ (attn); 3, $\mathcal{L}_{\text{score}}+\mathcal{L}_{\text{igrad}}$ (igrad); 4, $\mathcal{L}_{\text{score}}+\mathcal{L}_{\text{attn}}+\mathcal{L}_{\text{igrad}}$ (attn&igrad) as combinations of these loss functions.

4.5 Evaluation Metrics

In this experiment, we use QWK [4] as a measure of scoring accuracy. For explainability, we use the agreement of the justification cues [13] as plausibility evaluation and remove ratio [14,17] as faithfulness evaluation.[2] For each question, we trained the model with 10 different sets of seed values to produce 10 model instances and use the average result as the final result.

Agreement of Justification Cues (Plausibility). The agreement of the justification cues [13] is a plausibility measure that uses the f1 score as a measure of the degree of agreement between the gold and predicted justification cues. For item B in Fig. 3, we assume that the gold justification cue is "language is also an abstract symbol", and the model's output is "an abstract symbol, which blocks our direct". In this case, the true positive (TP) is 3 (an abstract symbol), the false positive (FP) is 4 (which blocks our direct), and the false negative (FN) is 3 (language is also). Therefore, the precision of this example is calculated as $\text{TP}/(\text{TP}+\text{FP}) = 0.43$, the recall is $\text{TP}/(\text{TP}+\text{FN}) = 0.5$, and the f1 score is $2 \times 0.43 \times 0.5/(0.43 + 0.50) = 0.46$.

[2] Although we also evaluated faithfulness using two other metrics, sufficiency and comprehensivenesss [5], we omit their results because they showed the same tendency.

Table 1. Scoring accuracy (QWK ↑)

unsup	attn	igrad	attn&igrad
0.766	0.801	0.803	**0.814**

Table 2. Results of explainability. AJC stands for agreement justification cues. Feature att. stands for feature attribution. Faith. stands for faithfulness.

Feature att.\Model	Plausibility (AJC ↑)				Faith. (Remove ratio ↓)			
	unsup	attn	igrad	attn&igrad	unsup	attn	igrad	attn&igrad
Attention Weight	0.467	**0.810**	0.548	0.805	0.238	0.232	0.165	0.157
Saliency Map	0.451	0.647	0.680	0.709	0.303	0.295	0.189	0.225
Input X Gradient	0.442	0.635	0.708	0.744	0.269	0.230	0.172	0.184
Integrated Gradients	0.443	0.551	0.771	0.772	**0.202**	**0.169**	**0.144**	**0.139**
Random	0.325	0.326	0.326	0.325	0.622	0.628	0.552	0.587

Remove Ratio (Faithfulness). Remove ratio is a faithfulness measure used by Serrano et al. [17] and Mohankumar et al. [14]. It masks words in descending order, starting with those with the highest feature attribution, and calculates the percentage of masked words when flipping the predicted label. If the feature attribution reflects the prediction process of the model, i.e., the words with the highest feature attribution match the words that the model considers important for a score prediction, then the prediction will change earlier because we can mask from the truly important word. The lower the value, higher the faithfulness.

In this study, we add scores to the answer if there is a basis statement for scoring in the answer. Therefore, masking words in a zero-score answer does not change the score, and we exclude the zero-score answer from the evaluation.

5 Results and Analysis

Scoring Accuracy. Table 1 shows the QWK results. Similar to previous studies [6,13,23], QWK was improved by attention supervision. The QWK was also improved by gradient norm supervision, and further improved by both attention and gradient norm supervision, compared with learning with only one method. Since we mainly focus on justification identification, we leave the interesting topic of QWK improvement in the attn&igrad model for our future work.

Plausibility. The left column of Table 2 shows the results of the agreement of justification cues. In the case of unsup, there is no significant difference in the values for any of the methods, but for attn, igrad, and attn&igrad, the f1 score was improved for all feature attributions except Random. A previous study has shown that gradient-based methods are not plausible [5], but we found that gradient norm supervision can also improve their plausibility. In addition, the

Table 3. Example of model output for Y14_2-1_2_3_D, showing the justification cues and remove ratio provided by the Attention Weight and Integrated Gradients in attn. Highlighted text spans are justification cues (gold or predicted).

Feature attributions	Answer	Remove ratio
Gold	テルは試合に出たのできっちり引退できたが、渡瀬は試合に出ていなく、 Teru could retire because he played in the match, but Watase didn't play in the match, 心のおさまりがつかないため、**やりきれない気持ち。** and he couldn't settle his mind, so **he felt unfulfilled.**	–
Attention Weight	テルは試合に出たのできっちり引退できたが、渡瀬は試合に出ていなく、 Teru could retire because he played in the match, but Watase didn't play in the match, 心のおさまりがつかないため、**やりきれない気持ち。** and he couldn't settle his mind, so **he felt unfulfilled.**	0.471
Integrated Gradients	テルは試合に出たのできっちり引退できたが、渡瀬は試合に**出ていなく、** Teru could retire because he played in the match, but Watase **didn't play** in the match, **心**のおさまりがつか**ないため、**やりきれない**気持ち**。 and **he couldn't** settle **his mind**, so he **felt** unfulfilled.	0.118

f1 score of Attention Weight in attn and Integrated Gradients in igrad were significantly improved, which indicate that one can improve the plausibility by aligning the feature attribution and learning methods. The highest f1 score was obtained with the combination of attn and Attention Weight. Attention Weight may be easier to learn than the norm of gradients because they are structured to directly solve the problem of sequential labeling, which predicts the label of each word. A more sophisticated analysis is a subject for future work.

Faithfulness. The right column of Table 2 shows the results for remove ratio. Among others, Integrated Gradients was the most faithful. Prior studies have shown that Attention Weight is sometimes inferior faithful to gradient-based methods [10, 14, 17]. In our experiment, we observed a similar tendency; namely, Attention Weight was inferior to the Integrated Gradients although, with a closer look, it was more faithful than the Saliency Map and Input X Gradient.

The difference in the remove ratio between Integrated Gradients and Attention Weight for attn was much larger than the difference between the two in the other settings. We analyzed the results of remove ratio in the attn model in detail. Table 3 shows a typical example for which the Attention Weight and Integrated Gradients behaved differently. The highlighted text is the gold or the predicted justification cues. Attention Weight is highly consistent with the gold, whereas the Integrated Gradients focus on words around the gold. Since the remove ratio of Integrated Gradients is lower than that of Attention Weight, the former faithfully reflects the model's behavior. Namely, the model predicts the score on the basis of surrounding words, not the gold justification cues that the model should focus on.

In SAS, training data are normally scarce and similar answers are likely to occur frequently, so extra information that is likely to co-occur with the statements that satisfy the rubric may be highly correlated with the score. In such cases, the model may learn to predict the score on the basis of extra information. We expect that Attention Weight supervision will suppress such pseudo-correlations; however, there may be cases where pseudo-correlations continue to exist as in the case shown in Fig. 3. In such cases, Attention Weight may be able to present a justification cue that is highly consistent with the gold justification cue through attention supervision, whereas the model may predict scores based on the basis of areas that are different from the gold due to pseudo-correlation, which causes a discrepancy in the behavior between the Attention Weight and the model.

6 Conclusion

In this study, we tackled the task of justification identification for SAS by the *explaining by feature attribution* approach that has emerged from the XAI research. Moreover, we investigated whether a method that achieves both plausibility and faithfulness in SAS exists. We found that Attention Weight is plausible and Integrated Gradients is faithful. We also found that it is possible to improve the plausibility of gradient-based methods by supervised learning. However, methods with high plausibility and high faithfulness are still different methods. The target audience in real applications is considered to be students and administrators (e.g., teachers). If the explanation of the score prediction is used to convince the student user of the scoring results or as a hint for review, the consistency with the rubric (i.e., plausibility) is considered to be important. On the other hand, if the explanation is used to allow the teacher to confirm the expected behavior of the scoring model, they are likely to place importance on faithfulness. Therefore, in SAS, it is crucial to choose the method of generating explanation according to whether the user considers plausibility or faithfulness more important.

In the field of education, it has not yet been explored as to how the presentation of explanations for score prediction by the model affects student users' learning. We plan to verify the system's effectiveness by using human subjects.

References

1. Riken: RIKEN Dataset for Short Answer Assessment. Informatics Research Data Repository (2020). https://doi.org/10.32130/rdata.3.1
2. Scoring short answer essays. ASAP short answer scoring competition system description (2013). https://www.kaggle.com/c/asap-sas/
3. Basu, S., Jacobs, C., Vanderwende, L.: Powergrading: a clustering approach to amplify human effort for short answer grading. In: TACL, pp. 391–402 (2013). https://doi.org/10.1162/tacl_a_00236
4. Cohen, J.: Weighted kappa: nominal scale agreement with provision for scaled disagreement or partial credit. Psychol. Bull. **70**, 213–220 (1968). https://doi.org/10.1037/h0026256
5. DeYoung, J., et al.: ERASER: a benchmark to evaluate rationalized NLP models. In: ACL, pp. 4443–4458 (2020). https://doi.org/10.18653/v1/2020.acl-main.408

6. Funayama, H., et al.: Preventing critical scoring errors in short answer scoring with confidence estimation. In: ACL, pp. 237–243 (2020). https://doi.org/10.18653/v1/2020.acl-srw.32
7. Guidotti, R., Monreale, A., Ruggieri, S., Turini, F., Giannotti, F., Pedreschi, D.: A survey of methods for explaining black box models. ACM Comput. Surv. **51**, 1–42 (2018). https://doi.org/10.1145/3236009
8. Herman, B.: The promise and peril of human evaluation for model interpretability, 6 p. arXiv:1711.07414 (2019)
9. Jacovi, A., Goldberg, Y.: Towards faithfully interpretable NLP systems: how should we define and evaluate faithfulness? In: ACL, pp. 4198–4205 (2020). https://doi.org/10.18653/v1/2020.acl-main.386
10. Jain, S., Wallace, B.C.: Attention is not explanation. In: EMNLP-IJCNLP, pp. 3543–3556 (2019). https://doi.org/10.18653/v1/N19-1357
11. Lage, I., et al.: An evaluation of the human-interpretability of explanation, 24 p. arXiv:1902.00006 (2019)
12. Mikolov, T., Sutskever, I., Chen, K., Corrado, G.S., Dean, J.: Distributed representations of words and phrases and their compositionality. In: NeurIPS, p. 9 (2013)
13. Mizumoto, T., et al.: Analytic score prediction and justification identification in automated short answer scoring. In: ACL, pp. 316–325 (2019). https://doi.org/10.18653/v1/W19-4433
14. Mohankumar, A.K., Nema, P., Narasimhan, S., Khapra, M.M., Srinivasan, B.V., Ravindran, B.: Towards transparent and explainable attention models. In: ACL, pp. 4206–4216 (2020). https://doi.org/10.18653/v1/2020.acl-main.387
15. Mohler, M., Bunescu, R., Mihalcea, R.: Learning to grade short answer questions using semantic similarity measures and dependency graph alignments. In: ACL, pp. 752–762 (2011)
16. Riordan, B., Horbach, A., Cahill, A., Zesch, T., Lee, C.M.: Investigating neural architectures for short answer scoring. In: BEA, pp. 159–168 (2017). https://doi.org/10.18653/v1/W17-5017
17. Serrano, S., Smith, N.A.: Is attention interpretable? In: ACL, pp. 2931–2951 (2019). https://doi.org/10.18653/v1/P19-1282
18. Shrikumar, A., Greenside, P., Shcherbina, A., Kundaje, A.: Not just a black box: learning important features through propagating activation differences, 6 p. arXiv:1605.01713 (2017)
19. Simonyan, K., Vedaldi, A., Zisserman, A.: Deep inside convolutional networks: visualising image classification models and saliency maps, 8 p. arXiv:1312.6034 (2014)
20. Sundararajan, M., Taly, A., Yan, Q.: Axiomatic attribution for deep networks. In: ICML, p. 3319–3328 (2017)
21. Teney, D., Abbasnejad, E., van den Hengel, A.: Learning what makes a difference from counterfactual examples and gradient supervision, 24 p. arXiv:2004.09034 (2020)
22. Tjoa, E., Guan, C.: A survey on Explainable Artificial Intelligence (XAI): toward medical XAI. IEEE Trans. Neural Netw. Learn. Syst. **32**, 4793–4813 (2021). https://doi.org/10.1109/TNNLS.2020.3027314
23. Wang, T., Inoue, N., Ouchi, H., Mizumoto, T., Inui, K.: Inject rubrics into short answer grading system. In: EMNLP-WS, pp. 175–182 (2019). https://doi.org/10.18653/v1/D19-6119
24. Zhong, R., Shao, S., McKeown, K.R.: Fine-grained sentiment analysis with faithful attention, 13 p. arXiv:1908.06870 (2019)

Experts' View on Challenges and Needs for Fairness in Artificial Intelligence for Education

Gianni Fenu, Roberta Galici(✉), and Mirko Marras

University of Cagliari, Cagliari, Italy
{fenu,roberta.galici}@unica.it, mirko.marras@acm.org

Abstract. In recent years, there has been a stimulating discussion on how artificial intelligence (AI) can support the science and engineering of intelligent educational applications. Many studies in the field are proposing actionable data mining pipelines and machine-learning models driven by learning-related data. The potential of these pipelines and models to amplify unfairness for certain categories of students is however receiving increasing attention. If AI applications are to have a positive impact on education, it is crucial that their design considers fairness at every step. Through anonymous surveys and interviews with experts (researchers and practitioners) who have published their research at top-tier educational conferences in the last year, we conducted the first expert-driven systematic investigation on the challenges and needs for addressing fairness throughout the development of educational systems based on AI. We identified common and diverging views about the challenges and the needs faced by educational technologies experts in practice, that lead the community to have a clear understanding on the main questions raising doubts in this topic. Based on these findings, we highlighted directions that will facilitate the ongoing research towards fairer AI for education.

Keywords: Education · Fairness · Data mining · Machine learning

1 Introduction

Educational systems that rely on artificial intelligence (AI) are increasingly influencing *the quality of the education we receive*. Notable examples of AI-based models integrated so far include early predictors of student success (e.g., [15]), clustering techniques for learner modelling (e.g., [2]), intelligent tutoring and scaffolding (e.g., [1]), agents for motivational diagnosis and feedback (e.g., [5]), and models for recommending peers or learning material (e.g., [20]). With this growth, the potential of AI to *amplify unfairness in educational applications* has received growing attention in both academia and industry as well as the press. Indeed, articles in mainstream media have reported systemic unfair behaviors of some AI-enabled educational systems. For example, an automated college enrolment system more likely to recommend enrolments from certain ethnic, gender,

M. M. Rodrigo et al. (Eds.): AIED 2022, LNCS 13355, pp. 243–255, 2022.
https://doi.org/10.1007/978-3-031-11644-5_20

or age groups [7], or a machine-learning system for evaluating PhD applicants in Computer Science that exacerbates current inequality in the field [17].

Concerted effort in the area of fairness in educational applications has mainly focused on the *design and operationalization of fairness definitions* [19] and *algorithmic methods to assess and mitigate undesirable biases* in relation to these definitions [9]. Other works have studied fairness in educational AI systems through a social and psychological lens [6]. As the field matures, literature reviews are collecting definitions, methods, and perspectives in a unified framework [10]. Specialized initiatives, such as workshops [3] and special issues [4], are also focusing on biases and unfairness in educational AI. While some fair AI studies are already being researched, they often represent isolated examples. If the resulting applications are to have a positive impact on education, however, *fairness-aware practices should become common for any person while creating an educational application that leverages AI.* Being informed about the actual challenges and needs for supporting the development of *fairer AI for education* is hence crucial.

Creating AI-based educational systems raises many unique challenges not commonly faced with intelligent systems in other domains [16]. In the broader AI field, several papers have dealt with fairness [8,11,13,14]. Despite this attention to unfairness, to the best of our knowledge, only two prior studies have investigated challenges and needs for supporting the creation of fairer AI by directly interrogating experts [10,18]. Unlike those studies, focused on the public-sector and commercial AI practitioners across a range of high-stakes contexts, our study focuses on educational researchers and practitioners (referred for convenience as experts) including AI in their workflow, who usually have experience in developing AI systems, but are relatively new to thinking about fairness. Considerations, beliefs, practices, motivations, and priorities in integrating fairness may be less clear within these contexts, applications, and cultures.

Our study in this paper investigates the challenges and needs faced by the educational community, whose products are going to affect the education of individuals, in *integrating and monitoring for unfairness and taking appropriate action.* Through an anonymous survey of 136 educational researchers and practitioners who have published their research at top-tier educational conferences in 2021 (e.g., AIED, EDM, and L@S), we analyze teams' existing opinions and experience around the development of fair educational AI, as well as their challenges and needs for support. To deeper the key themes of our survey, we also conducted semi-structured interviews with 29 of those researchers and practitioners. To our knowledge, this is the *first systematic investigation of experts' challenges and needs around fairness in educational AI.*

Through our investigation, we identify a range of real-world needs often not stated in the literature so far, as well as several common areas. For example differently from the broader AI field, large-scale data collection is not always considered as a solution, since biases are driven by complex reasons to be understood in the local context. Such concerns are also extended to research teams' own blind spots, since teams often struggled to anticipate the sub-populations and forms of unfairness they need to consider for specific kinds of applications. Moreover, though fair educational AI has overwhelmingly focused on data collection issues,

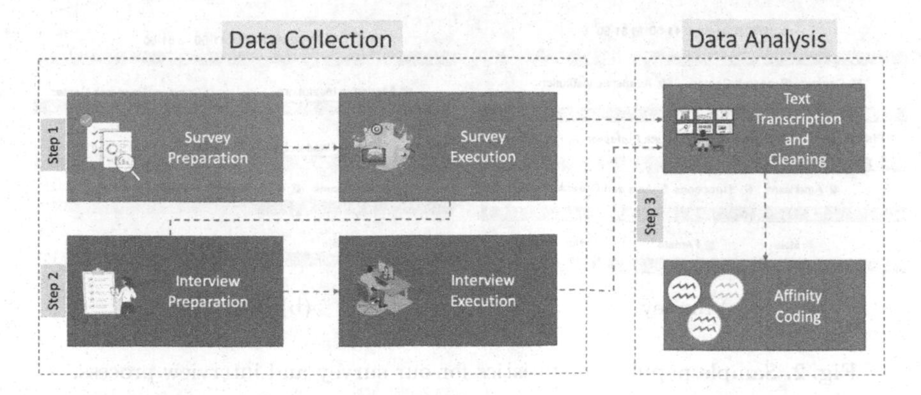

Fig. 1. We first conducted an anonymous survey (Step 1). To deeper the key themes from the survey, we then conducted semi-structured, one-on-one interviews (Step 2). We finally analyzed answers from both the surveys and the interviews (Step 3).

assessment and debiasing of unfairness are also crucial, e.g., by having continuous fairness assessment at all stages of the pipeline. Because fairness can be context and application dependent, domain-specific educational resources, metrics, processes, and tools are urgently needed, such as to open data and source code for public scrutiny and create participatory processes for fairness checking. Another identified area is the development of auditing processes and tools, to make fairness issues emerge. Based on these findings, we highlight opportunities to have a greater impact on the fair educational AI landscape.

2 Methodology

The goal of this study is to investigate the challenges and needs faced by the educational community, whose products are going to affect the education of individuals, in integrating and monitoring for unfairness and taking appropriate action. To this end, we adopted a two-step mixed approach, depicted in Fig. 1.

2.1 Survey Study Implementation

In a first step, to get a broad sense of challenges and needs for addressing fairness while developing educational AI, we conducted an anonymous online survey.

Participants Recruitment and Statistics. The survey participants were recruited using a systematic process, to make sure that our study was not based on opinions of an arbitrarily selected population. Specifically, we systematically scanned the 2021's proceedings of the top-tier educational conferences in a manual process, namely AIED, EAAI, EC-TEL, EDM, ICALT, ITS, LAK, and L@S, to identify the authors who had a paper accepted. We also considered the authors of papers in the special issues about fair educational AI in IJAIED. We then directly emailed the survey to those authors between Sep and Dec 2021, and

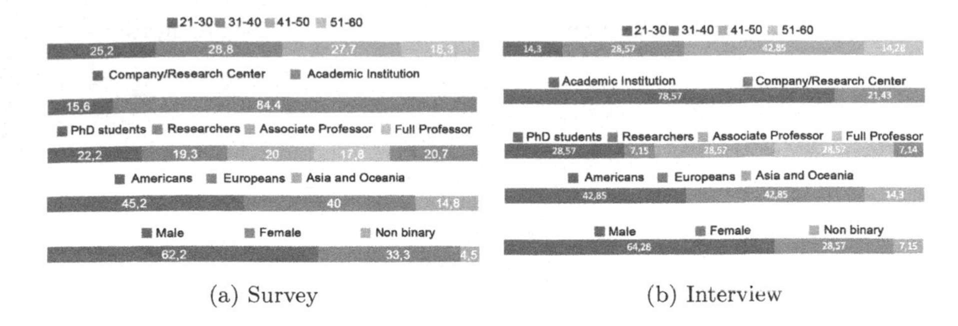

Fig. 2. Sample population statistics for our survey and interview process.

often invited them to pass the survey on to colleagues working on educational AI, within their organization. In total, we contacted 2,175 experts, and 136 out of them (6%) completed at least one section beyond demographics. A description of the population who answered to the survey is provided in Fig. 2a.

Survey Execution. We structured the survey as a Google Form and developed survey questions to investigate the prevalence and generality of emerging themes. First, we asked a set of demographic questions to understand our respondents' provenience and backgrounds, including their technology area(s) and role(s). In a branching sequence of survey sections, respondents were then asked about their opinions, challenges, and needs for support around fairness, with each section pertaining to one stage of the educational AI development pipeline[1]. For each of the latter questions, we provided open-ended response options that allowed respondents to elaborate on their arguments. We finally asked them to leave their email address in case there were willing to participate in a subsequent interview.

2.2 Interview Study Implementation

In a second step, to validate and deeper our findings from the previous survey answers, we then conducted a series of semi-structured, one-to-one interviews.

Participants Recruitment and Statistics. In Jan 2022, we involved the experts who answered the survey and gave their availability to participate in a follow-up interview by providing their e-mail addresses in the last question of the survey. In total, 29 out of 136 survey respondents (21%) were willing to participate in this second step, across as many research teams. Whenever possible, we tried to interview people in different roles on the same team to hear (potentially) different perspectives. Each interview lasted 30 min and was conducted remotely since involved people came from a diverse set of countries. Key statistics about our sample sub-population are depicted in Fig. 2b.

Interview Execution. Each participant was first reminded of the purpose of our research. Then, the interview focused on the participant awareness about

[1] A pdf copy of the survey questions is available at https://bit.ly/FairAIEdSurvey.

the debate and research on demographic fairness in educational AI and the most important challenges and open questions in the area in general (questions 7 and 8 of the survey). Each participant was then asked to better clarify the educational AI applications they are working on and who the target users of these applications are (question 8 of the survey). Interviewees were then asked whether fairness is regularly considered in their workflow and what it means for them to be fair in their applicative context (question 9 of the survey). Discussion points were often prompted from the survey answers.

We then provide survey questions about fairness at each stage in their educational AI development pipeline, from collecting data to designing datasets to assessing and potentially mitigating fairness issues (questions 10 to 13 of the survey). For each of these stages, interviewees were asked a broad opening question in line with the one reported in the corresponding question of the survey and follow-up questions based on the answers provided in the survey. This follow-up led interviewees to reflect more deeply on their practices.

2.3 Survey and Interview Data Analysis

We integrate the answers collected from each survey question with the corresponding interview counterpart. Interviews and surveys were identified with an ID, and the same ID was used if both come from the same participant. To synthesize findings using standard methodology from contextual design, we conducted interpretation sessions and adopted affinity diagramming (e.g., see also [12]). Specifically, we employed bottom-up affinity diagramming to iteratively generate codes for various individual texts and then group these codes into successively higher-level themes. The themes emerged from the data rather than being imposed. Key themes are presented in the following section.

3 Results and Discussion

In the following, we discuss current challenges and needs around fairness, organized by top-level themes associated to the survey questions and the interview-based deepening (the latter served as a confirmation and enrichment to the survey responses), granularly framed according to the resulting affinity diagram. These include needs and challenges on data collection and modelling (questions 9–10), unfairness detection and mitigation (question 11), fairness guarantees provision (question 12), holistic fairness auditing (question 13), paired with systemic aspects such as team composition, cross-organization collaborations, and educational AI maturity (question 7). Within each top-level theme, we present selected sub-themes. *It should be noted that our study aims more to uncover open questions than providing answers. The latter requires further discussion and work of the research community as whole*, also driven by our study.

3.1 Challenges and Needs in Data Collection and Grouping (Q9–10)

Cultural Dependencies in Demographic Representation. The majority of the participants recognizes that researchers, which are often not demographically representative of their societies, tend to involve people at their hand (their students), which are often not representative either. For instance, in question 10 of the survey, a participant reported that *"it is difficult to collect data enough representative of different contexts, such as countries, universities, and society, due to different culture, viewpoints and rules"*. The same participant, during the interview, further observed that *"most research focuses on certain countries, and is done in English, so findings are more representative of certain societies and educational systems"*. Overall, it was often pointed out that no dataset represents the diversity of the population, leaving always some people underrepresented.

Biases are Driven by Reasons to be Understood in the Local Context. Paired with the above point, differently from the broader AI field, several participants do not think that large-scale data collection will really get at the nuances of fairness in educational AI. For instance, in question 10 of the survey, a participant highlighted that *"unlike some AI applications where very broad groups are relevant (e.g., face recognition), biases in education are driven by complex reasons to be understood locally"*. This clearly calls for localized data collection paired with data sharing practices.

Hidden Relationships Between Demographics and Learning Variables. Our participants found that it is generally difficult to identify issues that really drive fairness. For example, during the interview, a participant reported that *"in some cases, ethnicity does not directly cause differences in how students interact with educational software or the data coming out of it, but rather students' life experiences that are correlated with ethnicity"* (e.g., feeling uncomfortable in class because discrimination). In general, it was observed that different demographic groups might have different ways of responding to psychological measures. A participant envisioned during the interview that *"educational AI models might need to be demographically stratified"*. Overall, challenges emerged on what the demographic attributes really proxy for and how experts can measure that.

Giving Individuals Continuous Control of Their Data. A vast segment of our participants acknowledges that individuals should always have complete access and control over their data as well as new data created about them. One participant, following up on what reported in the survey answer, suggested that *"access should be controlled in such a way that confidential information will not be inadvertently shared beyond their control"*. From our affinity diagram, it seems clear the need of supporting tools to inform the users about which data the system is using and for what. In this sense, another participant envisioned in the survey that *"in these tools, the user may turn on and off on which data they think the system needs to use"*. Overall, challenges and needs emerged on letting individuals have control on which of their data is used in any educational system.

3.2 Challenges and Needs of Fairness-Aware Technical Pipelines (Q11)

Continuous Fairness Assessment at all Stages of the Pipeline. Most of our participants emphasized that fairness should be taken into account from the beginning, and that all choices (data, optimization criteria, interventions) should be viewed from a fairness point of view. For instance, in question 11 of the survey, a participant envisioned that *"aims and objectives of the data collection have to be clearly defined and negotiated with the participants, through explicit discussions"*. Fairness is recognized as to be part of the design process from multiple lenses, from developing a team which can reach a high level of expertise in fairness until deploying the system. Overall, protocols and guidelines on how to include fairness through the pipeline are therefore needed.

Understanding and Acknowledging Weaknesses of the System. In designing educational AI systems, a segment of our participants recognizes as important to understand where systems work and the cases where they may suffer. In the survey, a participant suggested that *"this begins by understanding the scope and limitations of the data on which the systems are based, since it is often infeasible to achieve full transparency or explainability in regard to these systems"*. The way of working could also enable to keep the models as they are but inform users about the difference in accuracy between certain demographic groups, for instance. Overall, there is a need to acknowledge these aspects continuously to reach a better understanding of the strengths and limitations of the systems.

Reducing Frictions Between Model Effectiveness and Fairness. Our participants consistently mention that the main challenge is how to balance the accuracy of predictions with fairness. For example, a participant mentioned in the survey that *"we might get a good accuracy while directly using demographic features (e.g., gender), but the value of those features might be that they encode something else (e.g., prior experience)"*. This leads for instance to observe that models can achieve the same performance without using demographic features (e.g., prior experience). Moreover, if a model performs really well for one group and poorly for another, a participant found *"debatable whether the benefit should be withheld from the group it works well for"*. There may be alternative practices that still lead to fair uses, even if fairness is not incorporated into a model.

Creating Cross-Institutional Frameworks for Addressing Fairness. Some respondents highlighted the need of creating a consortium of organizations (e.g., universities and companies) from different countries and defining a unified framework for data collection and fairness-aware model evaluation applicable in all those universities. For instance, a participant in the interview said that *"there should be an increasing trust between government, institutions, researchers, and practitioners to access sensitive data, building on top of privacy regulations for de-identified data sharing in educational systems"* (e.g., by accessing data at institution level). However, some participants identified that leveraging data, even when anonymized, to improve educational systems is still challenging.

3.3 Challenges and Needs in Providing Fairness Guarantees (Q12)

Opening Data and Source Code for Public Scrutiny. In this line, a range of participants highlighted that people who develop educational AI systems should publish or release models and analyses for public scrutiny, especially when there are concerns about their models being unfair. For instance, a participant found important that *"sharing data, source code, and pre-trained models in open online repositories should be encouraged"*. Overall, this practice clearly goes in contrast to copyright, therefore creating guidelines and directives that regulate this sharing process is a major requirement to advance from these perspective.

Fairness Should not be a Property of the Model Only. In response to question 13 of the survey, several participants emphasized that a common practice is to try and offer as fair results as possible in AI models, but there might still be biases when using these models as a final service to the user (i.e., the way the model is deployed in the complex educational ecosystem). For instance, a participant reported that *"fairness should be therefore a property of the offered service as well at the end"*. Hence, there is a need, for a large segment, to envision guidelines and practices that embed fairness as a constraint or metric for the underlying predictive model as well as a key indicator for the service.

Showing Explicit Evidence the System's Potential Unfair Impacts. A segment of the participants believes that institutions adopting educational AI need data that supports a claim of the tool being fair. A participant believes that *"such aspect should not be explainable or transparent to students, since it is probably the case that drawing students' attention to demographics invokes stereotype threat, which might be in contrast with other participants ideas"*. For another segment, students should have the right to know how a system works and be informed on any shortcoming about fairness that the system might have. *"They should not feel that a system might guide them to take a decision because some fairness guarantees of the system are not met by the system"* is an example reported by another participant. Another interesting aspect identified by our participants is that the extent of transparency, accountability, and explainability should depend on the level of impact the system has. For example, a participant reported that *"decisions on weather a student must repeat a course should probably offer more accountability than small recommendations on a platform"*.

Creating Participatory Processes for Fairness Checking. A segment of our participants identified that, ideally, there should be an independent third-party entity that should be able to provide sample data to the educational AI service and then assess whether such service is fair. During the interview, a participant followed up on this point, envisioning that *"all aspects should be showcased to an ethical commission within the organization or that a learner advocate should be allowed to conduct exploratory research into how the system might have detrimental effects on later learner success"* (e.g., showing the system is fair by finding and correcting negative consequences). Who and how should be involved is an open question for the community and is expected to be defined by future advances.

Regulations for Defining Responsibilities Around Fairness Issues. Several participants found that, in fair educational AI systems, it is important to define who is the guarantor of the system's fairness and what are the consequences for not living up to the guarantee. As an example, a participant often compared this aspect to *"service level agreements provided by cloud services, where failing to live up to a guarantee might result in a financial penalty"*. Several of the participants emphasized that it should become mandatory to guarantee an overall fair treatment and to certify that the model is fair according to certain variables.

3.4 Challenges and Needs of a More Holistic Fairness Auditing (Q13)

Human-Centered Evaluation of Fairness. Most of the participants stated that the evaluation should be human-centered and cover different levels of analysis. A participant, while deepening question 13 of the survey during the interview, *"the evaluation should consider statistical metrics, expert audits regarding system design and training data sets, and meetings with stakeholders representing the most impacted groups"*. From the responses, it emerged that this multi-level evaluation gives the opportunity to make sure that the evaluation protocol is properly adapted to the specific applicative case. To make the protocol more efficient, some respondents also suggested that some parts of the evaluation protocol can be automated, but stakeholders must stay involved in any case.

Creation of Tools that Allow Stakeholders to Audit Models. Our participants highlighted making educational AI transparent to the end users is essential and envision a need of letting stakeholders analyze system data for fairness and outcomes, of course. A participant dug deeper into this aspect during the interview, saying that *"it would be necessary to have experts doing this, otherwise other stakeholders will have to spend a lot of time and resources"*. Overall, a common view is that the student (or the instructor) should understand why they are given a certain prediction so that they can reflect and react constructively. Some of the respondents observed that it might be tough to deeply explain AI-driven educational predictions to students, especially young students, though they should always have some high-level understanding of what the system is doing, also after some specific training on this task.

Contextualized and Application-Specific Properties to Inspect. The majority of the participants raised doubts regarding the extent to which the fairness metrics and protocols defined in the broader AI community can also work for educational AI systems. For instance, a participant expected that *"the fairness spectrum for AI in the educational field should be investigated, and a framework to be adapted to every context and application might be the outcome"*. Overall, tailored frameworks of processes and properties (also depending on the local data privacy and protection laws) should be better aligned to the specificity of the educational field, rather than being merely built upon those of other domains as black boxes. Indeed, education is recognized as a highly human-centered rather than data-centric field, and tailored frameworks become almost mandatory.

Long-Term Learning-Related Evaluation of Fairness. Our participants generally emphasized that it exists an overly computational definition of fairness in the broader AI field, that tends to evaluate the demographic differentials instead of identifying the strengths and weaknesses of each individual and with that help them to reach their full potential. Several participants identified that the concept is more complex than a metric or a protocol. For instance, during the interview, a participant emphasized that *"all the time is spent on optimising against one of these metrics, leading to unfair outcomes being declared fair just due to the model performance on the metric"*. Ideally, a common emerging view was that a system would be evaluated not just on the immediate intended effect (e..g, if an auto-generated hint helps students answer a question) but on broader educational goals (e.g., the student do well in classes).

3.5 Challenges and Needs in Team Blind Spots and Practices (Q7)

Support in the Selection of the Demographic Groups to Consider. Several participants think that the decision on which demographic attributes to consider in an analysis is challenging. In line with this, while answering to question 7 of the survey, a participant raised a doubt that *"there might not be a need to add gender data to certain problems as it should not be relevant or, on the other hand, it should be included to show it is irrelevant"*. Another participant emphasized that *"many studies do not consider social-economic characteristics of the students which may also bias the resulting models"*. In addition, interviews strengthened the view on the lack of datasets that do not just include all demographics, but also fair observations. For example, datasets containing salaries for people are probably still biased as salaries were decided by humans with their own biases. However, it is hard to define how an actual unbiased dataset should look like, and it is likely that all quantitative data have biases of some form or another. Overall, there is a need for standards on the demographic groups to consider.

Building Social and Multi-disciplinary Awareness in Teams. Several respondents highlighted that educational AI models are designed by technical people who often do not have the social science training to understand the socio-cultural implications of their algorithm designs. This might lead to prefer computational expediency over considerations of social justice. In this regard, a participant emphasized *"the need for equity training and understanding for developers and researchers"*. Overall, our respondents call for inclusivity and diversity in the teams in charge of developing educational AI. In current practices, several participants identified that most of the AI models are constructed with a single mindset from a specific field (generally Computer Science). Since data and models are about and impact on people, it would be important to include other perspectives from social sciences to understand why and when the variables to be collected are valid.

4 Conclusion and Future Directions

Though researchers and practitioners are already grappling with biases and unfairness in educational AI systems, research on this topic is rarely guided by a common understanding and view of the faced challenges and needs. In this work, we conducted the first systematic investigation of experts teams' challenges and needs for support in developing fairer educational AI. Even when experts are motivated to improve fairness in their educational applications, they often face technical and organizational barriers. We highlight a few emerged aspects below:

- Future research should also support experts in collecting and curating high-quality datasets, with an eye towards fairness in downstream AI models, reducing cultural dependencies in demographic representation. Moreover, large-scale data collection should be paired with an in-depth description of the local contexts, since biases are driven by complex reasons to be understood locally. Localized and causal data collection paired with data sharing practices are needed, posing attention in giving individuals control of their data.
- Though fair educational AI has mainly focused on data collection, assessment and debiasing of unfairness is also an important area of work. Challenges and needs in this area include having continuous fairness assessment at all stages of the pipeline, understanding and acknowledging the potential weaknesses of the system, reducing frictions between model effectiveness and fairness, and creating cross-institutional frameworks for addressing fairness.
- Domain-specific educational resources, metrics, processes, and tools are urgently needed. Challenges and needs in this perspective include, among others, practices for opening data and source code for public scrutiny, including fairness not only as a property of the AI model, showing explicit evidence the system's potential unfair impacts, creating participatory processes for fairness checking, and defining responsibilities around fairness issues.
- The development of processes and tools for fairness-focused auditing is also important, to surface fairness issues in complex, multi-component educational AI systems. Among others, challenges and needs include fostering a more focused human-centered evaluation of fairness, contextualized and application-specific tools for auditing, and long-term learning-related auditing of fairness.
- Finally, another area with several challenges and needs concern the teams working on educational AI. Among others, supporting the team in the selection of the demographic groups to consider and building multi-disciplinary awareness in teams are two of the more relevant aspects to work on.

The rapidly growing area of fairness in educational AI presents many challenges and needs. The resulting systems are increasingly widespread, with proved potential to amplify social inequities, or even to create new ones. As research in this area progresses, it is urgent that research agendas are aligned with the challenges and needs of those who affect and are affected by educational AI systems. We view the directions outlined in this paper as critical opportunities for the AI and the educational research communities to play more active, collaborative roles in making real-world educational AI systems fair.

Acknowledgments. Roberta Galici gratefully acknowledges the University of Cagliari for the financial support of her Ph.D. scholarship.

References

1. Abidi, S.M.R., Hussain, M., Xu, Y., Zhang, W.: Prediction of confusion attempting algebra homework in an intelligent tutoring system through machine learning techniques for educational sustainable development. Sustainability **11**(1), 105 (2019)
2. Abyaa, A., Khalidi Idrissi, M., Bennani, S.: Learner modelling: systematic review of the literature from the last 5 years. Educ. Tech. Res. Dev. **67**(5), 1105–1143 (2019)
3. Alameda-Pineda, X., Redi, M., Otterbacher, J., Sebe, N., Chang, S.F.: FATE/MM 20: 2nd international workshop on fairness, accountability, transparency and ethics in multimedia. In: Proceedings of the 28th ACM International Conference on Multimedia, pp. 4761–4762 (2020)
4. Alboaneen, D., Almelihi, M., Alsubaie, R., Alghamdi, R., Alshehri, L., Alharthi, R.: Development of a web-based prediction system for students' academic performance. Data **7**(2), 21 (2022)
5. Alsuliman, T., Humaidan, D., Sliman, L.: Ml and AI in the service of medicine: necessity or potentiality? Curr. Res. Transl. Med. **68**(4), 245–251 (2020)
6. Berendt, B., Littlejohn, A., Blakemore, M.: AI in education: learner choice and fundamental rights. Learn. Media Technol. **45**(3), 312–324 (2020)
7. Britto, J., Prabhu, S., Gawali, A., Jadhav, Y.: A machine learning based approach for recommending courses at graduate level. In: 2019 International Conference on Smart Systems and Inventive Technology (ICSSIT), pp. 117–121. IEEE (2019)
8. Caton, S., Haas, C.: Fairness in machine learning: a survey. arXiv preprint arXiv:2010.04053 (2020)
9. Hajian, S., Bonchi, F., Castillo, C.: Algorithmic bias: from discrimination discovery to fairness-aware data mining. In: Proceedings of the 22nd ACM SIGKDD International Conference on Knowledge Discovery and Data Mining, pp. 2125–2126 (2016)
10. Holstein, K., Wortman Vaughan, J., Daumé III, H., Dudik, M., Wallach, H.: Improving fairness in machine learning systems: what do industry practitioners need? In: Proceedings of the 2019 CHI Conference on Human Factors in Computer Systems, pp. 1–16 (2019)
11. Iosifidis, V., Fetahu, B., Ntoutsi, E.: FAE: a fairness-aware ensemble framework. In: 2019 IEEE International Conference on Big Data (Big Data), pp. 1375–1380. IEEE (2019)
12. Liu, J., Eagan, J.: ADQDA: a cross-device affinity diagramming tool for fluid and holistic qualitative data analysis. Proc. ACM HC Interact. **5**(ISS), 1–19 (2021)
13. Mehrabi, N., Morstatter, F., Saxena, N., Lerman, K., Galstyan, A.: A survey on bias and fairness in ML. ACM Comp. Surv. (CSUR) **54**(6), 1–35 (2021)

14. Mitchell, S., Potash, E., Barocas, S., D'Amour, A., Lum, K.: Prediction-based decisions and fairness: a catalogue of choices, assumptions, and definitions. arXiv preprint arXiv:1811.07867 (2018)
15. Rastrollo-Guerrero, J.L., Gomez-Pulido, J.A., Duran-Dom., A.: Analyzing and predicting students' performance by means of ml: a review. Appl. Sci. **10**(3), 1042 (2020)
16. Renz, A., Hilbig, R.: Prerequisites for artificial intelligence in further education: identification of drivers, barriers, and business models of educational technology companies. Int. J. Educ. Technol. High. Educ. **17**(1), 1–21 (2020). https://doi.org/10.1186/s41239-020-00193-3
17. Shahbazi, Z., Byun, Y.C.: Toward social media content recommendation integrated with data science and ml approach for e-learners. Symmetry **12**(11), 1798 (2020)
18. Veale, M., Van Kleek, M., Binns, R.: Fairness and accountability design needs for algorithmic support in high-stakes public sector decision-making. In: Proceedings of the 2018 CHI Conference on Human Factors in Computing Systems, pp. 1–14 (2018)
19. Verma, S., Rubin, J.: Fairness definitions explained. In: 2018 IEEE/ACM International Workshop on Software Fairness (FairWare), pp. 1–7. IEEE (2018)
20. Wang, J., Molina, M.D., Sundar, S.S.: When expert recommendation contradicts peer opinion: relative social influence of valence, group identity and artificial intelligence. Comput. Hum. Behav. **107**, 106278 (2020)

Balancing Fined-Tuned Machine Learning Models Between Continuous and Discrete Variables - A Comprehensive Analysis Using Educational Data

Efthyvoulos Drousiotis[1(✉)], Panagiotis Pentaliotis[1(✉)], Lei Shi[2], and Alexandra I. Cristea[2]

[1] Department of Electrical Engineering and Electronics, University of Liverpool, Liverpool, UK
{e.drousiotis,p.pentaliotis}@liverpool.ac.uk

[2] Department of Computer Science, Durham University, Durham, UK
{lei.shi,alexandra.i.cristea}@durham.ac.uk

Abstract. Along with the exponential increase of students enrolling in MOOCs [26] arises the problem of a high student dropout rate. Researchers worldwide are interested in predicting whether students will drop out of MOOCs to prevent it. This study explores and improves ways of handling notoriously challenging continuous variables datasets, to predict dropout. Importantly, we propose a *fair comparison methodology*: unlike prior studies and, for the first time, when comparing various models, we use algorithms with the dataset they are intended for, thus 'like for like.' We use a time-series dataset with algorithms suited for time-series, and a converted discrete-variables dataset, through feature engineering, with algorithms known to handle discrete variables well. Moreover, in terms of predictive ability, we examine the importance of finding the optimal hyperparameters for our algorithms, in combination with the most effective pre-processing techniques for the data. We show that these much lighter discrete models outperform the time-series models, enabling faster training and testing. This result also holds over fine-tuning of pre-processing and hyperparameter optimisation.

Keywords: Neural networks · Tree-based algorithms · Educational data mining · Feature engineering · MOOCs

1 Introduction

With the rapid development of the Internet and in combination with the growing training demands, the education industry has changed the way it operates. Massive Open Online Course (MOOC) platforms were introduced to the world, which has attracted millions of users [26]. This led to a revolution of big data in learning, with more resources and anonymised datasets for exploration. Over the years, an

M. M. Rodrigo et al. (Eds.): AIED 2022, LNCS 13355, pp. 256–268, 2022.
https://doi.org/10.1007/978-3-031-11644-5_21

undeniable challenge in online learning became to find ways to reduce and predict students' dropout rates, which fall roughly at 77%–87%. Many studies have been conducted to explore learning behavioural patterns, through statistical modelling and machine learning, towards predicting students' dropout [11,19]. Nonetheless, the majority of the studies, such as [30,31], used the same dataset and variables to implement predictive models, without taking into consideration the type of variables each model uses for maximising its performance. For example, [30] trained a time-series, Long Short-Term Memory (LSTM) model, using the same dataset that was used to train other non-time series machine learning models, including Logistic Regression, Random Forest, and Gradient Boosting Decision Tree (GBDT). The results showed that time-series models, such as LSTM, outperformed other machine learning models (i.e., Linear Regression and Decision Tree), and achieved higher accuracy, precision and recall when compared using data from their 'natural' environment (continuous/time-series variables). However, we argue that previous methods did not consider the functionality of the algorithms and how they could perform best, according to their nature. Some very preliminary previous research [10] has hinted that it may be a good practice to use sequential time-series 'as is', or first convert the dataset into discrete-variables, for obtaining enhanced metrics (precision, recall, f1-score, accuracy) on predicting students' dropout, when the models would be tuned, and the datasets would be pre-processed. The current paper aims to determine if traditional fine-tuning and optimisation methods can change this, or if the conversion into discrete variables is as robust as we assumed. We use the same application of predicting 'completers' and 'non-completers' with the same dataset to analyse this. We examine thus the following research question:

Can pre-processing, fine-tuning and hyperparameter optimisation change the balance between using time-series 'as is', or converting them into discrete variables?

The main contributions of this study are thus to perform, for the first time, a wide-scale analysis, showing, firstly, that discrete-feature methods outperform sequential time-series methods, on both discrete and sequential datasets. Secondly, we show that this result is further consistent, when performing model hyperparameter optimisations and optimal feature engineering. Our results, furthermore, outperform all other studies using the same dataset [15,17,20,21,24] in predicting dropouts. Moreover, as we compare several approaches, our work also shows that methods of capturing uncertainty outperform the others. This supports the more generic approach to converting the dataset, whenever possible, into the appropriate formats (in our case, time-series into discrete), which helps a different kind of predictive model than the default applied in previous studies, achieving faster training, testing, and predicting, as well as higher predictive accuracy and in general better performance compared to using multi-layer neural networks.

2 Related Work

Learning Analytics (LA) is the process of analysing and reporting multiple learners' data to understand and optimise their performance and the learning environment. Many recent studies focused on classifying students into 'completers' (i.e. students who completed the course) and 'non-completers'. Some of them [2,19] used traditional machine learning algorithms (e.g. Decision Tree, Logistic Regression, Random Forest), while others, such as [12,14], used more advanced algorithms (e.g. Neural Networks).

A few recent studies also [30,31] utilised both traditional machine learning algorithms and more advanced ones (Neural Networks). These studies [30,31] used the same dataset to train both Neural Networks and machine learning models (time-series), which showed that Neural Networks outperformed the other machine learning techniques.

However, Tensorflow[1]suggested that to train an LSTM, it is best to use a time-series dataset, while [13] suggested using discrete variables to train a tree-based algorithm (either categorical or continuous variables).

Moreover, some works [16,32] suggested that artificial Recurrent Neural Networks (RNN) with memory, such as Long-Short-Term-Memory (LSTM), are generally considered as superior models for time-series tasks, due to their nature - the way they operate and handle data. On the other hand, [28] suggested that traditional machine learning algorithms, such as Logistic Regression, Random Forest and GBDT, produce better results with discrete-variable data. The only study we could find that compared four benchmark models with their intended datasets [10], lacked, however, a thorough examination of possible outcomes after pre-processing and hyperparameter optimisation.

In our case, we convert the time-series data, through feature engineering, into discrete variables, and train each model on the type of data it can process best.

Furthermore, some current works used a combination of different types of data, when those were available, in order to obtain higher accuracy with the LSTM model. For example, [22,23] examined a combination of time-series data and other discrete data, which included features such as first quiz results, the number of playbacks of a video. The primary key aspect of our methodology is that we do not use any other data than video interactions, to *fairly* analyse different formats of this data for different algorithms.

Several studies, such as [15,17,24], conducted student dropout prediction (from MOOCs) on the same dataset that we use in this paper. Some of them [20,21,24] did not perform any hyperparameter optimisation on their predictive models, but they pre-processed their data; whilst others, such as [15], performed a basic feature engineering and hyperparameters tuning. Specifically, [24] used AME, a meta-embedding technique, through which they derived the optimal embedding for each sequence of object embedding vectors, and a temporal classification, which modelled the temporal nature of the data over multiple days. [20] filtered the logs, which contained unrelated events from users, rows with

[1] https://www.tensorflow.org/tutorials/structured_data/time_series.

missing values, and columns with no helpful information. [17] highlighted the unbalanced nature of the dataset, and to achieve better classifier performance, they applied the synthetic minority oversampling technique known as SMOTE on the training set. However, generating synthetic examples may increase the overlapping of classes and introduce additional noise, while the initial distribution of the dataset is compromised, such that it no longer corresponds to real-world data. Finally, [15] performed thorough hyperparameter tuning, while the data pre-processing techniques used were not to explore or enhance the data, but just to make the data compatible with their predictive model. As mentioned in Sect. 1, we explore further the results of [10] with optimised hyperparameters, and optimal pre-processing techniques.

Unlike prior work, this study shows, in a comprehensive way, that it is beneficial to convert time-series data into discrete variables, when testing several machine learning algorithms. Moreover, to ensure a fair comparison, we are using the same data for several testing cases, and we explore in depth the algorithmic performance, by hyperparameter optimising all the machine learning algorithms used.

3 Method

3.1 Dataset and Data Preparation

The dataset used in this study is comprised of 300,000 interactions performed by 2,000 unique students that were registered on XuetangX[2] (launched in October 2013, one of the largest MOOC platforms in the world). The dataset contains two modules delivered in 2014 in a Self-Paced Mode (SPM), where a student can have a more flexible schedule and study during the hours that suit them the best. In order to make a fair, controlled comparison and strengthen our claims, we trained all the models (with and without tuned hyperparameters, with and without pre-processed data) with the time-series dataset and discrete variable dataset. In particular, we converted the time-series dataset into a discrete-variables dataset. In the time-series dataset, the input variables include the type of actions and the time each action was performed for all the 300,000 interaction entries. For constructing the discrete-variables dataset, we used the time-series dataset and counted the number of unique actions for each student. The table is populated by the ID, the Truth (pass or dropout) and the 14 unique types of actions the students performed, namely, *click courseware, click forum, click info, click progress, click about, close courseware, create comment, load video, create thread, pause video, play video, problem get, seek video*, and *stop video*. In total, there are 14 unique types of actions, so we engineered 14 features for 14 input variables for our predictive models. Considering the LSTM model's preprocessing in preparation of the dataset, the actions performed by each student were sequentially grouped, according to the time they were performed. Thus, the essence of the time-series was preserved while still considering the unique

[2] http://moocdata.cn/data/user-activity.

actions performed. Afterwards, the actions were transformed into a sequence of binary numbers (see Table 1), to retain the categorical (nominal, i.e. no intrinsic ordering to the categories) nature of the actions.

3.2 Data Pre-processing/Features Engineering

Our feature engineering techniques aimed not to change the data distribution, but only to feed the data into the models in the most efficient way (Table 2).

Table 1. Time-series dataset

ID	Action	Time	Truth[a]
...	...	...	...
561867	pause_video	2015-10-25T10:52:06	0
561867	play_video	2015-10-25T10:52:09	0
561867	pause_video	2015-10-25T10:58:42	0
1368125	click_about	2015-10-05T15:43:55	1
1368125	click_info	2015-10-05T15:45:53	1
708122	pause_video	2015-10-04T21:41:30	1
708122	play_video	2015-10-04T21:24:40	1
...	...	...	...

[a] Truth value is stating if the particular student passes or withdraws the module. 1-states Pass, 0-states Withdrawn.

Table 2. Sample distribution per classification category

	Dropout	Continuing study
Sample number	76470	20059
Percentage of sample	79.22%	20.78%

For the tree-based algorithms, we adopted the Term Frequency-Inverted Document Frequency (TF-IDF) technique to feed the data into our models. TF-IDF is a statistical measure that estimates how important and relevant a word is to a document. Here, we examined the importance of each action compared to the total number of actions. Generally, an action's importance increases with the number of times an action appears in the current input, which, at the same time, is counterbalanced by the frequency of that action in general.

LSTMs use sequences of numerical values, so we transformed the sequences of actions performed by each student, which are identified by words, into numbers. Specifically, we gathered all the student's actions and ordered them according to their timestamps. By doing so, we retained the time-series nature of the action sequence. We then created a dictionary of the unique words from the actions and encoded them using One-Hot Encoding. One-Hot Encoding morphs the unique numbers into sequences of 0s and 1s, which maintain the sequences' categorical nature for the LSTM.

3.3 Building the Models

We implemented an LSTM model and several tree-based machine learning models, including Decision Tree, Random Forest, and BART. In this section, we introduce how these models were built.

LSTMs can process long sequences of data (in our case, sequences of actions). In the current study, we used LSTMs to train on the temporal sequence of actions performed by the students. For the discrete variables dataset training, we considered that all the inputs were from a single timestamp, meaning that all the actions had been executed at once (i.e., no temporal order) compared to the continuous variables dataset training, where each action had a different timestamp, adding a subtle continuous/temporal feature to the data.

A single standard LSTM unit is composed of a cell vector (c_t) Eq. (3), a hidden vector (h_t) Eq. (5), an input gate (i_t) Eq. (1), an output gate (o_t) Eq. (4) and a forget gate (f_t) Eq. (2). A cell remembers values over time intervals ($t-1$, t); and the three gates regulate the flow of information, by computing a series of functions for the cell vector and the hidden vector [18].

$$i_t = \sigma(W_{xi}x_t + W_{hi}h_{t-1} + W_{ci}c_{t-1} + b_i) \tag{1}$$

$$f_t = \sigma(W_{xf}x_t + W_{hf}h_{t-1} + W_{cf}c_{t-1} + b_f) \tag{2}$$

$$c_t = f_t c_{t-1} + i_t tanh(W_{xc}x_t + W_{hc}h_{t-1} + b_c \tag{3}$$

$$o_t = \sigma(W_{xo}x_t + W_{ho}h_{t-1} + W_{co}c_{t-1} + b_o) \tag{4}$$

$$h_t = o_t tanh(c_t) \tag{5}$$

Decision Tree is a non-parametric, supervised machine learning model, which learns simple decision rules inferred from the data variables. It can be used for both classification and regression tasks. Specifically, we used the CART model [6], which traverses the binary tree given a new input record, where the tree is trained by a greedy algorithm on the training data, to pick splits in the tree. The main reason of using CART is the algorithmic transparency provided, as it being proven to be a trustworthy baseline in prior researches [29].

Random Forest (RF) [5] is a supervised machine learning method formed on ensemble learning (the combination of different types of algorithms, or the same algorithms applied many times, to create a more precise predictive model). An RF receives the prediction from each sub-tree and chooses the solution that is in majority, by voting. Forests and specifically RF according to [9], tend to outperform the rest of the classification algorithms (based on a large-scale comparison of 179 classification algorithms emerging from 17 learning families over 121 datasets).

BART is a Bayesian version of a tree ensemble model, where the estimation is given by a sum of Bayesian CART trees. More information can be found in [7]. Bayesian methods are known to be better at modelling uncertainty [8], which addresses part of our aim in comparing these methods.

For the tree-based models we performed hyperparameter optimisation through Grid Search and Random Search techniques, as they are widely used in the state of the art research [3]. Moreover, we tested the model with 10-Fold Cross Validation, which is a widely used technique to ensure full usage of all available data.

The hyperparameter optimisation for the LSTM was performed with the help of an online open source application 'Weights & Biases'[3] [4]. The sweep method used to tune the model was 'Bayes' [27]. The 'Bayes' method uses a Gaussian Process (GP) Expected Improvement Markov Chain Monte Carlo (GP EI MCMC) technique to calculate the posterior distribution from a prior, and the 'expected improvement' of a parameter. A Gaussian Process distribution on prior functions, is chosen to express assumptions about the function being optimised [25,27]. The 'expected improvement' is the acquisition function (Eq. (6)), used to construct a utility function from the model posterior for our Bayesian optimisation, i.e. the main deciding factor for the MCMC [27]. If that parameter improves the F1-score of training, as we have requested from the MCMC search to monitor, the parameters are tuned, respectively.

$$a_{EI}(x; \{x_n, y_n\}, \theta) = \sigma(x; \{x_n, y_n\}, \theta)(\gamma(x)\Phi(\gamma(x)) + N(\gamma(x); 0, 1)) \quad (6)$$

where:

$a_{EI}(x; \{x_n, y_n\}, \theta)$ is the acquisition function that depends on the previous observations, and the GP hyperparameters;

$\Phi(\cdot)$ is the cumulative distribution function of the standard normal;

$N(\gamma(x); 0, 1)$ is the prior distribution with noise $(0, 1)$;

$\sigma(x; \{x_n, y_n\}, \theta)$ is the predictive variance function, and:

$$\gamma(x) = \frac{f(x_{best}) - \mu(x; \{x_n, y_n\}, \theta)}{\sigma(x; \{x_n, y_n\}, \theta)} \quad (7)$$

where:

$\mu(x; \{x_n, y_n\}, \theta)$ is the predictive mean function;

$x_{best} = argmin_{x_n} f(x_n)$ is the best current value.

Each training, with continuous or discrete variables, was performed 100 times. We used uniform distributions, as we had no prior belief information for the hyperparameters. The priors were either uniform or integer uniform, depending on the parameter checked. After monitoring tuning, the parameters providing the highest F1-score was chosen from 100 iterations. Then, to obtain the optimal hyperparameters, we run the resulting optimal algorithm 10-fold, to obtain the median ROC curve and the median testing results.

[3] https://wandb.ai/site.

The Random Forest and Decision Tree models were implemented using the scikit-learn version 1.0.1 in Python[4]. The BART model was implemented using the BART package in R[5]. The LSTM model was implemented using Keras version 2.2.5[6].

The overall purpose of our methodological setting was to:

- Compare the two datasets in their primitive forms, without any data pre-processing or hyperparameter optimisation (sequential time-series and discrete).
- Train and test the models using the two datasets applying hyperparameters optimisation and feature engineering techniques (sequential time-series and discrete).
- Compare and contrast all the above to find out if and when we should optimise model parameters and apply data pre-processing methods.

To evaluate our predictive models' performance, we utilised standard, comprehensive metrics: Precision, Recall, F-1 score and Accuracy. Moreover, we produced a ROC curve (receiver operating characteristic curve) for each model, i.e. the graph showing the performance of the classification models at all classification thresholds. This curve plots two parameters, the True Positive Rate and the False Positive Rate. This allowed us to also explore the Area under the ROC Curve (i.e., AUC) measure.

This way, we ensured a thorough, *fair* comparison of the algorithms under study.

4 Results and Discussions

We present the results of our study in Table 3, comprising of three tree-based models (Decision Tree, Random Forest, BART) and an LSTM model, for 4 test cases (Q1: Discrete dataset without hyperparameter optimisation and feature engineering, Q2: Discrete dataset with hyperparameter optimisation and feature engineering, Q3: Continuous dataset without hyperparameter optimisation and feature engineering, Q4: Continuous dataset with hyperparameter optimisation and feature engineering).

Firstly, we observed that BART is the most robust model - it maintains its high predictive accuracy for all 4 test cases (Table 3). Specifically, BART outperforms all the other models - and did not overfit in any of the test cases, achieving accuracies from 80% to 92% (see Table 3).

Secondly, for the LSTM for the 4 test cases, we observed overfitting (Table 3), which was caused by training on a (relatively) moderate amount of data. Decision Tree overfitted on the 3 out of 4 test cases (see Table 3), while hyperparameters optimisation and feature engineering did not enhance the model performance

[4] https://pypi.org/project/scikit-learn/.
[5] https://cran.r-project.org/web/packages/BART/BART.pdf.
[6] https://keras.io/api/layers/recurrent_layers/lstm/.

as expected. Random Forest overfitted as well on 2 out of 4 test cases, which indicates that it did not benefit from either feature engineering or data transformation. For all 4 test cases, BART showed an impressive ability to determine whether a student would pass or drop out, while hyperparameter optimisations and feature engineering improved its performance.

Thirdly, we used the Area Under the Curve (AUC) (see Figs. 1, 2, 3 and 4) as a criterion to measure the models' ability to discriminate the test cases. The closer the ROC curve to the upper left corner, the more efficient the test was. By taking into consideration the results (see Table 3) and comparing them with the ROC curves, we validated the fact that BART is the most consistent model, as it was not affected by neither the imbalanced nature of the data nor the low level of hyperparameters optimisation and it has an improved ability to discriminate the test values in comparison with the other four models (Decision Tree, Random Forest, BART and LSTM).

Fourthly, we observed the improved performance of Decision Tree and Random Forest models when they were trained on the dataset (discrete data) they are suited for, as they overfitted when trained on the 'unsuitable' dataset (sequential data). The LSTM model did not perform as well as the tree-based models, and especially not as expected for the continuous variables. That is possibly because LSTMs are known to require a large amount of data in order to be efficiently trained [1].

Table 3. Results: comparison of Decision Tree, Random Forest, BART on discrete and continuous data, with/out hyperparameter optimisation (4 test cases)

		DT	RF	BART	LSTM
Precision	Q1	0.64	0.77	**0.87**	0.40
	Q2	0.74	0.80	**0.88**	0.81
	Q3	0.60	0.67	**0.81**	0.41
	Q4	0.74	0.72	**0.87**	0.52
Recall	Q1	0.66	0.69	**0.96**	0.34
	Q2	0.63	0.71	**0.97**	0.66
	Q3	0.59	0.58	**0.98**	0.34
	Q4	0.65	0.66	**0.98**	0.51
F1	Q1	0.65	0.71	**0.91**	0.29
	Q2	0.66	0.75	**0.92**	0.68
	Q3	0.59	0.60	**0.89**	0.29
	Q4	0.68	0.68	**0.92**	0.50
Accuracy	Q1	0.77	0.85	**0.90**	0.32
	Q2	0.83	0.88	**0.92**	0.82
	Q3	0.82	0.86	**0.88**	0.32
	Q4	0.88	0.89	**0.89**	0.80

Our results suggest that, whenever possible, it would be beneficial to convert the time-series dataset into a discrete variables dataset and apply Bayesian methods, such as BART, as it is highly likely to produce better performance, especially when the time-series datasets are not populated enough.

Moreover, our results highlight the necessity of always finding the best hyperparameters for the models based on the data they are trained on, and the most efficient and effective data pre-processing techniques, as they can dramatically improve the models' performances and prevent overfitting. However, converting time-series datasets into discrete datasets can often be time-consuming.

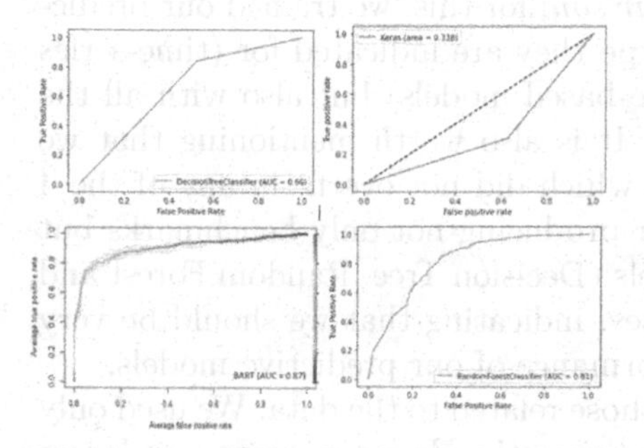

Fig. 1. Discrete data: no pre-processing, no hyperparameter optimisation (**for all ROC curves above:** upper left - Decision Tree, upper right - LSTM, bottom left - BART, bottom right - Random Forest)

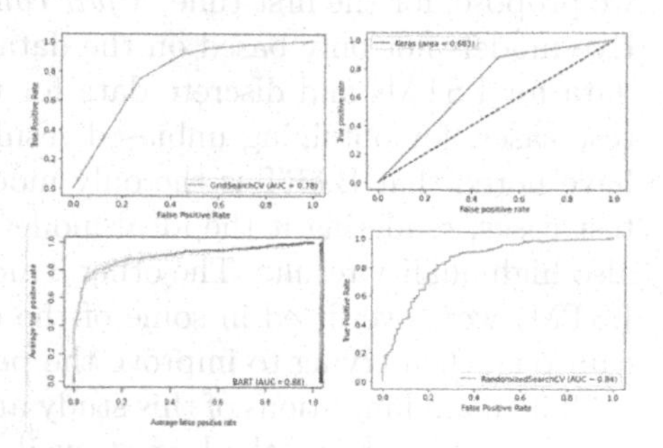

Fig. 2. Discrete data: hyperparameter optimisation, feature engineering (**for all ROC curves above:** upper left - Decision Tree, upper right - LSTM, bottom Left - BART, bottom right - Random Forest)

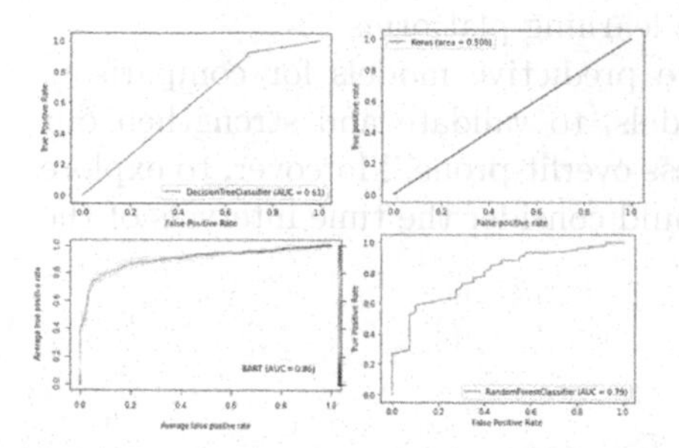

Fig. 3. Continuous data: no pre-processing, no hyperpar. optimisation (**for all ROC curves above:** upper left - Decision Tree, upper right - LSTM, bottom left - BART, bottom right - Random Forest)

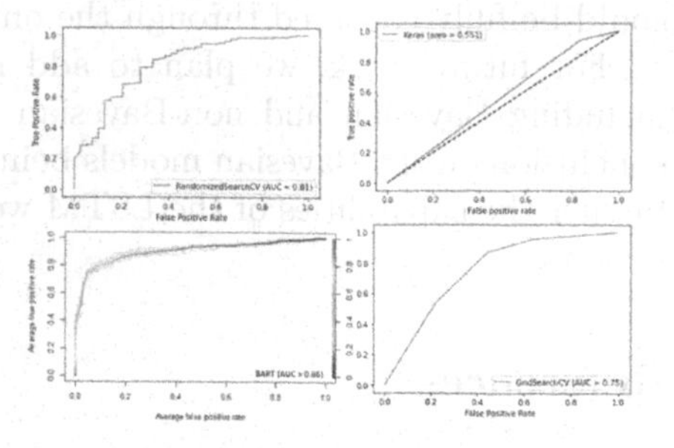

Fig. 4. Continuous data: hyperparameter optimisation, features engineering (**for all ROC curves above:** upper left - Decision Tree, upper right - LSTM, bottom left - BART, bottom right - Random Forest)

5 Conclusions, Limitations and Future Work

In summary, this paper presents the results of a comparison study with 4 test cases, swapping continuous and discrete datasets, as well as training with/without hyperparameter optimisation, on four different state-of-the-art algorithms (Decision Tree, LSTM, BART, Random Forest).

Our results conclude that researchers should transform data into suitable forms when feasible, and they should always try to identify the optimal data pre-processing techniques, as this can improve model performance. We have shown that this process assists different types of predictive models to obtain higher performance and enhanced learning ability. Different from other studies, we propose, for the first time, a *fair comparison*; for this, we trained our predictive models not only based on the data type they are indicated for (time-series data for LSTMs and discrete data for tree-based models) but also with all the test cases, for obtaining unbiased results. It is also worth mentioning that we have noted that BART is the only model which did not overfit in any of the 4 test cases, rendering it the ideal model for producing not only benchmarks but also high quality results. The other 3 models (Decision Tree, Random Forest and LSTM) were overfitted in some of the cases, indicating that we should be very cautious when trying to improve the performance of our predictive models.

The main limitations of this study are those related to the data. We used only one dataset, as being the largest available currently. However, more and larger datasets would be useful for further comparisons, as LSTM models especially tend to perform better when being trained on more data. Moreover, demographic or personal information of each student (unavailable in the dataset) might provide the models with more meaningful connections to variables for the classifications and thus allow better performance. It is also important to note that, as some students might prefer to download videos and watch them locally, not all the actions (i.e. interacting with a video player such as 'play' and 'pause') could be fully captured through the online learning platforms.

For future work, we plan to add more predictive models for comparison, including Bayesian and non-Bayesian models, to validate and strengthen our conclusion on the Bayesian models being less overfit-prone. Moreover, to explore further the capabilities of the LSTM we could consider the time intervals of the actions.

References

1. Adadi, A.: A survey on data-efficient algorithms in big data era. J. Big Data (2021). https://link.springer.com/article/10.1186/s40537-021-00419-9#citeas
2. Alamri, A., et al.: Predicting MOOCs dropout using only two easily obtainable features from the first week's activities. In: Coy, A., Hayashi, Y., Chang, M. (eds.) ITS 2019. LNCS, vol. 11528, pp. 163–173. Springer, Cham (2019). https://doi.org/10.1007/978-3-030-22244-4_20
3. Bergstra, J., Bengio, Y.: Random search for hyper-parameter optimization. J. Mach. Learn. Res. **13**(2), 281–305 (2012)

4. Biewald, L.: Experiment tracking with weights and biases (2020). software available from https://wandb.com
5. Breiman, L.: Random forests. Mach. Learn. **45**, 5–32 (2004)
6. Breiman, L., Friedman, J.H., Olshen, R.A., Stone, C.J.: Classification and Regression Trees (1983)
7. Chipman, H.A., George, E.I., McCulloch, R.E.: BART: Bayesian additive regression trees (October 2010). https://doi.org/10.1214/09-AOAS285
8. Clyde, M., George, E.I.: Model uncertainty. Stat. Sci. **19**(1), 81–94 (2004)
9. Delgado, M.F., Cernadas, E., Barro, S., Amorim, D.G.: Do we need hundreds of classifiers to solve real world classification problems? J. Mach. Learn. Res. **15**, 3133–3181 (2014)
10. Drousiotis, E., Pentaliotis, P., Shi, L., Cristea, A.I.: Capturing fairness and uncertainty in student dropout prediction – a comparison study. In: Roll, I., McNamara, D., Sosnovsky, S., Luckin, R., Dimitrova, V. (eds.) AIED 2021. LNCS (LNAI), vol. 12749, pp. 139–144. Springer, Cham (2021). https://doi.org/10.1007/978-3-030-78270-2_25
11. Drousiotis, E., Shi, L., Maskell, S.: Early predictor for student success based on behavioural and demographical indicators. In: Cristea, A.I., Troussas, C. (eds.) ITS 2021. LNCS, vol. 12677, pp. 161–172. Springer, Cham (2021). https://doi.org/10.1007/978-3-030-80421-3_19
12. Fei, M., Yeung, D.: Temporal models for predicting student dropout in massive open online courses. In: 2015 IEEE International Conference on Data Mining Workshop (ICDMW), pp. 256–263 (November 2015). ISSN 2375–9259
13. Freund, Y., Mason, L.: The alternating decision tree learning algorithm. In: Proceedings of the 16th International Conference on Machine Learning, ICML 1999, pp. 124–133. Morgan Kaufmann Publishers Inc., San Francisco (1999)
14. Gardner, J., Yang, Y.: Modeling and experimental design for MOOC dropout prediction: a replication perspective. In: Proceedings of The 12th International Conference on Educational Data Mining, EDM 2019, p. 10 (2019)
15. Goel, Y., Goyal, R.: On the effectiveness of self-training in MOOC dropout prediction. Open Comput. Sci. **10**, 246–258 (2020)
16. Hochreiter, S., Bengio, Y., Frasconi, P., Schmidhuber, J.: Gradient flow in recurrent nets: the difficulty of learning long-term dependencies (2001)
17. Hong, B., Wei, Z., Yang, Y.: Discovering learning behavior patterns to predict dropout in MOOC. In: 2017 12th International Conference on Computer Science and Education (ICCSE), pp. 700–704 (2017)
18. Huang, Z., Xu, W., Yu, K.: Bidirectional LSTM-CRF models for sequence tagging. CoRR abs/1508.01991 (2015). http://arxiv.org/abs/1508.01991
19. Jin, C.: MOOC student dropout prediction model based on learning behavior features and parameter optimization. Interact. Learn. Environ., 1–19 (2020). https://doi.org/10.1080/10494820.2020.1802300
20. Liang, J., Li, C., Zheng, L.: Machine learning application in MOOCs: dropout prediction. In: 2016 11th International Conference on Computer Science Education (ICCSE), pp. 52–57 (2016). https://doi.org/10.1109/ICCSE.2016.7581554
21. Liang, J., Yang, J., Wu, Y., Li, C., Zheng, L.: Big data application in education: dropout prediction in Edx MOOCs. In: 2016 IEEE 2nd International Conference on Multimedia Big Data (BigMM), pp. 440–443 (2016)
22. Liu, Z., Xiong, F., Zou, K., Wang, H.: Predicting learning status in MOOCs using LSTM (August 2018)
23. Mubarak, A.A., Cao, H., Ahmed, S.A.: Predictive learning analytics using deep learning model in MOOCs' courses videos (2021)

24. Pulikottil, S.C., Gupta, M.: ONet - a temporal meta embedding network for MOOC dropout prediction. In: 2020 IEEE International Conference on Big Data (Big Data), pp. 5209–5217 (2020). https://doi.org/10.1109/BigData50022.2020.9378001
25. Rasmussen, C.E., Williams, C.K.I.: Gaussian Processes for Machine Learning. Adaptive Computation and Machine Learning. The MIT Press (2005)
26. Rehfeldt, R.A., Jung, H.L., Aguirre, A., Nichols, J.L., Root, W.B.: Beginning the dialogue on the e-transformation: behavior analysis' first massive open online course (MOOC). Behav. Anal. Pract. **9**(1), 3–13 (2016)
27. Snoek, J., Larochelle, H., Adams, R.P.: Practical Bayesian optimization of machine learning algorithms. In: Proceedings of the 25th International Conference on Neural Information Processing Systems - Volume 2, NIPS 2012, pp. 2951–2959 (2012)
28. Song, Y., Lu, Y.: Decision tree methods: applications for classification and prediction. Shanghai Arch. Psychiatry **27**(2), 130–135 (2015)
29. Strecht, P., Cruz, L., Soares, C., Mendes-Moreira, J., et al.: A comparative study of classification and regression algorithms for modelling students' academic performance. International Educational Data Mining Society (2015)
30. Tang, C., Ouyang, Y., Rong, W., Zhang, J., Xiong, Z.: Time series model for predicting dropout in massive open online courses. In: Penstein Rosé, C. (ed.) AIED 2018. LNCS (LNAI), vol. 10948, pp. 353–357. Springer, Cham (2018). https://doi.org/10.1007/978-3-319-93846-2_66
31. Wang, L., Wang, H.: Learning behavior analysis and dropout rate prediction based on MOOCs data. In: 2019 10th International Conference on Information Technology in Medicine and Education (ITME), pp. 419–423 (August 2019)
32. Zhang, X., Liang, X., Zhiyuli, A., Zhang, S., Xu, R., Wu, B.: AT-LSTM: an attention-based LSTM model for financial time series prediction. IOP Conf. Ser. Mater. Sci. Eng. **569**, 052037 (2019)

"Teacher, Can You Say It Again?" Improving Automatic Speech Recognition Performance over Classroom Environments with Limited Data

Danner Schlotterbeck[1](✉), Abelino Jiménez[1], Roberto Araya[1], Daniela Caballero[1], Pablo Uribe[1], and Johan Van der Molen Moris[1,2]

[1] Center for Advanced Research in Education, University of Chile, Santiago, Chile
{danner.schmen,abelino.jimenez,roberto.araya,daniela.caballero, pablo.urible}@ciae.uchile.cl

[2] MRC Biostatistics Unit, University of Cambridge, Cambridge, UK
johan.vdmolen@mrc-bsu.cam.ac.uk

Abstract. Analyzing teachers' discourse plays a fundamental role in educational research and is a key component of Teaching Analytics. This usually involves transcribing lessons from audio recordings. As the number of recordings grows, Automatic Speech Recognition (ASR) systems gain popularity as a means for transcribing these recordings. However, most ASR systems are trained over very specific domains which usually involve read text and low environmental noise. This suggests common ASR systems available on the market may underperform over classroom recordings, as they present a unique type of environmental sound and spontaneous discourse, as opposed to the usual training domains. To address this challenge we present a system that automatically transcribes classroom discourse in a robust way with regard to classroom noise, which was trained over few annotated data. In particular, we used a state-of-the-art ASR model based on wav2vec 2.0 and fine-tuned it over a 6-h dataset of 4th to 8th grade Chilean lessons. We found that by leveraging its transformer-based architecture and changing the fine-tuning domain to classroom recordings, we can obtain a more accurate and robust transcriber for this source of audio which outperforms other popular cloud-based systems up to 35% and 59% in terms of Word and Character Error Rates, respectively. This work contributes by using state-of-the-art ASR techniques to develop a tool which is particularly adapted to classroom environments, making it robust and more reliable with regard to their environmental sound and the way teaching discourse is carried out.

Keywords: Automatic speech recognition · Classroom discourse · Teaching analytics · Transfer learning

Support from ANID/PIA/Basal Funds for Centers of Excellence FB0003 is gratefully acknowledged.

M. M. Rodrigo et al. (Eds.): AIED 2022, LNCS 13355, pp. 269–280, 2022.
https://doi.org/10.1007/978-3-031-11644-5_22

1 Introduction

The qualities and structures of teacher discourse have been proved to strongly affect students' learning [15,21]. Moreover, by analyzing them, one can obtain insights about aspects of teaching that could be improved which are not measurable through standardized tests or surveys [24]. Thus, teacher discourse analysis has become a critical part of teaching analytics and educational research in general. Such analyses regularly involve transcribing classroom speech from audio or video recordings in order to later detailedly analyze the textual information present in the lesson. However, due to the time and economic costs that come with manual transcription, it usually does not scale for educational research needs [10,26].

Therefore, to address this impediment, many researchers have started using Automatic Speech Recognition (ASR) technologies as part of their methods to obtain transcriptions at a larger scale. For example, [15] used ASR systems to develop a semi-automated method for analyzing teacher talk and used it to analyze teacher talk features over 127 secondary English Language Arts lessons. [17] used ASR to investigate automatic detection of teacher questions from audio recordings collected in live classrooms with the goal of providing automated feedback to teachers. [16,29], and [28] built systems for recognizing different teaching activities from automatic transcriptions of classroom audio centered on the teacher's discourse. These examples show how ASR systems can become a valuable tool for educational research and Teaching Analytics.

Nonetheless, these solutions are not exempt of concerns. First, most of these ASR systems are well tuned over minimal noise environments and are robust for transcribing voice assistant instructions. As a consequence, performance in noisy environments, such as a classroom, is oftentimes degraded [12]. Furthermore, there is evidence that the performance of speech recognizers can vary significantly depending on the audio domain [18]. Second, most times transcription is done via paid cloud-based systems, which can be an impediment when analyzing several classes over a long period of time for a large number of teachers. In addition, this type of application requires Internet access, limiting the use of systems in places with limited or no connectivity, which is common in rural areas and developing countries. Third, cloud usage may raise several concerns regarding recordings' privacy. Indeed, in many countries a recording of a teacher with his/her students can be considered confidential material, so storing and processing this information in an external party generates a series of privacy risks. Lastly, improvements in ASR performance over a specific domain, such as classroom recordings, usually depend on the quantity of annotated (manually transcribed) data which is hard to obtain in the educational field, particularly for non-english speakers. These considerations give rise to our research question:

- How can we improve the performance of ASR systems over classroom talk using limited data?

To answer this question, we proposed a methodology that uses transfer learning and minimal amounts of annotated data to develop a system which is robust to classroom environmental audio. In particular, we started by leveraging classroom audio recordings and transcribing a set of selected samples to generate a development dataset. Next, we fine-tuned a pretrained model based on wav2vec 2.0's architecture [11,20] over the selected samples. Subsequently, we improved the quality of the raw output transcripts using a spell-checking algorithm based on Levenshtein distance. Later, we assessed our model in terms of Word and Character Error Rates (WER and CER), two commonly used metrics for ASR evaluation, and compared it against base wav2vec 2.0 and two usually employed cloud-based systems. Finally, we posed a use-case scenario for our model consisting in content-related keyword counting through the transcripts and compared its performance in this task against the previously mentioned baselines.

We developed our model using a sample of 2121 $\sim$ 10 second segments and 4.6 h of audio from science and math lessons from several Chilean schools. In addition, we assessed our model over 920 segments from the first two sets and a third one containing Spanish language lessons. Results show that our model significantly outperforms the proposed baselines, surpassing the threshold of 40% WER and relatively reducing WER and CER up to 35% and 59%, respectively.

This way, our approach shows a method that leverages the advantages of transfer learning for developing an ASR system that is able to outperform other commonly used systems in this domain using limited annotated data. Thus, we expect this to be a valuable tool for educational researchers who might be able to improve their analyses by obtaining a more accurate transcription. Furthermore, as far as we are concerned, this is the first application of wav2vec 2.0 over a Spanish-speaking classroom environment, which could open the door for future development of better systems in this topic.

We start by making a brief review of previous studies that have inspired our research in Sect. 2. Next, Sect. 3 explains the procedures and details regarding our dataset, the preprocessing steps, and system development. Afterwards, we present the outcomes of model evaluation and benchmarking in Sect. 4. Finally, in Sect. 5 we discuss our findings and the implications, limitations and new questions raised by our research.

2 Related Work

Speech technologies have been widely used in different educational settings. One important area of their applications is in Intelligent Tutoring Systems, where speech processing and understanding can be critical. Some of the use cases explored include reading tutors [27], to teach and engage science [25,30] or to teach language and cultural skills [23]. In most of these systems, the solution is considered on a one-to-one human-computer interaction, typically conceived as a desktop application, having a controlled domain and reduced noise.

Furthermore, there are many different attempts at using speech recognition systems to analyze the content of teachers' talk and dialogues in a classroom. Blanchard et al. [13] compared the performance of 6 different speech recognition systems on audio recorded from noisy classroom environments, concluding that, despite the degrading performance (44% WER at best using Google Speech to Text), the obtained transcriptions from some cloud-based systems were a reasonable representation of the spoken dialogues. Donnelly et al. [17] built a system, based on both acoustic and linguistic features derived from automatic transcripts, to recognize authentic questions from audio recorded during the lessons. Moreover, within the different uses of ASR systems in classroom environments, one of the most outstanding is the development of feedback systems for teachers. For instance, Caballero et al. [14] designed a system that allows obtaining conceptual networks that relate key concepts used by the teacher. Jensen et al. [22] describe a system where the teacher must record classroom audio using a Wireless cardioid microphone connected to a laptop in order to receive feedback regarding eight key dimensions from teacher discourse.

Recently, wav2vec 2.0 [11], a pre-tranied model for speech recognition, has become very popular. This model is composed of multiple convolutional layers and self-attention layers. Convolutional layers reduce the dimensionality of the raw speech signal generating more compressed representations, whereas the self-attention layers build contextualized representations, capturing high level content. A remarkable result of wav2vec 2.0 is the low WER of 8.6% that can be obtained on noisy English speech using just 10 min of transcribed speech and 53K hours of unlabeled speech. According to the authors, this opens opportunities for speech recognition models in many more languages, dialects and domains. This has been explored successfully in different recent works [31–33], achieving below 30% WER in Mandarin, Japanese and German, and below 40% for Spanish and Arabic. As far as we know, our work constitutes the first attempt to use of wav2vec 2.0 to transcribe spanish-speaking recordings from classroom environments.

3 Methodology

3.1 Dataset

Data Collection. To answer our research question, we started by leveraging audio recordings from four different pilot projects. The first set consists of samples of 40 lessons from 24 schools obtained from a Randomized Control Trial carried out to measure the impact of using a technology guided platform for computer-assisted math lessons in fourth grades from low-performing schools in Chile. We refer to this dataset as MATH [8]. The second set consists of 9 recordings of first to fourth grade lessons carried out to measure computational thinking in a biological phenomenon and computational modeling of locomotion of bacteria [7]. The third set consists of a sample of 9 recordings from fourth to eight grade lessons on computational modeling of coronavirus infection dynamics and containment measures such as social distancing, face masks and mobility [9]. Due to the similarities between the contents, the second and third sets were merged into a new set which we called SCI. Finally, the fourth set contains 20 lessons from a project where first and second graders were taught to read and write [6], thus, we named this set LAN. During the pilots, lessons were recorded 44.1 kHz using Rode SmartLav+ for MATH and Swivl SW3322-C1 for SCI and

LAN. Microphones were pointed towards the teacher, therefore capturing mostly their voice instead of their students'.

Data Preprocessing. After data collection, lesson recordings were passed through a pre-trained neural Voice Activity Detection system [1] to obtain speech boundaries. Next, segments of around 10 s long were randomly sampled from every recording considering these boundaries to avoid cropping sentences in the middle. This way, our initial dataset consisted of 1200, 1600, and 400 segments from MATH, SCI, and LAN respectively, which added up to almost 9 h of audio.

Once the segments were selected, MATH and SCI sets were separated into train, validation, and test folds with proportions 60/20/20, and merged on each respective fold. This was considered as our development dataset. On the other hand, segments from LAN were left as a whole secondary test set in order to measure our model's performance over lessons that were outside the training domain. Moreover, this also allowed us to estimate the generalization capability of our system to different subjects, educational levels, and teacher voices considering the different conditions of each pilot and that lessons from LAN were taught from teachers who did not present lessons in the development set.

Afterward, the selected segments were manually transcribed by our team and the dataset was cleaned by removing the segments which were mostly noise or where the speech was unintelligible, leaving a total of 3041 segments and 6.6 h of audio. Table 1 shows the distribution of the final dataset in terms of number of segments and minutes of speech for every pilot. After the cleaning procedure, the respective test folds of the remaining segments were automatically transcribed using Google Speech and Microsoft Azure Cognitive Services API's [3,4], two of the most commonly used transcribers available in the market, to later compare our system's performance against these systems.

Table 1. Description of the datasets

Dataset	# Samples			Speech minutes		
	Train	Dev	Test	Train	Dev	Test
MATH	660	218	214	110	36	37
SCI	934	309	314	97	32	32
LAN	–	–	392	–	–	59

Finally, to prepare the datasets to train our model, we started by resampling every segment to a 16 kHz ratio and converting it to mono. Next, a sequence of steps was used to preprocess text in order to remove characters outside the model's vocabulary. Firstly, in MATH transcriptions we replaced mathematical symbols with their corresponding words. Secondly, we removed all punctuation marks and line end characters from the transcripts. Lastly, we converted all consecutive decimal characters to their number expressions in words. The previous steps were applied to the ground truth transcripts and to the ones obtanied by both automated systems. Thus, we finished with a total vocabulary of 35 characters, considering the accented vowels in Spanish and the blank space, which corresponds to one used by our base model.

3.2 Model Development and Evaluation

Fine-Tuning. As the base model for our system, we used a large version of wav2vec 2.0's architecture which was pre-trained over the Spanish subset of the common-voice dataset [20]. For an extended explanation of wav2vec 2.0's architecture the reader might refer to [11]. We chose this version as it presented the best metrics on the *paperswithcode* leader-board for Spanish ASR models over CommonVoice [2]. We fine-tuned this model over the training folds of MATH and SCI, while evaluating over the validation sets of the same pilots using the Trainer interface of the Hugging Face python library [5]. Neither the test folds from MATH and SCI nor the samples from the LAN pilot were considered for the development procedure. Moreover, teachers from LAN lessons were distinct from the teachers in MATH and SCI pilots to better assess the generalization capabilities of our system.

We tested different combinations of learning rate and weight decay to find a suitable combination of hyper-parameters for the model's architecture while trying to minimize the Connectionist Temporal Classification (CTC) [19] loss function over the validation dataset, obtaining an optima of 10^{-4} and 0.01 for the learning rate and weight decay, respectively. The model was trained with different hyper-parameter combinations for 15 epochs over batches of four samples and evaluated every 50 steps. Additionally, an early stopping callback with patience of 5 was used to save the best models once they started overfitting. In the end, the model that yielded the best CTC loss over the validation fold was considered for the rest of the development.

Spell-checking. When using neural-based transcribers, the transcription process is usually done at the character level, disregarding if the final output is a real word or not. Thus, the transcriptions from these systems are usually passed through a spell-checking algorithm or a language model before returning the final text. Considering this, we used a Python spell-checker library which bases on the Levenshtein distance between two words to assign possible corrections from a vocabulary to the words output by the neural-based transcriber before returning the final text. The algorithm was applied using a list of the 100,000 most common Spanish words as the vocabulary, which was developed and published by the Spanish Language Royal Academy.

4 Results

4.1 Model Evaluation and Benchmarking

After the model training and post-processing of the transcripts, we assessed model performances via Word and Character Error Rates (WER and CER, respectively). These metrics are based on the edit distance from the automated transcript to the ground truth transcription and are considered standard assessment measures for ASR tasks. We compared the results obtained by our model against three different baselines: Base wav2vec 2.0 without fine-tuning, Google Speech API, and Microsoft Azure Cognitive Services API over the three test folds. Furthermore, for every test set we ran a t-test for the mean of the WER difference from system to system to test for statistical significance regarding the difference in performance. Table 2 shows the results in terms of WER and CER for the four systems across the test folds from each pilot.

Table 2. Values of WER and CER for the different systems across different datasets. (*<0.05, **<0.01, ***<0.001)

System	WER			CER		
	MATH	SCI	LAN	MATH	SCI	LAN
Ours	38%	32%	38%	19%	14%	17%
Base wav2vec 2.0	45%***	49%***	43%***	21%***	27%***	20%**
Google	43%*	48%***	46%	29%***	34%***	34%***
Azure	40%**	33%*	38%	25%***	19%***	27%***

Results show that our model outperforms the three baselines over classroom environment domains, achieving below 40% WER and significant improvements in 2 of the 3 datasets, getting a significant relative WER reduction up to 5% and 33% respect with Azure and Google respectively. Moreover, we could also achieve values below 20% CER across the three test folds. Additionally, CER was relatively reduced between 15% and 58% with p-values below 0.01 in every case, which also reinforces the significance of the results. Table 3 shows some examples transcriptions obtained using the different systems.

4.2 A Use Case Scenario: Content-Related Keyword Counting

To test the impact of our model from an experimental standpoint, we decided to repeat a previous experiment similar to using the different transcribers. First, a content-related keyword list for the MATH set was made by the teachers who carried the pilot. The list included a total of 108 words which were related to fractions and decimal numbers, areas and perimeters, linear (in)equations, data representation and interpretation, basic probabilities, angles and polygons, and problem solving. Next, for every segment of the MATH test fold, we counted the keywords that appeared in the ground truth transcription. Subsequently, we repeated this process for the transcriptions obtained with the four systems. Lastly, we compared the keywords found by the systems with the ones from the ground truth in terms of precision and recall. Overall performance across the whole keyword list is reported for every system in Table 4, while Fig. 1 shows the precision and recall values for the 20 most frequent words across the dataset.

Table 3. Examples of transcriptions obtained using different ASR systems.

Ground truth text	Ours	Google	Azure
''próxima instrucción dice utiliza espacio para los miles pero s la mayoría no lee completamente la instrucción entonces por eso le estamos recordando que debe''	''ma instrucción dice utiliza espacio para los miles pero la mayoría no le completamente la instrucción dos por eso le estamos recordando que deb''	''introducción de se utiliza espacio para los miles pero la mayoría no le completamente la instrucción se puede solo estamos recordando quede''	''dice utiliza espacio para los miles pero la mayoría no lee completamente la instrucción entonces por eso le estamos recordando que debe''
''la segunda figura a cuánto equivale cuatro cuartos nuevamente cierto que es equivalente a un entero levantando''	''la segunda figura a cuánto equivale cuatro cuarto nuevamente cierto que es equivalente a un entero levantando''	''la segunda figura a cuánto equivale cuatro por cuarenta y cuatro nuevamente cierto que es equivalente a un entero levantando''	''la segunda figura a cuánto equivale cuatro cuartos nuevamente siento que es equivalente a un entero levanta''
''no por qué porque son iguales bien por lo tanto está en lo correcto la mamá de patricia no por lo tanto cómo se redacta la''	''no porque por que son iguales bien por lo tanto esta es lo correcto y lmaemás de patricia no porlo tanto cómo se reda''	''no porque porque son iguales bien por lo tanto está en lo correcto la matriz ya no tanto como se redacta''	''no porque porque son iguales bien por lo tanto estás en lo correcto la mamá de patricia elio por lo tanto cómo se redacta hola''

Table 4. Mean values of precision and recall across keywords for the four systems.

Model	Precision	Recall
Ours	0.88	0.85
Base wav2vec 2.0	0.73	0.72
Google	0.82	0.83
Azure	0.86	0.86

We found that in general, for the task of automatically recognizing content-related keywords, our model has better precision than the other three systems, as well as better recall than base wav2vec 2.0 and Google systems. Nevertheless, when testing for statistical significance between the assessment metric values across the MATH test fold samples, the results are significant just when we compare against base wav2vec 2.0 (p-values < 0.001). Therefore, we can conclude that for this task, our system has comparable results with respect to Azure and Google systems.

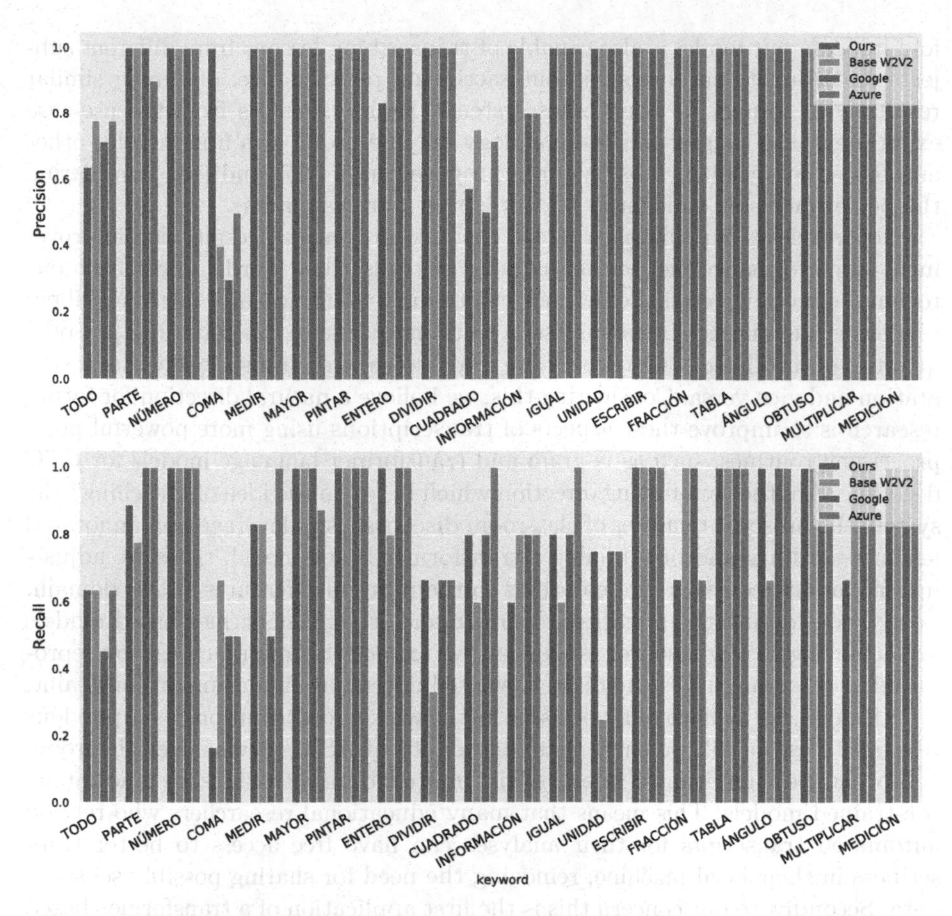

Fig. 1. Precision and recall values for the 20 most frequent words in the MATH test fold obtained by the different systems.

5 Discussion

We presented a methodology that leverages the goods of transfer learning and few annotated data to develop an ASR system which is robust with regard to classroom sound and its talk. We also compared our model to three commonly used machine transcribers across recordings from three pilots from different teachers, educational levels and subjects, helping us to estimate the generalization capability of our system more accurately. In addition, we also assessed the performance of the developed system against the three baselines in a use-case scenario by measuring the number of correctly identified content-related keywords across the transcripts.

We found our model can outperform the other systems in classroom audio, reducing significantly the relative error in the MATH and LAN test folds up to 35% and 59% WER and CER, respectively. In addition, results over LAN test fold indicate our model is also capable of transcribing lessons from different subjects and educational levels without sacrificing performance, obtaining similar results with respect to cloud-bases systems. Besides, results from the use-case experiment also suggest this methodology may be useful as a first step for other automated analyses such as the ones proposed in [14,17] and [16], considering the better precision to identify keywords that our system has.

Nevertheless, our approach is not exempt of caveats. Despite the improvement in performance metrics and number of transcribed words, words from our regional dialect were still sometimes erroneously identified and mistranscribed, rendering the sentences meaningless. Additionally, due to the particular prosodic qualities of the dialect, two consecutive words were often transcribed without separation and vice-versa. Considering this, we believe a natural direction for future research is to improve these aspects of transcriptions using more powerful post-processing routines, such as N-gram and transformer language models for CTC decoding. Another interesting direction which bases on the idea of "teaching" the system the prosodic qualities of classroom discourse is to leverage non-annotated lesson recordings and pre-training a transformer-based model, therefore adjusting the latent acoustic representations to the particular features of this domain. Lastly, considering the remarkable performance that transformer-based models are achieving in low-resource languages, we expect this approach can be reproduced in other languages and dialects where limited data is a common constraint.

All in all, we believe our work sets forth two key contributions over previous research. First of all, we show that the quality of ASR systems over classroom audio can be significantly improved in limited data situations by fine-tuning pre-trained models. This means that many educational researchers who rely on automated transcripts for their analyses can have free access to better transcribers in their local machine, removing the need for sharing possibly sensitive data. Secondly, to our concern this is the first application of a transformer-based ASR model for the Spanish language, and more remarkably, to a really strong Latin-American dialect in classroom environments. Thus, considering the several acoustic domains that these dialects generate and the difficulties implied in collecting large quantities of data for many of them, we believe this gives insights about the possibility of training dialect-specific systems to rapidly improve the transcription performance over those domains. Therefore, we expect our method to be useful for other educational researchers who want to improve ASR performance subject to data constraints, and to anyone else interested in automatically transcribing lesson recordings.

References

1. Neural speaker diarization with pyannote.audio. https://github.com/pyannote/pyannote-audio
2. Speech recognition on common voice 8.0 spanish. https://paperswithcode.com/sota/speech-recognition-on-common-voice-8-0-16
3. Speech to text: A speech service feature that accurately transcribes spoken audio to text; azure cognitive services. https://azure.microsoft.com/es-es/services/cognitive-services/speech-to-text/#features

4. Speech-to-text: Automatic speech recognition; google cloud. https://cloud.google.com/speech/
5. Huggingface trainer (2021). https://huggingface.co/docs/transformers/main_classes/trainer
6. Araya, R.: Early detection of gender differences in reading and writing from a smartphone-based performance support system for teachers. In: Vittorini, P., Di Mascio, T., Tarantino, L., Temperini, M., Gennari, R., De la Prieta, F. (eds.) MIS4TEL 2020. AISC, vol. 1241, pp. 137–146. Springer, Cham (2020). https://doi.org/10.1007/978-3-030-52538-5_15
7. Araya, R.: Enriching elementary school mathematical learning with the steepest descent algorithm. Mathematics **9**(11), 1197 (2021)
8. Araya, R., Diaz, K.: Implementing government elementary math exercises online: positive effects found in RCT under social turmoil in Chile. Educ. Sci. **10**(9), 244 (2020)
9. Araya, R., Isoda, M., van der Molen Moris, J.: Developing computational thinking teaching strategies to model pandemics and containment measures. Int. J. Environ. Res. Public Health **18**(23), 12520 (2021)
10. Araya, R., et al.: Estimation of teacher practices based on text transcripts of teacher speech using a support vector machine algorithm. Br. J. Edu. Technol. **43**(6), 837–846 (2012)
11. Baevski, A., Zhou, Y., Mohamed, A., Auli, M.: wav2vec 2.0: A framework for self-supervised learning of speech representations. In: Advances in Neural Information Processing Systems 33, 12449–12460 (2020)
12. Bhattacharjee, U., Gogoi, S., Sharma, R.: A statistical analysis on the impact of noise on MFCC features for speech recognition. In: 2016 International Conference on Recent Advances and Innovations in Engineering (ICRAIE), pp. 1–5. IEEE (2016)
13. Blanchard, N., et al.: A study of automatic speech recognition in noisy classroom environments for automated dialog analysis. In: Conati, C., Heffernan, N., Mitrovic, A., Verdejo, M.F. (eds.) AIED 2015. LNCS (LNAI), vol. 9112, pp. 23–33. Springer, Cham (2015). https://doi.org/10.1007/978-3-319-19773-9_3
14. Caballero, D., et al.: ASR in classroom today: automatic visualization of conceptual network in science classrooms. In: Lavoué, É., Drachsler, H., Verbert, K., Broisin, J., Pérez-Sanagustín, M. (eds.) EC-TEL 2017. LNCS, vol. 10474, pp. 541–544. Springer, Cham (2017). https://doi.org/10.1007/978-3-319-66610-5_58
15. Dale, M.E., Godley, A.J., Capello, S.A., Donnelly, P.J., D'Mello, S.K., Kelly, S.P.: Toward the automated analysis of teacher talk in secondary ELA classrooms. Teach. Teach. Educ. **110**, 103584 (2022)
16. Diosdado, D., Romero, A., Onaindia, E.: Recognition of teaching activities from university lecture transcriptions. In: Alba, E., et al. (eds.) CAEPIA 2021. LNCS (LNAI), vol. 12882, pp. 226–236. Springer, Cham (2021). https://doi.org/10.1007/978-3-030-85713-4_22
17. Donnelly, P.J., Blanchard, N., Olney, A.M., Kelly, S., Nystrand, M., D'Mello, S.K.: Words matter: automatic detection of teacher questions in live classroom discourse using linguistics, acoustics, and context. In: Proceedings of the Seventh International Learning Analytics & Knowledge Conference, pp. 218–227 (2017)
18. Georgila, K., Leuski, A., Yanov, V., Traum, D.: Evaluation of off-the-shelf speech recognizers across diverse dialogue domains. In: Proceedings of the 12th Language Resources and Evaluation Conference, pp. 6469–6476 (2020)

19. Graves, A., Fernández, S., Gomez, F., Schmidhuber, J.: Connectionist temporal classification: labelling unsegmented sequence data with recurrent neural networks. In: Proceedings of the 23rd International Conference on Machine Learning, pp. 369–376 (2006)
20. Grosman, J.: Xlsr wav2vec2 spanish by jonatas grosman (2021). https://huggingface.co/jonatasgrosman/wav2vec2-large-xlsr-53-spanish
21. Helaakoski, J., Viiri, J.: 6. content and content structure of physics lessons and students' learning gains: Comparing finland, germany and switzerland. Quality of Instruction in Physics: Comparing Finland, Switzerland and Germany, p. 93 (2014)
22. Jensen, E., et al.: Toward automated feedback on teacher discourse to enhance teacher learning. In: Proceedings of the 2020 CHI Conference on Human Factors in Computing Systems, pp. 1–13 (2020)
23. Johnson, W.L., Valente, A.: Tactical language and culture training systems: using AI to teach foreign languages and cultures. AI Mag. **30**(2), 72–72 (2009)
24. Kelly, S., Olney, A.M., Donnelly, P., Nystrand, M., D'Mello, S.K.: Automatically measuring question authenticity in real-world classrooms. Educ. Res. **47**(7), 451–464 (2018)
25. Litman, D., Silliman, S.: Itspoke: An intelligent tutoring spoken dialogue system. In: Demonstration papers at HLT-NAACL 2004, pp. 5–8 (2004)
26. Liu, J., Cohen, J.: Measuring teaching practices at scale: a novel application of text-as-data methods. Educ. Eval. Policy Anal. **43**(4), 587–614 (2021)
27. Mostow, J., et al.: Evaluating tutors that listen: An overview of project listen (2001)
28. Schlotterbeck, D., Uribe, P., Jiménez, A., Araya, R., van der Molen Moris, J., Caballero, D.: TARTA: teacher activity recognizer from transcriptions and audio. In: Roll, I., McNamara, D., Sosnovsky, S., Luckin, R., Dimitrova, V. (eds.) AIED 2021. LNCS (LNAI), vol. 12748, pp. 369–380. Springer, Cham (2021). https://doi.org/10.1007/978-3-030-78292-4_30
29. Slyman, E., Daw, C., Skrabut, M., Usenko, A., Hutchinson, B.: Fine-grained classroom activity detection from audio with neural networks. arXiv preprint arXiv:2107.14369 (2021)
30. Ward, W., Cole, R., Bolaños, D., Buchenroth-Martin, C., Svirsky, E., Weston, T.: My science tutor: a conversational multimedia virtual tutor. J. Educ. Psychol. **105**(4), 1115 (2013)
31. Yi, C., Wang, J., Cheng, N., Zhou, S., Xu, B.: Applying wav2vec2. 0 to speech recognition in various low-resource languages. arXiv preprint arXiv:2012.12121 (2020)
32. Yi, C., Wang, J., Cheng, N., Zhou, S., Xu, B.: Transfer ability of monolingual wav2vec2. 0 for low-resource speech recognition. In: 2021 International Joint Conference on Neural Networks (IJCNN), pp. 1–6. IEEE (2021)
33. Zouhair, T.: Automatic speech recognition for low-resource languages using wav2vec2: Modern standard Arabic (msa) as an example of a low-resource language (2021)

Deep Learning or Deep Ignorance? Comparing Untrained Recurrent Models in Educational Contexts

Anthony F. Botelho[1(✉)], Ethan Prihar[2], and Neil T. Heffernan[2]

[1] University of Florida, Gainesville, FL 32611, USA
abotelho@coe.ufl.edu
[2] Worcester Polytechnic Institute, Worcester, MA 01605, USA
{ebprihar,nth}@wpi.edu

Abstract. The development and application of deep learning methodologies has grown within educational contexts in recent years. Perhaps attributable, in part, to the large amount of data that is made available through the adoption of computer-based learning systems in classrooms and larger-scale MOOC platforms, many educational researchers are leveraging a wide range of emerging deep learning approaches to study learning and student behavior in various capacities. Variations of recurrent neural networks, for example, have been used to not only predict learning outcomes but also to study sequential and temporal trends in student data; it is commonly believed that they are able to learn high-dimensional representations of learning and behavioral constructs over time, such as the evolution of a students' knowledge state while working through assigned content. Recent works, however, have started to dispute this belief, instead finding that it may be the model's complexity that leads to improved performance in many prediction tasks and that these methods may not inherently learn these temporal representations through model training. In this work, we explore these claims further in the context of detectors of student affect as well as expanding on existing work that explored benchmarks in knowledge tracing. Specifically, we observe how well trained models perform compared to deep learning networks where training is applied only to the output layer. While the highest results of prior works utilizing trained recurrent models are found to be superior, the application of our untrained-versions perform comparably well, outperforming even previous non-deep learning approaches.

Keywords: Deep learning · LSTM · Echo state network · Affect · Knowledge tracing

1 Introduction

The availability of large-scale education datasets, often comprised of large numbers of interactions between learners and educational technologies over time, have coincided with an increase in applications of deep learning methodologies

M. M. Rodrigo et al. (Eds.): AIED 2022, LNCS 13355, pp. 281–293, 2022.
https://doi.org/10.1007/978-3-031-11644-5_23

to study various aspects of student learning. The data collected from massive open online courses (MOOCs), for example, researchers have been able to utilize large, complex models to study student learning strategies as well as unproductive behavior such as attrition and dropout [6,21,26]. Even beyond MOOCs, in K-12 classrooms, the adoption of educational technologies and learning platforms such as Cognitive tutor [20] and ASSISTments [13], among many others, have led to the recording and often public release of large datasets of anonymized student interaction logs. The application of deep learning, collectively referring to a growing variety of multi-layer neural network models, often require large amounts of data to learn from, assuming, of course, that these models are in fact well-structured to learn anything at all.

Due to the large number of learned parameters and often complex structure of many deep learning approaches, many researchers and developers attribute the success of these methods to their ability to learn rich high-dimensional representations of input data. While it is possible to interpret and visualize what is learned in some applications of deep learning, it is difficult to what is learned within certain deep learning model structures including, for example, recurrent neural networks (RNN) [25]; this also includes commonly applied variants of RNN such as long short term memory (LSTM) [14] and gated recurrent unit (GRU) [8] networks. These model structures are designed to learn dependencies within time-series data, which is common in educational contexts.

Prior research that suggests, contrary to initial assumptions, that many recurrent models are not learning rich representations of data. In fact, it was found that by randomizing network weights and only training the output layer (referred to in this paper as "untrained" models), such models performed nearly as well as their trained counterparts [9,24], as will be discussed further in the next section. This work seeks to build upon this prior research that has explored this phenomenon in the context of knowledge tracing [9], to compare trained and untrained recurrent models in another educational context, detecting student affect, where deep learning recurrent models have similarly been applied in recent years [3]. In addition, several related modeling approaches have been specifically designed to utilized randomized, untrained components. This work additionally explores the application of these approaches in educational contexts. Specifically, this work will address the following research questions:

1. How does the application of untrained recurrent models compare to similar trained models in detecting student affect?
2. Do the methods designed to utilize untrained components outperform other approaches in detecting student affect?
3. Do trained and untrained recurrent models exhibit an overlapping set of latent features within their hidden layers?

1.1 Representations Within Recurrent Networks

While the application of recurrent networks, or one of several common variants, has increased in recent years, it is important to examine the basic structure

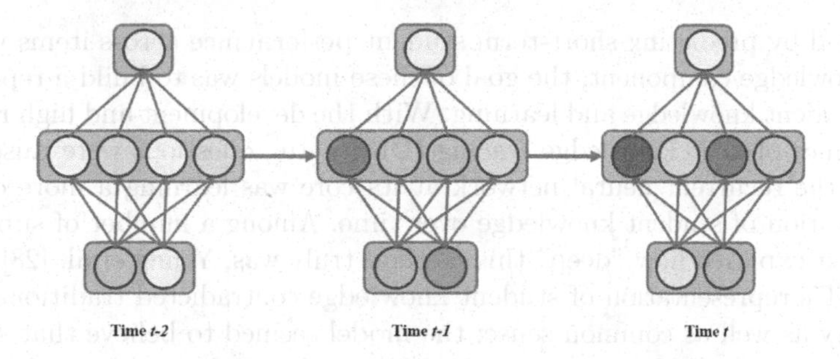

Fig. 1. This example illustrates how a recurrent network can build sequence representations within its hidden layer by combining new inputs with previous network states.

of these networks in order to better understand how such models could learn rich knowledge representations. Consider, for example, the simplified network representation depicted in Fig. 1. Like most "deep" models, recurrent networks are normally comprised of multiple layers of nodes representing values generated within the network structure. These values are calculated by multiplying the node values of earlier layers by learned weights that are traditionally fully-connected to all nodes in the subsequent layer. Unlike other network structures, recurrent models are designed to be applied to time-series data, where the values in the network's input layer (bottom layer in the figure) are combined with the previous hidden state (middle layer in the figure) for each time step. Intuitively, it is assumed that the model may learn how to combine new information with prior information within the series to make a more informed estimate for a given task; the network structure may learn how long to keep information, when to forget information, or certain conditions under which it should otherwise modify its understanding of the given sequence. The changing values within the hidden layer of the network contain and retain information from throughout the series.

How well these models are able to "understand" the given information, as represented by the values within the recurrent hidden layer, is a matter of recent speculation. It is precisely this question, as it applies to educational contexts, that is to be explored further within this work.

2 Background

In view of the application of recurrent models in education, it is important to better understand what applied models are learning from student data to understand how they can best be used to study various aspects of learning. Among the most well-known examples of using models to study learning, for example, is knowledge tracing [7]. In early models of knowledge tracing, interpretability was a primary goal; while later research has disputed how interpretable, or rather how identifiable traditional knowledge tracing models are [2,10], the structure of the models were built in alignment to learning theory. While the models themselves are trained by predicting short-term student performance across items within a given knowledge component, the goal of these models was to build a representation of student knowledge and learning. With the development and high reported performance of deep knowledge tracing (DKT) [19], questions were raised as to whether the recurrent neural network at its core was learning a more complex representation of student knowledge over time. Among a number of subsequent works that explored how "deep" this method truly was, Yeung et al. [28] showed that DKT's representation of student knowledge contradicted traditional learning theory as well as common sense; the model seemed to believe that students fluctuated frequently between states of knowledge and non-knowledge. While the authors proposed a fix to this problem using a form of regularization during model training, this work suggests that the model is able to perform well without a strong grasp of how learning is likely to occur.

These works, however, are not the first to question whether recurrent models are able to learn rich representations within sequential and temporal data. Wieting and Kiela [24] found that untrained recurrent models could perform comparably well to trained versions in natural language processing tasks. It was suggested that the applied models act as a type of sequence encoding, rather than by embedding deeper contextual information; the models are able to learn high-dimensional *encodings* of sequential data, but may be ignorant of the latent constructs and other factors that explain the data. The findings by Wieting and Kiela and subsequently Ding and Larson [9] who extended that work to further explore deep knowledge tracing, raise questions as to how useful these models are in educational contexts if they are not learning representations of deeper constructs; it is important to emphasize that these works did incorporate model training in the output layer of their compared models, so it is not the case that no training is required, only that the deeper layers of the networks may not learn composite features that align to latent factors in the data.

Prior work in applying recurrent models have explored how well such models are able to learn effective features in the context of detecting student affect and other learning behaviors [4]. In that work, the authors found that the use of expert-engineered features, developed in alignment to learning theory, led to higher predictive performance in a number of modeling tasks as compared to allowing a machine learning model to learn from the raw data logs used by the experts. The authors similarly identify an inherent difficulty within these networks to learn from the data.

This does not mean, however, that these models cannot still be useful in studying aspects of learning. Prior work has led to the development of sensor-free models of student affect [3] developed from student interaction logs paired with human-coded classroom observations [17,18]. Utilizing LSTM networks, these models have been successfully applied to study student affect even without the ability to interpret the learned representations within the model [5]; even if it is the case that many recurrent models are unable to learn deep representations of latent constructs, this does not mean that the estimates produced by these models cannot be useful to study learning (c.f. "discovery with models" [22].

3 Methodology[1]

Although there are many advantages in utilizing interpretable models to study learning, there are several practical benefits made possible if recurrent models are truly "ignorant" to deeper representations within data. As has been found in the prior works described in the previous section (i.e. [9,24]), untrained variants of recurrent models may perform comparably to similar trained models in several applications. If this finding holds in other contexts, these models may have increased potential for integration in a number of educational technologies and settings by significantly reducing training times or potentially even the amount of data needed to successfully fit such a model (needing to train only the output layer would be equivalent, in most cases, to training a linear or logistic regression model which traditionally requires fewer data samples to train).

In this work, we use student affect detection and knowledge tracing as two example cases of comparison for trained and untrained recurrent models on previously-published benchmark results ([4] and [27] for affect detection and knowledge tracing, respectively). While Ding and Larson [9] did explore applications of untrained models in the context of knowledge tracing, this work further expands upon this work by introducing two additional methods, Bag of Random Embeddings [24] and Echo State Networks [15,24], within educational contexts.

3.1 Affect and Knowledge Tracing Data

In this paper, we observe applications of these trained and untrained recurrent network models within two publicly-available datasets collected within ASSISTments [13]. ASSISTments is a free web-based learning platform used by primarily middle-school teachers and students for mathematics homework and classwork. Among a number of other features, the learning system allows teachers to assign traditional "complete all problems" assignments as well as mastery-based "skill builder" assignments. While working through problems, the system allows students to make multiple attempts to answer problems and offers supports in the form of on-demand hints and scaffolding problems. To support educational research, the system has also released a number of publicly-available datasets such as those utilized within this paper.

The first dataset observed in this work was released in [4], which was derived from several prior works focused on the development of sensor-free detectors of student affect (e.g. detectors that utilize only interaction logs without additional sensors such as video). ASSISTments data was used to develop affect detectors using expert-engineered features based on both theory and an iterative development process [18]. Additional works subsequently experimented with different features within a number of rule- and regression-based modeling methods [23] before recurrent deep learning methods were explored [3].

The dataset itself is comprised of student interaction logs paired with human-coded classroom observations of four states of student affect: engaged concentration, boredom, confusion and frustration. Following [4], the data exists in

[1] The code utilized by this work is made publicly available:https://osf.io/ubr2v/.

two forms, the first consisting of the 92 expert-engineered features used in prior works, and second consisting of the raw action-level logs that were used to build these features. While that work found that the use of expert-engineered features led to superior model performance as compared to the raw features, we explore both feature sets in this work utilizing untrained models to examine the performance benefits of training these models.

The second dataset observed in this work has been previously used to examine methods of knowledge tracing [27]. As described in Sect. 2, knowledge tracing is among the most widely studied problems in learner analytics, AI in education, and educational data mining communities. The original knowledge tracing (KT) model [7] and its bayesian implementation (BKT), attempt to model student latent knowledge using student performance metrics. The ASSISTments knowledge tracing dataset used in this work was made publicly available in [27] (specifically, the dataset refered to as "09–10 (c)" in that paper), after fixing several identified errors in the original version of that dataset used in Piech et al.'s original deep knowledge tracing paper [19]. This dataset is comprised of 275,459 math problems across 146 knowledge components answered by 4,217 students.

3.2 Leveraging Untrained Networks

This work explores the application of several untrained model structures. These model structures were applied across both the affect and knowledge tracing datasets, predicting the affect labels (as a multi-dimensional categorical outcome, as was done in prior works) and next problem correctness, respectively. As previously introduced, the terminology of "untrained" in the context of this work (in alignment with prior works [9,24]), refers to a partially-trained model. In most machine learning contexts, especially those observing deep learning approaches, models are typically trained by randomizing the initial values of a set of weights or coefficients that are then updated iteratively through an optimization procedure [16]. Considering deep learning models, this process is believed to help the model learn sets of features in lower layers of the network, with the final output layer (often functionally equivalent to a linear or logistic regression) then learning how to map those features to a set of outcomes. An "untrained" method effectively skips the optimization procedure, relying on the randomized weights to produce a large number of un-tuned features; in this process, a single regression model can be trained using these un-tuned features to map them to observed outcomes. This is an important distinction as this creates somewhat of a misnomer in that these methods still rely on some degree of training, but do not rely on training to "model" the data. These methods, as well as the application procedure, is described in this section.

Bag of Random Embeddings. The bag of random embeddings was the simplest untrained network. This method is used to simply project the time series data to a higher dimensional space. To create a bag of random embeddings for a time series of f features and t time-steps, the approach projects the time series into a n dimensional embedding by first creating a t by f matrix, referred to as the time-series matrix, where each row in the matrix is the full set of features from one time-step. Next, the approach generates an f by n matrix full of random values, referred to as the projection matrix. The time-series matrix is then multiplied by the projection matrix, resulting in a t by n matrix, referred to as the embedding matrix. Finally, a pooling operation is applied across all the time-steps in the embedding matrix, resulting in a final n dimensional vectorized embedding of the initial time series.

Following the advice of [24], the random numbers of the projection matrix were initialized between $\frac{-1}{\sqrt{f}}$ and $\frac{1}{\sqrt{f}}$. To find the best bag of random embeddings, all combinations of an n of 1, 2, 4, 8, 16, 32, 64, 128, 256, and 512, and both max and mean pooling, for five random seeds each were used to project the time-series data before using a 5-fold cross-validated logistic regression to classify affect or predict next problem correctness, depending on the dataset. The average performance of every fold of every random seed for each combination of hyper-parameters was used to determine the best values.

Long Short-Term Memory Networks. The Long Short-Term Memory Network (LSTM) [11,14] is a common recurrent network structure for modeling time-series data. The LSTM network is a form of recurrent neural network that in addition to utilizing information from its past state, is designed to learn when to incorporate new information into its state and when to forget previous information. In this context, the value of the LSTM network is often viewed as being in its internal state structure which incorporates a type of memory that is designed to capture long- and short-term dependencies within the series (thus its name). Even without training, the state of the LSTM network, if complex enough, can capture useful, predictive information from the time-series in certain contexts [24]. To determine if an untrained LSTM network would be capable of predicting either affect or next problem correctness, an LSTM network was created with all combinations of zero through four hidden layers (i.e. additional fully connected layers on top of the LSTM layer), and 1, 2, 4, 8, 16, 32, 64, 128, 256, and 512 nodes per layer, including the output layer, for five random seeds. Each network's output layer was given to a 5-fold cross-validated logistic regression and used to classify affect or predict next problem correctness. The average performance of every fold of every random seed for each combination of hyper-parameters was used to determine the best combination of hyper-parameters.

Echo State Networks. Echo State networks are similar to recurrent networks in that they have connections from forward nodes to their predecessors, but these networks usually lack the formality of layers. Instead, an echo state network has a reservoir of nodes that have many connections to many other nodes in the reservoir. The input layer connects to any subset, or all of the nodes in the reservoir, and the output layer receives the output from the reservoir nodes. The weights in the reservoir are never trained, but the weights of the output layer are [15]. The echo state network is designed to exploit the properties of a recurrent

Table 1. Comparison of trained and untrained models applied to the affect dataset

Model	Features	Best model	AUC	Kappa
Untrained models				
LSTM Network	Raw	$n = 512$, 0 *added hidden*	0.661	0.098
Bag of Random Emb.	Raw	$n = 64$, *max pooling*	0.631	0.066
Echo State Network	Raw	$n = 512$, 1 *added hidden*	0.673	0.121
LSTM Network	Expert	$n = 512$, 1 *added hidden*	0.701	0.152
Bag of Random Emb.	Expert	$n = 64$, *mean pooling*	0.741	0.128
Echo State Network	Expert	$n = 512$, 0 *added hidden*	0.694	0.127
Trained models				
LSTM (Botelho et al., 2019)	Raw		0.695	0.041
LSTM (Botelho et al., 2019)	Expert		0.760	0.172

network's state similarly to how the previous section uses the state of an LSTM network. Within the untrained weights of the reservoir lies the state of the echo state network. This state is designed to capture the latent information of the time-series data presented to it and when the output layer is trained.

To determine if an echo state network would be capable of predicting either affect or next problem correctness, the output of each of the random LSTM networks from the previous section was combined with the intermediate output from every node in the network, essentially converting the LSTM network to an echo state network. The outputs from every node were again used to classify affect or predict next problem correctness in a logistic regression, which functions as the output layer of the echo state network. The average performance of every fold of every random seed for each combination of hyper-parameters was used to determine the best combination of hyper-parameters.

4 Results

The results of our applied untrained models are compared to the results generated from trained models as reported in prior works utilizing the same respective datasets used here. For consistency, these results are compared using the same metrics as have been used in comparison in prior works; in regard to the affect data, the AUC measure is calculated using the multi-class categorical evaluation method as used in previous works [12].

The results of the untrained models applied in this work in comparison to the trained models described in [4] are reported in Table 1. The highest-performing of each model type is compared to the reported results of the prior work across measures of AUC and Kappa (in alignment to that prior work). In this table, it can be seen that the trained LSTM utilizing expert-engineered features exhibits the highest model performance across both metrics. However, the untrained LSTM and Bag of Random Embedding models each perform comparably close in regard to AUC and Kappa; these even outperform the trained LSTM model utilizing the raw dataset.

Table 2. Comparison of trained and untrained knowledge tracing models.

Model	Best model	AUC
Untrained models		
LSTM Network	$n = 512$, 0 *added hidden*	0.706
Bag of Random Emb.	$n = 512$, *mean pooling*	0.692
Echo State Network	$n = 512$, 0 *added hidden*	0.725
LSTM (Ding & Larson, 2019)		0.730
Trained models		
DKT (LSTM; Xiong et al., 2016)		0.750
BKT (Xiong et al., 2016)		0.630

Similarly, the results of the untrained models applied in this work in comparison to previous results are reported in Table 2. In this table, we also compare our untrained model results to the untrained model applied in [9]. Here, the trained DKT model does exhibit the highest AUC performance, but the untrained LSTM as reported in [9] and Echo State Network applied in this study perform comparably well. What is perhaps particularly worth noting, is that all untrained recurrent models outperformed the BKT model.

5 Exploring Latent Feature Overlap

We have seen over the previous set of analyses that the untrained models perform comparably well to their trained counterparts. This raises several questions including what, if anything, is being learned within the hidden layer of these trained recurrent models (i.e. is there an overlap of latent features utilized by these models). In addressing our third research question, we conduct a final analysis to explore the latent features represented by trained and untrained models in detecting student affect.

In this analysis, we compare an LSTM-based model architecture as presented in [3] as a basis of comparison. We train this method using one LSTM layer consisting of 200 nodes feeding to a dense output layer of 4 nodes corresponding to the four affective states, similar to those previously described. We train this model and then extract the hidden layer from the network. Similarly, we generate five untrained counterparts using the same model architecture differing only in the number of nodes used in the hidden layer (using 200, 400, 600, 800, and 1000). We similarly extract the hidden layers of these models corresponding with each sample of the affect detection dataset.

We conduct an exploratory factor analysis (EFA) to identify latent constructs represented by each set of hidden features. EFA is a common dimensionality reduction method that identifies latent factors, or features, that exist as the linear combination of other features [1]. With this, we want to observe whether the factors that emerge from the trained model overlap, or are meaningfully correlated, with the untrained model factors. If the trained model is not learning effectively from the data, we would expect that there would be a large overlap in factors when compared with the untrained models.

Table 3. Number of factors and overlap between untrained and trained models.

Model	EFA factors	N overlapping factors (Rho>0.6)
Trained LSTM (200)	31	—
Untrained LSTM (200)	35	5
Untrained LSTM (400)	50	1
Untrained LSTM (600)	74	5
Untrained LSTM (800)	91	4
Untrained LSTM (1000)	103	5

From our EFA, reported in Table 3, 31 features emerge from our trained model, with an increasing number of factors emerging from larger untrained dimensions (the number of factors were determined based on the number of factors with an eigenvalue greater than 1, following common practice). Using these features, we conducted a complete pair-wise comparison of untrained factors to trained model factors and computed a Spearman (Rho) ranked correlation for each pairing. We then simply counted the number of factor pairs that exhibited a Rho value greater than 0.6 as a measure of pseudo-overlapping feature sets. From the table, it can be seen that despite the increasing number of emerging factors, the number of overlapping factors remained relatively constant. This suggests that, while the untrained models constructed large feature sets, these were mostly uncorrelated with the trained model features.

6 Discussion

While it is surprising that the untrained models perform comparably well to their trained counterparts, the results of our analyses suggest that the trained models are learning effectively from the data; particularly from the EFA, we argue that the learned features are not simply random combinations of features due to the notable lack of overlap with the factors emerging from the untrained models. This lack of overlap is unexpected given the comparable model performance, suggesting that there are a small number of highly-predictive factors present.

In both affective and knowledge tracing contexts, the untrained models perform remarkably well, even outperforming other benchmarks (e.g. the trained LSTM using the raw data in Table 1 and the BKT model in Table 2). This work represents a step toward better understanding how deep learning models learn from given data. It is difficult to conclude that our findings will generalize to *all* recurrent models and applications, but the analyses conducted in this work in conjunction with those presented in prior works [9,24] have found similar results across multiple contexts. It is the goal that this work will lead to further work to better understand knowledge representations within deep learning models to either better utilize them in various contexts, or to improve them so that they may exhibit higher utility for the study of learning.

Following the results reported in this paper, it is important to clarify and emphasize the contribution and potential impact of our findings. First, as the untrained models were found to be comparable to prior results across both applications observed in this paper, this finding aligns with prior research that suggests that the trained recurrent models may not be learning deep representations. However, the lack of overlap between factors emerging from the trained and untrained models suggests that the trained model is learning a distinctive set of latent factors related to affect. This finding supports the use of such models to both detect affect, but also to better study the latent structures that indicate affect and other learning constructs (i.e. these features are not simply randomly generated or encoded features). With that said, the untrained models may additionally provide utility. As the models perform well above chance and other simple baselines, the estimates produced by these models may still highly correlate with outcomes of interest and may be used to study learning.

Acknowledgements. We would like to thank NSF (e.g., 2118725, 2118904, 1950683, 1917808, 1931523, 1940236, 1917713, 1903304, 1822830, 1759229, 1724889, 1636782, & 1535428), IES (e.g., R305N210049, R305D210031, R305A170137, R305A170243, R305A180401, & R305A120125), GAANN (e.g., P200A180088 & P200A150306), EIR (U411B190024), ONR (N00014-18-1-2768) and Schmidt Futures.

References

1. Bandalos, D.L., Finney, S.J.: Factor analysis: exploratory and confirmatory. In: The Reviewer's Guide to Quantitative Methods in the Social Sciences, pp. 98–122. Routledge, London (2018)
2. Beck, J.E., Chang, K.: Identifiability: a fundamental problem of student modeling. In: Conati, C., McCoy, K., Paliouras, G. (eds.) UM 2007. LNCS (LNAI), vol. 4511, pp. 137–146. Springer, Heidelberg (2007). https://doi.org/10.1007/978-3-540-73078-1_17
3. Botelho, A.F., Baker, R.S., Heffernan, N.T.: Improving sensor-free affect detection using deep learning. In: André, E., Baker, R., Hu, X., Rodrigo, M.M.T., du Boulay, B. (eds.) AIED 2017. LNCS (LNAI), vol. 10331, pp. 40–51. Springer, Cham (2017). https://doi.org/10.1007/978-3-319-61425-0_4
4. Botelho, A.F., Baker, R.S., Heffernan, N.T.: Machine-learned or expert-engineered features? exploring feature engineering methods in detectors of student behavior and affect. In: The 12th International Conference on Educational Data Mining (2019)
5. Botelho, A.F., Baker, R.S., Ocumpaugh, J., Heffernan, N.T.: Studying affect dynamics and chronometry using sensor-free detectors. Int. Educ. Data Min. Soc. (2018)

6. Chaplot, D.S., Rhim, E., Kim, J.: Predicting student attrition in MOOCs using sentiment analysis and neural networks. In: AIED Workshops, pp. 54–57 (2015)
7. Corbett, A.T., Anderson, J.R.: Knowledge tracing: modeling the acquisition of procedural knowledge. User Model. User-Adap. Inter. **4**(4), 253–278 (1994)
8. Dey, R., Salem, F.M.: Gate-variants of gated recurrent unit (GRU) neural networks. In: 2017 IEEE 60th International Midwest Symposium on Circuits and Systems (MWSCAS), pp. 1597–1600. IEEE (2017)
9. Ding, X., Larson, E.C.: Why deep knowledge tracing has less depth than anticipated. In: International Educational Data Mining Society (2019)
10. Doroudi, S., Brunskill, E.: The misidentified identifiability problem of Bayesian knowledge tracing. In: International Educational Data Mining Society (2017)
11. Gers, F.A., Schmidhuber, J., Cummins, F.: Learning to forget: continual prediction with LSTM (1999)
12. Hand, D.J., Till, R.J.: A simple generalisation of the area under the roc curve for multiple class classification problems. Mach. Learn. **45**(2), 171–186 (2001)
13. Heffernan, N.T., Heffernan, C.L.: The assistments ecosystem: building a platform that brings scientists and teachers together for minimally invasive research on human learning and teaching. Int. J. Artif. Intell. Educ. **24**(4), 470–497 (2014)
14. Hochreiter, S., Schmidhuber, J.: Long short-term memory. Neural Comput. **9**(8), 1735–1780 (1997)
15. Jaeger, H.: Echo state network. Scholarpedia **2**(9), 2330 (2007)
16. Le, Q.V., Ngiam, J., Coates, A., Lahiri, A., Prochnow, B., Ng, A.Y.: On optimization methods for deep learning. In: ICML (2011)
17. Ocumpaugh, J.: Baker Rodrigo Ocumpaugh monitoring protocol (BroMP) 2.0 technical and training manual. Technical report, Teachers College, Columbia University, New York, NY (2015)
18. Ocumpaugh, J., Baker, R., Gowda, S., Heffernan, N., Heffernan, C.: Population validity for educational data mining models: a case study in affect detection. Br. J. Edu. Technol. **45**(3), 487–501 (2014)
19. Piech, C., et al.: Deep knowledge tracing. In: Proceedings of the 28th International Conference on Neural Information Processing Systems, pp. 505–513 (2015)
20. Ritter, S., Anderson, J.R., Koedinger, K.R., Corbett, A.: Cognitive tutor: applied research in mathematics education. Psychonomic Bull. Rev. **14**(2), 249–255 (2007)
21. Rosé, C.P., et al.: Social factors that contribute to attrition in MOOCs. In: Proceedings of the 1st ACM Conference on Learning@ Scale Conference, pp. 197–198 (2014)
22. Siemens, G., Baker, R.S.d.: Learning analytics and educational data mining: towards communication and collaboration. In: Proceedings of the 2nd International Conference on Learning Analytics and Knowledge, pp. 252–254 (2012)
23. Wang, Y., Heffernan, N.T., Heffernan, C.: Towards better affect detectors: effect of missing skills, class features and common wrong answers. In: Proceedings of the 5th International Conference on Learning Analytics and Knowledge, pp. 31–35. ACM (2015)
24. Wieting, J., Kiela, D.: No training required: exploring random encoders for sentence classification. arXiv preprint arXiv:1901.10444 (2019)
25. Williams, R.J., Zipser, D.: A learning algorithm for continually running fully recurrent neural networks. Neural Comput. **1**(2), 270–280 (1989)
26. Xing, W., Chen, X., Stein, J., Marcinkowski, M.: Temporal predication of dropouts in MOOCs: reaching the low hanging fruit through stacking generalization. Comput. Hum. Behav. **58**, 119–129 (2016)

27. Xiong, X., Zhao, S., Van Inwegen, E.G., Beck, J.E.: Going deeper with deep knowledge tracing. In: International Educational Data Mining Society (2016)
28. Yeung, C.K., Yeung, D.Y.: Addressing two problems in deep knowledge tracing via prediction-consistent regularization. In: Proceedings of the 5th Annual ACM Conference on Learning at Scale, pp. 1–10 (2018)

Fine-grained Main Ideas Extraction and Clustering of Online Course Reviews

Chenghao Xiao(✉), Lei Shi, Alexandra Cristea, Zhaoxing Li, and Ziqi Pan

Department of Computer Science, Durham University, Durham, UK
{chenghao.xiao,lei.shi,alexandra.i.cristea,zhaoxing.li2, ziqi.pan2}@durham.ac.uk

Abstract. Online course reviews have been an essential way in which course providers could get insights into students' perceptions about the course quality, especially in the context of massive open online courses (MOOCs), where it is hard for both parties to get further interaction. Analyzing online course reviews is thus an inevitable part for course providers towards the improvement of course quality and the structuring of future courses. However, reading through the often-time thousands of comments and extracting key ideas is not efficient and will potentially incur non-coverage of some important ideas. In this work, we propose a *key idea extractor* that is based on fine-grained aspect-level semantic units from comments, powered by different variations of state-of-the-art pre-trained language models (PLMs). Our approach differs from both previous topic modeling and keyword extraction methods, which lies in: First, we aim to not only eliminate the heavy reliance on human intervention and statistical characteristics that traditional topic models like LDA are based on, but also to overcome the coarse granularity of state-of-the-art topic models like top2vec. Second, different from previous keyword extraction methods, we do not extract keywords to summarize each comment, which we argue is not necessarily helpful for human readers to grasp key ideas at the course level. Instead, we cluster the ideas and concerns that have been most expressed throughout the whole course, without relying on the verbatimness of students' wording. We show that this method provides high and stable *coverage* of students' ideas.

Keywords: MOOC · Key ideas extraction · Language models · Automated pipeline

1 Introduction

Identifying key ideas from course reviews is an essential way of obtaining insights into students' learning experience, especially in the context of massive open online courses (MOOCs), where it is hard for students and instructors to have further interaction [1,16].

M. M. Rodrigo et al. (Eds.): AIED 2022, LNCS 13355, pp. 294–306, 2022.
https://doi.org/10.1007/978-3-031-11644-5_24

However, reading through the often-time thousands of comments and extracting key ideas is not efficient and will potentially incur non-coverage of some important ideas [12]. This can be due to both aspects of feedback being forgotten throughout the reading due to readers' limited working memory [3] or even being ignored because of readers' perceptions and confirmation bias [8,10,30].

In this paper, we propose an *automated key idea extraction pipeline* that can be run with minimal human intervention and interpretation, with the purpose of efficiently covering as many as the most expressed ideas in massive corpus of students' reviews in online courses. While it is not necessarily feasible for course providers to sift through the often-time thousands of comments, it is advisable that they should attend to certain important aspects of concerns that have been most expressed in the comments [23]. We propose such an automated method, facilitated by state-of-the-art NLP algorithms. Moreover, we conduct experiments on the robustness of dimensionality reduction of text embeddings before applying hierarchical clustering, providing empirical and theoretical insights into the selection of this parameter and its impact on efficient *coverage* of ideas, for future users of this method. We also introduce a weighted centroid to select representative phrases for each cluster, and a flexible usage of a coefficient value to attend to under-represented ideas in a cluster.

We argue that for the efficient coverage of the most important aspect-level ideas expressed in massive corpus of online course comments, traditional keyword extraction and topic modeling methods might not work well, which is because the former only studies reducing the size of text instead of the number of documents [25,32], while the latter suffers from coarse granularity [14]. Facilitated by the state-of-the-art, our research provides a *fine-grained key idea extraction approach* to bridge this gap, while being wording-agnostic.

2 Related Work

Before static embedding methods such as word2vec [28] and contextualized language models such as BERT [9] and RoBERTa [20] were introduced, tasks of natural language processing (NLP) had been strongly relying on statistical characteristics extracted from language. For example, in the field of topic modelling, since Blei *et al.* [5] proposed Latent Dirichlet Allocation (LDA), this probabilistic model had been a major algorithm in topic modeling, whose limitations lie in both the unknown numbers of topic clusters that have to be decided by human through exhaustive experiments, and its statistical discrimination over rare but significant topic keywords - as topical words are not always frequently mentioned at the level of each document. On the other hand, recent development and deeper understanding in word embedding brings state-of-the-art topic modeling algorithms like top2vec [2] and BERTopic [11] to the table, which yield better results and require less human intervention than traditional topic models.

In the field of education, Miller [29] proposed leveraging BERT and k-means for extractive text summarization of lectures. They claimed that many approaches in the field used dated algorithms that produced sub-par results and relied on manual tuning. This provided a good pipeline to address similar extractive summarization scenarios, but we argue that the text representation that BERT itself produces, typically used in research by directly taking the [cls] (classification) token of BERT, is sub-optimal [31], and while being *de facto*, k-means is not necessarily a panacea for high-dimensional clustering. In this paper, we thus replace them by Sentence Transformers [31] and HDBSCAN [7]. Masala *et al.* [25] proposed extracting and clustering main ideas from student feedback based on a pipeline of KeyBERT-based keyword extraction and K-means context clustering. They first extracted top 10 keywords for each course, then clustered different contexts that mentioned these words. To the best of our knowledge, however, their keyword extraction component still relied on the verbatimness of the wording. This limitation is also addressed in our approach, through clustering directly on high-quality embeddings of fine-grained text.

We argue that for a course level analysis, starting from *top-n* verbatim keywords is not always a good approach as 1-gram keywords extracted might be mostly nouns which are over-general and hard to interpret on their own, while 2-gram or over 2-gram keywords strongly rely on the verbatimness of students' phrasing. For example, "well-organized" and "well-structured" convey close meanings that might otherwise be interpreted by course providers as one aspect. Considering two semantically similar words separately might affect the statistical significance of both of them, leading to both being ignored from *top-n*. By contrast, both being selected in *top-n* might affect the diversity of aspects included, as this prevents other important words from being selected. Therefore, we propose directly applying clustering on the level of fine-grained text, by breaking down each comment into chunks of long phrases or short sentences, which we argue is a good semantic unit that carries semantically interpretable meanings (as opposed to fragmented keywords), while mostly staying in only one aspect (as opposed to document level that covers different aspects which can twist the text embeddings and therefore affect the effectiveness of the clustering).

In line with our intuition, Luo and Litman [23] proposed summarizing students' responses at phrase level, and introduced *student coverage* as an evaluation of the method, based on the assumption that concepts mentioned by more students should receive more attention from the instructor, which chimes in with the purpose of our method. In this paper, we aim to realize these objectives with state-of-the-art algorithms. Moreover, on top of covering concepts that are semantically expressed the most, we also explore using outlier scores in a cluster, to 'listen' to under-represented phrases, as will be described in Sect. 4.1. In summary, we build upon the state of the art in text summarization and natural language processing to propose a novel pipeline, which also overcomes their limitations, and takes into account *readability*, *relevance*, and *coverage* [21].

3 Method

3.1 Corpus

We used the Coursera Course Reviews dataset[1], which comprises over 140k reviews of 1,835 courses, along with their corresponding ratings. For experiments

[1] https://www.kaggle.com/septa97/100k-courseras-course-reviews-dataset.

and demonstration of our proposed method, we focus on the field of machine learning and data science, which we filtered by the inclusion of either "machine", or both "data" and "science" in the course names, yielding 12 machine learning- and data science-related courses after we removed "machine design" which is irrelevant in this context. The filtering of data results in 9,980 unique comments with 246,290 tokens.

3.2 Pre-processing

What distinguishes our approach from other topic modeling methods is that we do not apply topic modeling at the entire document level, but instead at a fine-grained level, which requires that we first break each document into long phrases or short sentences. Although our method is mostly based on the state of the arts, the pre-processing step is inspired by a traditional method, RAKE [32], which observed that a document can be parsed into candidate keywords by breaking them down at delimiters and certain stopwords. We further customized our stopword list, removing as many useful words from the list as possible (e.g., opinionated ones like *don't, not, shouldn't*) to prevent them from being deleted during parsing. However, we find that what really matters is that a document is parsed into short sentences using delimiters. The stopwords that further break each short sentence into long phrases are less important, as a word that does not appear in a phrase will appear in the adjacent one anyway, preserving the information to be encoded and processed in later clustering.

3.3 Pipeline

We adopt similar pipeline described in [2], while making a few important adjustments to overcome its coarse granularity. First, as our method is based on fine-grained aspect-level linguistic units after pre-processing, the default doc2vec [18] would intuitively be insufficient to learn phrase embeddings that are semantically meaningful [17]. We thus replace this encoding method by two latest Sentence Transformers [31] models. Second, we find that the optimal number of embedding dimensions to reduce to at phrase level, before applying hierarchical clustering, is different from that on document level, and propose the method to empirically customize this hyperparameter through *coverage*. Lastly, we propose using a *local weighted centroid* to select the most representative phrase, so that readers can cover a big portion of the most important ideas through reading only a few phrases representing the largest clusters. Our pipeline is shown in Fig. 1.

Originally introduced using BERT as a backbone, Sentence Transformers (ST) have been shown to yield very effective representations of text when applied to similarity comparison, clustering, and information retrieval tasks [31], as opposed to previous approaches - taking [cls] token of BERT - which yielded sub-optimal semantic representation. It was not until recently that new ST methods that yielded significant performance boost based on newer Transformer models have been released. In this work, we empirically evaluate two ST models: one based on MPNet [33] that provides the best sentence embedding and semantic

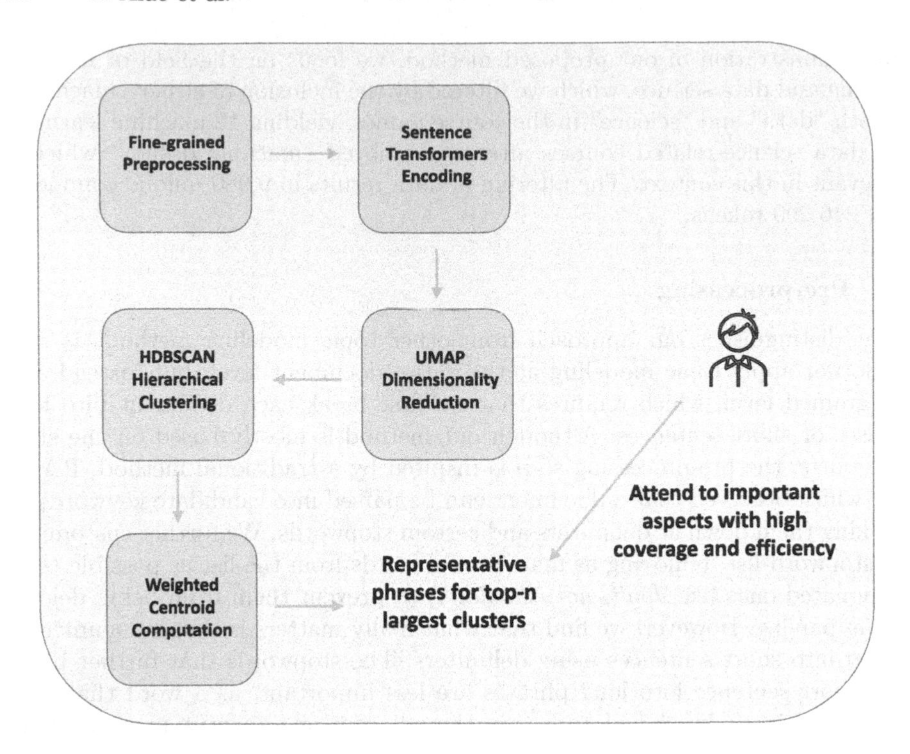

Fig. 1. Automated fine-grained key ideas extraction: pipeline and mechanism

search performance up to the date of this paper, *all-mpnet-base-v2*[2]; and the other being *all-MiniLM-L6-v2*[3], which is based on a distilled model MiniLM [35]. The latter achieves comparable results while being 5 times faster than the former. Thus we believe it is worth evaluating as an alternative for user cases in student feedback reading that require faster encoding of text embedding and output of following analysis results.

After encoding, the data further goes through dimensionality reduction and clustering. For dimensionality reduction, we use UMAP [27], which preserves better global structure of data [2] compared to t-SNE [24] as reflected in distances between clusters. For clustering, we use HDBSCAN [7,26], a robust hierarchical density-based clustering method which we use to replace the *de facto* k-means used in prior research, whose limitations lie in assumptions of inclusion of all instances and spherical shapes of clusters.

Equation (1) demonstrates the way we propose to find the centroid phrase CP in a cluster that shares the highest cosine similarity with our defined weighted centroid embedding CE as computed in Eq. (2).

[2] https://huggingface.co/sentence-transformers/all-mpnet-base-v2.
[3] https://huggingface.co/sentence-transformers/all-MiniLM-L6-v2.

$$CP = \arg\max_k \left(\frac{CE \cdot E_k}{\|CE\| \|E_k\|}\right), \quad (1)$$

where,

$$CE = \frac{\sum_1^K [(1 - \alpha O_k) E_k]}{K - \alpha \sum_1^K O_k}, \quad (2)$$

where K denotes the number of phrases in a cluster, k denotes the kth entry in that cluster, while O_k and E_k respectively denote its corresponding outlier score and embedding. We further introduce α as a flexible coefficient, to adjust the influence of the outlier score to the computation of the weighted centroid, where by default we set it to 1. However, we did find that α can be used flexibly, to output representative phrases that are far away from the vanilla centroid to obtain *unique ideas*, when it is set to a higher value. Although in [2], it is suggested that at document level, a weighted centroid will not make much of a difference to the vanilla centroid, we find that it does make a difference at the fine-grained aspect level, especially when applied to lower dimensional data after dimensionality reduction using UMAP.

4 Results and Discussions

4.1 Key Ideas Analysis

Results from our methods identified a range of important aspects that have been expressed in reviews of our selected machine learning- and data science-related courses (Table 1). We present the *top*-10 largest clusters from two results: respectively running with dimensionality reduction to 5 and 10 dimensions. [2] observed a best dimension reduction number of 5 for document level embedding clustering, while under our high-granularity context, we observed that it provides the best results when this hyperparameter is set to around 10, as will be further demonstrated through in-depth *coverage* study and empirical interpretation in Sect. 4.2. However, it is shown that our pipeline provides robust performance and a great overlap of topics under these two settings. Notably, we observed that the topics that are not overlapped in the top 10 clusters under these two results can be further found in the rest of their top 15 clusters.

Based on results shown in Table 1, we could easily get an overall idea that in machine learning- and data science-related courses, students express most concerns about: 1) programming language used in the courses, 2) math background required and covered, 3) content structured in the courses such as videos, quizzes and programming exercises, 4) the way and the degree to which instructors successfully convey the essence of the algorithms. These findings are in line with previous research using different methodologies [6,15,19,22,36]. In courses related to machine learning, students may encounter different kinds of difficulties [19]. For example, students lacking solid mathematical background struggled more with understanding math-related content in the course [15] and developing computational thinking [36]. Bolliger [6] suggested that an online course can be affected by various factors, such as instructor variables, technical issues, and interactivity, which can be interpreted through combinations of our identified topics as well. For example, identified clusters about Andrew Ng's way of teaching explain instructor variables, technical issues could involve codes debugging issues caused by difficulty of using programming languages like Octave, as expressed in comments, while interactivity could be supported by quizzes and exercises in this context. Lu *et al.* [22] found that flow experience significantly contributed to MOOC satisfaction, which relies on the students' not being distracted and frustrated during learning. In our case, we argue that ideas expressed in the topics identified, such as difficulty in using programming languages and following math intuition due to insufficient background knowledge, could account for no or low flow experience.

However, to acquire deeper insights into opinions in clusters, readers could further go into each cluster to see the well-represented and under-represented phrases in each cluster, through the usage of outlier scores. We provide the example of *Octave* to give a brief idea on how it works (see Table 2).

Table 1. Top-10 largest clusters by clustering on 5 and 10 dimensional embeddings, represented by their weighted centroid phrases. Cluster labels in bold show clusters that have been overlapped in the *top*-10 and thus *double-validated* by two outputs.

	5-d Clustering		10-d Clustering	
Top-n	Weighted centroid	Cluster label (interpreted)	Weighted centroid	Cluster label (interpreted)
1	The essence and purpose of the algorithms	**Algorithm**	The guts of the algorithms	**Algorithm**
2	The instructor uses Octave	**Octave**	best Coursera course I've ever taken	**Best Coursera course**
3	Made simple [..] understandable MATLAB	**Matlab**	Be afraid of using Octave	**Octave**
4	The best explanation of principles and ideas behind Machine Learning	Machine Learning	Matlab hands-on exercises permit a deeper understanding of the algorithms	**Matlab**
5	I enjoyed very much	**Enjoyed**	I enjoyed	**Enjoyed**
6	Do not want to watch the videos	Videos	The exercises	**Exercises**
7	Allows me to follow the quizzes	Quizzes	Very well constructed	Course structure
8	A statistics background [..]	**Math background**	A Complete online course	**Good MOOC**
9	Great MOOC	**Good MOOC**	Liked the way Andrew taught us the concept	Way of teaching
10	Great exercises	**Exercises**	Being proficient with Linear Algebra	**Math background**

Table 2. Sampled elements in an exemplar on the most representative cluster about *Octave* (the 3th largest cluster under 10-d clustering). The Example in bold is the centroid phrase computed by a weighted centroid. We present both well-represented and under-represented examples, showcased by outlier scores

Cluster examples	Outlier score
More chances to practice algorithm prototyping in Octave	0.0
Do not like Octave somehow and prefer the Python approach coming	0.0
Octave [...] instead of more modern languages	0.0
Force the students to use Octave	0.0
Be afraid of using Octave	0.0
In OCTAVE instead of popular languages like R	0.24
But I consistently felt unprepared for applying it in Octave	0.26
Along with a great introduction to Octave	0.26
Awesome assignment submission tool via Octave	0.34
I enjoyed learning Octave and performing the weekly homework	0.40

4.2 Robustness of Dimensionality Reduction

Recent research has indicated that the similarity measures of contextual word embeddings, as opposed to static word embeddings, have been dominated by a small number of what is referred to as the "rogue" dimensions [34]. Furthermore, the typically 768 or similar pre-defined dimensional space of BERT-facilitated embedding methods makes density-based clustering inefficient. It is also not difficult to intuitively picture how hierarchical density-based clustering will tend to only put phrases that are extremely close to one another into the same cluster, due to the vast geographical space created by high dimensions. This tends to, thereby, make the clusters no more than a collection of some almost semantically identical, or even worse, verbally identical phrases. Reducing dimensionality before apply clustering, however, greatly compresses the semantic space, making clusters that are otherwise separated in high dimensional space have to "accept" one another and merge to a large cluster.

We speculate that on a high granularity level, the number of embedding dimensions used in hierarchical clustering can be interpreted and utilized as a strong indicator of reader's tolerance towards semantic difference, and therefore the acceptance of larger cluster with semantically diverse, yet aspect-wise similar expressions. In general, a lower number of dimensions indicates higher tolerance, and leads to more otherwise separated clusters merged into one.

The results are demonstrated in Fig. 2, where we compare a few representative dimension numbers through their corresponding sizes of the biggest clusters, and average sizes of the *top*-10 clusters, which is extremely significant for measure of *coverage* of ideas. Coverage to more frequently expressed students' ideas through fewer *first* n clusters can lead to instructors' getting important opinion aspects from comments with higher efficiency. We empirically find that the

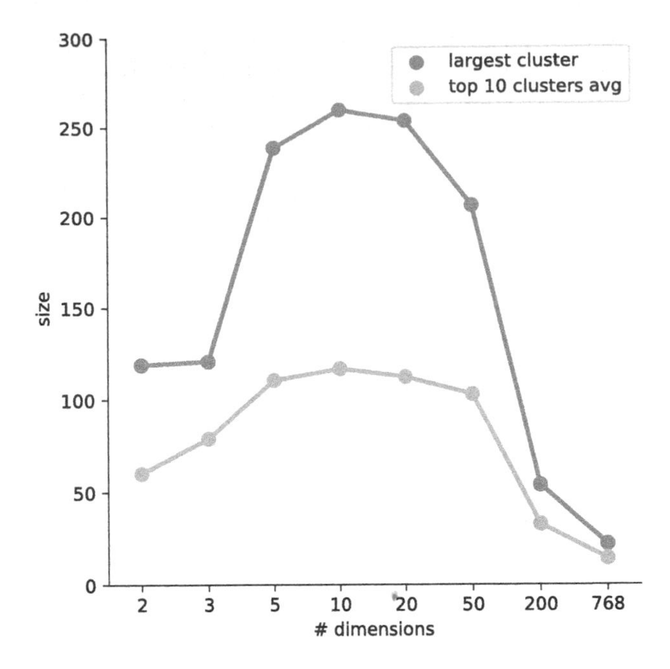

Fig. 2. Reducing to different numbers of dimensions before applying hierarchical clustering yields *coverage* of ideas through *top-n* clusters with different efficiency

coverage stays stable when dimension is reduced to around 5–20, yielding similar *top-n* cluster phrases and sizes, while 10 dimensions provide the best results in our case, indicated by both cluster quality and *coverage*. We suggest future studies to start from this range and find a customized value for their specific courses.

4.3 A Glimpse of the Algorithm's Wording-Agnostics

In this section, we briefly demonstrate the superiority of our algorithm in terms of how agnostic it is towards different wordings to convey similar meanings that are by their nature supposed to be clustered into one aspect.

Using the 7th largest cluster under 10-dimensional clustering as an example, we randomly select 13 phrases out of the 71 phrases in that cluster (see Table 3). It is clearly shown that wordings of reviews in this cluster are highly diverse, while our approach facilitates to understand them as conveying close meanings, albeit phrases in this cluster consist of no verbatim wordings.

Table 3. Diverse wordings of reviews in a sample cluster, whose weighted centroid phrase is computed as *Very well constructed*, while phrases in that cluster include almost no verbatim forms of the same wording

Cluster examples	Outlier score
Thoughtfully made	0.0
Very well put together	0.0
Carefully created	0.0
Very well constructed	0.0
Well constructed with good practicals	0.04
Really well crafted	0.27
Very well designed with a clear focus	0.43
Also well-built with a lot of warm-support and encouragement	0.52
Exceptionally well arranged	0.54
Very polished and it makes participating easy and smooth	0.54
Meticulously curated	0.56
Excellent selfcontained	0.57
Badly designed	0.58

5 Conclusion

In this work, we propose a novel pipeline for online course providers to receive insights into students' opinions, concerns and experience from online courses, and thus be able to attend to the most important aspects of comments efficiently. We empirically present the effectiveness of combining state-of-the-art embedding encoders, dimensionality reduction, and clustering algorithm on the *fine-grained aspect level*. We also present an empirical study on the robustness of embedding dimension selection that could optimize runtime without losing much semantic information of aspect-level linguistic units, and being more wording-agnostic for higher efficiency of student *coverage*.

Structured on the state-of-the-art, our proposed method contributes to achieving high coverage of important ideas, while being agnostic to students' wordings. We plan to deploy this pipeline in real-life classes and create teaching assistant-generated gold-standard summaries [23] for evaluation of algorithm-generated idea coverage against human readers' perceptions. Notably, our approach aims to extract information from course reviews, while we highlight that intrinsic biases in course evaluations do exist [4,13]. We encourage researchers in this field to build upon our method to detect and mitigate biases in course evaluations.

We also envision two possible directions of future work. First, we envision fine-tuning Sentence Transformer models with domain-specific text datasets, to make domain-specific aspects more positionally accurate in the semantic space, for facilitating better evaluations of courses in highly specialized domains. While in our case, reviews related to machine learning and data science courses might not be highly different from day-to-day writing, highly domain-specific ones like medical courses might require language models' deeper understanding about the field, to extract accurate clusters. Second, we envision efforts in human-AI interaction: if deploying our proposed method in industrial settings, we encourage to enable users (course providers, instructors, etc.) to accept, reject, and merge clusters. Such data can then be recorded and used to learn a feature-based activation layer [9, 31] for the system to provide more personalized cluster recommendations.

References

1. Anderson, A., Huttenlocher, D., Kleinberg, J., Leskovec, J.: Engaging with massive online courses. In: Proceedings of the 23rd International Conference on World Wide Web, pp. 687–698 (2014)
2. Angelov, D.: Top2vec: distributed representations of topics (2020). arXiv preprint arXiv:2008.09470
3. Baddeley, A.D.: Working memory and reading. In: Processing of Visible Language, pp. 355–370. Springer (1979). https://doi.org/10.1007/978-1-4684-0994-9_21
4. Baker, R., Dee, T., Evans, B., John, J.: Bias in online classes: evidence from a field experiment. Econ. Educ. Rev. **88**, 102259 (2022)
5. Blei, D.M., Ng, A.Y., Jordan, M.I.: Latent dirichlet allocation. J. Mach. Learn. Res. **3**, 993–1022 (2003)
6. Bolliger, D.U.: Key factors for determining student satisfaction in online courses. Int. J. E-learn. **3**(1), 61–67 (2004)
7. Campello, R.J., Moulavi, D., Zimek, A., Sander, J.: Hierarchical density estimates for data clustering, visualization, and outlier detection. ACM Trans. Knowl. Discov. Data (TKDD) **10**(1), 1–51 (2015)
8. Del Vicario, M., Scala, A., Caldarelli, G., Stanley, H.E., Quattrociocchi, W.: Modeling confirmation bias and polarization. Sci. Rep. **7**(1), 1–9 (2017)
9. Devlin, J., Chang, M.W., Lee, K., Toutanova, K.: Bert: Pre-training of deep bidirectional transformers for language understanding. In: Proceedings of the 2019 Conference of the North American Chapter of the Association for Computational Linguistics: Human Language Technologies, vol. 1 (long and short papers). pp. 4171–4186 (2019)
10. Frost, P., Casey, B., Griffin, K., Raymundo, L., Farrell, C., Carrigan, R.: The influence of confirmation bias on memory and source monitoring. J. Gen. Psychol. **142**(4), 238–252 (2015)
11. Grootendorst, M.: Bertopic: leveraging bert and c-tf-idf to create easily interpretable topics, vol. 4381785 (2020). https://doi.org/10.5281/zenodo
12. Hasan, K.S., Ng, V.: Automatic keyphrase extraction: A survey of the state of the art. In: Proceedings of the 52nd Annual Meeting of the Association for Computational Linguistics, vol. 1 (long papers), pp. 1262–1273 (2014)
13. Hassan, T.: On bias in social reviews of university courses. In: Companion Publication of the 10th ACM Conference on Web Science, pp. 11–14 (2019)
14. Jiang, D., Shi, L., Lian, R., Wu, H.: Latent topic embedding. In: Proceedings of COLING 2016, the 26th International Conference on Computational Linguistics: Technical Papers, pp. 2689–2698 (2016)

15. Kim, S.W.: Kepler vs Newton: teaching programming and math to almost all-majors in a single classroom. In: 2020 IEEE International Conference on Teaching, Assessment, and Learning for Engineering (TALE), pp. 956–957 (2020). https://doi.org/10.1109/TALE48869.2020.9368332
16. Kop, R.: The challenges to connectivist learning on open online networks: Learning experiences during a massive open online course. Int. Rev. Res. Open Distrib. Learn. **12**, 19–38 (2011)
17. Lau, J.H., Baldwin, T.: An empirical evaluation of doc2vec with practical insights into document embedding generation. arXiv preprint arXiv:1607.05368 (2016)
18. Le, Q., Mikolov, T.: Distributed representations of sentences and documents. In: International Conference on Machine Learning, PMLR, pp. 1188–1196 (2014)
19. Lishinski, A., Yadav, A., Enbody, R.: Students' emotional reactions to programming projects in introduction to programming: measurement approach and influence on learning outcomes. In: Proceedings of the 2017 ACM Conference on International Computing Education Research, pp. 30–38 (2017)
20. Liu, Y., et al.: Roberta: A robustly optimized BERT pretraining approach. arXiv preprint arXiv:1907.11692 (2019)
21. Liu, Z.: Research on Keyword Extraction Using Document Topical Structure. Tsinghua University, Beijing (2011)
22. Lu, Y., Wang, B., Lu, Y.: Understanding key drivers of MOOC satisfaction and continuance intention to use. J. Electron. Commer. Res. **20**(2), 105–117 (2019)
23. Luo, W., Litman, D.: Summarizing student responses to reflection prompts. In: Proceedings of the 2015 Conference on Empirical Methods in Natural Language Processing, pp. 1955–1960 (2015)
24. Van der Maaten, L., Hinton, G.: Visualizing data using t-sne. J. Mach. Learn. Res. **9**(11) 2579–2605 (2008)
25. Masala, M., Ruseti, S., Dascalu, M., Dobre, C.: Extracting and clustering main ideas from student feedback using language models. In: Roll, I., McNamara, D., Sosnovsky, S., Luckin, R., Dimitrova, V. (eds.) AIED 2021. LNCS (LNAI), vol. 12748, pp. 282–292. Springer, Cham (2021). https://doi.org/10.1007/978-3-030-78292-4_23
26. McInnes, L., Healy, J., Astels, S.: hdbscan: Hierarchical density based clustering. J. Open Source Softw. **2**(11), 205 (2017)
27. McInnes, L., Healy, J., Saul, N., Großberger, L.: UMAP: Uniform manifold approximation and projection. J. Open Source Softw. **3**(29), 861 (2018)
28. Mikolov, T., Sutskever, I., Chen, K., Corrado, G.S., Dean, J.: Distributed representations of words and phrases and their compositionality. In: Advances in Neural Information Processing Systems, pp. 3111–3119 (2013)
29. Miller, D.: Leveraging bert for extractive text summarization on lectures. arXiv preprint arXiv:1906.04165 (2019)
30. Oswald, M.E., Grosjean, S.: Confirmation bias. In: Cognitive illusions: a Handbook on Fallacies and Biases in Thinking, Judgement and Memory, vol. 79 (2004)
31. Reimers, N., Gurevych, I.: Sentence-bert: sentence embeddings using siamese bert-networks. In: Proceedings of the 2019 Conference on Empirical Methods in Natural Language Processing and the 9th International Joint Conference on Natural Language Processing (EMNLP-IJCNLP), pp. 3982–3992 (2019)
32. Rose, S., Engel, D., Cramer, N., Cowley, W.: Automatic keyword extraction from individual documents. Text Min. Appli. Theory **1**, 1–20 (2010)
33. Song, K., Tan, X., Qin, T., Lu, J., Liu, T.Y.: Mpnet: masked and permuted pre-training for language understanding. Adv. Neural. Inf. Process. Syst. **33**, 16857–16867 (2020)

34. Timkey, W., van Schijndel, M.: All bark and no bite: rogue dimensions in transformer language models obscure representational quality. In: Proceedings of the 2021 Conference on Empirical Methods in Natural Language Processing, pp. 4527–4546 (2021)
35. Wang, W., Wei, F., Dong, L., Bao, H., Yang, N., Zhou, M.: Minilm: deep self-attention distillation for task-agnostic compression of pre-trained transformers. Adv. Neural. Inf. Process. Syst. **33**, 5776–5788 (2020)
36. Weintrop, D., et al.: Defining computational thinking for mathematics and science classrooms. J. Sci. Educ. Technol. **25**(1), 127–147 (2016)

Assessing Readability by Filling Cloze Items with Transformers

Andrew M. Olney(✉)

University of Memphis, Memphis, Tennessee 38152, USA
aolney@memphis.edu

Abstract. Cloze items are a foundational approach to assessing readability. However, they require human data collection, thus making them impractical in automated metrics. The present study revisits the idea of assessing readability with cloze items and compares human cloze scores and readability judgments with predictions made by T5, a popular deep learning architecture, on three corpora. Across all corpora, T5 predictions significantly correlated with human cloze scores and readability judgments, and in predictive models, they could be used interchangeably with average word length, a common readability predictor. For two corpora, combining T5 and Flesch reading ease predictors improved model fit for human cloze scores and readability judgments.

Keywords: Readability · Assessment · Cloze · Transformers

1 Introduction

Cloze items, also known as fill-in-the-blank items, are widely used in education for assessment and some types of instruction (e.g. vocabulary instruction). However, cloze items also have a long history as a measure of readability, i.e. of text difficulty [16]. The standard approach to assessing readability with cloze items is called nth deletion, where every nth word in a text is deleted and replaced with a blank of fixed size. The task of the reader is to use their knowledge and context cues across the entire text to fill in the blanks.

It has long been known that a higher number of correct completions on nth deletion cloze tests is a strong indicator of higher readability (low difficulty) and aligns with well-known readability metrics like Flesch reading ease and Dale-Chall readability, aptitude tests, and standard comprehension questions [2,4,16,17]. Unlike comprehension test questions, which are difficult to create and confound the measurement of text difficulty with question difficulty, cloze items can be generated directly from text and have less measurement error.

Despite their effectiveness as a readability measure, cloze items are not a practical in most cases because they require human-subjects data collection. For this reason, practical readability metrics have been developed to have high correlation to measures like cloze, but otherwise use easily calculated characteristics of the text to determine a readability score. Common examples of such metrics

M. M. Rodrigo et al. (Eds.): AIED 2022, LNCS 13355, pp. 307–318, 2022.
https://doi.org/10.1007/978-3-031-11644-5_25

are Flesch reading ease [10], which uses average sentence length in words (ASL) and average word length in syllables (AWL), Dale-Chall readability, which uses the proportion of difficult words (defined by a word list) and ASL [5,7], and the Lexile measure, which uses word frequency and ASL [15].

These metrics are quite simple but also quite effective at assessing readability on a large scale - not because they are causally related to readability but rather because they are so strongly correlated with factors that influence readability. The seeming paradox that the simplest measures would be the best predictors of readability was addressed by Bormuth, who described it as a trade-off between face validity and predictive validity [3]: many linguistic variables correlate with readability, so a metric with face validity would include many linguistic variables; however, the measurement error associated with these variables means that a metric with fewer variables has better predictive power when applied to unseen texts. Thus, while there has been continued interest in creating better readability metrics, especially in the modern era (see [6] for a review), these simple metrics are a challenging baseline. For example, Martinc et al. found that their deep neural language models were not able to outperform an ASL baseline ($r = .906$) on the Newsela corpus in an unsupervised setting [12].

Modern deep learning methods, notably Transformers [19] offer a potential alternative to traditional readability metrics. As described above, the traditional approach is to use linguistic features to predict cloze item performance, which is a ground-truth measure of readability. In contrast, Transformers can be used to predict cloze difficulty directly because this is how they are trained in the first place - to predict masked tokens in their input. Deep learning methods based on masked language modeling have proven to be extremely effective in a variety of natural language processing (NLP) tasks [8,13], so presumably, they would function well for a task aligned with their pre-training objective. The idea of using Transformers to directly measure cloze difficulty was first investigated by Benzahra & Yvon, unfortunately without much success [1]. They used GPT-2, an autoregressive Transformer, to predict cloze completions on two corpora with experts-labeled grade levels and achieved overall correlations of .05 and .13, respectively. However, we argue that GPT-2 is the wrong model to use for this task because it is autoregressive and only allows leftward context to be used to predict the next word or words. In contrast, the nth deletion cloze task allows the use of both left and right context across the entire document. Therefore, additional study of Transformers to directly predict cloze difficulty is warranted.

The present investigation examines the application of Transformers to measuring both cloze difficulty and grade-level readability. Our primary research question is whether Transformer cloze scores correspond with these measures and standard readability metrics. The remainder of the paper is organized around three different studies with different corpora. The first corpus, the Bormuth passages [4], allow direct comparison to cloze item difficulty calculated from human subjects experiments, in addition to comparison to other relevant measures like comprehension tests. The second two corpora, the OneStopEnglish corpus (OSE) [18] and the Newsela corpus [20], allow comparison to expert-defined grade

<extra_id_0> of the Big Cats, <extra_id_1> well as the lesser <extra_id_2>, have wonderful eyes. They <extra_id_3> see clearly even on <extra_id_4> dark night. This is <extra_id_5> of the way they <extra_id_6> made. There is a <extra_id_7> of window in each <extra_id_8>. This window is called <extra_id_9> pupil. It is black <extra_id_10> is placed in the <extra_id_11> of the colored part <extra_id_12> the eye. The pupil <extra_id_13> light come in and <extra_id_14> a kind of mirror <extra_id_15> the back of each <extra_id_16>. These mirrors reflect everything <extra_id_17> is in front of <extra_id_18> eyes. Right away a <extra_id_19> nerve carries these reflected <extra_id_20> to the brain. Then <extra_id_21> brain sends a quick <extra_id_22> to all parts of <extra_id_23> body. This signal may <extra_id_24> to attack, hide, be <extra_id_25>

Fig. 1. A chunk representing one-half of a 250-word text submitted to T5. Each cloze word has been replaced by a highlighted T5 sentinel token, which is ordered sequentially using nth deletion, $n = 5$.

levels for each text. Across all corpora, additional comparisons will be made to standard readability metrics like Flesch reading ease.

2 Approach

All of the following studies use a Transformer called T5 [13] to measure both cloze difficulty and readability. T5 is a suitable model because it attends to both left and right contexts and because it is trained on a denoising objective that closely matches the cloze task. To match the method of Bormuth [4], only the first 250 words of each text are subjected to nth deletion ($n = 5$). Five clozed versions of each text are created by using different offsets for nth deletion, e.g. starting at words 1, 2, 3, 4, and 5, after which subsequent words have been deleted by a previous version. As a result, every word in the text is subjected to cloze in exactly one offset version. During development, it was discovered that the T5 model used[1] produces degenerate responses to cloze items after the 27th item[2]. Therefore, each text was split into two chunks, each representing 125 words and 25 cloze items, given the $n = 5$ nth deletion strategy, and the two chunks were submitted to T5 separately for each of the 5 offsets noted above The need to break the text into chunks for T5 is a notable departure from Bormuth's method because it creates less context for T5 to complete the task than what is afforded to humans, making the task more difficult. Otherwise, this task is broadly consistent with T5's unsupervised denoising training objective of predicting the randomly deleted 15% of tokens vs. predicting nth-deleted tokens, $n = 5$, or 20% of tokens. An example of a chunk input to T5 is shown in Fig. 1.

Several approaches to generating cloze predictions were explored during initial investigations, with the goal of generating multiple predictions for each cloze item. Multiple predictions are desirable because they allow partial credit

[1] https://huggingface.co/t5-large.
[2] https://github.com/huggingface/transformers/issues/8842.

for lower-ranked predictions using metrics like reciprocal rank, where a correct prediction at rank N receives a score of $1/N$. Multiple predictions are also desirable because they potentially reflect a distribution of predictions across human subjects, rather than a single prediction. Our investigations suggested that greedy beam search had desirable properties of being highly repeatable but the disadvantage of not producing much diversity when multiple predictions were requested, regardless of the number of beams and diversity penalties applied.

Therefore, for the top prediction, we used greedy beam search with one beam, and for the remaining predictions, we used sampling with both top-K [9] and top-p [11] approaches. Because sampling is stochastic, the sampling results are not highly repeatable, but because the cloze metrics are assessed per text, we consider these as being repeated 250 times, once for each word in the text. To avoid repetitions and multi-word predictions, which are impossible given the task, duplicate predictions were removed from lower ranks, and predictions that contained internal whitespace (as a separator between words) were excluded.

Two accuracy metrics were calculated for each cloze item using these predictions. Correct at rank 1 was defined by an exact match between the top prediction and the original word, normalized for case and leading/trailing whitespace. Correct at any rank was defined by a similar exact match on a prediction of any rank, weighted by reciprocal rank. In addition to the T5 cloze metrics, Flesch and Dale-Chall readability metrics were calculated for each text[3].

3 Study 1: Bormuth Passages

3.1 Data

The Bormuth passages were used in a major study of readability that incorporated cloze items (nth deletion; $n = 5$), reading rate, and pre/post comprehension questions [4]. To create these passages, Bormuth ranked 330 passages used in another study [3] by cloze difficulty, divided the difficulty range into 8 points, and selected the 4 passages closest to those 8 points, such that no more than 4 came from the same subject matter category and each text was at least 250 words in length. Thus the 32 passages represent 8 difficulty levels spanning from first grade to college. Each passage and corresponding measures were extracted from the Appendix [4] and manually checked for errors; passages were additionally submitted to Grammarly to catch any errors. Grammarly revealed that several passages (3213, 5226, 6441, 7151, and 8552) contained spelling mistakes. In order to prevent T5 from correctly predicting a word but not matching the original misspelled form, all spelling errors were corrected. Additionally, passage 6545 had no corresponding entry in the data tables and passage 6535 listed in the data tables had no corresponding text; these were assumed to refer to the same passage. Finally, Lexile scores were calculated using the Lexile Text Analyzer[4]. Because the Analyzer only allows 50 texts to be processed per month, Lexiles were only computed for this dataset.

[3] https://github.com/cdimascio/py-readability-metrics.

[4] https://hub.lexile.com/analyzer.

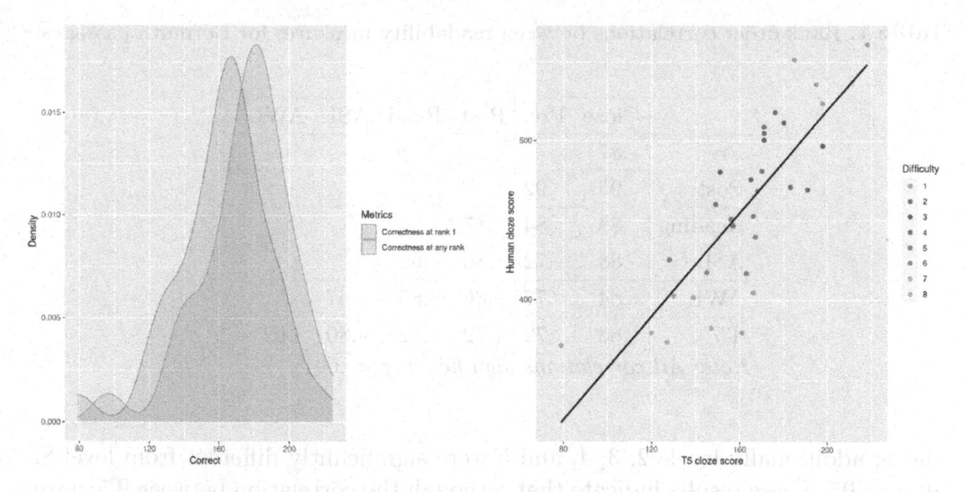

Fig. 2. Density plot for correctness at rank 1 and correctness at any rank.

Fig. 3. Scatterplot of T5 cloze correctness and human correctness. Regression lines show smoothed (blue) and linear (red) fits. (Color figure online)

3.2 Results

The primary questions of interest in this study are whether the T5 cloze scores correspond with the human cloze scores, as well as how these scores comparatively relate to other measures of readability. To address the first question, we examined the differences between the correctness at rank 1 metric and the correctness at any rank metric in order to determine which was the most appropriate measure for the following analyses. As shown in Fig. 2, the correctness at rank 1 ($M = 162.22, SD = 27.24$) and correctness at any rank ($M = 175.33, SD = 25.97$) are approximately equivalent in distribution, except correctness at any rank is slightly more lenient and therefore right-shifted. Because the difference seemed relatively negligible and correctness at rank 1 has better repeatability, we only report results for correctness at rank 1, which we will refer to as T5 cloze scores.

The Spearman rank order correlation between the T5 cloze scores and human cloze scores was significant, $r(30) = .86$, $p < .001$. A scatterplot between the two scores is shown in Fig. 3. The relationship is approximately linear, with visible separation of the eight difficulty levels along with visible overlap. Two ANOVA analyses were conducted to examine the discriminability of human and T5 cloze scores according to these levels. The human cloze score ANOVA was significant, $F(7, 24) = 93.19$, $p < .001$. Pairwise tests using Tukey's HSD revealed that every difficulty level was significantly different from the other, $p < .05$, except for levels 2 and 3, levels 4 and 5, and levels 7 and 8, i.e. there are effectively 5 levels of difficulty rather than 8 according to this measure. The T5 cloze score ANOVA was also significant, $F(7, 24) = 7.50$, $p < .001$. Pairwise tests using Tukey's HSD revealed that level 1 was significantly different from levels 6, 7,

Table 1. Rank order correlations between readability measures for Bormuth passages.

	Cloze	Pre	Post	Read	ASL	AWL
Pre	.87					
Post	.95	.92				
Reading	.88	.84	.87			
ASL	−.88	−.72	−.80	−.69		
AWL	−.84	−.77	−.86	−.85	.67	
T5	.86	.72	.72	.72	−.80	−.62

Note: All correlations significant, $p < .001$.

and 8; additionally levels 2, 3, 4, and 5 were significantly different from level 8, all $p < .05$. These results indicate that although the correlation between T5 cloze scores and human cloze scores is strong, the discriminability of T5 cloze scores with respect to the assigned difficulty levels is less than that of the human cloze scores. One possible reason for this is that the human cloze scores were based on students from grades 3 to 12, so the scores for difficult passages were drawn down by students from lower grades. In contrast, T5 is a single model with a single ability level.

Correlations with additional readability measures are shown in Table 1. The highest correlations were between the human measures in the upper left. The T5 cloze scores and the classic readability components, average sentence length in words (ASL) and average word length in syllables (AWL), have similar correlations to each of the human components, with the exception of post-test score. Since post-test score represents human performance after reading the text, a low correlation might be expected, but it is notable that AWL and ASL have a stronger correlation with post-test than pre-test, while T5 cloze scores have the same correlation with both. Surprisingly, the T5 cloze scores are more strongly correlated with ASL than AWL, suggesting that T5 is using linguistic information at the sentence level more than at the word level as it makes cloze predictions.

The results in Table 1 suggest that the T5 cloze scores could be combined with ASL, AWL, or both to create a model in the style of classic readability metrics like Flesch reading ease. To investigate this possibility and compare to the standard Flesch model, four models were constructed using combinations of these predictors. The models and their fits are reported in Table 2. The best-fitting model used all predictors, giving it a .06 improvement in fit over the Flesch reading ease model. However, this comparison is somewhat unfair as the Flesch model has only two predictors. The remaining models are two predictor contrasts to the Flesch model. The T5+ASL model has a fit .01 below the Flesch model, and the T5+AWL has a fit .03 below the Flesch model. Altogether, these models indicate that the T5 cloze scores potentially have some additive benefit to ASL and AWL and can be used almost interchangeably for this task.

Table 2. Linear models predicting human cloze scores for Bormuth passages.

Model	Coefficients			R^2
	T5	ASL	AWL	
All	.62	−4.56	−95.31	.94
Flesch		−6.09	−116.87	.88
T5+ASL	.79	−6.58		.87
T5+AWL	.95		−149.03	.85

Note: For all coefficients, $p < .001$.

4 Study 2: OneStopEnglish

4.1 Data

The OneStopEnglish corpus (OSE) is a balanced corpus consisting of 189 texts topics, each in three versions of difficulty, for a total of 567 texts [18]. Texts were collected from onestopenglish.com, a site for English language learners, and consisted of news stories that had been simplified by teachers for news-based lessons. The three levels of difficulty are thus aligned with pedagogical goals in ESL. Each difficulty level has a reported Flesch-Kinkaid Grade Level, 6.4 for Beginner, 8.2 for Intermediate, and 9.5 for Advanced. Unlike the Bormuth passages, OSE has no human-derived readability measures, so its primary utility for readability research stems from its expert-labeled difficulty levels. OSE is a popular corpus for readability research and was used in several studies mentioned in Sect. 1 [1,12].

4.2 Results

The primary research question for this study is the alignment of T5 cloze scores with expert difficulty and other measures of readability. To keep the results comparable with the last study, difficulty in these results is reverse scaled as ease. As in the previous study, we checked the distributions of the correct at rank 1 and the correct at any rank metrics. The distributions were comparable to Fig. 2 relative to each other, but the distributions of correct at rank 1 ($M = 156.82, SD = 11.28$) and correct at any rank ($M = 171.32, SD = 10.51$), were markedly narrower than in the last study, likely reflecting the smaller range of difficulty in OSE compared to the Bormuth passages. We again chose correct at rank 1 as our T5 cloze score metric in the following analyses.

The rank-order correlation between the T5 cloze scores and the three levels of ease was significant, $r(565) = .19$, $p < .001$, but notably smaller than in the last study. We additionally calculated Kendall's tau-b to compare to the previous work that used Transformers to predict cloze scores on this corpus, Benzahra and Yvon [1]. Our $\tau_b = .15$ for OSE versus their $\tau_b = .05$, a threefold improvement but a modest score nonetheless.

Table 3. Rank order correlations between readability measures for the OneStopEnglish corpus.

	Ease	ASL	AWL
ASL	-.58*		
AWL	-.37*	.29*	
T5	.19*	.02	-.07

Note: * $p < .001$.

Table 4. Linear models predicting reading ease for the OneStopEnglish corpus.

Model	Coefficients			R^2
	T5	ASL	AWL	
All	.01	−.09	−1.91	.41
Flesch		−.09	−2.05	.38
T5+ASL	.01	−.10		.37
T5+AWL	.01		−3.19	.16

Note: For all coefficients, $p < .001$.

Rank-order correlations with readability measures shown in Table 3 provide further insight into this low overall correlation between T5 cloze scores and levels of ease. In contrast to the previous study, the T5 cloze scores are not significantly correlated with either ASL or AWL. Additionally, the correlation between ASL and AWL is less than half what it was in the previous study. The cause of these changes in correlation is not clear, and possible explanations include the limited range of ease, the different genres (news text vs. informational text), and the mode of construction (artificially created vs. naturally occurring).

An ANOVA analysis was conducted to examine the discriminability of T5 cloze scores according to the levels of ease. The ANOVA was significant, $F(2, 564) = 14.47$, $p < .001$. Pairwise tests using Tukey's HSD revealed that Elementary texts ($M = 160.29, SD = 9.59$) were significantly easier than Intermediate texts ($M = 154.58, SD = 11.35$), $p < .001$, as well as Advanced texts ($M = 155.59, SD = 11.98$), $p < .001$. Intermediate and Advanced texts were not significantly different from each other, $p = .643$.

Although the correlations in Table 3 are lower than the previous study, each of the metrics is significantly correlated with the level of ease. Therefore additional regression models matching those in Table 2 were created, and the results are presented in Table 4. The rank order of model fit matches the previous study. The model containing all predictors had the best fit, followed by Flesch reading ease. The T5+ASL model has a fit .01 below the Flesch model, and the T5+AWL was markedly worse at .22 below the Flesch model. As before, models improve with the T5 predictor, and the T5 predictor is almost interchangeable with AWL. However, on OSE, the T5 predictor is not as interchangeable with ASL, as shown by the poor fit of the final model.

5 Study 3: Newsela

5.1 Data

The Newsela corpus[5] was introduced by Xu and colleagues [20] as a resource for text simplification research, but it has also been used for readability research [12]. Like OSE, the Newsela corpus contains multiple versions of the same text topic at different difficulty levels, and the text topics are drawn from the news. However, Newsela is different from OSE in a number of ways. Newsela has a greater number of versions for each text topic (typically 5) and spans a greater range of difficulty, grade 2 to grade 12. However, the distribution of grade-level text in the corpus is not balanced, and the number of texts at each grade level ranges from 2 to 2096. Newsela is designed to match English language learning needs of native, rather than ELL speakers. Finally, Newsela's grade levels are approximately aligned with Lexile, which increases its usefulness for readability research. All 9565 English texts of Newsela were used, consisting of 1911 text topics.

5.2 Results

The primary research question for this study is again the alignment of T5 cloze scores with expert difficulty and other measures of readability, and whether this alignment will be more consistent with the first or second study. To keep the results comparable, grade level is reverse scaled as ease. Distributions of the correct at rank 1 and the correct at any rank metrics were similar to the distributions in Sect. 4. The distribution of correct at rank 1 ($M = 159.00, SD = 12.89$) and correct at any rank ($M = 173.35, SD = 12.08$), were comparably narrow as the OSE distributions, suggesting that the narrowness of the distributions is not attributable to a restricted range of difficulty. To stay consistent with the other studies, correct at rank 1 was again chosen as our T5 cloze score metric in the following analyses.

The rank-order correlation between the T5 cloze scores and the 11 levels of ease was significant, $r(9563) = .33$, $p < .001$, was in between the correlations found in the previous studies. An ANOVA conducted to examine the discriminability of T5 cloze scores according to the levels of ease was significant, $F(10, 9554) = 126.91$, $p < .001$. Pairwise tests using Tukey's HSD revealed that texts from grade 2 were significantly easier than texts from grades 5–10 and 12; texts from grade 3 were significantly easier than texts from grades 4–10 and 12; texts from grade 4 were significantly easier than texts from grades 5–10 and 12; texts from grade 5 were significantly easier than texts from grades 6–10 and 12; texts from grade 6 were significantly easier than texts from grades 8, 10, and 12; texts from grade 7 were significantly easier than texts from grades 8, 10, and 12; texts from grade 8 were significantly easier than texts from grade 12; and texts from grade 9 were significantly easier than texts from grade 12,

[5] https://newsela.com/data/.

Table 5. Rank order correlations between readability measures for the Newsela corpus.

	Ease	ASL	AWL
ASL	−.95		
AWL	−.63	.62	
T5	.33	−.29	−.16

Note: For all r, p < .001.

Table 6. Linear models predicting reading ease for the Newsela corpus.

Model	Coefficients			R^2
	T5	ASL	AWL	
All	.02	−.56	−1.50	.83
Flesch		−.58	−1.49	.83
T5+ASL	.02	−.58		.83
T5+AWL	.05		−17.27	.40

Note: For all coefficients, p < .001.

all $p < .05$. Nonsignificant comparisons involving grades 10 and 11 are perhaps best explained by the small number of texts assigned to these levels, 22 total. Altogether, the ANOVA results indicate that T5 cloze scores afford a fair level of discriminability for Newsela grade levels.

Correlations with readability measures are shown in Table 5. The strength of the correlations again falls in between those of the previous studies. ASL and AWL are correlated comparably to the first study, but ASL is much more strongly correlated with ease than in the second study. Although the cause of these differences in correlation remains uncertain, it seems that genre can be ruled out as a cause, given that the corpora from the second and third studies are news corpora. These correlations provide additional evidence for another possible cause, which is the larger range of ease. A larger range of ease is a common characteristic between the first study and the third study and so may explain the similarities in correlation.

Regression models matching those used in the previous studies were created and results are presented in Table 6. The fits of the models follow a different pattern from the previous studies, with the first three models achieving the same fit. For the first time, the T5+ASL model has a fit equivalent to Flesch, providing additional evidence that T5 cloze scores are almost interchangeable with AWL. As in study 2, the poor fit of the T5+AWL indicates that T5 is not interchangeable with ASL.

6 Discussion

The focus of this work was to examine the use of T5 for predicting cloze item difficulty, a standard for readability, along with its analogous grade-level readability. A consistent pattern of results emerged across the three studies. In each case, T5 cloze scores significantly correlated with the outcome measures of interest, human cloze difficulty or expert-assigned grade level. Additionally, T5 cloze scores typically improved prediction of the outcome measures of interest when combined with the Flesch reading ease components of average sentence length (ASL) and average word length (AWL). In all studies, T5 cloze scores could be substituted for AWL in linear models and provide a fit almost as good, or as good, as Flesch reading ease.

However, there were some notable differences across the studies. The most striking difference is that T5 cloze scores were much more strongly correlated with human cloze scores (study 1) than with expert-assigned grade levels (studies 2 and 3). It seems unlikely that this difference can be explained by differences in genre or patterns of correlation across the studies, since studies 1 and 3 have similar patterns of correlation between the outcome measures, ASL, and AWL, but studies 2 and 3 shared the same genre, news. Neither can the differences be explained by the range of difficulties in the texts, since both studies 1 and 3 have approximately the same range of grades. Rather, the results across the studies suggest that T5 cloze scores are more aligned with human cloze scores than with expert-assigned grade levels, which is somewhat surprising because human cloze scores and expert-assigned grade levels themselves should be highly correlated [3,4]. Clearly, further research on this question is needed, focusing on naturalistic informational texts to replicate the strong findings found in study 1.

The question remains as to whether T5 cloze difficulty has the potential to improve readability measures that have been in place for many decades. After all, in our studies, T5 cloze scores at best replaced a component of Flesch reading ease. The primary reason that T5 might be useful going forward is that it encodes substantial knowledge about the world, and it makes cloze predictions using that knowledge. For example, T5 has been used for closed book trivia question answering without explicitly teaching it the knowledge involved [14]. This capability is analogous to a human reader bringing to bear background knowledge in order to understand a text, and it is something that isn't captured by word- or sentence-length metrics. Exactly how to manifest this capability in a readability model such that it consistently outperforms established metrics is a matter for future research.

Acknowledgments. This material is based upon work supported by the National Science Foundation under Grants 1918751 and 1934745 and by the Institute of Education Sciences under Grant R305A190448. Any opinions, findings, and conclusions or recommendations expressed in this material are those of the author(s) and do not necessarily reflect the views of the National Science Foundation or the Institute of Education Sciences.

References

1. Benzahra, M., Yvon, F.: Measuring text readability with machine comprehension: a pilot study. In: Proceedings of the Fourteenth Workshop on Innovative Use of NLP for Building Educational Applications, pp. 412–422. Association for Computational Linguistics, Florence, August 2019. https://doi.org/10.18653/v1/W19-4443
2. Bormuth, J.R.: Cloze test readability: criterion reference scores. J. Educ. Meas. **5**(3), 189–196 (1968)
3. Bormuth, J.R.: Development of readability analysis. Tech. Rep. ED 029 166, University of Chicago (1969). https://eric.ed.gov/?id=ED029166

4. Bormuth, J.R.: Development of standards of readability: Toward a rational criterion of passage performance. Final report. Tech. Rep. ED 054 233, University of Chicago (1971). https://eric.ed.gov/?id=ED054233
5. Chall, J., Dale, E.: Readability Revisited: The New Dale-Chall Readability Formula. Brookline Books (1995)
6. Collins-Thompson, K.: Computational assessment of text readability: a survey of current and future research. ITL - Int. J. Appli. Linguist. **165**(2), 97–135 (2014)
7. Dale, E., Chall, J.S.: A formula for predicting readability. Educ. Res. Bull. **27**(1), 11–28 (1948). http://www.jstor.org/stable/1473169
8. Devlin, J., Chang, M.W., Lee, K., Toutanova, K.: BERT: pre-training of deep bidirectional transformers for language understanding. In: Proceedings of the 2019 Conference of the North American Chapter of the Association for Computational Linguistics: Human Language Technologies, vol. 1, pp. 4171–4186. Association for Computational Linguistics, Minneapolis (2019)
9. Fan, A., Lewis, M., Dauphin, Y.N.: Hierarchical neural story generation. In: Gurevych, I., Miyao, Y. (eds.) Proceedings of the 56th Annual Meeting of the Association for Computational Linguistics, pp. 889–898. Association for Computational Linguistics (2018)
10. Flesch, R.: A new readability yardstick. J. Appl. Psychol. **32**(3), 221–233 (1948)
11. Holtzman, A., Buys, J., Du, L., Forbes, M., Choi, Y.: The curious case of neural text degeneration. In: 8th International Conference on Learning Representations. OpenReview.net (2020). https://openreview.net/forum?id=rygGQyrFvH
12. Martinc, M., Pollak, S., Robnik-Šikonja, M.: Supervised and unsupervised neural approaches to text readability. Comput. Linguist. **47**(1), 141–179 (2021)
13. Raffel, C., et al.: Exploring the limits of transfer learning with a unified text-to-text transformer. J. Mach. Learn. Res. **21**(140), 1–67 (2020)
14. Roberts, A., Raffel, C., Shazeer, N.: How much knowledge can you pack into the parameters of a language model? In: Proceedings of the 2020 Conference on Empirical Methods in Natural Language Processing (EMNLP), pp. 5418–5426. Association for Computational Linguistics, November 2020
15. Stenner, A.J., Sanford-Moore, E.E., Burdick, D.: The Lexile Framework for Reading technical report. MetaMetrics, Inc., Tech. rep. (2007)
16. Taylor, W.L.: Cloze procedure: a new tool for measuring readability. J. Quart. **30**(4), 415–433 (1953)
17. Taylor, W.L.: Cloze readability scores as indices of individual differences in comprehension and aptitude. J. Appli. Psychol. **41**(1), 19–27 (1957)
18. Vajjala, S., Lučić, I.: OneStopEnglish corpus: a new corpus for automatic readability assessment and text simplification. In: Proceedings of the Thirteenth Workshop on Innovative Use of NLP for Building Educational Applications, pp. 297–304. Association for Computational Linguistics, New Orleans, Louisiana, June 2018
19. Vaswani, A., et al.: Attention is all you need. In: Guyon, I., et al., (eds.) Proceedings of the Thirty-first Annual Conference on Neural Information Processing Systems, pp. 5998–6008 (2017)
20. Xu, W., Callison-Burch, C., Napoles, C.: Problems in current text simplification research: new data can help. Trans. Assoc. Comput. Linguist. **3**, 283–297 (2015)

CurriculumTutor: An Adaptive Algorithm for Mastering a Curriculum

K. M. Shabana[1](✉), Chandrashekar Lakshminarayanan[2], and Jude K. Anil[1]

[1] Indian Institute of Technology Palakkad, Palakkad, India
shabana.meethian@gmail.com
[2] Indian Institute of Technology Madras, Chennai, India
chandrashekar@cse.iitm.ac.in

Abstract. An important problem in an intelligent tutoring system (ITS) is that of adaptive sequencing of learning activities in a personalised manner so as to improve learning gains. In this paper, we consider intelligent tutoring in the *learning by doing* (LbD) setting, wherein the *concepts* to be learned along with their inter-dependencies are available as a *curriculum graph*, and a given concept is learned by performing an activity related to that concept (such as solving/answering a problem/question). For this setting, recent works have proposed algorithms based on multi-armed bandits (MAB), where activities are adaptively sequenced using the student response to those activities as a direct feedback. In this paper, we propose CurriculumTutor, a novel technique that combines a MAB algorithm and a *change point detection* algorithm for the problem of adaptive activity sequencing. Our algorithm improves upon prior MAB algorithms for the LbD setting by (i) providing better learning gains, and (ii) reducing hyper-parameters thereby improving personalisation. We show that our tutoring algorithm significantly outperforms prior approaches in the benchmark domain of two operand addition up to a maximum of four digits.

Keywords: Adaptive sequencing · Multi-armed bandits · Change-point detection · Personalisation

1 Introduction

Personalised learning approaches tailored to address the individual needs, skills, and interests of each student, have been found to cause a significant improvement in learning gains for the students, apart from providing an engaging learning experience. Intelligent Tutoring Systems (ITS) have been effective in delivering personalised learning to students in an automatic manner. An ITS consists of three important components namely: (i) *domain model* that captures the relations or dependencies between the various concepts to be learned, (ii) *student model* that represents the student's current knowledge level and how it changes as the tutoring progresses and (iii) *tutoring model* that decides the sequence of

M. M. Rodrigo et al. (Eds.): AIED 2022, LNCS 13355, pp. 319–331, 2022.
https://doi.org/10.1007/978-3-031-11644-5_26

learning activities presented to students. The tutoring model performs *adaptive sequencing of activities*, wherein learning activities to be presented to students are selected based on student competence levels estimated through their interactions with the system. This leads to better personalisation of learning and hence higher learning gains.

Clement et al. [4] proposed a tutoring algorithm called ZPDES that uses a *curriculum graph* capturing the concept inter-dependencies as a domain model. The key highlight of ZPDES is the use of ideas from multi-armed bandits to perform adaptive assignment of learning activities based on student responses. Later Brunskill et al. [10], used ZPDES to propose an adaptive sequence of activities to advance students through a curriculum graph generated based on an algorithmic representation of the concepts for the domain of two operand addition upto a maximum of 4 digits.

Our Contribution: In this paper, we consider the *learning by doing* (LbD) setting used in [10], wherein the concepts to be mastered are available in the form of a curriculum graph and the students learn a concept by performing activities related to the concept. Even though ZPDES is less reliant on the underlying student learning model, it has its own set of hyper-parameters which adversely affect personalisation. Our main contribution is a novel tutoring algorithm called CurriculumTutor for the LbD setting. CurriculumTutor combines the ideas of multi-armed bandits and *change point detection.* Here, the change point detector separately tracks the mastery level of each activity and an upper confidence based MAB algorithm manages the exploration vs exploitation tradeoff. We show that our CurriculumTutor significantly outperforms ZPDES in the benchmark domain of two operand addition up to a maximum of four digits, especially in scenarios where the learner faces difficulty in learning a subset of concepts in the curriculum graph.

2 Background

In this section, we first describe the learning by doing setting and then explain how prior works have used ideas from multi-armed bandits in the tutoring model.

Learning by Doing (Fig. 1): A *concept* is a representational unit of educational content which is learned by performing an activity related to that concept (such as solving/answering a problem/question). A *concept/curriculum graph* has concepts as nodes and the edges represent dependency/prerequisite relations between concepts. Several methods have been proposed in the literature to construct concept graphs from text corpus [6,13]. In this paper, we assume access to a concept graph for the curriculum to be mastered. *Zone of Proximal Development (ZPD)* is a set of concepts on the boundary of the student's knowledge (Fig. 1). It is based on an idea from classical psychology and education research which states that learning is the fastest and most engaging when practicing on material slightly beyond the current abilities of the student [3]. A *Question bank*

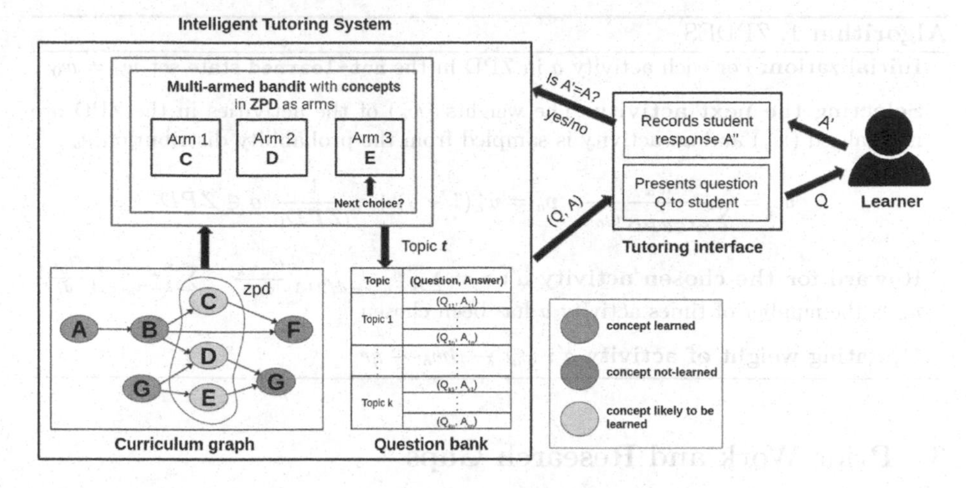

Fig. 1. Multi-armed bandit framework for adaptive activity sequencing in an ITS

contains a collection of question answer pairs associated with each concept. It is further assumed that all the questions associated with a concept are of the same difficulty level.

Multi-armed bandit (MAB) problem consists of a fixed set of actions (arms), and an unknown reward distribution associated with each arm. The system has to select actions (arms) such that its long term cumulative reward is maximized. The key challenge in this problem is to balance exploiting the information that has already been gained about the effectiveness of each action and exploring actions where the estimates about their value are still relatively uncertain, referred to as the *exploration-exploitation* tradeoff. Over time the system learns which actions are more effective and can earn larger rewards.

MAB Applied to Tutoring Model: In the multi-armed bandit framework for adaptive sequencing, the activities/topics in the ZPD form the arms and rewards are computed based on student responses, as seen in Fig. 1. The system maintains an estimate of the learning progress associated with each concept in the ZPD, based on which it decides the next activity to be offered to the student. In the process of maximizing cumulative rewards, the MAB framework would pick activities from the concepts that the student tends to learn faster, thereby generating a personalized activity sequence. This formulation is less reliant on the underlying model of student learning and can therefore better adapt to different individual learning behaviors.

Algorithm 1. ZPDES

Initialization: For each activity a in ZPD in the `not-learned` state set $w_a = w_0$

Selecting the next activity: The weights (w_a) of the activities in the ZPD are normalized (w'_a) and an activity is sampled from the probability distribution p_a

$$w'_a = \frac{w_a}{\sum_{a' \in ZPD} w_{a'}}, \quad p_a = w'_a(1-\gamma) + \gamma \frac{1}{|ZPD|} \quad a \in ZPD$$

Reward for the chosen activity a': $r = \sum_{k=n_{a'}-d/2+1}^{n_{a'}} \frac{C_{a',k}}{d/2} - \sum_{k=n_{a'}-d}^{n_{a'}-d/2} \frac{C_{a',k}}{d/2}$
n_a is the number of times activity a has been chosen

Updating weight of activity a': $w_{a'} \leftarrow \beta w_{a'} + \eta r$

3 Prior Work and Research Gaps

In this section, we discuss the ZPDES algorithm proposed by Clement et al. [4] and then look at the research gaps in ZPDES, in particular about how the hyper-parameters in ZPDES adversely affect personalization.

Zone of Proximal Development and Empirical Success (ZPDES) [4] (see Algorithm 1) is based on the multi-armed bandits framework for adaptive activity sequencing where pre-conditions between activities are provided in the form of an expert defined curriculum graph. At each time-step, ZPDES selects the next activity a' based on the normalized weights of activities (w'_a) in the ZPD as well as an exploration factor γ, $0 \leq \gamma \leq 1$. The correctness of student response to the activity a' at the ith attempt, $C_{a',i}$, is a binary random variable - it takes a value of 1 for a correct response and 0 for incorrect response. The reward is computed by taking the difference of the number of successes in the last $d/2$ samples with the $d/2$ previous samples, where d is the window-size. The reward provides an empirical measure of how the success rate is increasing. The reward becomes close to zero when either a concept has been mastered or the student struggles to master it. A high positive reward indicates good learning progress, whereas a negative reward implies stalled learning. These rewards are then used to update the activity weights ($w_{a'}$) based on the hyper-parameters η and β. ZPDES doesn't propose a method to determine if a concept has been mastered or not.

Later Brunskill et al. [10] proposed a novel system that combines automatic curriculum ordering [1] with ZPDES for automatically and adaptively advancing a student through a curriculum. A 'sliding windowed' average was used in this framework for inferring mastery, where a concept is determined to be *mastered* when the accuracy over the past d attempts reaches above a specified threshold t.

Research Gap: The major drawback of ZPDES is the use of many hyper-parameters such as d, w_0, γ, β and η, which are critical for its deployment, and eventual success. Using fixed values for these hyper-parameters would have an adverse effect on personalisation since the same set of hyper-parameters may not be effective for the entire population. On the contrary, tuning these hyper-parameters for different individuals is an additional computational overhead and a time consuming process.

4 Our Approach: Breaking Adaptive Activity Sequencing into Sub-problems

Our algorithm `CurriculumTutor` addresses the research gap in the previous section. We describe our approach in this section and present the algorithm in the next section. The main reason for the presence of hyper-parameters in ZPDES is that it tries to address together both the exploration vs exploitation tradeoff as well as the non-stationarity of the student responses. On the contrary, our approach is to decompose the adaptive activity sequencing problem into the following two sub-problems namely (i) **activity selection** to take care of the exploration vs exploitation tradeoff, and (ii) **detection of mastery** to take care of the non-stationarity of the student responses. We now describe these in detail.

4.1 Activity Selection

The tutoring algorithm at each time step has to select an activity that offers higher learning progress to the student. In the multi-armed bandit framework for activity selection, the concepts in the ZPD form the arms while rewards are computed based on student responses. We consider the case where the rewards are binary, i.e., a reward of one is obtained if an activity related to the selected concept is correctly performed and otherwise zero. The problem can now be modelled as a *Bernoulli multi-armed bandit*, where a reward of 1 is obtained with probability p that is dependent on the knowledge level of the associated concept(arm).

4.2 Detection of Mastery

The tutoring algorithm has to infer whether a concept has been mastered based on the sequence of student responses. Determining whether a concept has been mastered helps to avoid over-practicing which leads to a drop in the learning gains, and under-practicing that results in the concept not being learned.

Each concept is assumed to be associated with two states - *learned* and *not-learned*. Let r_i denote the correctness of the student response at the ith attempt. When the concept has not been mastered, $r_i \sim P_{\text{not-learned}}$, where

$$P_{\text{not-learned}}(r=1) = p_{\text{guess}} \quad P_{\text{not-learned}}(r=0) = 1 - p_{\text{guess}} \tag{1}$$

When the concept has been mastered $r_i \sim P_{\text{learned}}$, where

$$P_{\text{learned}}(r=1) = 1 - p_{\text{slip}} \quad P_{\text{not-learned}}(r=0) = p_{\text{slip}} \tag{2}$$

The reward distribution is now *piecewise stationary* i.e., remains constant for a certain period, and shifts at some unknown time step referred to as a change-point, that corresponds to the transition from the *not-learned* to *learned* state. Detection of mastery of a concept can now be modelled as a *change-point detection problem*, which can be described as follows [12]: Given a sequence of observations $\{X_n\}_{n\geq 1}$, the observed random variables $X_1, X_2, ...$ have distribution function p_0 until a change occurs at an unknown point in time $\lambda, \lambda \in \{1, 2, ...\}$, after which the observations have another distribution p_1, $p_0 \neq p_1$. A sequential change-point detection procedure is identified with a stopping time τ for an observed sequence $\{X_n\}_{n\geq 1}$, i.e., the time of alarm τ at which it is declared that a change has occurred, which is a random variable depending on the observations. The *average detection delay (ADD)* and *false alarm rate (FAR)* of a detection procedure is defined as follows:

$$ADD_\lambda(\tau) = E_\lambda(\tau - \lambda | \tau \geq \lambda) \tag{3}$$

$$FAR(\tau) = \frac{1}{E_0(\tau)} \tag{4}$$

where $E_0(\tau)$ denotes the expectation of the sequence $\{X_n\}_{n\geq 1}$ when there is no change, i.e., $\lambda = \infty$. A good detection procedure should have small values of average detection delay with a low FAR.

5 Our Algorithm: CurriculumTutor

We now describe CurriculumTutor, a novel tutoring algorithm that combines a Bernoulli multi-armed bandit algorithm with a change point detection technique for adaptive activity sequencing to master a curriculum. The algorithm maintains a zpd at each time step, based on which activities are presented to the student. A concept is removed from the ZPD when it is *determined* to have been transitioned into the learned state. The ZPD is then updated with concepts in the not-learned state, all of whose prerequisites are in the learned state. This continues until the student has mastered all the concepts in the curriculum graph. The *CurriculumTutor* algorithm has been described in Algorithm 2.

CurriculumTutor uses the *Kullback-Leibler Upper Confidence Bound (KL-UCB)* algorithm to perform activity selection, as it is shown to have better performance bounds than UCB and its variants [8]. The change-point detection algorithm used by CurriculumTutor is *Cumulative Sum (CUSUM)*. Here the log-likelihood ratio (LLR) is used to test the hypothesis that a change occurred at the point λ and that there is no change at all ($\lambda = \infty$) which is defined as:

$$Z_{n,\lambda} = \sum_{k=\lambda}^{n} log \frac{p_1(X_k | X_1, ..., X_{k-1})}{p_0(X_k | X_1, ..., X_{k-1})}, \quad n \geq \lambda \tag{5}$$

Here, p_0 and p_1 are the pre-change and post-change probability density functions respectively. The maximum LLR statistic $U_n = max_{1\leq\lambda\leq n} Z_{n,\lambda}$ is compared with

Algorithm 2. CurriculumTutor

1: Initialize zpd = Minimal elements of the partial ordering curriculum, P
2: Add arms corresponding to elements in zpd
3: $t = 1$
4: **repeat**
5: c_t = SELECT-ACTIVITY(zpd)
6: Present an activity of type c_t and observe correctness of response r_t
7: Update estimates associated with c_t based on r_t
8: DETECT-MASTERY(c_t)
9: $t \leftarrow t + 1$
10: **until** zpd is empty

11: **procedure** SELECT-ACTIVITY(zpd)
12: **return** concept selected by KL-UCB
13: **end procedure**

14: **procedure** DETECT-MASTERY(c)
15: Update CUSUM estimate for c
16: **if** change-point detected for c **then**
17: Mark c as *learned*
18: Remove c from zpd
19: $zpd \leftarrow zpd \cup \{q' \in P \mid (c \leq q') \wedge (\nexists\ p \in P$ such that $p \leq q' \wedge$ p is not-learned$)\}$
20: Add arms corresponding to new elements in zpd
21: **end if**
22: **end procedure**

a threshold h and a change is detected when the value of U_n exceeds h. When $h > 0$ and the observations are independent and identically distributed (i.i.d.), U_n can be replaced by the statistic $\tilde{U}_n$ which obeys the recursion:

$$\tilde{U}_n = max\left\{0,\ \tilde{U}_{n-1} + log\ \frac{p_1(X_n)}{p_0(X_n)}\right\} \tag{6}$$

with the initial condition $\tilde{U}_0 = 0$. It has been proved [9] that in the i.i.d. case CUSUM minimizes the worst case average detection delay among all the detection algorithms for which the FAR is fixed at a given level $\overline{FAR}$. With p_{guess} and p_{slip} defined by the user, the pre and post change distributions are as follows:

$$p_0(k) = p_{guess}{}^k(1 - p_{guess})^{(1-k)} \tag{7}$$

$$p_1(k) = (1 - p_{slip})^k(p_{slip})^{(1-k)} \quad \text{for } k \in \{0, 1\} \tag{8}$$

CurriculumTutor uses the $\tilde{U}_n$ statistic for change-point detection, as the correctness of student responses are independent random variables given the state of each concept.

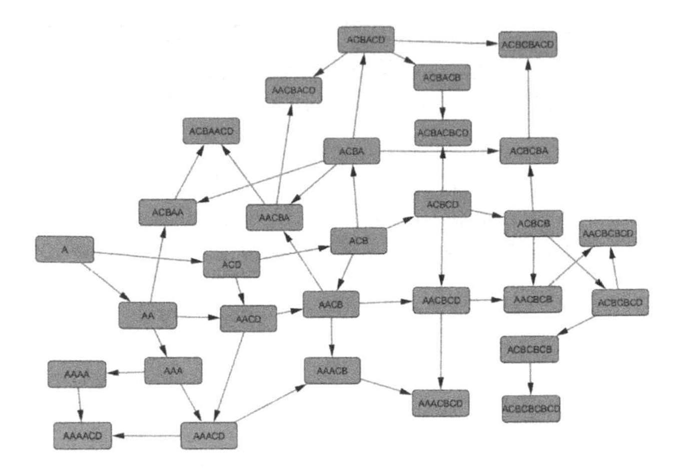

Fig. 2. Curriculum graph for two operand addition up to a maximum of 4 digits. We test our algorithm in two scenarios, *Scenario 1*: no struggle with any concept and *Scenario 2*: struggling with the concepts AAACD and AAAACD which are colored in orange. (see the Results section for more details)

6 Experiment Setup

We will now describe the experimental setup to evaluate our algorithm in the domain of two operand addition with simulated students. The curriculum graph generation for the domain, and models for simulated students are described below. While the curriculum graph is used by our algorithm, the simulated students are used only to evaluate our algorithm, and not by the algorithm itself.

6.1 Generating the Curriculum Graph

The trace based framework proposed by Andersen et al. [1] was used to generate a curriculum graph for the domain of two operand addition up to a maximum of four digits. Four basic operations were identified to be required to solve an integer addition problem: one digit addition without a carry (A), one-digit addition with a carry (B), writing a carry (C), and bringing down a final carry (D). For instance, problems can be decomposed into basic skills as shown in Table 1:

These traces can then be ordered by complexity based on the N-gram-based partial ordering as follows: *Let n be any positive integer. A trace $T1$ is said to be at least as complex as trace $T2$ if every n-gram of trace $T2$ is also present in trace $T1$.* The curriculum graph generated with n=3 has been given in Fig. 2.

Table 1. Traces generated for addition problems

Problem	2 + 3	15 + 18	93 + 15	298 + 865
Trace	A	ACB	AACD	ACBCBCD

6.2 Student Models for Simulation

Students were simulated using the following popular knowledge tracing models. For each student model, separate knowledge components were defined corresponding to addition of different lengths, presence of carry and overflow. Prerequisites were imposed on knowledge components and each problem is associated with exactly one knowledge component. All activities are assumed to be of the same difficulty level.

Bayesian Knowledge Tracing (BKT). Here a student's knowledge state is modelled as a two-state Hidden Markov Model(HMM) [5], one per knowledge component/skill, where the skill is either mastered by the student or not. Prerequisites between knowledge components were enforced by varying the transition probability between states $p(T)$, depending on whether the prerequisites were learned or not.

Learning Factors Analysis (LFA). In LFA [2], the probability p of a student performing correctly on an activity is modeled as a logistic function with parameters: n_i - number of times knowledge component i has been practiced, α - student learning parameter, β_i - coefficient of knowledge component i, γ - difficulty level of the activity

$$ln(\frac{p}{1-p}) = \frac{\alpha}{|P_{kc}|} \sum_{i \in P_{kc}} n_i \beta_i - \gamma \tag{9}$$

where kc is the knowledge component associated with the activity and P_{kc} is the set containing kc along with its prerequisite knowledge components

Performance Factor Analysis (PFA). In PFA [11], the probability p of a student performing correctly on an activity is modeled as a logistic function with parameters: s_i - number of times knowledge component i has been used correctly, f_i - number of times knowledge component i has been used incorrectly, α - student learning parameter, β_i - coefficient for the success count of knowledge component i, η_i - coefficient for the failure count of knowledge component i, γ - difficulty level of the activity

$$ln(\frac{p}{1-p}) = \frac{\alpha}{|P_{kc}|} \sum_{i \in P_{kc}} (s_i \beta_i + f_i \eta_i) - \gamma \tag{10}$$

Integrating Knowledge Tracing and Item Response Theory (KT-IRT). KT-IRT [7] combines the BKT model with item response theory. Here the student's knowledge is represented as a HMM with binary states. The probability of performing correctly on an activity based on knowledge component k at the ith attempt is given by:

$$p_i = \begin{cases} p_g(\frac{1}{1+e^{-(\beta_k n_k + c_{k,0})}}) & \text{when } state_i = \text{not-learned} \\ (1-p_s)(\frac{1}{1+e^{-(\beta_k n_k + c_{k,1})}}) & \text{when } state_i = \text{learned} \end{cases} \quad (11)$$

6.3 Parameters

Hyper-parameters for ZPDES: The best value of hyper-parameters such as η, β, γ and w_0 were found using grid search separately for each student model.

Change-Point Detection: For selecting the window size d and accuracy threshold t for ZPDES, and the threshold h for CUSUM, the FAR was fixed as 5e−05. To find the FAR of the sliding window based method, a markov chain was constructed with the states corresponding to the number of ones to be obtained to reach the accuracy threshold t. With zero set as the absorbing state, the mean time to absorption for the chain was computed, which corresponds to $E_0(\tau)$. Similarly for CUSUM, a markov chain was constructed with the states corresponding to the possible values taken by the statistic $\tilde{U}_n$ and states with values greater than h were designated as the absorbing states. Again, the mean time to absorption was calculated to compute the FAR.

For the sliding window, the window size was set as 8 and the accuracy threshold as 0.7. Thus, out of the last 8 activities the student has to get at least 6 of them correct in order to infer that the concept has been mastered. Similarly the threshold h for CUSUM was set as $ln(1/0.0004)$. For CUSUM p_{guess} and p_{slip} for each concept were sampled from $Beta(20, 40)$.

7 Results

In this section we present the experimental results that compare the performance of three tutoring algorithms namely *CurriculumTutor*, *ZPDES* and *Blocking* (a basic scheduling scheme where activities associated with the same concept are presented to a student until the concept is mastered, uses a sliding window to infer mastery). For these three algorithms, the following two scenarios were simulated using the four different student models:

Scenario 1. *Student not struggling with any of the concepts* Here similar model parameters (transition probabilities, learning coefficients, intercept values, etc.) were used for knowledge components (KC) of similar complexity and the parameter values were made smaller with increasing complexity. For instance, the KC parameters associated with two digit addition were lower as compared to those associated with one-digit addition.

Scenario 2. *Student struggling with a few concepts in the intermediate phase* This scenario was simulated by using a lower transition probability as well as learning parameters for the student model for one of the KCs - *three digit addition without carry and overflow*, as compared to Scenario 1. This causes the students to struggle on the concepts $AAACD$ and $AAAACD$ in Fig. 2.

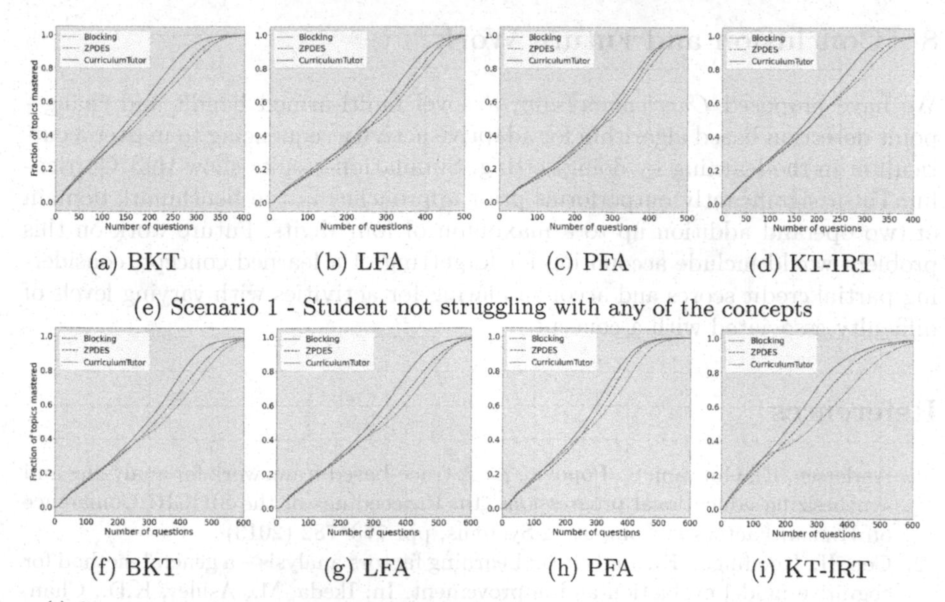

(a) BKT (b) LFA (c) PFA (d) KT-IRT

(e) Scenario 1 - Student not struggling with any of the concepts

(f) BKT (g) LFA (h) PFA (i) KT-IRT

(j) Scenario 2 - Student struggling with a few concepts in the intermediate phase

Fig. 3. Simulation results

Learning gains were measured by computing the fraction of concepts in the curriculum graph that was mastered by the student over a given number of questions. The average curves for 500 runs were plotted for each student model, as seen in Fig. 3. It could be observed that for all the student models in both the scenarios, the learning curves of *CurriculumTutor* lie above those of ZPDES and Blocking, implying higher learning gains. In scenario 2, it could be observed that the learning curves of both the bandit algorithms, ZPDES and CurriculumTutor, are quite higher than that of Blocking (Fig. 3j). This is because while Blocking continuously tries to present activities related to a concept that the student is struggling with, the bandit algorithms performs exploration to find another concept that gives better rewards and thus contributes to higher learning progress. Thus **CurriculumTutor performs better than ZPDES and Blocking** in both the scenarios, irrespective of the student models chosen.

Limitations of the Approach: CurriculumTutor requires a curriculum graph that models the prerequisite relations between the concepts to be learned. When the curriculum graph has a chain-like structure, the algorithm cannot perform much better than Blocking as the scope for exploration becomes limited. Also the algorithm requires the user to define p_{guess} and p_{slip} for each concept.

8 Conclusion and Future Work

We have proposed *CurriculumTutor*, a novel multi-armed bandit and change-point detection based algorithm for adaptive activity sequencing to master a curriculum in the learning by doing setting. Simulation results show that CurriculumTutor significantly outperforms prior approaches in the benchmark domain of two operand addition up to a maximum of four digits. Future work on this problem would include accounting for forgetting of a learned concept, considering partial credit scores and accommodating for activities with varying levels of difficulty associated with a concept.

References

1. Andersen, E., Gulwani, S., Popovic, Z.: A trace-based framework for analyzing and synthesizing educational progressions. In: Proceedings of the SIGCHI Conference on Human Factors in Computing Systems, pp. 773–782 (2013)
2. Cen, H., Koedinger, K., Junker, B.: Learning factors analysis – a general method for cognitive model evaluation and improvement. In: Ikeda, M., Ashley, K.D., Chan, T.-W. (eds.) ITS 2006. LNCS, vol. 4053, pp. 164–175. Springer, Heidelberg (2006). https://doi.org/10.1007/11774303_17
3. Chaiklin, S.: The zone of proximal development in Vygotsky's analysis of learning and instruction. In: Vygotsky's Educational Theory in Cultural Context, vol. 1, pp. 39–64 (2003)
4. Clement, B., Roy, D., Oudeyer, P.Y., Lopes, M.: Multi-armed bandits for intelligent tutoring systems. arXiv preprint arXiv:1310.3174 (2013)
5. Corbett, A.T., Anderson, J.R.: Knowledge tracing: modeling the acquisition of procedural knowledge. User Model. User-adapt. Interact. **4**(4), 253–278 (1994). https://doi.org/10.1007/BF01099821
6. Gordon, J., Zhu, L., Galstyan, A., Natarajan, P., Burns, G.: Modeling concept dependencies in a scientific corpus. In: Proceedings of the 54th Annual Meeting of the Association for Computational Linguistics (Volume 1: Long Papers), pp. 866–875 (2016)
7. Khajah, M.M., Huang, Y., González-Brenes, J.P., Mozer, M.C., Brusilovsky, P.: Integrating knowledge tracing and item response theory: a tale of two frameworks. In: CEUR Workshop Proceedings, vol. 1181, pp. 7–15. University of Pittsburgh (2014)
8. Lattimore, T., Szepesvári, C.: Bandit Algorithms. Cambridge University Press, Cambridge (2020)
9. Lorden, G., et al.: Procedures for reacting to a change in distribution. Ann. Math. Stat. **42**(6), 1897–1908 (1971)
10. Mu, T., Goel, K., Brunskill, E.: Program2Tutor: combining automatic curriculum generation with multi-armed bandits for intelligent tutoring systems. In: Conference on Neural Information Processing Systems (2017)
11. Pavlik, P.I., Jr., Cen, H., Koedinger, K.R.: Performance factors analysis-a new alternative to knowledge tracing. Online Submission (2009)

12. Tartakovsky, A.G., Rozovskii, B.L., Blazek, R.B., Kim, H.: A novel approach to detection of intrusions in computer networks via adaptive sequential and batch-sequential change-point detection methods. IEEE Trans. Signal Process. **54**(9), 3372–3382 (2006)
13. Wang, S., et al.: Using prerequisites to extract concept maps from textbooks. In: Proceedings of the 25th ACM International on Conference on Information and Knowledge Management, pp. 317–326 (2016)

Pedagogical Agent Support and Its Relationship to Learners' Self-regulated Learning Strategy Use with an Intelligent Tutoring System

Daryn A. Dever(✉), Nathan A. Sonnenfeld, Megan D. Wiedbusch, and Roger Azevedo

School of Modeling, Simulation, and Training, University of Central Florida, Orlando, FL 32816, USA
{daryn.dever,nathan.sonnenfeld,megan.wiedbusch, roger.azevedo}@ucf.edu

Abstract. Oftentimes learners are unable to engage in effective self-regulated learning (SRL) strategies while learning about complex topics. To combat this, intelligent tutoring systems (ITSs) incorporate pedagogical agents to guide learners in understanding how to engage in several SRL strategies effectively and efficiently throughout learning. To identify how ITSs can best support learners' SRL strategy usage, data from 105 undergraduate students across several North American public universities were collected as they learned with MetaTutor, a hypermedia-based ITS about the human circulatory system. Participants were randomly assigned to two conditions – a prompt and feedback condition in which pedagogical agents prompted learners to engage in specific cognitive and metacognitive SRL strategies and provided feedback to performance in addition to learners' self-initiated SRL strategy usage, and the control condition in which learners were not prompted nor were provided feedback on their performance of self-initiated SRL strategies. Results found that learners receiving external support from pedagogical agents had greater learning gains and deployed a greater number of both cognitive and metacognitive SRL strategies than learners who only self-initiated strategies. While probabilities obtained from a Markov model did not find differences between conditions in learners' sequential transitions between SRL strategies, metrics from auto-recurrence quantification analysis found that learners receiving external support enacted less repetitive interactions of SRL strategies throughout their entire time interacting with MetaTutor. Implications of these results encourage the use of pedagogical agents in prompting more novel SRL strategies to increase learning within ITSs.

Keywords: Self-regulated learning · Pedagogical agents · Intelligent tutoring systems · Auto-recurrence quantification analysis

1 Self-regulation in Intelligent Tutoring Systems

Self-regulated learning (SRL) refers to learners' ability to monitor and regulate cognitive, affective, metacognitive, motivational, and social processes continuously throughout

M. M. Rodrigo et al. (Eds.): AIED 2022, LNCS 13355, pp. 332–343, 2022.
https://doi.org/10.1007/978-3-031-11644-5_27

their learning experiences [1]. SRL primarily encompasses cognitive and metacognitive strategies that can be deployed by learners to efficiently and accurately identify and integrate instructional content essential to learning outcomes into existing mental model structures [1, 2]. However, learners typically demonstrate an inability to engage in and deploy SRL strategies while learning about a complex topic [3]. This may be due to several factors including learners' constrained metacognitive knowledge [1, 4].

Many intelligent tutoring systems (ITSs) embed pedagogical agents to externally support learners' self-regulatory behaviors [5–8]. MetaTutor exemplifies a hypermedia-based ITS which incorporates multiple pedagogical agents to identify when learners are using SRL strategies and which strategies are used, didactically prompts learners to engage in several types of cognitive and metacognitive SRL strategies, and provides feedback evaluating the accuracy of executed strategies [9]. These virtual pedagogical agents can act as teachers by providing an external scaffold that enhances the learning experience [1, 5, 6, 10]. The goal of this study is to use MetaTutor to examine how embedded pedagogical agents within an ITS support learners' use and sequences of SRL strategies in relation to overall learning gains. Specifically, this study examines how pedagogical agent support is related to learning gains in the way in which learners deploy their cognitive and metacognitive strategies while learning with MetaTutor.

1.1 Pedagogical Agents Within ITSs

Due to the complexity and context-specific nature of self-regulatory skills and abilities, teachers may promote and foster regulatory development through modeling within the classroom [8]. In the absence of physical teachers during learning with a virtual ITS, pedagogical agents can serve as a source of external support through their human-like interactions with learners [11, 12]. Pedagogical agents within ITSs use interactions, such as dialogues, with the learners to enhance learning, increasing learners' knowledge of SRL strategies [5, 13, 14]. These agent-learner interactions are further promoted using dialogue between the agent and learner in which the agent prompts learners to engage in SRL strategies and provides feedback to validate or correct learners' SRL strategy.

For example, MetaTutor, as previously mentioned, features pedagogical agents as teachers acting to prompt learners' use of self-regulatory strategies and support their knowledge acquisition pertaining to the human circulatory system [15]. This interaction allows for pedagogical agents to externally scaffold learners' use of SRL strategies, immediately supporting learners' conceptual understanding, and eventually growing learners' knowledge of different SRL strategies.

1.2 MetaTutor: Relevant Works

Several studies have examined how learners engage in SRL with MetaTutor across varying contexts of emotions [16–20], motivation [21, 22], and SRL strategy deployment [23–25]. For this paper, we review three papers that examine patterns of how learners deploy SRL strategies and their relationship with pedagogical agent support.

Taub and Azevedo [24] examined how learners' prior knowledge on the human circulatory system was related to how they deployed SRL strategies during learning with MetaTutor. Results from this study found that learners with high prior knowledge had greater frequencies of strategy usage than learners with low prior knowledge. Similarly, Wiedbusch et al. [25] examined how learners differed in their SRL strategy usage, but specifically the role of pedagogical agent support. This study utilized principle component analyses on learners' log-file data as they learned with MetaTutor to examine how learners' differing in their pedagogical agent support utilized SRL strategies. Results showed that strategy usage was driven by two phases of SRL - defining the task and goal setting/planning - according to Winne [1]. Further, Wiedbusch and colleagues [25] found that learners tended to deploy strategies that were more familiar (i.e., summarization) than those that were not (i.e., judgments of learning).

Dever et al. [23] also examined how pedagogical agents supported learners' SRL strategy deployment, but focused on metacognitive monitoring strategies (e.g., content evaluations). This study, using hierarchical clustering, found that learners who had a greater frequency of engaging in content evaluations and feelings of knowing had greater learning gains than learners who engaged more in monitoring goal progress.

1.3 Theoretical Framework

This study is grounded within Kramarski and Heaysman's [26] model of teachers' triple SRL-SRT processes which describes the impact of teachers' activation of students' SRL (i.e., student-focused self-regulated teaching (SRT)) on outcomes related to learning, metacognition, and learners' use of metacognitive strategies over time. According to this model, the traditional dichotomy between teachers' roles in SRL and SRT (e.g., [27]) was expanded into three processes: teachers' regulation of their own learning (teacher-focused SRL), teachers' regulation of their teaching practices' (teacher-focused SRT), and teachers' activation of student' SRL (student-focused SRT). This model posits these processes have an impact upon student outcomes including SRL and achievement [26]. We extend this model to include pedagogical agents acting as teachers and narrow our focus to the impact of student-focused SRT processes on student outcomes by using a series of pedagogical agents as a proxy for the teachers' different roles during instruction. The different pedagogical agents take on the role of teachers as they engage in student-focused SRT by prompting learners to engage in SRL strategies and providing feedback on their success.

1.4 Current Study

The primary goal of this current study was to use MetaTutor, a hypermedia-based ITS, to examine how embedded pedagogical agents fulfill the role of a virtual teacher of SRL. The secondary goal of this study was to add to current literature by: (1) identifying how SRL strategy use differs between learners who receive external support and those who do not; (2) examining how learners' sequences and repetitive SRL strategy use is related to pedagogical agent prompting; and (3) transforming current and new theoretical grounding to personify pedagogical agents as teachers of SRL. We achieve these goals by proposing three questions:

Research Question 1: How do learning gains differ between conditions? Based on Kramarski and Heaysman's [26] theoretical framework, we hypothesize that learners who receive prompts from pedagogical agents to engage in SRL strategies as well as feedback on their accuracy will show greater learning gains than learners who only self-initiate SRL strategy usage.

Research Question 2: How are the frequencies of SRL strategies different between conditions? Based on prior literature [23, 25], we hypothesize that learners who are prompted to engage in SRL strategy usage will demonstrate a greater frequency of SRL strategies overall. However, we do hypothesize that learners receiving prompts to engage in SRL strategies will have a lower frequency of self-initiated SRL strategies, demonstrating an over-reliance on system aid [28, 29].

Research Question 3: How does pedagogical agent external-regulation relate to learners' deployment of SRL strategies? According to Kramarski and Heaysman's [26] theoretical framework, pedagogical agents may relate to how learners deploy SRL strategies over time. As such, we hypothesize that learners differing in their pedagogical agent support will vary in terms of which SRL strategies learners deploy and in what order they are deployed.

2 Method

2.1 Participants

Undergraduate students from several large universities across North America were recruited for a two-day laboratory study. From our current dataset of 105 students ($M\_age = 20.5$; $SD\_age = 3.31$; 53% female), participants were randomly assigned to one of two conditions, the Control Condition ($N = 47$) or the Prompt & Feedback (P&F) Condition ($N = 58$). In the Control Condition, learners initiated SRL strategy interactions and were not prompted by the system. In the P&F Condition, the system initiated SRL strategy interactions by having pedagogical agents prompt learners to engage in SRL strategies. The pedagogical agents used were diegetically appropriate to the type of strategy which was prompted (see section Pedagogical Agents). After concluding the experimental protocol (detailed below in Procedure), participants were monetarily compensated ($10/h, up to $40).

2.2 MetaTutor

General Architecture. MetaTutor is a hypermedia-based ITS with 47 pages of instructional content related to the human circulatory system (e.g., anatomy, function, composition; see Fig. 1). MetaTutor was designed to support SRL strategy use through several features. A table of contents affords learners the opportunity to select the most appropriate pages relevant to their learning goals and subgoals. A progress bar and a timer allow learners to monitor progress toward learning goals. Text and diagrams afford learners the opportunity to acquire knowledge and coordinate information. An SRL palette offers learners the opportunity to identify and select SRL strategies while exploring instructional content. Finally, one of four pedagogical agents may interact with learners depending on the triggered production rules, each supporting a specific component of SRL.

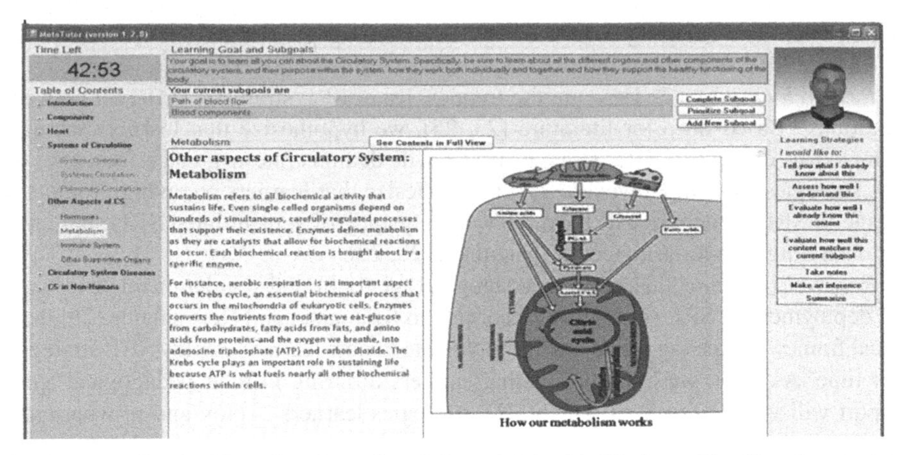

Fig. 1. Example MetaTutor interface (from left to right) with (1) timer (2) table of contents to jump between content; (3) goal setting; (4) set subgoals and progress bar; (5) textual content; (6) supporting biology diagram; (7) pedagogical agent; and (8) SRL palette.

Conditions. In the P&F Condition, the system detected and supported learners' SRL by prompting strategies during learning [9]. Production rules were triggered based on specific actions or after a specified time period and could respond to prompt learners to enact various cognitive strategies (i.e., prior knowledge activation, planning, summarizing, note-taking) or metacognitive strategies (i.e., feelings of knowing, judgements of learning, CEs, or progress monitoring). When these production rules are triggered, pedagogical agents are used to prompt specific SRL strategies and provide learners with feedback on their performance of that strategy. In the Control Condition, the pedagogical agent appears static and non-interactive. Regardless of condition, participants are still able to self-initiate SRL strategies without being prompted by agents.

Pedagogical Agents. Each pedagogical agent within MetaTutor is responsible for diegetically appropriate cognitive and metacognitive SRL processes [31]. Gavin the Guide supports learners' navigation of the MetaTutor interface and learning environment. Pam the Planner supports learners' planning (PLAN) and prior knowledge activation (PKA) strategies. Sam the Strategizer supports taking notes (TN) and summarizing (SUMM) strategies. Finally, Mary the Monitor supports multiple strategies including monitoring progress towards goals (MPTGs), content evaluations (CEs), judgments of learning (JOLs), and feelings of knowing (FOKs). Operational definitions for each of these SRL strategies has been provided below in Table 1 [25, 30].

2.3 Experimental Procedure

On Day 1, participants gave informed consent and completed a demographics questionnaire along with several self-report measures. Participants then completed a 30-item multiple-choice pretest on instructional content covered within MetaTutor (e.g., human circulatory system knowledge items). On Day 2, researchers briefed participants on

Table 1. SRL strategies deployed in MetaTutor and brief operational definitions [25, 30].

Strategy type	Strategy code	Operational definitions
Cognitive	PKA	Recall of relevant prior knowledge prior to or during task performance
	PLAN	Coordination and selection of operators for learners' execution of behaviors conditionally resulting in a state-transition from the current state along a hierarchy of goal/subgoal states
	SUMM	Provision of brief statements about main concepts recently read, inspected, or heard in the learning environment
	TN	Copying and/or extending of textual information from the learning environment
Metacognitive	CE	Monitoring of content and its quality relative to goals/subgoals
	FOK	Awareness of familiarity with concepts and concurrent inability to recall that information on demand
	JOL	Self-assessed self-efficacy with respect to learned concepts
	MPTG	Assessment of goal/subgoal attainment

the learning task, learning environment, and learning goals, before using MetaTutor to engage in SRL over the course of a 60-min session. Instructional activities within MetaTutor involved reading texts, inspecting diagrams, and completing quizzes. Regardless of condition, participants could indicate their deployment of SRL strategies using the SRL palette within MetaTutor. After the learning phase, participants completed a final set of questionnaires, a 30-item multiple-choice post-test on human circulatory system knowledge, and were debriefed by investigators.

2.4 Apparatus

Multimodal multichannel data were collected using several apparati including self-report questionnaires, log-files, eye-trackers, etc. For this study, only participants' collected log-file data were used. Specifically, log-file data of which SRL strategy was used, in what order these strategies were used, and the initiator of the strategy (i.e., participant vs. pedagogical agent) were used to identify how participants engaged in SRL strategy use during their time learning with MetaTutor.

2.5 Coding and Scoring

Learning gain was calculated using the differences in learners' scores on the content pre- and post-test about the human circulatory system while accounting for prior knowledge. This was calculated using Marx and Cumming's [32] series of equations which assigned a different calculation to each learner based on if their score increased after MetaTutor, decreased, or was stagnant.

Sequences were identified using learners' log files with a categorical representation of strategies. For example, a learner could have a sequence of—CE, PKA, CE, TN.

Because we used a categorical representation of time series data, the sequences of SRL strategies were periodically broken by other behaviors such as moving to a different content page. As such, sequences do not account for the time in between SRL strategies used, nor for how long those strategies were used.

Novelty in Research Question 3 is identified through learners' log-file sequences of SRL strategies. Novelty refers to the use of a broad range of SRL strategies over time whereas repetitive, or recurrent, behavior is identified by learners' sequential repeated use of the same SRL strategy. For example, a learner whose sequence of SRL strategies is SUMM, TN, CE, JOL demonstrates novel SRL strategy usage. Conversely, a learner whose sequence is SUMM, SUMM, TN demonstrates more repetitive behaviors.

3 Results

3.1 Research Question 1: How Do Learning Gains Differ Between Conditions?

An independent t-test was run to examine how learning gains differed between condition. Learners who received the external prompts from pedagogical agents had significantly greater learning gains ($M = 0.39$, $SD = 0.16$) than learners who only engaged in self-initiated SRL ($M = 0.10$, $SD = 0.19$; $t(91.4) = 8.50$, $p < .01$).

3.2 Research Question 2: How Are the Frequencies of SRL Strategies Different Between Conditions?

Given that learners who received external prompts from pedagogical agents to engage in self-regulatory behaviors had greater learning gain, this research question examines how learners between these two conditions differ in their strategy use. A one-way MANOVA found a significant main effect where learners in the P&F Condition ($M = 48.0$, $SD = 17.2$) had significantly greater overall strategies than learners in the Control Condition ($M = 24.9$, $SD = 13.5$; $F(8, 97) = 60.8$, $p < .01$). Using post-hoc analyses, there are significant differences in learners' CE, JOL, PKA, PLAN, and SUMM strategy frequencies between conditions where, with the exception of PLAN, learners in the P&F Condition had significantly greater frequencies than those in the Control Condition (see Table 2).

Table 2. Descriptive statistics of significant overall SRL frequencies between condition.

Strategy	Condition means (SDs)		F-statistic
	P&F	Control	
CE	7.31 (4.68)	0.38 (0.77)	$F(1, 104) = 100.5$
JOL	6.81 (7.71)	2.09 (5.13)	$F(1, 104) = 13.1$
PKA	2.44 (1.86)	0.38 (0.97)	$F(1, 104) = 47.3$
PLAN	2.07 (2.13)	6.15 (3.51)	$F(1, 104) = 54.6$
SUMM	9.88 (6.08)	0.23 (0.48)	$F(1, 104) = 117.5$

Another one-way MANOVA was used to identify how learners differing in their assigned condition vary in the frequency of *self-initiated* SRL strategies. Results found that there are significant overall differences between conditions in the frequency of self-initiated strategy use where learners in the P&F Condition ($M = 29.6$, SD $= 17.84$) self-initiated strategies a greater number of times than learners in the Control Condition ($M = 22.06$, $SD = 14.1$; $F(8, 97) = 6.31, p < .01$). When examining post-hoc comparisons, learners between conditions differed in their JOL, PLAN, and SUMM strategy usage (see Table 3). From this, we infer that the prompting of pedagogical agents encouraged the use and self-initiation of JOLs and SUMMs.

Table 3. Descriptive statistics of significant self-initiated SRL frequencies between condition.

Strategy	Condition means (SDs)		*F*-statistic
	P&F	Control	
JOL	5.53 (8.15)	2.09 (5.13)	$F(1, 104) = 6.37$
PLAN	1.14 (1.75)	3.28 (2.28)	$F(1, 104) = 22.8$
SUMM	3.34 (5.51)	0.23 (0.48)	$F(1, 104) = 14.8$

3.3 Research Question 3: How Does Pedagogical Agent External-Regulation Relate to Learners' Deployment of SRL Strategies?

Each use of an SRL strategy was recorded across participants. Using this sequencing, each state of strategy usage was mapped to a unique state in a Markov chain. Probabilities of state transitions were then calculated for each participant and for each possible interaction. For example, if a participant had state transitions TN, TN, CE, this participant had a probability of transitioning from TN to CE of 33.3%. For state transitions that did not occur, the probability was 0. These values were then inputted into a one-way MANOVA with participants from each condition being compared across all possible state transitions where a significant difference between conditions was not found ($F(64, 41) = 0.75, p > .05$).

Because we know that learners receiving prompts have greater learning gains (RQ1) and typically have greater frequencies of strategy usage (RQ2), the lack of significance in learners' sequential state transitions (RQ3) encourages a comprehensive approach to understanding strategy use. We used auto-Recurrence Quantification Analysis (aRQA; [33, 34]), a method to calculate the degree of repetitive behaviors, or recurrence, within a learner's time series. One of the metric outputs from this analysis, recurrence rate, is the proportion of the frequency of observed recurrent behaviors in relation to the total number of time series points. As such, this metric identifies the degree of novel behaviors throughout a learner's entire interaction with MetaTutor.

A t-test found a significant difference where learners in the P&F Condition displayed greater novelty in deploying SRL strategies ($M = 35.2$, $SD = 12.2$) than learners in the Control Condition ($M = 60.9$, $SD = 14.3$; $t(90.3) = -9.82, p < .01$). These results suggest that learners who only self-initiate SRL strategies without any external aid from pedagogical agents tend to reuse strategies rather than engage in novel behaviors.

4 Discussion and Future Directions for SRL and ITSs

The goal of this study was to eliminate gaps in current literature on pedagogical agents in ITSs by examining how agents promote learners' use of SRL strategies. Learners who received external support from pedagogical agents had greater overall learning gains than learners who did not have support (RQ1). Kramarski and Heaysman's [26] theory supports AIEd design research on pedagogical agents acting as SRL teachers, which our work empirically supports. We can imagine future pedagogical agents designed to engage in self-learning via teacher-focused SRT to better engage in interactions with learners to guide their use of SRL strategies (student-focused SRT).

For Research Question 2, results found that learners who had external regulation of their SRL strategies engaged in CEs, JOLs, PKA, and SUMM more often than learners who only self-initiated SRL strategies. However, these learners had greater frequencies of PLAN strategies. When examining only self-initiated strategies between learners differing in their pedagogical agent support, we found that learners who received support had greater JOLs and SUMM whereas learners without support had greater PLAN strategies. From Research Question 2 analyses, we find that pedagogical agents drive learners' CE and PKA strategies. This result provides implications for the design of pedagogical agents in that agents need to prompt greater metacognitive strategies. Additionally, this result further provides evidence for pedagogical agents as teachers since they promote the use of strategies required for learning with an ITS (see [35]).

For Research Question 3, we examined learners' SRL strategy sequences. Sequential transitions of learners' SRL strategy states across all strategies did not significantly differ between conditions; however, learners who were prompted to engage in SRL strategies had a greater range of SRL strategy usage where their behaviors were significantly more novel. For the design of pedagogical agents in ITSs, agents should prompt SRL strategies that have not been used recently or interrupt learners who try to engage in more repetitious behaviors of SRL strategy usage. Our findings raise other important questions for the field of AIEd such as–Is there an optimal transition between repetitive to novel SRL strategy use that reflect the development of SRL competencies? Can individual differences, learning activities, timing, sequencing, and accuracy of SRL behaviors, etc. influence when and how pedagogical agents should intervene?

5 Conclusion

Learners who are externally-regulated by pedagogical agents within an ITS demonstrate a greater use of SRL strategies across several different types of strategies, relating to greater learning gains. However, learners who are not supported while learning a complex scientific concept within an ITS demonstrate more repetitive SRL strategy usage with smaller learning gains. From these results, using a wide variety of SRL strategies, rather than the repetitive use of a singular strategy, contributes to greater learning gains. This is important within SRL literature as models should reflect that the constant use of a range of SRL strategies benefits learning. Applied implications from these findings include the increased use of pedagogical agents as virtual teachers to promote the use of novel sequences of SRL strategies using a new model of SRL and SRL that infuses and balances self-regulated learning with self-regulated teaching.

Acknowledgement. This research was funded by the National Science Foundation (DRL #1916417). The authors would like to thanks the members of UCF's SMART Lab for their numerous contributions.

References

1. Winne, P.H.: Cognition and metacognition within self-regulated learning. In: Schunk, D.H., Greene, J.A. (eds.) Educational Psychology Handbook Series, Handbook of Self-Regulation of Learning and Performance, pp. 36–48. Routledge/Taylor & Francis Group (2018)
2. Azevedo, R., Dever, D.: Multimedia learning and metacognitive strategies. In: Mayer, R.E., Fiorella, L. (eds.) The Cambridge Handbook of Multimedia Learning, vol. 3. Cambridge University Press, Cambridge (in press)
3. Munshi, A., Biswas, G.: Personalization in OELEs: developing a data-driven framework to model and scaffold SRL processes. In: Isotani, S., Millán, E., Ogan, A., Hastings, P., McLaren, B., Luckin, R. (eds.) Artificial Intelligence in Education. Lecture Notes in Computer Science, vol. 11626, pp. 354–358. Springer, Cham (2019)
4. Schunk, D.H., Greene, J.A.: Handbook of Self-regulation of Learning and Performance, 2nd edn. Routledge/Taylor & Francis Group, New York (2018)
5. Aguirre, H.R.O.: Pedagogical agents as virtual tutors: applications and future trends in intelligent tutoring systems and virtual learning environments. In: Negrón, A.P.P, López, G.L., Aguirre, H.R.O. (eds.) Virtual Reality Designs, 1st edn., pp. 118–150. CRC Press, Boca Raton (2020)
6. Baker, R.S.: Stupid tutoring systems, intelligent humans. Int. J. Artif. Intell. Educ. **26**(2), 600–614 (2016)
7. Greene, J.A., Azevedo, R.: A theoretical review of Winne and Hadwin's model of self-regulated learning: new perspectives and directions. Rev. Educ. Res. **77**(3), 334–372 (2007)
8. Johnson, W.L., Lester, J.C.: Pedagogical agents: back to the future. AI Mag. **39**(2), 33–44 (2018)
9. Azevedo, R., Taub, M., Mudrick, N.V.: Understanding and reasoning about real- time cognitive, affective, and metacognitive processes to foster self-regulation with advanced learning technologies. In: Alexander, P.A., Schunk, D.H., Greene, J.A. (eds.) Handbook of Self-Regulation of Learning and Performance, 2nd edn., pp. 254–270. Routledge, New York (2018)
10. Azevedo, R., Martin, S.A., Taub, M., Mudrick, N.V., Millar, G.C., Grafsgaard, J.F.: Are pedagogical agents' external regulation effective in fostering learning with intelligent tutoring systems? In: Micarelli, A., Stamper, J., Panourgia, K. (eds.) ITS 2016. LNCS, vol. 9684, pp. 197–207. Springer, Cham (2016). https://doi.org/10.1007/978-3-319-39583-8_19
11. Lippert, A., Shubeck, K., Morgan, B., Hampton, A., Graesser, A.: Multiple agent designs in conversational intelligent tutoring systems. Technol. Knowl. Learn. **25**, 443–463 (2020)
12. Biswas, G., Segedy, J.R., Bunchongchit, K.: From design to implementation to practice a learning by teaching system: Betty's Brain. Int. J. Artif. Intell. Educ. **26**, 350–364 (2016)
13. Desai, S., Chin, J.: An explorative analysis of the feasibility of implementing metacognitive strategies in self-regulated learning with the conversational agents. In: Proceedings of the Human Factors and Ergonomics Society Annual Meeting, vol. 64, no. 1, pp. 495–499. SAGE Publications, Los Angeles (2020)
14. Graesser, A., McNamara, D.: Self-regulated learning in learning environments with pedagogical agents that interact in natural language. Educ. Psychol. **45**(4), 234–244 (2010)

15. Dever, D.A., Amon, M.J., Vrzakova, H., Wiedbusch, M.D., Cloude, E.B., Azevedo, R.: Capturing sequences of learners' self-regulatory interactions with instructional material during game-based learning using auto-recurrence quantification analysis. Frontiers (2022)
16. Bouchet, F., Harley, J.M., Azevedo, R.: Evaluating adaptive pedagogical agents' prompting strategies effect on students' emotions. In: Nkambou, R., Azevedo, R., Vassileva, J. (eds.) ITS 2018. LNCS, vol. 10858, pp. 33–43. Springer, Cham (2018). https://doi.org/10.1007/978-3-319-91464-0_4
17. Cloude, E.B., Wortha, F., Dever, D.A., Azevedo, R.: Negative emotional dynamics shape cognition and performance with MetaTutor: toward building affect-aware systems. In: 9th International Conference on Affective Computing and Intelligent Interaction (ACII), pp. 1–8. IEEE (2021)
18. Harley, J.M., Bouchet, F., Hussain, S., Azevedo, R., Calvo, R.: A multi-componential analysis of emotions during complex learning with an intelligent multi-agent system. Comput. Hum. Behav. **48**, 615–625 (2015)
19. Lallé, S., Murali, R., Conati, C., Azevedo, R.: Predicting co-occurring emotions from eye-tracking and interaction data in MetaTutor. In: Roll, I., McNamara, D., Sosnovsky, S., Luckin, R., Dimitrova, V. (eds.) AIED 2021. LNCS (LNAI), vol. 12748, pp. 241–254. Springer, Cham (2021). https://doi.org/10.1007/978-3-030-78292-4_20
20. Sinclair, J., Jang, E.E., Azevedo, R., Lau, C., Taub, M., Mudrick, N.V.: Changes in emotion and their relationship with learning gains in the context of MetaTutor. In: Nkambou, R., Azevedo, R., Vassileva, J. (eds.) ITS 2018. LNCS, vol. 10858, pp. 202–211. Springer, Cham (2018). https://doi.org/10.1007/978-3-319-91464-0_20
21. Duffy, M.C., Azevedo, R.: Motivation matters: interactions between achievement goals and agent scaffolding for self-regulated learning within an intelligent tutoring system. Comput. Hum. Behav. **52**, 338–348 (2015)
22. Martha, A.S.D., Santoso, H.B.: Investigation of motivation theory on pedagogical agents design in the online learning environment. In: Proceedings of the 10th International Conference on Education Technology and Computers (ICETC 2018), pp. 217–222. Association for Computing Machinery, New York, October 2018
23. Dever, D.A., Wortha, F., Wiedbusch, M.D., Azevedo, R.: Effectiveness of system-facilitated monitoring strategies on learning in an intelligent tutoring system. In: Zaphiris, P., Ioannou, A. (eds.) HCII 2021. LNCS, vol. 12784, pp. 250–263. Springer, Cham (2021). https://doi.org/10.1007/978-3-030-77889-7_17
24. Taub, M., Azevedo, R.: How does prior knowledge influence eye fixations and sequences of cognitive and metacognitive SRL processes during learning with an intelligent tutoring system? Int. J. Artif. Intell. Educ. **29**(1), 1–28 (2018). https://doi.org/10.1007/s40593-018-0165-4
25. Wiedbusch, M., Dever, D., Wortha, F., Cloude, E.B., Azevedo, R.: Revealing data feature differences between system- and learner-initiated self-regulated learning processes within hypermedia. In: Sottilare, R.A., Schwarz, J. (eds.) HCII 2021. LNCS, vol. 12792, pp. 481–495. Springer, Cham (2021). https://doi.org/10.1007/978-3-030-77857-6_34
26. Kramarski, B., Heaysman, O.: A conceptual framework and a professional development model for supporting teachers "triple SRL–SRT processes" and promoting students' academic outcomes. Educ. Psychol. **56**(4), 298–311 (2021)
27. Karlen, Y., Hertel, S., Hirt, C.: Teachers' professional competences in self-regulated learning: an approach to integrate teachers' competences as self-regulated learners and as agents of self-regulated learning in a holistic manner. Front. Educ. **5**, 159 (2020)
28. Johnson, A.M., Azevedo, R., D'Mello, S.K.: The temporal and dynamic nature of self-regulatory processes during independent and externally assisted hypermedia learning. Cogn. Instr. **29**(4), 471–504 (2011)

29. Lindquist, T.M., Olsen, L.M.: How much help, is too much help? An experimental investigation of the use of check figures and completed solutions in teaching intermediate accounting. J. Account. Educ. **25**, 103–117 (2007)
30. Greene, J.A., Azevedo, R.: A macro-level analysis of SRL processes and their relations to the acquisition of a sophisticated mental model of a complex system. Contemp. Educ. Psychol. **34**(1), 18–29 (2009)
31. Mudrick, N., Azevedo, R., Taub, M., Bouchet, F.: Does the frequency of pedagogical agent intervention relate to learners' self-reported boredom while using multi-agent intelligent tutoring systems? In: Noelle, D.C., et al. (eds.) 37th Annual Meeting of the Cognitive Science Society, pp. 1661–1666. Cognitive Science Society, Austin, July 2015
32. Marx, J.D., Cummings, K.: Normalized change. Am. J. Phys. **75**, 87–91 (2007)
33. Webber, C.L., Jr., Zbilut, J.P.: Recurrence quantification analysis of nonlinear dynamical systems. In: Riley, M.A., Van Orden, G.C. (eds.) Tutorials in Contemporary Nonlinear Methods for the Behavioral Sciences, vol. 94, pp. 26–94 (2005)
34. Wallot, S., Roepstorff, A., Mønster, D.: Multidimensional recurrence quantification analysis (MdRQA) for the analysis of multidimensional time-series: a software implementation in MATLAB and its application to group-level data in joint action. Front. Psychol. **7**, 1835 (2016)
35. du Boulay, B., Luckin, R.: Modelling human teaching tactics and strategies for tutoring systems: 14 years on. Int. J. Artif. Intell. Educ. **26**, 393–404 (2016)

Evaluating AI-Generated Questions: A Mixed-Methods Analysis Using Question Data and Student Perceptions

Rachel Van Campenhout[1] (✉), Martha Hubertz[2], and Benny G. Johnson[1]

[1] VitalSource Technologies, Pittsburgh, PA 15218, USA
rachel.vancampenhout@vitalsource.com
[2] University of Central Florida, Orlando, FL 32816, USA

Abstract. Advances in artificial intelligence (AI) have made it possible to generate courseware and formative practice questions from textbooks. Courseware applies a learn by doing approach by integrating formative practice with text, a method proven to increase learning gains for students. By using AI for automatic question generation, the learn by doing method of courseware can be made available for nearly any textbook subject. As the generated questions are a primary learning feature in this environment, it is necessary to ensure they function as well for students as those written by humans. In this paper, we will use student data from an AI-generated Psychology courseware used in an online course at the University of Central Florida. The courseware has both generated questions and human-authored questions, allowing for a unique comparison of question engagement, difficulty, and persistence using student data from a natural learning context. The evaluation of quality metrics is critical in automatic question generation research, yet on its own is not comprehensive of students' experience. Student perception is a meaningful qualitative metric, as student perceptions can inform behavior and decisions. Therefore, student perceptions of the courseware and questions were also solicited via survey. Combining question data analysis with student perception feedback gives a more comprehensive evaluation of the quality of AI-generated questions used in a natural learning context.

Keywords: Artificial intelligence · Automatic question generation · Automatically generated questions · Student perception · Courseware · Mixed-method

1 Introduction

Automatic question generation (AQG) is a process with tremendous potential for impact in education as its application is both practical and varied. As outlined in a recent systematic review by Kurdi et al. [11], AQG has been developed for diverse subjects, learners, and purposes. While the advances in AQG have been ever increasing, Kurdi et al., were unable to identify a "gold standard" regarding the performance of automatically generated (AG) questions. This is in part due to the heterogeneity in the measurement of

M. M. Rodrigo et al. (Eds.): AIED 2022, LNCS 13355, pp. 344–353, 2022.
https://doi.org/10.1007/978-3-031-11644-5_28

quality and the limited research on AG questions using student data from natural contexts. Of the 93 papers included in the systematic review, only one evaluated them in the classroom and only 14 evaluated question difficulty. Expert evaluation of AG questions is imperative during the development and validation phases of AQG, but the ultimate—and necessary—test of these questions is how well they perform in natural learning contexts.

While some research on AQG focuses on high-stakes assessment items, a different avenue for AQG is formative practice. Formative practice is a well-established method of learning across domains and for learners of all ages, and while formative practice can raise achievement for all students, it helps low-performing students most of all [4]. In an evaluation of study techniques, practice testing (which includes no- or low-stakes formative practice) and distributed practice received a high utility assessment because they benefit learners of different ages and abilities and in varied educational contexts [6]. More specifically, formative practice has been studied in a courseware learning environment where it is delivered along with the text content. This learn by doing approach was found to have about six times the effect size on learning than reading alone [10]. Additionally, this learn by doing method was found to be causal to learning in varying natural learning contexts [10, 12]. It is clear, then, that AQG could expedite the availability of this method of combining formative practice with learning content to more students.

In this paper, we analyze AG questions that were created as part of an artificial intelligence (AI) process called SmartStart [5] that generates courseware from e-textbooks. The SmartStart process uses machine learning (ML) and natural language processing (NLP) to read the textbook, identify learning objectives, divide the content into short lessons aligned to learning objectives, and generate practice questions for each lesson [8]. This application reduces the high cost and long timeframe of developing courseware and writing hundreds to thousands of formative questions. Yet, it is imperative to ensure the AG questions in this generated courseware provide the same learning benefit to students as their traditional human-authored (HA) counterparts. Analyzing AG questions in relation to HA questions using student data can answer one of the most fundamental questions regarding AQG: are these generated questions as good as those written by subject matter experts? Kurdi et al. [11] identified that mixing AG and HA questions to evaluate quality has been a focus of other AQG studies, though the evaluation was done through expert review. Furthermore, [11] noted that standard quality metrics such as difficulty or discrimination provide a statistical approach to evaluating performance and quality of AQ questions. Previous research studied six SmartStart courses with over 750,000 student-question interactions with both AG and HA questions [13]. An analysis of engagement, difficulty, and persistence performance metrics revealed that AG questions did not perform any differently than the HA questions in the course, but rather found differences primarily by the cognitive process dimension of the question type (recognition or recall). Similar trends were identified in an initial analysis of AG and HA question discrimination [9].

These initial findings from the SmartStart courseware's AG questions advances the study of AQG both in the scale of natural student data used as well as the different performance metrics studied, showcasing the benefits of the type and scale of data collected via the courseware platform as well as the insights gained through learning analytics.

Yet, this type of analysis is on data collected solely from student actions. It is also beneficial to understand student perceptions of the AG questions, which cannot be interpreted from platform data. Therefore, there is a need for research that combines a quantitative analysis of question data with qualitative feedback data from those same students who completed the questions. In this study, we evaluate AI-generated courseware for a Psychology of Sex and Gender course at the University of Central Florida (UCF) by analyzing the performance metrics of AG and HA questions as well as student perceptions of the courseware and questions.

2 Methods

The Psychology of Sex and Gender course is offered online every spring semester at UCF. There were 122 students enrolled in the courseware during the spring 2021 semester. According to student self-reporting, more than 80% of students were juniors or higher. UCF has a large population of transfer and first-generation students, reflected in this course with about 60% transfer students, and 30% first-generation college students. In this course, 95% of students reported having taken online courses previously (unsurprising given the number of online courses at UCF in addition to the online shift from COVID-19).

The Psychology of Sex and Gender textbook [3] was an ideal fit for the SmartStart process as it is a specialized subject that does not have custom publisher courseware or enhanced learning resources. The ML and NLP methods in SmartStart analyzed the text to identify learning objectives and determine a new structure with shorter lessons. The text was also the corpus for AQG, which works to identify important content that is suitable for questions. Kurdi et al. [11] put forth an AQG categorization that includes two parts: level(s) of understanding and procedure(s) of transformation. The AQG employed here includes both syntactic and semantic levels of understanding, and the procedure of transformation is primarily rule-based (for more details on the AQG process, see [13]). Two types of questions were generated for this courseware: matching questions that require students to drag terms to their correct location in a sentence (a recognition cognitive process on Bloom's hierarchy [1]), and fill-in-the-blank (FITB) questions that require students to type in the correct term (a recall process type on Bloom's). Both recognition and recall questions have been long studied for their learning benefits [2]. In total, 607 AG FITB and AG matching questions were delivered in the Psychology of Sex and Gender courseware.

In addition to the generated questions, 48 additional HA questions were added by the instructor and instructional designer. The purpose of the additional questions was to take advantage of additional features available in the courseware platform, such as predictive learning estimates and adaptive activities. Multiple choice (MC) questions from the textbook ancillary materials were used for this additional practice, including some that were implemented as a two-option true/false question.

During the semester, the courseware platform collected clickstream data that were used for analyzing the performance metrics of the questions. These data provide insights into how students chose to engage with different question types, how questions compared in difficulty, and how often students persisted until they reach the correct response. Given that there were a mix of AG and HA questions on lesson pages, question types could be directly compared on each metric using a mixed effects logistic regression model. At the end of the semester students were sent a survey that included student demographics, characteristics, prior experience with online learning and resources, perception of the courseware and the questions within, and overall course feedback. Students were incentivized to take the survey with a small number of bonus points, but could request an alternate activity (though none did). Student identities were not reported in the results to add a level of anonymity to encourage honest responses. In total, 67 students (54.9%) participated in the survey.

3 Results and Discussion

3.1 Performance Metrics

A data visualization called an engagement graph shows how students engaged with the courseware by plotting the number of students who engaged with the courseware's content elements on each page. A perfect engagement graph would be a horizontal line across the top, showing that every student read and did practice on every page. Some attrition is always expected at both the beginning of the course (as some students may drop the course) and over time as students begin to lose motivation to complete the reading and homework.

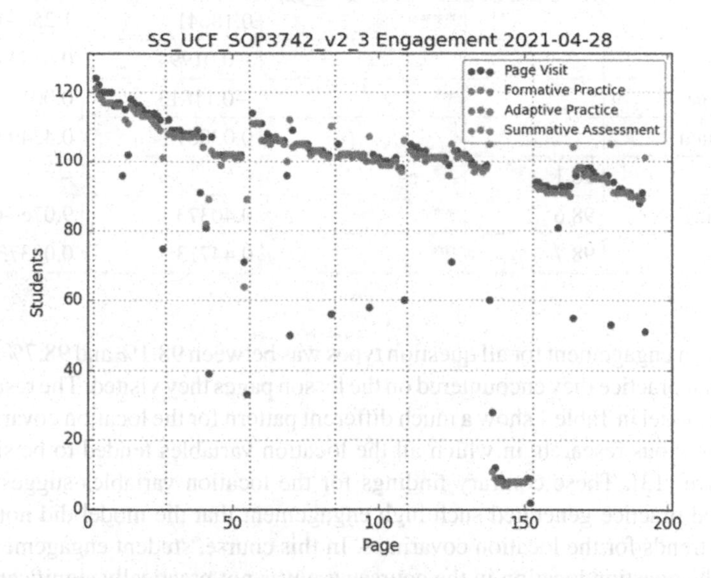

Fig. 1. Engagement graph for psychology of sex and gender.

While not perfect, this engagement graph (Fig. 1) is very close, in part due to assigning the formative practice for a percentage of the course grade [7]. Over 100 students consistently read the content and did the practice on the pages. The dotted vertical lines indicate unit boundaries, and the small jumps in engagement show typical—though in this case, minor—attrition within the unit. Figure 1 shows the red formative practice dots nearly on top of the blue reading dots, indicating nearly all students who read a page also did the practice. However, there are 5 to 7 students who only enter the courseware to do the summative assessments (which also contributed to the grade) but did not read or practice.

We can already tell that students had high engagement with the courseware, but were there any differences in how they chose to answer the AG questions compared to the HA questions? Following [13], an engagement data set was constructed as the set of all opportunities students had to engage with the practice questions in the course. Engagement opportunities were taken as all student-question pairs on course pages that the student visited (very short page visits of under 5 s were excluded); an answered question was recorded as an outcome of 1 and an unanswered question was recorded as −0. The resulting data set of 66,939 student-question observations was then analyzed using a mixed effects logistic regression model, which controls for question placement in the course to account for naturally occurring attrition.

Table 1. Engagement results for AG and HA questions.

Fixed effects	Mean	Significance	Estimate	*p*
Intercept		***	4.96956	<2e−16
Course page		***	0.18641	1.25e−06
Unit page			−0.01994	0.61734
Module page		**	−0.11813	0.00158
Page question			0.03081	0.43492
AG FITB	98.1	–	–	–
AG matching	98.6	***	0.40373	9.07e−08
HA MC	98.7	**	0.44713	0.00373

The mean engagement for all question types was between 98.1% and 98.7%; students did nearly all practice they encountered on the lesson pages they visited. The results of the regression model in Table 1 show a much different pattern for the location covariates than found in previous research, in which all the location variables tended to be significant and negative [13]. These contrary findings for the location variables suggest that the incentivized practice generated such high engagement that the model did not find any consistent trends for the location covariates. In this course, student engagement was so high that the question location in the courseware was not practically significant.

The comparison between question types revealed that both the AG matching and HA MC were more likely to be answered than the AG FITB (the baseline for the question type variable), which is consistent with previous findings. Both AG matching and HA MC are recognition type questions with mean engagement within 0.1% of each other. Even in a course where engagement was incentivized, recognition type questions have higher mean engagement and are more likely to be answered than recall questions. However, when we consider that the mean engagement of the AG FITB was within 0.6% of the recognition questions, it is clear that in this case the significance has little practical effect.

To evaluate the relative difficulty of questions, we narrow the data set to only the questions students chose to answer, resulting in 65,859 student-question observations. The same mixed effects model was used with this difficulty data set, using correctness of the student's first attempt as the outcome. The results in Table 2 show the AG matching were the least difficult (89.3%), AG FITB a few points behind (86.3%), and HA MC were the most difficult by almost 30 points (60.3%). The results of the model mirror the differences in mean scores. (Location variable results are omitted for brevity, as they continued not to be significant.) Students were more likely to get AG matching correct than AG FITB, but less likely to get HA MC correct than AG FITB. However, when we consider that most of the HA MC questions had a two-option true/false format, these results may indicate that these HA MC questions were either difficult, or possibly that a high proportion of students simply guessed. While previous research found students were more likely to get recognition questions correct [13], these results show question characteristics can create exceptions to these trends.

Table 2. Difficulty results for AG and HA questions.

Fixed effects	Mean	Significance	Estimate	*p*
AG FITB	86.3	–	–	–
AG matching	89.3	***	0.58409	1.4e−10
HA MC	60.3	**	−1.53956	<2e−16

Both the AG and HA questions analyzed here were formative in nature, so students received immediate feedback and could continue to answer as many times as they chose. Persistence is the rate at which students who are incorrect on their first response continue to answer a question until they get the correct response. Therefore, the persistence data set is a subset of the difficulty data, including only those student responses that were incorrect the first try (a total of 9,302 observations). The results of the regression model in Table 3 show students nearly always persisted on the AG matching (99.6%), also showing students were more likely to persist in this question type than AG FITB. Interestingly, AG FITB and HA MC had the same mean persistence (97.1%) and were not statistically different from one another. The HA MC had a lower mean difficulty score, yet students persisted at the same rate as HA FITB. The most reasonable explanation is the cognitive process dimension of the question types: it takes less effort to answer a recognition question than a recall question—and even more so if the recognition type only has two options to choose from. The most important finding in persistence, however, is the overall high rate. While engagement was required for points, we see no patterns that indicate students simply answered carelessly to meet the requirement and moved on. Nearly all students who were incorrect continued to try until they reached the correct answer.

Table 3. Persistence results for AG and HA questions.

Fixed effects	Mean	Significance	Estimate	*p*
AG FITB	97.1	–	–	–
AG matching	99.6	***	2.21753	1.96e−10
HA MC	97.1		−0.32718	0.2964

Significance codes: *** $p < .001$, ** $p < .01$, * $p < .05$

3.2 Student Perceptions

While evaluating question data provides a clear understanding of what students did, it is unable to give insights into what students think. Student perception of AG questions in natural learning contexts is a valuable addition to AQG research, as student beliefs can influence student behavior. In this research, students were unaware of which questions were generated through AI and which were authored by subject matter experts to avoid preconceived bias. Due to self-selection, it is possible that students who completed the survey ($n = 67$) were not representative of the class. To investigate this, we can use the question, "How often did you do the practice questions located at the bottom of the page?" to gauge the distribution of engagement. Students responded always (67%), most of the time (18%), about half the time (3%), sometimes (9%), and never (3%). While the Likert scale response categories for question engagement are qualitative rather than quantitative and therefore prohibit a detailed statistical comparison of the engagement data and survey response distributions, a qualitative comparison to the engagement graph (Fig. 1) shows similarities. Most students did all (or nearly all) the practice, while a very small percentage did little to none. This provides some confidence that this sample of students is representative of the class, and that students were honest in their self-reporting.

Before soliciting feedback on the questions, students were asked about their previous experience with adaptive/interactive platforms, as UCF uses a variety of platforms and resources. About 84% of students reported having previously used an adaptive platform such as Connect, LaunchPad, Reveal, Realizeit, ALEKS, etc. Of the students who had used adaptive platforms, 26 students liked them a great deal, 23 liked them somewhat, 3 were neutral, 3 disliked them somewhat, and 1 disliked them a great deal. This baseline indicates that while most students enjoy learning with digital platforms, a small percentage do not. Students were then asked, "Compared to other adaptive courseware you have used, was your experience with Acrobatiq this semester:" far above average (34%), somewhat above average (39%), average (21%), somewhat below average (7%), far below average (0%). This is positive feedback for AI-generated courseware; most students preferred it or found it to be equivalent to other adaptive platforms they had used.

Students were then asked a series of questions regarding the formative practice in general. This began with the question, "If you answered the practice questions, how helpful did you find them?" Students answered: extremely useful (32%), very useful (37%), somewhat useful (20%), slightly useful (5%), not at all useful (3%). No distinction was made between the AG and HA questions, but given the high proportion of

AG questions, this reveals that most students had positive reactions to them. To better understand how students perceived the learning benefit, students were asked, "Do you think that the Acrobatiq features helped you learn the material?" Students responded: definitely yes (45%), probably yes (40%), might or might not (4%), probably not (4%), definitely not (6%). Students were also asked, "Do you think that the practice questions increased your test scores?" Students responded: yes (69%), maybe (24%), no (7%). We see that the majority of students felt that the practice questions helped them learn the material better as well as increased their test scores. A small number of students did not, however, but we should note a similarly small number of students also disliked adaptive/interactive platforms in general and rarely or never answered the practice questions. It is to be expected that not all students prefer this style of learning resource, but it is encouraging that this group is proportionally very small. Further investigation of this group's preferences and rationale would be beneficial for future research.

Students were also prompted to provide elaboration for their thoughts on the practice questions. "If you did not answer the practice questions often or at all, what were the reasons you chose not to?" In total, 19 students responded, but 13 of those responses were to indicate this question was not applicable—they did answer them. The remaining 6 responses can be grouped into three categories: students didn't realize the questions were assigned or beneficial at the time, students had limited time and had to prioritize, and students felt confident with the material and so bypassed them. No students responded that they didn't care for the questions or thought they were of poor quality; all responses were related to student circumstances.

To know more about student preference for individual question types, students were asked, "Were there any question types that you liked or disliked more than others?" Forty students responded, with 21 answering no, they didn't have a preference. Six students reported liking the fill-in-the-blank the best ("I liked the fill in the blank questions because it requires you to recall information"), four liked the matching ("the ones where they gave you the options and you had to fill them in were my favourite"), and one liked the true/false. One student replied that they liked "definition" questions, and three students responded with additional course feedback. Four students responded that they disliked the true/false questions—which were the HA MC questions. This was the only question type students identified disliking and could suggest a relationship to the low mean score for this question type.

Finally, students were asked if there were any other types of study features they would like to have in the courseware. Of the 35 students who responded, 22 said no, with two elaborating, "No, I thought the questions and quizzes were enough," and "not really, I liked the courseware a lot." Videos, vocabulary, immediate feedback on quizzes, and flashcards were all suggested by one student each and one answer was also not readable. Two students requested note-taking and three requested a study guide ("A study guide for the exams would be helpful"). Three students requested the ability to get more/random questions ("Yes an option for new questions to keep practicing or more questions"). This request for more formative practice is especially encouraging as it identifies the value students place on practice as a study feature in general, and indicates that the generated practice was perceived on par with other human-authored resources.

4 Conclusion

Measurable performance metrics are key to evaluating the quality of AG questions and advancing toward standards for AQG at large. Quality metrics such as difficulty and cognitive level are standard measures that can be evaluated through automatic scoring and statistical procedures, yet these are less common in the AQG literature [11]. In this study, we contribute to the research reporting question difficulty, but also include engagement and persistence to provide a more thorough understanding of how students interact with AG questions. Furthermore, using student data from natural learning contexts expands the external validity of AG question performance.

While more research is needed in AQG that uses student data for evaluation of performance metrics, it is notable that no research is mentioned that investigates student perceptions of the AG questions as a learning feature [11]. Student perception is particularly relevant to this courseware environment because the formative practice questions are the primary learning method that engages the doer effect [12]. Should students perceive the AG questions poorly, this could affect their use of them or how seriously the students regard the value of their learning benefit. While the AG question performance metrics do not show evidence of negative student perception (i.e., low engagement, unusual difficulty patterns, low persistence), combining this with student feedback creates a more comprehensive evaluation. Survey results showed that students were generally experienced with adaptive/interactive platforms and—apart from a handful of students—generally like using them for learning. The comparative value students found in the AI-generated courseware is notable, especially given the minimal time and effort required from the instructor and designer to create it. Nearly all students thought the practice questions were useful, helped them learn the material, and helped increase their test scores. Research on the doer effect has repeatedly shown that doing formative practice while reading is causal to learning [10, 12], but it is also beneficial for students to believe that this learn by doing method is helping them. For students to have such positive perceptions of questions generated through artificial intelligence is a significant outcome for AG questions as well as the overall learning method they were intended for. While there are many avenues of future research remaining for AG questions, these results point to an optimistic future for the application of AQG in formative learning contexts.

References

1. Anderson, L.W., et al.: A Taxonomy for Learning, Teaching, and Assessing: A Revision of Bloom's Taxonomy of Educational Objectives (Complete Edition). Longman, New York (2001)
2. Andrew, D.M., Bird, C.: A comparison of two new-type questions: recall and recognition. J. Educ. Psychol. **29**(3), 175–193 (1938). https://doi.org/10.1037/h0062394
3. Bosson, J.K., Vendello, J.A., Buckner, C.V.: The Psychology of Sex and Gender, 1st edn. SAGE Publications, Thousand Oaks (2018)
4. Black, P., William, D.: Inside the black box: raising standards through classroom assessment. Phi Delta Kappan **92**(1), 81–90 (2010). https://doi.org/10.1177/003172171009200119
5. Dittel, J.S., et al.: SmartStart: Artificial Intelligence Technology for Automated Textbook-to-Courseware Transformation, Version 1.0. VitalSource Technologies, Raleigh (2019)

6. Dunlosky, J., Rawson, K., Marsh, E., Nathan, M., Willingham, D.: Improving students' learning with effective learning techniques: promising directions from cognitive and educational psychology. Psychol. Sci. Public Interest **14**(1), 4–58 (2013). https://doi.org/10.1177/1529100612453266
7. Hubertz, M., Van Campenhout, R.: Teaching and iterative improvement: the impact of instructor implementation of courseware on student outcomes. In: IICE 2022: The 7th IAFOR International Conference on Education, Honolulu, Hawaii (2022). https://iafor.org/archives/conference-programmes/iice/iice-programme-2022.pdf
8. Jerome, B., Van Campenhout, R., Johnson, B.G.: Automatic question generation and the SmartStart application. In: Proceedings of the Eighth ACM Conference on Learning@Scale, pp. 365–366 (2021). https://doi.org/10.1145/3430895.3460878
9. Johnson, B.G., Dittel, J.S., Van Campenhout, R., Jerome, B.: Discrimination of automatically generated questions used as formative practice. In: Proceedings of the Ninth ACM Conference on Learning@Scale (2022). https://doi.org/10.1145/3491140.3528323
10. Koedinger, K., McLaughlin, E., Jia, J., Bier, N.: Is the doer effect a causal relationship? How can we tell and why it's important. In: Proceedings of the Sixth International Conference on Learning Analytics & Knowledge, Edinburgh, United Kingdom, pp. 388–397 (2016). https://doi.org/10.1145/2883851.2883957
11. Kurdi, G., Leo, J., Parsia, B., Sattler, U., Al-Emari, S.: A systematic review of automatic question generation for educational purposes. Int. J. Artif. Intell. Educ. **30**(1), 121–204 (2019). https://doi.org/10.1007/s40593-019-00186-y
12. Van Campenhout, R., Johnson, B.G., Olsen, J.A.: The doer effect: replicating findings that doing causes learning. In: Proceedings of eLmL 2021: The Thirteenth International Conference on Mobile, Hybrid, and On-line Learning, pp. 1–6 (2021). https://www.thinkmind.org/index.php?view=article&articleid=elml_2021_1_10_58001. ISSN 2308-4367
13. Van Campenhout, R., Dittel, J.S., Jerome, B., Johnson, B.G.: Transforming textbooks into learning by doing environments: an evaluation of textbook-based automatic question generation. In: Third Workshop on Intelligent Textbooks at the 22nd International Conference on Artificial Intelligence in Education. CEUR Workshop Proceedings, pp. 60–73 (2021). http://ceur-ws.org/Vol-2895/paper06.pdf. ISSN 1613-0073

Representing Scoring Rubrics as Graphs for Automatic Short Answer Grading

Aubrey Condor(✉), Zachary Pardos(✉), and Marcia Linn(✉)

University of California Berkeley, Berkeley, CA 94720, USA
aubrey_condor@berkeley.edu, pardos@berkeley.edu, mclinn@berkeley.edu

Abstract. To score open ended responses, researchers often design a scoring rubric. Rubrics can help produce more consistent ratings and reduce bias. This project explores whether an automated short answer grading model can learn information from a scoring rubric to produce ratings closer to that of a human. We explore the impact of adding an additional transformer encoder layer to a BERT model and training the weights of this extra layer with only the scoring rubric text. Additionally, we experiment with using Node2Vec sampling to capture the graph-like ordinal structure in the rubric text to further pre-train the model. Results show superior model performance when further pre-training with the scoring rubric text. Specifically, questions that elicit a very simple rubric structure show the most improvement from incorporating rubric text. Using Node2Vec to capture the structure of the text had an inconclusive impact.

Keywords: Automatic short answer grading · BERT · Node2Vec

1 Introduction

Automatic Short Answer Grading (ASAG) is an emerging field of research, as the education community has started to embrace the use of machine learning to assist both students and education professionals. ASAG refers to the machine scoring of open-ended (OE) items that are shorter than essay length - often a sentence or two. It has been shown that the use of open-ended questions helps facilitate learning by self-explanation [9], or information recall [5]. The "generation effect" has been studied extensively and findings show that subjects who generate information are able to remember the information better than material they have simply read [5]. Additionally, open-ended items may be adapted to assess student knowledge and understandings across many content areas and have been shown to enhance learning [1,19]. However, educators are often deterred from the use of OE items because grading requires much more time than that for multiple choice items [19]. The use of an automated system for grading short-answer questions could decrease the amount of time teachers spend grading, while allowing students to grow in their knowledge through self-explanation and information recall. However, limitations do exist when considering the practical

M. M. Rodrigo et al. (Eds.): AIED 2022, LNCS 13355, pp. 354–365, 2022.
https://doi.org/10.1007/978-3-031-11644-5_29

use of ASAG, including the resource-heavy need to label (grade) many student responses, and that ASAG models do not seem to generalize well to questions outside of those included in the training set [10].

Many assessments are created via collaboration between teachers, researchers, and test developers where scoring rubrics for each item are carefully designed. Rubrics can provide raters with both exemplar student answers as well as competency descriptions corresponding to each grading category. Rubrics are created to increase both reliability and validity of scores [32]. When a rubric clearly and rigidly defines scoring criteria, inter and intra rating consistency can be increased [31]. There are, however, grading scenarios in which rubrics are not created such as contract grading where evaluation is solely based on completion of a task [11], or intuitive grading where instructors rely on their experience to grade [14].

In order to exploit the advantages of using scoring rubrics, we propose to incorporate item-specific scoring rubrics into an ASAG model. We do so by further training a model called BERT (Bidirectional Encoder Representations of Transformers) using the scoring rubric text and structure. BERT is a type of language model called a transformer, which has been pre-trained on the English language and can be later trained for downstream tasks such as classification [12]. BERT, as-is, encapsulates latent relationships in natural language from the text on which it has been pre-trained. However, we want it to capture such relationships in the scoring rubric text specifically. Before we train it on our classification task of grading, we continue the model's pre-training with the scoring rubrics. We hypothesize that this will improve the ASAG accuracy beyond that of similar models not trained on rubrics. Additionally, we look at model performance at the question level to examine if certain types of questions and their corresponding rubrics exhibit varied results.

The original BERT model is a neural network consisting of 12 transformer encoder layers with pre-trained weights. To provide the model with the rubric text, we add one more (13th) transformer encoder layer to the model to train with the masked language model (MLM) task, using the rubric text as input. Thus, the model will also incorporate relationships in the text specific to the rubric as it classifies student responses. Additionally, because the ordinality, of the rubric is important - namely that its text would not mean much to a human grader if they could not interpret that a level 3 score is ranked "higher" than a level 2 score - we incorporate a sampling method called Node2Vec [17] for creating a vector representation of the rubric text. If we conceptualize the rubric as a graph network to capture the ordinality in the scoring levels, Node2Vec will learn a latent representation of the structure of the rubric.

The novelty of this research is the modification of BERT to incorporate a rubric for each item. The purpose is to increase an automated grading system's performance for practical use in terms of producing scores that match that of human graders. Using BERT for downstream classification tasks is common for applied machine learning research, as well as for ASAG-specific tasks. Modification of BERT to incorporate a rubric has not yet been studied.

We hypothesize that pre-training BERT on rubric text, termed "further pre-training" will improve the model's ability to match human ratings of the student responses. However, we do anticipate that this effect may differ depending on the type of questions and their corresponding rubrics.

2 Related Work

We believe that this work will contribute to the growing research for, and improvement of, ASAG models. A systematic review of trends in ASAG [6] illustrates an increasing interest in the field of automatic grading for education; researchers have used a broad range of NLP methods.

Unsupervised methods have been explored such as concept mapping, semantic similarity, and clustering to assign ratings [2,21,25]. In addition, many types of supervised classification methods have been utilized for ASAG [4,20,24]. More recent ASAG research exploits advances in deep learning methods [23,26] [27,30]. Further, much of the newest ASAG work makes use of state-of-the-art transformer based language models such as BERT, GPT and ELMo [7,15].

More specifically related to our project, researchers have started to focus on using extra, question-relevant information within ASAG models. Sung et al. [28, 29] examined the effectiveness of pre-training BERT on relevant domain texts, as a function of the size of training data and generalizability across domains. Our research expands on that of Sung by investigating pre-training BERT but on a much different corpus of text (rubrics) and accounting for the structure of the text as well. Wang et al. [33] used an attention based method to extract key phrases from student answers that are highly related to phrases and answers from a scoring rubric. Condor et al. [10] experimented with concatenating different pieces of information to student responses such as question text, or rubric text, as input to a grading model. Chen and Li [8] make use of prior knowledge to enrich features of student responses by incorporating question text information within an extra forward propagation training step.

3 Background

3.1 BERT

BERT [12] was the first language model to successfully learn relationships within sequences of words by pre-training with a bi-directional prediction task, masked language modeling (MLM). It does so by randomly masking 15% of tokens, or words, throughout the text during training and uses a self-attention mechanism to predict the masked texts. Additionally, BERT learns representations between sequences of sentences by its pre-training on a next sentence prediction (NSP) task, where the model is given two sentences and predicts whether one sentence should come directly after the other. The standard BERT model was pre-trained on large amounts of natural language text from Wikipedia and BooksCorpus, and can be fine-tuned for different tasks, such as classification, by adding only one additional layer to the existing neural network [12]. We use a compressed version of the original BERT model called BERT-base, consisting of 12 encoder layers and a total of 110 million parameters. With limited computational resources, the base version is faster to train yet still retains much of the impressive performance on standard benchmark tasks.

3.2 Node2Vec

Node2Vec is an approach for learning latent representations of vertices in a network such that graph structures can be represented as a continuous vector space [17]. By modeling random walks over the graph structure, Node2Vec creates input sequences conceptually similar to that for a sequence of words, however the sequences capture the structure of the graph.

The algorithm works as follows: firstly, a random walk generator takes in a graph structure, and uniformly samples a node as the root of the walk. Then, the walk uniformly samples from the direct neighbors of the last visited node, and this process continues until a maximum length has been reached. This sequence of nodes would represent one input sequence, and as many sequences can be generated as specified. In addition to the number of walks and the length of walks, two other parameters exist which specify how likely it is for the walk to wander far from the starting node (breadth of the search), and how likely it is to return to the start node (depth of the search).

4 Methods

In this section, we describe the data used for this project, outline the methods used to alter the architecture of the BERT-base model, explain the Node2Vec sampling approach for creating input sequences to further pre-train the extra layer of the model, and detail the process of training the classifier.

4.1 The Data Sets

We used two data sets for this project, consisting of short responses to science questions. The first data set we used was sourced from a 2019 field test of an assessment [3] created at the Berkeley Evaluation and Assessment Research (BEAR) center. The data consists of 5,550 student responses from 558 distinct students to 31 different items. The mean number of responses per question is 179 with the minimum being 128 and the maximum being 313. Items all were administered in four different test forms. We refer to this data set as 'BEAR'. Responses were rated from 0 (incorrect) to 4 (fully correct) by multiple subject-matter experts at the BEAR center. The quality and consistency of ratings were evaluated by an inter-rater reliability score, and when a high percentage of rating mismatches between raters existed (more than 15%), incongruous ratings were discussed until a consensus was reached by the raters.

To exemplify the type of questions included, for one item labelled 'Crude Oil', students are presented with two images relating to oil production - one being a line graph and the other, a table of values. Respondents are to choose between the graph or the table for which would be better to represent the historical patterns of oil production on a poster for a class presentation. The correct answer for this question is the graph, and students are expected to provide reasons why this is the right choice. An example of a student response to the Crude Oil item rated as a 4 (fully correct) is: *"The graph easily displays patterns over time whereas the table and equations require more analyzing."* In contrast, a student response to the same question rated as a 0 (incorrect) is: *"The table is more clear, the information is seen in the table."*

We also show results with an open-source data set called the Automatic Student Assessment Prize (ASAP) Short Answer Scoring (SAS) data. The data was used in a 2012 Kaggle competition sponsored by the Hewlett Foundation, and consists of almost 13,000 short answer responses to 10 science and English questions. We used only the 5 science questions since the domain is the same as the BEAR data set. The questions were scored from 0 (incorrect) to 3 (fully correct), and each question includes a scoring rubric. We refer to this data set as ASAP. An example of a question from the ASAP data set is: *"Starting with mRNA leaving the nucleus, list and describe four major steps involved in protein synthesis.* The rubric consists of a list of key elements of an exemplary response such as *"mRNA exits the nucleus via nuclear pore"*, and specifies how many key elements need to be included for the answer to be rated at a given level.

Standard pre-processing of both the data sets included removing null responses, adding a period to the end of sentences, and concatenating a numeric question identifier to the beginning of student responses. We split the data sets into training, validation and testing sets with a 70/15/15% split.

4.2 Architecture Modification

To further pre-train the model on the scoring rubric text, we add a 13th encoder transformer layer to the original BERT-base model. The weights of the 13th layer are randomly initialized, and the weights of the other 12 encoder layers are frozen during the additional pre-training. Thus, the original 12 layers maintain the weights that capture latent relationships in general English language from the model's original pre-training, and the new weights of the 13th layer start as essentially a blank slate. We then train the 13th layer with the MLM task using our scoring rubric text as input. Standard neural network parameters were tuned to minimize the MLM loss, in addition to considering the best configuration for our computational limitations. Optimal parameters include a batch size of 8, a learning rate of 4e−4, and 12 training epochs.

4.3 Sampling Methods

To produce input sequences that capture the ordinal structure of the scoring rubric for pre-training the BERT model, we use the Node2Vec sampling method.

We create a graph representation of the structure of the scoring rubric for each question in the data set. An example of a scoring rubric is included in Fig. 3, as well as an image of the graph representation of the scoring rubric. In order to conceptualize the rubric as a graph, we make each level of the scoring rubric a node in the graph, and undirected edges connect each level to both the level below and above it. The description of each level as well as the example responses are included as features of the level nodes. We use the python package, 'networkx' [18] to create the graph structure (Fig. 1).

Level	Description	Example Response
4	Student provides a fully correct positive and negative justification for ...	"The graph best illustrates the visual trend of oil production. The table or the equation won't provide an overall ..."
3	Student provides a fully correct positive justification for selected representation ...	"The table cannot show a trend in the oil production effectively. "
2	Student provides a partial/general positive justification for selected ...	"The graph helped me with more questions."
1	Student provides incorrect justification for selected representation.	"Because the graph gives us better projections."
0	Student makes no attempt to provide a justification for selected representation.	"I am not sure why."

L4
L3
L2
L1
L0

Fig. 1. Example of a scoring rubric (left) and corresponding graph structure (right)

Parameters to be tuned for Node2Vec include the walk length, number of walks, and the p and q parameters that indicate how far the random walk strays from the original node, and how likely it is to return to the previous node. The best set of parameters were chosen based on the performance with our downstream classification task and consisted of a walk length of 10, number of walks being 50, a p of 0.5 and a q of 2.

Included as features of each node of our scoring rubric graph is both the description of the grading level as well as exemplary student responses. When the random walk lands on a node, the method randomly samples one piece of text from the node (rubric level). For example, if the algorithm starts on the level 2 node, we grab either the level description text, or one of the exemplary responses for the input sequence. If, during the random walk, we return to the same node, the algorithm will similarly randomly choose one of the pieces of text. The full input sequence for each walk is created by concatenating the text from each "step" of the walk in the order they are sampled. The sampling process is repeated for each question included in the data, as each question has a unique corresponding scoring rubric. Resulting input sequences for each question are combined to create all the input text. All of the input sequences are used to pre-train the weights of the 13th encoder layer of our model.

4.4 Fine-Tuning Experiments

To test our hypothesis that providing the BERT model with information about the scoring rubric will increase the performance of the classifier for ASAG, we performed four different experiments, with four separately trained models. We choose the particular experiments to help us clarify the implication of our results such that we can distinguish between whether the text itself or the latent graph structure is helping the ASAG model.

Each model is fine-tuned for our multi-class classification task - i.e. we add a classification head to the model after its encoder layers (regardless if it has 12 or 13 encoder layers) and train the classification head with the labeled data. We performed a standard parameter search and achieved the best validation set results with a batch size of 8, 4 training epochs, and a learning rate of 5e−4.

We include two standard evaluation metrics for both datasets (BEAR and ASAP) in our results table: Cohen's Kappa (CK) and weighted F1 score (F1). The F1 score includes both precision and recall and accounts for class imbalance, and CK reports the agreeability between two scores beyond random chance - a more robust measure than accuracy. Additionally, we provide a majority class classifier for the F1 score results as a baseline comparison. The CK results have an 'n/a' for the majority class classifier because a majority class model is the same as classification by random chance for multi-class classification, which would always result in a CK metric of 0.

A 5x2 cross validation (CV) paired t-test was used to evaluate the statistical significance of the difference in models. The 5x2 CV paired t-test is based on five iterations of twofold cross-validation, and is presented in [13] as the recommended approximate statistical test for whether one machine learning model outperforms another because of it's more acceptable type I error, and stronger statistical power than other methods such as McNemar's test, or a paired t-test based on 10-fold CV. We used the CK metric for the 5x2 CV paired t-tests (with 5 degrees of freedom). For each of the 'Rubric' models, there is a corresponding t-test comparing the CK metric for the BERT-base model with the given model. The null hypothesis for each t-test states that the CK metric for the specified model is no different than the CK metric for the BERT-base model.

5 Results

We present our results below, in Table 1, for the four models we wish to compare. The first model, 'BERT-base', is a standard BERT-base classification model (no extra 13th layer). This model serves as a baseline with which to compare our modified architecture ('Rubric') models. The second model, 'Node2Vec Rubric', incorporates the Node2Vec sampling method in the further pre-training of the 13th layer. This is the model we hypothesize will perform best because it captures both the text and structure of the scoring rubric in the added 13th layer. The third model, 'Top-level Rubric', also has an added 13th encoder layer but the extra layer is trained with only the top level of scoring rubric text (i.e., explanations and examples of the most correct responses) and the text is randomly sampled instead of using Node2Vec Sampling. In this experiment, we still sample 10 pieces of text for each input sequence, and 50 sequences total for each question for consistency in the amount of training data. Finally, the fourth model, 'Random Rubric', we add the 13th encoder layer and train it with all of the scoring rubric text, randomly sampled (no Node2Vec sampling). Thus we can compare a regular BERT-base model to three other models that have been altered in structure with a new 13th layer to examine: if just the rubric text itself with no structure is helpful (Random Rubric), if only including exemplar level descriptions and text is helpful (Top-level Rubric), or if providing the model a signal about the structure of all the rubric text, along with the text itself is helpful (Node2Vec Rubric) in terms of the performance of our ASAG model.

Table 1. Test set weighted F1 score and Cohen's Kappa

	BEAR F1	BEAR CK	ASAP F1	ASAP CK
BERT-base	0.6742	0.6831	0.7755	0.8677
Graph Rubric	0.6946	0.7361	0.8001	0.8808
Top Rubric	0.6932	0.7170	0.7879	0.8739
Random Rubric	0.6815	0.6961	0.7972	0.8825
Majority class	0.2082	n/a	0.2737	n/a

Results of the 5x2 CV paired t-test are shown in Table 2. For a paired one-tailed t-test, the critical values for an alpha, or probability of a type I error, of 0.05, 0.10 are 2.015 and 1.476 respectively. In Table 2, statistical significance at the 0.05 level is represented with an appended *.

Table 2. 5x2 Cross Validation Paired one-tailed t-test (5 df)

BERT-base versus:	BEAR t(5)	BEAR p-val	ASAP t(5)	ASAP p-val
Graph Rubric	1.814	0.0647	3.098*	0.0135
Top Rubric	1.880	0.0594	3.191*	0.0121
Random Rubric	1.778	0.0678	4.110*	0.0046

From Table 1, we see that all of the models, including BERT-base, on both the BEAR and ASAP data, perform much better than the majority class classifier. Additionally, across both data sets and considering both metrics, the 'BERT-base' model performs slightly worse than all the others. For the BEAR data, the 'Node2Vec Rubric' model performs best and the difference between this model and the 'BERT-base' model is considerable, especially in terms of the CK metric.

The 'Top-level Rubric' model still seems to be an improvement to the 'BERT-base' model, and the 'Random Rubric' model shows only slight improvements from the 'BERT-base' model. With the ASAP data, we see similar results when comparing the 'BERT-base' model and the 'Node2Vec Rubric' model - we get at least two percentage points improvement in terms of both F1 and CK metrics. Additionally similar, we get improved performance, but to a lesser extent, with the 'Top-level Rubric' model. However, one considerable difference is that we see about the same performance improvement with the 'Random Rubric' model as with the 'Node2Vec Rubric' model, when comparing against 'BERT-base'.

In Table 2, We further validated our results with the 5x2 CV t-test for statistical significance for whether each of the three models with an added 13th layer model performs superior to the 'BERT-base' model in terms of the CK metric. With the BEAR data, results are statistically significant at the 0.10 type I error level, but not quite at the 0.05 level as shown by the p-values for each of the three 'Rubric' models. However, with the open-source ASAP data, all three models show statistically significant improvement over 'BERT-base' at the 0.05 type I error level.

Additionally, since the ASAP data showed more promising results for all three 'Rubric' models in terms of statistical significance, we broke down the results for this data set by question in order to see if our method is more useful when autograding a particular type of question or with a certain rubric type. We saw the biggest performance increase between the 'BERT-base' model and the 'Node2Vec Rubric' model with one specific question that elicited a very straightforward rubric. The question context presents an experimental procedure, and asks students to describe what other pieces of information are needed to replicate the experiment. The rubric is simply broken down into the number of pieces of information that the student provides; to achieve the highest score, a student must provide three additional pieces of information, and to achieve the second highest score, two additional pieces of information. In contrast, the least performance increase (actually, a slight performance decrease) was observed for a question that had a much more open-ended nature and abstract rubric. The question context presents an experimental design and corresponding results, and asks students to describe two ways to improve the experimental design and/or the validity of the results. Rubric guidance is vague, including statements such as: *"The response draws a valid conclusion supported by the student's data but fails to describe, or incorrectly describes, how the student could have improved the experimental design and/or the validity of the results"*.

6 Discussion and Conclusion

Most evident with the ASAP data set results, adding an extra 13th layer to the model, further pre-trained with some version of the rubric text improves the classification performance of the ASAG model notably (from the t-test, we see a statistically significant improvement in performance, at a significance of 0.05, when comparing all three models to the BERT-base model with no 13th layer). Less evident, but also noteworthy, is the performance increase for all three 'Rubric' models with the BEAR data (from the CV t-test, we see a statistically significant improvement in performance, at a significance of 0.10, when comparing all three 'Rubric' models to the BERT-base model with no 13th layer). Thus, we conclude that incorporating rubric text to BERT by further pre-training an extra encoder layer adds value to the ASAG model.

However, we cannot conclude that using Node2Vec sampling to capture the structure of the rubric text prior to further pre-training BERT adds value. We do not see consistently lower p-values in the 5x2 CV t-test results for the 'Node2Vec Rubric' model than the other 'Rubric' models ('Top-level Rubric' and 'Random Rubric'), nor do we observe a noticeable increase in performance for the results in Table 1 when comparing the 'Node2Vec Rubric' model with the other 'Rubric' models. So, although it appears that allowing the BERT model to learn about the text in a rubric is beneficial for the model's performance, the structure of the rubric seems not as important as we hypothesized. It may be the case that the model benefits simply from ingesting the subject-specific jargon in the rubric, versus the more general English text on which the other 12 encoder layers have been pre-trained. Further, there are methods outside of Node2Vec that may better capture the rubric structure and further research may attempt to include other methods to represent structured text for language models.

Finally, after investigating the results broken down by question, we conclude that our method of further pre-training a BERT model using rubric text may be useful only for specific types of questions that elicit a very straightforward rubric structure. Questions with rubrics that call for more abstract interpretation or subjective judgment may not be a good application for our proposed method. Further investigation on different short answer question types and rubric structures is necessary to further generalize this conclusion.

The outlook of automatic grading continues to improve with advances in machine learning and natural language processing. Incorporating human-created, domain related text such as a scoring rubric may be one way to ensure that ASAG models reflect human judgment, even more than with supervised classification (i.e. training the model to match human ratings), while still improving model performance. However, if it is the case that our methods are only applicable to questions of a simplistic nature, future research may need to focus on how we can automatically grade questions that elicit more complex responses. For example, the knowledge integration perspective on science inquiry supports that students can maintain multiple ideas, sometimes conflicting, about a particular topic [22]. Automatically identifying a student's various, and potentially contradicting, ideas within a response may be an important step in creating quality autograders for more complex science questions [16].

References

1. Al-Absi, M.: The effect of open-ended tasks-as an assessment tool-on fourth graders' mathematics achievement, and assessing students' perspectives about it. Jordan J. Educ. Sci. **9**(3), 345–351 (2013)

2. Anandakrishnan, P., Raj, N., Nair, M.S., Sreekumar, A., Narayanan, J.: Automated short-answer scoring using latent semantic indexing. In: Smart Computing, pp. 298–307 (2021). https://doi.org/10.1201/9781003167488-35
3. Arneson, A., Wihardini, D., Wilson, M.: Assessing college-ready data-based reasoning. In: Quantitative Measures of Mathematical Knowledge, pp. 93–120 (2019)
4. Bailey, S., Meurers, D.: Diagnosing meaning errors in short answers to reading comprehension questions. In: Proceedings of the Third Workshop on Innovative Use of NLP for Building Educational Applications, pp. 107–115 (2008)
5. Bertsch, S., Pesta, B.J., Wiscott, R., McDaniel, M.A.: The generation effect: a meta-analytic review. Memory Cogn. **35**(2), 201–210 (2007)
6. Burrows, S., Gurevych, I., Stein, B.: The eras and trends of automatic short answer grading. Int. J. Artif. Intell. Educ. **25**(1), 60–117 (2015)
7. Camus, L., Filighera, A.: Investigating transformers for automatic short answer grading. In: Bittencourt, I.I., Cukurova, M., Muldner, K., Luckin, R., Millán, E. (eds.) AIED 2020. LNCS (LNAI), vol. 12164, pp. 43–48. Springer, Cham (2020). https://doi.org/10.1007/978-3-030-52240-7_8
8. Chen, S., Li, L.: Incorporating question information to enhance the performance of automatic short answer grading. In: Qiu, H., Zhang, C., Fei, Z., Qiu, M., Kung, S.-Y. (eds.) KSEM 2021. LNCS (LNAI), vol. 12817, pp. 124–136. Springer, Cham (2021). https://doi.org/10.1007/978-3-030-82153-1_11
9. Chi, M.T., De Leeuw, N., Chiu, M.H., LaVancher, C.: Eliciting self-explanations improves understanding. Cogn. Sci. **18**(3), 439–477 (1994)
10. Condor, A., Litster, M., Pardos, Z.: Automatic short answer grading with SBERT on out-of-sample questions. In: Proceedings of the 14th International Conference on Educational Data Mining (2021)
11. Danielewicz, J., Elbow, P.: A unilateral grading contract to improve learning and teaching. Coll. Compos. Commun. **61**, 244–268 (2009)
12. Devlin, J., Chang, M.W., Lee, K., Toutanova, K.: BERT: pre-training of deep bidirectional transformers for language understanding. arXiv preprint arXiv:1810.04805 (2018)
13. Dietterich, T.G.: Approximate statistical tests for comparing supervised classification learning algorithms. Neural Comput. **10**(7), 1895–1923 (1998). https://doi.org/10.1162/089976698300017197
14. Ecclestone, K.: 'i know a 2: 1 when i see it': understanding criteria for degree classifications in franchised university programmes. J. Further High. Educ. **25**(3), 301–313 (2001)
15. Gaddipati, S.K., Nair, D., Plöger, P.G.: Comparative evaluation of pretrained transfer learning models on automatic short answer grading. arXiv preprint arXiv:2009.01303 (2020)
16. Gerard, L.F., Ryoo, K., McElhaney, K.W., Liu, O.L., Rafferty, A.N., Linn, M.C.: Automated guidance for student inquiry. J. Educ. Psychol. **108**(1), 60 (2016)
17. Grover, A., Leskovec, J.: node2vec: scalable feature learning for networks. In: Proceedings of the 22nd ACM SIGKDD International Conference on Knowledge Discovery and Data Mining, pp. 855–864 (2016)
18. Hagberg, A., Swart, P., Chult, D.S.: Exploring network structure, dynamics, and function using networkX. Technical report, Los Alamos National Lab (LANL), Los Alamos, NM (United States) (2008)
19. Hancock, C.L.: Implementing the assessment standards for school mathematics: enhancing mathematics learning with open-ended questions. Math. Teach. **88**(6), 496–499 (1995)

20. Hou, W.J., Tsao, J.H.: Automatic assessment of students' free-text answers with different levels. Int. J. Artif. Intell. Tools **20**(02), 327–347 (2011)
21. Jordan, S.: Investigating the use of short free text questions in online assessment. Final project report, Centre for the Open Learning of Mathematics, Science, Computing and Technology, The Open University, Milton Keynes, United Kingdom (2009)
22. Linn, M.C., Gerard, L., Ryoo, K., McElhaney, K., Liu, O.L., Rafferty, A.N.: Computer-guided inquiry to improve science learning. Science **344**(6180), 155–156 (2014)
23. Liu, T., Ding, W., Wang, Z., Tang, J., Huang, G.Y., Liu, Z.: Automatic short answer grading via multiway attention networks. In: Isotani, S., Millán, E., Ogan, A., Hastings, P., McLaren, B., Luckin, R. (eds.) AIED 2019. LNCS (LNAI), vol. 11626, pp. 169–173. Springer, Cham (2019). https://doi.org/10.1007/978-3-030-23207-8_32
24. Madnani, N., Burstein, J., Sabatini, J., O'Reilly, T.: Automated scoring of summary-writing tasks designed to measure reading comprehension. Grantee Submission (2013)
25. Mohler, M., Mihalcea, R.: Text-to-text semantic similarity for automatic short answer grading. In: Proceedings of the 12th Conference of the European Chapter of the ACL (EACL 2009), pp. 567–575 (2009)
26. Qi, H., Wang, Y., Dai, J., Li, J., Di, X.: Attention-based hybrid model for automatic short answer scoring. In: Song, H., Jiang, D. (eds.) SIMUtools 2019. LNICST, vol. 295, pp. 385–394. Springer, Cham (2019). https://doi.org/10.1007/978-3-030-32216-8_37
27. Rajagede, R.A., Hastuti, R.P.: Stacking neural network models for automatic short answer scoring. IOP Conf. Ser. Mater. Sci. Eng. **1077**(1), 012013 (2021). https://doi.org/10.1088/1757-899x/1077/1/012013
28. Sung, C., Dhamecha, T., Saha, S., Ma, T., Reddy, V., Arora, R.: Pre-training BERT on domain resources for short answer grading. In: Proceedings of the 2019 Conference on Empirical Methods in Natural Language Processing and the 9th International Joint Conference on Natural Language Processing (EMNLP-IJCNLP), pp. 6071–6075 (2019)
29. Sung, C., Dhamecha, T.I., Mukhi, N.: Improving short answer grading using transformer-based pre-training. In: Isotani, S., Millán, E., Ogan, A., Hastings, P., McLaren, B., Luckin, R. (eds.) AIED 2019. LNCS (LNAI), vol. 11625, pp. 469–481. Springer, Cham (2019). https://doi.org/10.1007/978-3-030-23204-7_39
30. Tulu, C.N., Ozkaya, O., Orhan, U.: Automatic short answer grading with SemSpace sense vectors and MaLSTM. IEEE Access **9**, 19270–19280 (2021)
31. Wainer, H.: Measurement problems. J. Educ. Meas. **30**(1), 1–21 (1993)
32. Waltman, K., Kahn, A., Koency, G.: Alternative approaches to scoring: the effects of using different scoring methods on the validity of scores from a performance assessment. Center for the Study of Evaluation, National Center for Research on Evaluation, Standards, and Student Testing (1998)
33. Wang, T., Funayama, H., Ouchi, H., Inui, K.: Data augmentation by rubrics for short answer grading. J. Nat. Lang. Process. **28**(1), 183–205 (2021). https://doi.org/10.5715/jnlp.28.183

Educational Equity Through Combined Human-AI Personalization: A Propensity Matching Evaluation

Danielle R. Chine[1(✉)], Cassandra Brentley[2], Carmen Thomas-Browne[2], J. Elizabeth Richey[2], Abdulmenaf Gul[1], Paulo F. Carvalho[1], Lee Branstetter[1(✉)], and Kenneth R. Koedinger[1(✉)]

[1] Carnegie Mellon University, Pittsburgh, PA 15213, USA
{dchine,pcarvalh,branstet,kk1u}@andrew.cmu.edu,
menafgul@gmail.com

[2] University of Pittsburgh, Pittsburgh, PA 15260, USA
{cassandrabrentley,cgt9}@pitt.edu, jelizabethrichey@gmail.com

Abstract. Recent developments in combined human-computer tutoring systems show promise in narrowing math achievement gaps among marginalized students. We present an evaluation of the use of the Personalized Learning^2, a hybrid tutoring approach whereby human mentoring *and* AI tutoring are combined to personalize learning with respect to students' motivational and cognitive needs. The approach assumes achievement gaps emerge from differences in learning opportunities and seeks to increase such opportunities for marginalized students through after-school programs, such as the Ready to Learn program. This program engaged diverse middle school students from three schools in an urban district. We compared achievement growth of 70 treatment students in this program with a control group of 380 students from the same district selected by propensity matching to have similar demographics and prior achievement. Based on standardized math assessments (NWEA Measures of Academic Progress) given one year apart, we found the gain of treatment students (6.8 points) was nearly double the gain of the control group (3.6 points). Further supporting the inference that greater learning was caused by the math-focused treatment and not by some selection bias, we found no significant differences in reading achievement between treatment and control participants. These results show promise that greater educational equity can be achieved at reasonable costs through after-school programs that combine the use of low-cost paraprofessional mentors and computer-based tutoring.

Keywords: Personalized learning · Cognitive tutoring · Design-based research

1 Introduction

The impact of combined human mentoring and AI-driven computer-based tutoring on student performance is encouraging, with an expanding stream of research showing promise in improving learning gains, especially in mathematics [8, 24]. AIED technologies involving human-computer teaming can lower the financial cost of personalized

M. M. Rodrigo et al. (Eds.): AIED 2022, LNCS 13355, pp. 366–377, 2022.
https://doi.org/10.1007/978-3-031-11644-5_30

tutoring and increase student learning [3, 19, 24]. The human mentors in these teams generally require less professional and on-the-job training than classroom teachers [8], which keeps human resource costs low. However, these mentors need additional support in providing personalized resources and skills development to assist with specific student's needs. The use of human-computer teaming, particularly in the wake of the COVID-19 pandemic, gives mentors access to individualized resources using AI-software based on students' existing math learning software and mentor input and feedback. We present results (i.e., EdTech usage, math and reading learning gains and outcomes) from the deployment of an after-school learning support system that integrates human mentors and AI tutoring (e.g., [17, 23]), with the aim of substantially reducing income and racial gaps in learning opportunities and outcomes. The Personalized Learning (PL^2) approach intends to maximize the synergies between the motivational capability of human mentors and the ability of computer-aided learning systems to provide low-cost personalized learning in pursuit of more equitable educational outcomes.

1.1 Related Work

Narrowing the Opportunity Gap. Marginalized students lack the means to access quality instructional services and experience lesser opportunities for learning [24] creating an opportunity gap. We define marginalized learners as, "students systematically denied equitable access to the same opportunities theoretically available to all students (p. 216)" often due to socioeconomic status, disability, or racial minoritization among other factors [12]. Racial and economic learning gaps are preventing millions of American students from realizing their potential which perpetuates inequalities of income and opportunity across generations [2]. Recently, the COVID-19 pandemic has exacerbated these inequalities with lower student achievement at the start of the 2021–22 school year (9 to 11 percentile points on standardized achievement assessments) than previous years hitting marginalized groups the hardest—minority students experiencing high-poverty [15]. Although achievement was lower across all groups, the achievement gap is present now more than ever with higher achieving and non-marginalized students making gains consistent with projected normative growth and marginalized, often under-achieving, students lagging behind further exacerbating the learning gap [15]. While these are recent and long-standing problems, researchers have struggled to identify effective solutions. Recent research undertaken in the Chicago Public Schools in some of the city's highest-poverty neighborhoods, provides new grounds for hope [3, 7, 11]. Using a randomized control trial consisting of 2,700 students of whom 95% were Black or Hispanic, they demonstrate that just one year of intensive, personalized tutoring can narrow racial achievement gaps in mathematics by as much as one third. These gains come at a substantial cost. With one tutor providing instruction to just two students per class period, the cost exceeds the threshold of political feasibility in many districts, despite its proven efficacy.

Offering Low-Cost Tutoring. Advances in computer-aided learning provide a method of substantially lowering the cost of personalized tutoring, while maintaining the magnitude of the learning gains. Research on AI-driven computer-based tutoring has shown computer tutors can substantially accelerate student learning, especially in mathematics. In one recent large-scale randomized control trial, this technology was shown to double the rate of math learning [21]. Setren [25] showed that the use of another commercially available tool (eSparks), has a positive effect on learning gains for all students and can contribute to reducing inequality. Similarly, Muralidharan et al. [19] showed that a personalized technology-aided after-school program was successful in generating large learning gains among under-achieving students in a developing country. Despite positive findings, many students do not partake in the practice opportunities provided. We propose human mentoring to help motivate students to participate and to round out their learning experiences.

Supporting the Whole Child. While computer tutors can often provide effective support for student thinking and learning, these systems do not provide human support for social motivational development such as self-efficacy building [26], feelings of belonging [31], growth mindset [32], and valuing utility of STEM [10]. Using the last as an example, motivational support for students and parents to better appreciate the value of STEM learning produced about 50% greater achievement and future course enrollment, especially for low-performing and underserved students. Our proposed intervention supports the whole child similar to Guryan et al. [7] in attending to the social-motivational needs and relationship building which is particularly important in middle school years. Milner [18] posits that to foster excellence, a learning environment should center on building and cultivating relationships with students. The synergy of human and computers has been studied in a similar fashion with the use of trialogues (the interaction of two agents with a human student) to address pedagogical goals and student's emotional state [6] and the use of tutorial dialogue agents to increase learning gains [14]. Similarly, peer-to-peer interaction within intelligent tutoring systems to scaffold learning has been researched via adaptive collaborative learning supports for both improving content learning and collaboration [30].

2 The Personalized Learning^2 Approach

Introduced in Schaldenbrand et al. [24], PL^2 is a learning app that syncs with students' existing math learning software and mentor input and feedback to improve students' math achievement.[1] This paper presents an evaluation of the general PL^2 approach, as both a learning app and tutoring method, to human-computer teaming for motivational mentoring and content tutoring. By combining research-driven mentor training with AI-powered software, the PL^2 approach improves mentoring efficiency by connecting mentors to personalized resources with the click of a button. This connection is achieved by a web app used by mentors and mentor supervisors. The PL^2 approach serves out-of-school tutoring programs, which choose a computer-based math tutoring system for students to use. The data from student interactions is passed to the PL^2 web app to power mentor decisions. Mentors make post-session reflections based on reports of student effort and progress. Mentors work with students to set or modify intermediate

[1] http://personalizedlearning2.org/.

effort goals, much like the 10,000 step goals in physical fitness apps, such as doing 40 min of math practice a week. When students are missing effort or progress goals, the PL^2 app provides suggestions for resources that the mentor can use themselves or with students to enhance student motivation, cognition, or metacognition.

PL^2 has been integrated with several math EdTech systems [24]. The two used in the evaluation we report on were MATHia and ALEKS. MATHia (formerly Cognitive Tutor) uses a cognitive model of student problem solving to implement the model tracing and knowledge tracing algorithms for personalized tutoring [23]. It has been demonstrated to improve student learning in large-scale randomized field trials (e.g., [21]). ALEKS is an intelligent tutoring system based on knowledge space theory and it too has been demonstrated to improve student learning [17].

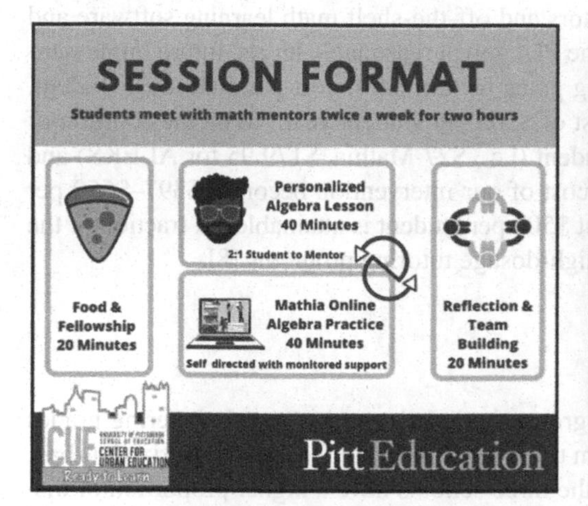

Fig. 1. The Ready to Learn program 2-h in-person format commences with fellowship, followed by rotations between AI and in-person tutoring (40 min. each), and ends with reflection and team building. Students shown working with an in-person mentor and AI tutors.

2.1 Description of the Program

The Ready to Learn (RtL) program is offered by the Center for Urban Education at the University of Pittsburgh (Pitt). The program is a combination tutoring-mentoring initiative that connects students from Pitt and Carnegie Mellon University (CMU) with middle school students in select urban public schools to provide mentoring and math tutoring in an out-of-school context. The overarching goal of the RtL program is to provide students, especially students from disadvantaged contexts (i.e., living in poverty, experiencing racial bias, being part of a marginalized group) with experiences to support their academic improvement in mathematics. Since 2019, CMU researchers have partnered with Pitt's Center for Urban Education to provide personalized math mentoring. Small scale pilot versions of RtL in the spring and summer of 2019 paved the way for full implementation in the 2019–2020 academic year and following summer. RtL combines small group lessons, individual mentoring with trained CMU and Pitt undergraduate students, student engagement with adaptive AI-driven learning software, and use of the Personalized Learning[2] app to help mentors work together with technology to customize the learning experience for each student. The in-person session format (see Fig. 1) consisted of both a human personalized math lesson and personalized computer-based tutoring provided by the MATHia intelligent tutoring system. Because of the COVID pandemic, the program moved online in the summer of 2020. At the same time for logistical reasons, the computer-based tutoring was changed to ALEKS. RtL builds student math confidence and competence at no cost to students or to partnering schools. In the evaluation we describe below we included students that participated in an RtL program between the available assessments, that is, during the spring and summer of 2020.

By relying on undergraduate mentors and off-the-shelf math learning software and keeping the price of subscriptions to the PL[2] app at reasonable levels, future implementations may be able to deliver learning gains for a modest cost per additional student. Our calculations suggest a mentor cost of $360 per student/year.[2] With the addition of an annual EdTech license cost per student (i.e., $27 Mathia, $179.95 for ALEKS) and an annual PL[2] student fee ($10), the cost of our intervention becomes $397–$550 per student. Thus, a marginal cost of about $500 per student is attainable—a fraction of the $3,500–$4,300 per student for other high-dosage tutoring programs [8].

3 Method

Participants in treatment and control groups included students mostly entering grades 6–7 at the 2019–2020 school year from three schools located in a medium-sized, urban, Pennsylvania school district. Two of the three schools have a higher proportion of disadvantaged students compared to the district aligning with the goal of RtL of reaching marginalized groups. The majority of students were in 6th (57%) and 7th grade (33%) grade. Students' demographics are summarized in Table 1. Among the treatment, approximately 74% were Black with an approximately equal gender distribution (48% female). Most of the participants were eligible for free or reduced lunch (71%) and 20% were receiving an Individualized Education Program (IEP), which is a special education service.

Students' achievement was measured by the NWEA Measures of Academic Progress (MAP) assessment which the district administered a few times per year assessing students' math and reading achievement. In our evaluation, MAP scores for fall and winter of 2019–20 were used as pretest scores and MAP scores for fall and winter of 2020–21 were used as posttest scores. We used all four test scores to maximize the number of students for which at least one pre and one post score was available (see missing data discussion below). We note concerns that majority-based norming of standardized tests can create cultural bias and may exaggerate achievement gaps [13]. At the same time,

[2] Estimated mentor cost: $15/hr.*2 h./session*2 sessions/week*24 weeks/4 students per mentor = $360 per student/year.

we note efforts to reduce bias in standardized testing in general [22] and that the questions on this test are representative of important learning goals for students (e.g., using rational numbers to solve real-world problems).

Table 1. Demographic group percent distribution (and number) demonstrates about 3/4 of participants belong to marginalized groups (i.e., Black and low SES)

Group	Demographic	Treatment (n)	Control
Gender	Female	48% (34)	52% (199)
Race	Black	74% (52)	81% (306)
	White	20% (14)	11% (43)
	Hispanic	3% (2)	5% (19)
Free/Reduced lunch	Yes	71% (50)	79% (301)
IEP status	IEP	20% (14)	17% (64)

Control Group Creation via Propensity Matching. Toward our goal to evaluate whether extra learning opportunities provided by the PL^2 approach enhanced student learning, we created a matched control group of similar students who did not receive these opportunities. The district provided anonymized score and demographic data from a total of 20,628 students across all grades for academic years 2019–20 and 2020–21. This data provided scores and demographics for the 72 students that participated in the PL^2 treatment. These treatment demographics and pre-test scores were used as input into a rigorous propensity matching process to select a set of students as a control who were as similar as possible in demographics, grade level, and pre-test scores. An optimal full matching method [9, 27] was used to match each treatment student with multiple matching control students. Initially, all demographic factors were used to match students. However, gender and socioeconomic status (determined by free and reduced lunch designation) were found to be non-significant factors to balance groups and were removed from matching criteria. In the final matching, 70 students out of 72 in the treatment group were matched with 380 control students based on grade level, race, IEP (Individualized Education Program) status, and pretest math scores. The two students dropped from the treatment had a combination of these features for which there was no adequate match. Grade level was defined as an exact matching factor and a clipper value was defined for pretest math scores to ensure that matched units are close enough in terms of pretest math scores. In addition to propensity score matching, manual matching using exact matching of grade level, gender, race, socioeconomic status, IEP group, and math score within a 3-point range was used. Manual matching replicated all significant differences reported below with similar magnitudes.

Pandemic, Missing Data, and Imputation. Especially because of the pandemic, some participants completed only one of the two pre or posttests, mostly in fall 2020. We started with a total of 13,554 students in the district with at least one pretest and posttest

test score. The test with the highest missing rate was fall 2020 (14.9% control, 15.3% treatment). Missing data for all other tests ranged between 0% and 8.3%. Missing data were imputed using a single deterministic imputation model based on linear regression using the MICE package in R [29]. In this method, all demographic factors were entered as predictors and plausible synthetic values were generated for incomplete test scores. Incomplete math and reading scores were imputed separately. In addition to the single imputation method, multiple imputation based on stochastic regression with random error added to predicted values was tested. This method was tested with 20 samples. The results of both single and multiple imputation were identical with respect to all statistical threshold judgements. Since multiple imputation produces different matched samples, we present subsequent results using the single matched sample resulting from the single imputation method.

4 Results

Math Learning Gains. Figure 2a summarizes pre-to-post learning gains for the treatment group (rightmost bar) and a matched demographic control (leftmost bar) as well as two other reference points, a national average gain and a non-matched grade level control. On average, students in the treatment group grew 6.8 Rasch Interval Unit (RIT) points from pre to posttest, compared to 3.6 points for students in the matched control. NWEA MAP reports a typical one-year average growth as 5.5 RIT points [20].

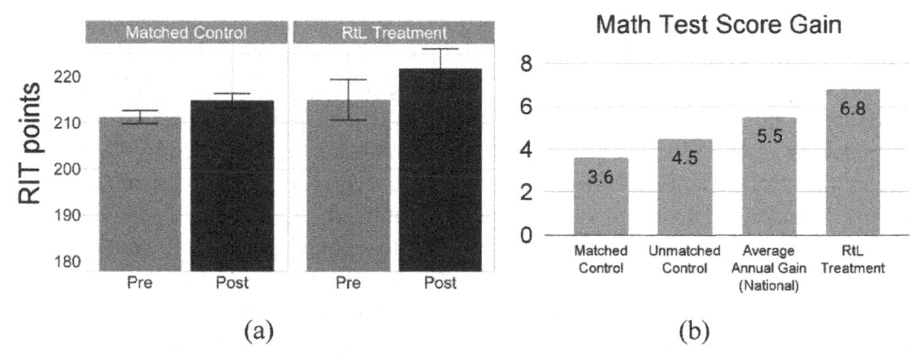

Fig. 2. a) Math test score gain comparison. b) Mean pre and posttest results. Both (a) and (b) illustrate the substantial gain for RtL participants compared to the matched control.

Further contextualizing this difference, we calculated the average growth in the same period for a non-matched control group, which included all students in the same grade without accounting for demographics. The non-matched control group showed a growth of 4.5 points in the same period, higher than the matched control, suggesting that the RtL program met its goal of working with disadvantaged students. In addition, the results may indicate evidence of the "COVID slide" among the control group given that the 3.6-point gain and grade level, "non-demographic" group gain of 4.5 points, are significantly lower than the average MAP expected annual growth. This aligns with the pandemic-related lag in mean growth evidenced by Lewis et al. [16] with RIT scores ranging approximately 2–3 points lower than pre-pandemic years dependent upon grade level and test administration. Past performance data of the participating schools in comparison with national growth gains are needed to confirm evidence of "COVID slide." The most striking finding was the substantial growth differences among RtL participants with students performing significantly over the normative one year's growth.

Figure 2b provides average pre and posttest results for the matched control and RtL treatment. To assess whether the RtL treatment had a statistically greater pre-posttest gain than the matched control, we performed a repeated measures regression analysis. The model predicted MAP scores using group (experimental vs. control), test-time (pre vs. post) and the interaction between the two variables. As can be seen in Table 2, all students performed better in the posttest tests than the pretest tests, $\beta = 3.60$, $t\,(448) = 8.30$, $p < .0001$. Importantly, the interaction between type of test and group was also statistically significant, $\beta = 3.21$, $t\,(448) = 2.92$, $p = .0004$. MAP scores for students in the treatment group improved considerably more than for students in the control group (effect size of $d = 0.40$). An analysis of density plots of pre and posttest performance suggest the treatment raised student posttest performance across nearly the entire range of student pretest performance levels.

Table 2. ANOVA of math score differences between treatment and control groups

Variable	Estimate	*SE*	*df*	*t*	*p*
(Intercept)	211.16	0.79	520.30	267.17	**0.000*** **
Treatment	3.82	2.00	520.30	1.90	0.057
TestTime = PostTest	3.60	0.43	448.00	8.30	**0.000*** **
Treatment × TestTime	3.21	1.10	448.00	2.92	**0.004** **

*** $p < 0.001$, ** $p < 0.01$, * $p < 0.05$

Reading Learning Gains. To ensure the observed differences in math performance among treatment and control participants were not driven by selection bias or an overall "mentoring effect," similar analysis was conducted using nonequivalent-groups design comparing students' reading test score gains [28]. Unlike what we saw for math MAPs scores, on average students in both the control and treatment group showed similar growth in reading over the span of a year (2.1 RIT score points for the control group and 2.7 RIT score points for the treatment group). As we saw before, this growth is below the national average (4.5 RIT points).

Using statistical analyses similar to that used with math scores, we saw an overall positive growth in reading scores from pre to posttest, $\beta = 2.05$, $t\,(417) = 4.20$, $p < .0001$, but no overall effect of treatment, $\beta = 1.43$, $t\,(492.02) = 0.67$, $p = .051$, nor evidence of an interaction between the treatment group and time of the test, $\beta = 0.61$, $t\,(417) = 0.50$, $p = .619$. Overall, these findings suggest that there is no overall "mentoring effect", or improvement in student performance due to the noncognitive effects of mentoring (i.e., social-motivational supports, relationship building) that extends beyond the targeted math learning gains. In addition, the results support the exclusion of a selection effect or the possibility that the RtL treatment students are generally better, more motivated students. The differences between control and treatment groups on math MAPs scores are likely due to the RtL treatment.

EdTech Usage and Outcomes. We investigated the role of the computer tutoring element of the PL^2 approach to combined human-computer mentorship by analyzing the relationship between students' EdTech usage during the program and their MAP scores. We combined data from both EdTech sources and used a measure common to both: total time in the program. Fifty-four students used MATHia and 16 used ALEKS. On average, students spent a median 102 min in the educational technology system during the RtL program ($M = 150.70$, $SD = 201.00$). We used a multiple regression model predicting student pre-post change, using pretest, EdTech usage (mins) and their interaction, as well as the type of EdTech used and its interaction with EdTech usage as predictors. The type of EdTech used did not have a significant relationship with score growth, $\beta = 3.16$, t $(64) = 0.72$, $p = .48$, and did not vary depending on amount of EdTech usage, $\beta = 0.20$, t $(64) = -1.10$, $p = .28$. Pretest scores also did not have a significant relationship with score growth either, $\beta = 0.03$, t $(64) = -0.32$, $p = .75$. Importantly, there was a positive relationship between minutes spent on EdTech and score growth,, $\beta = 0.18$, t $(64) = 2.22$, $p = .03$, indicating that spending more time on EdTech during the RtL program was associated with higher learning growth. Moreover, the relationship between EdTech usage and learning growth was moderated by pretest scores, $\beta = -0.001$, t $(64) = -2.10$, $p = .04$.

5 Discussion and Conclusion

The combined human-computer personalized approach of PL^2 is based on the following hypotheses, which we posit as explanations for the demonstrated enhanced learning gains of the RtL participants in this study. Many marginalized students are not given sufficient learning opportunities [24]; thus, they do not get the deliberate practice they need to achieve success [5]. Educational technologies can provide such deliberate practice, which is one piece of PL^2, but only if the technology is used. Thus, the second piece of PL^2 is the notion that human mentors provide needed social-motivational support that help students engage in rigorous deliberate practice. Afterall, deliberate practice is motivationally challenging [4]. PL^2 helps human mentors not only personalize their math content tutoring, but also personalize motivational support. It gives mentors strategies for relationship building, which is foundational to student learning outcomes [18] and helps them personalize whether a student's effort could be enhanced by one or another motivational intervention, including growth mindset [32], valuing utility value of STEM [10], and self-efficacy building [26].

There is limited flexibility in schools to add extra learning opportunities and there is good evidence that out-of-school learning opportunities are a major source of opportunity gaps. For example, evidence of "summer slide" indicates that racial achievement gaps widen over the summer and implicate greater learning opportunities for privileged students than for marginalized students [1]. Recently, the "COVID slide" has exacerbated such opportunity gaps among marginalized students, particularly in math [15]. No amount of improvement during the school day will address this opportunity gap.

Given our goal to increase student opportunities beyond those available schools, we did not seek or create a control group that was matched for opportunities. Some students in the matched control may have attended other out-of-school programs. To be sure, out-of-school opportunities intended to support academic learning do not necessarily do so. Our results demonstrate that our out-of-school activities, which mix human and computer tutoring, do enhance student learning and quite substantially.

In addition, we hypothesize one of the reasons for the positive academic impact evidenced from the PL^2 approach comes from the intentionally designed culturally relevant training and tutoring practices within the RtL program format. Guryan et al. [8] reports similar success of "high impact" tutoring, however, occurring during the school day taking away from instructional time for other content. In two randomized control trials (RCTs), Guryan et al. [8] states increased math scores of 0.16 and 0.37 standard deviations, respectively, with evidence of impacts existing over time. Our results indicate a larger effect size ($d = 0.40$), however without the rigor that comes with RCTs. Guryan et al. [8] reports a cost per participant per year of $3,500 to $4,300. Our research team's calculations suggest that marginal costs on the order of $500 per student appear feasible within a few years, perhaps yielding stronger academic impact without sacrificing valued school time.

While RCTs are the gold standard experimental research method, they are not always practical. Quasi-experimental methods are especially valuable early in a project lifecycle to determine whether the costs of a full RCT are justified. We illustrate cost-effective use of two quasi-experimental methods, propensity score matching [9, 27] and a nonequivalent dependent variables (NEDV) design [28]. Propensity score matching removes the costs of random assignment and can be straightforwardly employed when school partners can provide student-level demographic data. Similarly, school partnership can make employing the NEDV design simple when the school can provide two kinds of test results: one test that is aligned with the content of your instructional intervention (a math test in our case) and one test that is not (a reading test in our case).

While we presented evidence for the benefits of EdTech learning opportunities, we were not able to similarly investigate the role of the mentor. We hope to explore whether students learn more if they have mentors that provide more learning opportunities or that use PL^2 more often or more effectively. We are also interested in determining the role mentor and mentee matching based on demographics and socioeconomics plays in learning gains. Further research assessing the impact of the online Ready to Learn (RtL) program will be analyzed to determine the impact of the PL^2 system in a virtual environment with hope of increasing scalability. A fully online version was developed and implemented during the 2020–2021 school year with a RtL virtual session format consisting of virtual personalized instruction in conjunction with student self-directed use of ALEKS in a 4:1 student to mentor ratio.

Our results demonstrate progress in human-computer teaming in mentoring and tutoring providing needed out-of-school learning opportunities and producing substantial and significant learning. More generally, this work supports the idea that greater educational equity can be achieved at reasonable costs by supporting after-school programs with technology that improves mentoring and student learning.

Acknowledgements. This work is supported with funding from the Chan Zuckerberg Initiative (Grant #2018-193694), Richard King Mellon Foundation (Grant #10851), Bill and Melinda Gates Foundation, and the Heinz Endowments (E6291). Any opinions and conclusions expressed in this material are those of the authors

References

1. Alexander, K., Pitcock, S., Boulay, M.C. (eds.): The Summer Slide: What We Know and Can Do About Summer Learning Loss. Teachers College Press (2016)
2. Autor, D.: Skills, education, and the rise of earnings inequality among the "other 99 percent." Science **344**(6186), 843–851 (2014)
3. Cook, P.: Not too late: improving academic outcomes for disadvantaged youth. Northwestern University Institute for Policy Research. Working Paper No. 15-01 (2015)
4. Duckworth, A.L., Kirby, T.A., Tsukayama, E., Berstein, H., Ericsson, K.A.: Deliberate practice spells success: why grittier competitors triumph at the National Spelling Bee. Soc. Psychol. Pers. Sci. **2**(2), 174–181 (2011)
5. Ericsson, K.A., Krampe, R.T., Tesch-Römer, C.: The role of deliberate practice in the acquisition of expert performance. Psychol. Rev. **100**(3), 363 (1993)
6. Graesser, A.C., Forsyth, C.M., Lehman, B.A.: Two heads may be better than one: learning from computer agents in conversational trialogues. Teachers Coll. Board **119**(3), 1–20 (2017)
7. Guryan, J.: The effect of mentoring on school attendance and academic outcomes: a randomized evaluation of the check & connect program. Institute for Policy Research Working Paper Series, WP-16-18. Northwestern University (2017)
8. Guryan, J.: Not too late: improving academic outcomes among adolescents (No. w28531). National Bureau of Economic Research (2021)
9. Hansen, B.B.: Full matching in an observational study of coaching for the SAT. J. Am. Stat. Assoc. **99**(1), 609–619 (2004)
10. Harackiewicz, J., Rozek, C., Hulleman, C., Hyde, J.: Helping parents to motivate adolescents in mathematics and science: an experimental test of a utility-value intervention. Psychol. Sci. **23**(8), 899–906 (2012)
11. Heller, S.B., Shah, A.K., Guryan, J., Ludwig, J., Mullainathan, S., Pollack, H.A.: Thinking, fast and slow? Some field experiments to reduce crime and dropout in Chicago. Q. J. Econ. **132**(1), 1–54 (2017)
12. Hutson, K.M.: Missing faces: making the case for equitable student representation in advanced middle school courses. In: Promoting Positive Learning Experiences in Middle School Education, pp. 200–216. IGI Global (2021)
13. Kim, K.H., Zabelina, D.: Cultural bias in assessment: can creativity assessment help? Int. J. Crit. Pedagogy **6**(2), 129–148 (2015)
14. Kumar, R., Rosé, C.P., Wang, Y.C., Joshi, M., Robinson, A.: Tutorial dialogue as adaptive collaborative learning support. Front. Artif. Intell. Appl. **158**, 383 (2007)
15. Lewis, K., Kuhfeld, M.: Learning during COVID-19: an update on student achievement and growth at the start of the 2021–22 school year. NWEA (2021)
16. Lewis, K., Kuhfeld, M., Ruzek, E., McEachin, A.: Learning during COVID-19: reading and math achievement in the 2020–21 school year. NWEA (2021)

17. Matayoshi, J., Uzun, H., Cosyn, E.: Studying retrieval practice in an intelligent tutoring system. In: Proceedings of the Seventh ACM Conference on Learning@ Scale, pp. 51–62. Springer, August 2020
18. Milner, H.R.: Start Where You Are, But Don't Stay There: Understanding Diversity, Opportunity Gaps, and Teaching in Today's Classrooms, 2nd edn. Harvard Education Press, Cambridge (2020)
19. Muralidharan, K., Singh, A., Ganimian, A.J.: Disrupting education? Experimental evidence on technology-aided instruction in India. Am. Econ. Rev. **109**(4), 1426–1460 (2019)
20. NWEA. 2020 MAP growth normative data overview, March 2020. https://teach.mapnwea.org/impl/MAPGrowthNormativeDataOverview.pdf
21. Pane, J.F., Griffin, B.A., McCaffrey, D.F., Karam, R.: Effectiveness of cognitive tutor algebra I at scale. Educ. Eval. Policy Anal. **36**(2), 127–144 (2014)
22. Popham, W.J.: Assessment Bias: How to Banish It. Routledge, New York (2006)
23. Ritter, S., Anderson, J.R., Koedinger, K.R., Corbett, A.: Cognitive tutor: applied research in mathematics education. Psychon. Bull. Rev. **14**(2), 249–255 (2007). https://doi.org/10.3758/BF03194060
24. Schaldenbrand, P.: Computer-supported human mentoring for personalized and equitable math learning. In: Roll, I., McNamara, D., Sosnovsky, S., Luckin, R., Dimitrova, V. (eds.) International Conference on Artificial Intelligence in Education. AIED 2021. LNCS, vol. 12749, pp. 308–313. Springer, Cham (2021). https://doi.org/10.1007/978-3-030-78270-2_55
25. Setren, E.M.: Essays on the economics of education. (Unpublished doctoral dissertation). Massachusetts Institute of Technology (2017)
26. Siegle, D., McCoach, D.B.: Increasing student mathematics self-efficacy through teacher training. J. Adv. Acad. **18**(2), 278–312 (2007)
27. Stuart, E.A., Green, K.M.: Using full matching to estimate causal effects in nonexperimental studies: examining the relationship between adolescent marijuana use and adult outcomes. Dev. Psychol. **44**(2), 395–406 (2008). https://doi.org/10.1037/0012-1649.44.2.395
28. Trochim, W.M., Donnelly, J.P.: The Research Methods Knowledge Base, 3rd edn. Cengage, Ohio (2007)
29. Van Buuren, S., Groothuis-Oudshoorn, K.: MICE: multivariate imputation by chained equations in R. J. Stat. Softw. **45**, 1–67 (2011)
30. Walker, E.: Automated adaptive support for peer tutoring. Doctoral dissertation, Carnegie Mellon University (2011)
31. Walton, G.M., Cohen, G.L.: A brief social-belonging intervention improves academic and health outcomes of minority students. Science **331**(6023), 1447–1451 (2011)
32. Yeager, D.S., Dweck, C.S.: Mindsets that promote resilience: when students believe that personal characteristics can be developed. Educ. Psychol. **47**(4), 302–314 (2012)

Eye to Eye: Gaze Patterns Predict Remote Collaborative Problem Solving Behaviors in Triads

Angelina Abitino(✉), Samuel L. Pugh, Candace E. Peacock, and Sidney K. D'Mello

University of Colorado Boulder, Boulder, CO 80309, USA
{angelina.abitino,samuel.pugh,sidney.dmello}@colorado.edu

Abstract. We investigated the feasibility of using eye gaze to model collaborative problem solving (CPS) behaviors in 96 triads (N = 288) who used videoconferencing to remotely collaborate on an educational physics game. Trained human raters coded spoken utterances based on a theoretically grounded framework consisting of three CPS facets: constructing shared knowledge, negotiation/coordination, and maintaining team function. We then trained random forest classifiers to identify each of the CPS facets using eye gaze features pertaining to each individual (e.g., number of fixations) and/or shared across individuals (e.g., eye gaze distance between collaborators) in conjunction with information about the unfolding task context. We found that the individual gaze features outperformed the shared features, and together yielded between 6% to 8% improvements in classification accuracy above task-context baseline models, using a cross-validation scheme that generalized across teams. We discuss how our findings support CPS theories and the development of real-time intervention systems that provide actionable feedback to improve collaboration.

Keywords: Collaborative problem solving · Eye tracking · Machine learning

1 Introduction

Collaborative problem solving (CPS) occurs when two or more people work together to construct and execute a solution to a problem [1]. CPS is a crucial 21st century skill that students need to develop in order to contribute to a successful workforce and to address critical issues such as global pandemics, climate change, and childhood poverty [2]. At the same time, students need to adapt to new ways of collaborating through remote working and learning [1, 3]. In particular, since the onset of the COVID-19 pandemic, teachers have had to migrate their classrooms to online platforms (e.g., Zoom) where students might work together in smaller groups of "breakout rooms" where a teacher is not present to help students and instead, students must rely on available resources and their own collaboration skills to succeed. Whereas many classrooms have since returned to in-person learning, the workplace of the future may have been changed forever with remote work and especially remote teamwork expected to be here to stay

M. M. Rodrigo et al. (Eds.): AIED 2022, LNCS 13355, pp. 378–389, 2022.
https://doi.org/10.1007/978-3-031-11644-5_31

[4]. The pertinent question is whether students will have the requisite collaboration skills to succeed in a workforce increasingly driven by remote teamwork.

Unfortunately, this may not be the case. A 2015 study conducted by the OECD Programme for International Student Assessment (PISA) found significant deficiencies in students' CPS skills, with less than 10% of students exhibiting the highest level of proficiency [1]. The learned ability to problem solve with others begins with students practicing in the classroom, but it is not an easy task for teachers to measure and scaffold students' CPS skills, especially since the measurement of CPS itself is a nascent endeavor [2]. Can technology help? We think indeed this is the case. One way that technology can support student collaboration is via intelligent *team* tutoring systems (ITTSs) where the computer tutor interacts with two or more learners working on a shared task with the same or different roles [5]. Another is for the computer to play the role of a guide by monitoring and encouraging particular conversation moves, called academic productive talk, during collaborative discussions [6]. Rather than intervene in real-time, yet another possibility is for the computer to monitor CPS as it unfolds and provide after-action feedback on improving those skills.

A key component that underlies these technologies is the ability to measure CPS skills, which may entail analyses of multimodal data characterizing collaborators' interactions [7]. As reviewed next, researchers have investigated a variety of modalities for this task. The present work focuses on eye gaze as a novel (i.e., relatively unexplored) modality to automatically measure collaboration skills. As we elaborate below, eye gaze provides an index into social visual attention [8], as well as individuals' cognitive states [9], making it a particularly promising modality for modeling CPS, which entails both social (collaboration) and cognitive (problem solving) processes.

1.1 Background and Related Work

Collaboration is a social endeavor, and verbal communication is one of the key products of CPS [10]. Accordingly, researchers have successfully used linguistic information to identify CPS skills such as negotiation [10, 11], information sharing [10, 11], and task regulation [12]. Verbal communication has also been used to predict CPS outcomes such as task performance [13] and learning gains [14]. Whereas previous work used conversation gathered from typed chat transcripts [10] or human transcription of audio recordings [14], recent studies have shown that transcripts generated via automatic speech recognition (ASR) can also be used to model CPS skills [11, 15]. However, the accuracy of such models is greatly reduced in real-world learning environments due to ASR errors (e.g., as much as 20% compared to human transcripts [11]).

In addition to these modalities, gaze information is thought to be a strong complementary signal to speech because it can provide insights into social visual attention, including mutual gaze (individuals look at each other), joint attention (individuals focus on the same subject), and gaze aversion (an individual looks at their partner, who is looking elsewhere) [16]. Accordingly, gaze has been used in some CPS studies, often focusing on dyadic interaction (e.g., dual eye-tracking). For instance, Olsen et al. [17] analyzed individual and combined gaze activity during dyadic CPS, and found that gaze predicted learning gains. Other studies have used eye gaze data (often focusing on measures of social visual attention such as joint attention) to predict task performance [18], learning gains [19], and CPS skills [20].

1.2 Current Work: Contributions and Novelty

We aim to contribute to recent work on modeling CPS behaviors and skills by investigating the potential of eye gaze, a relatively unexplored modality, for this task. Our first research question examined the accuracy of gaze-based machine learning models at predicting three CPS facets (constructing shared knowledge, negotiation/coordination, and maintaining team function) from a validated CPS framework [21]. A second question examined whether gaze features pertaining to the individual (e.g., number of fixations, fixation dispersion) or shared across individuals (e.g., eye gaze distance between collaborators) were better predictors of CPS facets. Next, based upon work that suggests it might be beneficial to differentiate signals coming from different teammates [22], we also examined whether it is favorable to model gaze based upon a given teammate's fixed role (i.e., controller, observer) or dynamically based on the ongoing conversation (i.e., speakers' vs. listeners' gaze). Finally, we examined which gaze features are most predictive of the CPS facets by analyzing our models' feature importances.

To our knowledge, the present study is the first attempt at a fully automated approach to detect CPS behaviors using eye gaze in a manner that generalizes across teams. Additionally, while some studies have examined eye gaze during dyadic CPS, we extend this work by investigating the role of gaze in triadic collaboration, which raises pertinent issues on incorporating gaze signals in multiparty (i.e., beyond dyads) CPS.

2 Method

2.1 Data Collection

The dataset was collected for a larger study on remote CPS [23]; only details pertinent to the present study are discussed here.

Participants. Participants (N = 288, average age = 22, 56% female) were students from two large public universities in the Western US (111 from School 1 and 177 from School 2). Participants were assigned to 96 triads based on scheduling constraints and were compensated with a $50 Amazon gift card (95.8%) or course credit (4.2%). To participate, students had to meet three inclusion criteria: (1) they spoke English, (2) they had no significant uncorrected vision impairments, and (3) they had no prior experience with the physics game.

CPS Task and Procedure. Participants first completed at-home tasks consisting of a Qualtrics survey to assess individual difference measures, followed by a tutorial to learn how to play Physics Playground (see Fig. 1) [22]. Next, there was an in-lab session where each participant was assigned to a computer-enabled workstation in separate rooms (School 1) or partitioned with dividers (School 2), and fitted with webcams, headsets, and Tobii 4C eye trackers (for which licenses to record data were purchased).

All interactions occurred over Zoom (https://zoom.us) with video and screen sharing. Gaze data was recorded at 90 Hz.

Teams of three were tasked with collaboratively playing Physics Playground (PP) [24], an educational computer game for learning Newtonian physics concepts (see Fig. 1). The goal is to guide a ball to a red balloon separated by obstacles by drawing objects on the screen (e.g., ramps, levers, pendulums, springboards, weights). Everything in the game obeys the laws of physics. Teams participated in four 15-min blocks, where the first three blocks involved the PP task, and the fourth block involved a task irrelevant to this study; therefore, we only used data from the first three blocks. For each block, a team member was randomly assigned to control the game environment, while the other two teammates were able to communicate their ideas through audio and video. The controller's screen was shared with all teammates and each teammate was the controller once during the three blocks. Participants could begin, restart, and exit levels at any time. No hints or support mechanisms were provided, except a tutorial on game mechanics which could be viewed at any time.

2.2 Human Coding of CPS Facets

Using the IBM Watson ASR, we obtained automated transcripts of each participant's utterances. These were then coded alongside videos of the CPS interaction (see Fig. 1) using a validated CPS framework [21] consisting of three CPS facets: (1) constructing shared knowledge, (2) negotiation/coordination, and (3) maintaining team function. These facets were coded from 14 verbal indicators, such as "shares specific solutions" (constructing shared knowledge), "responds to others' questions/ideas" (negotiation/coordination), and "asks for suggestions" (maintaining team function). Three trained raters coded a pseudorandom 90 s from the first, second, and third five minutes of each block, resulting in 30% of the utterances coded. Coder agreement on the indicators ranged from .88 to 1.00 (Gwet's AC1 metric [25]) on ten 90-s video samples consisting of 406 utterances. After achieving adequate reliability, videos were randomly assigned to the three coders for independent coding. There could be multiple indicators per utterance, but this rarely occurred (<3%). Therefore, consistent with previous work [15], we created binary labels for each facet by assigning a 1 if any of the indicators were present for the facet, else 0. We also included a fourth binary label, "no facet", when none of the indicators were present.

2.3 Data Processing and Feature Computing

Role- vs. Conversation-Based Identities. In the case of multiparty interactions, where data from multiple individuals are used for modeling, it is essential to assign meaningful identities to each individual [22]. Accordingly, we experimented with two forms of identification. For role-based identities, participants were identified as the controller and observers, the latter subdivided into the more- vs. less-verbose observer for that block. For conversation-based identities, participants were identified as the current speaker, previous speaker, and other teammate. Whereas the role-based identities were consistent within each block, the conversation-based identities varied across utterances.

Gaze Features. The raw eye gaze data was first fixation filtered using PyGaze Analyser [26]. Fixations were defined as points where gaze was maintained on a location (within 25 pixels) for at least 50 ms and trimmed at 1 s [22] because gaze features were computed based on the fixations in non-overlapping 1 s windows (see Fig. 2). We focused on a small set of eye gaze features that are ostensibly generalizable to other task environments as an alternative to features specific to PP (i.e., AOI features). We explored two sets of gaze features: individual and shared (see Fig. 1). The individual gaze features ($n = 6$ per individual) included partial gaze validity (proportion of samples in a given second where at least one eye was successfully tracked), fixation dispersion (mean Euclidean distance of each raw gaze point in a 1 s window from the centroid), pupil size (diameter in mm), mean saccade amplitude (mean pixel distance between two subsequent fixations), fixation count, and mean fixation duration. The shared gaze features ($n = 3$ per team) were the pairwise Euclidean distances between the centroids of each participant, with low values indicating joint attention [27, 28], where the participants are fixated on similar areas on the screen. All "individual" gaze features were z-scored within-participant to account for individual differences.

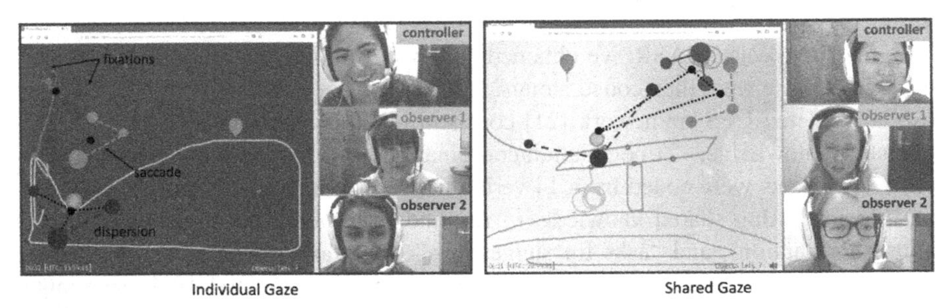

Fig. 1. Visualizations of individual (left) and shared (right) gaze features from gaze scan paths. The controller's gaze is red, one observer's gaze is green, and the other observer's gaze is blue. The colored circles represent fixations, and the circle size indicates the fixation duration. The black circles represent the centroid of each participant's gaze. (Color figure online)

Data Alignment. Because eye gaze features vary as a function of time (e.g., more fixations for longer utterances), we first segmented utterances into 1 s intervals, a time window of sufficient length to accommodate the short utterance durations (median of 1.4s). We then aligned the 1 s gaze feature vectors to the corresponding utterance and averaged the features within each utterance's time frame to produce one feature vector per utterance (Fig. 2). Fixation counts were converted to fixation rates by summing across each utterance and dividing by the utterance duration.

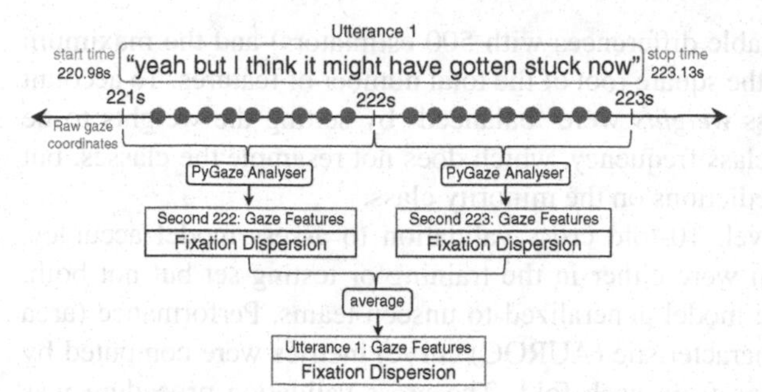

First, utterance start/stop times were rounded to the nearest second. Then, raw gaze coordinates (blue circles) were fixation filtered and aligned to the 1s intervals. Features were computed per 1s interval and averaged across the intervals corresponding to the utterance.

Fig. 2. Example of gaze to utterance alignment for one gaze feature (fixation dispersion).

Task Context Features. Eye gaze patterns are constrained within a given task context, hence, we computed a set of context features as a high-level representation of how teams interacted on the CPS task [22]. We focused on high-level task context without encoding task-specific information for generalizability. The main context features were utterance duration (in seconds) and changes in the Physics Playground area of the screen, computed using a validated motion tracking algorithm [29] to distinguish moments of action and inaction during gameplay. Screen motion was computed on individual video frames and averaged to the utterance-level like the gaze features (see Fig. 2). Additional context features included speaker shift (whether the speaker changed from the previous utterance), the block (warmup, block 1, or block 2) and the speaker's role (controller, more-verbose observer, or less-verbose observer), resulting in five context features. For the conversation-based identities, we also included the previous speaker's role and the other teammate's role (i.e., controller, more- or less-verbose observer), resulting in seven context features. If the same teammate spoke consecutive utterances (12% of the time), the previous speaker identity was assigned to the teammate who spoke immediately before the current speaker.

Data Exclusion and Missing Value Imputation. Two teams' transcripts were unavailable and an additional two teams' data were excluded due to technical issues, resulting in 23,277 utterances from 92 (out of 96) teams for use in the study. For each feature, we calculated what percent of examples had a missing value for the given feature. On average, feature values were missing for 7.66% (range = .00% to 18.83%) of cases, which we replaced (post alignment) with their median values for the team. There were four cases where gaze data was not available for an entire block, but we included them for robustness; removing these blocks did not impact the results.

2.4 Machine Learning Models

We trained random forest classifiers (RFCs) because they support nonlinearity, interactivity, and generalizability. We separately trained four RFCs for binary classification of each CPS facet as well as the "no facet" cases. Separate models were trained because the facets were not mutually exclusive, but their joint occurrence was too infrequent (2.9%) to warrant multilabel learning. RFCs were implemented using Scikit Learn with 100 estimators (there were no notable differences with 500 estimators) and the maximum number of features was set to the square root of the total number of features. To account for class imbalances, the class *weights* were 'balanced' by setting the weights to be inversely proportional to the class frequency, which does not resample the classes, but instead penalizes incorrect predictions on the minority class.

We implemented team-level, 10-fold cross validation to assess model accuracy, where utterances from a team were either in the training or testing set but not both, allowing us to assess how the model generalized to unseen teams. Performance (area under the receiver operator characteristic (AUROC) curve) metrics were computed by pooling the test set predictions from each fold. The cross-validation procedure was repeated for 25 iterations, using a different random assignment of teams in each fold.

We fit separate models for context + gaze and context-only to ascertain the additional value of gaze features over the context baseline. We did not fit separate gaze-only models because they interact with the task context (as noted above). We also fit shuffled baseline models by randomly shuffling the feature vectors within each team to break the temporal dependency between gaze and CPS facets while retaining the same values.

3 Results

Table 1 lists the main results. Across the 25 iterations, the standard deviation of AUROCs ranged from .001 to .003. Therefore, we determined the variance was small enough to report the AUROC and 95% confidence interval from the iteration with the *median* accuracy only. We used the roc.test function from the pROC package in R [30] to compare correlated pairs of ROC curves using the bootstrap test with 2,000 iterations, and report p-values using a false discovery rate (FDR) adjustment for four comparisons (corresponding to the four classes).

Gaze Predicts the CPS Facets. We focus on the role-based models for the main analyses (comparisons to conversation-based identification are presented next). Overall, the shuffled role-based models yielded chance performance, as expected. We tested whether there was added value to using gaze relative to context, finding that that all four context + gaze models performed better than the context-only models (all FDR-corrected $ps < .001$; Table 1), although the improvement was modest (average improvement of 7.29%). Importantly, the improvements were highest for negotiation/coordination and maintaining team function (about 8%) compared to constructing shared knowledge (6%). This is likely because the context-only model performed better for constructing shared knowledge (AUROC = .72) than the other two facets (.56 and .63). This difference may be due to the fact that constructing shared knowledge was generally characterized by longer utterance durations (average of 3.5 s, 2.5 s, and 2.9 s for constructing shared knowledge, negotiation/coordination, and maintaining team function, respectively), which is a feature in the models, or because it was more frequent, enabling better prediction due to more training examples.

Table 1. AUROC with 95% CIs for median performing role-based models across 25 iterations.

Facet (Base rate)	Shuffled	Context only	Context + Gaze	Percent improvement
Construct. Knowldg. (25.45%)	.49 [.48, .50]	.72 [.72, .73]	.76 [.76, .77]	5.69%
Neg./Cord. (14.29%)	.50 [.49, .51]	.56 [.55, .57]	.61 [.60, .62]	8.05%
Maintain. Team. Fn. (10.22%)	.50 [.48, .51]	.63 [.62, .64]	.68 [.67, .70]	8.39%
No Facet (53.07%)	.50 [.49, .51]	.67 [.66, .68]	.72 [.72, .73]	7.74%
Weighted average		.66	.71	7.29%

Comparison of Different Feature Types and Representations. We explored whether the individual gaze features were more predictive than the shared gaze features, finding that this indeed was the case (*ps* < .006; Table 2), though the advantage was minor. This may be because our shared features (which measure joint attention) were not as comprehensive as the individual features in characterizing the collaboration activity. We next investigated whether gaze was better characterized in a role- or conversation-based fashion, finding no significant difference (all *ps* > .1; Table 2).

Table 2. Median AUROC with 95% CIs comparing individual versus shared features and role-based versus conversation-based models. Context features are included in all models.

Facet	Feature comparison		Identity comparison	
	Individual	Shared	Role	Conversation
Constuct. Knowldg.	.77 [.76, .77]	.76 [.75, .77]	.76 [.76, .77]	.76 [.76, .77]
Neg./Cord.	.61 [.60, .62]	.59 [.58, .60]	.61 [.60, .62]	.60 [.59, .61]
Maintain. Team. Fn.	.69 [.68, .70]	.67 [.66, .68]	.68 [.67, .70]	.69 [.68, .70]
No Facet	.72 [.72, .73]	.71 [.71, .72]	.72 [.71, .73]	.72 [.71, .73]
Weighted Average	.71	.70	.71	.71

Feature Analysis. We used Shapley Additive exPlanations (SHAP) [31] on the median-performing role-based model for each facet to examine feature importances. SHAP bases feature importance on how much and in what direction features influenced the outcome. We first correlated SHAP scores across the three facets and found them to be correlated (*rs* from .78 to .84) suggesting similar patterns of feature importances for all facets. Beyond utterance duration, which had the highest SHAP scores, the controller's fixation duration was one of the three most important gaze features for constructing shared knowledge and negotiation/coordination, while the fixation rate of the controller and less-verbose observer was important for maintaining team function.

Follow Up Experiments. We conducted additional experiments to assess if the following modifications would improve performance. First, we no longer computed features per second and instead computed features directly based on the fixations that occurred during the utterance. Second, we computed features for the duration of silence after an utterance, up to a maximum of 2s, to capture gaze patterns in response to the utterance. Third, we included these three additional shared features: the convex hull area (in pixels) of all participants' fixation locations and the proportion of fixations that were in the PP and partners' view AOIs (see Fig. 1). However, none of these modifications improved overall performance from the results reported.

4 Discussion

Main Findings. The main goal of the present study was to explore whether eye gaze could predict collaborative problem solving (CPS) behaviors in triads. The results demonstrated that there were generalizable patterns of gaze that were able to predict our three CPS facets. However, our models were only moderately accurate, with gaze information yielding an average of 7.29% improvement over a context-only baseline.

We further demonstrated that individual gaze features were better predictors of CPS than were our shared gaze features (which measured joint attention), although there were fewer of the latter type. In related CPS studies, shared gaze has been identified as an indication of productive CPS because it demonstrates a "meeting of the minds", in which teammates must focus their attention on the same area to work towards the same goal [20, 32]. Whereas joint attention has been demonstrated to reflect coordination towards CPS goals, it has also been shown to obscure imbalances in team dynamics which could be due to the "free rider" effect or a dominating partner [18]. A dominant partner might frequently gesture to areas on the screen to direct teammates' attention. If partners passively follow along, Schneider et al. [18] found that collaboration outcomes were negative, while if partners challenged and discussed different ideas, shared gaze was more dispersed, but outcomes were more positive.

Furthermore, the improved performance of individual gaze features over shared features suggests that modeling successful collaboration involves more than capturing synchrony and shared attention between group members, and that individual features, which can provide information about teammates' cognitive states, are informative in identifying CPS behaviors. The individual gaze features in our study are likely reflecting how on task each individual is and whether they are processing task-relevant information [33], since pupil size, saccade amplitude, fixation count, and fixation duration can index cognitive load, and dispersion measures how focused the gaze is [17]. However, it is evident that our shared features (the distance between centroids of each partner pair), are missing additional context (e.g., leader–follower patterns), and performance might improve with additional shared features that capture this information.

Finally, there was no difference between a role- and conversation-based model of gaze, suggesting that little information was added by considering teammate identity. Indeed, [22] found that individual signals weighted by assigned role or behavior did not produce significantly different results, which is consistent with the present findings.

Applications. There are several potential applications, such as intelligent computer interfaces that can monitor and support CPS processes by providing personalized feedback to support CPS skill acquisition. For example, if the model predicts that a group is not engaging in shared knowledge construction, the system could intervene (in real time) and encourage teammates to build on each other's ideas or review the constraints of the task together. Similarly, if the system detects that a particular team member is not engaging in negotiation with their teammates, this insight could be conveyed to that individual as part of an after-task review, along with actionable suggestions to improve their negotiation skills. Furthermore, our gaze-based models could be particularly valuable in noisy classroom environments, where accurate speech recognition is difficult (see Introduction), and additional modalities might be needed to model CPS processes. Eye gaze can also serve as a complementary signal when speech quality is too poor.

Limitations and Future Work. Like any study, ours has limitations. First, our strategy for feature aggregation (averaging values across the duration of an utterance) resulted in the loss of fine-grained temporal and visuospatial information from the eye tracker, which outputs high spatial/temporal resolution data at 90 Hz. Thus, future work should investigate whether predictive accuracy can be improved by operating on a more fine-grained representation of gaze activity (i.e., time series). Next, our shared features representing joint attention could benefit from additional information to characterize the nature of shared gaze, so future work should include additional shared features that can control for imbalanced team participation, such as the difference in a group's "visual leadership", proposed by Schneider et al. [18], which measures the balance of leader–follower gaze patterns. Another limitation is that we only considered a single modality (eye gaze), and did not explore how gaze could be combined with other modalities that have been shown to predict CPS facets (e.g., speech [15, 34], facial expressions [34]). In future work, we plan to investigate multimodal approaches for incorporating gaze in models of CPS facets. In particular, we hypothesize that gaze information may be a useful complement to language-based models of CPS, as eye gaze and speech production are tightly coupled [16]. Finally, we collected our data in a controlled lab setting to obtain relatively accurate gaze tracking, but future research in more ecologically valid settings is warranted.

Conclusion. Remote collaborative problem solving is a critical 21st century skill that is important across many domains, such as the classroom and remote work. Our results suggest that how people move their eyes during remote collaboration is related to how well they construct shared knowledge, negotiate, and maintain team function. In addition to its theoretical relevance, the ability to model CPS skills from gaze has exciting implications for intelligent systems that aim to improve collaboration in small groups.

Acknowledgments. This research was supported by the NSF National AI Institute for Student-AI Teaming (iSAT) (DRL 2019805) and NSF DUE 1745442/1660877. The opinions expressed are those of the authors and do not represent views of the NSF.

References

1. OECD: PISA 2015 Results (Volume V): Collaborative Problem Solving. Organisation for Economic Co-operation and Development, Paris (2015)

2. Fiore, S.M., Graesser, A., Greiff, S.: Collaborative problem-solving education for the twenty-first-century workforce. Nat Hum Behav. **2**, 367–369 (2018)
3. Franken, E., et al.: Forced flexibility and remote working: opportunities and challenges in the new normal. J. Manag. Org. 1–19 (2021)
4. Kniffin, K.M., et al.: COVID-19 and the workplace: implications, issues, and insights for future research and action. Am. Psychol. **76**, 63–77 (2021)
5. Sottilare, R.A., Graesser, A.C., Hu, X., Sinatra, A.M.: Design Recommendations for Intelligent Tutoring Systems: Volume 6 - Team Tutoring. US Army Research Laboratory (2018)
6. Dyke, G., Adamson, D., Howley, I., Penstein Rosé, C.: Towards academically productive talk supported by conversational agents. In: Cerri, S.A., Clancey, W.J., Papadourakis, G., Panourgia, K. (eds.) Intelligent Tutoring Systems, vol. 7315, pp. 531–540. Springer, Heidelberg (2012). https://doi.org/10.1007/978-3-642-30950-2_69
7. Schneider, B., Dowell, N., Thompson, K.: Collaboration analytics—current state and potential futures. J. Learn. Anal. **8**, 1–12 (2021)
8. Shockley, K., Richardson, D.C., Dale, R.: Conversation and coordinative structures. Top. Cogn. Sci. **1**, 305–319 (2009)
9. Duchowski, A.T.: Visual attention. In: Duchowski, A.T. (ed.) Eye Tracking Methodology: Theory and Practice, pp. 3–15. Springer, London (2003). https://doi.org/10.1007/978-1-84628-609-4_1
10. Hao, J., Chen, L., Flor, M., Liu, L., von Davier, A.A.: CPS-Rater: automated sequential annotation for conversations in collaborative problem-solving activities. ETS Res. Rep. Ser. **2017**, 1–9 (2017)
11. Pugh, S., Subburaj, S.K., Rao, A.R., Stewart, A., Andrews-Todd, J., D'Mello, S.: Say what? Automatic modeling of collaborative problem solving skills from student speech in the wild. In: Proceedings of the 14th Educational Data Mining Conference (2021)
12. Emara, M., Hutchins, N., Grover, S., Snyder, C., Biswas, G.: Examining student regulation of collaborative, computational, problem-solving processes in open-ended learning environments. J. Learn. Anal. **8**, 49–74 (2021)
13. Chopade, P., Edwards, D., Khan, S.M., Andrade, A., Pu, S.: CPSX: using AI-machine learning for mapping human-human interaction and measurement of CPS teamwork skills. In: 2019 IEEE International Symposium on Technologies for Homeland Security (HST), pp. 1–6 (2019)
14. Reilly, J.M., Schneider, B.: Predicting the quality of collaborative problem solving through linguistic analysis of discourse. International Educational Data Mining Society (2019)
15. Pugh, S.L., Rao, A., Stewart, A.E.B., D'Mello, S.K.: Do speech-based collaboration analytics generalize across task contexts? In: LAK 2022: 12th International Learning Analytics and Knowledge Conference, pp. 208–218. Association for Computing Machinery, NewYork, (2022)
16. Jermann, P., Sharma, K.: Gaze as a proxy for cognition and communication. In: 2018 IEEE 18th International Conference on Advanced Learning Technologies (ICALT), pp. 152–154 (2018)
17. Olsen, J., Sharma, K., Rummel, N., Aleven, V.: Temporal analysis of multimodal data to predict collaborative learning outcomes. Br. J. Educ. Technol. **51** (2020)
18. Schneider, B., Sharma, K., Cuendet, S., Zufferey, G., Dillenbourg, P., Pea, R.: Leveraging mobile eye-trackers to capture joint visual attention in co-located collaborative learning groups. Int. J. Comput.-Support. Collab. Learn. **13**(3), 241–261 (2018). https://doi.org/10.1007/s11412-018-9281-2
19. Celepkolu, M., Boyer, K.E.: Predicting student performance based on eye gaze during collaborative problem solving. In: Proceedings of the Group Interaction Frontiers in Technology, pp. 1–8. Association for Computing Machinery, New York (2018)

20. Vrzakova, H., Amon, M.J., Stewart, A.E.B., D'Mello, S.K.: Dynamics of visual attention in multiparty collaborative problem solving using multidimensional recurrence quantification analysis. In: Proceedings of the 2019 CHI Conference on Human Factors in Computing Systems, Glasgow, Scotland, UK, pp. 1–14. ACM (2019)
21. Sun, C., Shute, V.J., Stewart, A., Yonehiro, J., Duran, N., D'Mello, S.: Towards a generalized competency model of collaborative problem solving. Comput. Educ. **143**, 103672 (2020)
22. Subburaj, S.K., Stewart, A.E.B., Ramesh Rao, A., D'Mello, S.K.: Multimodal, multiparty modeling of collaborative problem solving performance. In: Proceedings of the 2020 International Conference on Multimodal Interaction, pp. 423–432. Association for Computing Machinery, New York (2020)
23. Stewart, A.E.B., Amon, M.J., Duran, N.D., D'Mello, S.K.: Beyond team makeup: diversity in teams predicts valued outcomes in computer-mediated collaborations. In: Proceedings of the 2020 CHI Conference on Human Factors in Computing Systems, pp. 1–13. Association for Computing Machinery, New York (2020)
24. Shute, V.J., Ventura, M., Kim, Y.J.: Assessment and learning of qualitative physics in Newton's playground. J. Educ. Res. **106**, 423–430 (2013)
25. Gwet, K.L.: Handbook of inter-rater reliability: the definitive guide to measuring the extent of agreement among raters (2014)
26. Dalmaijer, E.S., Mathôt, S., Van der Stigchel, S.: PyGaze: An open-source, cross-platform toolbox for minimal-effort programming of eyetracking experiments. Behav. Res. Methods **46**(4), 913–921 (2013). https://doi.org/10.3758/s13428-013-0422-2
27. Richardson, D.C., Dale, R., Tomlinson, J.M.: Conversation, Gaze coordination, and beliefs about visual context. Cognit. Sci. **33**, 1468–1482 (2009)
28. Vrzakova, H., Amon, M.J., Rees, M., Faber, M., D'Mello, S.: Looking for a deal? Visual social attention during negotiations via mixed media videoconferencing. Proc. ACM Hum.-Comput. Interact. **4**, 260:1–260:35 (2021)
29. Westlund, J.K., D'Mello, S.K., Olney, A.M.: Motion tracker: camera-based monitoring of bodily movements using motion Silhouettes. PLoS ONE **10**, e0130293 (2015)
30. Robin, X., et al.: pROC: an open-source package for R and S+ to analyze and compare ROC curves. BMC Bioinform. **12**, 77 (2011)
31. Lundberg, S.M., Lee, S.-I.: A unified approach to interpreting model predictions. in: advances in neural information processing systems. Curran Associates, Inc. (2017)
32. Pöysä-Tarhonen, J., Awwal, N., Häkkinen, P., Otieno, S.: Joint attention behaviour in remote collaborative problem solving: exploring different attentional levels in dyadic interaction. Res. Pract. Technol. Enhanc. Learn. **16**(1), 1–24 (2021). https://doi.org/10.1186/s41039-021-00160-0
33. Peacock, C.E., et al.: Gaze dynamics are sensitive to target orienting for working memory encoding in virtual reality. J. Vis. **22**, 2 (2022)
34. Stewart, A.E.B., Keirn, Z., D'Mello, S.K.: Multimodal modeling of collaborative problem-solving facets in triads. User Model. User-Adap. Inter. **31**(4), 713–751 (2021). https://doi.org/10.1007/s11257-021-09290-y

Improving Automated Evaluation of Formative Assessments with Text Data Augmentation

Keith Cochran[1(✉)], Clayton Cohn[2], Nicole Hutchins[2], Gautam Biswas[2], and Peter Hastings[1]

[1] DePaul University, Chicago, IL 60604, USA
kcochr11@depaul.edu
[2] Vanderbilt University, Nashville, TN 37240, USA

Abstract. Formative assessments are an important component of instruction and pedagogy, as they provide students and teachers with insights on how students are progressing in their learning and problem-solving tasks. Most formative assessments are now coded and graded manually, impeding timely interventions that help students overcome difficulties. Automated evaluation of these assessments can facilitate more effective and timely interventions by teachers, allowing them to dynamically discern individual and class trends that they may otherwise miss. State-of-the-art BERT-based models dominate the NLP landscape but require large amounts of training data to attain sufficient classification accuracy and robustness. Unfortunately, educational data sets are often small and unbalanced, limiting any benefits that BERT-like approaches might provide. In this paper, we examine methods for balancing and augmenting training data consisting of students' textual answers from formative assessments, then analyze the impacts in order to improve the accuracy of BERT-based automated evaluations. Our empirical studies show that these techniques consistently outperform models trained on unbalanced and unaugmented data.

Keywords: Data augmentation · Text augmentation · BERT · Formative assessments · Imbalanced data sets · Educational texts · Natural language processing

1 Introduction

The current generation of intelligent learning environments (ILEs) for K-12 students focuses on inquiry, problem-based, game-based, and open-ended learning [10,14,16, for example]. Working on open-ended tasks provides students with

The assessment project described in this article was funded, in part, by the NSF Award # 2017000. The opinions expressed are those of the authors and do not represent views of NSF.

M. M. Rodrigo et al. (Eds.): AIED 2022, LNCS 13355, pp. 390–401, 2022.
https://doi.org/10.1007/978-3-031-11644-5_32

choices in how they develop and pursue their learning and problem-solving processes [22]. Research on ILEs has demonstrated the challenges in framing adaptive support in the context of the specific difficulties that students face as they work on their learning and problem-solving tasks, and develop productive learning strategies [2,21]. Formative assessments have been employed in the learning sciences and education research as interventions that (1) help students learn components of knowledge they need to build and solve larger problem-solving and learning tasks [3] and (2) communicate conceptual understanding of the target domain for self-reflection as well as teacher and environment pedagogical support that aids students' achievement in the context of their current learning [3]. Therefore, formative assessments can help students develop their conceptual understanding of the domain, while supporting their self-assessment and self-regulated learning skills [6,11].

However, formative assessments are often time-consuming to grade [12], limiting the ability to leverage them for *in-time* pedagogical adjustments and feedback. Our long-term goal is to develop robust deep learning-based, natural language processing (NLP) approaches to support rich, in-time formative feedback to students' responses to short answer questions. Formative assessments often go beyond statement-of-fact conceptual knowledge applications, requiring students to reason about causal relations between concepts, explain a scientific process or phenomena, or construct an argument that justifies or negates a particular statement. Simple text-processing methods like keyword matching and templates are often insufficient to uncover the nuanced reasoning in students' short answers to formative assessment questions [13]. Advances in NLP allow us to dive deeper into students' knowledge and reasoning applications, and help students understand the difficulties they face with the instructional material they are being taught. In parallel, they also support teachers in understanding and responding to student difficulties soon after they occur, and before they move on to teach new content. However, issues such as data insufficiency, data imbalance, and lack of variation in student responses limit our ability to apply these advances in a robust and reliable way.

Our approach develops automated text assessments that shed light on students' conceptual knowledge. Educational data sets present several difficulties in NLP because studies typically conducted in classroom environments generate rich data, but the data collection is often limited to about a 100 students at a time. In this paper, we address increasing the effectiveness of NLP evaluation when there is limited and unbalanced training data, as is often the case in educational contexts. We do this by augmenting the training data with generated sentences that share characteristics of the original data. In the rest of this paper, we summarize related work, present our research questions and hypotheses describing the educational context of the formative assessments, and conclude with findings and future work.

2 Background and Research Questions

Transformer-based NLP architectures, such as BERT [8] and GPT-3 [4], are now the industry standard for modeling many NLP tasks. They leverage language knowledge from massive corpora of unlabeled texts via unsupervised pretraining, and they can be fine-tuned on a downstream task with only a fraction of the training instances that would otherwise be required to train a neural network from scratch. However, despite the prevalence of transformer models, many data sets are still too small to effectively fine-tune a model out-of-the-box. There are few areas where this is more apparent than with educational texts in general, and educational assessments in particular.

These texts are also domain-specific, focusing on a wide variety of general areas and specific examples within them. Domain-specific subject matter often includes esoteric jargon that is not well-represented in the canonical corpora that these large transformer models are pretrained on, and there can often be performance degradation when these models are applied to texts whose vocabularies differ considerably from their own [7]. In addition to the issue of educational data sets being non-canonical semantically, they are often non-canonical syntactically as well. Wikipedia, which is used to pre-train both BERT and GPT, is written using proper language syntax. Conversely, many educational texts, such as answers to formative and summative assessments written by children or adolescents, use informal syntax and are written in a much more colloquial manner. This type of text is often incompatible with pre-trained models derived from canonical corpora, as model performance is affected by the quality of data used for training. For example, middle school short answer questions typically use a shallower vocabulary, and this has to be factored into the augmentation techniques used. It is possible to further pre-train BERT with domain-specific corpora, but this also requires large quantities of data. As such, the only practical approach is to select a base model and fine-tune it using labeled data to improve the model's performance.

One salient solution to mitigate the aforementioned issues is *data augmentation*. Data sets once small, imbalanced, and sparsely populated can be made robust by adding instances that are similar in both syntax and semantics. However, hand-crafting these instances can be extremely tedious, so automated approaches are preferable. Data augmentation has been used in areas such as image processing with great effectiveness, e.g., by translating or shifting the images. However in NLP, data augmentation techniques are more complex. A newly generated sentence must retain the same semantic intent as the original sentence. Issues arise when augmented data stray away from the label they are intended to augment. This has led some researchers to assess and label augmented data using experts to ensure correct labeling.

One textual data augmentation technique adds noise in the form of substitution or deletion of words or characters [20]. Another approach uses "back translation" where the data is translated into another language, then translated back, producing alternate ways of saying the same thing [15]. Other forms of data augmentation introduce noise by adding a random character in a word, avoiding the first and last characters of the word. Some methods use random synonym replacements in the form of hypernym (more general) and hyponym (more specific) word replacements using WordNet. Hypernyms have been shown to outperform hyponyms because generalizing a sentence is more likely to preserve the same meaning [9]. BERT uses a masking feature, where a word in the sentence is masked with a special token, and the model tries to predict the masked word. This can serve as another form of augmentation, where a model can be further trained to generate more semantically similar sentences.

In this paper, we examine the benefits of textual data augmentation for evaluating middle school formative assessments (short-answer questions), especially in cases of data scarcity and data imbalance. We compare four different data augmentation techniques: (1) masking using BERT, (2) noise injection, (3) hyponym/hypernym replacement, and (4) oversampling using the existing data. The goal, as discussed, is to provide accurate, timely feedback to students and their teachers. Accordingly, we formulate three research questions:

RQ 1: Does data augmentation improve the classification of student answers? Furthermore, if augmentation is beneficial, is that primarily due to increasing the amount of data, improving the balance between classes, or some combination of both? Our first hypothesis (**H1**) is that both more balanced data and larger amounts of data will improve classification accuracy.

RQ 2: How does the method used for generating new texts affect augmentation performance? Our hypothesis **H2** is that the masking technique will be most effective due to its alignment with BERT. Our expectations for the other three are mixed. In principle, WordNet should provide semantically related substitutes, but its knowledge base is so broad that it may bring in words far outside the learning context.

RQ 3: Do characteristics of the questions and answers affect the effectiveness of data augmentation? For example, some questions may call for fact-based answers. Others may call for descriptions of processes or for causal reasoning that requires the answers to adopt a meaningful structure to produce a correct answer. **H3** proposes that augmenting the data with *wrong answers* will reduce performance because there are such a wide variety of wrong answers for any question. **H4** proposes that augmenting the data with sentences generated from a very small set of examples will also hurt performance due to the limited variability of the samples.

3 The SPICE Curriculum

The formative assessments analyzed in this paper are part of the SPICE (Science Projects Integrating Computation and Engineering) curriculum [23]. This is a three-week, NGSS-aligned unit that challenges students to redesign their schoolyard using appropriate surface materials that meet design constraints and minimize the amount of water runoff after heavy rainfall.

The curriculum (Fig. 1) includes a conceptual modeling unit, where students construct conceptual models of the water runoff phenomenon; then translate it to a computational model of water runoff; and then use the model to solve an engineering design challenge problem, where students construct a playground

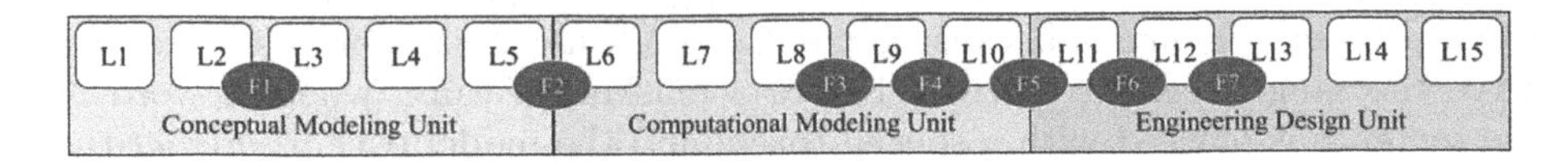

Fig. 1. SPICE curriculum overview.

that adheres to specified constraints [17]. Formative assessments (identified in red) are integrated throughout the curriculum to evaluate students' conceptual understanding in science, computing, and engineering. For this paper, we focus on formative assessment F1 in the conceptual modeling phase.

We leverage evidence-centered design (ECD; [18]) as the overarching framework for assessment development. This process supports our analysis of knowledge construction and problem-solving skill development in the integrated science, computing, and engineering curriculum by linking components of the curriculum and assessments to evidence of students' proficiency with the target knowledge and skills [17]. For instance, students are presented with an incorrect conceptual model and are tasked with (1) identifying and correcting errors and (2) describing positive information presented by the model. These tasks can be linked to key science and engineering practices as described by NGSS [19], including engaging in argument from evidence, and developing and using models, and allow us to evaluate students' science knowledge through its application in model evaluation. In-time analysis of these assessments may allow us to provide key evidence-based, formative feedback to better support students' construction, debugging, and evaluation of their own conceptual models during the curriculum.

4 Methods

Our exploratory analysis leverages student data collected from a classroom study with 99 6th-grade students in the southeastern United States. The study, conducted in Fall 2019, was led by two experienced science teachers with three university researchers providing additional support in the classroom.

The data set for this study consists of student responses to three separate questions that are based on a fictitious student-constructed visual model shown in Fig. 2. Each question had 95 student responses.[1] The concepts the students must identify for each question are enumerated in Table 1.

1. *What do you think the different sized arrows in Libby's model could mean?* This question has one correct response: the size of the arrows indicates the amount of water. There is only one concept, which evaluates students' understanding of the model representation.
2. *What are two things that you would change about Libby's model to explain where the water goes?*

[1] While there were 99 students in the study, not all students answered each question.

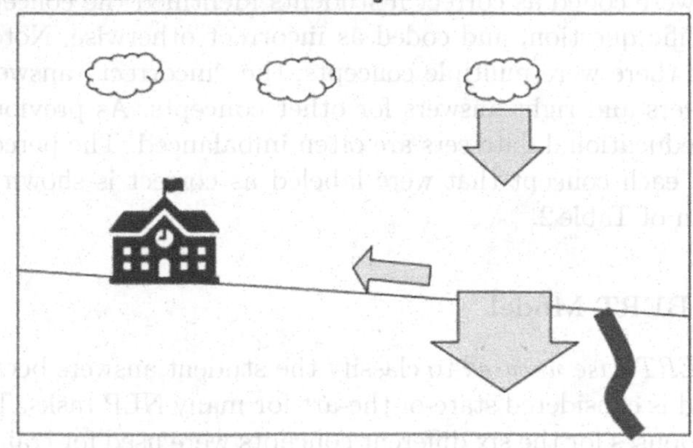

Fig. 2. Libby's model demonstrating where water goes after precipitation.

The focus of this question is on finding errors in the model, explaining the errors, and providing the correct answer. It includes two concepts: the size of the runoff and absorption arrows should sum to the size of the rainfall arrow (conservation of matter), and the direction of the runoff arrow should be pointing downhill. This question evaluates students' knowledge of scientific concepts rather than model representation.

3. *What are two things that Libby's model does a good job of explaining?* Extending the previous question, this question also targets students' ability to observe and evaluate a science model. In this case, although Libby's model contained errors (previous question), the model (1) demonstrates rainfall either is absorbed or becomes runoff, (2) illustrates where water is coming from, and (3) uses arrow size to indicate water amounts. Students are tasked with listing two of these positive model elements and assesses students' knowledge of the scientific concepts as well as their understanding of the model representation.

Table 1. Concepts present in each question.

Question	Concept	Description
1	C1	Arrow size indicates amount of water
2	C2a	Size of runoff and absorption arrows should sum to size of rainfall arrow
2	C2b	Direction of runoff arrow should be pointing downhill
3	C3a	Model demonstrates rainfall either absorbed or becomes runoff
3	C3b	Model illustrates where water is coming from
3	C3c	Model uses arrow size to indicate water amount

Each of the six concepts described above (correct responses for each assessment question) was modeled individually as a binary classification task.

Responses were coded as correct if students identified the concept(s) associated with a specific question, and coded as incorrect otherwise. Note that for questions where there were multiple concepts, the "incorrect" answers include both wrong answers and right answers for other concepts. As previously mentioned, such small educational data sets are often imbalanced. The percentage of the 95 answers for each concept that were labeled as correct is shown in the leftmost data column of Table 2.

4.1 The BERT Model

We used *BERT-base uncased* to classify the student answers because it is widely adopted and is considered state-of-the-art for many NLP tasks. The three sets of student responses for the six different concepts were used for training, validating, and testing the models. For each concept, a separate BERT model was fine-tuned for classification on the training data by adding a single feed-forward layer. We used the micro-F_1 metric as the performance measurement.

In all experiments, the models were trained and evaluated 10 times, with each training iteration using a different seed for the random number generator, which partitions the training and testing instances. During training, the following hyperparameters were used: learning rate 9e-5, batch 12, epochs 2, max sequence 128, train/test split 80/20. Devlin et al. [8] recommend learning rates of 2e-5 to 5e-5, and batch size of 16 or 32, but we chose different values due to data scarcity.

4.2 Baseline Evaluation

For each concept, we evaluated two different baseline models without augmentation or balancing. The *a priori* model simply chose the majority classification for each concept. For our *unaugmented* baseline, we applied BERT in the prototypical way, without data augmentation.

4.3 Augmentation Approach

We chose four textual alteration methods for augmenting the data sets because they are among the leading modern methods at both the word and character level. This gave us a wider sample range to compare and contrast different augmentation methods [1]. Techniques were chosen to minimize the risk of changing semantic intent. WordPiece-level masking is cited as the best augmentation method for classification tasks by Chen et al. [5]. Therefore, our first approach used masking to mask a word in the sentence, then used the BERT model to generate a substitute for the masked word. The second method, noise injection, randomly inserted, deleted or changed a character in the original sentence [9]. The hypo/hypernym method generated sentences by selecting a keyword in the given sentence, and replacing it with both types of related word to generate new candidate sentences [9]. Last, an oversampling method using multiple copies of each instance in the data set was used for augmentation.

The majority label quantity in Table 2 became the **majority quantity of reference** for that particular data set. We first left the data unbalanced from 0x to 1x, then augmented the minority class only by adding $N = (Maj - Min)/5$ sentences[2] at a time until parity was reached. Next, we performed another test by forcing the data to be balanced by removing majority label responses to match the minority level, ensuring parity at each level of augmentation up to 1x. After the data reached 1x, all data sets thereafter were balanced and were augmented in multiples of the majority quantity from 1x to 20x. Initial tests with imbalanced data showed inconsistent results as more augmentation was applied. Additionally, we found empirically that model performance decreased when higher augmentation levels were used over 20x.

5 Results

The high-level view of our results is presented in Table 2. Each row corresponds to a concept. The leftmost data column shows the percentage of the answers for each concept that were originally marked as correct. The next two columns present the baseline results. On the right are the maximum F_1 scores for each concept using one form of data augmentation, and indicating what augmentation quantity level reached that maximum. The highest performance achieved for each concept are shown in bold.

Table 2. Performance (micro-F_1) of baseline vs all augmented models

Concept	%	Baseline		Max Performance	
	Correct	*a priori*	Unaug.	F_1	Aug. Level
C1	89	**0.940**	0.735	0.936	0.6x
C2a	73	0.850	0.757	**0.995**	5x
C2b	33	0.670	0.000	**0.958**	5x
C3a	54	0.700	0.399	**0.873**	8x
C3b	23	0.770	0.000	**0.979**	3x
C3c	40	0.600	0.098	**0.900**	8x

Figure 3 illustrates how balanced and unbalanced data sets affect performance during augmentation. As augmentation increases, the balanced approach shows worse initial performance, however, as augmentation was applied, the balanced approach had a more stable rise in performance. The "balanced" approach (shown by the solid line) forced equality by increasing the minority label as before, but this time, diminishing the majority label such that both had equal representation in the data set.

[2] Here, Maj and Min refer to the number of available sentences from the majority and minority classes.

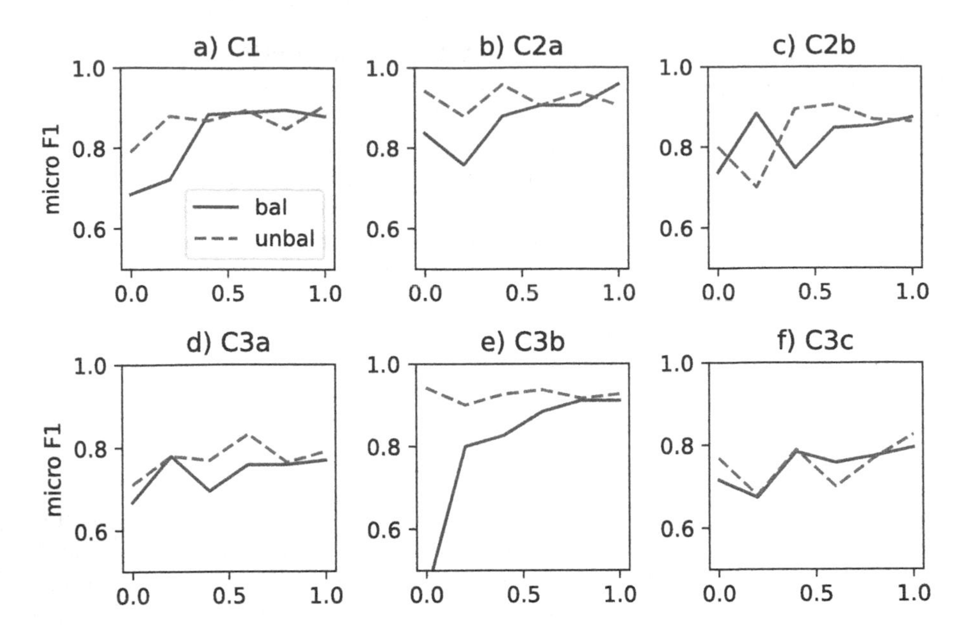

Fig. 3. Model Performance for Balanced vs Unbalanced Data with Augmentation less than 1x the reference majority class.

Figure 4 shows how the performance varied with different augmentation types, masked ("mask"), noise injection ("noise"), hypo-hypernym ("hyp"), and oversampling ("over") and the amount of augmentation for each student response. For each of the concepts, performance improved when augmentation levels from 3x-8x were applied, but tended to fall off slightly with additional augmentation.

6 Discussion

Recall **RQ 1** which asked whether the performance could be improved with an augmented data set. Table 2 shows that data augmentation does improve classification performance over the *a priori* baseline in five of the six concepts, and improves on the unaugmented model baseline in every case. **H1** states that the effect of a balanced data set and larger amounts of data improve classification accuracy. Our results show that balancing the data is vital, and augmentation additionally improves performance. However, there is a limit to how much augmentation can be applied before the model levels off and begins to degrade in performance. Also, for questions with a high percentage of majority label quantities (>90%), guessing the majority label outperforms any model used in our testing. Therefore, **H1** was almost completely confirmed. The only exception was for concept 1, where the majority label represented over 90% of the given responses.

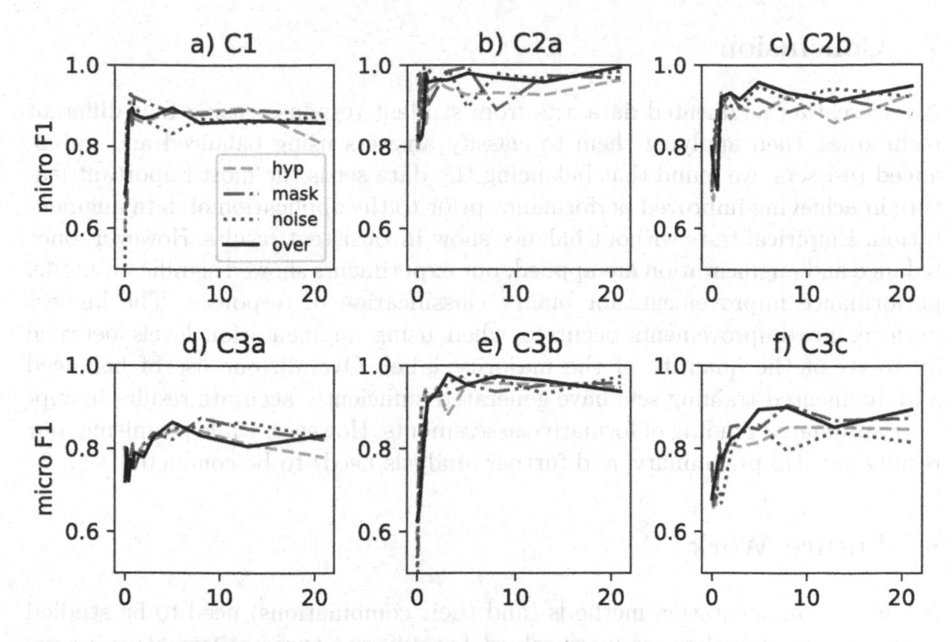

Fig. 4. Model Performance as a Function of Augmentation Level. The x-axis shows the amount of augmentation applied from 0x to 20x.

RQ 2 addresses the four augmentation methods. **H2** predicted that masking would be a clear winner in making the model perform better because of its relation to BERT. Our results in Fig. 4 revealed that performance varies with augmentation method as well as the characteristics of the questions and student responses. No clear winner was evident but a combination of methods may produce better results (currently only an empirical observation). Although **H2** was not supported, stability of performance did vary among the different types of augmentation, so more types of augmentation should be investigated.

RQ 3 speculated that the characteristics of the questions and answers affects model performance. The concept characteristics in our study are that concepts C1, C2a, and C3a are more fact-based answers. The remaining concepts required causal reasoning by the student in order to provide a correct answer. Fact-based concepts do not show a significantly different performance than those requiring causal reasoning. This result means **H3** was not supported.

Figure 3 shows increasing data balance best improves classification accuracy. Once followed up by additional augmentation, significant model improvements can be gained. Some interesting outcomes arose from studying this phenomenon. The model performance improvements in our study varied based on initial data set balance. Data sets that are already close to having an even balance (50%-65%) start out performing in an acceptable range, then increase slightly with applied augmentation to about 8x before falling off. For concepts that originally have close to a two-thirds majority (C2a and C3b), model performance peaked at around 8x.

7 Conclusion

After creating augmented data sets from student responses using four different techniques, then applying them to classify answers using balanced and unbalanced test sets, we found that balancing the data set is the most important feature in achieving improved performance prior to the application of data augmentation. Empirical tests without balance show inconsistent results. However, once balance and augmentation are applied, our experiments showed significant model performance improvements for binary classification of responses. The highest performance improvements occurred when using augmentation levels between 3x to 8x of the quantity of the majority label. Overall, our use of balanced and augmented training sets have generated sufficiently accurate results to support automated grading of formative assessments. However, while promising, our results are still preliminary, and further analysis needs to be conducted.

8 Future Work

Additional augmentation methods (and their combinations) need to be studied to develop models that are more robust for different types of formative assessment questions. After seeing initial ratios of data balance around two-thirds boost performance the most, further investigation is needed. Finally, teachers and education researchers may determine some questions require multiple levels of grading. To grade such answers, we need to train multi-class classifiers or construct hierarchical grading models.

References

1. Bayer, M., Kaufhold, M.A., Reuter, C.: A survey on data augmentation for text classification. arXiv preprint arXiv:2107.03158 (2021)
2. Biswas, G., Segedy, J.R., Bunchongchit, K.: From design to implementation to practice a learning by teaching system: Betty's brain. Int. J. Artif. Intell. Educ. **26**(1), 350–364 (2016)
3. Black, P., Wiliam, D.: Developing the theory of formative assessment. Educ. Assessm. Evaluat. Accountab. **21**, 5–31 (2009)
4. Brown, T.B., et al.: Language models are few-shot learners. arXiv preprint arXiv:2005.14165 (2020)
5. Chen, J., Tam, D., Raffel, C., Bansal, M., Yang, D.: An empirical survey of data augmentation for limited data learning in NLP. arXiv preprint arXiv:2106.07499 (2021)
6. Clark, I.: Formative assessment: assessment is for self-regulated learning. Educ. Psychol. Rev. **24**, 205–249 (2012). https://doi.org/10.1007/s10648-011-9191-6
7. Cohn, C.: BERT Efficacy on Scientific and Medical Datasets: A Systematic Literature Review. DePaul University (2020)
8. Devlin, J., Chang, M.W., Lee, K., Toutanova, K.: BERT: pre-training of deep bidirectional transformers for language understanding. arXiv preprint arXiv:1810.04805 (2018)

9. Feng, S.Y., Gangal, V., Kang, D., Mitamura, T., Hovy, E.: GenAug: data augmentation for finetuning text generators. arXiv preprint arXiv:2010.01794 (2020)
10. Geden, M., Emerson, A., Carpenter, D., Rowe, J., Azevedo, R., Lester, J.: Predictive student modeling in game-based learning environments with word embedding representations of reflection. Int. J. Artif. Intell. Educ. **31**(1), 1–23 (2020). https://doi.org/10.1007/s40593-020-00220-4
11. Hattie, J., Timperley, H.: The power of feedback. Rev. Educ. Res. **77**(1), 81–112 (2007). https://doi.org/10.3102/003465430298487
12. Higgins, M., Grant, F., Thompson, P.: Formative assessment: balancing educational effectiveness and resource efficiency. J. Educ. Built Environ. **5**(2), 4–24 (2010). https://doi.org/10.11120/jebe.2010.05020004 https://doi.org/10.11120/jebe.2010.05020004 https://doi.org/10.11120/jebe.2010.05020004
13. Hughes, S.: Automatic Inference of Causal Reasoning Chains from Student Essays. Ph.D. thesis, DePaul University, Chicago (2019). https://via.library.depaul.edu/cdm_etd/19/
14. Käser, T., Schwartz, D.L.: Modeling and analyzing inquiry strategies in open-ended learning environments. Int. J. Artif. Intell. Educ. **30**(3), 504–535 (2020)
15. Liu, P., Wang, X., Xiang, C., Meng, W.: A survey of text data augmentation. In: 2020 International Conference on Computer Communication and Network Security (CCNS), pp. 191–195. IEEE (2020)
16. Luckin, R., du Boulay, B.: Reflections on the Ecolab and the zone of proximal development. Int. J. Artif. Intell. Educ. **26**(1), 416–430 (2015). https://doi.org/10.1007/s40593-015-0072-x
17. McElhaney, K.W., Zhang, N., Basu, S., McBride, E., Biswas, G., Chiu, J.: Using computational modeling to integrate science and engineering curricular activities. In: Gresalfi, M., Horn, I.S. (Eds.). The Interdisciplinarity of the Learning Sciences, 14th International Conference of the Learning Sciences (ICLS) 2020, vol. 3 (2020)
18. Mislevy, R.J., Haertel, G.D.: Implications of evidence-centered design for educational testing. Educational Measurement: Issu. Pract. **25**(4), 6–20 (2006) https://doi.org/10.1111/j.1745-3992.2006.00075.x
19. NGSS: Next Generation Science Standards. For States, By States. The National Academies Press (2013)
20. Wei, J., Zou, K.: EDA: Easy data augmentation techniques for boosting performance on text classification tasks. arXiv preprint arXiv:1901.11196 (2019)
21. Winne, Philip H.., Hadwin, Allyson F..: nStudy: tracing and supporting self-regulated learning in the Internet. In: Azevedo, Roger, Aleven, Vincent (eds.) International Handbook of Metacognition and Learning Technologies. SIHE, vol. 28, pp. 293–308. Springer, New York (2013). https://doi.org/10.1007/978-1-4419-5546-3_20
22. Zhang, N., Biswas, G., Hutchins, N.: Measuring and analyzing students' strategic learning behaviors in open-ended learning environments. Int. J. Artif. Intell. Educ. (2021). https://doi.org/10.1007/s40593-021-00275-x
23. Zhang, N., et al.: Studying the interactions between science, engineering, and computational thinking in a learning-by-modeling environment. In: Bittencourt, I.I., Cukurova, M., Muldner, K., Luckin, R., Millán, E. (eds.) AIED 2020. LNCS (LNAI), vol. 12163, pp. 598–609. Springer, Cham (2020). https://doi.org/10.1007/978-3-030-52237-7_48

Clustering Learner's Metacognitive Judgment Accuracy and Bias to Explore Learning with AIEd Systems

Megan D. Wiedbusch(✉), Nathan Sonnenfeld, Daryn Dever, and Roger Azevedo

School of Modeling, Simulation, and Training, University of Central Florida, Orlando, FL 32816, USA
{megan.wiedbusch,nathan.sonnenfeld,daryn.dever,roger.azevedo}@ucf.edu

Abstract. Metacognitive monitoring and regulation are key dynamic psychological processes in predicting learning, reasoning, and problem solving across AIEd systems. They are impacted by one's metacognitive knowledge, skills, and experiences. Understanding the dynamical processes underlying metacognition are critical to design intelligent adaptive scaffolding. Metacognition may be assumed to function at hierarchical levels of abstraction including a local level (i.e., isolated metacognitive judgements) and global level (i.e., self-beliefs). However, metacognitive research for complex learning has traditionally disregarded the concept of a general metacognitive ability due to local level fluctuations in temporal metacognitive accuracy. In our study, we shift our analyses to reflect a global-level approach to study metacognitive judgment ability by measuring both accuracy and bias across a series of metacognitive judgments. Using hierarchical clustering on undergraduates' (n = 58) metacognitive judgments' accuracy and bias while learning about nine human biological systems with MetaTutor-IVH, a multimedia-based learning environment, we show that some learners show patterns of global-level metacognitive ability. Specifically, we find that learners who tended to have low metacognitive accuracy across all judgment types performed worse overall on learning outcomes. Examining the bias of these learners, we found they tended to be under-confident across all judgment types. Our work suggests that considering multiple metrics of local-level metacognitive judgments' accuracy and bias, can be aggregated to depict a global-level metacognitive ability of learners that is correlated to learning outcomes. We further discuss the impact of our findings for the design of adaptive scaffolding of AIEd systems to foster metacognition.

Keywords: Metacognition · Self-regulated learning · Empirical study

The research presented in this paper is supported by funding from the National Science Foundation (DRL 1431532).

M. M. Rodrigo et al. (Eds.): AIED 2022, LNCS 13355, pp. 402–413, 2022.
https://doi.org/10.1007/978-3-031-11644-5_33

1 Metacognition

Metacognition is a key psychological process in predicting learning, reasoning, and problem solving across AIEd systems [1]. Metacognition is a learner's thinking about their cognition's processes and one's prior knowledge [2–5]. This occurs across the entire dynamic timeline of cognitive events and includes the (in)accurate monitoring (i.e., meta-level) and regulation (i.e., object-level) of those events [6]. Of particular importance to AIEd research and design, is whether or not metacognition is domain-specific or domain-general. Domain general meta-cognition suggests that metacognitive skills and abilities learned under one set of conditions or tasks could then be transferred across contexts [7].

Metacognition can be abstracted into two hierarchical levels- a local level which includes confidence in isolated judgments and a global level which encompasses more stable self-beliefs about our abilities and skills [8]. Local level metacognition occurs throughout the learning session [6] beginning with Ease-of-Learning (EOL) judgements, or the preliminary evaluation of how difficult learning will be made prior to instruction or content [9]. Within many laboratory studies, these judgments are often thought to be poor indicators of learning outcomes [10–12]. However, they have been shown to help direct learner attention and effort allocation with varying degrees of accuracy [9]. Some research has begun to emerge, however, that indicate by controlling for task elements such as grading criterion, task type, and item presentation timing, EOLs could be highly informative about learning outcomes [9].

After learning has occurred, learners can also reflect on their performance and ability to accurately recall information using Retrospective Confidence Judgments (RCJs; [14,15]. RCJs have been consistently shown to be better predictors of learning outcomes than judgments made prior to (i.e., EOLs) or during learning [16–18]. While research on metacognitive judgements has been dominated by cognitive, developmental, and educational psychologists, much of the work has been applied to several learning technologies, including AIEd systems [1,13,19].

Traditionally, research has focused on the local level metacognitive monitoring by calculating a variety of metrics about metacognitive sensitivity (i.e., discriminating correct from incorrect) and bias (confidence irrespective of performance) on EOLs [20,21], FOKs [22], and retrospective confidence judgments [14,15]. The inconsistent individual differences in metacognitive sensitivity, specifically absolute and relative accuracy, have been used as evidence against a general metacognitive ability [23]. However, as other researchers have pointed out, this could be due to differences in learning tasks and complexity. Research that uses multiple tasks have proven inconclusive in their defense of general versus specific metacognition [24–28]. The assumptions underlying these metacognitive processes regarding their timing, nature, role, function, and impact on learning, reasoning, and problem solving in complex tasks while using AIED systems has not been explored.

1.1 Metacognition in AIEd Systems

The relationship between traditional metacognitive research and AIEd systems is one that is symbiotic in nature. Metacognitive research has contributed to the design, development, and research of AIEd systems by informing features and addressing assumptions of interventions. Examples of these systems include the Help-Seeking Tutor, Betty's Brain, Crystal Island, and SimStudent. Additionally, AIEd systems built with these theoretical and conceptual models of metacognition then in turn inform our understanding of metacognition outside of the laboratory and during complex learning. Azevedo and colleagues' (2018) [29] MetaTutor intelligent tutoring system, production rules (whose use and deployment are informed by Winne's 2018 [5] information processing model) prompt and then provide feedback to support learners as they study the human circulatory system. Research using MetaTutor has in turn led to research contributing to our understanding of individual differences [30], internal and external conditions [31], feedback mechanisms [32], etc. In general, evidence of metacognition within AIEd systems tend to be collected through self-report measures embedded within the environments (e.g., using pull-down menus, pedagogical agents), and in some cases, the systems use those measures to infer metacognitive states or processes to inform interventions or feature deployment [32]. Despite current attempts at measuring metacognition with contemporary AIEd systems, our aim is to build on this work and extend it by providing evidence of the complexity of metacognitive judgements and discuss implications for the design of adaptive scaffolding for future AIED systems.

1.2 Current Study

The argument against general metacognitive ability does not hold in the AIEd field. Specifically, this argument has primarily been supported through laboratory based study in which certain assumptions must be made (i.e., no prior-knowledge). Additionally, studies that use metacognitive theory, such as Nelson & Naren's (1990)'s [6] framework do not attempt to capture a global-level abstraction of metacognition either through aggregation of multiple local levels or global components such as self-belief or concepts. This study addresses these gaps by aggregating multiple local-level measurements of metacognition (i.e., metacognitive confidence judgements) to create a global-level measure of metacognition (i.e., metacognitive ability) as they relate to learning outcomes. This study addresses the following research questions:

(1) Can metacognitive accuracy on multiple local-level instances (i.e., judgements) be used to profile learners for a global-level abstraction (i.e., ability)? (2) Can metacognitive bias on multiple local-level instances (i.e., judgments) be used to profile learners for a global-level abstraction (i.e., ability)? And (3) Are there differences in learning outcomes between learner profiles?

We hypothesize that both accuracy and bias can be used to profile learners who differ in learning outcomes. We expect that learners who show high metacognitive accuracy across all judgments will outperform their peers.

2 Methods

2.1 Participants and Materials

Following IRB approval, undergraduate students from a large North American university were recruited for our study (N = 58; 63% female; ages 18–29, M = 20.21, SD = 2.19). All participants were financially compensated for their involvement ($10/hr, up to $30).

Questionnaires. Participants responded to a series of questionnaires (not used in this current analysis), including a demographics questionnaire, the Achievement Emotions Questionnaire (AEQ [33]), the Emotion Regulation Questionnaire (ERQ [34]), the Perceived Affect Utility Scale (PAUSe [35]), and an 18-item multiple-choice pretest on human physiological systems.

MetaTutor-IVH. MetaTutor-IVH (Intelligent Virtual Human) is a linearly-structured multimedia-based learning environment with content pertaining to nine human physiological systems (e.g., cardiovascular, musculoskeletal, nervous, etc.). Embedded within the environment is a virtual pedagogical agent designed to represent a peer learner that provides a non-verbal metacognitive judgment on the relevance of the instructional content via a facial expression. Developed with the assistance of a human biology expert, the instructional content consists of multimedia content slides each which present three text paragraphs and a diagram illustrating the physiological concept described within the text (Flesch-Kincaid text readability score: M = 10.5). The contextual relevance of the instructional content to the target comprehension question is modified, providing learners with the opportunity to consult the IVH and associated metacognitive judgment prompts to support their own metacognitive processing. Previous work with this environment has examined the degree to which learners attended to content [36].

Apparatuses. Multimodal data, not used in this analysis, were collected from participants including eye movements, affect, electrodermal activity (EDA), and log files of participants' interactions within MetaTutor-IVH. See [36] for eye-tracking and EDA apparatus specifics.

2.2 Experimental Design

Our study involved a 3 (content relevance) × 3 (agent congruency) × 2 (inferential question type) within-subjects design (for a total of 18 trials). Content relevance varied among three levels over the series of experimental trials: fully relevant (text and diagram contained relevant information), text somewhat relevant (text contained relevant information), and diagram somewhat relevant (diagram contained relevant information). Regardless of the combination of independent variables and the source of relevant information, instructional content always provided sufficient information to answer each target comprehension question. Agent congruency varied among three levels during trials: neutral facial expression (no change to baseline expression), congruent facial expression (agent expresses joy regarding fully relevant content and confusion regarding somewhat relevant content), and incongruent facial expression (agent expresses confusion regarding fully relevant content and joy regarding somewhat relevant content). Finally, the target comprehension questions focused on either human body functions (e.g., "Please explain how cortisol travels in the body") or human body malfunctions (e.g., "Please explain what would happen if the thyroid hormone were to diffuse freely from the thyroid all the time").

2.3 Procedure

After being briefed and providing informed consent, participants were calibrated to multimodal data collection apparatuses. Participants then filled out a demographic questionnaire, a series of questionnaires assessing emotions and motivation, and a pretest on human physiological systems (see section Materials). Following the pretest, participants completed eighteen trials in MetaTutor-IVH. After the experimental trials, participants completed a second round of emotion and motivation questionnaires, attended debriefing, and were compensated for their participation.

Participants engage in a series of eighteen identically-structured trials which are linear, counterbalanced, randomized, and self-paced. Beginning each trial, participants are presented with an inferential comprehension question related to the content domain, and asked to submit an EOL judgment (i.e., "How easy do you think it will be to learn the information needed to answer this question?"), responding on a discrete unit scale (0–100). Next, participants are presented with multimedia instructional content within the default MetaTutor interface. After thirty seconds, participants are prompted to perform a CE judgment (i.e., "Do you feel the content (text/diagram) on this page is relevant to the question?"), using a three-point Likert-style rating scale (i.e., "text/diagram is relevant"; "text/diagram is somewhat relevant"; "text/diagram is not relevant"). Subsequently, the IVH expresses their judgment (i.e., joy, confusion, neutral) about content relevance for ten seconds. When the participant is finished reading the slide, they are presented with a multiple (four) choice question associated with the learning objective for the trial. After submitting their answer, participants are asked to perform an immediate RCJ (i.e., "How confident are you that the answer you provided is correct?"). Following their immediate RCJ (RCJ1), participants are asked to provide a justification for their answer (open response), then are prompted to perform a delayed RCJ (RCJ2).

2.4 Coding and Scoring

Judgment sensitivity was calculated using the absolute accuracy index (AAI [37]). It is important to emphasize that this score is the discrepancy between a participant's confidence and their performance. This means the smaller the AAI, the higher their metacognitive accuracy. **Judgment bias** was calculated using the bias index [37]. These scores remove the squaring of the difference between a participant's confidence and performance so that direction of confidence may be established. That is, when confidence is high but performance is low, a participant is considered over-confident. When confidence is low but performance is high, a participant is considered under-confident. The distance from zero (i.e., the magnitude of this discrepancy) is a measurement of error severity. **Learning performance** was calculated as the average number of the multiple choice questions answered correctly throughout the learning session (i.e., out of 18 total trials). We choose not to examine CEs for this analysis as there is currently no validated method for examining bias for CEs, as CEs are not confidence judgements. Specifically, CEs are about the relevancy of content to an overall goal where as the other metacognitive judgements reflect confidence in one's ability and responses.

3 Results

3.1 RQ1: Can Metacognitive Accuracy on Multiple Local-Level Instances (i.e., Judgements) Be Used to Profile Learners for a Global-Level Abstraction (i.e., Ability)?

We use learners' average EOL, RCJ1, and RCJ2 AAI across all trials as variables for hierarchical clustering both across judgment type and learners. General descriptives are reported in Table 1. Using this method, we identified 3 clusters of participants and 2 clusters of judgment type. Results are visualized using a heatmap that have been sorted using the resulting clustering dendrograms (see Fig. 1).

Table 1. AAI & bias general descriptives

Judgment type	AAI		Bias	
	Mean (SD)	Min, Max	*Mean (SD)*	Min, Max
EOL	0.34 (0.08)	[0.18, 0.52]	−0.22 (0.20)	[−0.52, 0.38]
RCJ1	0.42 (0.08)	[0.25, 0.59]	−0.41 (0.13)	[−0.68, 0.00]
RCJ2	0.41 (0.09)	[0.23, 0.63]	−0.40 (013)	[−0.66, 0.00]

Metacognitive judgments can be seen in two main clusters- (1) EOLs and (2) RCJs. This is unsurprising as these judgements occur at two separate times in the learning session (prior to and post learning respectively). Participants have 3 emerging groups (more groups were not considered due to sample size limitations). Cluster 1 (N = 24) is composed of participants who have consistently higher accuracy in their metacognitive judgments across judgment types. Cluster 2 (N = 19) have roughly average or mixed accuracy in their judgments.

Cluster 3 (N = 15) are learners with low metacognitive accuracy across RCJs and a split between high and low EOL accuracy.

These results suggest that learners can have identifiable patterns in their accuracy across judgments that are made across the learning session. That is, learners who tend to have high EOL judgment accuracy additionally have high RCJ judgment accuracy. We have also identified a subset of learners who have a shift in their accuracy from highly accurate prior to learning to less accurate after learning (see top half of cluster 3).

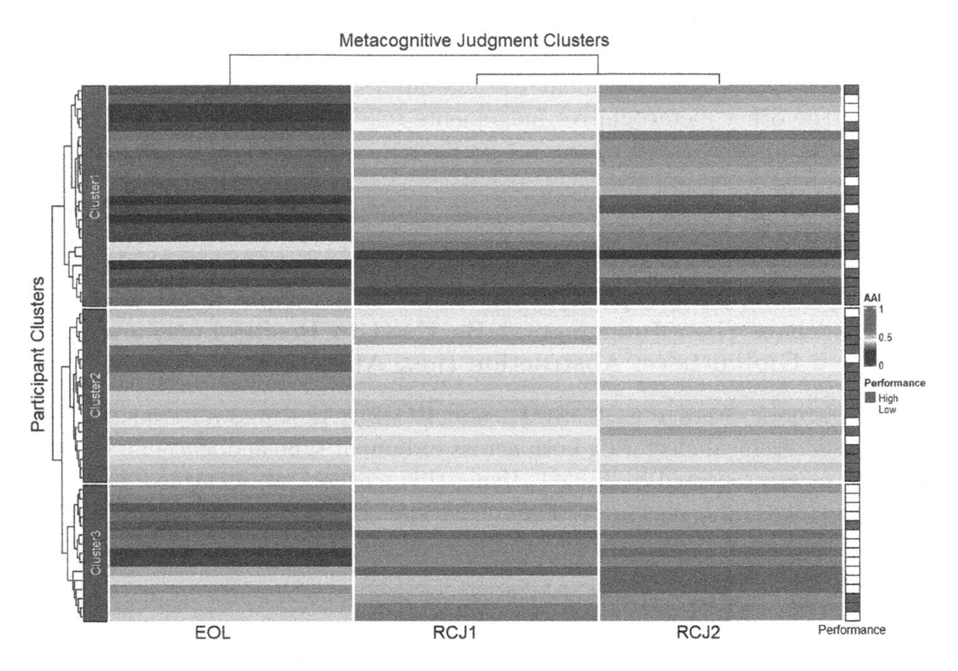

Fig. 1. Heatmap of Metacognitive Judgment Accuracy - Each row represents a unique participant and each column represents a metacognitive judgment. Each cell is shaded on a spectrum from a small AAI (blue; high metacognitive accuracy) to a larger AAI (red; low metacognitive accuracy). The green and white bar annotation along the right-hand size shows high and low performance (See RQ3). (Color figure online)

3.2 RQ2: Can Metacognitive Bias on Multiple Local-Level Instances (i.e., Judgements) Be Used to Profile Learners for a Global-Level Abstraction (i.e., Ability)?

We use learner's average EOL, RCJ1, and RCJ2 bias indices across all trials as variables for hierarchical clustering both across judgment type and learners. General descriptives are reported in Table 1. Results are visualized using a heatmap that have been sorted using the resulting clustering dendrograms (see Fig. 2)

Using this method, we are able to further examine the clusters identified in RQ1 (see the annotation bar on the right hand side of the graph). Specifically, we see that for Cluster 3 (those with low metacognitive accuracy), the majority of participants were under confident across all judgments. This under-confidence was also greater (i.e., more saturation seen in the heatmap) post-learning. The majority of participants in Cluster 1 (those with high metacognitive accuracy) are split in their bias. One group of these participants start the learning sessions over-confident, but their confidence reaches an appropriate level once learning has concluded and they have attempted to answer the posed question and justify that answer. Another group of these participants become somewhat under-confident for these later judgements, suggesting the aggregation method of averaging across all judgements may be losing some of the nuanced temporal fluctuations. That is, students that appear to have high metacognitive accuracy across the task actually have both inaccurate over- and under-confident judgements.

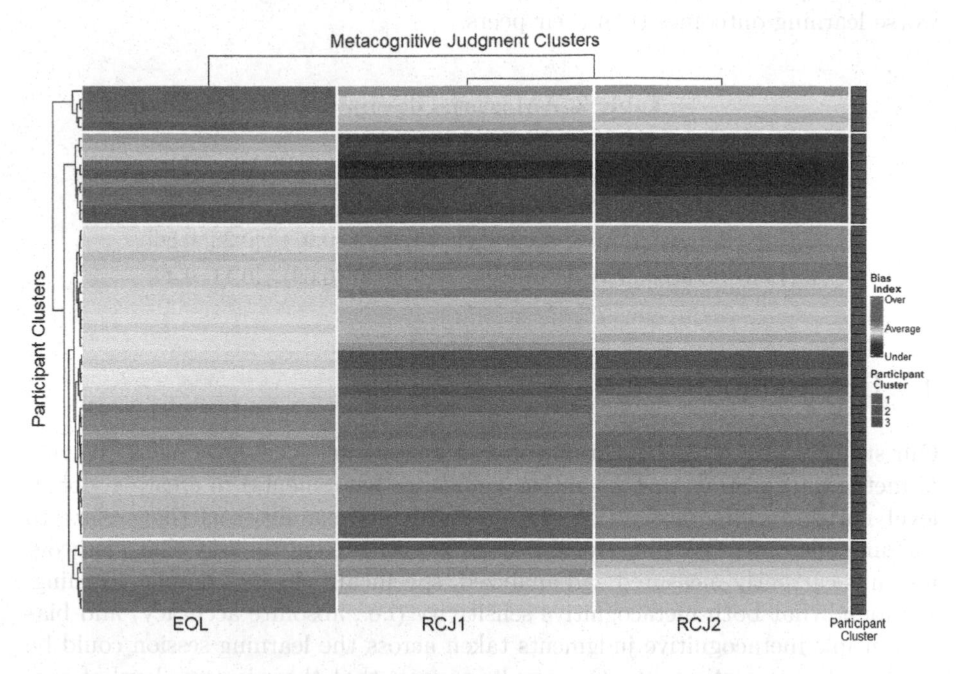

Fig. 2. Heatmap of Metacognitive Judgment Bias - Each row represents a unique participant and each column represents a metacognitive judgment. Each cell is shaded on a spectrum from under-confident (blue) to over-confident (red). The green, pink and purple bar annotation along the right-hand size shows the clusters from RQ1. (Color figure online)

3.3 Are There Differences in Learning Outcomes Between Learner Profiles?

Participants answered 39% (SD = 0.11) of the multiple choice questions correctly on average (roughly 7/18 questions). Table 2 provides general descriptives about learning performance by clusters identified in RQ1.

A one-way ANOVA was performed to compare this learning outcome across the identified clusters of participants. We found a statistically significant difference in learning performance between at least two of the clusters ($F(2,55) = 9.355$, $p < 0.005$). Post-hoc Tukey's HSD Test for multiple comparisons found the mean value of learning performance was significantly different between Cluster 1 (high metacognitive accurate participants) and Cluster 3 (low metacognitive accurate participants) ($p = 0.001$, 95% CI = [−0.194, −0.042]). Additionally, learning performance was significantly different between Cluster 2 (mixed accuracy participants) and Cluster 3 (low metacognitive accurate participants) ($p < 0.005$, 95% CI = [−0.210, −0.051]). There was no significant difference in learning performance between Cluster 1 (high metacognitive accurate participants) and Cluster 2 (mixed accuracy participants) ($p = 0.905$). These results suggest that participants in Cluster 3 (i.e., those with low metacognitive accuracy) had worse learning outcomes than their peers.

Table 2. AAI general descriptives

Cluster	N	Mean (SD)	Min, Max
1 (High metacognitive accuracy)	24	0.41 (0.11)	[0.17, 0.56]
2 (Mixed metacognitive accuracy)	19	0.43 (0.07)	[0.28, 0.56]
3 (Low metacognitive accuracy)	15	0.30 (0.09)	[0.11, 0.44]

4 Discussion

Our study explored the benefit of aggregating multiple local-level measurements of metacognition (i.e., metacognitive confidence judgements) to create a global-level measure of metacognition (i.e., metacognitive ability) as they relate to learning outcomes. We directly addressed several gaps in the way that metacognition is currently measured and analyzed, specifically during complex learning. We found that both metacognitive sensitivity (i.e., absolute accuracy) and bias of multiple metacognitive judgments taken across the learning session could be used to cluster participants. Our results suggest that there is some level of consistency in regards to the metacognitive accuracy of multiple types of judgments made throughout the learning session. That is, participants' accuracy did not fluctuate drastically based on when they provided the judgment (either prior to or post learning). These results fully support our original hypothesis. We also found learners who showed low metacognitive accuracy tended to be underconfident for all of their judgments. Learners with high metacognitive accuracy showed some over-confidence in their EOL judgments while some then moved to a more appropriate level of confidence once learning had concluded in their RCJs while a small subset of these learners became slightly underconfident. Finally, we found that learners who showed low metacognitive accuracy across all judgments under performed compared to their peers. However, learners who showed high metacognitive accuracy across all judgments did not outperform their peers. These results somewhat support our original hypothesis that metacognitive ability would show differences in learning outcomes.

5 Conclusion and Future Directions

Our work has begun to explore how we may use multiple local-level metrics of metacognitive sensitivity and bias and aggregate them into a global-level measurement of learner profiles that supports the concept of general metacognitive ability. While we did find significant learning outcome differences, our work suggests that aggregation through averaging judgements may not be the best approach. Specifically, we appear to lose temporal nuance using such a method, that other more sophisticated statistical approaches could help reveal (e.g., latent growth modeling, weighted modeling, etc.). Additionally, we have only begun to consider this approach with similar confident-based judgments. Future work should begin to address how other components of metacognition (e.g., metacomprehension evaluations) could be introduced into this approach and incorporated in AIEd systems. Similarly, we should also begin to consider more online measures, such as eye-tracking, facial expressions, etc., to help infer metacognitive states and judgment. Metacognitive research for complex learning is of great importance to the AIEd community in that it can help inform the development, design, and integration of new interventions and features. This type of research can specifically help answer questions about when system-deployed interventions and interruptions should occur to better support and foster student learning. For example, if we see that a student is overconfident based on a metacognitive ability profile from previous learning sessions, it might be more prudent to provide more context and activate prior knowledge before jumping straight into the new content. This would help students re-evaluate how easy or difficult they might believe the new task to be, and therefore reevaluate where and how much attention to allocate moving forward. The accurate identification of (in)accurate metacognitive processes is imperative to help improve the quality and type of scaffolding that are built into future AIEd systems that are both individualized and adaptive.

References

1. Azevedo, R., Wiedbusch, M.: Theories of metacognition and pedagogy applied in aied systems. In: du Boulay, B., Mitrovic, A., Yacef, K. (eds.) Handbook of Artificial Intelligence in Education. (In Press)
2. Flavell, J.H.: Metacognition and cognitive monitoring: a new area of cognitive-developmental inquiry. Am. Psychol. **34**(10), 906 (1979)
3. Koriat, A.: When two heads are better than one and when they can be worse: the amplification hypothesis. J. Exp. Psychol. Gen. **144**(5), 934 (2015)
4. Tarricone, P.: The Taxonomy of Metacognition. Psychology Press, New York (2011)

5. Winne, P.H.: Cognition and metacognition within self-regulated learning. In: Schunk, D., Greene, J. (eds.) Handbook of Self-regulation of Learning and Performance, pp. 36–48. Routledge, New York (2018)
6. Nelson, T.O., Narens, L.: Metamemory: A theoretical framework and new findings. Psychology of Learning and Motivation, p. 125 (1990)
7. Veenman, M.V., Hout-Wolters, V., Bernadette, H., Afflerbach, P.: Metacognition and learning: conceptual and methodological considerations. Metacogn. Learn. **1**(1), 3–14 (2006)
8. Seow, T.X., Rouault, M., Gillan, C.M., Fleming, S.M.: How local and global metacognition shape mental health. Biol. Psychiat. **90**(7), 436–446 (2021)
9. Jemstedt, A., Kubik, V., Jönsson, F.U.: What moderates the accuracy of ease of learning judgments? Metacogn. Learn. **12**(3), 337–355 (2017). https://doi.org/10.1007/s11409-017-9172-3
10. Leonesio, R.J., Nelson, T.O.: Do different metamemory judgments tap the same underlying aspects of memory? J. Exp. Psychol. Learn. Mem. Cogn. **16**(3), 464 (1990)
11. Mazzoni, G., Cornoldi, C., Tomat, L., Vecchi, T.: Remembering the grocery shopping list: A study on metacognitive biases. Applied Cognitive Psychology: The Official Journal of the Society for Applied Research in Memory and Cognition **11**(3), 253–267 (1997)
12. McCarley, J.S., Gosney, J.: Metacognitive judgments in a simulated luggage screening task. In: Proceedings of the human factors and ergonomics society annual meeting. vol. 49, pp. 1620–1624. SAGE Publications Sage CA: Los Angeles, CA (2005)
13. Dever, D.A., Wiedbusch, M., Azevedo, R.: Learners' gaze behaviors and metacognitive judgments with an agent-based multimedia environment. In: Isotani, S., Millán, E., Ogan, A., Hastings, P., McLaren, B., Luckin, R. (eds.) AIED 2019. LNCS (LNAI), vol. 11626, pp. 58–61. Springer, Cham (2019). https://doi.org/10.1007/978-3-030-23207-8_11
14. Brewer, W.F., Sampaio, C.: The metamemory approach to confidence: a test using semantic memory. J. Mem. Lang. **67**(1), 59–77 (2012)
15. Chua, E.F., Schacter, D.L., Rand-Giovannetti, E., Sperling, R.A.: Understanding metamemory: neural correlates of the cognitive process and subjective level of confidence in recognition memory. Neuroimage **229**(4), 1150–1160 (2006)
16. Dougherty, M.R., Scheck, P., Nelson, T.O., Narens, L.: Using the past to predict the future. Memory Cogn. **33**(6), 1096–1115 (2005)
17. Hines, J.C., Touron, D.R., Hertzog, C.: Metacognitive influences on study time allocation in an associative recognition task: an analysis of adult age differences. Psychol. Aging **24**(2), 462 (2009)
18. Ryals, A.J., Rogers, L.M., Gross, E.Z., Polnaszek, K.L., Voss, J.L.: Associative recognition memory awareness improved by theta-burst stimulation of frontopolar cortex. Cereb. Cortex **26**(3), 1200–1210 (2016)
19. Biswas, G., Segedy, J.R., Bunchongchit, K.: From design to implementation to practice a learning by teaching system: betty's brain. Int. J. Artif. Intell. Educ. **26**(1), 350–364 (2016)
20. Britton, B.K., Van Dusen, L., Gülgöz, S., Glynn, S.M., Sharp, L.: Accuracy of learnability judgments for instructional texts. J. Educ. Psychol. **83**(1), 43 (1991)
21. Jönsson, F.U., Kerimi, N.: An investigation of students' knowledge of the delayed judgements of learning effect. J. Cogn. Psychol. **23**(3), 358–373 (2011)
22. Koriat, A.: How do we know that we know? the accessibility model of the feeling of knowing. Psychol. Rev. **100**(4), 609 (1993)

23. Kelemen, W.L., Frost, P.J., Weaver, C.A.: Individual differences in metacognition: evidence against a general metacognitive ability. Memory Cogn. **28**(1), 92–107 (2000)
24. Schraw, G., Dunkle, M.E., Bendixen, L.D., Roedel, T.D.: Does a general monitoring skill exist? J. Educ. Psychol. **87**(3), 433 (1995)
25. Schraw, G., Nietfeld, J.: A further test of the general monitoring skill hypothesis. J. Educ. Psychol. **90**(2), 236 (1998)
26. Glaser, R., Schauble, L., Raghavan, K., Zeitz, C.: Scientific reasoning across different domains. In: De Corte, E., Linn, M.C., Mandl, H., Verschaffel, L. (eds.) Computer-Based Learning Environments and Problem Solving. NATO ASI Series, vol. 84, pp. 345–371. Springer, Heidelberg (1992). https://doi.org/10.1007/978-3-642-77228-3_16
27. Veenman, M.V., Wilhelm, P., Beishuizen, J.J.: The relation between intellectual and metacognitive skills from a developmental perspective. Learn. Instr. **14**(1), 89–109 (2004)
28. Veenman, M.V., Prins, F.J., Verheij, J.: Learning styles: self-reports versus thinking-aloud measures. Br. J. Educ. Psychol. **73**(3), 357–372 (2003)
29. Azevedo, R., Taub, M., Mudrick, N.V., Martin, S., Grafsgaard, J.: Using multi-channel trace data to infer and foster self-regulated learning between humans and advanced learning technologies. In: Handbook of self-regulation of learning and performance, vol. 2. Routledge New York (2018)
30. Dever, D.A., Wortha, F., Wiedbusch, M.D., Azevedo, R.: Effectiveness of system-facilitated monitoring strategies on learning in an intelligent tutoring system. In: Zaphiris, P., Ioannou, A. (eds.) HCII 2021. LNCS, vol. 12784, pp. 250–263. Springer, Cham (2021). https://doi.org/10.1007/978-3-030-77889-7_17
31. Taub, M., Azevedo, R.: How does prior knowledge influence eye fixations and sequences of cognitive and metacognitive SRL processes during learning with an intelligent tutoring system? Int. J. Artif. Intell. Educ. **29**(1), 1–28 (2019)
32. Wiedbusch, M., Dever, D., Wortha, F., Cloude, E.B., Azevedo, R.: Revealing data feature differences between system- and learner-initiated self-regulated learning processes within hypermedia. In: Sottilare, R.A., Schwarz, J. (eds.) HCII 2021. LNCS, vol. 12792, pp. 481–495. Springer, Cham (2021). https://doi.org/10.1007/978-3-030-77857-6_34
33. Pekrun, R., Goetz, T., Frenzel, A.C., Barchfeld, P., Perry, R.P.: Measuring emotions in students' learning and performance: the achievement emotions questionnaire (AEQ). Contemp. Educ. Psychol. **36**(1), 36–48 (2011)
34. Gross, J.J., John, O.P.: Individual differences in two emotion regulation processes: implications for affect, relationships, and well-being. J. Pers. Soc. Psychol. **85**(2), 348 (2003)
35. Chow, P.I., Berenbaum, H.: Perceived utility of emotion: the structure and construct validity of the perceived affect utility scale in a cross-ethnic sample. Cultur. Divers. Ethnic Minor. Psychol. **18**(1), 55 (2012)
36. Wiedbusch, M. D., Azevedo, R.: Modeling metacomprehension monitoring accuracy with eye gaze on informational content in a multimedia learning environment. In: ACM Symposium on Eye Tracking Research and Applications, pp. 1–9 (2020)
37. Schraw, G.: A conceptual analysis of five measures of metacognitive monitoring. Metacogn. Learn. **4**(1), 33–45 (2009)

Towards an Inclusive and Socially Committed Community in Artificial Intelligence in Education: A Social Network Analysis of the Evolution of Authorship and Research Topics over 8 Years and 2509 Papers

Yipu Zheng(✉), Zhuqian Zhou, and Paulo Blikstein

Teachers College, Columbia University, New York, NY 10027, USA
{yz3204,zz2404,paulob}@tc.columbia.edu

Abstract. This paper presents an overview of the last decade of research on Artificial Intelligence in Education by conducting keyword and social network analysis on the time-evolving co-authorship networks in four major research conferences: the International Conference on Artificial Intelligence in Education, the International Conference on Educational Data Mining, the International Conference on Learning Analytics and Knowledge, and the ACM Conference on Learning at Scale. Time-evolving co-authorship networks were used as a proxy for the collaboration dynamic in the field, while keyword analysis was conducted to supplement the social network analysis in order to pinpoint foci of individuals and cliques. Recent research foci and the level of openness of the four research communities were examined to inform strategies on how to promote diverse ideas and further collaborations within the field of AI in Education.

Keywords: Co-authorship network · Social network analysis · Keyword analysis · Bibliometric analysis

1 Introduction

With the growing attention to Artificial Intelligence (AI) in recent years, the application of AI technologies to education has grown considerably, developing from a niche field to a major interdisciplinary area [7]. Propelled by the global COVID-19 pandemic and remote learning, online and AI technologies (such as AI-powered content delivery systems, learning management software, and adaptive testing platforms) have impacted millions of students, teachers and schools and might continue in future years. Such significant implications of AI in Education (AIEd) leave research communities with greater responsibility than ever to examine its strengths, weaknesses, risks and opportunities. There is also growing concern about AIEd in terms of ethics, data privacy, and pedagogy [3]. As more EdTech companies (e.g., Cognii, Quizlet) claim that they are approaching the implementation of AI systems at a large scale, the AIEd research community needs

M. M. Rodrigo et al. (Eds.): AIED 2022, LNCS 13355, pp. 414–426, 2022.
https://doi.org/10.1007/978-3-031-11644-5_34

to not only recommend effective systems from a technical or cognitive perspective, but also shed light on potential social, ethical, cultural or even political issues that might come with the AI "hype" [3]. For example, how should students and parents deal with data privacy and security concerns? How should school districts and teachers react to biased algorithms and inequity in access? Who would control the content and pedagogy in public educational system in which decisions are outsourced to private AIEd companies? In order to foster meaningful discussions to address these emerging questions, AIEd research communities need to create intellectual spaces that welcome and foster new ideas.

Given such a context, this study conducts keyword analysis and social network analysis on the time-evolving co-authorship networks on the four major research conferences in the field – the International Conference on Artificial Intelligence in Education (IAIED), the International Conference on Educational Data Mining (EDM), the International Conference on Learning Analytics and Knowledge (LAK), and the ACM Conference on Learning at Scale (L@S) – from 2013 to 2020 to examine recent research foci and the level of openness of each research community. The analysis is conducted on three levels. On the macro-level, we investigated the most frequently used keywords and measures of connectedness and centralization over time in each of the four research communities. On the meso-level, we explored the formation of different cliques within each research community. In particular, we inspected the evolution of the largest connected components in each research community over time. On the micro-level, we focused on the demographics of the core authors in the field and their research topics. Overall, this paper investigates the major topics and collaboration patterns over time within the four major research communities to examine if they align with the societal needs and concerns around AIEd, and to inform strategies to promote diverse ideas and further collaborations in AIEd.

2 Related Work

Co-authorship networks have been acknowledged as an important indicator of the effectiveness of a research field because they visualize the social aspect of academic research that citation networks often overlook [5, 10]. In addition, longitudinal studies on co-authorship networks have been used to examine collaboration patterns and trends that evolve over time in a research community [1]. Building upon existing methods of social network analysis of co-authorship networks, this paper combines keyword analysis [8] and social network analysis to conduct a detailed inspection of current research trends and investigate how those are carried out within and among different cliques. Keywords of scholarly publications represent the authors' opinion of their articles and are good indicators of research trends [9]. Thus, we overlaid keyword analysis on top of our co-authorship network in order to garner a more comprehensive view of the focus of distinctive cliques and the collaboration dynamics within the four research communities.

There is an ongoing effort reviewing the research in the field of AIEd. Luckin et al. [8] provided a comprehensive view of AIEd and described existing applications in education in three categories: a) personal tutors, b) intelligent support for collaborative learning, and c) intelligent virtual reality. Zawacki-Richter et al. [13] explored the landscape of AIEd research in higher education and identified four potential AI use cases, namely "profiling and prediction", "intelligent tutoring systems", "assessment and evaluation", and "adaptive systems and personalisation". Feng and Kirkley [7] assessed disciplinary diversity in AIEd research collaboration at the individual, dyadic, and team levels for research on AIEd, and reported that disciplinary diversity was reflected by the diverse research experiences of individual researchers rather than diversity within groups. Berendt et al. [3] examined the benefits and risks of Artificial Intelligence (AI) in education in relation to fundamental human rights, and pointed out a need to balance the cost and reward as AI tools are developed, marketed, and deployed. This study contributes to this ongoing conversation by comparing how different research foci and collaboration dynamics were carried out within and among the major four research communities in the field over time, and providing insights on how to promote diverse ideas and further collaborations within the field. Worth noting, our discussion and findings are also framed within a larger debate about surveillance capitalism and the monetization of personal data [14], which could be made more deleterious if applied to schoolchildren with the increasingly significant presence of "big-tech" companies implementing AIEd solutions in public schools systems, especially in the developing world [11].

3 Methods

3.1 Data Collection and Analysis

To examine research foci and collaboration patterns in the field of AIEd, we first identified the major conferences (i.e., IAIED, EDM, LAK, and L@S) related to AIEd in consultation with expert researchers in the field. We then retrieved metadata of full conference papers from these four conferences from 2013 to 2020 from Springer, Scopus, and the ACM website. In total, there are 776 papers from IAIED, 560 papers from EDM, 719 papers from LAK, and 454 papers from L@S[1]. We specified the links of the web pages of target papers, automatically accessed the website's source code with *requests* and *selenium* libraries in Python, stored the source code as text files, and extracted data from the source with the *BeautifulSoup* library in Python. The dataset includes the following fields: article title, author names and affiliations, publication year, keywords, abstract, and references. Since the L@S community does not have a research journal, journal publications were not included to allow fair comparison across the four communities. In our data processing, we used the stemming (*PorterStemmer*) packages in NLTK library to group keywords with similar meanings (e.g., Tutor and Tutoring). We also created a unique identifier for each author in the dataset. Since some of the authors may use their names with variations (e.g., abbreviating first names or removing middle names) when publishing their papers, we wrote Python code to detect authors with slightly different names in the dataset, double-checked those names manually, and unified different variations into one unique identifier for each author. In total, there are 1632 authors from IAIED, 1377 authors from EDM, 1592 authors from LAK, and 1152 authors from L@S.

[1] The IAIED conference was held bi-annually up to 2017 and annually later, so our dataset on IAIED covers full conference papers in 2013, 2015, and 2017–2020. Additionally, the first L@S conference was held in 2014, so there are no papers from L@S in 2013.

Our data analysis was organized into three levels – the macro-level analysis displaying the structural and topical comparison of the four research communities, the meso-level analysis presenting the evolution of cliques within each research community, and the micro-level analysis revealing the demographics and distinctive research foci of core authors of the four research communities. Based on the collected dataset, we examined the overall trends of topics of each community through the most frequently used keywords, ascending keywords, and descending keywords over time. Additionally, we built time-evolving co-authorship networks with a two-year sliding window and utilized them as a proxy for the evolving collaboration patterns of the four communities. Core members of each community were identified using degree centrality. Cliques were detected using the modularity-optimization method offered in the R package, *igraph*, which maximizes the proportion of edges within communities relative to the expected proportion of such if all edges were placed randomly. Keywords produced by core members and major cliques in the co-authorship network were further classified into distinct conceptual groups to reveal different ideas and research foci of those major contributors. We used both node-level metrics (e.g. degree centrality) and network-level metrics (e.g. average degree) of social network analysis to reveal the co-authorship dynamics of the four communities.

3.2 Network Metrics

For the undirected co-authorship networks in this study, each node represents one author. Two nodes are connected to each other with an unweighted link if the two authors have co-authored at least one paper. The degree centrality of a node was used to measure how widely an author collaborates with other authors in paper publications, which equals the number of edges that this node possesses.

To measure the network-level properties, four metrics were used in this paper – the diameter, the average degree, the percentage of authors in the largest connected component, and the degree centralization, which are defined as follows.

The diameter is the shortest distance between the two most distant nodes in the network. The larger the diameter, the less connected the network is. It is defined below:

$$dia = \max\{min_{ij}\{p_{ij}\}\} \tag{1}$$

where p_{ij} is the number of edges between node i and node j.

The average degree is the average number of edges per node in the network. The larger the average degree, the more connected the network is. It is defined below:

$$d.avg = \frac{1}{2n}\sum\nolimits_{i=1}^{n} d_i \tag{2}$$

where n is the total number of nodes of the network, and d_i is the degree centrality of node i.

The LCC, i.e. the largest connected component, is a maximal set of nodes such that each pair of nodes is connected by a path. The larger the percentage of authors in the LCC, the more centralized the network is. It is defined below:

$$LCC\% = \frac{1}{n}|LCC| \tag{3}$$

where n is the total number of nodes on the network.

The degree centralization is the distribution of degree centrality among the nodes of a network. The closer the degree centralization is to 1, the more centralized the network is. The closer the degree centralization is to 0, the more equal distribution of degree centrality among nodes. It is defined below:

$$d.cen = \frac{\sum_{i=1}^{n}(\max\{d_j\} - d_i)}{(n-1)(n-2)} \tag{4}$$

where n is the total number of nodes of the network, d_i is the degree centrality of node i, $max\{d_i\}$ is the highest degree centrality among all nodes of the network. The numerator is the sum of the difference between the highest node-level degree centrality on the network in question and the degree centrality of each node on this network. The denominator is the largest value that the numerator can possibly achieve on a network with n nodes.

4 Findings

4.1 Macro-level Analysis: Frequently Used Keywords, Connectedness, and Centralization

On the macro-level, an analysis of the most frequently used keywords provided us with the first impression of the four research communities around AIEd (see Table 1). *Learning Analytics*, *Educational Data Mining*, *Machine Learning*, *Natural Language Processing*, *Intelligent Tutoring Systems (ITSs)*, and *Massive Open Online Courses (MOOCs)* are the top six keywords shared by all four research communities. Apart from the commonality, the four research communities also possessed different high-frequency keywords that distinguish itself from the other three (see Table 1). *Affect* and *Metacognition* highlighted the IAIED's focus on the cognitive aspect of learning while *Higher Education* and *Self-Regulated Learning* demonstrated the LAK's special attention to post-secondary education and matured learners. The EDM's distinctive keywords, *Deep Learning* and *Clustering*, stressed the community's emphasis on particular data mining techniques while the L@S's *Online Learning*, *Online Education*, and *Distance Learning* implied its strong interest in online learning environments.

Furthermore, among a host of network metrics and statistics, we found that connectedness and centralization were two meaningful dimensions to compare the four communities. The assumption that the oldest community, IAIED, whose first conference was in 1989, would have a more connected co-authorship network than the youngest community, L@S, whose first conference was in 2014, was validated via two social network metrics of connectedness, i.e. the diameter of the network (the shortest distance between the two most distant nodes in the network, *dia(IAIED)* $= 13 <$ *dia(L@S)* $= 16$) and the average degree (the average number of edges' per node in the network, *d.avg(IAIED)* $= 5.314 >$ *d.avg(L@S)* $= 4.696$). In addition to being more connected, the IAIED community was also more centralized than the L@S community as reflected by degree centralization (the distribution of degree centrality among the nodes of a network, *d.cen(IAIED)* $= 0.047 >$ *d.cen(L@S)* $= 0.034$) and the percentage of authors in the Largest Connected Components (the percentage of a maximal set of nodes such that each pair of nodes is connected by a path over the total number of nodes, *LCC%(IAIED) = 54.3% > LCC%(L@S) = 32.1%*).

Table 1. Ten most frequently used keywords of four AIEd research communities in 2013–2020.

IAIED	EDM	LAK	L@S
Intelligent tutoring system (113)	Massive open online courses (46)	Learning analytics (249)	Massive open online courses (168)
Natural language processing (36)	Educational data mining (33)	Massive open online courses (65)	Learning analytics (33)
Machine learning (28)	Intelligent tutoring system (28)	**Higher education (32)**	**Online learning (30)**
Student model (26)	Student model (27)	Educational data mining (25)	**Online education (26)**
Educational data mining (23)	Learning analytics (26)	Natural language processing (24)	**Education (23)**
Collaborative learning (20)	Knowledge tracing (20)	**Visualizations (24)**	**Assessment (17)**
Game-based learning (19)	Machine learning (17)	**Data mining (22)**	Machine learning (16)
Affect (19)	**Deep learning (16)**	Intelligent tutoring systems (18)	**Distance learning (13)**
Learning analytics (19)	Natural language processing (13)	Machine learning (18)	Intelligent tutoring systems (13)
Metacognition (19)	**Clustering (11)**	**Self-regulated learning (16)**	Knowledge tracing (13)

Note. The frequency of the keywords is presented in parentheses. Bolded keywords are the keywords unique to their community's top ten keyword list

Table 2. Four network metrics of four AIEd-related research communities in 2013–2020.

Dimensions	Network metrics	IAIED	EDM	LAK	L@S
–	Year of 1st Conference	1989	2008	2011	2014
Connectedness	Diameter (*dia*)	13	17	12	16
	Average Degree (*d.avg*)	5.314	4.593	5.175	4.696
Centralization	Percentage of Authors in the Largest Connected Component (*LCC%*)	0.543	0.466	0.563	0.321
	Degree Centralization (*d.cen*)	0.047	0.032	0.042	0.034

However, the LAK community is an exception. Although its first conference was held later than the IAIED and the EDM in 2011, two out of its four metrics (i.e. *dia* and *LCC%*) showed that the LAK community could be the most connected and the most centralized community in some aspects among the four (see Table 2).

4.2 Meso-level Analysis: Development of Sub-groups Around Different Topics Overtime

On the meso-level, we explored research topics of different co-authorship cliques and their changes over time within the four research communities. Co-authorship cliques were ranked by their sizes, and their research foci were identified by classifying their frequently used keywords. We saw a similar pattern to that found in the macro-level keyword analysis. The field is largely dominated by research around *Intelligent Tutoring Systems (ITSs)*, *Massive Open Online Courses (MOOCs)*, *Online Learning*, and *Game-Based Learning* but with a decreasing percentage over time. We visualized annual changes of the LCCs of each research community to study the evolution of the field. Due to the space limit, Table 3 only presents two LCC graphs from IAIED and LAK at the beginning and the end of the time window studied as examples. Denoted by different colors of nodes, the LCCs highlighted the research topics that were central to the research community in the corresponding year. One notable finding was that, despite the increasing societal emphasis on ethics, equity, privacy, and social justice issues related to AI in Education, we didn't see related keywords shown in the LCCs or in the major cliques in each community. Additionally, it's worth noting that the evolution of the LCC networks all started as a less connected chain-like network, then gradually grew more branches of collaboration around the central authors. The LAK community stood out once again by the well-connected mesh structure and ring structure of the LAK's LCC, which indicated vibrant endogenous growth of collaborations happening throughout the LAK's LCC apart from the increasing collaborations with the central authors.

To zoom in on research related to the increasing societal emphasis on ethics and equity issues related to AIEd, we analyzed out how many papers included such related keywords in each research community over time. The result is striking – only 19 papers address ethical concerns related to AIEd in all four research communities in the eight years.

Another unexpected finding from our analysis is the use of synonymous but distinct terminologies to describe the same concepts in different cliques. This phenomenon may, arguably, hinder inter-group communication since multiple terminologies need to be learned and used accordingly. Here is a typical example from the IAIED community: *Adaptive Learning* as a keyword is used by five co-authorship cliques, while another ontologically similar keyword, *Personalized Learning*, is used by another six co-authorship cliques. Although they are studying similar topics, *there are no overlaps or collaborations between these two sets of authors*. Other examples are *Student Model* versus *Learner Model* versus *User Model*, and *Automated Writing Evaluation* versus *Automated Writing Assessment* versus *Automatic Essay Assessment* versus *Automated Essay Scoring*. Even though more experienced researchers might be familiar with all synonyms and easily navigate the terminology, newcomers or external audiences might have issues understanding their differences. Take the example of *Adaptive* vs. *Personalized*

Learning: the two terms might be seen as equivalent in some research communities, but to an outside audience *personalized* learning carries a connotation of student-centered, constructivist learning—which does not happen to *adaptive* learning. These differences are consequential, and the language we use to describe different AIEd learning projects have deep implications for the way policymakers perceive and implement them, as we have shown in previous work [4].

Table 3. Largest Connected Components (LCCs) of IAIED and LAK in 2013–2014 and 2019–2020, colored by sub-communities and their respective most frequently used keywords.

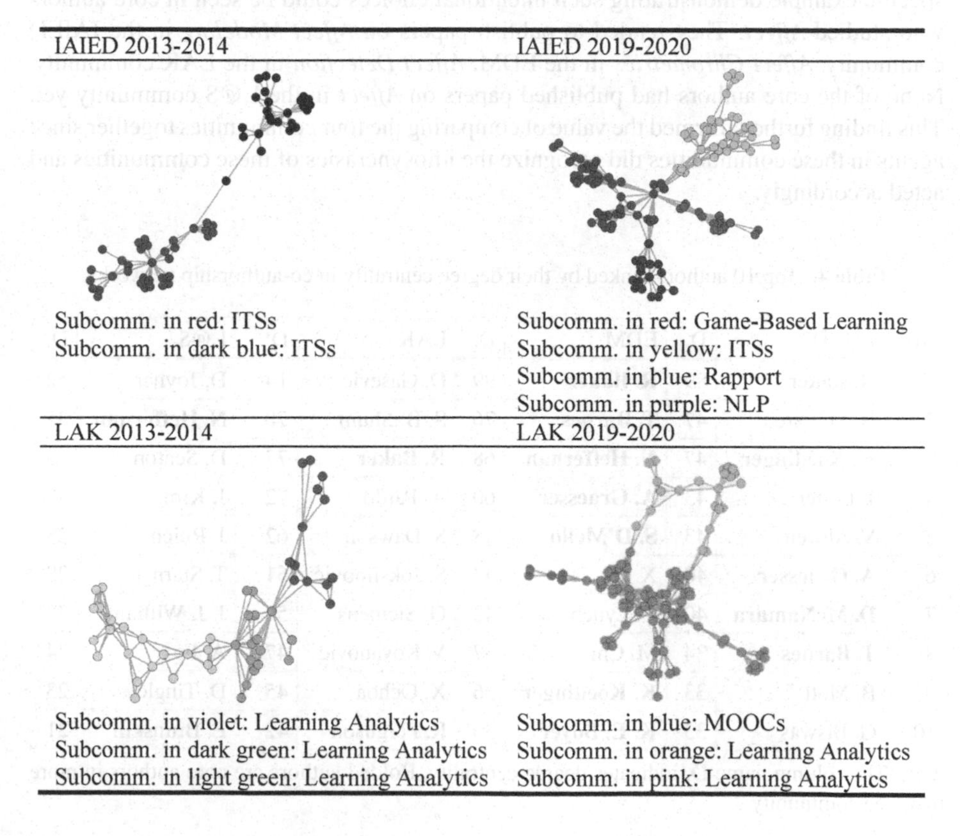

IAIED 2013-2014	IAIED 2019-2020
Subcomm. in red: ITSs Subcomm. in dark blue: ITSs	Subcomm. in red: Game-Based Learning Subcomm. in yellow: ITSs Subcomm. in blue: Rapport Subcomm. in purple: NLP
LAK 2013-2014	**LAK 2019-2020**
Subcomm. in violet: Learning Analytics Subcomm. in dark green: Learning Analytics Subcomm. in light green: Learning Analytics	Subcomm. in blue: MOOCs Subcomm. in orange: Learning Analytics Subcomm. in pink: Learning Analytics

4.3 Micro-level Analysis: Demographics and Research Foci of Core Authors in Different Communities

On the micro-level, we looked into authors who were central in the co-authorship networks and had co-authored with more than 20 people in this community from 2013 to 2020 (i.e. degree centrality greater than 20). According to this definition, there were in total 66 core authors from the four communities. More than 70% of them had doctoral degrees in Computer Science or Electrical Engineering.

Additionally, nine out of the 66 core authors served as the core for more than one community (see Table 4 for the top 10 core authors). It was interesting to see that these shared core authors intentionally chose publication venues to emphasize different aspects of their work. Consistent with the general impression and the macro-level analysis in this article, these authors included more analytical or technical terms as keywords, e.g. *Coh-Matrix*, *Bayesian Knowledge Tracing*, *A/B Testing*, and *Crowdsourcing*, in their EDM or L@S publications while more educational goals and learning constructs as keywords in their IAIED or LAK publications, e.g. *Computational Thinking*, *Hybrid Human-System Intelligence*, *Self-Regulated Learning*, and *Engaged Concentration*. A specific example demonstrating such intentional choices could be seen in core authors who studied *Affect*. They tended to publish papers on *Affect Modeling* in the IAIED community, *Affect Chrometrics* in the EDM, *Affect Detection* in the LAK community. None of the core authors had published papers on *Affect* in the L@S community yet. This finding further affirmed the value of comparing the four communities together since agents in these communities did recognize the idiosyncrasies of these communities and acted accordingly.

Table 4. Top 10 authors ranked by their degree centrality in co-authorship networks.

No.	IAIED	D.	EDM	D.	LAK	D.	L@S	D.
1	**R. Baker**	89	**R. Baker**	99	D. Gasevic	146	D. Joyner	52
2	B. McLaren	47	**T. Barnes**	76	S. B. Shum	78	**N. Heffernan**	35
3	**K. Koedinger**	47	**N. Heffernan**	68	**R. Baker**	77	D. Seaton	33
4	J. Lester	43	**A. Graesser**	60	A. Pardo	72	J. Kim	30
5	V. Aleven	43	**S. D'Mello**	48	S. Dawson	62	J. Reich	28
6	**A. Graesser**	40	X. Hu	43	S. Joksimović	51	T. Starner	28
7	**D. McNamara**	40	C. Lynch	42	G. Siemens	50	J. J. Williams	27
8	**T. Barnes**	34	M. Chi	37	V. Kovanovic	47	A. Fox	24
9	B. Mott	33	**K. Koedinger**	36	X. Ochoa	45	D. Tingley	23
10	G. Biswas	33	**K. E. Boyer**	33	R. Ferguson	42	E. Brunskill	21

Note. The column name D. indicates degree centrality. Bolded authors are core authors in more than one community

The LAK community was distinct from the other three communities again in terms of its core authors in two aspects. On the one hand, while 90% of core authors of the IAIED (22 out of 24), the EDM (18 out of 20), and the L@S (9 out 10) worked for institutions in the United States, the percentage flipped for the LAK community with 80% (19 out of 24) of its core authors worked outside of the United States, including eleven in Australia, two in China, one each in Japan, Belgium, Germany, Netherland, UK, and Canada. How the LAK community managed to achieve high connectedness and centralization while having such a diverse core author body is a question that deserves more investigation in the future. On the other hand, the LAK's core authors, though primarily had computer science backgrounds as their peers from the other three communities, studied Data Ownerships, Ethics, Legal Rights, Privacy, and Surveillance as well.

Apart from degree centrality, we also listed below the top 10 authors in each research community ranked by their betweenness centrality in co-authorship networks (see Table 5). Both centrality measures gave us a similar set of top-ranked authors, while the betweenness metric highlighted authors who served as bridges to connect different parts in the networks, which are worth noting from a collaboration dynamics perspective.

Table 5. Top 10 authors ranked by their betweenness centrality in co-authorship networks.

No.	IAIED	EDM	LAK	L@S
1	**R. Baker**	**R. Baker**	A. Pardo	J. Reich
2	N. Heffernan	C. Studer	X. Ochoa	R. Kizilcec
3	J. Beck	**T. Barnes**	S. B. Shum	S. Halawa
4	Y. Wang	A. Lan	**R. Baker**	J. Ruipérez-Valiente
5	V. Aleven	X. Hu	S. Dawson	G. Davis
6	G. Biswas	**S. D'Mello**	S. Crossley	D. Joyner
7	B. McLaren	L. Paquette	R. Ferguson	E. Glassman
8	**T. Barnes**	T. Yang	C. Mills	D. Seaton
9	V. Shute	C. Lynch	G. Siemens	Y. Rosen
10	S. Bull	C. Piech	**S. D'Mello**	J. Kim

Note. Bolded authors are top authors in more than one community

5 Discussion

By focusing on a recent dataset including conference full papers published in major conferences (IAIED, EDM, LAK, and L@S) in the field of AIEd, this paper provides insights on recent research foci and the level of openness of research communities within AIEd, and formulates strategies on how to possibly promote diverse ideas and further collaborations in the field.

Absence of Research on Ethics and Social Justice-Related Topics. Our analysis shows that only a minuscule number of publications focus on ethical or social justice concerns in all four conferences around AIEd. This is particularly striking given the overwhelming exposure that such topics have had in the popular media and in academia [2, 14]. In other words, there has been enough time for the AIEd communities to investigate these topics and make them a significant part of the community's research, but that has not yet taken place. This might indicate that the communities still see themselves as more technical and less engaged in issues of ethics, or that those issues do not warrant acceptance of papers into the conferences. Even though our data does not allow us to know the exact cause of such gap, we see this stance as problematic given the negative shift in the nature of the public discourse around AI in the larger society, the exploitation and monetization of people's data, and issues around algorithmic bias. In addition, in education, there have been central discussions about the privacy of children's data and the intrinsic limitations of AI in classrooms [12]. To further investigate this issue and take action, we recommend focusing on authors who did publish papers on ethics and social justice in AIEd, analyzing their position within the co-authorship network, and collecting new data about why their work was not taken more enthusiastically by the community. It would also be significant to investigate if core (or other) authors are involved in ethics-related work in other venues, and if they find obstacles in bringing such work to AIEd communities. Finally, research communities have ways to promote and incentivize research in new areas. The AIEd community could, thus, convene panels on ethics at conferences, create special issues of journals, give awards, create special conference tracks, or invite keynote speakers researching such issues.

Limited Global Participation. 90% of the core authors in the co-authorship networks of the IAIED, the EDM, and the L@S work in US institutions, while the 80% of the LAK core authors work outside of the US, but very few in the Global South. This is a phenomenon that requires further exploration. But with the global impact of AIEd technologies (especially after the COVID-19 pandemic), an international conversation around the topic is very much needed. In particular, the implications of AIEd technologies need to be framed within a larger debate about the monetization of personal (and children's) data [14], which could be made more harmful if applied to public school systems, especially in the developing world [11]. These communities are, so far, worryingly absent from academic discussions on the matter. In those countries, we could take action to incentivize collaborative research and elevate local scholars, creating mechanisms to bring them more to the center of the international community, thus empowering them in their local contexts as well.

Looking into LAK's Model. Our analysis on all three levels indicated LAK was doing a relatively good job in fostering greater participation and maintaining openness to diverse ideas. More research could illuminate how LAK achieves such progress by interviewing core authors and leaders of the community. In addition, the fact that LAK is more international and more interdisciplinary might provide important clues. We recognize, however, that not all conferences and communities should have the same goals, so the "lessons" of LAK should be taken critically.

Synonymous Keywords and Inconsistent Nomenclature as Potential Barriers for Collaboration. The use of synonymous but distinct terminologies to describe the same concepts in different cliques may hinder inter-group communication. However, two other plausible hypotheses should also be considered. First, perhaps some cliques simply work in different countries or institutions. Second, small cliques may not be crossing over because they are fleeting members of the community and have fewer ties with authors in general. However, over half of these authors have five or more co-author connections. Still, this finding suggests little overlap amongst many researchers who study almost identical phenomena, which could be caused by a lack of communication or of a defined terminology within AIEd. It is also the case, as we have discussed, that some terms such as "personalized learning" have become part of the commercial branding of many companies, making its more precise definition imperative for researchers. Finally, with the growth of the field and its four conferences, it is to be expected that many groups would generate their own terms and academic jargon, so an intentional effort to better define keywords and their use within the whole AIEd community is also timely and needed.

6 Conclusion

The last few years solidified AIEd as one of the core areas in educational research, as online, remote, hybrid, game-based, constructionist, and other types of technology-enabled learning entered the educational mainstream--especially during the COVID-19 pandemic. Now, and in the near future, AIEd will impact millions of students, and it is more important than ever to ascertain where the community is going, if we are balancing technical development and ethical considerations, and if we are welcoming new ideas and authors. This paper is a contribution to examining these topics, to build a more inclusive, socially committed, and robust research community.

References

1. Abbasi, A., Hossain, L., Uddin, S., Rasmussen, K.J.R.: Evolutionary dynamics of scientific collaboration networks: multi-levels and cross-time analysis. Scientometrics **89**(2), 687–710 (2011)
2. Benjamin, R.: Race After Technology: Abolitionist Tools for the New Jim Code. Polity, Cambridge (2019)
3. Berendt, B., Littlejohn, A., Blakemore, M.: AI in education: learner choice and fundamental rights. Learn. Media Technol. **45**(3), 312–324 (2020)
4. Blikstein, P., Blikstein, I.: Do educational technologies have politics? A semiotic analysis of the discourse of educational technologies and artificial intelligence in education. In: Algorithmic Rights and Protections for Children. PubPub (2021)
5. Cheong, F., Corbitt, B.: A social network analysis of the co-authorship network of the Australasian conference of information systems from 1990 to 2006. In: Proceedings of 17th European Conference on Information Systems (ECIS 2009), Verona, Italy (2009)
6. Clauset, A., Newman, M.E., Moore, C.: Finding community structure in very large networks. Phys. Rev. E **70**(6), 066111 (2004)
7. Feng, S., Kirkley, A.: Mixing patterns in interdisciplinary co-authorship networks at multiple scales. Sci. Rep. **10**(1), 1–11 (2020)
8. Luckin, R., Holmes, W., Griffiths, M., Forcier, L.B.: Intelligence unleashed: an argument for AI in Education (2016)
9. Pesta, B., Fuerst, J., Kirkegaard, E.O.: Bibliometric keyword analysis across seventeen years (2000–2016) of intelligence articles. J. Intell. **6**(4), 46 (2018)
10. Sarigöl, E., Pfitzner, R., Scholtes, I., Garas, A., Schweitzer, F.: Predicting scientific success based on coauthorship networks. EPJ Data Sci. **3**(1), 1–16 (2014). https://doi.org/10.1140/epjds/s13688-014-0009-x
11. Selwyn, N., et al.: What's next for Ed-Tech? Critical hopes and concerns for the 2020s. Learn. Media Technol. **45**(1), 1–6 (2020)
12. Taddeo, M., Floridi, L.: How AI can be a force for good. Science **361**(6404), 751–752 (2018)

13. Zawacki-Richter, O., Marín, V.I., Bond, M., Gouverneur, F.: Systematic review of research on artificial intelligence applications in higher education–where are the educators? Int. J. Educ. Technol. High. Educ. **16**(1), 39 (2019)
14. Zuboff, S.: Big other: surveillance capitalism and the prospects of an information civilization. J. Inf. Technol. **30**(1), 75–89 (2015)

Scaling Mixed-Methods Formative Assessments (mixFA) in Classrooms: A Clustering Pipeline to Identify Student Knowledge

Xinyue Chen(✉) and Xu Wang

University of Michigan, Ann Arbor, MI 48109, USA
{xinyuech,xwanghci}@umich.edu

Abstract. Formative assessments provide valuable data for teachers to make instructional decisions and help students actively manage their progress and learning. Multiple-choice questions (MCQ) and free-text open-ended questions are typically employed as formative assessments. While MCQs have the benefit of ease of grading and visualizing student answers, they lack capabilities in revealing diverse student ideas and reasoning beyond the options. On the other hand, open-ended tasks and free-text submissions may elicit students' perspectives more comprehensively, though it requires laborious work for instructors to analyze such responses. In this work, we explore the use of mixed-methods formative assessments in a college-level CS class, in which we assign MCQs and ask students to explain their answers. We propose a clustering pipeline to categorize students' free-text explanations leveraging the meta-data the original MCQs provide. We find that using students' choices in MCQs to resolve co-reference in their explanations and adding students' choices as features significantly improve clustering performance. Moreover, our work demonstrates that providing structures in the data collection process improves the clustering of free-text responses without making changes to the algorithm.

Keywords: Formative assessments · Self-explanations · Clustering pipeline

1 Introduction

College classes witness high enrollment in recent years [13]. Especially in computer science, students in introductory courses are from increasingly diverse backgrounds [1]. This introduces difficulties for instructors to accurately and efficiently predict students' knowledge and monitor student progress to plan for and adjust their instruction[14,22]. In-class formative assessments, in the format of multiple-choice questions (MCQ) or open-ended questions (OEQ), are often employed by instructors to identify students' strengths and weaknesses.

M. M. Rodrigo et al. (Eds.): AIED 2022, LNCS 13355, pp. 427–439, 2022.
https://doi.org/10.1007/978-3-031-11644-5_35

As an example, during a lecture, an instructor may use MCQs to probe into students' understanding of concepts and visualize student options in real-time [36]. In other cases, instructors may use OEQs and walk around the classroom to sample students' answers and prompt the class to discuss further [7].

Although MCQs have the benefit of ease of grading and help instructors quickly visualize student answers, prior work has raised concerns on whether the options experts designed could correctly and comprehensively reflect students' understanding and misconceptions [19,21]. Some studies have shown that learners may be over-tested by MCQ because they can select the right answer even when they are not able to complete the task [10]. Moreover, instructors could gain little insights into the reasoning behind students' choices [21]. On the other hand, OEQs have the benefit of revealing students' ideas and reasoning behind a problem [21]. However, using OEQs as formative assessments lacks the immediacy for instructors to identify students' weaknesses and monitor their progress, as analyzing a large amount of textual data is laborious [28,32]. To solve this problem, researchers have explored Natural Language Processing (NLP) methods to detect the common misconceptions in students' textual responses [26,29,33], however, it remains a challenging task for several reasons. First, it is difficult to parse the contextual information in students' answers, e.g. domain-specific terms and abbreviations, and incomplete sentences; Second, students' answers often have nuanced differences in the meaning they convey, but existing studies focus on detecting right answers from wrong rather than capture diverse students' perspectives. Third, although we have seen well-performed domain-specific classification models in short-answer grading, the generalizability across question topics and disciplines is unsatisfactory [29].

In this work, we explore the use of *mixed-methods formative assessments (mixFA)* to identify students' knowledge. Specifically, in a college-level user interface development class with 373 students, we assigned MCQs and ask students to explain their answers. We created a mixFA dataset with labels of students' ideas as shown in their explanations. We propose a clustering pipeline to categorize students' free-text explanations leveraging the meta-data the original MCQs provide. We see several benefits of using mixFA. First, mixFA elicits in-depth student reasoning and diverse student ideas compared to using MCQs alone. Second, the clustering pipeline can quickly and effectively cluster students' free-text explanations. We find using students' choices in MCQs to resolve co-reference in their explanations and adding students' choices as features significantly improve clustering performance.

We present a case study where providing structures in the data collection process improves the clustering of free-text student responses without making changes to the algorithm. We discuss the implications on collecting meta-data and improving feature representation as our community makes improvements on short answer clustering and classification problems. Through a qualitative error analysis of the clustering outcome, we surface the need to give instructors more control over the clustering setup, e.g., providing input for the algorithm to improve and being able to explore and rectify clustering results. We discuss implications on building human-in-the-loop interfaces to invite instructor input and allow for more versatile NLP-powered short answer clustering and classification pipelines. We suggest that mixFA could support instructors in identifying students' knowledge and monitoring student progress in a way that achieves quality and scale at the same time.

2 Related Work

In this section, we discuss relevant literature on how formative assessments can be used in classrooms to help instructors with decision making and prior machine learning-powered methods to identify students' knowledge and ideas.

2.1 Formative Assessments in Supporting Teaching and Learning

Decades of research has shown the benefit of using formative assessments to facilitate student learning and help instructors identify students' strengths and weaknesses to adjust their teaching [4,24]. In the process, it is critical for instructors to analyze student responses and translate the insights to help with their instructional decision-making [16]. Commonly used formative assessments could take the form of multiple-choice questions (MCQ) or open-ended questions (OEQ). While MCQs have the benefit of ease of grading, they may not elicit students' prior knowledge and ideas comprehensively [12]. On the other hand, while OEQs are better positioned to capture diverse student ideas, they do not provide immediacy for instructors to visualize student answers [4]. Research has shown that integrating self-explanations into MCQs could improve students' learning of complex concepts and skills, and help students develop meta-cognitive skills [5]. In our study, we explore the use of *mixFA*, combining MCQs with self-explanation prompts. We investigate whether mixFA could elicit diverse student reasoning and ideas beyond the options in MCQ and offer opportunities for automatically clustering student ideas and reasoning. This can support downstream educational applications, including supporting automatic short answer grading [27], crowdsourcing explanations for future students [34], and generating high quality questions leveraging natural student mistakes [32,33].

2.2 Automatic Methods for Identifying Students' Prior Knowledge and Misconceptions

Prior work has explored machine learning techniques to detect students' prior knowledge in short-answer textual responses [8,17,26]. Michalenko et al. developed a probabilistic model to differentiate students' correct and wrong answers [17]. Other work proposed NLP models to cluster students' short answers, with a focus on programming tasks that provide more structured features than free text [25,26]. More recent work applied pre-trained language models, such as BERT on short answer grading [6,18,30]. They found that transformers improved the accuracy of automatic grading results. We summarize the following limitations in prior work: 1) Most existing techniques in the space of classifying and clustering student free-text responses have a focus on detecting correct and incorrect answers. However, in a formative assessment setting, instructors have the desire to identify diverse student ideas and reasoning to plan for and adjust their teaching [4,24]. 2) Although we have seen successes with recent short answer grading techniques, the performance remains inconsistent across data sets [9]. Domain-specific models require substantial efforts on data annotation, whereas domain-general models also require abundant data input [35]. Nuanced meanings conveyed in students' short answers are hard to be captured with existing approaches [26]. In this work, through collecting the mixFA response dataset, we investigate whether structured student responses allow for the development of novel clustering pipelines to help instructors identify students' knowledge and misconceptions using mixFA.

3 MixFA Response Dataset

3.1 Data Collection

We explored the use of *mixFA* and collected a response dataset in a college-level introductory Human-Computer Interaction course with 373 students at the University of Michigan. Through discussion with two of the course instructors, we designed 7 Multiple-choice questions (MCQ) on topics including ideation, prototyping, think-aloud protocols, and universal and accessible design principles. Each MCQ offers 4–5 different options for students to choose from. The options were designed based on both instructors' predictions of students' prior knowledge and past students' mistakes. The MCQs were used in mixFA, in which students were also asked to explain their answers. The class was offered in the fall of 2021. The study was IRB approved, and 373 students in the class consented to have their data collected. At the end of the class, we collected 987 mixFA responses (with student MCQ choices and explanations).

3.2 Data Preparation

Since our goal is to investigate whether mixFA can elicit student reasoning and misconceptions behind their choices, we developed a coding scheme for each question to annotate unique student ideas or misconceptions emerging from their mixFA explanations. For each of the 7 questions, one author did the initial coding of the mixFA explanations. In the first step, answers that did not contain explanatory information (e.g., "refer to the slides"; "In lecture"; "Yes/No") were coded as "Non-informative". There were 284 student responses labeled as "non-informative" and excluded for further analysis. Two authors then did axial coding based on the initial codes. In this process, we made sure all initial codes with similar meanings were merged and determined whether a code is a correct understanding or a misconception. We then developed a codebook for each question which showed unique student ideas emerging from the mixFA explanations.

Two authors used the codebook to code 10% of data for each question in the mixFA dataset and achieved an average Cohen's kappa [15] of 0.91. One author then coded the whole mixFA dataset. Table 1 displays examples with initial and final codes after merging.

Table 1. An example of merging similar initial codes to a final code.

Student explanations	Initial code	Final code
You want to get a lot of ideas first and then judge them	Get idea first and then judge them later	Get ideas first (without judgment) and then evaluate/narrow them down later
Evaluation of quality is certainly more required for the next step in the process	Evaluation of the quality in the next step	Get ideas first (without judgment) and then evaluate/narrow them down later

Following the data annotation process, we built the mixFA response dataset with each student explanation labeled with a code representing a unique idea. The labels are used as the ground truth for our subsequent clustering and classification experiments. One thing to highlight here is that our ultimate goal is to support instructors using mixFA to identify diverse students' knowledge and misconceptions. So we tried our best to retain the meaning in students' explanations and made nuanced distinctions between codes in our coding process. Some codes may share common keywords but they demonstrate different specificity and levels of understanding from the students. For example, *"Block-based programming is easier because dragging is easier than typing for people with motor disabilities"* and *"Block-based programming is easier or more accessible"* are treated as different codes, since the former one displays extra reasoning, and both codes are misconceptions. The dataset includes 703 annotated free-text self-explanations in response to 7 multiple-choice questions. The dataset and the coding manual can be downloaded at this link[1].

4 Methods: A Clustering Pipeline for Identifying Students' Knowledge

To help instructors identify diverse student ideas and reasoning from students' self-explanations in mixFA, we develop a clustering pipeline. The novelty of the clustering pipeline lies in applying meta-level data that mixFA responses provide. Specifically, we use the original MCQ options to resolve the co-reference in students' explanations and use the MCQ answer as an additional feature.

[1] https://github.com/UM-Lifelong-Learning-Lab/AIED2022-MixFA-dataset.

4.1 Co-reference Resolution

One challenge presented in short answer grading is that incomplete sentences are common [9]. Similarly, in our mixFA dataset, student explanations often rely on contextual information in the question itself. For example, students may use pronouns or abbreviations to refer to the entities in the original MCQ options. Thus we applied co-reference resolution to contextualize students' explanations. Specifically, we used the NeuralCoref pipeline in SpaCy to resolve co-references with the combined option and explanation as input [11]. We then split the output to extract the resolved explanations. Here we present an example, before co-reference resolution: "It is an iterative process.", and after co-reference resolution: "The transition between lo-fi and high-fi prototyping is an iterative process." More examples are shown in Table 4.

4.2 Data Representation

We used sentence-BERT [23] to represent the textual data. Sentence-BERT is a state-of-the-art method for sentence embeddings. It utilized siamese and triplet network structures to derive semantically meaningful sentence embeddings. Prior work showed that sentence-BERT performed well on short answer grading tasks in an educational context [6,20], with better performance on clustering tasks than alternative GloVe and BERT embeddings [23]. We also tried Word2Vec, GloVe, BERT, sentence-BERT to represent student text answers and found sentence-BERT to be the best by comparing the clustering resutls with manual labels.

We extracted students' answers in the corresponding MCQ as an additional feature since students' MCQ answers represent their prior knowledge [21]. For example, for Question 3 as shown in Table 3, students explain the transition between low-fidelity and high-fidelity prototypes when they select option A while focusing on the benefit of the low-fidelity prototype when they select option B. In our dataset, students can have up to 14 different combinations of option selection since some MCQs used were select all that apply questions. To construct the feature space, we combined the feature column of students' MCQ answers with the vectorized explanation using sentence-BERT.

4.3 Clustering

We used the agglomerative clustering method with euclidean distance measure and average linkage provided in Scikit-Learn to cluster students' explanations leveraging the feature representation presented above. Agglomerative clustering is a bottom-up algorithm that treats each data point as a singleton cluster at the outset and then successively agglomerates pairs of clusters until all clusters have been merged into a single cluster that contains all data. We adopted this approach since it resembles instructors' natural process of discovering and merging different student ideas. In this study, we examine whether providing structure in the data collection process could improve the clustering of free-text student responses. We set the number of clusters to be the same as the number of codes extracted from the annotation process. We then evaluate the clustering outcome by comparing the results with our manual labels using Adjusted Mutual Information Score [31] Adjusted mutual information score (AMI) is a commonly used metric for comparing clustering outcomes and it corrects the effect of the agreement solely due to chance between clustering algorithms [31]. We also use the Silhouette Coefficient score to evaluate the density of the clusters [3].

5 Findings

In this section, we report the performance of the clustering pipeline with baseline models. We also report findings on an in-depth error analysis of the clustering outcome to suggest future pathways for more effective clustering of students' free-text answers.

5.1 Experiment Results

We use the mixFA dataset, as shown in the public link (see footnote 1). We run the clustering algorithm separately for each of the 7 questions. There are four experimental setups with different feature representations: 1) Sentence-BERT only; 2) Sentence-BERT applied after co-reference resolution (Resolved-SBERT); 3) Sentence-BERT plus MCQ options as a column feature (Option-SBERT); 4) Sentence-BERT applied after co-reference resolution plus MCQ options as a column feature (Resolved-Option-SBERT).

Meta-level Data from MCQ Improves the Clustering Outcome. Table 2 shows the AMI scores for the four experimental setups for each of the 7 questions. We applied Anova one-way analysis with Dunn's posthoc pairwise test. The column "p" shows the p-value of each model compared with the baseline model in Dunn's test. We see marginally significant improvement in Adjusted Mutual Information score with Resolved-SBERT (Ave. AMI $= 0.30$, $p < 0.1$) and significant improvement with Option-SBERT conditions (Ave. AMI $= 0.34$, $p < 0.05$). Maximum improvement is obtained when using both resolved explanations and the MCQ options as a column feature (Ave. AMI $= 0.42$, $p < 0.01$). This suggests that our proposed feature space with co-reference resolution and MCQ options improves the data representation.

Table 2. Adjusted Mutual Information score (AMI) for the four experimental setups with different feature representations. AMI score improved significantly in Resolved-SBERT, Option-SBERT, and Resolved-Option-SBERT compared with the Baseline.

Model	Questions	Q1	Q2	Q3	Q4	Q5	Q6	Q7	***Ave.***	***p***
	Clusters	18	22	14	16	17	18	13		
Baseline	AMI	0.14	0.30	0.19	0.11	0.3	0.16	0.24	0.21	
Resolved-SBERT	AMI	0.23	0.40	0.32	0.19	0.37	0.27	0.39	0.30	0.08*
Option-SBERT	AMI	0.18	0.41	0.36	0.21	0.33	0.30	0.51	0.34	0.04**
Resolved-Option-SBERT	AMI	0.29	0.44	0.38	0.29	0.41	0.36	**0.61**	**0.41**	0.003***

Tradeoff Between Capturing Nuanced Differences in Student Answers and Achieving Better Clustering Outcomes. In the experiments, we set the number of clusters to be the same as the number of manual labels shown in our dataset (see footnote 1), which gives us a relatively large number of clusters (average clusters = 17) under each question. Therefore, the lower distance-thresholds for hierarchical clustering will increase the possibility that student explanations with a certain degree of similarity are not merged, causing the errors. We present evidence that reducing the number of clusters may increase the AMI score and the Silhouette score. However, that will leave some unique student ideas, and nuances between student explanations uncaptured. Figure 1 shows the average AMI score across 7 questions when changing the number of clusters. We can see that for the Resolved-Option-BERT setup, the AMI score is peaked when N (number of clusters) = 13. Figure 2 shows the average Silhouette score across 7 questions when changing the number of clusters. A general trend is that the silhouette score is higher when there are fewer clusters. This is understandable because student answers may appear linguistically similar but convey different meanings.

This set of experiments demonstrates that if our goal is to capture diverse student ideas in a formative assessment scenario for instructors to understand student's knowledge and reasoning, especially the subtle differences in student answers, optimizing for existing ML metrics (such as AMI or Silhouette score) may not be sufficient.

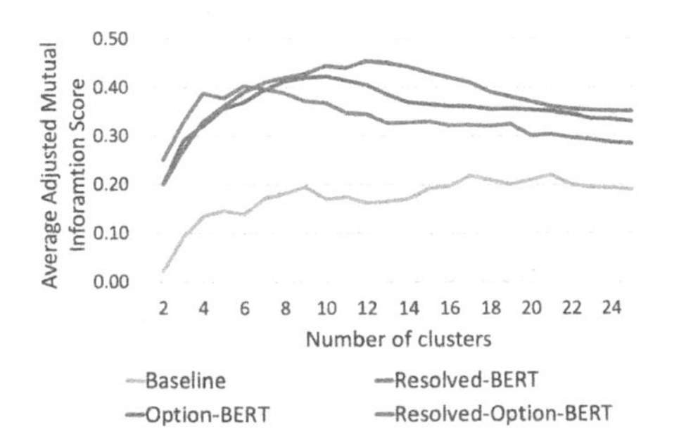

Fig. 1. Average AMI changes with the number of clusters. For the Resolved-Option-BERT setup, AMI peaks when N= 13, <the number of manual labels.

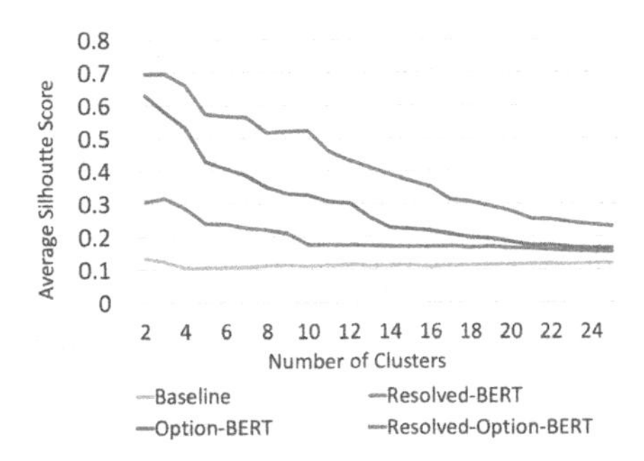

Fig. 2. The average Silhouette score increases as the number of clusters decreases. Resolved-Option-BERT setup has the overall best performance.

5.2 Qualitative Assessment

We performed an in-depth qualitative assessment of the clustering outcome to see how the new feature representation influenced the result and where the errors came from. We summarize the drawbacks of the clustering pipeline and propose future improvement ideas. We use Question 3 as an example, as shown in Table 3.

Table 3. An example question in the mixFA dataset (Question 3). Students are provided with a text field following this question to explain their answers.

Which of the following is NOT correct about the relationship between low-fidelity versus high-fidelity prototypes? Select all that apply (Correct Answers: B, C)
A. It is always better to first do a low-fidelity prototype versus a high fidelity prototype because we need to know the basics of user interaction
B. Lo-fi prototypes, if done well, could give us everything we need to understand user interactions with the system
C. The transition between lo-fi and high-fi prototyping is a linear process
D. Lo-fi prototypes could provide us with valuable data and help us evaluate high-level characteristics of the system that could inform us on how to build a high-fi prototype

Benefits of the New Data Representation with Co-reference Resolution and MCQ Option Column Feature. The co-reference resolution step successfully helps complete students' sentences. Table 4 shows examples where entities in students' explanations are successfully replaced and enriched with co-reference resolution. We found that adding the MCQ options as a column feature had mixed effects on the clustering outcome. On the one hand, the option feature helps when student explanations are aligned with the original options, e.g., the second example shown in Table 4. However, when student explanations are widely disparate, the option feature is distracting, e.g., the first example shown in Table 4.

Error Analysis. One source of error we observed was that two labels were clustered together. This was often due to the fact that student answers present similar linguistic features, however, when we analyze them qualitatively, they demonstrate subtle differences in student understanding. For example, students' explanations "Low-fi are important for an initial part of the prototyping process" and "Do Lo-fi at first helps gather data to build hi-fi" were grouped in one cluster as they were similar to some extent. However, in our manual coding, we take these as two different students' perspectives, "It is helpful to first do lo-fi first." and "Do Lo-fi first could help with hi-fi.", because the latter one is more specific about the relationship between lo-fi and hi-fi. On the contrary, another type of error is that students' explanations in one label were distributed into two clusters due to the length or quality of the explanations. For example, for the label "Lo-fi can not represent everything", students' simple answers such as *"Not everything"* were placed into one cluster, while other explanations with higher specificity such as *"Lo-fi prototypes intentionally exclude some of the details about how the app works"* were placed into a different cluster.

Table 4. Example of successful Co-reference resolution. The subject in the student's explanation was correctly replaced with domain-specific keywords.

	Options that students chose	Explanations	Resolved explanations
1	The transition between lo-fi and high-fi prototyping is a linear process	It is an iterative process	The transition between lo-fi and high-fi prototyping is an iterative process
2	Lo-fi prototypes, if done well, could give us everything we need to understand user interactions with the system	It wouldn't give us everything we need to know	Lo-fi prototypes wouldn't give us everything us need to know

Another main source of error was that incorrect and correct explanations with similar linguistic features were wrongly clustered together, e.g., students' explanation *"It is always better to do low-fi first"* and *"It's not necessarily true that it's always better to do low-fi first"* were wrongly clustered together. Since it is critical to recognize the polarity and sentiment in student answers, future work could incorporate additional features to highlight such tendencies during data representation. Besides, the linguistic distance between MCQ options, and the level of student knowledge they represent could serve as additional features. For example, some options are partially incorrect, whereas others are completely wrong. We also observe cases where co-reference resolution doesn't work well. This may happen when the option sentence has complex structures with multiple entities.

These errors point to design ideas for giving instructors more control in the process and interaction with the clustering or classification algorithm. First, in the mixFA dataset, the explanations for different questions possess varying properties, e.g., to what extent student explanations target the options in the MCQ. We can give instructors more control to decide what data representations to use, adjust the number of clusters, and determine the threshold for clustering depending on the nuanced level they want to get at. Second, instructor input on keywords, synonyms, and opposing arguments could help correct many of the errors we have seen in the experiments. Lastly, when clustering outcomes are not ideal, instructors need to have the freedom to freely explore and rectify the clustering result.

5.3 Validation of the New Feature Representation

We applied a supervised learning approach to examine how the new data representation supports classification compared to existing approaches on short answer grading. Specifically, with the feature representation of the Resolved-Option-BERT setup, we trained logistic regression classifiers and evaluated the classifiers through 10-fold cross-validation. The results are shown in Table 5. In comparison to a recent study [6] which uses SBERT for student answer classification (SBERT accuracy, 0.621), our setup reaches a higher level of accuracy (Resolved-option-SBERT, 0.661). This offers triangulation that the meta-level data provided by mixFA improves the data representation in students' free-text explanations.

Table 5. Accuracy of the classifiers built on the mixFA dataset with a 10-fold cross validation. This offers triangulation that the meta-level data provided by mixFA improves the data representation in students' free-text explanations.

	SBERT-baseline	Resolved-SBERT	Option-SBERT	Resolved-Option-SBERT	Condor, 2021 [6]
Accuracy	0.587	0.612	0.629	**0.661**	0.621
AMI	0.245	0.298	0.358	0.422	—

6 Discussion and Conclusion

In this work, we contribute the mixFA dataset which contains students' answers to MCQ questions and their free-text explanations. We then propose a clustering pipeline that improves the vectorization of students' free-text explanations using the meta-level data the corresponding MCQs provide. Our findings show that MCQ options could be used to resolve co-references in students' free-text answers and their MCQ choices provide additional context for clustering. We demonstrate that the clustering pipeline with co-reference resolution and the choice information significantly outperforms the baseline setup with sentence-BERT only. We show a case study where providing structures in the data collection process improves the clustering of free-text student responses without making changes to the algorithm. Besides, our findings show the trade-offs between capturing nuanced differences in students answers and optimizing for metrics such as the AMI and the Sillhoutte scores. Future studies in the space need to devise and use metrics that are aligned with instructional goals.

We present a qualitative error analysis which points to failure cases of the proposed clustering pipeline. We discuss the design implications on building a human-in-the-loop interface where instructors control the clustering setup and provide input to improve the outcomes [2]. For example, instructors may experiment with alternative data representations, choose when and how to use meta-level data, provide keywords and synonyms, specify opposing arguments, and rectify clustering mistakes.

In conclusion, our study shows that mixFA is a viable approach for eliciting diverse and nuanced student ideas and reasoning, while at the same time instructors can use the clustering pipeline to quickly examine students' knowledge.

References

1. Alhazmi, S., Hamilton, M., Thevathayan, C.: CS for all: catering to diversity of master's students through assignment choices. In: Proceedings of the 49th ACM Technical Symposium on Computer Science Education, pp. 38–43 (2018)
2. Amershi, S., Cakmak, M., Knox, W.B., Kulesza, T.: Power to the people: the role of humans in interactive machine learning. AI Mag. **35**(4), 105–120 (2014)
3. Aranganayagi, S., Thangavel, K.: Clustering categorical data using silhouette coefficient as a relocating measure. In: International Conference on Computational Intelligence and Multimedia Applications (ICCIMA 2007), vol. 2, pp. 13–17 (2007)

4. Bennett, R.E.: Formative assessment: a critical review. Assess. Educ. Principles Policy Pract. **18**(1), 5–25 (2011)
5. Chung, C.-Y., Hsiao, I.-H.: Examining the effect of self-explanations in distributed self-assessment. In: De Laet, T., Klemke, R., Alario-Hoyos, C., Hilliger, I., Ortega-Arranz, A. (eds.) EC-TEL 2021. LNCS, vol. 12884, pp. 149–162. Springer, Cham (2021). https://doi.org/10.1007/978-3-030-86436-1_12
6. Condor, A., Litster, M., Pardos, Z.: Automatic short answer grading with SBERT on out-of-sample questions. International Educational Data Mining Society (2021)
7. Crouch, C.H., Mazur, E.: Peer instruction: ten years of experience and results. Am. J. Phys. **69**(9), 970–977 (2001)
8. Feldman, M.Q., Cho, J.Y., Ong, M., Gulwani, S., Popović, Z., Andersen, E.: Automatic diagnosis of students' misconceptions in K-8 mathematics. In: Proceedings of the 2018 CHI Conference on Human Factors in Computing Systems (2018)
9. Galhardi, L.B., Brancher, J.D.: Machine learning approach for automatic short answer grading: a systematic review. In: Simari, G.R., Fermé, E., Gutiérrez Segura, F., Rodríguez Melquiades, J.A. (eds.) IBERAMIA 2018. LNCS (LNAI), vol. 11238, pp. 380–391. Springer, Cham (2018). https://doi.org/10.1007/978-3-030-03928-8_31
10. Harrison, C.J., Könings, K.D., Schuwirth, L.W., Wass, V., Van der Vleuten, C.P.: Changing the culture of assessment: the dominance of the summative assessment paradigm. BMC Med. Educ. **17**(1), 1–14 (2017)
11. Huggingface: Huggingface/neuralcoref: fast coreference resolution in spacy with neural networks. https://github.com/huggingface/neuralcoref
12. Kanli, U.: Using a two-tier test to analyse students' and teachers' alternative concepts in astronomy. Sci. Educ. Int. **26**(2), 148–165 (2015)
13. Kara, E., Tonin, M., Vlassopoulos, M.: Class size effects in higher education: differences across stem and non-stem fields. Econ. Educ. Rev. **82**, 102104 (2021)
14. Karataş, P., Karaman, A.C.: Challenges faced by novice language teachers: support, identity, and pedagogy in the initial years of teaching. Int. J. Res. Teach. Educ. **4**(3), 10–23 (2013)
15. Landis, J.R., Koch, G.G.: The measurement of observer agreement for categorical data. Biometrics **33**, 159–174 (1977)
16. Mandinach, E.B., Gummer, E.S., Muller, R.D.: The Complexities of Integrating Data-Driven Decision Making into Professional Preparation in Schools of Education: It's Harder Than You Think. CNA Analysis & Solutions, Alexandria (2011)
17. Michalenko, J.J., Lan, A.S., Baraniuk, R.G.: Data-mining textual responses to uncover misconception patterns. In: Proceedings of the Fourth ACM Conference on Learning @ Scale, L@S 2017, New York, NY, USA (2017)
18. Nandini, V., Maheswari, P.U.: Automatic assessment of descriptive answers in online examination system using semantic relational features. J. Supercomput. **76**(6), 4430–4448 (2020)
19. Nathan, M.J., Petrosino, A.: Expert blind spot among preservice teachers. Am. Educ. Res. J. **40**(4), 905–928 (2003)
20. Ndukwe, I.G., Amadi, C.E., Nkomo, L.M., Daniel, B.K.: Automatic grading system using sentence-BERT network. In: Bittencourt, I.I., Cukurova, M., Muldner, K., Luckin, R., Millán, E. (eds.) AIED 2020. LNCS (LNAI), vol. 12164, pp. 224–227. Springer, Cham (2020). https://doi.org/10.1007/978-3-030-52240-7_41
21. Polat, M.: Analysis of multiple-choice versus open-ended questions in language tests according to different cognitive domain levels. Novitas-ROYAL (Res. Youth Lang.) **14**(2), 76–96 (2020)

22. Qian, Y., Lehman, J.: Students' misconceptions and other difficulties in introductory programming: a literature review. ACM Trans. Comput. Educ. (TOCE) **18**(1), 1–24 (2017)
23. Reimers, N., Gurevych, I.: Sentence-BERT: sentence embeddings using Siamese BERT-networks. arXiv preprint arXiv:1908.10084 (2019)
24. Schildkamp, K., van der Kleij, F.M., Heitink, M.C., Kippers, W.B., Veldkamp, B.P.: Formative assessment: a systematic review of critical teacher prerequisites for classroom practice. Int. J. Educ. Res. **103**, 101602 (2020)
25. Shi, Y., Mao, T., Barnes, T., Chi, M., Price, T.W.: More with less: exploring how to use deep learning effectively through semi-supervised learning for automatic bug detection in student code. In: Proceedings of the 14th International Conference on Educational Data Mining (EDM) 2021 (2021)
26. Shi, Y., Shah, K., Wang, W., Marwan, S., Penmetsa, P., Price, T.: Toward semi-automatic misconception discovery using code embeddings. In: LAK21: 11th International Learning Analytics and Knowledge Conference, pp. 606–612 (2021)
27. Singh, A., Karayev, S., Gutowski, K., Abbeel, P.: GradeScope: a fast, flexible, and fair system for scalable assessment of handwritten work. In: Proceedings of the Fourth ACM Conference on Learning@ Scale, pp. 81–88 (2017)
28. Sirkiä, T., Sorva, J.: Exploring programming misconceptions: an analysis of student mistakes in visual program simulation exercises. In: Proceedings of the 12th International Conference on Computing Education Research, pp. 19–28 (2012)
29. Sung, C., Dhamecha, T.I., Mukhi, N.: Improving short answer grading using transformer-based pre-training. In: Isotani, S., Millán, E., Ogan, A., Hastings, P., McLaren, B., Luckin, R. (eds.) AIED 2019. LNCS (LNAI), vol. 11625, pp. 469–481. Springer, Cham (2019). https://doi.org/10.1007/978-3-030-23204-7_39
30. Uto, M., Uchida, Y.: Automated short-answer grading using deep neural networks and item response theory. In: Bittencourt, I.I., Cukurova, M., Muldner, K., Luckin, R., Millán, E. (eds.) AIED 2020. LNCS (LNAI), vol. 12164, pp. 334–339. Springer, Cham (2020). https://doi.org/10.1007/978-3-030-52240-7_61
31. Vinh, N.X., Epps, J., Bailey, J.: Information theoretic measures for clusterings comparison: variants, properties, normalization and correction for chance. J. Mach. Learn. Res. **11**, 2837–2854 (2010)
32. Wang, X., Rose, C., Koedinger, K.: Seeing beyond expert blind spots: online learning design for scale and quality. In: Proceedings of the 2021 CHI Conference on Human Factors in Computing Systems, pp. 1–14 (2021)
33. Wang, X., Talluri, S.T., Rose, C., Koedinger, K.: Upgrade: sourcing student open-ended solutions to create scalable learning opportunities. In: Proceedings of the Sixth ACM Conference on Learning@ Scale, pp. 1–10 (2019)
34. Williams, J.J., et al.: Axis: generating explanations at scale with learnersourcing and machine learning. In: Proceedings of the Third (2016) ACM Conference on Learning@ Scale, pp. 379–388 (2016)
35. Zhang, L., Huang, Y., Yang, X., Yu, S., Zhuang, F.: An automatic short-answer grading model for semi-open-ended questions. Interact. Learn. Environ. **30**(1), 177–190 (2022)
36. Zou, D., Xie, H.: Flipping an English writing class with technology-enhanced just-in-time teaching and peer instruction. Interact. Learn. Environ. **27**, 1127–1142 (2019)

Student-Tutor Mixed-Initiative Decision-Making Supported by Deep Reinforcement Learning

Song Ju(✉), Xi Yang, Tiffany Barnes, and Min Chi

Department of Computer Science, North Carolina State University, Raleigh, NC 27695, USA
{sju2,yxi2,tmbarnes,mchi}@ncsu.edu

Abstract. One fundamental goal of education is to enable students to *act independently* in the world by continuously adapting and learning. Certain learners are less sensitive to learning environments and can always perform well, while others are more sensitive to variations in learning environments and may fail to learn. We refer to the former as *high performers* and the latter as *low performers*. Previous research showed that low performers benefit more from tutor-driven Intelligent Tutoring Systems (ITSs), in which the tutor makes pedagogical decisions, while the high ones often prefer to take control of their own learning by making decisions by themselves. We propose a *student-tutor mixed-initiative (ST-MI)* decision-making framework which balances allowing students some control over their own learning while ensuring effective pedagogical interventions. In an empirical study, ST-MI significantly improved student learning gains than an Expert-designed, tutor-driven pedagogical policy on an ITS. Furthermore, our ST-MI framework was found to offer low performers *the same benefits as* the Expert policy, while that for high performers was *significantly greater* than the Expert policy.

Keywords: Critical decisions · Reinforcement learning · Student choice

1 Introduction

One fundamental purpose of education is to enable students to act independently in the world—to make good decisions and to adapt and continue to learn. On one hand, students who are more actively involved in deciding what and how to learn will benefit from the sense of control, such as becoming more engaged, motivated, and persistent [4,6,9]. On the other hand, not all students are adept at making decisions. Prior research has shown that *low performing learners* may not always have the necessary metacognitive skills to make effective pedagogical decisions [1,19]. As a result, most Intelligent Tutoring Systems (ITSs) are tutor-driven in that *the tutor* decides what to do in the next step. For example, the tutor can *elicit* the subsequent step from the student, either with prompting and support or without. When a student enters a step, the ITS records its success or failure and

M. M. Rodrigo et al. (Eds.): AIED 2022, LNCS 13355, pp. 440–452, 2022.
https://doi.org/10.1007/978-3-031-11644-5_36

may give feedback (e.g., correct/incorrect markings) and/or hints. Alternatively, the tutor can choose to *tell* them the next step directly, or provide a partially-worked step [11]. Each of these decisions affects the student's successive actions and performance. *Pedagogical policies* are used for the agent (i.e., tutor) to decide what action to take next among several alternatives.

In this work, we present a generalizable ***student-tutor mixed-initiative (ST-MI) decision-making*** framework which balances allowing students some control over their own learning while ensuring effective pedagogical interventions. More specifically, our framework is supported by a general Critical Deep Reinforcement Learning (Critical-DRL) approach, which uses Long-Short Term Rewards (LSTRs) and Critical Deep Q-Network (Critical-DQN). In the ST-MI framework, the tutor would take over decision-making ***only when students fail to make the optimal choice at critical moments***.

Figure 1 illustrates that our ST-MI framework consists of two loops with two agents: a student agent (SA) and a pedagogical agent (PA). The SA interacts with the environment in the inner loop (dashed area in Fig. 1), whereas the PA interacts with the inner loop in the outer loop. Here the SA is the front-end decision-maker, and our PA is the back-end. *If the SA makes a sub-optimal choice on a critical decision, the PA will intervene by taking an alternative choice and explain why it is better; Otherwise, the SA's decision is carried out.* To identify critical decisions, we proposed and developed a *Critical-DRL* approach using Long-Short Term Rewards and Critical Deep Q-Network described in 3.1.

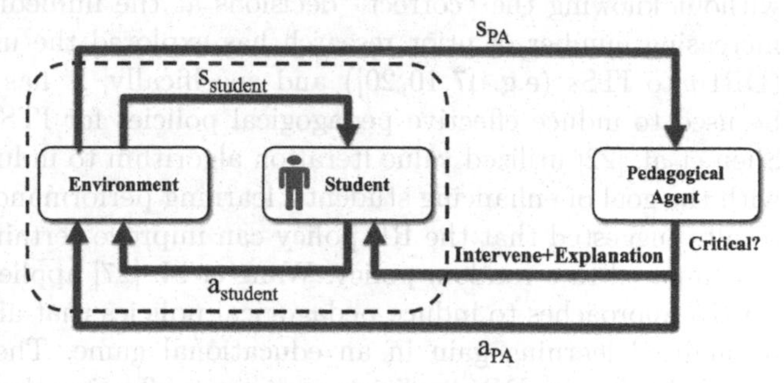

Fig. 1. Our ST-MI Decision-making Framework

The effectiveness of the ST-MI framework is empirically compared against an Expert-designed policy, referred to as the *Expert policy*, where the tutor makes all pedagogical decisions. In this study, we focused on the decisions on whether to present the next problem as a Worked Example (WE), a Problem Solving (PS), or a faded worked example (FWE). In WE, students were given a detailed example showing how the tutor solves a problem; in PS, by contrast, students were tasked with solving the same problem on their own on the ITS; in FWEs, the students and the tutor *co-construct* in that their solutions are intertwined. Our results showed that the ST-MI students achieved significantly higher learning gains than the Expert peers. Further, we separated students based on their incoming competencies, i.e., pretest scores, and examined the impact of the ST-MI framework on the Aptitude-Treatment Interaction (ATI). For low incoming competence students, in particular, prior research has shown that they are less likely to benefit from making pedagogical decisions on their own [30], our ST-MI framework was found to offer the low performers the same benefits as the Expert policy. While previous research has shown that high incoming competence students are just as effective at learning as those who make their own decisions or follow the Expert policy [30], our findings showed that the ST-MI framework can significantly improve their learning over the Expert policy.

2 Related Work

Applying RL to ITSs: In ITSs, the student-agent interactions can be described as sequential decision-making problems under uncertainty, which can be formulated as problems of RL, a learning paradigm that depends on long-term rewards without knowing the "correct" decisions at the immediate time-steps [24]. An increasing number of prior research has explored the use of RL and Deep RL (DRL) to ITSs (e.g. [7,10,20]) and specifically, it has showed that they can be used to induce effective pedagogical policies for ITSs [10,27]. For example, Shen et al. [22] utilized value iteration algorithm to induce a pedagogical policy with the goal of enhancing students' learning performance. Empirical evaluation results suggested that the RL policy can improve certain learners' performance as compared to a random policy. Wang et al. [27] applied a variety of Deep RL (DRL) approaches to induce pedagogical policies that aim to improve students' normalized learning gain in an educational game. The simulation evaluation revealed that the DRL policies were more effective than a linear model-based RL policy. Recently, Zhou et al. [29] applied offline Hierarchical Reinforcement Learning (HRL) to induce a pedagogical policy to improve students' normalized learning gain. In a classroom study, the HRL policy was significantly more effective than the other two flat-RL baseline policies. In summary, prior studies suggest that RL-induced pedagogical policies can enhance the effectiveness of tutor-driven ITS where tutors are the ones making pedagogical choices. As far as we know, none of the prior work has attempted to employ RL for an ST-MI-like framework that would allow both students and tutors to make pedagogical decisions, and none of them has examined the effectiveness of the ST-MI framework on student learning.

Identifying Critical Decisions: The advances of computational neuroscience allow researchers to treat the brain as a supercomputing machine to understand the learning and decision-making process in animals and humans [15,18,23]. A lot of studies have shown that RL-like signals and decision-making processes exist in humans/animals and we humans use immediate reward and Q-value to make decisions [12]. In RL, the Q-value is defined as the expected cumulative reward for taking an action at a state and following the policy until the end of the episode. Therefore, the difference in Q-values between two actions for a given state reflects the magnitude of the difference in the final outcomes. Motivated by research in human and animal behaviors, lots of RL work has applied Q-value difference as a heuristic measurement for the importance of a state and decide when to give advice in a simulated environment called the "Student-Teacher" framework [5,25,32]. Their research question is when to provide an advice and their results showed that the Q-value difference was an effective heuristic function to estimate the importance of a state.

Student Decisions: Much of prior research has shown that while students can benefit from making their own decisions during learning [4,21], they are not always good at making effective pedagogical decisions. For example, Mitrovic et al. showed that even college students often make poor problem selections [13]. Aleven & Koedinger found that students often do not use hints effectively in that they tended to wait too long before asking for hints [1]. Wood et al. found that students with low prior knowledge exhibit ineffective help-seeking behaviors than those with high prior knowledge [28].

WE, PS, and FWE: Many studies have examined the effectiveness of WE, PS, and FWE, as well as their different combinations [16,17,26]. Renkl et al. [17] compared WE-FWE-PS with WE-PS pairs and the results showed that WE-FWE-PS condition significantly outperformed WE-PS condition on posttest scores. Similarly, Najar et al. [16] compared adaptive WE/FWE/PS with WE-PS pairs and found that the former is significantly more effective than the latter on improving student learning. Overall, it is demonstrated that adaptively alternating amongst WE, PS, and FWE is more effective than hand-coded expert rules in terms of improving student learning. However, when students making decisions among WE, PS, and FWE, there's no significant difference with tutor making decisions on students' learning performance [30]. As far as we know, no prior research has explored how to combine students' decision-making with RL-induced policy's decision-making to facilitate learning.

3 Method

3.1 Long-Short Term Rewards

To determine whether a state is critical, Critical-DRL considers both short-term reward (ShortTR) and long-term reward (LongTR) [8]. For the ShortTR, it considers the *immediate rewards* over all possible actions to determine the criticality of a state. A primary challenge, however, is that in most ITSs we only have delayed rewards, and immediate rewards are often not available. Specifically, in ITSs, student's learning performance is the most appropriate reward, but it is typically not available until the entire learning trajectory has been completed. Due to the complex nature of learning, it is difficult to assess students' knowledge level moment by moment, and more importantly, many instructional interventions that boost short-term performance may not be effective over the long term. To tackle this issue, we apply a Deep Neural Network-based approach called InferNet, which infers the immediate rewards from delayed rewards. Prior work showed that the InferNet-learned immediate rewards can be as effective as real immediate rewards [2]. Here we employ the InferNet to infer the ShortTR for each state-action pair. Furthermore, to determine whether a state is critical or not, we calculate two thresholds by applying the elbow method to the inferred immediate rewards distribution: one is a *positive reward threshold* above which the agent should pursue, and the other is a *negative reward threshold* below which the agent should avoid. A state is critical if any action on it can lead to an inferred immediate reward either higher than the positive threshold or lower than the negative threshold.

For the LongTR, *Q-value difference* is used to measure the criticality of a state. Q-values are the expected cumulative reward for an agent to take an action a at state s and follow the policy to the end. In theory, if all the actions for a given state have the same Q-value, which one should be taken doesn't matter because they all lead to the same reward. Conversely, if the Q-values of various actions differ widely, taking the wrong action could result in a significant loss of rewards. We define the LongTR of a state s, then, as the difference between its minimum and maximum Q-values: $LongTR(s) = \max_a Q(s,a) - \min_{a'} Q(s,a')$. In general, the higher the LongTR, the more important the state should be.

3.2 Critical Deep Q-Network

In order to determine LongTR, we developed a Critical-DRL approach using Deep Q-Networks (DQN) because of its great success in handling complicated tasks, such as robot control and video game playing [14]. DQN approximates the Q-value function using deep neural networks following the Bellman equation. In the original Bellman equation, the Q-values are calculated assuming that the agent takes the optimal action in every state. In our ST-MI framework, however, optimal actions are taken in critical states, and any action can be taken in non-critical states. Thus, we used the modified Bellman equation as:

$$Q(s,a) = \begin{cases} r + \gamma * max(Q(s',a')) & \text{s' is critical} \\ r + \gamma * average(Q(s',a')) & \text{s' is non-critical.} \end{cases} \quad (1)$$

For a state s and an action a, $Q(s,a)$ follows the original Bellman equation (top) if the next state s' is critical; otherwise we use the average Q-value over all the available actions for s' to update $Q(s,a)$ (bottom). To induce the Critical-DQN policy, we first apply the ShortTR threshold to identify a fixed set of critical states. Then, during each iteration in training, our Critical-DQN algorithm first calculates the Q-value difference $\Delta(Q)$ for all states in the training dataset. Then the median of the Q-value differences is defined as a threshold. If the $\Delta(Q)$ of a state is greater than the threshold, it is critical; otherwise, it is non-critical. The critical states are the union of the two sets identified by the ShortTR and the LongTR, respectively. After the critical states have been determined, the algorithm follows Eq. 1 to update the Q-values. Then in the next iteration, the updated Q-values are applied to determine a new median threshold to update the critical states recursively. This process will repeat until convergence. Once the Critical-DQN policy is induced, for any given state, we calculate its Q-value difference and compare it with the corresponding median threshold. If the Q-value difference is larger than the threshold, the state is critical.

3.3 Hierarchical RL Policy Induction

Our ITS first makes the problem-level decisions (WE/PS/FWE) and if a FWE is selected, step-level decisions (elicit/tell) will be made. With the two levels of decisions, we extended the existing flat-RL algorithm to Hierarchical RL (HRL), which aims to induce an optimal policy to make decisions at different levels. Most HRL algorithms are based upon an extension of MDPs called Discrete Semi-Markov Decision Processes (SMDPs). Different from MDPs, SMDPs have an additional set of complex activities or options, each of which can invoke other activities recursively, thus allowing the hierarchical policy to function [3]. The complex activities are distinct from the primitive actions in that a complex activity may contain multiple primitive actions. In our applications, WE, PS, and FWE are complex activities, while elicit and tell are primitive actions. For HRL, learning occurs at multiple levels. A global learning generates a policy for the complex level decisions and local learning generates a policy for the primitive level decisions in each complex activity. More importantly, the goal of local learning is not inducing the optimal policy for the overall task but the optimal policy for the corresponding complex activity. Therefore, our HRL approach learns a global problem-level policy to make decisions on WE/PS/FWE and learns a local step-level policy for each problem to choose between elicit/tell.

4 Policy Induction

Training Corpus: Our training dataset contains a total of 1,307 students' interaction logs collected over seven semesters' classroom studies (2016 Fall to 2020 Spring). During the studies, all students used the same tutor, followed the same general procedure, studied the same training materials, and worked through the same training problems. The training corpus provides us with the state representation, action, and reward information for policy induction. **State:** We extracted 142 features that might impact student learning from the student-system interaction logs. More specifically, these state features can be categorized into the following five groups: *Autonomy:* the amount of work done by the student; *Temporal Situation:* the time-related information about the work process; *Problem-Solving:* information about the current problem-solving context; *Performance:* information about the student's performance during problem-solving; *Student Action:* the statistical measurement of student's behavior. **Action:** Our tutor makes decisions at two levels of granularity: problem and step. In the problem-level, there are three actions WE/PS/FWE. In the step-level, there are two actions elicit/tell. **Reward:** There's no immediate reward during tutoring, and the delayed reward is the students' Normalized Learning Gain (NLG), which measures their learning gain irrespective of their incoming competence. NLG is defined as $\frac{posttest-pretest}{\sqrt{1-pretest}}$, where 1 is maximum score for both pre- and post-test.

Two Policies: Our **ST-MI** follows the *Critical-DRL model* in that in non-critical states, the student's decision is always carried out, while in critical states, if the student's choice aligns with the ST-MI policy's optimal action, the tutor executes it; otherwise, *the tutor executes the policy's choice and explains it to the student why it is better.* Each type of pedagogical actions has multiple explanation messages, and the tutor will select a message at random to display to students. Because of space constraints, we only include one example message per type of intervention in Table 1. The explanation is intended to smooth the student-system interactions. In our prior work, we found that adding these explanations to RL-induced policies does not improve their effectiveness while adding them to a policy that is not effective does not harm it [31]. The **Expert policy** is designed by an instructor with more than 20 years of experience on the subject. Based on our ITS and prior instructional experience, the Expert policy consists of alternating between elicit and tell at step-level, which was shown to be more effective than other baselines [30].

Table 1. Examples of Explanation Messages in Problem-Level

Student	*ST-MI*	*Explanation messages*
WE	FWE	*"We are good on time. Let's work together on this problem."*
WE/FWE	PS	*"We are good on time. Try to solve this one yourself."*
PS	FWE	*"To learn more efficiently, let's solve this together."*
PS/FWE	WE	*"You performed pretty well so far. Let me solve this problem."*

5 Experiment Setup

Participants: This study was given to students as a homework assignment in an undergraduate Computer Science class in the Fall of 2020. Students were told to complete the study in one week, and they will be graded based on demonstrated effort rather than learning performance. 153 students were *randomly assigned* into the two conditions: $N = 65$ for Expert and $N = 88$ for ST-MI. *It is important to note that the difference in size between the two conditions is due to the fact that we prioritized having a sufficient number of participants in the ST-MI condition to perform a meaningful analysis of the ATI effect.* Due to preparation for final exams and the length of study, 117 students completed the study. In addition, 12 students were excluded from our subsequent statistical analysis due to the perfect performance in the pre-test. The final group sizes were $N = 47$ for Expert and $N = 58$ for ST-MI. A Chi-square test on the relationship between students' condition and their completion rate found no significant difference between the two conditions: $\chi^2(1) = 2.4335$, $p = 0.12$.

Pyrenees Tutor: Our tutor is a web-based ITS to teach students probability and covers 10 major principles, such as the Complement Theorem, Bayes' Rule, etc. It provides step-by-step instruction and immediate feedback. As with other systems, Pyrenees provides students with help via a sequence of increasingly specific hints, which prompts them with what they should do next. The last hint in the sequence, i.e., the bottom-out hint, tells the student exactly what to do.

Experiment Procedure & Grading: Both conditions went through the same four phases: 1) textbook, 2) pre-test, 3) training on the ITS, and 4) post-test. The only difference among them was how the pedagogical decisions were made. During **textbook**, all students read a general description of each principle, reviewed some examples, and solved some training problems. The students then took a **pre-test** which contained a total of 14 single- and multiple-principle problems. Students were not given feedback on their answers, nor were they allowed to go back to earlier questions (this was also true for the post-test). During **training**, both conditions received the same 12 problems in the same order. Each domain principle was applied at least twice. Finally, all students took the 20-problem **post-test**: 14 of the problems were isomorphic to the pre-test, and the remainders were non-isomorphic multiple-principle problems. All of the tests were graded in a double-blind manner by a single experienced grader. For comparison purposes, all test scores were normalized to the range of $[0, 1]$.

6 Results

6.1 ST-MI vs. Expert

Pre-test Score: No significant difference was found between the Expert condition ($M = 0.77, SD = 0.13$) and the ST-MI condition ($M = 0.73, SD = 0.22$) on the pre-test scores: $t(103) = 1.18$, $p = 0.23$, $d = 0.23$. It suggests that the two conditions are balanced in terms of incoming competence.

Improvement Through Training: A repeated measures analysis using test type (pre-test vs. isomorphic post-test) as a factor and test score as the dependent measure showed a main effect for test type for both conditions in that students scored significantly higher in the isomorphic post-test than in the pre-test: $F(1, 46) = 10.6$, $p = .0016$, $\eta = 0.319$ for Expert and $F(1, 57) = 13.64$, $p = .0003$, $\eta = 0.315$ for ST-MI respectively. In details, the isomorphic post-test scores in the ST-MI condition is ($M = 0.86, SD = 0.19$) while the Expert condition is ($M = 0.85, SD = 0.11$). It shows that both conditions learned significantly from training on our tutor.

Learning Performance & Training Time: In comparing students' learning performance between the two conditions, we compared their isomorphic posttest and full posttest scores, as well as their isomorphic and full NLGs. The goal of the isomorphic posttest is to assess the learning gain and whether or not the tutor is helpful, while the purpose of the full posttest is to determine whether the intervention makes a difference in student learning. There was no significant difference between the two conditions on either isomorphic posttest or posttest. For example, the ST-MI students had higher post-test scores ($M = 0.81, SD = 0.21$) than

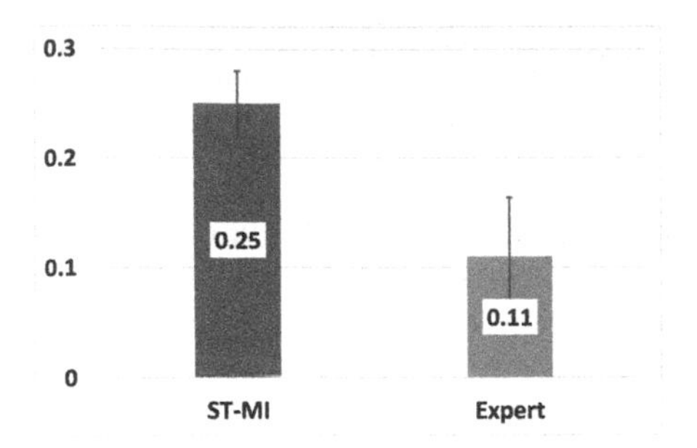

Fig. 2. Isomorphic NLG

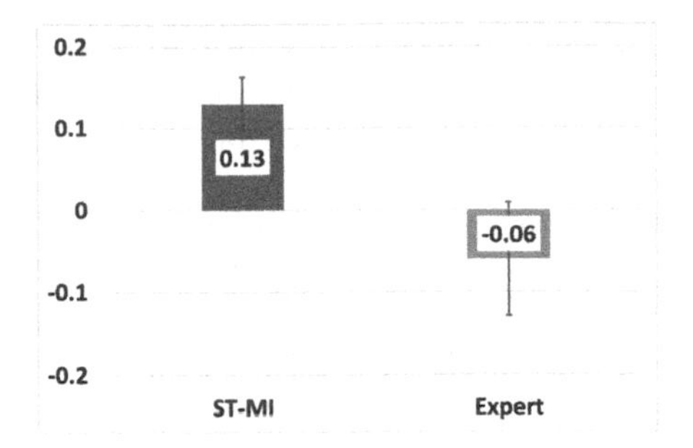

Fig. 3. Full NLG

the Expert students ($M = 0.78, SD = 0.15$) but such difference is not significant: $t(103) = 0.83, p = 0.411, d = 0.16$. Our most important interest, however, is in student performance improvement from pre- to posttest, so we focus on isomorphic NLG and NLG. ***NLGs of both types demonstrate how beneficial our ITS actually are, as well as their role as reward functions in our Critical DRL framework***. The ST-MI condition scored significantly higher than the Expert condition on both the isomorphic NLG: $t(103) = 2.35$, $p = 0.021$, $d = 0.46$ and the full NLG: $t(103) = 2.72$, $p = .008$, $d = 0.53$. In Fig. 2, the isomorphic NLG for the ST-MI condition is ($M = 0.25, SD = 0.23$) and the Expert condition is ($M = 0.11, SD = 0.36$). Similarly, in Fig. 3, the full NLG for the ST-MI condition is ($M = 0.13, SD = 0.25$) while the Expert condition is ($M = -0.06, SD = 0.47$). Finally, on training time the ST-MI condition spend less time (measured in minutes, $M = 109.6, SD = 38.1$) than the Expert condition ($M = 123.2, SD = 47.1$) during the training on the tutor but the difference is not significantly: $t(103) = -1.63$, $p = 0.106$, $d = 0.32$. In short, our results indeed show that the ST-MI policy significantly improves students' learning gains with less time cost than the Expert policy.

6.2 The Impact of ST-MI on ATI Effect

In order to measure ATI, we further divided students into High vs. Low groups by a median split on their pretest scores, also known as incoming competence. Thus, we had four groups based upon their pretest scores and policies: High-ST-MI (n=28), Low-ST-MI (n=30), High-Expert (n=21), Low-Expert (n=26). No significant difference was found among the two conditions on the distribution of High vs. Low students: $\chi^2(1) = 0.0291$, $p = 0.86$. Table 2 presents the comparison between the policies {ST-MI, Expert} and incoming competence {High, Low} in terms of learning performance. As expected, in both conditions the high group significantly outperformed their low peers in the pretest: $t(45) = 6.07$, $p < 0.001$, $d = 1.10$ for Expert and $t(56) = 9.03$, $p < 0.001$, $d = 1.10$ for ST-MI. Moreover, while no significant difference was found between the High-Expert and High-ST-MI ones: $t(47) = 0.33$, $p = 0.74$, $d = 1.10$, the Low-Expert significantly out-performed the Low-ST-MI ones: $t(54) = 2.57$, $p = 0.012$, $d = 1.10$.

Table 2. Learning performance for four groups

Group	*Pre*	*Iso Post*	*Post*	*Iso NLG*	*NLG*	*Time*
Low-Expert	0.67 (0.08)	0.80 (0.11)	0.71 (0.15)	0.23 (0.20)	0.06 (0.27)	129.9 (52)
Low-ST-MI	0.58 (0.21)	0.78 (0.23)	0.70 (0.24)	0.31 (0.22)	0.18 (0.24)	108.6 (44)
High-Expert	0.90 (0.05)	0.91 (0.08)	0.86 (0.11)	−0.02 (0.46)	−0.21 (0.61)	114.8 (40)
High-ST-MI	0.88 (0.06)	0.96 (0.06)	0.92 (0.07)	0.19 (0.23)	0.07 (0.26)	111.8 (31)

Table 2 shows that the test score results are consistent with our hypothesis. Despite their significantly lower pre-test scores, the Low-ST-MI students catch up with their Low-Expert peers on the following four performance measures in that no significant difference was found between them on Iso-Post, Post, Iso NLG, and full NLG. According to [30], the low incoming competence students are less likely to benefit from making pedagogical decisions on their own, but our results showed that our ST-MI framework with ST-MI policy could make them catch up to their peers in the Expert condition. As for the two High groups, both scored high for the isomorphic and full posttests, and the High-ST-MI group outperformed the High-Expert group on both the Iso NLG: $t(47) = 2.59, p = 0.011, d = 0.41$, and NLG: $t(47) = 2.78, p = 0.007, d = 0.40$. While previous research has shown that high incoming students are just as effective at learning as those who make their own decisions or follow the Expert policy [30], our findings showed that despite having a high score in the Iso-post and Posttest scores, the High-Expert group does not seem to benefit from the tutor, as their average NLG is negative. In contrast, ST-MI policy can significantly enhance the High performers' learning gains when compared with Expert policy.

In summary, our findings confirm that ST-MI can benefit both High and Low performers. More specifically, low performers who are more sensitive to learning environments can parallel their Expert peers with our framework, while high performers, who are less sensitive to learning environments and always perform well, can further boost their learning gains with our ST-MI framework.

6.3 Log Analysis

Table 3. Problem-Level Critical Decisions in ST-MI

Decisions	*High*	*Low*	*T-test result*
Critical decision	8.2 (1.8)	6.1 (2.9)	$\mathbf{t(56) = 3.36, p = 0.001^*, d = 0.88}$
Correct critical choice	3.4 (2.0)	2.5 (2.3)	$t(56) = 1.46$, $p = 0.150$, $d = 0.38$
Intervention	4.9 (2.6)	3.5 (1.9)	$\mathbf{t(56) = 2.21, p = 0.031^*, d = 0.58}$

Next, we analyze the pedagogical decision behaviors between the High and Low groups in the ST-MI condition. Table 3 shows the average number of different types of critical decisions students received in the problem-level. In Table 3, there are three types of critical decisions: 'Critical Decision' means the decision state is identified as critical by our ST-MI policy; 'Correct Critical Choice' means the students select the optimal actions (same as our policy's choice) in the critical decision; 'Intervention' means the students select the sub-optimal actions (different from our policy's choice) in the critical decision. By definition, correct critical choices and intervention are exclusive and they are subsets of critical decisions. First, the High students experienced significantly more critical decisions than the Low students. Then, by facing more critical moments, not only were the High students able to make more correct critical choices (not significant), but also they received more interventions (significant) to achieve their goals. Additionally, there's no significant difference between High vs. Low on all three types of critical decisions in the step-level. In summary, the results showed that the High students experienced more interventions than the Low group students, and as a result, the intervention could help the High students experience more critical optimal actions, which can lead to better learning performance.

7 Conclusion

In the classroom study, we evaluated the effectiveness of the ST-MI framework by comparing the ST-MI policy with a baseline Expert policy. In the ST-MI condition, students could control their own learning process by making decisions on what type of questions they want, and in the meantime, the RL-induced policy would intervene when they make sub-optimal choices in critical decisions and give dedicated explanations. The results show that the students in the ST-MI condition significantly outperform the students in the Expert condition in terms of learning performance. Additionally, a log analysis suggests that the students with high incoming competence received more interventions than the students with low incoming competence. The reason is that the RL-induced policy aims to maximize NLG, and the high students usually have lower NLG due to little room to improve. As a result, the RL-induced policy would intervene more on the high students to improve their NLG. Finally, we observe a trend that giving students control over their learning could make the learning more efficient. Overall, the empirical study demonstrates that our proposed ST-MI framework could improve students' learning without the trivial tutor-driven step decisions.

Acknowledgements. This research was supported by the NSF Grants: 1660878, 1651909, 1726550 and 2013502.

References

1. Aleven, V., Koedinger, K.R.: Limitations of student control: do students know when they need help? In: Intelligent Tutoring Systems, pp. 292–303 (2000)
2. Ausin, M.S., Maniktala, M., Barnes, T., Chi, M.: Tackling the credit assignment problem in reinforcement learning-induced pedagogical policies with neural networks. In: AIED (2021)

3. Barto, A.G., Mahadevan, S.: Recent advances in hierarchical reinforcement learning. Discrete Event Dyn. Syst. **13**(1–2), 41–77 (2003)
4. Cordova, D.I., Lepper, M.R.: Intrinsic motivation and the process of learning: Beneficial effects of contextualization, personalization, and choice. J. Educ. Psychol. **88**(4), 715 (1996)
5. Fachantidis, A., Taylor, M.E., Vlahavas, I.P.: Learning to teach reinforcement learning agents. Mach. Learn. Knowl, Extra. (2017)
6. Flowerday, T., Schraw, G., Stevens, J.: The role of choice and interest in reader engagement. J. Exp. Educ. **72**(2), 93–114 (2004)
7. Ju, S., Zhou, G., Abdelshiheed, M., Barnes, T., Chi: M.: Evaluating critical reinforcement learning framework in the field. In: AIED, pp. 215–227 (2021)
8. Ju, S., Zhou, G., Barnes, T., Chi, M.: Pick the moment: Identifying critical pedagogical decisions using long-short term rewards. In: EDM (2020)
9. Kinzie, M.B., Sullivan, H.J.: Continuing motivation, learner control, and CAI. Education Tech. Research Dev. **37**(2), 5–14 (1989)
10. Mandel, T., Liu, Y.E., Levine, S., Brunskill, E., Popovic, Z.: Offline policy evaluation across representations with applications to educational games. In: AAMAS, pp. 1077–1084 (2014)
11. Maniktala, M., Cody, C., Barnes, T., Chi, M.: Avoiding help avoidance: Using interface design changes to promote unsolicited hint usage in an intelligent tutor. Int. J. Artif. Intell. Educ. **30**(4), 637–667 (2020)
12. McClure, S.M., Laibson, D.I., Loewenstein, G., Cohen, J.D.: Separate neural systems value immediate and delayed monetary rewards. Science **306**(5695), 503–507 (2004)
13. Mitrovic, A., Martin, B.: Scaffolding and fading problem selection in sql-tutor. In: AIED, pp. 479–481 (2003)
14. Mnih, V., et al.: Human-level control through deep reinforcement learning. Nature **518**, 529–533 (2015)
15. Morris, G., Nevet, A., Arkadir, D., Vaadia, E., Bergman, H.: Midbrain dopamine neurons encode decisions for future action. NatureNeuro **9**(8), 1057–1063 (2006)
16. Najar, A.S., Mitrovic, A., McLaren, B.M.: Adaptive support versus alternating worked examples and tutored problems: which leads to better learning? In: Dimitrova, V., Kuflik, T., Chin, D., Ricci, F., Dolog, P., Houben, G.-J. (eds.) UMAP 2014. LNCS, vol. 8538, pp. 171–182. Springer, Cham (2014). https://doi.org/10.1007/978-3-319-08786-3_15
17. Renkl, A., Atkinson, R.K., Maier, U.H., Staley, R.: From example study to problem solving: smooth transitions help learning. J. Exp. Educ. **70**(4), 293–315 (2002)
18. Roesch, M.R., Calu, D.J., Schoenbaum, G.: Dopamine neurons encode the better option in rats deciding between different delayed or sized rewards. Nat. Neurosci. **10**(12), 1615–1624 (2007)
19. Roll, I., Wiese, E.S., Long, Y., Aleven, V., Koedinger, K.R.: Tutoring self-and co-regulation with intelligent tutoring systems to help students acquire better learning skills. Design Recomm. Intell. Tutoring Syst. **2**, 169–182 (2014)
20. Rowe, J.P., Lester, J.C.: Improving student problem solving in narrative-centered learning environments: a modular reinforcement Learning framework. In: Conati, C., Heffernan, N., Mitrovic, A., Verdejo, M.F. (eds.) AIED 2015. LNCS (LNAI), vol. 9112, pp. 419–428. Springer, Cham (2015). https://doi.org/10.1007/978-3-319-19773-9_42
21. Schneider, S., Nebel, S., Beege, M., Rey, G.D.: The autonomy-enhancing effects of choice on cognitive load, motivation and learning with digital media. Learn. Instr. **58**, 161–172 (2018)

22. Shen, S., Chi, M.: Reinforcement learning: the sooner the better, or the later the better? In: UMAP, pp. 37–44. ACM (2016)
23. Sul, J.H., Jo, S., Lee, D., Jung, M.W.: Role of rodent secondary motor cortex in value-based action selection. Nat. Neurosci. **14**(9), 1202–1208 (2011)
24. Sutton, R., Barto, A.: Reinforcement Learning: An Introduction. MIT Press (2018)
25. Torrey, L., Taylor, M.E.: Teaching on a budget: agents advising agents in reinforcement learning. In: AAMAS, pp. 1053–1060 (2013)
26. Van Gog, T., Kester, L., Paas, F.: Effects of worked examples, example-problem, and problem-example pairs on novices' learning. Contemp. Educ. Psychol. **36**(3), 212–218 (2011)
27. Wang, P., Rowe, J., Min, W., Mott, B., Lester, J.: Interactive narrative personalization with deep reinforcement learning. In: IJCAI (2017)
28. Wood, H., Wood, D.: Help seeking, learning and contingent tutoring. Comput. Educ. **33**(2), 153–169 (1999)
29. Zhou, G., Azizsoltani, H., Ausin, M.S., Barnes, T., Chi, M.: Hierarchical reinforcement learning for pedagogical policy induction. In: AIED, pp. 544–556 (2019)
30. Zhou, G., Chi, M.: The impact of decision agency & granularity on aptitude treatment interaction in tutoring. In: CogSci, pp. 3652–3657 (2017)
31. Zhou, G., Yang, X., Azizsoltani, H., Barnes, T., Chi, M.: Improving student-tutor interaction through data-driven explanation of hierarchical reinforcement induced pedagogical policies. In: UMAP. ACM (2020)
32. Zimmer, M., Viappiani, P., Weng, P.: Teacher-student framework: A reinforcement learning approach. In: AAMAS Workshop (2013)

An Intelligent Interactive Support System for Word Usage Learning in Second Languages

Yo Ehara(✉)

Tokyo Gakugei University, Koganei, Tokyo 1848501, Japan
ehara@u-gakugei.ac.jp

Abstract. Second-language learners typically encounter difficulty in learning how to use words properly. One reason for this is that words that express multiple meanings - that is, polysemous words - vary from language to language. For example, the word “figure” can mean either a number, an image, or a person. However, this is not the case in all languages. Although the word “figure” is often used with both meanings, some words have only rare usage cases. Thus, ideally, a second language vocabulary learner would benefit from the following learning aids. For words all meanings of which are frequently used, all of the senses of the word should be presented with a high learning priority, whereas senses of a word that are rarely used should be given a low priority. Furthermore, learners should be able to visually understand the semantic proximity of word senses. In this study, we propose an intelligent, interactive user interface to support learning such word usages in second languages. The proposed interface estimates the difficulty of the usage example of the word by estimating its exceptionality. Our method measures semantic closeness with contextualized word embeddings. The model also incorporates a deep anomaly detection model to measure the exceptionality of each usage example. Using our interface, learners of English as a second language (ESL) learners can learn about the semantic closeness between word usage examples and the exceptionality of the examples.

Keywords: Second language learning · Interactive visualization · Word usage examples

1 Introduction

For the practical applications of language learning support systems, it would be desirable to be able to not only recommend words to be learned, but also to recommend which senses of a word should be learned. Since many words that learners learn initially are frequent and polysemous, there is a crucial need for a system to be able to discern which meanings of a word are already known to a learner. However, previous assessment methods have not been able to extract such fine-grained information from a quick vocabulary test.

M. M. Rodrigo et al. (Eds.): AIED 2022, LNCS 13355, pp. 453–464, 2022.
https://doi.org/10.1007/978-3-031-11644-5_37

This paper proposes a polysemy-aware method for investigating the vocabulary of a second language learner. In addition to estimating which word the learner knows, our method can also estimate which meaning of a polysemous word the language learner knows *without imposing a heavy burden on learners.*

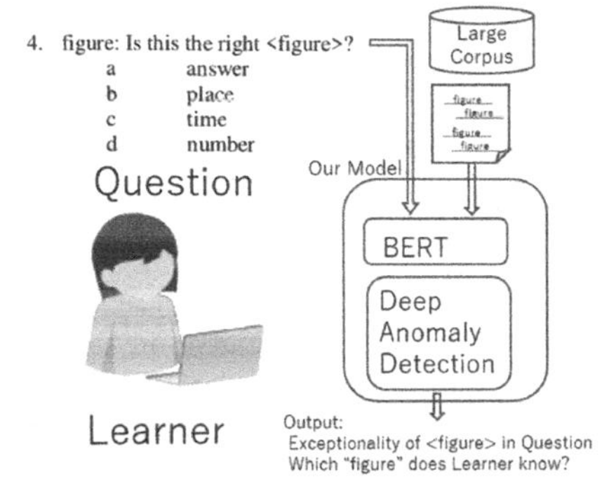

Fig. 1. Overview of the proposed model

4. figure: Is this the right <figure>?

a	answer
b	place
c	time
d	number

Fig. 2. Example of a vocabulary test

Figure 1 presents an overview of our framework. Our input is the same as that of previous vocabulary assessment methods: a half-hour vocabulary test for each language learner. "Question" in Fig. 1 shows an example of a quick vocabulary question. Given that a learner responded to the questions, our goal is to estimate which meaning of the word "figure" the learner knows. As a source of the other meanings of the word "figure", we use *no* annotated data because a corpus of meanings of words that are manually annotated is not always obtainable. We use a large *raw* corpus and extract the occurrences of the word "figure" from the large corpus. Given the occurrences of the word "figure" embedded in sentences, our model first obtains the contextualized word embeddings via bidirectional encoder representations from transformers (BERT) [6]. Pre-trained on a large corpus written by native speakers, BERT captures semantic proximities among various usages of a word. If a word has different meanings in different sentences, BERT can capture these differences.

Our key idea is to employ a deep outlier detection model called DAGMM [18], which can detect outliers of high-dimensional vectors in an unsupervised manner. Given a set of contextualized word embeddings, DAGMM calculates the *exceptionality* of each occurrence of the word. Our model employs BERT and DAGMM to identify semantic outliers in each meaning of a word. Our model automatically determines the threshold that separates exceptional from non-exceptional using vocabulary test results so that the number of non-exceptional occurrences of a word fits to the test results. Hence, our model does not necessarily require a test-taker response to exceptional occurrences to capture exceptionality, which is one of the model's strengths. Experimental results show that our model achieves competitive accuracy in predicting the test-taker responses of vocabulary test results.

Our contributions are summarized as follows:

1. We propose a cost-efficient, polysemy-aware method for personalized assessment of a language learner's vocabulary.
2. Our method employs unsupervised anomaly detection models to capture the typical meaning of a word in order to measure the typical usage of each word without imposing a heavy testing burden on learners.
3. We built a dataset to evaluate this novel task. The experimental results show competitive predictive accuracies, suggesting that our model can capture learners' semantic vocabulary knowledge without imposing a heavy testing burden. We plan to make our dataset openly available on http://yoehara.com/ or http://readability.jp/.

2 Related Work

Figure 2 shows an example of a question on a typical vocabulary test, taken from the Vocabulary Size Test [3], which is widely used in applied linguistics as a placement test. From the multiple choices in the test, learners are asked to choose the option with the closest meaning to the underlined word in the sentence. All options, including the incorrect answers, are grammatically correct when used to replace the underlined word. The options are designed in this manner to help prevent learners from guessing the correct option based on grammar. A representative vocabulary test might comprise some 100 questions, each of which tests whether a learner knows the meaning of a word embedded in a sentence. Each question is designed to test a different word. Hence, no duplicate words are included. Typically, learners take 30 to 40 min to finish answering all questions. In Fig. 2, a question is asked on the word "figure" in the sense of a number. Although the word "figure" has multiple alternative meanings, such as a person or a picture, generally only a single meaning is measured on a given test, because the number of questions that can reasonably be asked are limited.

If a language learner correctly answers a question like that *in* Fig. 2, *does the learner need to understand "figure" as a person as well?* To measure this, we used the semantic exceptionality of each occurrence of the word "figure". If the two senses are used with a similar frequency in the texts of a corpus, both are non-exceptional, and the learner needs to understand both *for reading such texts.* This is the case for "figure". However, there are some cases in which one meaning is exceptional in most corpora, for example, the use of "period" as a direction of time compared to "period" in the sense of a menstrual cycle. In this case, the latter is usually infrequent and exceptional in the texts of a corpus, and the learner does not need to understand both *for reading such texts.* Thus, what kind of texts the learner is expected to encounter is important: we used a general corpus in this paper.

Language tutoring system studies typically focus on learning languages via dialogues [1,10]. Learning words and their usages *incidentally* in reading, dialogue, conversations, and other natural activities is called *incidental learning.*

In applied linguistics, however, incidental learning has been shown to be ineffective in terms of increasing language learners' vocabulary knowledge compared to *intentional learning*, in which learners intentionally learn vocabulary knowledge [13,15,16]. In this study, we have focused on supporting the intentional learning of vocabulary knowledge. Among intentional learning strategies, spaced repetitions have been extensively studied [5,17] in language learning. However, to the best of our knowledge, we are the first to practically identify important word usages for language learning via automatic deep anomaly detection models.

3 Deep Anomaly Detection and Proposed Method

DAGMM [18] is a recent method designed to perform deep anomaly detection. It is a deepened version of the well-known Gaussian mixture model (GMM), which is a clustering method that can also perform anomaly detection. After compressing high-dimensional vectors and clustering them in a low-dimensional representation based on the GMM, DAGMM can compute each vector's *energy value* that can be intuitively understood as the sum of distances from each cluster center, and DAGMM detects points far from any cluster center as anomalies.

Regarding word sense disambiguation, a method of clustering contextualized word embedding expressions and grouping them together by treating each cluster as a word sense has been proposed [11]. As GMM is a widely-used method of clustering, we considered the affinity and interpretability of existing studies on word sense disambiguation. In contrast, as DAGMM has been rarely applied in natural language processing (NLP) [14], to the best of our knowledge, it has never been applied to the field of second-language learning.

In this section, we explain the DAGMM model. Then, we explain the proposed neural network model, which contains a DAGMM model as a block. DAGMM is a deep learning model that converts the input vector $\boldsymbol{x}$ into a low-dimensional representation $\boldsymbol{z}$ using an autoencoder and reconstructs $\boldsymbol{x}$ from $\boldsymbol{z}$. Let the reconstructed vector be $\boldsymbol{x}' = g(\boldsymbol{z}_c; \theta_d)$ and the low-dimensional representation be $\boldsymbol{z}_c = h(\boldsymbol{x}; \theta_e)$. Let $\boldsymbol{z}_r = f(\boldsymbol{x}, \boldsymbol{x}')$ be a function measuring the proximity of the reconstructed vector to the original input; several such functions are available. DAGMM uses $\boldsymbol{z} = [\boldsymbol{z}_c, \boldsymbol{z}_r]$ as the final latent representation. $\boldsymbol{z}$ includes the reconstruction error $\boldsymbol{z}_r$, which penalizes poor reconstruction in the latent space.

The clustering of latent representations is defined in terms of Eq. 1, following the typical GMM notation, where K is the number of clusters, N is the number of data, and MLN denotes the multilayer network. The mixture coefficients of cluster k are given by Eq. 2. The mean and covariance matrices of a cluster k are given by Eq. 3.

$$\mathbf{p} = MLN(\mathbf{z}; \theta_m), \hat{\gamma} = \text{softmax}(\mathbf{p}) \tag{1}$$

$$\hat{\phi_k} = \sum_{i=1}^{N} \frac{\hat{\gamma}_{ik}}{N}, \tag{2}$$

$$\hat{\mu_k} = \frac{\sum_{i=1}^{N} \hat{\gamma}_{ik} \boldsymbol{z}_i}{\sum_{i=1}^{N} \hat{\gamma}_{ik}}, \hat{\boldsymbol{\Sigma}}_k = \frac{\sum_{i=1}^{N} \hat{\gamma}_{ik} \left(\boldsymbol{z}_i - \hat{\mu}_k\right) \left(\boldsymbol{z}_i - \hat{\mu}_k\right)^{\top}}{\sum_{i=1}^{N} \hat{\gamma}_{ik}} \tag{3}$$

For a latent representation $\boldsymbol{z}$ of input $\boldsymbol{x}$, the degree of anomaly is represented by the value of the energy function of Eq. 4. The energy function of point $\boldsymbol{z}$ can be interpreted intuitively as the sum of the distances between $\boldsymbol{z}$ and the center of each cluster $\hat{\mu}_k$ in terms of the measured $\boldsymbol{\Sigma}_k$. The threshold of detection for anomalies depends on the data. For example, in [18], the points with the top 20% energy values were simply identified as anomalies.

$$E(\boldsymbol{z}) = -\log \left(\sum_{k=1}^{K} \hat{\phi}_k \frac{\exp\left(-\frac{1}{2} \left(\boldsymbol{z} - \hat{\mu}_k\right)^{\top} \boldsymbol{\Sigma}_k^{-1} \left(\boldsymbol{z} - \hat{\mu}_k\right)\right)}{\sqrt{\left|2\pi \hat{\boldsymbol{\Sigma}}_k\right|}} \right) \tag{4}$$

We can train DAGMM by minimizing the following objective functions for neural network parameters: θ_e, θ_d, and θ_m; here, L is a loss function for vector reconstruction, P is a penalty term, and λ is a hyperparameter.

$$J(\theta_e, \theta_d, \theta_m) = \frac{1}{N} \sum_{i=1}^{N} L(\boldsymbol{x}_i, \boldsymbol{x}_i') + \frac{\lambda_1}{N} \sum_{i=1}^{N} E(\boldsymbol{z}_i) + \lambda_2 P(\hat{\Sigma}) \tag{5}$$

3.1 Proposed Model

This section explains how the DAGMM model relates to vocabulary tests in the proposed model. i is an index of the occurrences of the word. Here, we aim to consider multiple words; let k be the index of the words to be considered. Let y_{jk} be the result of the vocabulary test; then, y_{jk} is 1 if learner j correctly answered a question on word k.

The Rasch model [2] is typically used to model vocabulary tests. Let σ denote the logistic sigmoid function, i.e., $\sigma(x) := \frac{1}{1+\exp(-x)}$. Thus, the Rasch model can be expressed as follows.

$$P(y_{jk} = 1) = \sigma(a_j - d_k) \tag{6}$$

In Eq. 6, we have two parameters tuned by training data y_{jk}: a_j and d_k. a_j denotes the ability of learner j and d_k denotes the difficulty of word k. Equation 6 can be seen as a special case of logistic regression.

The difficulty of word k, namely d_k, is the key that links Eq. 6 to the DAGMM model. Within the occurrences of word k, we can regard the count of non-exceptional occurrences as the count of occurrences important to language learners. Because the logarithm of word counts has been shown to be to be a good estimate of the word difficulty parameter d_k [2], we modeled d_k as follows.

$$d_k = -\log \left(\sum_{i=1}^{N_k} I\left[E(\boldsymbol{z}_{ik}) < t\right] \right) \tag{7}$$

Fig. 3. Non-exceptional usage of "period".

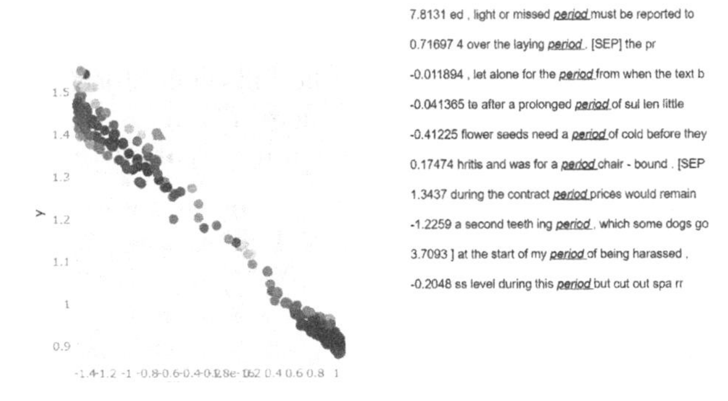

Fig. 4. Exceptional use of "period". The text corresponding to this green point is displayed in the top-right corner.

In Eq. 7, t denotes the threshold parameter and $\boldsymbol{z}_{ik}$ is the latent DAGMM vector of occurrence i of word k, while N_k denotes the number of raw occurrences of word k. Here, I denotes the indicator function that takes 1 if its argument holds true and 0 otherwise.

During training, all parameters are trained jointly, including, the DAGMM parameters, a_j, and threshold t, by using the Adam optimization algorithm [12].

Intuitive Explanation of Equations: Intuitively, words that frequently appear in a given corpus may be considered easy to learn; Eq. 7 follows this intuition. Equation 7 counts the number of occurrences of a word k inside the $-\log$ function. The larger the count, the lower d_k, and the lower the difficulty of word k. However, simple word counting cannot distinguish exceptional from non-exceptional word usage. Hence, we count only non-exceptional occurrences of word k. $E(\boldsymbol{z}_{ik})$ denotes the degree of exceptionality; it denotes how exceptional the i-th occurrence of word k compared to the other occurrences of word k. In Eq. 7, t is the threshold that determines whether an occurrence is sufficiently non-exceptional to count.

4 Experimental Results

4.1 BERT Vectors

As a balanced corpus of English by native speakers of British English, we applied BERT [6] to 100,000 sentences in the British national corpus (BNC) [4], and retrieved contextualized word embedding vectors from the highest layer (the layer closest to the output). As the BERT model, we utilized the **bert-base-uncased.**[1] The number of dimensions of the contextualized word embedding vector was 768. For a given inputted word, all occurrences of the word and their corresponding contextualized word embeddings were retrieved.

4.2 Experimental Setup and Interface

We used a PyTorch implementation of DAGMM made by a third party.[2] We used the same training hyperparameters as in the previous work in this implementation, except for the number of dimensions. In particular, the number of clusters in DAGMM was set to 4. We discuss the word "period" as an example. The latent expression $\boldsymbol{z}$ of the example by the DAGMM can be displayed in three dimensions. Using the first two dimensions of $\boldsymbol{z}$, we show examples of the usage of the word "period" in Fig. 3 and Fig. 4.

The word "period" appeared 376 times in the 100,000 sentences. Each point is a two-dimensional representation of the contextualized word vector of each occurrence of "period" in two-dimensional coordinates, corresponding to each occurrence. The color of each point is an energy value represented by Eq. 4. The higher the value, the more exceptional; that is, higher the green values indicate more exceptional usages. In contrast, the red area is considered a non-exceptional example, which and can be interpreted intuitively as a heatmap.

The horizontal and vertical axes represent the first and second dimensions of the latent-space representation of DAGMM $\boldsymbol{z}$, respectively. The gray triangle is the reference point, and the ten points arranged in order of proximity to this point on the figure, along with the corresponding ten texts, are presented on the right-hand side as examples. The actual calculated energy values are shown on the left side of the example. The reference point can be moved by dragging the mouse, and the students can see examples by moving the reference point near the point of interest, namely $\boldsymbol{z}$.

Existing unsupervised clustering technology for word usage in word senses is not fully accurate. Hence, it is difficult to use this technology for educational purposes because it may show learners incorrect clustering results. However, word senses defined by linguists are typically quite sensitive to subtle differences. For example, "period" in the "Jurassic period" and in "Picasso's blue period" are considered to have different senses in the WordNet dictionary [9]. This is the case because the former indicates a geological period, which is defined as

[1] https://github.com/huggingface/transformers.
[2] https://github.com/danieltan07/dagmm.

Table 1. Counts of non-outlier/Outlier occurrences.

Word	Raw Freq.	Non-outlier Freq.	Outlier Freq.
time	2,863	2,753	110
see	1,359	1,288	71
period	376	368	8
poor	275	269	6
deficit	137	136	1
restore	53	52	1
olive	43	41	2
ubiquitous	13	13	0
retro	13	13	0
weep	8	8	0

a subdivision of a geological era. In this way, for language education, our goal is to measure the exceptionality of the meanings rather than distinguish the subtle differences between the meanings. In contrast, clustering errors between such subtly different senses can be important in understanding the meaning of a given word when it appears in texts.

The colors shown in Fig. 3 and Fig. 4 exactly measure importance in the context of language education. The red points are *non-exceptional* points and are expected to have high priority in language learning. In contrast, the green points indicate the opposite; they are exceptional and are expected to have a low priority in language learning. In Fig. 3, regardless of the cluster to which each point belongs, it may be observed that the central points of the cluster are marked in red and the edge points of the cluster, i.e., the outliers, are shown in green. In Fig. 3, each round dot corresponds to an example of each use of "period" (each occurrence in the corpus). The color of each point indicates the degree of anomaly (energy value), with green being exceptional and red being non-exceptional. The gray triangle in the lower right-hand corner represents the probe point around which the majority of red dots are concentrated. The 10 occurrences nearest to the probe point are shown on the right-hand side as text. The numbers in front of the text are the actual energy values for each example. All text examples in this study were obtained from BNC [4]. In Fig. 3, we illustrate an example in which the probe point is placed in the center of the cluster. The examples of the widely known meaning of the word "period" around the reference point are listed on the right. Thus, deep anomaly detection can present words with a low anomaly degree as non-exceptional usages of the word.

An example of exceptional usage marked in green is shown in Fig. 4. The example sentence on the right side of Fig. 4 shows a use of the phrase "light or missed period", in which the word "period" has the meaning of "menstruation," rather than "a span of time". We can see that the use of "period" to mean "menstruation" is judged as exceptional because it is shown in green. This implies that the use of the word in this meaning exhibited a low frequency in the corpus used for pre-training, which was Wikipedia in our experiments. Notably, "period" in the sense of menstruation is a noun, as is "period" in the sense of a span of time. Therefore, capturing the exceptionality of "period" in the sense of menstruation using other natural language processing technologies such as part-of-speech tagging is considered to be difficult.

4.3 Experiments Using Vocabulary Test Results

To evaluate the proposed approach, we used a publicly available dataset containing the results of vocabulary tests taken by language learners [7]. In this dataset, we used 23 words $\times$ 100 people, i.e., $2,300$ cases, to training the proposed model, and 10 words $\times$ 100 people, i.e., $1,000$ cases, to perform testing.

Table 1 shows the counts of exceptional and non-exceptional occurrences identified by our model for each word, where **Raw Freq.** is the number of raw occurrences of each word. **Non-Outlier Freq.** is the number of occurrences identified as non-exceptional, that is, those that met $E(\boldsymbol{z}) < t$ for each word. **Outlier Freq.** is the number of occurrences that satisfy $E(\boldsymbol{z}) \geq t$ for each word. The sum of **Non-Outlier Freq.** and **Outlier Freq.** is the **Raw Freq**. For the word "time," it may be observed that approximately 3.8% of the occurrences of the word were detected as exceptional usages (i.e., outliers). Table 1 shows that no usages were identified as outliers when the value of **Raw Freq.** was small. This result shows that we cannot distinguish exceptional from primary usages of rarely-used words, which is in accordance with our intuition.

If the non-exceptional usages identified by our model are inaccurate, predicting language learners' vocabulary test results accurately using the frequency of the identified non-exceptional usages is difficult. To evaluate this problem, we simply compared two machine-learning classifiers constructed to predict learners' vocabulary test results. One of the two used **Non-Outlier Freq.** as its feature, and the other used **Raw Freq.** In Eq. 6, word difficulty d_k is determined by the frequency of word k; hence, we compared the accuracies in predicting each learner's performance on each word in the vocabulary test dataset by changing the frequency value that determines d_k to **Non-Outlier Freq.** and to **Raw Freq**. Consequently, both classifiers achieved the same accuracy: 0.75.

The competitive accuracy of the proposed model in predicting each language learner's vocabulary knowledge implies that our model does not contradict the vocabulary test result dataset. The number of non-exceptional occurrences was a good predictor of vocabulary test results, as was the number of raw occurrences. The competitive accuracy implies that our model was able to capture occurrences that the learners may or may not have known by using only a dataset of language learners' vocabulary test results. Being mostly unsupervised, our model performed efficiently in terms of cost; in particular, because it operates without costly manual annotations of word senses, our model is practical in that it is applicable to realistic situations in which a heavy testing burden cannot be imposed on language learners. These results show that the proposed model may be considered promising for vocabulary tutoring systems, providing much more visual information for learners' second-language vocabulary.

She had a missed _____.
time period hour duration

Fig. 5. A question for testing another sense of a word.

4.4 Evaluation Dataset

Fairly assessing multiple senses of a word is not straightforward. For example, let us consider another multiple-choice question that tests a learner's knowledge of the word "figure" in the sense of a person like Fig. 2. If we ask a learner two questions, they can easily guess that the word "figure" has two senses and answer the question based on this guess. This affects their response patterns to the two questions. For example, a distractor (i.e., incorrect option) of one question may influence the response patterns of another question.

To avoid such problems, we developed a test to measure whether learners were aware of other meanings of a word with as little influence as possible on their answer patterns. To prevent the two questions from interfering with each other, before the start of the typical vocabulary test, we conducted a special *preceding test* to determine whether the students knew the other meanings of the words in the vocabulary test.

Figure 5 shows the preceding test conducted before the typical vocabulary test. While the learners were asked to choose the correct option, we designed the preceding test as a fill-in-the-blank test to prevent the interference problem. This enabled us to avoid revealing which option had multiple meanings in the vocabulary test. The preceding test consisted of 13 questions, each of which was checked by 2 native English speakers who were English teachers. The correct answers to each question were sourced from the typical vocabulary test (VST). The typical vocabulary test included 70 questions.[3] In total, we asked 83 questions to 270 test-takers. The test-takers were crowdworkers contracted by the Lancers crowdsourcing service; most participants were Japanese. To guarantee that test-takers had learned English to some extent, we limited participants to those who had taken the Test of English for International Communication (TOEIC), a commercial test developed by the English Testing Service. Because it costs approximately 70 USD to take this test, those who have done so at least once are likely to have learned English to some degree.

4.5 Energy-Value-Based Difficulty

In Eq. 7, we obtained the polysemy-aware overall difficulty of a word k using the frequency of word k's usage examples whose energy value (i.e. the degree of being a semantic outlier) is below a threshold t. Additionally, by directly using the energy value in $E(\boldsymbol{z}_{ik})$, can we obtain the difficulty of the i-th usage of

[3] Although the original VST consists of 100 questions because most test-takers are unlikely to know the answers to the difficult questions, we omitted 30 difficult questions to avoid the influence of the fatigue of test-takers.

Table 2. Accuracy of predicting learners' responses.

	Test of typical meanings	Test of exceptional meanings
Baseline (Size-based)	62.8%	44.4%
Baseline (LR)	63.1%	47.9%
Ours (Outlier-based)	63.7% (*)	48.5% (*)

word k, namely d_{ik}? To answer this research question, we conducted experiments using a special dataset in which test-takers answered for both exceptional and typical meanings of each word, as described in Sect. 4.4. This dataset contains 58 typical vocabulary questions and 12 pairs of exceptional and typical word usage questions for 12 words. We used 235 test-taker learners' responses to the 58 typical vocabulary questions and 12 pairs of exceptional/typical word questions as the training and test data, respectively.

Table 2 shows the results of our method compared with previous methods. **Size-based** is a vocabulary-size-based method widely used in the field of applied linguistics [3]. This method first measured the vocabulary size of the learner and assumed that the learner knows all the meanings of all the words for which the frequency is ranked higher than the vocabulary size, and vice versa. **LR** is a logistic-regression-based method. We used the COCA (https://www.english-corpora.org/coca/) and BNC (http://www.natcorp.ox.ac.uk/) corpora for the general corpus, as in [8] The **LR** can consider each learner's ability by introducing a one-hot feature vector representing the learner ID, as explained in [8]. **Ours** is ours: we added the DAGMM energy value of the target word within the question sentence as a feature for **LR** in addition to the corpora features. Note that this experimental setting is quite challenging for the classifiers because, both non-exceptional and exceptional usages are in the test data, the classifiers must predict the responses for words that do not appear in the training data.

Table 2 shows that our method outperforms the previous methods in both methods in terms of predicting the response of second language learners. This is because our method can consider the semantic outlierness of word usage by using the energy value of each word usage, whereas the others assume that exceptional word usage has the same difficulty as typical word usage. "(*)" denotes that the difference between **Ours** and **Baseline (LR)** was statistically significant (Wilcoxon, $p < 0.01$).

5 Conclusion

In this study, to estimate the semantic knowledge of words of language learners, we have proposed a novel method based on unsupervised deep anomaly detection. Our method was capable of distinguishing exceptional and non-exceptional word usage cases with the same accuracy and recognizing non-exceptional and exceptional cases in a qualitative evaluation. Our future work includes analyzing word usages that are non-exceptional but not known by language learners.

Acknowledgements. This work was supported by JST ACT-X Grant Number JPMJAX2006, Japan. We used ABCI of AIST and miniRaiden of RIKEN for the computational resources. We appreciate the anonymous reviewers for their valuable and insightful comments.

References

1. Al Emran, M., Shaalan, K.: A survey of intelligent language tutoring systems. In: Proceedings of ICACCI, pp. 393–399 (2014)
2. Beglar, D.: A Rasch-based validation of the vocabulary size test. Lang. Test. **27**(1), 101–118 (2010)
3. Beglar, D., Nation, P.: A vocabulary size test. Lang. Teach. **31**(7), 9–13 (2007)
4. BNC Consortium: The British National Corpus (2007)
5. Choffin, B., Popineau, F., Bourda, Y., Vie, J.J.: DAS3H: modeling student learning and forgetting for optimally scheduling distributed practice of skills. In: Proceedings of EDM 2019. arXiv:1905.06873, May 2019
6. Devlin, J., Chang, M.W., Lee, K., Toutanova, K.: BERT: pre-training of deep bidirectional transformers for language understanding. In: Proceedings of NAACL (2019)
7. Ehara, Y.: Building an English vocabulary knowledge dataset of Japanese English-as-a-second-language learners using crowdsourcing. In: Proceedings of LREC, May 2018
8. Ehara, Y., Sato, I., Oiwa, H., Nakagawa, H.: Mining words in the minds of second language learners for learner-specific word difficulty. J. Inf. Process. **26**, 267–275 (2018). https://doi.org/10.2197/ipsjjip.26.267
9. Fellbaum, C.: WordNet. In: Theory and Applications of Ontology: Computer Applications, pp. 231–243. Springer, Cham (2010). https://doi.org/10.1007/978-90-481-8847-5_10
10. Graesser, A., Chipman, P., Haynes, B., Olney, A.: AutoTutor: an intelligent tutoring system with mixed-initiative dialogue. IEEE Trans. Educ. **48**(4), 612–618 (2005)
11. Huang, L., Sun, C., Qiu, X., Huang, X.: GlossBERT: BERT for word sense disambiguation with gloss knowledge. arXiv preprint arXiv:1908.07245 (2019)
12. Kingma, D.P., Ba, J.: Adam: a method for stochastic optimization. In: Proceedings of ICLR (2015)
13. Laufer, B., Hulstijn, J.: Incidental vocabulary acquisition in a second language: the construct of task-induced involvement. Appl. Linguist. **22**(1), 1–26 (2001)
14. Luo, Y., Zhao, H., Zhan, J.: Named entity recognition only from word embeddings (2019)
15. Nation, I.: How large a vocabulary is needed for reading and listening? Can. Mod. Lang. Rev. **63**(1), 59–82 (2006)
16. Nation, I.S.P., Waring, R.: Teaching Extensive Reading in Another Language. Routledge, November 2019. google-Books-ID: xRu_DwAAQBAJ
17. Settles, B., Meeder, B.: A trainable spaced repetition model for language learning. In: Proceedings of ACL, Berlin, Germany, pp. 1848–1858 (2016)
18. Zong, B., et al.: Deep autoencoding Gaussian mixture model for unsupervised anomaly detection. In: Proceedings of ICLR (2018). https://openreview.net/forum?id=BJJLHbb0-

Balancing Cost and Quality: An Exploration of Human-in-the-Loop Frameworks for Automated Short Answer Scoring

Hiroaki Funayama[1,2(✉)], Tasuku Sato[1,2], Yuichiroh Matsubayashi[1,2], Tomoya Mizumoto[2,3], Jun Suzuki[1,2], and Kentaro Inui[1,2]

[1] Tohoku University, Sendai, Japan
{h.funa,tasuku.sato.p6}@dc.tohoku.ac.jp,
{y.m,jun.suzuki,inui}@tohoku.ac.jp
[2] RIKEN, Tokyo, Japan
[3] Future Corporation, Tokyo, Japan
t.mizumoto.yb@future.co.jp

Abstract. Short answer scoring (SAS) is the task of grading short text written by a learner. In recent years, deep-learning-based approaches have substantially improved the performance of SAS models, but how to guarantee high-quality predictions still remains a critical issue when applying such models to the education field. Towards guaranteeing high-quality predictions, we present the first study of exploring the use of *human-in-the-loop* framework for minimizing the grading cost while guaranteeing the grading quality by allowing a SAS model to share the grading task with a human grader. Specifically, by introducing a confidence estimation method for indicating the reliability of the model predictions, one can guarantee the scoring quality by utilizing only predictions with high reliability for the scoring results and casting predictions with low reliability to human graders. In our experiments, we investigate the feasibility of the proposed framework using multiple confidence estimation methods and multiple SAS datasets. We find that our human-in-the-loop framework allows automatic scoring models and human graders to achieve the target scoring quality.

Keywords: Neural network · Natural language processing · Automated short answer scoring · Confidence estimation

1 Introduction

Short answer scoring (SAS) is a task used to evaluate a short text written as input by a learner based on grading criteria for each question (henceforth, prompt). Figure 1 gives examples of rubrics and student's answers. Automatic SAS systems have attracted considerable attention owing to their abilities to provide fair

M. M. Rodrigo et al. (Eds.): AIED 2022, LNCS 13355, pp. 465–476, 2022.
https://doi.org/10.1007/978-3-031-11644-5_38

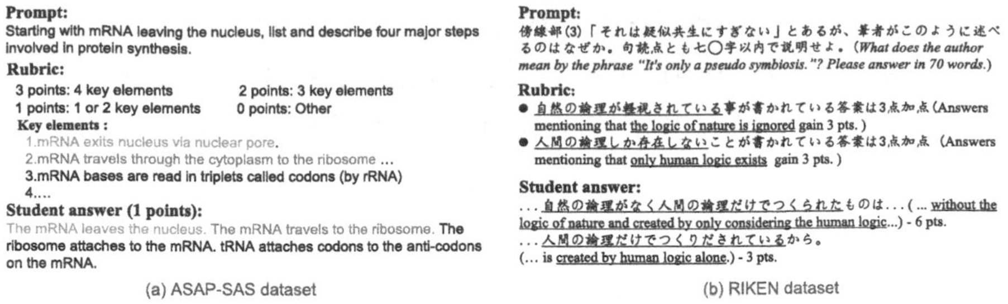

Prompt:
Starting with mRNA leaving the nucleus, list and describe four major steps involved in protein synthesis.
Rubric:
3 points: 4 key elements 2 points: 3 key elements
1 points: 1 or 2 key elements 0 points: Other
Key elements :
1.mRNA exits nucleus via nuclear pore.
2.mRNA travels through the cytoplasm to the ribosome ...
3.mRNA bases are read in triplets called codons (by rRNA)
4....
Student answer (1 points):
The mRNA leaves the nucleus. The mRNA travels to the ribosome. The ribosome attaches to the mRNA. tRNA attaches codons to the anti-codons on the mRNA.

(a) ASAP-SAS dataset

Prompt:
傍線部(3)「それは疑似共生にすぎない」とあるが、筆者がこのように述べるのはなぜか。句読点とも七〇字以内で説明せよ。(*What does the author mean by the phrase "It's only a pseudo symbiosis."? Please answer in 70 words.*)
Rubric:
- 自然の論理が軽視されている事が書かれている答案は3点加点 (Answers mentioning that the logic of nature is ignored gain 3 pts.)
- 人間の論理しか存在しないことが書かれている答案は3点加点 (Answers mentioning that only human logic exists gain 3 pts.)

Student answer:
. . . 自然の論理がなく人間の論理だけでつくられたものは. . . (... without the logic of nature and created by only considering the human logic...) - 6 pts.
. . . 人間の論理だけでつくりだされているから。
(... is created by human logic alone.) - 3 pts.

(b) RIKEN dataset

Fig. 1. (a) Examples of rubrics and student's answers of prompt 5 excerpted from kaggle Automated Student Assessment Prize Short Answer Scoring (ASAP-SAS) dataset, and (b) example of a prompt, scoring rubric, and student's answers excerpted from RIKEN dataset [15], translated from Japanese to English. For space reasons, some parts of the rubrics and answers are omitted.

and low-cost grading in large-scale examinations and to support learning in educational settings [13,14]. In recent years, deep -learning-based approaches have improved the performance of automated scoring models [17]. However, the possibility of errors produced by an automatic grading system cannot be completely eliminated, and such errors may interfere with the learner's learning process [18]. Owing to a concern for such automatic scoring errors, current automatic grading systems are often used as references for human graders to detect grading errors [1]. To further utilize automatic scoring systems in the education field, it is critical to guarantee high-quality grading.

To tackle this challenge, we propose a human-in-the-loop automatic scoring framework in which a human grader and an automatic scoring system share the grading in order to minimize the grading cost while guaranteeing the grading quality. In this framework, we attempt to guarantee the scoring quality using confidence estimation methods by adopting only the highly reliable answers among the automatic scoring results and cast the remaining answers to human graders (see Fig. 2). Specifically, we perform the following two-step procedure: (1) estimating the threshold of confidence scores to achieve the desired scoring quality in the development set, and (2) filtering the test set using the determined threshold. The achievement of the entire framework is evaluated in two aspects: the percentage of automatically scored answers and how close the scoring quality is to the predetermined scoring quality. In the experiments, we simulated the feasibility of the proposed framework using two types of dataset with a general scoring model and several confidence estimation methods. The experimental result verifies the feasibility of the framework.

The contributions of our study are as follows. (i) We are the first, to the best of our knowledge, to provide a realistic framework for minimizing scoring costs while aiming to ensure scoring quality in automated SAS. (ii) We validated the feasibility of our framework through cross-lingual experiments with a general scoring model and multiple confidence estimation methods. (iii) We gained promising results showing that our framework enabled the control of the scoring quality while minimizing the human grading load, and the framework worked well even for prompts for which the agreement between human graders is relatively low. The code for our experiments and all experimental setting information is publicly available.[1]

2 Previous Research

In recent years, SAS has attracted considerable attention owing to its ability to provide a fair and low-cost scoring in large-scale exams [2]. A central challenge in the use of SAS has focused primarily on improving the performance of scoring models [20]. With the recent advancement of models using deep learning, the performance of scoring models has also significantly improved [13,17]. Towards realizing the practical use of SAS systems in the real world, several researchers have explored outputting useful feedback for an input response [21], utilizing rubrics for scoring [19], and investigating adversarial input in SAS [4], to name a few. In addition, research on various languages has also been reported in recent years, including Indonesian [7], Korean [9], and Japanese [15].

To the best of our knowledge, the only study in which the use of confidence estimation in SAS was investigated by Funayama et al. [5]. They introduced the concept of unacceptable critical scoring errors (CSEs). Subsequently, they proposed a new task formulation and its evaluation for SAS, in which automatic scoring is performed by filtering out unreliable predictions using confidence scores to eliminate CSEs as much as possible.

In this study, we extend the work of Funayma et al. [5], and propose a new framework for minimizing human scoring costs while controlling the overall scoring quality of the combining human scoring and automated scoring. We also conducted cross-lingual experiments using a Japanese SAS dataset, as well as the ASAP dataset commonly used in the SAS field. The human-in-the-loop approach in SAS was also used in a previous work to apply active learning to the task [8]; however, our study is unique in that we have attempted to minimize scoring costs while focusing on ensuring overall scoring quality.

3 Short Answer Scoring: Preliminaries

3.1 Task Definition

Suppose $\mathcal{X}$ represents a set of all possible student's answers of a given prompt, and $\mathbf{x} \in \mathcal{X}$ is an answer. The prompt has an integer score range from 0 to N, which is basically defined in rubrics. Namely, the score for each answer is selected from one of the integer set $S = \{0, ..., N\}$. Therefore, we can define the SAS task as assigning one of the scores $s \in S$ for each given input $\mathbf{x} \in \mathcal{X}$. Moreover, to construct an automatic SAS model means to construct a mapping function m from every input of student answer $\mathbf{x} \in \mathcal{X}$ to a score $s \in S$, that is, $m : \mathcal{X} \to S$.

[1] https://github.com/hiro819/HITL_framework_for_ASAS.

3.2 Scoring Model

A typical, recent approach to constructing a mapping function m is the use of newly developed deep neural networks (DNNs). As discussed a priori, the set of scores S consists of several consecutive integers $\{0, \ldots, N\}$. We often consider each discrete number as one class so that the SAS task can be treated as a simple $N+1$-class classification task. This means that each integer in S is considered as a class label, not a consecutive integer. In this case, a SAS model is often constructed as a probabilistic model. Suppose $\mathcal{D}$ is training data that consist of a set of actually obtained student's answers $\mathbf{x}$ and its corresponding human annotated score s pairs, that is, $\mathcal{D} = ((\mathbf{x}_i, s_i))_{i=1}^{I}$, where I is the number of training data. To train the model m, we try to minimize the empirical loss on training data $L_m(\mathcal{D})$ that consist of the sum of negative log-likelihood for each training data calculated using model m. Therefore, we can write the training process of the SAS model as the following minimization problem:

$$m = \underset{m'}{\operatorname{argmin}} \left\{ L_{m'}(\mathcal{D}) \right\}, \quad L_m(\mathcal{D}) = - \sum_{(\mathbf{x},s) \in \mathcal{D}} \log \left(p_m(s \mid \mathbf{x}) \right), \tag{1}$$

where $p_m(s \mid \mathbf{x})$ represents the probability of class s given input $\mathbf{x}$ calculated using model m. Once m is obtained, we can predict the score $\widehat{s}$ of any input (student answer) by using trained model m. We often use the argmax function for determining the most likely score $\widehat{s}$ given $\mathbf{x}$ as

$$\widehat{s} = \underset{s}{\operatorname{argmax}} \left\{ p_m(s \mid \mathbf{x}) \right\}. \tag{2}$$

4 Proposed Framework

Guaranteeing a high-quality scoring framework is crucial, especially when applying it to the education field in actual use. To address this issue, we propose a novel human-in-the-loop framework that can minimize the grading cost of human graders while guaranteeing the overall grading quality at a certain level.

4.1 Scoring Framework Overview

The basic idea of our scoring framework is intuitive and straightforward. We first score all the student answers by the method described in Sect. 3.2, and simultaneously estimate the confidence of the predicted scores. We then ask trained human graders to reevaluate the student answers when their confidence scores are below the predefined threshold. Figure 2(a) shows an overview of our scoring framework. Our human-in-the-loop framework is based on the assumption that automated SAS systems cannot achieve zero-error, while trained human graders can achieve it. Therefore, the unreliable scores obtained are reevaluated by human graders to realize high-quality scoring.

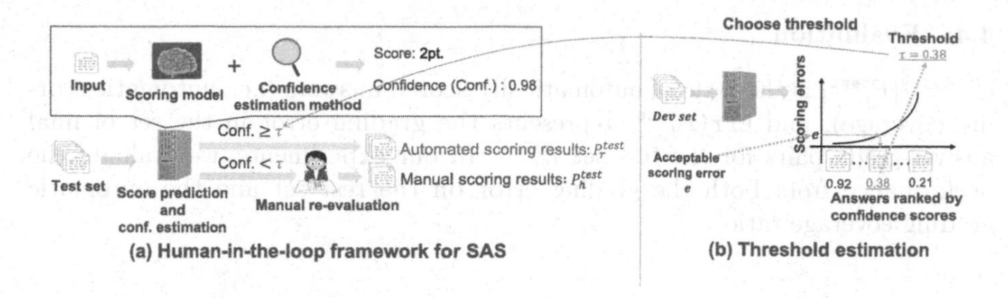

Fig. 2. Overview of the proposed human-in-the-loop framework for SAS.

The goal of our framework is to minimize the cost of human graders while maintaining erroneous scoring given by automated SAS systems to a minimum. Thus, the critical components of our framework are the methods o estimating the confidence of scoring and detemining the threshold of reasonable confidence level in advance.

4.2 Utilizing Confidence Estimation for SAS

We define confidence C as a function that is determined using three factors: model m, input student answer $\mathbf{x}$, and score $\widehat{s}$ predicted from m, that is, $C(\mathbf{x}, \hat{s}, m)$. Suppose we have ℓ unique answers to grade $(\mathbf{x}_1, ..., \mathbf{x}_\ell)$. Let P be the set of all the student answers and its corresponding score pairs. Let P_r be a subset of P, whose confidence is above the predefined threshold τ. Additionally, $\overline{P}_r$ is the complementary set of P_r. Namely,

$$P = \{(\mathbf{x}_i, \hat{s}_i)\}_{i=1}^{\ell}, \quad P_r = \{(\mathbf{x}, \hat{s}) \in P \mid C(\mathbf{x}, \hat{s}, m) \geq \tau\}, \quad \text{and} \quad \overline{P}_r = P \backslash P_r. \quad (3)$$

We treat P_r and $\overline{P}_r$ as the reliable and unreliable scoring results, respectively. In this study, we consider that the answers in $\overline{P}_r$ need rescoring by trained human graders. The set of pairs of the answers in $\overline{P}_r$ and their corresponding regraded score by human graders is denoted by P_h. Finally, we combine P_r and P_h as the set of final scoring results, P_f. The relation can be written as $P_h = P_f \backslash P_r$.

4.3 Threshold Estimation

It is difficult to determine the optimal confidence threshold τ in advance. To find a reasonable one, we use the development set, a distinct dataset from the training data used for obtaining the model m. Let P^{dev}, P_r^{dev} and P_f^{dev} be P, P_r and P_f, respectively, based on the development data. For a given, acceptable scoring error e, we find the confidence threshold τ such that the errors of scoring results in the development set do not exceed e[2].

$$\tau = \underset{\tau'}{\text{argmax}} \left\{ |P_{r'}^{\text{dev}}| \right\} \quad \text{subject to } \texttt{Err}(P_f^{\text{dev}}) \leq e, \quad (4)$$

where $\texttt{Err}(P_f^{\text{dev}})$ represents the sum of scoring errors occurring within P_f^{dev}.

[2] We assume that the acceptable scoring error is determined by test administrators.

4.4 Evaluation

$|P_r^{\mathtt{test}}|/|P_f^{\mathtt{test}}|$ is the ratio of automatically scored answers (i.e., automatic scoring coverage), and $\mathtt{Err}(P_f^{\mathtt{test}})$ represents the grading error in the set of final answer-grade pairs for the test set $P_f^{\mathtt{test}}$. In our experiments, we evaluate the performance from both the grading error on the test set and the automatic grading coverage ratio.

5 Experiments

In the experiments, we verify the feasibility of the proposed framework using three confidence estimation methods as our test cases. We publish the code and instance IDs used for training, development and test data.[3]

5.1 Base Scoring Model

We selected the BERT [3] based classifier as the base scoring model because this is one of the most promising approaches for many tasks in text classification. We can also expect to achieve near state-of-the-art performance in our experiments.

The model first converts the input student answer $\mathbf{x}$ into a feature representation $\mathbf{Z} \in \mathbb{R}^{d_h \times n}$ using the BERT encoder $\mathtt{enc}(\cdot)$, that is, $\mathbf{Z} = \mathtt{enc}(\mathbf{x})$, where d_h is the vector dimension and n is the number of words (tokens) in $\mathbf{x}$. The model then calculates the probability $p_m(s \mid \mathbf{x})$ using the vector of CLS token $\mathbf{h}^{(\mathrm{CLS})}$, which is the first token in the BERT model, in $\mathbf{Z}$ with the following standard Affine transformation and softmax operation:

$$p_m(s \mid \mathbf{x}) = \mathbf{o}_s{}^{\top}\mathtt{softmax}\left(\mathbf{W}\mathbf{h}^{(\mathrm{CLS})} + \mathbf{b}\right), \tag{5}$$

where $\mathbf{W}$ and $\mathbf{b}$ are the trainable model parameters of $(N+1) \times d_h$ and $(N+1) \times 1$ matrices, respectively. $\mathbf{o}_s$ is the one-hot vector, where 1 is located at score s in the d_h-vector. Note that $\mathbf{o}_s$ is a kind of selector of a score in the vector representation generated from the softmax function. Finally, we used the standard operation, Eq. (2), to obtain the predicted score $\widehat{s}$.

5.2 Confidence Estimation Methods

Posterior Probability. The most straightforward way to estimate the confidence score in our classification model is to consider the prediction probability (posterior) calculated Eq. (5):

$$C_{\mathtt{prob}}(\mathbf{x}, \hat{s}, m) = p_m(\widehat{s} \mid \mathbf{x}). \tag{6}$$

[3] https://github.com/hiro819/HITL_framework_for_ASAS.

Trust Score. Funayama et al. [5] used the *trust score* [10] to estimate confidence in automatic scoring of written answers. Here, we also consider using the *trust score* as one of the confidence estimation methods.

We calculate the trust score as follows. The first step is to input the training data $\{(\mathbf{x}_1, s_1), ..., (\mathbf{x}_k, s_k)\}$ into the trained automatic scoring model m and obtain a set of feature representations $H = \{\mathbf{h}_1, ..., \mathbf{h}_k\}$ corresponding to each training instance. In addition, the feature representations are kept as clusters $H_s = \{\mathbf{h}_i \in H \mid s_i = s\}$ for each score label s. We then obtain the feature representation $\mathbf{h}$ and predicted score $\hat{s}$ for an unseen input $\mathbf{x}$ from the test data. The *trust score* $C_{\mathtt{trust}}(\mathbf{x}, \hat{s}, m)$ for the scoring result $(\mathbf{x}, \hat{s})$ is calculated as

$$C_{\mathtt{trust}}(\mathbf{x}, \hat{s}, m) = \frac{d_c(\mathbf{x}, H)}{d_p(\mathbf{x}, H) + d_c(\mathbf{x}, H)}, \tag{7}$$

where $d_p(\mathbf{x}, H) = \min_{\mathbf{h}' \in H_{\hat{s}}} d(\mathbf{h}, \mathbf{h}')$ and $d_c(\mathbf{x}, H) = \min_{\mathbf{h}' \in (H \setminus H_{\hat{s}})} d(\mathbf{h}, \mathbf{h}')$, and $d(\mathbf{h}, \mathbf{h}')$ denotes the Euclidean distance from $\mathbf{h}$ to $\mathbf{h}'$. Note that H can be obtained by using m and $\mathbf{x}$. To normalize the values in the range from 0 to 1, d_c is newly added to the denominator. We note here that the normalization does not change the original order without normalization.

Gaussian Process. Previous works have shown that the performance of regression models is superior to that of classification models in SAS since the regression models treat the score as a consecutive integer, not a discrete category [11,17].

Gaussian Process Regression (GPR) [16] is a regression model that can estimate a variance of the predicted score. We utilize this value as the confidence of its prediction and use GPR as a representative of regression models that are compatible with our framework. We use a publicly available GPR implementation [6] and train the GPR model on the feature representations output by the trained encoder shown in Sect. 5.1.

5.3 Dataset

We conduct cross-lingual experiments using the Automated Student Assessment Prize Short Answer Scoring (ASAP-SAS) dataset, which is commonly used in the field of SAS, and the RIKEN dataset, which is the only publicly available SAS dataset in Japanese.

ASAP-SAS Dataset. The ASAP-SAS dataset is introduced in kaggle's ASAP-SAS contest.[4] It consists of 10 prompts and answers on academic topics such as biology and science. The answers are graded by two scorers. The experiment was conducted on the basis of the score given by the first scorer according to the rules of kaggle's ASAP contest. In our experiment, we used the official training data and test data. We use 20% of the training data as development data.

[4] https://www.kaggle.com/c/asap-sas.

RIKEN Dataset. We also used the publicly available Japanese SAS dataset[5] provided in [15]. The RIKEN dataset consists of 17 prompts in total. Each prompt has its own rubric, student answers, and corresponding scores. All contents were collected from examinations conducted by a Japanese education company, Yoyogi Seminar. In this dataset, the scoring rubric of each prompt consists of three or four elements (referred to as *analytic criteria* in [15]), and each answer is manually assessed on the basis of each analytic criterion (i.e., analytic score). For example, the scoring rubric shown in Fig. 1 is the analytic criterion excerpted from Item C in prompt Y14_2-2_1_4 and gives an analytic score to each student answer. Since each analytic score is given independently from each other, we treat the prediction of each individual analytic score for a given input answer as an independent task. For our experiment, we divided the data into 250 training instances and 250 test instances. The training data were further divided into five sets, four sets for training (200 instances), and one set for development (50 instances).

5.4 Setting

As described in Sect. 5.1, we used the pretrained BERT [3] as the encoder for the automatic scoring model and use the vectors of CLS tokens as feature vectors for predicting answers[6,7]. The root mean square error (RMSE) is adopted as the function Err representing the scoring error. To stabilize the experimental results using a small development set, we estimated the threshold value using 250 data sets, which were integrated with five development datasets.

5.5 Results

Correlation Between Confidence Scores and Scoring Accuracy. First, we investigate the relationship between confidence estimation and scoring quality. Figure 3 shows changes in RMSE for the top n% confident predictions in the test set using the ASAP-SAS dataset and the RIKEN dataset. All methods reduced the RMSE for the higher confident predictions, indicating that there is a correlation between confidence and scoring quality. We also observed that the scoring error is reduced more for the RIKEN dataset than for the ASAP-SAS dataset.

Feasibility of our Proposed Framework. Next, we applied the proposed framework to SAS data and confirmed its feasibility. Figure 4 shows the percentage of answers automatically scored (automatic scoring coverage [%]) and the RMSE for the combined scoring result P_f^{test} for human scoring and automatic

[5] https://aip-nlu.gitlab.io/resources/sas-japanese.

[6] We used pretrained BERTs from https://huggingface.co/bert-base-uncased for English and https://github.com/cl-tohoku/bert-japanese for Japanese.

[7] Quadratic weghted kappa (QWK) of our model is 0.722 for the ASAP-SAS dataset, which is comparable to previous studies [12,17].

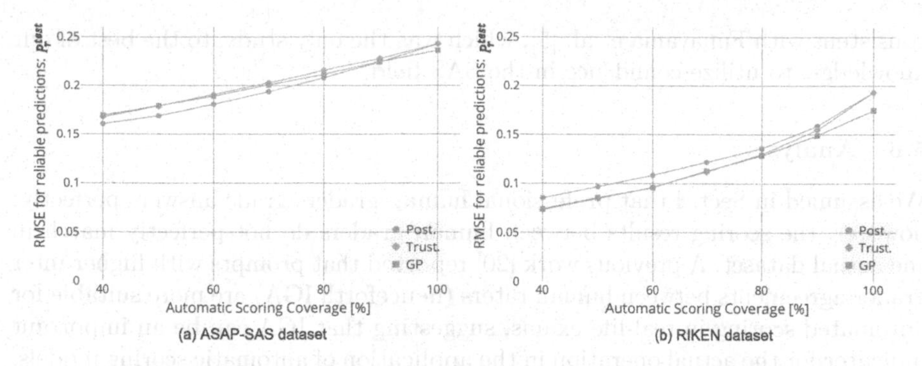

Fig. 3. Changes of RMSE when the prediction with the top n% confidence score is adopted as automatic scoring result in the test set. Green markers represent posterior (Post.), red ones the trust score (Trust.), and blue ones, the Gaussian process (GP). The markers represent mean values.

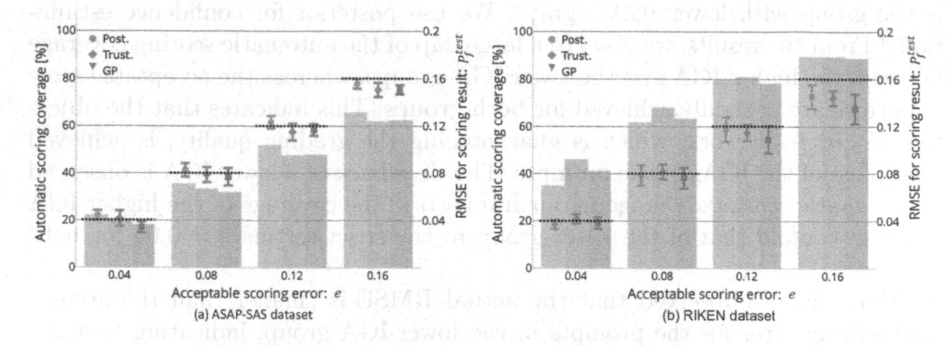

Fig. 4. Scoring errors for the test data after applying our proposed framework using ASAP-SAS and RIKEN datasets. The bars represent automatic scoring coverage [%], and the markers and error bars represent RMSE for test set after applying our proposed framework. Green, red, and blue marks represent the results of Post., Trust., and GP, respectively. The error bars represents standard deviation of the RMSE. The dotted line represents the acceptable scoring error e.

scoring when the acceptable scoring error was varied from 0.04, 0.08, 0.12, to 0.16 on RMSE.

From Fig. 4, we can see that we are successfully able to control scoring errors around the acceptable scoring error in most settings for both datasets. In terms of the ability to achieve the acceptable scoring error, there is no significant difference in performance between the confidence estimation methods. The standard deviation indicated by the error bars also shows that our proposed framework performs well with the three confidence estimation methods. For automatic scoring coverage, the posterior tends to be slightly dominant in ASAP-SAS, whereas the trust score is slightly dominant in the RIKEN dataset. The slightly higher performance of the trust score for the RIKEN dataset is consistent with Funayama et al. [5], which was the only study, to the best of our knowledge, to utilize confidence in the SAS field.

5.6 Analysis

We assumed in Sect. 4 that professional human graders grade answers perfectly; however, the scoring results between human graders do not perfectly match in the actual dataset. A previous work [20] reported that prompts with higher inter grader agreements between human raters (henceforth IGA) are more suitable for automated scoring in real-life exams, suggesting that IGA can be an important indicator for the actual operation in the application of automatic scoring models. In this section, we analyze the relevance of the inter grader agreement between two human graders for our framework.

Figure 5 shows the results of applying our proposed framework to two groups: one is the group with higher IGAs than the average IGA (green) and the other is the group with lower IGAs (gray). We use posterior for confidence estimation.[8] From the results, we observe a large gap of the automatic scoring coverage between the higher IGA and the lower IGA groups, whereas the acceptable scoring errors are generally achieved for both groups. This indicates that the objective of our framework, which is guaranteeing the grading quality, is achieved regardless of the IGAs of the prompts. The drawback of a lower IGA is observed in automatic scoring coverage in our framework; the coverage of the higher IGA group is twofold that of the lower group in the strict setting $e = 0.04$ for both datasets.

Moreover, we observed that the actual RMSE is smaller than the acceptable scoring error for the prompts in the lower-IGA group, indicating that the framework provides more 'cautious' automatic scoring for this group. For the

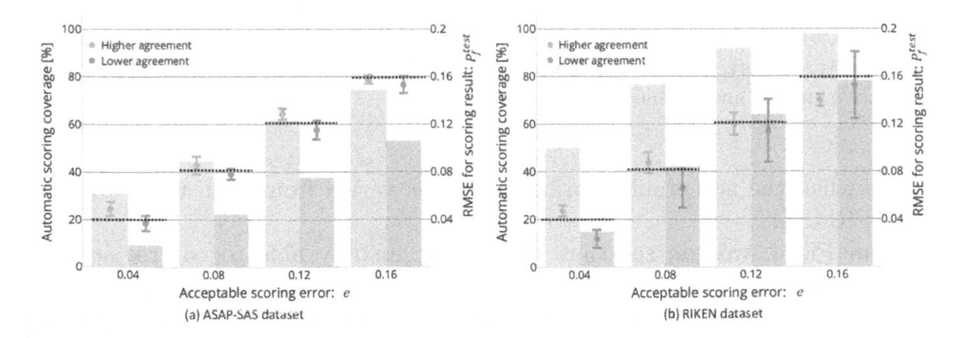

Fig. 5. The automatic scoring coverage (bars) and RMSE for P_f^{test} (markers and error bars) when we apply our proposed framework to the two groups: prompts with higher inter grader agreement and prompts with lower inter grader agreement.

[8] Posterior is used because there is no significant difference in performance among the three methods and the most widely used way to estimate confidence score is posterior.

noisy training data, an instance with a high prediction confidence tends to be judged as incorrect even though the model predicts the correct score. As a result, the scoring quality of the model for predictions with high confidence is underestimated compared with the true scoring quality of the model. Then, a strict threshold is selected in the development set, which may lead to such 'cautious' scoring. Indeed, in the actual situation, when estimating the threshold value in the development set, a human grader is expected to reconfirm and correct the ground truth score for a training instance that has a scoring error despite its high confidence. Thus, a negative impact of such human scoring errors is expected to be insignificant when using our framework that combines human grading and the confidence of automatic scoring models.

6 Conclusion

In recent years, the accuracy of automatic grading systems has considerably improved. Ensuring the scoring quality of SAS models is crucial challenge in the promotion of further applications of SAS models in the education field. In this paper, we presented a framework for ensuring the scoring quality of SAS systems by allowing such systems to share the grading task with human graders. In our experiments, we discovered that the desired scoring quality can be achieved by using our proposed framework. Our framework is designed to control the overall scoring quality (i.e., RMSE) and is therefore expected to work complementarily with Funayama's work [5], which aims to reduce the number of critical scoring errors on individual answers.

Acknowledgments. This work was supported by JSPS KAKENHI Grant Number JP22H00524, JP19K12112, and JP21H04901.

References

1. Attali, Y., Burstein, J.: Automated essay scoring with e-rater®v.2. J. Technol. Learn. Assess. **4**(3) (2006)
2. Burrows, S., Gurevych, I., Stein, B.: The eras and trends of automatic short answer grading. Int. J. Artif. Intell. Educ. **25**(1), 60–117 (2015)
3. Devlin, J., Chang, M.W., Lee, K., Toutanova, K.: BERT: pre-training of deep bidirectional transformers for language understanding. In: NAACL-HLT, pp. 4171–4186, June 2019. https://doi.org/10.18653/v1/N19-1423
4. Ding, Y., Riordan, B., Horbach, A., Cahill, A., Zesch, T.: Don't take "nswvtnvakgxpm" for an answer -the surprising vulnerability of automatic content scoring systems to adversarial input. In: COLING, pp. 882–892. International Committee on Computational Linguistics, December 2020. https://doi.org/10.18653/v1/2020.coling-main.76
5. Funayama, H., et al.: Preventing critical scoring errors in short answer scoring with confidence estimation. In: ACL-SRW, pp. 237–243. Association for Computational Linguistics, July 2020. https://doi.org/10.18653/v1/2020.acl-srw.32
6. Gardner, J.R., Pleiss, G., Bindel, D., Weinberger, K.Q., Wilson, A.G.: Gpytorch: Blackbox matrix-matrix gaussian process inference with GPU acceleration (2021)

7. Herwanto, G., Sari, Y., Prastowo, B., Riasetiawan, M., Bustoni, I.A., Hidayatulloh, I.: Ukara: a fast and simple automatic short answer scoring system for bahasa indonesia. In: ICEAP Proceeding Book, vol. 2, pp. 48–53, December 2018
8. Horbach, A., Palmer, A.: Investigating active learning for short-answer scoring. In: BEA. Association for Computational Linguistics, San Diego, June 2016
9. Jang, E.S., Kang, S., Noh, E.H., Kim, M.H., Sung, K.H., Seong, T.J.: Kass: korean automatic scoring system for short-answer questions. CSEDU **2014**(2), 226–230 (2014)
10. Jiang, H., Kim, B., Guan, M.Y., Gupta, M.R.: To trust or not to trust a classifier. In: NIPS, pp. 5546–5557 (2018)
11. Johan Berggren, S., Rama, T., Øvrelid, L.: Regression or classification? automated essay scoring for Norwegian. In: BEA, pp. 92–102. Association for Computational Linguistics, Florence, August 2019. https://doi.org/10.18653/v1/W19-4409
12. Krishnamurthy, S., Gayakwad, E., Kailasanathan, N.: Deep learning for short answer scoring. Int. J. Rec.Technol. Eng. **7**, 1712–1715 (2019)
13. Kumar, Y., Aggarwal, S., Mahata, D., Shah, R.R., Kumaraguru, P., Zimmermann, R.: Get it scored using autosas - an automated system for scoring short answers. In: AAAI/IAAI/EAAI. AAAI Press (2019). https://doi.org/10.1609/aaai.v33i01.33019662
14. Leacock, C., Chodorow, M.: C-rater: automated scoring of short-answer questions. Comput. Human. **37**(4), 389–405 (2003). https://doi.org/10.1023/A:1025779619903
15. Mizumoto, T., et al.: Analytic score prediction and justification identification in automated short answer scoring. In: BEA, pp. 316–325 (2019). https://doi.org/10.18653/v1/W19-4433
16. Rasmussen, C.E.: Gaussian processes in machine learning. In: Bousquet, O., von Luxburg, U., Rätsch, G. (eds.) ML -2003. LNCS (LNAI), vol. 3176, pp. 63–71. Springer, Heidelberg (2004). https://doi.org/10.1007/978-3-540-28650-9_4
17. Riordan, B., Horbach, A., Cahill, A., Zesch, T., Lee, C.M.: Investigating neural architectures for short answer scoring. In: BEA, pp. 159–168 (2017). https://doi.org/10.18653/v1/W17-5017
18. Sychev, O., Anikin, A., Prokudin, A.: Automatic grading and hinting in open-ended text questions. Cogn. Syst. Res. **59**, 264–272 (2020)
19. Wang, T., Inoue, N., Ouchi, H., Mizumoto, T., Inui, K.: Inject rubrics into short answer grading system. In: DeepLo, pp. 175–182, November 2019. https://doi.org/10.18653/v1/D19-6119
20. Williamson, D.M., Xi, X., Breyer, F.J.: A framework for evaluation and use of automated scoring. Educ. Meas. Issues Pract. **31**(1), 2–13 (2012). https://doi.org/10.1111/j.1745-3992.2011.00223.x
21. Woods, B., Adamson, D., Miel, S., Mayfield, E.: Formative essay feedback using predictive scoring models. In: KDD 2017, pp. 2071–2080. Association for Computing Machinery (2017). https://doi.org/10.1145/3097983.3098160

Auxiliary Task Guided Interactive Attention Model for Question Difficulty Prediction

V. Venktesh(✉), Md. Shad Akhtar, Mukesh Mohania, and Vikram Goyal

Indraprastha Institute of Information Technology, Delhi, India
{venkteshv,shad.akhtar,mukesh,vikram}@iiitd.ac.in

Abstract. Online learning platforms conduct exams to evaluate the learners in a monotonous way, where the questions in the database may be classified into Bloom's Taxonomy as varying levels in complexity from basic knowledge to advanced evaluation. The questions asked in these exams to all learners are very much static. It becomes important to ask new questions with different difficulty levels to each learner to provide a personalized learning experience. In this paper, we propose a multi-task method with an interactive attention mechanism, Qdiff, for jointly predicting Bloom's Taxonomy and difficulty levels of academic questions. We model the interaction between the predicted bloom taxonomy representations and the input representations using an attention mechanism to aid in difficulty prediction. The proposed learning method would help learn representations that capture the relationship between Bloom's taxonomy and difficulty labels. The proposed multi-task method learns a good input representation by leveraging the relationship between the related tasks and can be used in similar settings where the tasks are related. The results demonstrate that the proposed method performs better than training only on difficulty prediction. However, Bloom's labels may not always be given for some datasets. Hence we soft label another dataset with a model fine-tuned to predict Bloom's labels to demonstrate the applicability of our method to datasets with only difficulty labels.

Keywords: Question difficulty prediction · Transformers · Multi-task learning

1 Introduction

The academic questions in online learning platforms help the learner evaluate his understanding of concepts. However, serving a static set of questions to all the users is not desirable as all the users do not have the same learning abilities. Hence, there is a need to dynamically adapt to the user's learning profile and accordingly select a question. This would require accurate prediction of the difficulty level of each question so that the system or the academician can choose appropriate questions for the exams. A system for labelling the new questions

M. M. Rodrigo et al. (Eds.): AIED 2022, LNCS 13355, pp. 477–489, 2022.
https://doi.org/10.1007/978-3-031-11644-5_39

Table 1. Some samples from our QC-Science dataset.

Question text	Difficulty	Bloom's taxonomy
The value of electron gain enthalpy of chlorine is more than that of fluorine. Give reasons	Difficult	Understanding
What are artificial sweetening agents?	Easy	Remembering
Explain the concept of rotation of Earth	Medium	Understanding

with appropriate difficulty levels would obviate the need for manual intervention. When the questions are automatically labelled with difficulty levels, it helps in designing adaptive tests where questions in tests are dynamically changed at test time according to the performance in previous questions, tests, or as per the users' capability. For instance, a student who answers the questions in a certain topic like 'calculus' correctly is presented with a question of increasing difficulty as the test progresses. This strategy is adopted in online platforms that offer practice for standardized tests like GRE[1].

Given the advantages of automated difficulty prediction, we propose, `QDiff`, a method for predicting the difficulty label of a question that is derived from the difficulty levels denoted as *'easy'*, *'medium'*, or *'difficult'*. We collect a set of academic questions in the Science domain from a leading e-learning platform[2]. Bloom's taxonomy provides a mechanism for describing the learning outcomes. The different levels in Bloom's taxonomy as observed in our dataset are *'remembering'*, *'understanding'*, *'applying'*, and *'analyzing'* which form the Bloom's labels. The questions are tagged with an appropriate level in Bloom's taxonomy [5] and a difficulty level. Some samples from our dataset, named QC-Science, are shown in Table 1. From the collected QC-Science data, we observe that the difficulty level is related to the levels in Bloom's taxonomy as shown in Table 2. For instance, in Table 2 most of the questions tagged with the *'remembering'* level of Bloom's taxonomy are categorized as being *'easy'* questions. To verify the strength of association between Bloom's taxonomy and the difficulty levels, we use the *Cramer's V* test since it is best suited for a large sample size. We compute the value V using the formula, $V = \sqrt{\frac{\chi^2}{n(min(r-1, c-1))}}$ where χ^2 is the chi-squared statistic, n is the total sample size, r is the number of rows and c is the number of columns. We obtain a value of **0.51** for V, indicating that there is a strong association between Bloom's taxonomy and the difficulty labels. Therefore, the Bloom's taxonomy labels can serve as a strong indicator for the difficulty labels and could help in the question difficulty prediction task.

As mentioned in the previous section, we observe a strong association between Bloom's taxonomy labels and difficulty labels. Hence, we propose an interactive attention model to predict the difficulty level and Bloom's taxonomy level jointly for the questions collected for classes VI to XII in the K12[3] education system.

[1] https://www.prepscholar.com/gre/blog/how-is-the-gre-scored/.
[2] We don't disclose the identity of the source due to the anonymity requirement.
[3] Kinder-garden to grade-12.

Table 2. Distribution of samples across Bloom's taxonomy levels and difficulty levels (contingency table)

		Bloom's taxonomy			
		Analyzing	Applying	Remembering	Understanding
Difficulty	Easy	756	1488	7146	4505
	Difficult	2089	2529	2518	7010
	Medium	585	980	1712	2242

The Bloom's taxonomy level prediction is considered as an auxiliary task, and the attention weights are computed between the vector representations of the predicted Bloom's taxonomy labels and the input vector representation through the interactive attention mechanism. We use the *hard parameter sharing* approach [12] where the backbone is a Transformer-based [15] model (like BERT [3]) with task specific output layers on top of the backbone network. The conventional multi-task learning methods do not explicitly model the interactions between the task labels and the input. Hence we propose the interactive attention mechanism to explicitly model the interaction between Bloom's taxonomy label and the input, which enables to model the relationship between the tasks better. We observe that `QDiff` outperforms the existing baselines as measured by the macro-averaged and weighted-average F1-scores.
Following are the core technical contributions of our work:

- We propose a multi-task learning and interactive attention based approach, `QDiff` for difficulty prediction, where Bloom's taxonomy predictions are used to determine the input representations using an attention mechanism.
- We evaluate the proposed method on the QC-Science dataset. We also evaluate the proposed method on another dataset QA-data [13] which consists of only difficulty labels. We show that our method trained on QC-Science dataset can be used to soft-label the QA-data dataset with Bloom's taxonomy levels. The experiments demonstrate that our method can be extrapolated to new datasets with only difficulty labels.

Code and data are at https://github.com/VenkteshV/QDIFF_AIED_2022.

2 Related Work

Question difficulty prediction is an intriguing NLP problem, but it has not been explored to the extent it deserves. A few recent works in difficulty prediction focus on evaluating the readability of the content [4,18]. Authors in [4] proposed to combine classical features like Flesch readability score with non-classical features derived automatically. In [18], the authors observed that combining a large generic corpus with a small population specific corpus improves the performance of difficulty prediction.

Another application where the task of question difficulty prediction has been discussed is automated question generation [6]. In the work [6], the authors propose to estimate the difficulty measure as the semantic similarity between the question and the answer options. Similarly, in the work [1], the authors propose a similarity based theory for controlling the difficulty of the questions.

Prior work has also focused on estimating the difficulty of questions in community question-answering (QA) platforms like StackOverflow [8,16]. Other studies like [9,10] explore automated grading by estimating the difficulty of academic questions from the Science domain. In the work [9], the authors propose to infer the difficulty of the questions by first mapping them to the concepts. In the work [17], the authors propose fine-tuning the ELMo [11] model on 18,000 MCQ type questions from the medical domain for predicting the question difficulty. More recently, in the work [2], the authors propose fine-tuning transformer based language models like BERT for the task of difficulty estimation. However, the discussed works do not exploit information from related tasks.

3 Methodology

In this section, we describe the proposed approach `QDiff` for the question difficulty prediction as the primary task and Bloom's taxonomy prediction as the secondary task. The input to `QDiff` is a corpus of questions, $C = \{q_1, q_2, ..., q_n\}$ where each q_i corresponds to a question along with a difficulty label and Bloom's taxonomy label (skill levels). Since most questions are short texts, we augment the question with the answer as auxiliary information to obtain more semantic information. Hence, we refer to the augmented question as a *'question-answer'* pair in the remainder of the paper. We show that the performance of various methods improves when using the question along with the answer rather than the question text alone. We obtain contextualized representations for the inputs using BERT [3] followed by task specific layers, H^{diff} and H^{bloom} where each tasks specific layer comprises of two linear layers with a non-linearity in between the two layers.

3.1 Contextualized Input Representations - BERT

The academic questions also have **polysemy** terms that refer to different semantics depending on the context of their occurrence in the input sentence. To tackle the mentioned problem, we use BERT, a transformer-based masked language model, for projecting the input text to the embedding space.

Self-attention is the core of Transformer [15] model, and BERT utilizes it to obtain better representations. Self-attention encodes each word in the sentence using **Q**uery, **K**ey, and **V**alue vectors to obtain attention scores, which determines how much attention to pay to each word when generating an embedding for the current word. Mathematically, it is defined as:

$$Attention(Q, K, V) = \frac{Softmax(Q * K^T)}{\sqrt{d_k}} * V \tag{1}$$

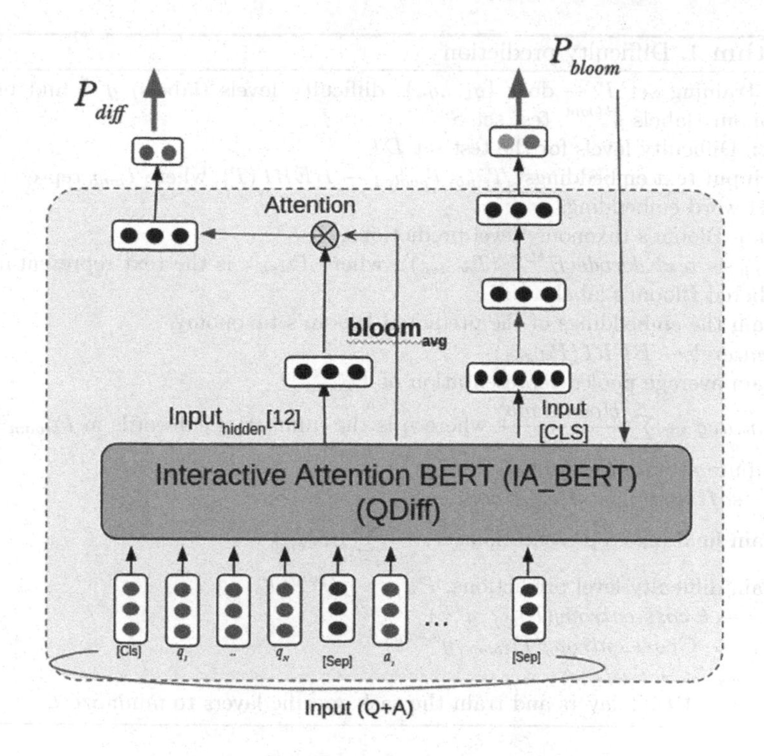

Fig. 1. `QDiff` network architecture (middle).

where, d_k is the dimension of query, key, and value vectors and is used to scale the attention scores, and K^T denotes the transpose of the **K**ey vector.

3.2 Auxiliary Task Guided Interactive Attention Model

Based on the strength of association between Bloom's labels and the difficulty labels verified through Cramer's V test (V = **0.51**), we hypothesize that leveraging Bloom's taxonomy representations to compute input representations using an attention mechanism would lead to better performance in difficulty prediction. The proposed approach would help capture the relationship between the words in the input question and Bloom's taxonomy level, leading to better representations for the task of difficulty prediction as Bloom's taxonomy and difficulty levels are related. Our method also jointly learns to predict both Bloom's taxonomy and the difficulty level, obviating the need for providing Bloom's taxonomy labels at inference time. Figure 1 shows the architecture of the proposed approach `QDiff`.

During training, as shown in Algorithm 1, the question-answer pair is first passed through a transformer based language model BERT, to obtain contextualized word representations (T_{emb}) and the pooled representation T_{pooled} (step 1). Then the representations are passed to the task specific output layer H^{bloom}

Algorithm 1. Difficulty prediction

Input: Training set $T \leftarrow$ docs $\{q_1, ..q_n\}$, difficulty levels (labels) y^{diff} and bloom's taxonomy labels y^{bloom}, test set S
Output: Difficulty levels for the test set DT
1: Get input text embeddings , $T_{emb}, T_{pooled} \leftarrow BERT(T)$, where T_{emb} represents the set of word embeddings
2: Obtain Bloom's taxonomy level predictions,
$P_{bloom} \leftarrow text\_decode(H^{bloom}(T_{pooled}))$, where P_{bloom} is the text representation of predicted Bloom's label.
3: Obtain the embeddings of the predicted Bloom's taxonomy,
$bloom\_emb \leftarrow BERT(P_{bloom})$
4: Obtain average pooled representation of P_{bloom},
$bloom\_avg \leftarrow \sum_{i=1}^{n} \frac{bloom\_emb^i}{n}$ where n is the number of subwords in P_{bloom}.
5: Compute attention weights,
$\alpha_i \leftarrow softmax(f_{attn}(T^i_{emb}, bloom\_avg))$
6: Obtain final text representations, $T_r \leftarrow \sum_{i=1}^{n} \alpha_i T^i_{emb}$
7: Obtain difficulty level predictions, $P_{diff} \leftarrow H^{diff}(T_r)$
8: $\mathcal{L}_{diff} \leftarrow Cross\_entropy(P_{diff}, y^{diff})$
9: $\mathcal{L}_{bloom} \leftarrow Cross\_entropy(P_{bloom}, y^{bloom})$
10: $\mathcal{L} \leftarrow \mathcal{L}_{bloom} + \mathcal{L}_{diff}$
11: Fine-tune BERT layers and train the task specific layers to minimize $\mathcal{L}$

to obtain Bloom's taxonomy level predictions (step 2). The vector representations for Bloom's taxonomy prediction are obtained using the **same BERT** model (step 3). Then the representations of subwords in P_{bloom} are averaged to obtain a fixed 768 dimensional vector representation (step 4). With the input vector representations T_{emb}, the attention mechanism generates the attention vector α_i using Bloom's taxonomy representations $bloom\_avg$ (step 5) by

$$\alpha_i = \frac{exp(f_{attn}(T^i_{emb}, bloom\_avg))}{\sum_i^N exp(f_{attn}(T^i_{emb}, bloom\_avg))} \tag{2}$$

$$f_{Attn} = \tanh(T^i_{emb}.W_a.bloom\_avg^T + b_a) \tag{3}$$

where, tanh is a non-linear activation, W_a and b_a are the weight matrix and bias, respectively.

Then the final input representations are obtained using the attention weights α_i (step 6). The difficulty predictions are then obtained by passing the final input representation T_r through the task specific layer H^{diff} (step 7). The BERT model acts as the backbone as its parameters are shared between the tasks. The loss function is a combination of the loss for difficulty prediction $\mathcal{L}_{diff}$ and the loss for Bloom's taxonomy level prediction $\mathcal{L}_{bloom}$ (steps 8, 9 and 10). The BERT layers are fine-tuned and the task specific layers are trained to minimize the combined loss (step 11).

During the *inference* phase, the contextualized representations are obtained for the test set question-answer pairs as described above. Subsequently, the softmax probability distributions of the difficulty labels are obtained by passing the contextualized representations through the H^{diff} layer. Since the BERT layers are shared between the tasks during training, the contextualized representations obtained at test time improve the performance on the task of difficulty prediction. Since we use an interaction layer with attention where the interaction between the input representations and Bloom's taxonomy label representations are captured, we also call our method as **IA_BERT** (Interactive Attention BERT).

4 Experiments

In this section, we discuss the experimental setup and the datasets used.

4.1 Dataset

QC-Science: We compile the QC-Science dataset from a leading e-learning platform. It contains 45766 question-answer pairs belonging to the science domain tagged with Bloom's taxonomy levels and the difficulty levels. We split the dataset into 39129 samples for training, 2060 samples for validation, and 4577 samples for testing. Some samples are shown in Table 1. The average number of words per question is 37.14, and per answer, it is 32.01.

QA-data: We demonstrate the performance of the proposed method on another dataset [13]. The dataset is labeled only with difficulty labels. We soft-label the dataset with Bloom's taxonomy levels using the Bloom's taxonomy prediction model trained on the **QC-Science** dataset. We demonstrate that the proposed model, when fine-tuned on a large enough dataset like QC-Science with labels for both the tasks, can be extrapolated to datasets without Bloom's taxonomy labels. The dataset consists of 2768, 308 and 342 train, validation and test samples, respectively. The average number of words per question is 8.71, and per answer, it is 3.96.

4.2 Baselines

We compare with baselines like *LDA + SVM* and *TF-IDF + SVM* [14], *ELMo fine-tuning* [17]. We also propose a new baseline.

- *TF-IDF + Bloom verb weights (BW)*: In this method the samples are first grouped according to difficulty levels. Then, we extract the Bloom verbs from each sentence using following POS tag patterns: 'VB.*', 'WP', 'WRB'. Once the Bloom verbs are extracted, we obtain Bloom verb weights as follows:

$$Bl\_W = freq(verb, label) * \frac{no.\ of\ labels}{n_l}$$

where, $freq(verb, label)$ indicates the number of times a verb appears in a label, and n_l indicates the number of labels that contain the verb. The above operation assigns higher relevance to rare verbs. Then the TF-IDF vector representation of each sentence is multiplied by the weights of the Bloom verbs contained in the sentence. Then for each difficulty level, we obtain the mean of the vector representations of the sentences (centroid). At inference time, the difficulty label whose centroid vector representations is closest to test sentence representation is obtained as output.

We also explore the following deep learning approaches.

- *Bi-LSTM with attention* and *Bi-GRU with attention and concat pooling* [7].
- *BERT cascade*: In this method, a BERT (base) model is first fine-tuned on Bloom's taxonomy level prediction followed by the task of difficulty prediction.

We also conducted several ablation studies for the proposed architecture `QDiff`.

- **Multi-Task BERT**: In this method, the interaction layer in IA_BERT is removed, and the model is trained to jointly predict Bloom's taxonomy and difficulty labels.
- **IA_BERT (B/D) (bloom/difficulty label given)**: In this method, Bloom's taxonomy label is not predicted but rather assumed to be given even at inference time. The model is trained on the objective of difficulty prediction (loss $\mathcal{L}_{diff}$) only. The Bloom's taxonomy label is considered as given when difficulty is predicted and difficulty label is considered as given when the task is Bloom's taxonomy label prediction.
- **IA_BERT (PB) (pre-trained Bloom model)**: In this method first, a BERT model is fine-tuned for predicting Bloom's taxonomy labels alone, given the input. Then the model's weights are frozen and are used along with another BERT model, which is fine-tuned to predict the difficulty labels.

The interaction layer is the same as in IA_BERT for the last two methods mentioned above. The HuggingFace library (https://huggingface.co/) was used to train the models. All the BERT models were fine-tuned for 20 epochs with the ADAM optimizer, with learning rate (lr) of 2e−5 [3]. The LSTM and GRU based models are trained using ADAM optimizer and with lr of 0.003.

5 Results and Discussion

The performance comparison of various methods is shown in Table 3. We use macro-average and weighted-average Precision, Recall, and F1-scores as metrics for evaluation. From Table 3, we can observe that most of the deep learning based methods outperform classical ML based methods like TF-IDF + SVM and LDA + SVM. However, we observe that TF-IDF + SVM method outperforms the ELMo baseline [17] on the QC-Science dataset. It is also evident that the transformer based methods significantly outperform the 'TF-IDF + Bloom verbs' baseline. This demonstrates that the contextualized vector representations obtained have more representational power than carefully hand-crafted features for the task of difficulty prediction.

5.1 Results Analysis

We provide a detailed analysis of results in this section.

Table 3. Performance comparison for the difficulty prediction and Bloom's taxonomy prediction tasks. † indicates significance at 0.05 level (over Multi-task BERT). D1 - QC-Science, D2 - QA-data

	Method	Difficulty prediction						Bloom's level prediction					
		Macro			Weighted			Macro			Weighted		
		P	R	F1	P	R	F1	P	R	F1	P	R	F1
D1	LDA+SVM [14]	0.319	0.354	0.336	0.406	0.492	0.445	0.279	0.266	0.267	0.320	0.360	0.338
	TF-IDF + SVM [14]	0.471	0.415	0.440	0.510	0.532	0.520	0.421	0.341	0.377	0.413	0.410	0.411
	ELMo [17]	0.466	0.403	0.432	0.495	0.503	0.499	0.407	0.367	0.386	0.429	0.429	0.429
	TF-IDF + BW	0.426	0.437	0.431	0.486	0.429	0.456	0.330	0.346	0.338	0.364	0.344	0.342
	Simple rule baseline	0.359	0.402	0.379	0.454	0.542	0.494	0.138	0.239	0.175	0.199	0.344	0.252
	Bi-LSTM with attention	0.491	0.407	0.445	0.518	0.529	0.523	0.487	0.419	0.450	0.497	0.481	0.489
	Bi-GRU with attention	0.438	0.369	0.400	0.476	0.499	0.488	0.505	0.359	0.420	0.503	0.441	0.470
	BERT (base) [2]	0.499	0.450	0.473	0.530	0.550	0.539	0.484	0.459	0.471	0.494	0.502	0.498
	BERT cascade	0.494	0.454	0.473	0.530	0.550	0.539	0.470	0.441	0.455	0.486	0.486	0.486
	Multi-task BERT (ours)	0.518	0.441	0.476	0.538	0.556	0.547	0.490	0.439	0.463	0.497	0.499	0.498
	IA_BERT (QDiff) (ours)	**0.544**	**0.447**	**0.491†**	**0.556**	**0.564**	**0.560†**	**0.497**	0.447	**0.471**	**0.502**	**0.506**	**0.504†**
D2	LDA+SVM [14]	0.356	0.359	0.357	0.370	0.409	0.388	0.232	0.205	0.218	0.580	0.655	0.615
	TF-IDF + SVM [14]	0.518	0.487	0.502	0.517	0.518	0.517	0.514	0.319	0.394	0.796	0.795	0.796
	ELMo [17]	0.635	0.623	0.629	0.654	0.658	0.656	0.450	0.386	0.415	0.779	0.784	0.781
	TF-IDF + BW	0.580	0.573	0.576	0.608	0.581	0.594	0.312	0.338	0.324	0.705	0.646	0.674
	Simple rule baseline	0.236	0.341	0.279	0.233	0.392	0.292	0.131	0.200	0.158	0.429	0.655	0.518
	Bi-LSTM with attention	0.628	0.605	0.616	0.644	0.655	0.650	0.430	0.357	0.390	0.765	0.766	0.766
	Bi-GRU with attention	0.524	0.534	0.529	0.563	0.591	0.577	0.402	0.295	0.340	0.753	0.722	0.737
	BERT (base) [2]	0.640	0.638	0.639	0.661	0.660	0.661	0.455	0.404	0.428	0.814	0.822	0.818
	BERT cascade	0.662	0.666	0.664	0.683	0.681	0.682	0.437	0.401	0.418	0.816	0.827	0.821
	Multi-task BERT (ours)	0.664	0.644	0.654	0.681	0.687	0.684	0.389	0.365	0.377	0.799	0.804	0.802
	IA_BERT (QDiff) (ours)	**0.684**	**0.682**	**0.683†**	**0.702**	**0.708**	**0.705†**	**0.494**	**0.420**	**0.454†**	**0.841**	**0.830**	**0.836†**

Is the Simple Rule-Based Baseline Enough?

We implement a simple rule-based baseline (Table 3) where the question or answer content is not considered and the difficulty label is predicted based on co-occurrence with the corresponding bloom's label alone. We form a dictionary recording the co-occurrence counts of the bloom's labels and difficulty labels in the training samples as shown in Table 2. For each test sample, we look up into the dictionary the entries for the corresponding bloom's taxonomy label of the test sample. Then the difficulty label with maximum co-occurrence count is chosen. We observe that this baseline performs poorly when compared to even other ML baselines from Table 3. This baseline performance demonstrates the need for learning based methods to analyze the given content.

Is the Proposed Approach Better than Baselines?

We observe that QDiff (IA_BERT), which jointly learns to predict Bloom's taxonomy level and the difficulty level, outperforms all the deep learning (DL) and

Table 4. Ablation studies for `QDiff` (IA_BERT). D1 - QC-Science, D2 - QA-data

	Method	Difficulty prediction						Bloom's level prediction					
		Macro			Weighted			Macro			Weighted		
		P	R	F1	P	R	F1	P	R	F1	P	R	F1
D1	IA_BERT (`QDiff`)	0.544	0.447	0.491	0.556	0.564	0.560	0.497	0.447	0.471	0.502	0.506	0.504
	IA_BERT (B/D)	0.523	0.486	**0.503**	0.554	**0.569**	0.561	0.476	0.473	0.475	0.499	0.506	0.503
	IA_BERT (PB)	0.491	0.487	0.489	0.535	0.530	0.533	0.482	0.463	0.472	0.496	0.501	0.498
D2	IA_BERT (`QDiff`)	0.684	0.682	0.683	0.702	0.708	**0.705**	0.494	0.420	0.454	0.841	0.830	0.836
	IA_BERT (B/D)	0.682	0.688	0.685	0.702	0.696	0.699	0.458	0.429	0.443	0.837	0.842	0.839
	IA_BERT (PB)	0.642	0.641	0.641	0.667	0.652	0.659	0.449	0.424	0.436	0.825	0.825	0.825

machine learning (ML) based baselines on both the datasets. It also performs better than BERT (base) [2] on the ask of difficulty prediction by advancing weighted F1-score from 0.539 to 0.560 (+3.89%). It also outperforms Multi-task BERT by **2.37%** (weighted F1-score) which can be considered as ablation of the proposed method without the interactive attention mechanism. This demonstrates that in addition to jointly learning to predict labels for both the tasks, the interactive attention mechanism yields better representations leading to improved performance. We also observe that on the QA-data dataset `QDiff` (IA_BERT) outperforms other methods as measured by macro and weighted F1 scores. For Bloom's label prediction, we observe that the IA_BERT (`QDiff`) leads to good results (from Table 3) as Bloom's labels are jointly learned and used as signal in attention mechanism. In addition, we also perform an experiment where we use the jointly predicted difficulty labels as signal in the interactive attention mechanism in IA_BERT for Bloom's label prediction task. We observe that the macro F1-score increases to **0.479** from 0.471 for QC-Science and also increases the weighted F1-score on QA-data to **0.852** from 0.836. This demonstrates that both tasks can benefit from each other through the interactive attention mechanism in addition to joint learning. We do not tabulate this result as our focus is difficulty prediction and mention it here for completion.

Are the Results Statistically Significant?

We perform statistical significance test (*t-test*) on obtained outputs. We observe that the results obtained using `QDiff` (IA_BERT) are significant with p-value = 0.000154 (weighted-F1) and p-value = 0.011893 (macro-F1) for difficulty prediction on QC-Science dataset. We also observe that results are statistically significant on QA-data with p-value = 0.003813 (weighted-F1) and p-value = 0.02248 (macro-F1) for difficulty prediction.

Does Augmenting Question with Answer Lead to Better Performance?

From Table 3, it is evident that augmenting the question with the answer provides better performance when compared to using the question text alone. When we evaluated on the QC-Science dataset using the question text alone, we observed a drop in performance. For instance, QDiff only yielded a macro-F1 score of 0.455 when using question text alone. The baselines also show a decline in performance, demonstrating that the answer helps provide some context.

What if Bloom's Taxonomy Labels Are Randomly Labeled for QA-Data?
We also perform an **ablation study** on the QA-data dataset by randomly labeling the dataset with Bloom's levels instead of the proposed soft labeling method. We observe that this lowers the performance on the task of difficulty prediction. For instance, QDiff yields a macro F1 score of 0.656 and a weighted F1 score of 0.674 on the difficulty prediction task using the random Bloom's labels. This ablation study supports the significance of the proposed soft labeling method and the interactive attention mechanism as random labels lead to erroneous predictions of difficulty labels.

How Does the Ablations of IA_BERT QDiff Perform?

We also perform several ablations studies by varying the components of the proposed method. The results are as shown in Table 4. We observe that using a pre-trained model for Bloom's taxonomy prediction performs poorly as it is not jointly trained on the two related tasks resulting in the error from Bloom's label prediction model propagating to difficulty prediction task through the interactive attention mechanism. Additionally, we observe that directly feeding the Bloom's taxonomy label for the task of difficulty prediction provides gain in performance in certain scenarios as demonstrated by the second ablation in Table 4 for datasets D1 and D2. However, this setting is not possible in real-time scenarios as during inference, the incoming content would not be labeled with Bloom's taxonomy labels. We also observe that the performance of the original IA_BERT (QDiff) method is very close the mentioned ablation study. This demonstrates that the error propagation from Bloom's label prediction is mitigated in our approach.

6 Conclusion

In this paper, we proposed a novel method for predicting the difficulty level of the questions. The proposed method, QDiff, leverages an interactive attention mechanism to model the relation between bloom's taxonomy labels and the input text. We observe that QDiff and the proposed ensemble construction approach outperforms existing methods. The results also confirm the hypothesis that modeling the interaction between the input and the task labels through an attention mechanism performs better than implicit interactions captured using only multi-task learning. Though question difficulty estimation is subjective, we observe that modeling the interaction between related tasks improves the performance.

Acknowledgements. The authors acknowledge the support of Extramarks Education India Pvt. Ltd., SERB, FICCI (PM fellowship), Infosys Centre for AI and TiH Anubhuti (IIITD).

References

1. Alsubait, T., Parsia, B., Sattler, U.: A similarity-based theory of controlling MCQ difficulty. In: ICEEE 2013, pp. 283–288 (2013)
2. Benedetto, L., Aradelli, G., Cremonesi, P., Cappelli, A., Giussani, A., Turrin, R.: On the application of transformers for estimating the difficulty of multiple-choice questions from text. In: Proceedings of the 16th Workshop on Innovative Use of NLP for Building Educational Applications, Online, pp. 147–157. ACL (April 2021)
3. Devlin, J., Chang, M., Lee, K., Toutanova, K.: BERT: pre-training of deep bidirectional transformers for language understanding. CoRR abs/1810.04805 (2018)
4. François, T., Miltsakaki, E.: Do NLP and machine learning improve traditional readability formulas? In: Proceedings of the 1st Workshop on Predicting and Improving Text Readability for Target Reader Populations, Montréal, Canada. ACL (June 2012)
5. Gogus, A.: Bloom's taxonomy of learning objectives, pp. 469–473 (2012)
6. Ha, L.A., Yaneva, V.: Automatic distractor suggestion for multiple-choice tests using concept embeddings and information retrieval. In: BEA (June 2018)
7. Howard, J., Ruder, S.: Universal language model fine-tuning for text classification. arXiv preprint arXiv:1801.06146 (2018)
8. Liu, J., Wang, Q., Lin, C.Y., Hon, H.W.: Question difficulty estimation in community question answering services. In: Proceedings of the 2013 EMNLP, Seattle, Washington, USA. ACL (October 2013)
9. Nadeem, F., Ostendorf, M.: Language based mapping of science assessment items to skills. In: Proceedings of the 12th Workshop on Innovative Use of NLP for Building Educational Applications, Copenhagen, Denmark. ACL (September 2017)
10. Padó, U.: Question difficulty - how to estimate without norming, how to use for automated grading. In: BEA, Copenhagen, Denmark. ACL (September 2017)
11. Peters, M., et al.: Deep contextualized word representations. In: Proceedings of the 2018 Conference of the NAACL, Volume 1 (Long Papers). ACL (June 2018)
12. Ruder, S.: An overview of multi-task learning in deep neural networks. arXiv preprint arXiv:1706.05098 (2017)
13. Smith, N.A., Heilman, M., Hwa, R.: Question generation as a competitive undergraduate course project. In: Proceedings of the NSF Workshop on the Question Generation Shared Task and Evaluation Challenge, pp. 4–6 (2008)
14. Supraja, S., Hartman, K., Tatinati, S., Khong, A.W.H.: Toward the automatic labeling of course questions for ensuring their alignment with learning outcomes. In: EDM (2017)
15. Vaswani, A., et al.: Attention is all you need. In: Advances in Neural Information Processing Systems, pp. 5998–6008 (2017)
16. Wang, Q., Liu, J., Wang, B., Guo, L.: A regularized competition model for question difficulty estimation in community question answering services. In: EMNLP 2014, pp. 1115–1126 (2014)

17. Xue, K., Yaneva, V., Runyon, C., Baldwin, P.: Predicting the difficulty and response time of multiple choice questions using transfer learning. In: BEA. ACL (July 2020)
18. Yaneva, V., Orăsan, C., Evans, R., Rohanian, O.: Combining multiple corpora for readability assessment for people with cognitive disabilities. In: BEA (September 2017)

A Basic Study on Educational Growth Indicators Based on Quantitative Evaluation of Strokes Quality in Drawing Works

Kanato Sugii[1], Takashi Nagai[2], and Mizue Kayama[1(✉)]

[1] Shinshu University, Wakasato 4-17-1, Nagano, Nagano, Japan
kayama@shinshu-u.ac.jp

[2] Institute of Technologists, Maeya 333, Gyoda, Saitama, Japan
t_nagai@iot.ac.jp

Abstract. This study develops a set of indicators for the quantitative evaluation of drawing work in the study of drawing. In basic drawing classes, students are taught to simplify the strokes that they use. To simplify a stroke means to use only straight lines and simple curves. In this study, we focus on the shapes of individual strokes in drawing work. We have been conducting a remote drawing learning support system at a Japanese art school since 2012. At this school, a digital drawing instruction using a digital pen is offered for 3 months at the beginning of each school year. Through drawing with a digital pen, each stroke that the learner applies is recorded together with the relevant temporal and geometrical information. In this paper, we describe the abstract stroke information collected and classify it using a self-organizing feature map, a machine learning method.

Keywords: Art education · Drawing · Stroke · Machine learning · Quality · Growth indicators

1 Introduction

Since the mid-2000s, a rapid increase has been seen in the publication of academic papers related to the digitization of art education. In these papers, both theoretical proposals and practice-based discussions have been reported. These have included a framework to support art education, an approach to mining art course grades, the presentation of examples of specified motifs by GAN, the emotional analysis of art-works, the automatic evaluation of students' artworks based on GLCM and color moments, and the automatic evaluation of ink painting artworks by MVPD-CNN [1–7]. However, most of these studies were focused on completed works. By contrast, we have explored learning support methods for the process of constructing artworks. We have focused on professional art education for students seeking to enter an art university.

M. M. Rodrigo et al. (Eds.): AIED 2022, LNCS 13355, pp. 490–501, 2022.
https://doi.org/10.1007/978-3-031-11644-5_40

Drawing is a fundamental technique in professional art education, and it is generally considered to be the first skill that students should learn [8,9]. Particularly, in the early stages of learning drawing, learners are instructed to increase their number of strokes and draw motifs using simple lines. Quantitative evaluation using the number of strokes was achieved using a previous study by Sakimoto et al. [10]. In this study, we focus on the shape of the strokes and explore the possibility of supporting drawing learning by evaluating the quality of the shape of the stroke.

In this paper, we report the results of stroke shape classification using a self-organizing map (SOM), an unsupervised learning method. The classification features of the map created by the learner's strokes are compared between what the learner and the instructor produce, and the possibility of classification as a growth indicator is discussed.

2 Research Goal

This study is conducted to develop growth indicators based on the stroke shapes used in drawing. To achieve this goal, we adopt a machine learning method to process the drawing data (strokes) that are stored in the drawing learning support system.

This basic study was conducted to verify the applicability of the machine learning method. We classified the stroke data of three learners and one instructor. We com-pared the classification results among multiple drawings of the same learner, among multiple learners, and between learners and instructors. These results present the possibility that classification by stroke shape can be used as an indicator to confirm the development of students' drawing.

3 Method

3.1 Drawing Learning Support System and Drawing Process Data

It is difficult to recognize one's own habits and weaknesses when learning to draw by oneself. For this reason, students who hope to pursue art professionally often study at art school. However, in one-to-many classes, it is difficult for instructors to grasp all the drawings of each student. In response, Nagai et al. developed a remote drawing learning support system for art learners [1,10,11]. Figure 1 shows the interfaces of this system. Using the system, learners can obtain advice and evaluation from their instructors without being constrained by time or place. In this system, a digital pen is a writing tool for learning drawing [12]. Using a digital pen, the system records the geometric information of all the lines (strokes) that the learner uses in the process of drawing.

In this study, process data from drawing stored in the drawing learning support system are analyzed. These data were generated in a digital drawing class at a Japanese art school from 2012 to 2022. In a digital drawing class, students draw on a specified motif with a digital pen within a specified time. The specified time is 20 min, and the specified motifs are a paper box in the first half

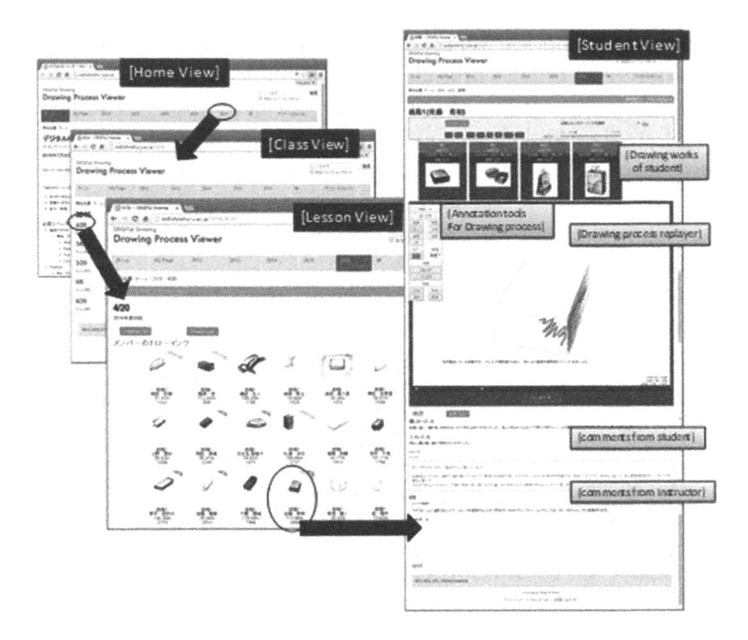

Fig. 1. Overview of Nagai's remote drawing learning system [1,10,11].

Fig. 2. Paper box drawing with digital pen in the digital drawing class.

of the class (three lessons) and a paper bag in the second half (three lessons). Figure 2 shows the classroom scene. Learners in this figure engaged in the paper box drawing with digital pens. We analyze each stroke applied in this drawing work.

3.2 Machine Learning Method for Stroke Classification

In this study, we classify strokes using convolutional NN. The input data consist of two patterns: the stroke image and the feature vector of the stroke. The line types identified by the rule base are used as teacher data in training. However, the system is unable to classify the strokes correctly in either case. All such strokes are classified as complex lines. This may be due to the inclusion relation of the stroke images. For example, a point is included in a straight line, a curve, and a complex line. This suggests that machine learning based on images and features has limitations. We there-fore attempted to classify strokes based on abstract data that maintain the shape information of the strokes.

An SOM is a neural network proposed by T. Kohonen [13]. It is a topology-preserving map that performs unsupervised learning. The basic function of an SOM is to map a high-dimensional data set in a low-dimensional space while preserving the topological structure of the data distribution. For mapping to a two-dimensional (2D) space, the data distribution is visualized as a topographic map. This map is used for data mining. The procedure of map generation using an SOM is shown below.

1. An initial map of a specified size is generated. A vector with the same number of elements as the input data is assigned to each unit in the map. The initial value of the vectors for each unit in the map is given randomly.
2. The data are input to the SOM.
3. One unit with the closest vector in Euclidean distance to the input data is selected.
4. The input data are reflected to the vectors of the unit selected in step 3 and to multiple units in the neighborhood of the selected unit at a certain learning rate.
5. Repeat steps 2–4 for the specified number of times, decreasing the learning rate and the range of the neighborhood.
6. A learned map is generated.

The unit selected in step 3 is called the best match unit (BMU). An SOM is used in this study for the following reason.

- Where data with similar geometrical features are output at units near to each other, it can identify the coherence of the strokes.
- The results are less complicated, even when the number of data is large and there are a lot of similar data because similar elements are aggregated in the nearby units.
- The algorithm is intuitive and simple, which makes it suitable for verifying its applicability.

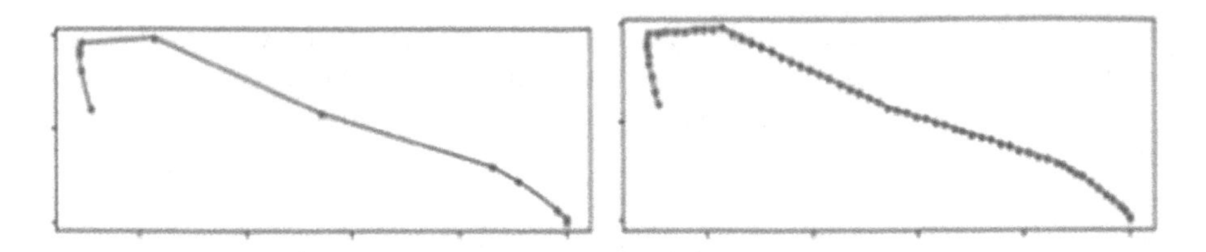

Fig. 3. Before (left) and after (right) our completion calculation. In this figure, the number of coordinates complemented is smaller than the number of coordinates originally complemented to enable visibility.

In this study, we map the 2D coordinate data of a stroke to a 2D SOM. Following that, we assign the set of strokes contained in each drawing in the generated map. Then, we observe the classification of the strokes in the drawings.

4 Data Mining of the Strokes in Drawing

4.1 Formation of Input Data for SOM

To analyze the stroke data using SOM, it is necessary to reshape these data into a unified format. The formation method adopted in this study has four steps: rotation, completion, translation and scaling, and binarization.

Rotation. The digital pen paper has an origin. Coordinate data based on the origin are recorded in the digital pen. In the digital drawing class, the orientation of the paper is decided by the learner, so the coordinates of the strokes stored in the learning support system may be different for each drawing. Therefore, the orientation of the stroke coordinates was unified so that the motif faces downward toward the origin of the special paper. In this process, we used affine transformation.

Completion. Each stroke in a drawing is represented by a set of geometric coordinates. To input them into the SOM, the number of coordinates of each stroke should be unified. Using the distribution of the number of coordinates of the strokes recorded in the drawing learning support system, we set the number of vectors to be input to the SOM as 1.5 times its IQR. In this case, we set the number of vectors to 943. Drawings with strokes having more coordinates than this value were excluded from the analysis. We set the coordinate data such that the total number of strokes for all is 943. The interpolation was done linearly, accounting for the distance between each coordinate of the stroke. In other words, more interpolated coordinates were assigned to the more distant coordinates, so that the distance between the coordinates after completion would be equal. Figure 3 shows an image of coordinate completion.

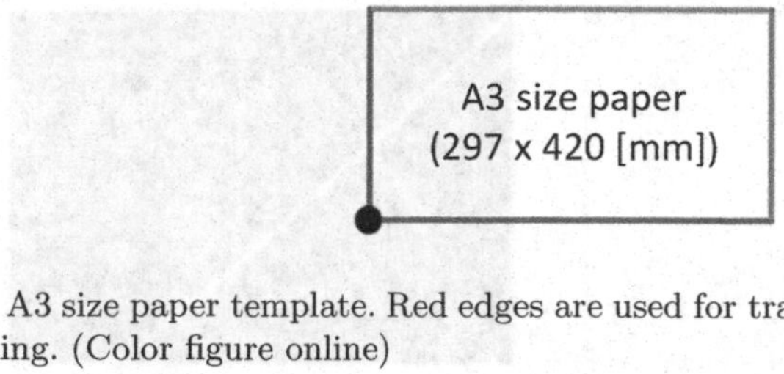

Fig. 4. A3 size paper template. Red edges are used for translation. Blue edges are used for scaling. (Color figure online)

Translation and Scaling. In this study, we consider strokes of the same shape but with different orientations and sizes to be nevertheless the same shape. For this purpose, the translation, scaling, and rotation of the data are necessary. For this, we adopt data translation and scaling. The procedure is shown as follows:

- The bounding box of the stroke is computed.
- The bounding box is translated so that it touches both red sides of the canvas in the orientation shown in Fig. 4.
- The bounding box is scaled so that it touches one of the blue edges in Fig. 4. The ratio of the height and width of the bounding box to the canvas is found, and the larger ratio is used as the scaling factor.

Binarization. To make the SOM classification more effective, we binarize the stroke data while maintaining the shape information of the stroke. The procedure runs as follows:

- The paper area is divided into 3 mm squares and makes a paper matrix of 99 × 140.
- A 99 × 140 quadratic zero matrix (paper matrix) is generated.
- The coordinate data of the stroke are divided by 3 to obtain the quotient. The correspondence between these values and the paper matrix is found.

In the paper matrix, the value of the element corresponding to the stroke coordinate is replaced with 1. The width of the paper is set to 3 mm because, in interviews with the instructor, it was judged that strokes smaller than 3 mm square could be regarded as points in the drawing. Figure 5 presents an example of the stroke data before and after binarization.

4.2 Input for SOM

The input data to the SOM is a one-dimensional (1D) vector. Therefore, we transform the 99 × 140 2D matrix generated by the binarization into a 1D vector of length 13,860. We set the SOM to learn strokes as a 100 × 100 2D map. We assign a vector of length 13,860 to each unit of the initial SOM map. The initial values of the vector elements are randomly assigned binary values (0 and 1).

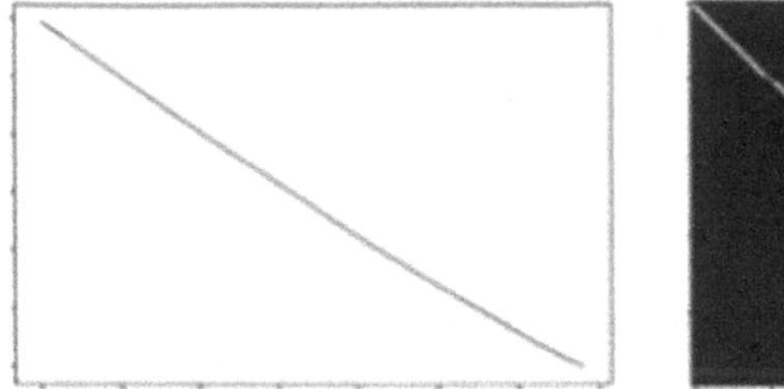
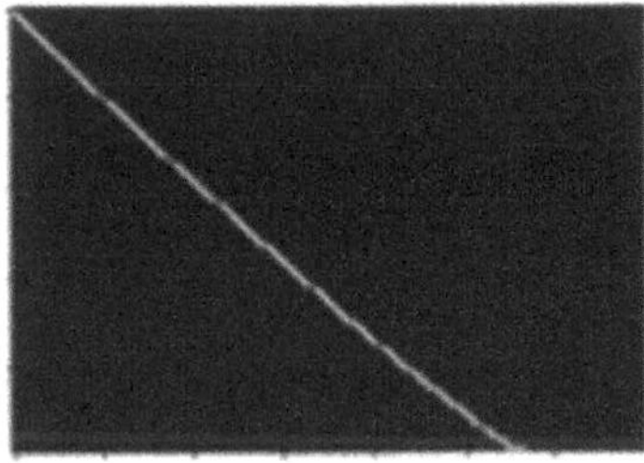

Fig. 5. Before (left) and after (right) our binarization calculation. In this figure, at the visualization after binarization (right), the 0s in the matrix are displayed in black and the 1s are in white.

As a basic analysis of stroke classification by SOM, three learners (called Subject_A, Subject_B, and Subject_C) and one instructor are selected for the study. These three learners were selected because they were judged by the drawing instructor to have shown a high degree of growth in the digital drawing class. We analyze three drawings made by each learner for each motif of the paper box and paper bag. The first drawing of the paper box is the initial learning datum, and the third drawing of the paper bag is the final learning datum. As Subject_C has only two drawings of the paper bag, only the paper box data are used in that case. The data for the instructor are one drawing for each motif. Figure 6 shows all drawings performed using this analysis.

Figure 7 shows an overview of the stroke analysis in an SOM. To generate a learned map, we input all stroke data of the three drawings of each motif for each learner. This process is repeated for each learner and each motif. Then, for each learned map, the stroke data of the learner were used to generate that map, and the unlearned instructor was input on a per drawing, and the BMU of each stroke was recorded.

4.3 Results

Trained SOM. Figure 8 shows the results of visualizing the vectors of the units of the trained SOM. These are the vectors assigned to the 24 units (from $[50, 48]$ to $[53, 53]$) in the 100×100 map, where the origin is $[0, 0]$. Clearly, the shape of the strokes used in the drawing is learned in the vectors assigned to each unit. We can also see that vectors with similar shapes are located in nearby units in the 2D map. In all the learned SOMs generated in this study, the stroke shape was examined in all regions. We confirmed that the similar vectors are clustered in a 2D neighborhood.

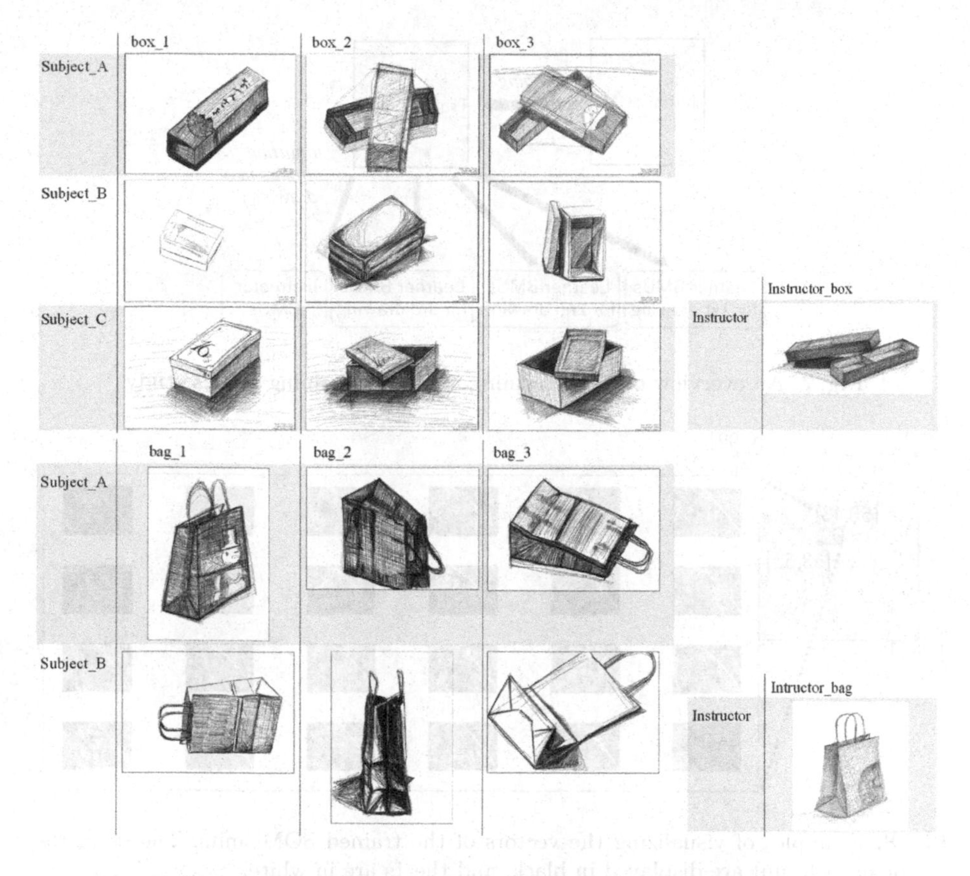

Fig. 6. Drawings by three subjects and an instructor using this analysis.

Data Mining. We analyze the BMUs obtained by inputting the strokes of each drawing into a trained SOM. We define the percentage of BMUs in each drawing as the variation rate (VR) of the strokes used in that drawing. The lower the VR, the more strokes of a similar shape are used. The numerator of the VR is the number of BMU types (excluding duplicated BMUs), and the denominator is the total number of strokes in the drawing. Figure 9 shows some of the VRs obtained in this analysis. These VRs are calculated by the BMUs inputted of each drawing shown on the horizontal axis in the paper box SOM trained by three drawings of each subject. For example, the points ,with the legend of "Subject_A" are taken from the trained SOM with all the strokes in Subject_A's three paper box drawings. At most eight drawings are inputted to each subject's trained SOM. They are three paper box drawings created by each subject, one paper box drawing created by an instructor, three paper bag drawings created by each subject, and one paper bag drawing created by an instructor (see Fig. 6). The box_1 of Subject_A shown in Fig. 9 indicates that all the strokes of the first drawing of the paper box are input to this trained map, and the VR is 85%.

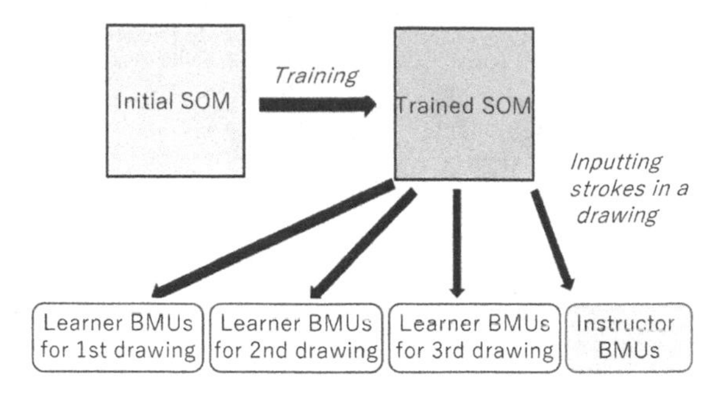

Fig. 7. An overview of SOM training and stroke mining in this study.

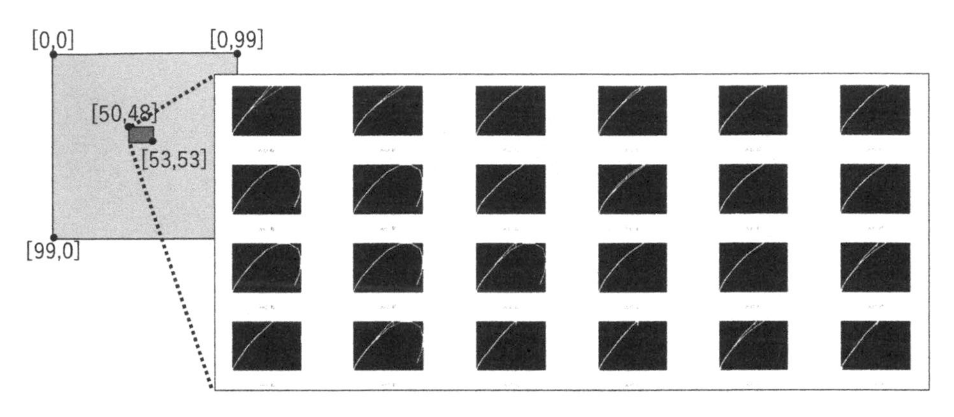

Fig. 8. Examples of visualizing the vectors of the trained SOM units. The 0s in the vector in each unit are displayed in black, and the 1s are in white.

Table 1 shows the VRs obtained from the paper bag trained. This table shows the VRs calculated from three learners' drawings for each row unit. The Subject_A row shows the result of training all of the strokes in Subject_A's three paper bag drawings.

4.4 Considerations

Figure 9 shows that the VRs in the third paper box drawings were decreased more than the first time for all subjects. The motif changed to a paper box after the paper bag was learned. For Subject_A and Subject_B, the VR for paper bag drawing was lower than that for paper box drawing. The reason for the less pronounced decrease in VR values for the paper bag drawings, as seen in Fig. 9 and Table 1, may be that the skill in drawing had progressed, and the shape of the strokes used in each drawing is within a narrower range than that of the paper box.

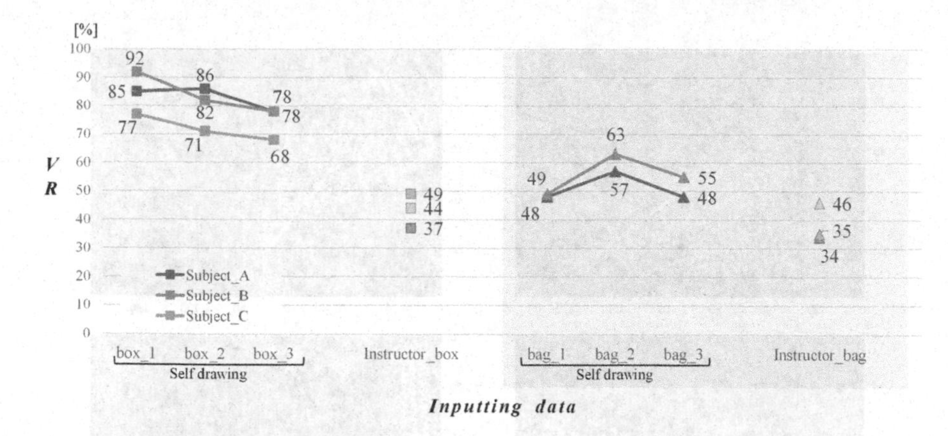

Fig. 9. VRs of each subject's drawings and an instructor's drawings.

Table 1. VRs for paper bag motif.

Trained SOM	Inputted data			
	bag_1	bag_2	bag_3	Instructor
Sublect_A	77%	80%	85%	36%
Sublect_B	75%	88%	72%	33%
Sublect_C	-	-	-	-

The VR of the instructor was calculated by inputting the instructor's strokes of the paper box and the paper bag into each learner's paper box trained SOM (see Fig. 9). The instructor's VRs are lower than those of the learners. This suggests that the stroke shapes used by the instructor in drawing are more similar in shape than those used by the learners.

We also compared the BMU of the learner strokes (learner BMU) with the BMU of the instructor's strokes (instructor BMU) in the learner's trained SOM. The maximum number of coordinates for the instructor's strokes was 183. Among the learners' strokes that were assigned to only a learner's BMU, we identified stroke shapes with stroke coordinate numbers greater than 183. Figure 10 shows four examples of these stroke shapes. In the context of learning drawing, a simple line is a straight line or a smooth curve. The examples in Fig. 10 are generally not simple lines. From interviews with drawing instructors, it was found that the strokes in Fig. 10 are inappropriate shapes for drawing learning and should not be repeated as learners develop.

Meanwhile, some of the learner strokes that were assigned only to learner BMUs were judged to be simple lines. These strokes were considered to have the same BMUs as other learners' strokes and as the instructor's strokes when they were rotated. In this basic analysis, we did not apply any rotation to the stroke

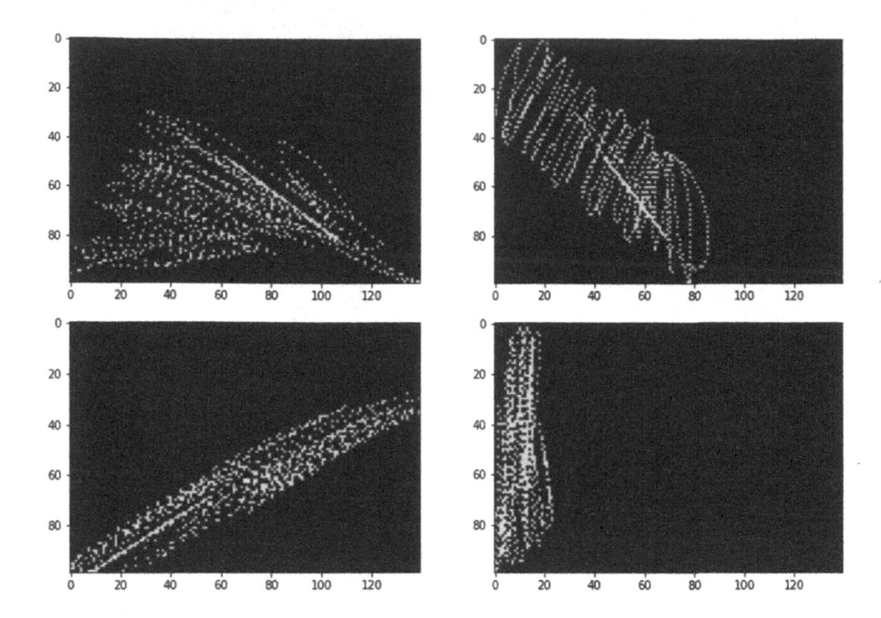

Fig. 10. Examples of stroke shapes assigned to only a learner BMU, and with stroke coordinate numbers greater than 183. These examples are included in the first drawing of Subject_A.

data. In the future, we will add rotation to the formatting process to make it clearer to identify the target strokes for instruction, as shown in Fig. 10.

We summarize our findings in this analysis as follows.

- VR decreases as learning progresses for drawings in the early stages of learning.
- For trained SOMs that have learned the learner's drawing style, the instructor's VR is lower than the learner's VR.
- Learner strokes assigned only to learner BMUs tend not to be simple lines.

These findings indicate the following two points as possible indicators for drawing growth in stroke analysis by SOM.

- Decreasing VR.
- Decreased number of strokes allocated only to the learner's BMU.

5 Conclusion

This study aims to develop a concrete index for quantitative evaluation of the results of drawing learning. In this paper, we first identified the characteristics of strokes in drawing work and described the preprocessing for machine learning with SOM. Next, the characteristics of machine learning based on the SOM method are described, and the results of its application to stroke classification are shown. We showed the results of classifying the strokes of learners and an instructor and discussed its potential as a growth indicator.

On the basis of the considerations described in this paper, we seek to improve our mining method. Following this, we will merge the method into an auto assessment function in a remote drawing learning support system.

References

1. Nagai, T., Kayama, M., Itoh, K.: A drawing learning support system based on the drawing process mode. Interact. Technol. Smart Educ. **11**(2), 146–164 (2014)
2. Wen, Z., Shankar, A., Antonidoss, A.: Modern art education and teaching based on artificial intelligence. J. Interconnect. Netw. (2021). https://doi.org/10.1142/S021926592141005X
3. Long, Y.: Research on art innovation teaching platform based on data mining algorithm. Clust. Comput. **22**, S14943–S14949 (2019)
4. Elalfi, A.E.E., Elatawy, M.F., Mahmoud, N.M.: Using artificial intelligence techniques for evaluating practical art products for the students in art education. Int. J. Comput. Appl. **182**(15), 19–26 (2018)
5. Yuxi, J.Y., Li, P., Wang, W., et al.: GAN-based pencil drawing learning system for art education on large-scale image datasets with learning analytics. Interact. Learn. Environ. (2019). https://doi.org/10.1080/10494820.2019.1636827
6. Gang, L., Weishang, G.: The effectiveness of pictorial aesthetics based on multiview parallel neural networks in art-oriented teaching. Comput. Intell. Neurosci. (2021). https://doi.org/10.1155/2021/3735104
7. Sartori, A., Yanulevskaya, V., Salah, A.A., et al.: Affective analysis of professional and amateur abstract paintings using statistical analysis and art theory. ACM Trans. Interact. Intell. Syst. **5**(2), 1–27 (2015)
8. Sato, K.: Developmental trial in drawing instruction at art/design universities. Shizuoka Univ. Art Cult. Bull. **4**, 153–162 (2004)
9. Sekine, E.: A trial to develop the ART SYSTEM. Art Educ. **6**, 89–100 (1984)
10. Sakimoto, T., Nagai, T., Kayama, M., et al.: A consideration for the classification algorithm for hatching-like strokes in complex line set for effectiveness improvement of auto drawing assessment function in digital drawing learning support system, SIG in "Knowledge, Skill, and Technology succession" in Japanese Society of Artificial Intelligence, SIG-KST-030-02, pp. 1–6 (2017)
11. Nagai, T., Sakimoto, T., Kayama, M.: DEGITal drawing: an innovative challenge for drawing skill development in art education, e-learning excellence awards. In: Academic Conference and Publishing International, UK England, pp. 133–154 (2016)
12. Anoto Group AB: Anoto Digital pen. https://www.anoto.com/cases/anoto-digital-pen/. Accessed 3 May 2022
13. Kohonen, T.: Self-organized formation of topologically correct feature maps. Biol. Cybern. **43**(1), 59–69 (1982)

Short Papers

What Is Relevant for Learning? Approximating Readers' Intuition Using Neural Content Selection

Tim Steuer(✉), Anna Filighera, Gianluca Zimmer, and Thomas Tregel

Technical University of Darmstadt, Darmstadt, Germany
{tim.steuer,anna.filighera,thomas.tregel}@kom.tu-darmstadt.de

Abstract. Experienced readers intuitively mark text passages containing central concepts as learning-relevant when reading actively. Although this intuitive process of marking important information is sometimes imperfect, it fosters comprehension. It would be beneficial to approximate this intuition by automatically detecting potential learning-relevant content. It is a building block for various upstream tasks such as automatic self-assessment or intelligent author assistance. This work argues that learners often apply heuristics based on different sentence types to determine the learning-relevant contents in texts. We show that such heuristics can be approximated using neural sentence classifiers and implement two neural sentence classifiers detecting causal and definitory sentences. We evaluate the classifiers' ability to detect learning-relevant information in an empirical study (N = 37). Furthermore, a system performance evaluation compares the proposed classifiers with unsupervised summarization systems. We find evidence for a small but reliable association between the chosen automatically detectable sentence types (definition/causal) and the learners' perception of content relevance. Additionally, the classifiers outperform most other relevant content selection techniques in our experiments. Interestingly, other simple heuristics based on sentence position or length also exhibit strong performance.

Keywords: Content selection · Information extraction · Natural language processing · Education

1 Introduction

Reading is an essential activity in education. While reading, learners construct mental models which depend on various factors such as the readers' attention, prior knowledge, the text coherence or the text difficulty [2,8]. However, model construction is imperfect, and readers sometimes miss crucial points or misunderstand fundamental ideas. Thus, supporting these construction processes may be achieved by educational assistance systems. They may detect the texts' essential segments to scaffold the reading comprehension accordingly.

M. M. Rodrigo et al. (Eds.): AIED 2022, LNCS 13355, pp. 505–511, 2022.
https://doi.org/10.1007/978-3-031-11644-5_41

However, what learners perceive as learning-relevant is an intuitive and multifaceted concept challenging to define. Hence, this work explores texts' characteristics for the detection of learning-relevant information. The objective is not to perfectly capture learners' intuition of what is relevant, but to determine if text characteristics, particularly definition and causal sentences, provide a reasonable approximation of the intuition. We pose the following research question:

RQ To what extent are automatically detected definitions and causal sentences perceived as learning-relevant?

The underlying assumption behind the research question is that if the respective sentence types are important for learning, learners should also recognize them as relevant. We investigate the research question with an empirical reading comprehension study with $N = 37$ German-speaking learners. We furthermore conduct a system evaluation of the proposed detection algorithms.

2 Related Work

There is a considerable amount of research applying unsupervised machine learning algorithms for learning-relevant content selection (e.g. [3,14]). Most of this work assumes that text summarization algorithms transfer well to the problem of learning-relevant content selection. However, this assumption does not hold in general, and position and length heuristics sometimes outperform all other algorithms [3]. Besides, supervised approaches extracting relevant content from texts exist. The models aim to either extract keyphrases [1] or whole sentences [6,9,15]. These approaches use different machine learning techniques to learn what constitutes learning-relevant information and apply a wide range of different algorithms (e.g. [1,6]). Finally, some works focus on detecting definitions or causal sentences to solve their upstream tasks. Steuer et al. [13] detect definitions in texts to generate questions about their contents. Additionally, Stasaski et al. [12] apply causal sentences selection to generate valuable questions about texts.

3 Definition and Causal Sentence Extraction Models

This work focuses on the direct evaluation of the content selection task in an educational scenario with actual learners. We investigate the research question stated in the introduction with an empirical study comprising German learners. Consequently, we collect German corpora to train the respective sentence classification tasks. The collected corpus contains 5641 definitions and 1460 causal sentences from textbooks. We train a BERT-based classifier [5] for both sentence types. We model the classification task as a binary task with two separately trained BERT classifiers for definitions and causal sentences. The amount of negative labels sampled from the corpus equals the positive label count. Each model trains for five epochs using *Hugging Face*'s default BERT configuration and the *bert-base-german-cased* snapshot. We do not perform any hyperparameter tuning and select the last model checkpoint after five epochs for the study. The models' performance is evaluated using a 70% train and 30% test split. The final performance of the trained classifiers on the test set is $F1_{macro} = 0.8$ for definitions and $F1_{macro} = 0.78$ for causal sentences with precision and recall values identical to the F1 score.

4 Experiments

We address the research question by conducting an empirical study ($N = 37$) in which readers select text sentences they deem important. We run experiments from a learner and system perspective on this data. From the learners' perspective, we seek evidence of whether automatically detected definitions and causal sentences are perceived differently than other sentences. We test the hypothesis:

H1 There is an association between the type of a sentence (definition/causal) and its selection probability.

Moreover, the system perspective seeks to find out if the selection criteria perform strong enough to build a competitive, learning-relevant sentence selection system. It is guided by the question:

E1Which extraction algorithms perform best for learning-relevant sentence extraction?

For the data collection a study design similar to Dee-Lucas and Larkin [4] is adapted ($N = 37$). Participants read five short science texts between 30 and 50 lines. The texts concern introductory concepts in biology, physics, and psychology. The chosen concepts are purposefully selected such that the sample readership will most likely have limited prior knowledge. We sample our readership via voluntary sampling, mainly from the local university. The mean age of participants is $M = 26.8$, $SD = 9.64$ with 23 female and 14 male participants.

While reading, participants selected six sentences they perceived the most learning-relevant for imaginary exam preparation. Every participant selected and rated 30 sentences from 182 sentences in every response. The full dataset thus comprises $182 * 37 = 6734$ selection decisions.

4.1 Learners' Perspective: Characteristics of Selected Sentences

We test the given hypotheses with a logistical regression model (significance level $\alpha = 0.05$). The model operates on the sentence level with the dependent variable sentence selection probability and four independent variables. The first two binary variables describe the classifiers' selection decision (S_{def} and S_{cause}). The third variable S_{pos} is computed by dividing the sentences index by the text's sentence count. It is included due to its observed relationship with the perceived sentence importance [3]. The fourth variable S_{len} divides the sentence length by the maximum sentence length of the text. It is an intuitive heuristic: the longer a sentence, the more information is comprised. The fitted model approximates the data with $pseudo - R^2 = 0.09$. It is significantly better than the null model ($LL - Null = -3014.1 \quad LLR - p = 0.0$). All coefficients are significant. The average marginal effects are $S_{def} = 0.06$, $S_{cause} = 0.02$, $S_{pos} = -0.19$ and $S_{len} = 0.28$. Thus, the significant model indicates a relationship between the independent variables and the participants' selection. However, the model approximates the data only moderately ($pseudo - R^2 = 0.09$). For comparison, a regularly used heuristic states an $pseudo - R^2 > 0.2$ as excellent model fit [10]. Yet, a moderate approximation is expected as we rely solely on text characteristics. Thus, the given four-variable model will not thoroughly explain learners' decision processes. Nevertheless, it shows a reliable association between the independent variables and perceived learning relevancy. Furthermore, all independent variables' coefficients are significant, providing evidence of the usefulness of definition and causal classification for learning-relevant content selection. Note, that the other indepedent variables are also significant and have higher marginal effect sizes.

4.2 Systems' Perspective: Algorithm Comparison

A gold standard dataset is necessary to compare the different systems' performances. Thus, we transform the collected dataset labelling sentence in the fourth quartile (eight or more selections) as relevant and sentences in the first quartile (zero or one selection) irrelevant. The transformed corpus consists of 101 data points, with 51 learning-relevant and 50 irrelevant data points. We report precision, recall, and macro F1 score and compare seven algorithms.

The first comparison algorithm is SumBasic [11]. We apply German word stemming and stop word removal, and for every text, the algorithm extracts $N = 10$ sentences. Second, the LexRank algorithm [7] with TF-IDF matrices initialized from the 10kGNAD[1] dataset is used. It selects $N = \{10, 20\}$ sentences. Moreover, a sentence length heuristic selecting sentences longer than $p = \{30, 40, 50\}$ percent of the longest sentence in the text is applied. Furthermore, a position-based heuristic selecting all sentences in a given text's initial $p = \{20, 25, 30\}$ percent is applied. The best-performing length and position heuristics are combined by selecting a sentence classified by at least one heuristic. Finally, the definition and causal classifier from the previous experiments are used. They are also combined using a sentence if it is classified at least once.

Results. The evaluation results can be seen in Table 1. According to F1 scores, the best algorithms are the combined classifiers and combined heuristics ($F1 = 0.77$). Although the combined heuristics perform well, it is unknown which length and position parameters should be chosen under real-world conditions. The best standalone heuristics perform decently, and a length heuristic

[1] https://tblock.github.io/10kGNAD/.

Table 1. Classification results on the gold standard dataset.

Classifier	Precision	Recall	F1	# Pred.
Cause	0.65	0.55	0.60	43
Definition	0.79	0.67	0.72	43
Def-cause-combined	0.69	0.86	**0.77**	64
Sumbasic [10]	0.48	0.27	0.35	29
Lexrank [10]	0.74	0.45	0.56	31
Lexrank [20]	0.60	0.71	0.65	60
Position [20%]	**0.88**	0.45	0.60	26
Position [25%]	0.84	0.53	0.65	32
Position [30%]	0.78	0.57	0.66	37
Length [30%]	0.65	**0.90**	0.75	71
Length [40%]	0.76	0.73	0.74	49
Length [50%]	0.87	0.53	0.66	31
Length-pos-combined	0.64	0.96	**0.77**	68

is the best standalone approach. Among the classifiers, the definition classifier performs best slightly worse than some length heuristics. SumBasic scores $F1 = 0.35$ and the best LexRank result scores $F1 = 0.65$. Yet, this ranking is not showing the complete picture. We have to consider the operation characteristics of upstream systems. Tasks such as automatic question generation must not necessarily recall all learning-relevant content before being useful. However, if they identify many irrelevant sentences as relevant, users will probably lose trust. Thus, we believe that classification precision is more critical than recall for most upstream tasks. Hence, if the F1 score is similar, we theorize it is sensible to prefer precision-oriented systems. In contrast to all other highly ranked systems, the definition classifier has a high precision ($p = 0.79$) and reasonable recall. We hence assume it to be a solid approach for many upstream tasks.

5 Conclusion

The conducted experiments show that neural sentence classification can be helpful in predicting learning-relevant sentences. The learner-centric view shows that both classifications were weakly but reliably associated with increased perceived learning relevancy. Therefore, the sentence type seems to influence what is perceived as important in texts. From the system-centric view, combining both classification heuristics or the length and position heuristics performs best. However, the best parameters for the length and position heuristics in a real-world scenario are unknown. Thus, We interpret the system-centric results as evidence that the proposed classifiers approximate some of the learners' intuition of learning relevancy. In summary, there is no single algorithm that clearly identifies learning-relevant information. This is not surprising because learning relevance is a complex, multifaceted concept. However, the proposed approaches, while not perfect, have a reliable association with learning relevancy and outperform unsupervised systems on the given dataset.

References

1. Becker, L., Basu, S., Vanderwende, L.: Mind the gap: learning to choose gaps for question generation. In: Proceedings of the 2012 Conference of the North American Chapter of the Association for Computational Linguistics: Human Language Technologies, pp. 742–751 (2012)
2. Best, R.M., Rowe, M., Ozuru, Y., McNamara, D.S.: Deep-level comprehension of science texts: the role of the reader and the text. Top. Lang. Disord. **25**(1), 65–83 (2005)
3. Chen, G., Yang, J., Gasevic, D.: A comparative study on question-worthy sentence selection strategies for educational question generation. In: Isotani, S., Millán, E., Ogan, A., Hastings, P., McLaren, B., Luckin, R. (eds.) AIED 2019. LNCS (LNAI), vol. 11625, pp. 59–70. Springer, Cham (2019). https://doi.org/10.1007/978-3-030-23204-7_6
4. Dee-Lucas, D., Larkin, J.H.: Novice strategies for processing scientific texts. Discourse Process. **9**(3), 329–354 (1986)
5. Devlin, J., Chang, M.W., Lee, K., Toutanova, K.: Bert: Pre-training of deep bidirectional transformers for language understanding. In: NAACL-HLT (1) (2019)
6. Du, X., Cardie, C.: Identifying where to focus in reading comprehension for neural question generation. In: Proceedings of the 2017 Conference on Empirical Methods in Natural Language Processing, pp. 2067–2073 (2017)
7. Erkan, G., Radev, D.R.: Lexrank: graph-based lexical centrality as salience in text summarization. J. Arti. Intell. Res. **22**, 457–479 (2004)
8. Filighera, A., Steuer, T., Rensing, C.: Automatic text difficulty estimation using embeddings and neural networks. In: Scheffel, M., Broisin, J., Pammer-Schindler, V., Ioannou, A., Schneider, J. (eds.) EC-TEL 2019. LNCS, vol. 11722, pp. 335–348. Springer, Cham (2019). https://doi.org/10.1007/978-3-030-29736-7_25
9. Mahdavi, S., An, A., Davoudi, H., Delpisheh, M., Gohari, E.: Question-worthy sentence selection for question generation. In: Canadian Conference on AI, pp. 388–400 (2020)
10. McFadden, D., et al.: Conditional logit analysis of qualitative choice behavior (1973)
11. Nenkova, A., Vanderwende, L., McKeown, K.: A compositional context sensitive multi-document summarizer: exploring the factors that influence summarization. In: Proceedings of the 29th annual international ACM SIGIR Conference on Research and Development in Information Retrieval, pp. 573–580 (2006)
12. Stasaski, K., Rathod, M., Tu, T., Xiao, Y., Hearst, M.A.: Automatically generating cause-and-effect questions from passages. In: Proceedings of the 16th Workshop on Innovative Use of NLP for Building Educational Applications, pp. 158–170 (2021)
13. Steuer, T., Filighera, A., Meuser, T., Rensing, C.: I do not understand what i cannot define: Automatic question generation with pedagogically-driven content selection (2021). arXiv preprint arXiv:2110.04123

14. Steuer, T., Filighera, A., Rensing, C.: Remember the facts? investigating answer-aware neural question generation for text comprehension. In: Bittencourt, I.I., Cukurova, M., Muldner, K., Luckin, R., Millán, E. (eds.) Artificial Intelligence in Education, pp. 512–523. Springer International Publishing, Cham (2020)
15. Willis, A., Davis, G., Ruan, S., Manoharan, L., Landay, J., Brunskill, E.: Key phrase extraction for generating educational question-answer pairs. In: Proceedings of the Sixth (2019) ACM Conference on Learning@ Scale, pp. 1–10 (2019)

Computer-Aided Response-to-Intervention for Reading Comprehension Based on Recommender System

Ming-Chi Liu[1], Wei-Yang Lin[1], and Chia-Ling Tsai[2](✉)

[1] Chung Cheng University, Chiayi, Taiwan
[2] Queens College, CUNY, Queens, NY 11367, USA
ctsai@qc.cuny.edu

Abstract. In 2019, New York State Education Department announced 54.6% of all students in grades 3 to 8 not meeting the standard of reading proficiency. Motivated by the need for a more efficient intervention model, we propose a recommender system to leverage the technology in machine learning to recommend suitable reading materials for effective intervention. The recommendation is based on the student's prior reading comprehension assessments and also assessments of other students at the same grade level using collaborative filtering. No other prior academic or demographic information of students is available. Two main challenges are lack of explicit ratings of reading passages by students and the small data size. Both are addressed in this paper. BERT is applied to determine the textual evidence of a question, and linguistic properties are extracted to generate a continuous rating for a question answered by a student to reflect the skill level of the student. The difficulty level of a passage is determined by the associated multiple-choice questions. The system is trained with a collection of fourth grade New York English Language Arts assessments. The training dataset is augmented with synthetic data using SMOTE for better generalizability. Our system achieves 75.7% in accuracy and 59.23% in F1-score.

Keywords: Reading comprehension · Intervention program · Recommender system · Collaborative filtering

1 Introduction

In New York State, USA, students in grades 3 to 8 take the State English Language Arts (ELA) test each spring. An ELA test contains multiple-choice questions and open-ended questions based on short passages in the test. To do well, students should be able to read the text closely for textual evidence and to make logical inferences from it. In 2019, New York State Education Department reported that 54.6% of all students in grades 3 to 8 do not meet the standard

M. M. Rodrigo et al. (Eds.): AIED 2022, LNCS 13355, pp. 512–518, 2022.
https://doi.org/10.1007/978-3-031-11644-5_42

of proficiency [2]. To drive the changes in students who are at some level of risk for not meeting academic expectation, schools arrange academic intervention service for all students who are well below or partially proficient, based on the ELA score. However, a single performance score is not instrumental in explaining the lack of specific language knowledge and skills typically demonstrated at that grade level.

Our work is motivated by the need for a more effective Response-to-Intervention model that continuously assess the need for changes in instruction and goals, driven by students' progress data [6]. We develop a machine learning (ML) system for recommending instruction materials for reading comprehension, based on the prior reading assessments of an individual student and also of students in the same grade level. A recommender system for e-learning comes in various formats, depending on the data availability. A common approach is to make recommendations of courses or predictions of student performance based on known student and course characteristics [7–9]. Such problem can be easily formulated as a classification or regression problem. Thai-Nghe et al. [10] proposed a recommender system for math assessment based on matrix factorization where student factors and some of the problem factors are known. Our work is most similar to [10], but performs recommendation of reading passages based on the predicted ratings of associated multiple-choice questions by a given student, without any prior information regarding either the student or the assessment material.

There are two main challenges to be addressed. First, the multiple-choice questions have only dichotomous ratings (correctly or incorrectly answered) from students. Second, the performance of a recommender system is limited by the prior data for training, but our dataset is relatively small, comparing to the growing dataset of millions of records for a commercial recommender system. We explore linguistic properties of reading passages to address the first challenge and employ data augmentation for the second challenge.

2 Dataset

The dataset consists of two parts. The first part is the set of six fourth grade New York State mock ELA examinations from year 2005 to 2010. Only multiple-choice questions are considered. In each examination there are five passages, each having five to six associated questions, for a total of 28 questions. Question and choice statements are typically short. The second part is the set of student assessments, involving a total of 378 randomly-selected fourth grade students with various levels of reading proficiency from 17 reading intervention classes. Every participant was assigned an identification number and participated in most three mock examinations. Each student answered from 4 to 84 questions in total. No other background information, such as prior academic performance or demographic data, was collected to protect the privacy of the participants.

3 Methodology

A recommender system can be understood as a ML problem trained with a set of 3-tuples: {[user ID, product ID, rating]}. Our input is a tuple of [student ID, question ID, student answer]. As shown in Fig. 1, the weights of features associated with the j-th student and with the i-th question should be estimated jointly by the recommender system, and the rating of a new student-question pair can be predicted using the estimated features. The fitness level of a reading passage can be computed as the average of the predicted ratings of associated multiple-choice questions by a given student. The system makes recommendation by choosing passages with the rating in the range $[0 - \epsilon, 0]$—a difficulty level slightly surpassing the recent strength of a student—to promote learning.

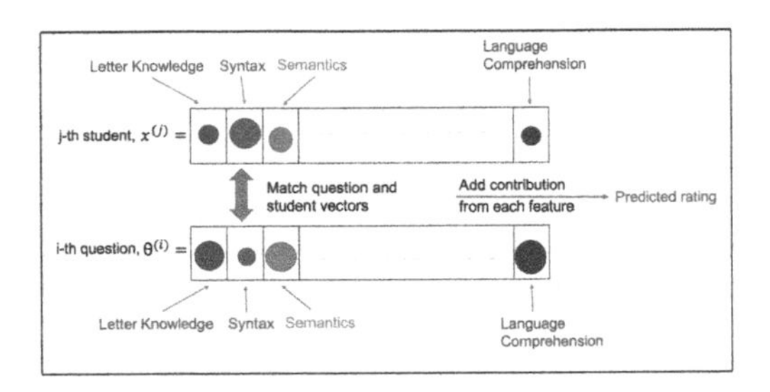

Fig. 1. General concept of a recommender system demonstrated for e-learning. Student features and question features are estimated from known ratings, and a new rating is predicted as a combination of the estimated features.

To generate the ratings for student-question pairs, neither the nominal answer choices nor the dichotomous scoring outcome of 0/1 is appropriate for a recommendation system. It is necessary to convert the student's answer choice of a question to a continuous rating, reflecting the skill level of a student for answering a given question. Data augmentation is applied to the dataset to increase the data variation for better generalizability of the model. Collaborative filtering is adopted as the filtering technique of the recommendation system, which simultaneously computes the student features and the question features using all available ratings in the training pool.

3.1 Rating Transformation

To transform the rating from either nominal or dichotomous to continuous for better discrimination, we explore ML techniques in natural language processing to connect the questions to the passage with the textual evidence, and to extract linguistic properties of the questions. A rating computed from the linguistic properties of a question should reflect the underlying language skills needed to correctly answer the given question.

To locate the textual evidence of a question in the passage, Bidirectional Encoder Representation from Transformer (BERT) [5] is applied to identify connection between a question and a sentence in the passage. The pre-trained version used in this study is BERT-base-uncased, which is further fine-tuned with OneStopQA dataset [3] for reading comprehension with multiple-choice questions.

Combining with the textual evidence, linguistic features are identified for each answer choice of a question, because linguistic properties have been shown to be important indicators for readability of a passage. We explore the Suit for Automatic Linguistic Analysis Tools (SALAT) [1] for the social sciences. There is a total of 3908 features generated for each answer choice (combined with the textual evidence) of a question. Values of a feature are normalized to Z-scores using the mean and standard deviation of the feature.

To generate the rating of a student-question pair, the 3908-tuple feature vector of the student's answer choice is converted into a scalar value using the magnitude of the vector, which is normalized to a Z-score again and scaled to the range of $[0, 1]$, using a sigmoid function for linear scaling mainly in the clustered sections closer to 0. Let d_f be the magnitude of the feature vector f in Z-score, the sigmoid score $s(d_f) \in [0, 1]$ is computed as $s(d_f) = \frac{1}{1+\exp(c*(-d_f))}$, where $c = 2.5$ determined empirically. The final rating $r = s(d_f)$ is set for a correct answer choice and $r = s(d_f) - 1$ for an incorrect answer choice, so all correct answers have ratings greater than 0 and all incorrect answers have ratings less than 0. $r \in [-1, 1]$ with 1 representing exceptional and –1 for well below proficient.

3.2 Data Augmentation

Data augmentation is common in ML to increase the amount of data to deal with the problem of class imbalance or to improve generalizability of the model. To add synthetic data that are slightly modified of existing data, we adopt SMOTE (Synthetic Minority Oversampling Technique) [4] to generate additional samples for each student. Given a question that a student completed, the feature vector f is computed as the average of the feature vectors of the 4 answer choices. Based on the cosine similarity, its K (=3) nearest neighbors of the same student are located, and one neighbor f' is randomly decided to determine the direction of perturbation. The new feature $\bar{f}$ is f perturbed with a random portion η between [0,1] of the difference between f and f': $\bar{f} = f + \eta(f' - f)$. $\bar{f}$ is marked as correctly or incorrectly answered question the same as f, and is converted to the rating following the same steps described in Sect. 3.1.

3.3 Collaborative Filtering

Collaborative filtering implemented using matrix factorization aims to estimate two types of feature vectors: $x^{(i)} \in \Re^n$ representing i-th question and $\theta^{(j)} \in \Re^n$ representing j-th student, where $n = 10$ set empirically. There are n_m questions and n_u students. $x^{(i)}$ is constrained by ratings of all students who answered question i and $\theta^{(j)}$ is constrained by ratings of all questions that student j answered. When both $[x^{(1)}, \ldots, x^{(n_m)}]$ and $[\theta^{(1)}, \ldots, \theta^{(n_u)}]$ are unknown, they can be estimated jointly by minimizing the following cost function:

$$J\left(x^{(1)}, .., x^{(n_m)}, \theta^{(1)}, .., \theta^{(n_u)}\right) = \frac{1}{2} \sum_{(i,j):r(i,j)=1} \left(\left(\theta^{(j)}\right)^T x^{(i)} - y^{(i,j)}\right)^2 + \frac{\lambda}{2}\sum_{i=1}^{n_m}\sum_{k=1}^{n}\left(x_k^{(i)}\right)^2 + \frac{\lambda}{2}\sum_{j=1}^{n_u}\sum_{k=1}^{n}\left(\theta_k^{(j)}\right)^2, \tag{1}$$

where $y^{(i,j)}$ is the rating for student j and question i, and $r(i,j) = 1$ indicates valid rating for student j and question i. The last two terms in Eq. 1 are for regularization to avoid overfitting. The minimization process is initialized with random values for all vectors and alternates the estimations of $[x^{(1)}, \ldots, x^{(n_m)}]$ and $[\theta^{(1)}, \ldots, \theta^{(n_u)}]$ by fixing another vector until the process converges.

To predict whether student j will correctly answer question i, the rating r is estimated as $r = (\theta^{(j)})^T x^{(i)}$. To map r back to the answer choices, ratings of the choices are computed and the choice with the rating closest to r is the predicted choice of student j if given question i.

4 Experiments and Results

There is a total of 27731 answered records with 67.5% correctly answered. Data augmentation was applied to add another 35773 records for training—168 questions per student in total. Leave-one-out validation was performed on only the original dataset of 27731 records. As a limitation of our current study, we were able to only validate the performance prediction of a student on a question, not the effectiveness of the recommendation for Response-to-Intervention.

We assessed the performance of the system with accuracy and F1-score. Since incorrectly answered records are considered the minority for the classification problem, they are considered the positive class, whereas the correctly answered records are the negative class. A prediction is considered correct if the student scored or not scored the question and system predicts the same outcome, regardless of the answer choice picked. The F1-score provides better insight to a problem with imbalance classes since it ignores the correct predictions of the majority class (i.e. true negatives) by considering only the precision and recall of the minority class. Our system achieves an F1-score of 55.93% and accuracy of 72.83% without data augmentation, and 59.23% and 75.7% with data augmentation. We also compared our system with content-based filtering as the engine of the recommender system [11]—the rating given by a student for a new question is computed from the ratings of K (=3) nearest questions answered by the same student. Collaborative filtering outperforms content-based filtering by 2.45% in accuracy and 6.34% in F1-score. It shows the importance of automatic determination of feature representation of questions from the data using all ratings available by the cohort.

If the process of rating transformation is completely removed and the continuous rating is replaced with 0/1 rating, i.e. 1 for a correctly answered question and 0 for an incorrectly answered question, the F1-score degrades substantially from 55.93% to 27.78%. The use of BERT for evidence identification only improves the F1-score by 1.59%. An explanation for the very minor improvement from BERT is lack of discrimination of strong evidence supporting a question; on average, close to 54% of a passage is considered as the evidence for an answer choice of a question, but most statements are irrelevant. As a result, 3908-tuple feature vectors for 4 answer choices of a question can be very close in the feature space and the transformed ratings are less dispersed.

5 Conclusions and Discussion

The proposed system supports intervention for reading comprehension by recommending reading passages of difficulty level slightly surpassing the recent strength of a student to promote learning. Our proposed rating transformation scheme doubles the F1-score by converting the binary score to a continuous value using linguistic properties of a question with its supporting evidence from the reading passage. Data augmentation further boosts the performance by 3.3%. Our model can be easily generalized for other formats of reading comprehension, such as short-answer questions, if linguistic properties can be reliably computed from the associated question(s) with the supporting textual evidence identified.

Acknowledgments. This work was partially supported by grants from NSF-CHS Award 1543639, Taiwan MOST Award 109-2221-E-194-040 and PSC-CUNY Research Award 65406-00-53.

References

1. Suit for Automatic Linguistic Analysis Tools. https://www.linguisticanalysistools.org/. Accessed Dec 2021
2. New York SED: 3–8 assessment database (2019). http://www.nysed.gov/news/2019/state-education-department-releases-spring-2019-grades-3-8-ela-math-assessment-resuls. Accessed Dec 2021
3. Berzak, Y., Malmaud, J., Levy, R.: STARC: structured annotations for reading comprehension. In: Proceedings of 58th Annual Meeting of the Association for Computational Linguistics, pp. A:567–576 (2020)
4. Chawla, N.V., Bowyer, K.W., Hall, L.O., Kegelmeyer, W.P.: SMOTE: synthetic minority over-sampling technique. J. Artif. Intell. Res. **16**, 321–357 (2002)
5. Devlin, J., Chang, M.W., Lee, K., Toutanova, K.: BERT: pre-training of deep bidirectional transformers for language understanding. ArXiv abs/1810.04805 (2019)
6. Fuchs, D., Fuchs, L.: Responsiveness-to-intervention: a blueprint for practitioners, policymakers, and parents. Teach. Except. Child. **38**(1), 57–61 (2001)
7. Goga, M., Kuyoro, S., Goga, N.: A recommender for improving the student academic performance. Procedia Soc. Behav. Scoi. **180**, 1481–1488 (2015)

8. Kurniadi, D., Abdurachman, E., Warnars, H., Suparta, W.: A proposed framework in an intelligent recommender system for the college student. J. Phys. Conf. Ser. **1402**(6), 066100 (2019). https://doi.org/10.1088/1742-6596/1402/6/066100
9. Sweeney, M., Rangwala, H., Lester, J., Johri, A.: Next-term student performance prediction: a recommender systems approach. J. Educ. Data Mining **8**(1), 22–50 (2016)
10. Thai-Nghe, N., Drumond, L., Krohn-Grimberghe, A., Schmidt-Thieme, L.: Recommender system for predicting student performance. Procedia Comput. Sci. **1**(2), 2811–2819 (2010)
11. Thorat, P.B., Goudar, R.M., Barve, S.S.: Survey on collaborative filtering, content-based filtering and hybrid recommendation system. Int. J. Comput. Appl. **110**(4), 31–36 (2015)

Improving the Quality of Students' Written Reflections Using Natural Language Processing: Model Design and Classroom Evaluation

Ahmed Magooda[1](✉), Diane Litman[1](✉), Ahmed Ashraf[2], and Muhsin Menekse[2]

[1] University of Pittsburgh, Pittsburgh, USA
{aem132,dlitman}@pitt.edu

[2] Purdue University, West Lafayette, USA
{butt5,menekse}@purdue.edu

Abstract. Having students write reflections has been shown to help teachers improve their instruction and students improve their learning outcomes. With the aid of Natural Language Processing (NLP), real-time educational applications that can assess and provide feedback on reflection quality can be deployed. In this work, we first evaluate various NLP approaches for developing a reflection quality prediction model, aiming to find a configuration that balances model simplicity and generalizability across courses. Second, using the model that best balances runtime performance and predictive accuracy, we evaluate the impact of using this model to trigger real-time feedback regarding reflection quality in a mobile application currently being deployed in multiple courses across universities. Analysis of students' long-term (semester-level) and short-term (reflection writing level) changes in reflection quality across multiple classes demonstrate the utility of the deployed model in encouraging students to submit reflections with higher quality.

Keywords: Reflections · NLP · Quality prediction · Feedback

1 Introduction

Enabling students to write *free-text responses* to *reflection prompts* has been shown to improve learning gain and teaching quality [10]. Prior computational work has largely focused on reflection quality assessment [3,7,9,12,13], but has typically considered data from only single course domains and evaluated models for accuracy without regard to runtime performance. Moreover, while reflection quality modeling has been used to understand learning outcomes, its potential

The research reported here was supported by a grant from the Institute of Education Sciences (R305A180477). The opinions expressed do not represent the views of the U.S. Department of Education.

M. M. Rodrigo et al. (Eds.): AIED 2022, LNCS 13355, pp. 519–525, 2022.
https://doi.org/10.1007/978-3-031-11644-5_43

for adaptive reflection scaffolding largely remains an area for future research [2] or only studied in the lab [5]. Expanding on this prior literature, we perform research in two stages to provide students with real-time reflection quality assessment and feedback. **In the first stage**, we design new *quality prediction models* using recent transformer-based NLP techniques, and investigate model performance along two dimensions: 1) accuracy within and across conditions common to classroom use cases (e.g., differing courses), and 2) run-time (e.g., to determine which models can be integrated into a real-time application). **In the second stage**, we incorporate the best model into the *CourseMIRROR* mobile app, with the goal of providing students with *real-time feedback*. An in-the-wild evaluation of the technology deployment across multiple college classes demonstrates how providing feedback improves the quality of submitted reflections.

2 First Stage: Reflection Quality Prediction

Data for Model Development. We use the publicly available *CourseMIRROR (CM)* corpus[1] [5,9] which contains student reflections collected from 4 undergraduate classes (Chemistry (Chm), Statistics (ST), and Material Science (MSG1, MSG2)) at the end of each lecture. Each reflection was scored for quality in terms of specificity by trained raters on a 4 point scale, according to the guidelines in [9]. Table 1 summarizes the reflection distributions in the corpus.

Model Design. Following prior work, we implement a *feature extractor module* to encode reflections, followed by a *prediction module* to predict the reflection quality. While early models used handcrafted predictive features [7,9,12], recent research used neural network (NN) encoders to automatically extract features [3,6,13]. We similarly use a NN to extract all features, but unlike prior work, we use recent BERT-based transformer encoders as they have achieved better performance on many downstream tasks compared to earlier NN encoders (e.g., word2vec, GloVe, etc.) [1]. We integrate and compare two sentence encoders within our model. **DistilBERT** [11] is a distilled version of the BERT transformer-based encoder [4]. **RoBERTa** [8] is an optimized version of the original BERT, where model hyperparameters were tuned to achieve better performance compared to BERT. We predict quality using **classification** (following CourseMIRROR [5,9]) and report results using support vector machines (SVM).

Model and Data Configurations. We experimented with three different configurations of RoBERTa to observe the impact of model encoder size: RoBERTa (base, and large), and DistilRoBERTa. The encoder parameters are kept fixed during model training. For evaluation, we use **leave-one-out** to split the data, where reflections in each testing fold come from a held-out course not used during model training. This corresponds to the use case for the second stage of our

[1] https://engineering.purdue.edu/coursemirror/download/reflections-quality-data/.

research, where the model trained at the conclusion of stage one will be used to predict the quality of reflections from new courses in stage two.

Evaluation Results. With respect to *predictive performance*, Table 1 shows that while all transformer models perform very closely, the best QWK (Quadratic Weighted Kappa) score is achieved by DistilBERT. As predicted, all four BERT-based transformer encoders outperform a baseline model using a GloVe NN encoder (which was used to encode reflections in [6]). With respect to *runtime*, Table 1 shows that RoBERTa large takes on average 6 times and 3 times the time to embed compared to DistilBERT and DistilRoBERTa, respectively. We decided to choose the DistilBERT model for our real-time deployment as it is slightly faster than DistilRoBERTa and achieves best predictive performance.

Table 1. CourseMIRROR (CM) corpus reflections distribution and model performance (QWK) and runtime (in seconds) results (best in **bold**).

CM data distribution				Model	QWK	Reflection embedding time		
Courses		Scores				Max	Avg	Min
ST	1769	1	1354	GloVe (Baseline)	0.66	NA		
MSG1	395	2	2035	RoBERTa base	0.77	0.24	0.13	0.11
Chm	1034	3	2377	RoBERTa large	0.78	0.9	0.35	0.26
MSG2	3626	4	1058	DistilBERT	**0.79**	**0.13**	**0.06**	**0.05**
Total = 6824				DistilRoBERTa	0.77	0.16	0.10	0.09

3 Second Stage: Improving Reflection Quality

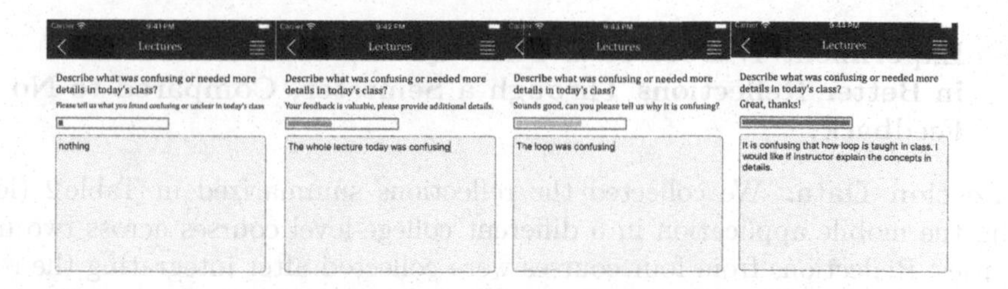

Fig. 1. Real-time quality feedback as a reflection is being written.

We now turn to presenting real-time feedback to students while writing reflections in a mobile application. First, we hosted the DistilBERT model from Sect. 2 on a server and provided an API to communicate with the hosted quality prediction model. Second, we integrated communication into the *CourseMIRROR* mobile application that students used to write and submit reflections, to enable

CourseMIRROR to provide a real-time indicator of the predicted reflection quality while students are actively writing before submission. Figure 1 shows the interface of the reflection submission mobile application. To avoid flooding the server with requests, we decided not to call the API for each student's change within the typing session as we didn't expect the quality score to change with every character change. Instead, we performed API calls whenever the number of words became odd, e.g., number of words is 1, 3, 5, 3, 5, 7, etc.

Table 2. Data used for feedback evaluation: across semester analysis and within session analysis. ARPL refers to average number of reflections per lecture.

Across semester analysis					Within session analysis			
	Course	# Lecs.	# Students	ARPL	Course	# Lecs.	# Students	ARPL
+ Feedback	PHYS1	32	143	72	CS1	26	33	14.19
	PHYS2	18	123	47	CS2	28	30	12.14
	PHYS3	9	92	17	CS3	26	46	12.34
	ENGR1	20	90	53	IS1	16	54	20.9
No feedback	ENGR2	26	124	64	CS4	27	19	4.5
					Avg # logs per reflection			9.06

Table 3. *Percentage of students* with last lecture's reflection score less than first lecture's reflection score and vice versa (left), and *average reflection quality score* for submitted reflections for the first and last lecture of the semester (right).

		Last < First	Last >= First	First Lec mean	Last Lec mean
Avg. of with feedback		27.5%	**72.5%**	2.89	2.85
No feedback	ENGR2	**65%**	35%	3.2	2.5

3.1 Experiment 1: Does Real-Time Quality feedback result in Better Reflections Through a Semester, Compared to No Feedback?

Reflection Data. We collected the reflections summarized in Table 2 (left) using the mobile application in 5 different college-level courses across two universities. Reflections from four courses were collected after integrating the real-time feedback algorithm during the Spring 2021 semester. Reflections from the remaining course were used as a control group,[2] as they were collected before the feedback algorithm was integrated into the mobile application. ***We performed human annotation of reflection quality for data from all courses and carried out the analysis using these human scores.*** Three annotators evaluated the reflections based on the annotation guidelines [9].

[2] We didn't randomly assign students to feedback and control groups, as the data collection happened in two different semesters.

Evaluation Results. Table 3 compares average reflection quality change for the courses with real-time quality feedback versus the course without. At the *reflection level*, the last two columns show the average score of submitted reflections for the first and last lecture of the semester.[3] For the average of the four courses with feedback, score of the first and last lectures are very close, with a 0.04 difference. For the course with no feedback, the scores show a 0.7 degradation, which is around a seventeen times larger difference than the feedback course difference. At the *student level*, the first two columns show the percentage of students where their last lecture's reflection score was less than their first lecture's score and vice-versa. With feedback, the percentage of students submitting equal or higher-quality reflections for the last lecture is greater (bolded) than those who submit lower-quality reflections. When no feedback is presented, the majority of students (bolded) tend to submit lower-quality reflections at the semester's end. In sum, our results support feedback utility.

Table 4. Score improvement within sessions.

	Final score vs first score (Endpoint category)			Score change direction (Trend category)		
	Improved	Constant	Decreased	Increasing	Constant	Other
Avg of 5 courses	54.1%	39.9%	5.9%	35.6%	39.9%	24.4%

3.2 Experiment 2: Do Students Keep Writing a Reflection Until It Is of High Quality, Each Time They Submit a Reflection?

Reflection Data. For Experiment 2, we logged the changes in reflection quality provided by the deployed model while students were typing the reflections. This can help us observe if the feedback provided helped students improve their submission quality within a writing session. We used data from 5 different college level courses that used the application after within session logging was incorporated into the application. ***Logs were collected from typing sessions and contained scores for each partial reflection***. Table 2 (right) summarizes data size and the average number of logged scores.

Evaluation Results. We first categorize each series of logs using one of three *trend categories*, based on the pattern when considering all logged scores for a given reflection. **Increasing** series are monotonically increasing, **constant** series have constant value, while **other** series are neither monotonically increasing nor constant. We also categorize each series using one of three *endpoint categories*, based on comparing only the starting and ending values. **Improved** series have a final value higher than the starting value, **constant** series have a final value

[3] We performed additional experiments comparing the mean of the first/last quarters of lectures instead of the first/last lecture only, and we observed similar findings.

equal to the starting value, while **decreased** series have a final value less than the starting value. Table 4 shows the distribution of these categories for average of all courses using the application after within-session API logging was implemented (Table 2). For trends (right columns), on average, more than 75% of series are either improving or constant, with around 35% improving. This shows that students often keep improving their reflections until they submit, supporting the utility of real-time feedback. Similarly, comparing the last score to the first score in the series (left columns) shows that in most cases (54%), students end the writing session with higher quality reflections than what they started with. Only around 6% of series end with lower quality reflections than what they started with, suggesting that even when a score drops during a writing session (the "other" trend category), most students recover or improve the quality by the end of the session. In sum, our results again support feedback utility.

4 Summary

Our first stage experiments in model development focused on balancing accuracy and efficiency when predicting reflection quality. Our results suggested DistilBERT as the most promising model for deployment in a real-time application. Our second stage experiments showed that using the model to provide real-time quality feedback did indeed help students submit higher-quality reflections within a reflection writing session and over the semester. For future work, we plan to tackle a few limitations of our current research. First, the feedback generated consists of a color corresponding to an ordinal value and a static message. We would like to explore generating dynamic messages tailored to the reflection content. Additionally, we plan to investigate the utility of generating more personalized feedback that integrates multiple dimensions in addition to specificity.

References

1. Bommasani, R., Davis, K., Cardie, C.: Interpreting pretrained contextualized representations via reductions to static embeddings. In: Proceedings of ACL (2020)
2. Carpenter, D., Cloude, E., Rowe, J., Azevedo, R., Lester, J.: Investigating student reflection during game-based learning in middle grades science. In: LAK21: 11th International Learning Analytics and Knowledge Conference, pp. 280–291 (2021)
3. Carpenter, D., Geden, M., Rowe, J., Azevedo, R., Lester, J.: Automated analysis of middle school students' written reflections during game-based learning. In: Bittencourt, I.I., Cukurova, M., Muldner, K., Luckin, R., Millán, E. (eds.) AIED 2020. LNCS (LNAI), vol. 12163, pp. 67–78. Springer, Cham (2020). https://doi.org/10.1007/978-3-030-52237-7_6
4. Devlin, J., Chang, M.W., Lee, K., Toutanova, K.: BERT: pre-training of deep bidirectional transformers for language understanding. arXiv:1810.04805 (2018)
5. Fan, X., Luo, W., Menekse, M., Litman, D., Wang, J.: Scaling reflection prompts in large classrooms via mobile interfaces and natural language processing. In: Proceedings of 22nd International Conference on Intelligent User Interfaces, pp. 363–374 (2017)

6. Geden, M., Emerson, A., Carpenter, D., Rowe, J., Azevedo, R., Lester, J.: Predictive student modeling in game-based learning environments with word embedding representations of reflection. Int. J. AI Educ. **31**(1), 1–23 (2021)
7. Kovanović, V., et al.: Understand students' self-reflections through learning analytics. In: Proceedings of 8th International Conference on Learning Analytics and Knowledge, pp. 389–398 (2018)
8. Liu, Y., et al.: RoBERTa: a robustly optimized BERT pretraining approach. arXiv preprint arXiv:1907.11692 (2019)
9. Luo, W., Litman, D.: Determining the quality of a student reflective response. In: The Twenty-Ninth International FLAIRS Conference (2016)
10. Menekse, M., Stump, G., Krause, S., Chi, M.: The effectiveness of students' daily reflections on learning in engineering context. In: ASEE Conference & Exposition (2011)
11. Sanh, V., Debut, L., Chaumond, J., Wolf, T.: DistilBERT, a distilled version of BERT: smaller, faster, cheaper and lighter. arXiv preprint arXiv:1910.01108 (2019)
12. Ullmann, T.D.: Automated analysis of reflection in writing: validating machine learning approaches. Int. J. AI Educ. **29**(2), 217–257 (2019). https://doi.org/10.1007/s40593-019-00174-2
13. Wulff, P., et al.: Computer-based classification of preservice physics teachers' written reflections. J. Sci. Educ. Technol. **30**(1), 1–15 (2020). https://doi.org/10.1007/s10956-020-09865-1

Providing Insights for Open-Response Surveys via End-to-End Context-Aware Clustering

Soheil Esmaeilzadeh(✉), Brian Williams, Davood Shamsi, and Onar Vikingstad

Apple, Cupertino, CA, USA
{sesmaeilzadeh,brian_d_willaims,davood,vikingstad}@apple.com

Abstract. Teachers often conduct surveys in their classes to gain insights into topics of interest. When analyzing surveys with open-ended responses, a teacher traditionally has to read the responses one by one, which is a labor-intensive and time-consuming process. We present a novel end-to-end context-aware framework that extracts, aggregates, and abbreviates embedded semantic patterns in open-response survey data. Our framework uses a pre-trained natural language model to encode the textual data into semantic vectors. The encoded vectors then get clustered either into an optimally tuned number of groups or into a set of groups with pre-specified titles. We provide context-aware word-clouds that demonstrate the semantically prominent keywords within each group. Honoring user privacy, we have successfully built the on-device implementation of our framework suitable for real-time analysis on mobile devices and have tested it on a synthetic dataset. Our framework reduces the costs at-scale by automating the process of extracting the most insightful information pieces from survey data.

Keywords: Surveys · Context-aware · Clustering · Natural language model

1 Introduction

Formative assessment refers to a set of activities undertaken by teachers to gather information about the learning progress of students. Surveys are a commonly used formative assessment method in classrooms. Formative assessment using surveys usually includes four steps, namely (i) creation (ii) collection (iii) analysis, and (iv) action. Teachers often use four types of questions in surveys, namely (i) multiple-choice (ii) rating scale (iii) likert scale, and (iv) open-response. In the analysis step, analyzing the responses to open-response questions is not straightforward since they can include a wide range of topics and concepts. For analyzing the open-responses teachers commonly go through the responses one by one and find out the key themes. This analysis approach is extremely time-consuming, inefficient, challenging, and often biased. There have been efforts to automate

M. M. Rodrigo et al. (Eds.): AIED 2022, LNCS 13355, pp. 526–532, 2022.
https://doi.org/10.1007/978-3-031-11644-5_44

survey analysis to save time and cost and be able to use the outcome of such analysis to increase user satisfaction and improve services. A common approach for analyzing open-ended survey responses is the use of topic modeling [9]. The main challenge behind using topic modeling techniques is that they do not account for contextual information and merely rely on the word-level frequencies. Moreover, they require careful pre-processing steps, are vulnerable to not generalizing well on unseen samples, need extensive hyperparameter tuning, and cannot capture implicit semantics such as sarcasm, anger, metaphors, and figurative languages [1]. In order to overcome such shortcomings, in this work, we present an end-to-end framework for context-aware analysis of open-response survey data.

2 Methodology

In this work, we use pre-trained natural language models in order to extract the contextual semantic patterns in a collection of open-responses. Pre-trained neural models are commonly used to capture the semantics at the word as well as sentence levels in a wide range of tasks such as text generation, building dialogue systems, text classification, hate speech detection, sentiment analysis, named entity recognition, question answering, and text summarization. There are two main categories of such models for capturing the semantics, namely (i) word-level models and (ii) sentence-level models. Pre-trained word-level models such as Word2Vec [4] and GloVe [5] are used to encode words into so-called *embedding* vectors. One major limitation of word-level embedding vectors obtained by Word2Vec and GloVe is that they do not capture the context of the words in sentences. Context-aware word-level embeddings such as ELMo [6] and BERT [2] on the other hand attempt to address that shortcoming by accounting for the context of a word within a sentence. Recently, pre-trained language models have been used for capturing the contexts beyond word-level in order to encode sentences into embedding vectors. In this work, we use the Sentence-BERT (SBERT) [7] model provided by HuggingFace [8] to get the embedding vectors of words as well as sentences.

2.1 Context-Aware Clustering

In the clustering task, the inputs are the raw open-responses gathered from the surveys. Using the SBERT pre-trained language model we first tokenize and then extract the embedding vectors for each input sample. The SBERT model maps sentences and paragraphs to a 384 dimensional dense vector space. Once the embedding vectors are created we use the k-means algorithm to cluster the input samples. In the k-means algorithm, the number of clusters k is unknown and needs to be tuned for and provided as an input. We use the silhouette score to find the best number of clusters k^*. For doing that we calculate the silhouette score values for multiple number of clusters between 2 and an upper bound of k_{max} and choose k^* as the number of clusters where the silhouette score obtains its maximum value as $k^* = \arg\max_k SS^{(k)}$ for $k \in [2, k_{max}]$, where $SS^{(k)}$ is the silhouette score for a given clustering configuration with k number of clusters. Finally, we annotate each cluster with the prominent keywords of its samples, and we generate wordclouds for each cluster separately as well as a unified wordcloud for all the clusters together.

2.2 Context-Aware Cluster Assignment

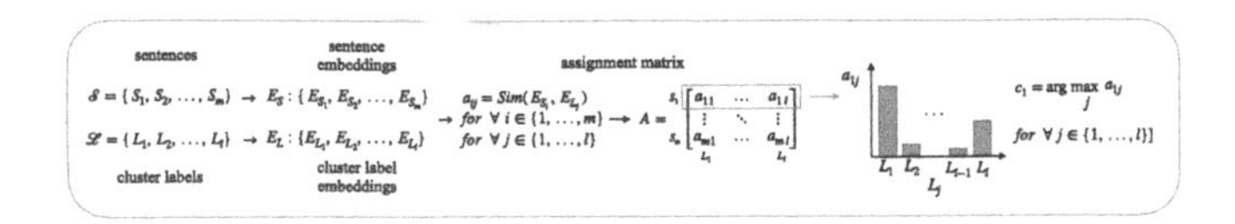

Fig. 1. Overview of the cluster assignment approach.

In the clustering assignment approach, we have two groups of inputs, namely, the raw open-responses ($\mathcal{S}$) and the labels of the clusters ($\mathcal{L}$). Using the SBERT pre-trained language model we tokenize and extract the embedding vectors E_S and E_L, respectively for each group of inputs. Next, we calculate the assignment matrix A with its elements being the pairwise cosine similarity between the sentence embeddings and the cluster label embeddings as $A = a_{ij} = Sim(E_{S_i}, E_{L_j})$ for $\forall i \in \{1, \ldots, m\}$, for $\forall j \in \{1, \ldots, l\}$, where m and l are the number of input open-responses and input cluster labels, respectively. $Sim(\,)$ represents the cosine similarity function. In this work, the length of the embedding vectors generated by the SBERT models is $V = 384$. Once we build the assignment matrix A, the corresponding assigned label c_i for sentence i can be found as $c_i = \arg\max_j a_{ij}$ for $\forall j \in \{1, \ldots, l\}$ and $\forall i \in \{1, \ldots, m\}$, where each sentence is assigned the label with the highest cosine similarity in the embedding space. Figure 1 illustrates the steps involved in the cluster assignment approach for calculating the assignment matrix A and finding the assigned labels. It is worth noting that since the cluster labels are provided as an input in the cluster assignment approach, different from clustering in Sect. 2.1, we do not annotate the clusters. The input titles are considered as the labels of the clusters.

2.3 Context-Aware Insights

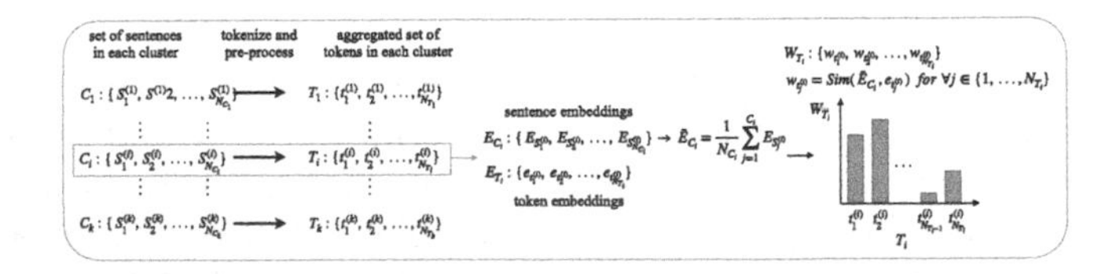

Fig. 2. Overview of the cluster annotation (labeling) approach.

Cluster Annotation: In the clustering task, the output of the clustering function is a set of grouped open-responses. Assuming that the cluster C_i has N_{C_i} number of sentences, we tokenize all the sentences and pre-process (*e.g.*, removing the stop-words, lemmatization and/or stemming) the tokens. We then gather the set of tokens T_i corresponding to cluster C_i. Next, using the SBERT pre-trained language model we extract the embedding vectors for the set of sentences (E_{C_i}) and tokens (E_{T_i}). Afterward, we calculate the average of sentence embeddings as $\bar{E}_{C_i}$, which is the centroid of a cluster in the embedding space. Next, for each cluster C_i, we calculate the weight value w_t of each token as the cosine similarity between $\bar{E}_{C_i}$ and the token embedding vector e_t. We then sort the tokens in each cluster with respect to w_t values in descending order. Finally, we use the top 5 prominent tokens (*i.e.*, with the largest w_t values) for annotating each cluster. Figure 2 illustrates an overview of the described cluster annotation approach.

Wordcloud Generation: Wordcloud as a visual representation of text data is commonly used to depict keywords where the sizes of the words represent their frequency or importance level. In this work, we present an approach for creating context-aware wordclouds. We consider (i) cluster-level wordclouds for each cluster, and (ii) unified wordcloud for all the input open-responses. The cluster-level wordclouds include the top prominent tokens with their sizes accounting for weight values. The unified wordcloud shows the prominent keywords across all clusters. To create the unified wordcloud, we use the words in the cluster-level wordclouds together with their corresponding weight values w. We then scale w with a density coefficient ρ that accounts for the relative number of samples in each cluster. The density coefficient for cluster C_i is defined as $\rho_i = N_{C_i}/m$ where $i \in \{1, \ldots, k\}$, where N_{C_i} is the number of samples in cluster C_i, m is the total number of input open-responses, and k is the number of clusters. Accordingly, the scaled weight values of the prominent tokens (w^*) for the cluster C_i become $w^*_{t_j^{(i)}} = \rho_i w_{t_j^{(i)}}$ for $\forall j \in \{1, \ldots, 5\}$ and $\forall i \in \{1, \ldots, k\}$, where k is the number of clusters. The scaled weight values (w^*) not only take into account the importance of each keyword at the cluster level (*i.e.*, through w), but also account for the relative importance of each cluster depending on how many samples a cluster entails (*i.e.*, through ρ).

3 Dataset

As the dataset for this work, the authors of this manuscript have manually written a plausible set of responses to an open-response survey question for a chemistry class. In this survey, the teacher asks about one topic that each student would like to be reviewed before their upcoming exam.

4 Results and Discussion

4.1 Context-Aware Clustering

Upon our investigation, $k^* = 6$ as the number of clusters achieved the maximum silhouette score of $SS = 0.299$. Accordingly, in context-aware clustering, we consider the number of clusters to be six. Table 1 shows the list of clustered responses. In Table 1, we see that the responses that have fallen under each group discuss identical topics. For instance, cluster C_1 entails responses that mostly talk about atomic interactions, whereas cluster C_6 is mostly related to different forms of unit conversion.

Figure 3(a) illustrates the UMAP [3] projection of the sentence embeddings where the color represents different clusters. We can see that through the UMAP projection in the embeddings space the clusters get properly segregated and form six distinct groups. Next, we annotate each cluster by following the steps presented earlier. Upon calculating the importance values of all the tokens in each cluster (W_{T_i}), we found that the top five tokens for clusters C_1 to C_6 respectively are: {ionic, bonding, covalent, bond, atom}, {proton, neutron, electron, atomic, atom}, {enthalpy, entropy, thermodynamic, explain, difference}, {acid, chemical, chemistry, reaction, compound}, {periodic, chemical, table, reaction, element}, {unit, kilogram, conversion, meter, convert}. Figure 3(b) shows the wordclouds for the prominent tokens of each cluster where the size of words represents the importance values of tokens W_{T_i}. In Fig. 3(b) the middle wordcloud

Table 1. List of clustered responses distributed across the six clusters C_1 to C_6.

Cluster ID	Clustered Responses
C_1	About the differences between ionic bonding & covalent bonding.
	About the ionic bonding and its properties when reacting with other substances.
	About the way atoms join together through ionic and covalent bonding.
	I have a hard time understanding how in ionic bonding, atoms transfer electrons to each other?
	could you please explain why Ionic bonds form between a metal and a nonmetal mostly?
	please clarify about the covalent and ionic bonding and how they are different and similar.
C_2	About the differences between proton and neutron.
	About the similarities between protons and neutrons with electrons.
	it would be great if you could explain more about the atomic structure and neutron, proton, electron.
	please explain more about the composition of atoms such as electron, neutron and proton.
	please explain what are the common properties of protons and neutrons.
C_3	About the differences between entropy and enthalpy, and also their similarities.
	Could you please explain how entropy can get transformed in to enthalpy and vice versa.
	Explain the entropy and enthalpy concepts that we learned in the beginning of semester.
	Why entropy and enthalpy are important and how they are used in thermodynamic.
C_4	About acids & bases that we learned in the last lecture.
	About the difference between bases and acids in their chemical formula.
	About the use cases of both acids and bases in industry.
	Please elaborate on the reactions of acids and bases with inert compounds.
	Please explain the applications of acids in chemistry.
C_5	About how we can use periodic table to identify reactions.
	About the total number of elements in the periodic table that we studied.
	Regarding the periodic table and the order of chemical elements in each column.
C_6	about how distance unit in foot can get converted to distance unit in meter.
	about the unit conversion in SI system and how that differs with UK system.
	please explain about units and how unit conversion works.
	please explain how we can transform pounds unit to kilograms.
	please explain unit conversion again with a few more examples.

shows the unified wordcloud where the size of the words represents the scaled weight values of the prominent tokens w^*. The unified wordcloud enables a quick understanding of the prominent tokens across all the clusters by accounting for their relative importance within each cluster as well as the number of samples that each cluster entails.

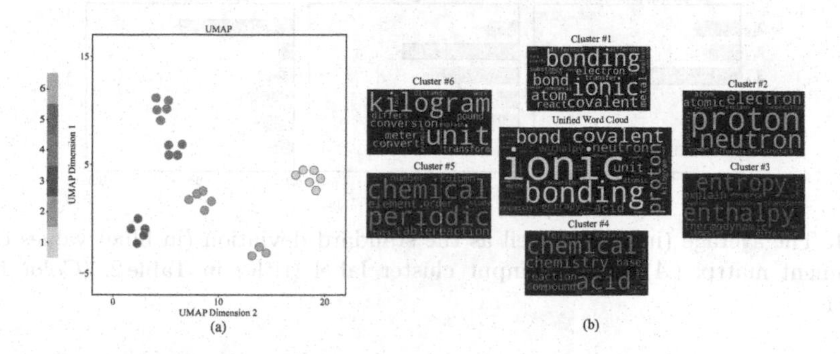

Fig. 3. (a) UMAP representation of sentence embeddings. (b) Wordclouds of the prominent tokens of the clusters.

4.2 Context-Aware Cluster Assignment

Table 2 presents the list of a plausible set of input titles for the response categories of an open-response survey question for a chemistry class that the authors of this manuscript have manually written. The main difference between the cluster assignment task in this section and the clustering task in the previous section is that here the teacher has a set of bucket titles in mind where he/she wants to group the responses with respect to. Here, in addition to the input responses we also extract the embeddings of the input titles in Table 2 using the pre-trained language model. Upon building the assignment matrix A we assign to each input response the title with the highest cosine similarity in the embedding space. Figure 4 shows the average (in red) as well as the standard deviation (in blue) values of the assignment matrix A for each cluster label (title). Even though in Fig. 4 we show the average values of the assignment matrix (A) for each input cluster label, the assignment happens at the input response and input title level where we assign to each input response the title with the highest cosine similarity in the embedding space.

Table 2. List of input titles for the open-response survey question.

ID	Input Titles
#1	The chapter on molecular ionic and covalent bonds
#2	The chapter on atomic subparticles such as proton, electron, neutron
#3	The chapter on thermodynamic concepts such as enthalpy and entropy
#4	The chapter on acid and base reactions
#5	The chapter on periodic table layout
#6	The chapter on converting different units

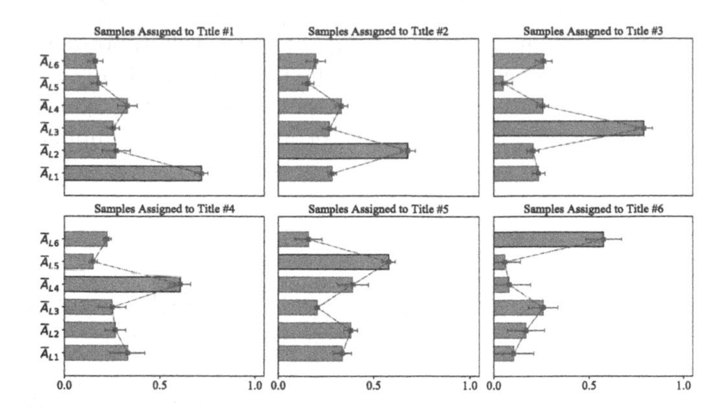

Fig. 4. The average (in red) as well as the standard deviation (in blue) values of the assignment matrix (A) for each input cluster label (title) in Table 2. (Color figure online)

5 Conclusion and Future Work

In this work, we presented a novel end-to-end context-aware framework to extract, aggregate, and annotate embedded semantic patterns in open-response survey data. As a future direction, we intend to investigate extending our proposed framework to non-English languages using multi-language pre-trained models.

References

1. Buenano-Fernandez, D., Gonzalez, M., Gil, D., Lujan-Mora, S.: Text mining of open-ended questions in self-assessment of university teachers: an LDA topic modeling approach. IEEE Access **8**, 35318–35330 (2020)
2. Devlin, J., Chang, M.W., Lee, K., Toutanova, K.: BERT: pre-training of deep bidirectional transformers for language understanding (2019)
3. McInnes, L., Healy, J., Melville, J.: UMAP: Uniform Manifold Approximation and Projection for Dimension Reduction (2018)
4. Mikolov, T., Chen, K., Corrado, G., Dean, J.: Efficient estimation of word representations in vector space. In: 1st International Conference on Learning Representations, ICLR 2013 - Workshop Track Proceedings, pp. 1–12 (2013)
5. Pennington, J., Socher, R., Manning, C.D.: GloVe: global vectors for word representation. In: EMNLP (2014)
6. Peters, M.E., Neumann, M., Iyyer, M., Gardner, M., Clark, C., Lee, K., Zettlemoyer, L.: Deep contextualized word representations. In: NAACL HLT, pp. 2227–2237 (2018)
7. Reimers, N., Gurevych, I.: Sentence-BERT: Sentence embeddings using siamese BERT-networks. In: EMNLP-IJCNLP, pp. 3982–3992 (2019)
8. Reimers, N., Gurevych, I.: Sentence Transformers Trained on the MiniLM Paraphrase Corpus (2019)
9. Vayansky, I., Kumar, S.A.: A review of topic modeling methods. Inf. Syst. **94**, 101582 (2020)

Towards Aligning Slides and Video Snippets: Mitigating Sequence and Content Mismatches

Ziyuan Liu(✉) and Hady W. Lauw(✉)

School of Computing and Information Systems, Singapore Management University, Singapore, Singapore
{ziyuan.liu.2018,hadywlauw}@smu.edu.sg

Abstract. Slides are important form of teaching materials used in various courses at academic institutions. Due to their compactness, slides on their own may not stand as complete reference materials. To aid students' understanding, it would be useful to supplement slides with other materials such as online videos. Given a deck of slides and a related video, we seek to align each slide in the deck to a relevant video snippet, if any. While this problem could be formulated as aligning two time series (each involving a sequence of text contents), we anticipate challenges in generating matches arising from differences in content coverage and sequence of content between slide deck-video pairs. To mitigate these challenges, we propose a two-stage algorithm that builds on time series alignment to filter out irrelevant content and to align out-of-sequence slide deck and video pairs. We experiment with real-world datasets from openly available lectures, which have been manually annotated with start and end times of each slide in the videos to facilitate the evaluation of matches.

Keywords: Slide to video alignment · Dynamic time warping · Sequence mismatch · Content mismatch

1 Introduction

Many instructors use slides as teaching aid, and often make these available to students as reference material. The compact and terse nature may render slides, on their own, inadequate for the latter function of reference materials. Students may need to rely on additional outside materials, such as videos that can be found in course webpages, massive open online courses, or video sites.

Some works attempt to augment academic or educational materials with additional content [3,8]. Adamson et al. [1] set out a means of automatically generating questions to support instruction and learning, while others seek to support teaching by generating answers to questions [2,13].

We envision a system where a student who is reviewing a deck of slides can be pointed to a snippet of a video that is relevant to the slide currently being viewed. Given a video relevant to a deck of lecture slides, we seek to align each

M. M. Rodrigo et al. (Eds.): AIED 2022, LNCS 13355, pp. 533–539, 2022.
https://doi.org/10.1007/978-3-031-11644-5_45

slide to a snippet within the video. This also involves detecting when there is no snippet within the video relevant to the slide. The technical challenge here concerns aligning two collections of different modalities, e.g., slides and videos. Our approach treats both slides and videos as time series of text contents.

As the first contribution in this paper, we propose SEQUENTIALIGN (see Sect. 2), a methodology for aligning a slide deck and a video that mitigates sequence mismatch and content mismatch. As a second contribution, we build an annotated dataset of aligning slides to video snippets from pairings of educational slides and videos from computer science topics such as artificial intelligence, operating systems, and systems programming. As a third contribution, we empirically validate the approach (see Sect. 3) on the afore-mentioned annotated data against comparable baselines.

2 Methodology: Sequentialign

Data. Our data consists of slide deck-video pairs, each consisting of a slide deck and a related video. Within each pair, at least one slide in the deck matches one snippet in the video. We work with the textual contents of the slides and the videos (i.e., transcripts). Each slide deck $\mathbf{s}$ consists of a number of slides, given by a sequence of vectors $\mathbf{s} = \{\boldsymbol{s}_1, \boldsymbol{s}_2, \boldsymbol{s}_3 ... \boldsymbol{s}_m\}$, each representing the textual context of a single slide. These vectors could be based on bag-of-words representation such as tf-idf or word embeddings [9]. In turn, each video $\mathbf{v}$ is divided into video snippets of a specified equal duration,[1] given by a sequence of vectors $\mathbf{v} = \{\boldsymbol{v}_1, \boldsymbol{v}_2, \boldsymbol{v}_3 ... \boldsymbol{v}_n\}$ each representing the transcribed content of a single snippet.

Problem. For each slide deck-video pair, our task is to find a set of matches (an alignment) between slides and video snippets, such as $(s_i \rightarrow [v_{j1}, v_{j2} ... v_{jn}])$, where $[v_{j1}, v_{j2} ... v_{jn}]$ is a set of video snippets matched to s_i. Each slide may be assigned to 0 or more snippets. Each snippet can be assigned to 0 or 1 slide.

Dynamic Time Warping. We can view a deck of slides as a time series whereby each slide is a time point. Similarly, each snippet is a time point within a video time series. Among techniques for measuring the similarity between two time series [10], Dynamic Time Warping (DTW) is known as a robust way to measure similarity between two time series that vary in speed [6]. It also produces an alignment of time points between the two respective time series. Without losing generality, we build our proposed algorithm using DTW as a building block. However, DTW has a couple of constraints that render it unsuitable for direct use. The monotonicity property requires that indices of successive matches on either sequence should be monotonically increasing, thereby forcing false matches when the two time series are out of sequence. The continuity property requires matched indices on each sequence to increase one at a time, thereby continuing

[1] In our experiments, each basic unit of video snippet is of 30-s duration. The last snippet in a video may be shorter.

false matches over periods of irrelevant snippets. To counter these, we propose SEQUENTIALIGN that addresses the sequence and content mismatches.

2.1 Mitigating Sequence Mismatch

Consider a slide deck-video pair which cover a similar set of topics. If sequence were not informative, we may consider matching using a distance measure alone. A naive means of doing this would be to divide the slides or the videos into blocks, and perform minimum weight bipartite matching [4]. Between each of the blocks, we calculate a distance and find a matching that minimizes the total distance.

While the overall sequence of a slide deck and video might be different, there may be common local subsequences in which the flow of topics follow a similar, logical order. It may be useful to use DTW as a distance measure locally, while relying on bipartite matching globally. This forms the basis for the alignment subroutine of SEQUENTIALIGN. Our alignment subroutine divides both the slides and the video snippets into a number of blocks set in a 2-dimensional grid (given by the grid factor, g), each containing an equal number of slides or snippets, as the case may be. For every cell in the grid, each representing a possible match between a slide block and a video snippet block, we run DTW locally, giving each cell a warping distance. The Hungarian algorithm [7] is used to find a minimum-weight bipartite matching between the 2 axes on the grid, to identify a set of cells representing one-to-one matches with the lowest total distance measure.

While the initial grids enforce equal-sized blocks, to model more natural alignment that may involve different-sized blocks, after the alignment subroutine obtains the matches given by the bipartite matching, it runs the DTW algorithm again on adjacent slide blocks in the matches, and the warping path returned is used to adjust the boundaries between their matched video snippet blocks, while leaving the slide block boundaries unchanged. The intuition is that this process will break through the rigidity of the uniform length of blocks, and allow snippets on the cell boundaries to be assigned to the correct slides, while the bipartite matching between slide blocks and video blocks makes it possible for common subsequences to be matched together out-of-sequence between a slide deck-video pair, even as the monotonicity constraint of DTW is respected locally.

What remains is the determination of the value of g, which we consider a hyperparameter. Our approach is to search from 1 to two-thirds of number of slides, and pick the value of g which yields the minimum distance measure.

2.2 Mitigating Content Mismatch

To mitigate the content mismatch, we identify irrelevant slides and video snippets and remove them before alignment. For one naive approach to identify an irrelevant slide, we can consider its minimum distance to any snippet and impose a maximum threshold. For another, we can let each video identify its closest slide, and remove any slide not picked by any video. Analogously, we can attempt to identify an irrelevant video snippet. Both look at each slide (resp. snippet) independently of any other slide in the deck (resp. snippet in the video).

We postulate that to identify whether a slide is relevant, we need to consider the neighbouring slides, and whether as a group of slides they may match a sequence of video snippets as well. To allow the consideration of multiple target window sizes, we introduce the concept of *relevance score*. For video snippets, the primary component of this score is the number of best-match windows it is part of. For slides, the primary component of this score is the number of video snippets it is matched to across all queries. The raw scores are adjusted for the distance between each query and its best-match window, and the distance between each video snippet and query slide pair within best-match windows.

Drawing from [12], our subsequence search subroutine uses DTW as a subroutine. For each slide, it constructs a 'query' by taking the slide itself and a number of subsequent slides, given by the length of the query, r. DTW is run between the query and target windows of video snippets of length q, with starting index incremented by 1 for each successive window. For each query, we match the first slide in the query with the sequence of snippets starting with the first snippet of the best-match window, and the snippet immediately before that identified by the path as the starting point of the second slide in the query in the window. If the first snippet matched to second slide is the same as that of the first, we do not match the first slide with any snippet, and return an empty set.

To filter out irrelevant content, we run the subsequence search subroutine multiple times, using varying window sizes for the target windows of video snippets. For each query q, we take note of the best match window, the video snippets matched to each slide in the query $[v_1, v_2 ... v_n]$, the total distance (cost) between the query and best match window, and the cosine distances (distance) between each slide and the video snippets it is matched to. Having calculated relevance scores for all slides and video segments, we set a percentile threshold for determining relevance, and remove slides and video segments with a relevance score below the relevance score value at the percentile threshold. For instance, if the 25th percentile of slide relevance scores is 75.5, slides with relevance scores below 75.5 are labelled as irrelevant and removed. We name filtering subroutines (and the SEQUENTIALIGN implementation it is used in) according to the percentile threshold of relevance scores used to identify irrelevant slides. For example, SEQUENTIALIGN-33 combines the filtering subroutine with 33rd percentile threshold with the alignment algorithm described in the previous section. The use of percentile threshold, instead of absolute threshold, is to guard against different levels of text similarities across domains.

3 Experiments

Our objective is to evaluate efficacy of various methods at producing alignments between video and slides on real-world datasets.

Datasets. We annotate 6 datasets containing slide-deck video pairs from publicly available lectures, as summarized in Table 1, covering subjects such as Artificial Intelligence, Operating Systems, and Systems Programming in C/C++.

Each slide is labelled with start and end times corresponding to the video portion which in the opinion of the annotator best matches the slide content. The datasets have content and sequence mismatch between slide decks and videos.

Baselines. We compare our SEQUENTIALIGN algorithm with these baselines:

Table 1. Summary of datasets

	Video course	Slides course	Pairs	Slide count		Video duration (s)	
				Mean	Median	Mean	Median
BERKELEYSTANFORD-AI	Berkeley CS188	Stanford CS221	8	39.5	36.5	4735.8	4863.0
STANFORDBERKELEY-AI	Stanford CS221	Berkeley CS188	8	42.3	42.0	4149.6	4140.0
BERKELEYVIRGINIA-OS	Berkeley CS162	UVirginia CS4414	8	90.5	87.0	5203.3	5233.5
VIRGINIABERKELEY-OS	UVirginia CS4414	Berkeley CS162	8	66.1	61.0	4551.9	4489.0
CMUCORNELL-C	CMU 15213	Cornell CS4414	5	44.8	49.0	3305.6	2980.0
CORNELLCMU-C	Cornell CS4414	CMU 15213	5	56.4	56.0	4295.0	4613.0

Table 2. Performance on various slide deck-video pairs from different sources

	Artificial intelligence				Operating systems				Systems programming in C			
	BERKELEYSTANFORD		STANFORDBERKELEY		BERKELEYVIRGINIA		VIRGINIABERKELEY		CMUCORNELL		CORNELLCMU	
	Acc	IoU	Acc	IoU	Acc	IoU	Acc	IoU	Acc	IoU	Acc	IoU
Random	0.004	0.001	0.005	0.003	0.006	0.004	0.017	0.004	0.001	0.002	0.004	0.001
DTW	0.027	0.011	0.051	0.022	0.010	0.007	0.048	0.011	0.087	0.040	0.137	0.061
HMM+IBM1	0.003	0.001	0.017	0.007	0.010	0.007	0.012	0.002	0.030	0.016	0.008	0.003
SEQUENTIALIGN-25	0.109	0.126	0.189	0.172	0.234	0.168	0.236	0.167	0.125	0.211	0.179	0.198
SEQUENTIALIGN-33	0.211	0.190	0.253	0.224	0.305	0.230	0.300	0.230	0.160	0.260	0.241	0.276
SEQUENTIALIGN-50	0.407	0.314	0.389	0.343	0.444	0.365	0.406	0.365	0.414	0.331	0.366	0.405

HMM+IBM1. The closest related work in terms of task is the HMM+IBM1 [11]. We align video snippets with slides using a window of jump probabilities [−2, 2]. It mainly targets sequence alignment without targeting content mismatch.

Dynamic Time Warping (DTW). To evaluate the performance without the mitigation of the sequence and content mismatch provided by SEQUENTIALIGN over the base alignment algorithm, we compare to the vanilla DTW.

Random. We split the video snippets into as many segments as there are slides, and assign each segment randomly to a slide.

Metrics. We use the following metrics that are commonly associated with multimedia retrieval or alignment:

Accuracy (Acc). Accuracy is the number of seconds in the video with true positive and true negative alignment outcomes, over the duration of the video in seconds. True positive is defined as seconds correctly aligned to the right slide. True negative is defined as seconds which are irrelevant to any slide and correctly identified. We average accuracy across all slide deck-video pairs.

Intersection over Union (IoU). Following [5], for each slide, we measure the intersecting duration between the predicted and the ground truth video spans and divide this by the union. For true negative hits, the IoU value is taken to be 1. We then average this IoU over the slides in a deck, and over the decks.

Empirical Results. In Sect. 2.2, we describe dealing with content mismatch by filtering out irrelevant content that involves specifying a percentile threshold, yielding the various SEQUENTIALIGN variants (at 25th, 33rd, and 50th percentiles). The results are shown in Table 2. The SEQUENTIALIGN variants tend to outperform over the baselines across all the datasets here. DTW and HMM+IBM1 perform rather poorly due to the considerable content and sequence mismatch in these datasets. The performance of SEQUENTIALIGN steadily improves as we remove more irrelevant content.

4 Conclusion

In conclusion, we have proposed a framework for the generation of matches between slide deck-video pairs. To mitigate the content mismatch and sequence mismatch problems which can cause an unmodified DTW algorithm to be less suitable for the task of generating matches, we propose a 2-step solution, by first identifying probable irrelevant slides using a subsequence search approach, and then focusing on finding good matches despite the sequence mismatch problem, using the alignment subroutine. Experiments on slides and videos from real courses show promise. We identify several directions for future work. In our experiments, we produce alignments for slide decks with a single video. We could run SEQUENTIALIGN across several videos to find more matches for a given slide. Being more aggressive with content filtering may achieve higher quality matches with smaller quantity from each video but higher quantity across videos.

Acknowledgements. This research is supported by the Centre for Teaching Excellence at Singapore Management University under its Educational Research Fellowship Programme.

References

1. Adamson, D., Bhartiya, D., Gujral, B., Kedia, R., Singh, A., Rosé, C.P.: Automatically generating discussion questions. In: Lane, H.C., Yacef, K., Mostow, J., Pavlik, P. (eds.) AIED 2013. LNCS (LNAI), vol. 7926, pp. 81–90. Springer, Heidelberg (2013). https://doi.org/10.1007/978-3-642-39112-5_9
2. Atapattu, T., Falkner, K., Falkner, N.: Educational question answering motivated by question-specific concept maps. In: Conati, C., Heffernan, N., Mitrovic, A., Verdejo, M.F. (eds.) AIED 2015. LNCS (LNAI), vol. 9112, pp. 13–22. Springer, Cham (2015). https://doi.org/10.1007/978-3-319-19773-9_2
3. Csomai, A., Mihalcea, R.: Linking educational materials to encyclopedic knowledge. Front. Artif. Intell. Appl. **158**, 557 (2007)

4. Duan, R., Pettie, S.: Linear-time approximation for maximum weight matching. J. ACM (JACM) **61**(1), 1–23 (2014)
5. Gao, J., Sun, C., Yang, Z., Nevatia, R.: TALL: temporal activity localization via language query. In: ICCV, pp. 5267–5275 (2017)
6. Keogh, E.J., Pazzani, M.J.: Scaling up dynamic time warping for datamining applications. In: Proceedings of the Sixth ACM SIGKDD International Conference on Knowledge Discovery and Data Mining, pp. 285–289 (2000)
7. Kuhn, H.W., Yaw, B.: The Hungarian method for the assignment problem. Nav. Res. Logist. Q. **2**, 83–97 (1955)
8. Labhishetty, S., Bhavya, Pei, K., Boughoula, A., Zhai, C.: Web of slides: automatic linking of lecture slides to facilitate navigation. In: ACM L@S (2019)
9. Le, Q., Mikolov, T.: Distributed representations of sentences and documents. In: International Conference on Machine Learning, pp. 1188–1196. PMLR (2014)
10. Morse, M.D., Patel, J.M.: An efficient and accurate method for evaluating time series similarity. In: SIGMOD, pp. 569–580 (2007)
11. Naim, I., Song, Y.C., Liu, Q., Kautz, H., Luo, J., Gildea, D.: Unsupervised alignment of natural language instructions with video segments. In: AAAI (2014)
12. Rakthanmanon, T., et al.: Searching and mining trillions of time series subsequences under dynamic time warping. In: KDD (2012)
13. Zylich, B., Viola, A., Toggerson, B., Al-Hariri, L., Lan, A.: Exploring automated question answering methods for teaching assistance. In: Bittencourt, I.I., Cukurova, M., Muldner, K., Luckin, R., Millán, E. (eds.) AIED 2020. LNCS (LNAI), vol. 12163, pp. 610–622. Springer, Cham (2020). https://doi.org/10.1007/978-3-030-52237-7_49

Student Behavior Models in Ill-Structured Problem-Solving Environment

Deepti Reddy(✉), Vedant Balasubramaniam, Salman Shaikh, and Shreyas Trapasia

SIES Graduate School of Technology, Nerul, Navi Mumbai, India
deeptir@sies.edu.in, {vedant.balasubramaniam17,salman.shaikh17, shreyas.trapasia17}@siesgst.ac.in

Abstract. Identifying the various cognitive processes that learners engage while solving an ill-structured problem online learning environment will help provide improved learning experiences and outcomes. This work aims to build a student model and analyze student behaviors in our technology-enhanced learning environment named Fathom used for teaching-learning of ill-structures problem-solving skills in the context of solving software design. Students' interactions on the system, captured in log files represent their performance in applying the skills towards understanding the problem as a whole and formulating it into sub-problems, generating alternative designs, and selecting the optimal solution. We discuss methods for analyzing student behaviors and linking them to student performance. The approach used is a hidden Markov model methodology that builds students' behavior models from data collected in the log files.

Keywords: Ill-structured problem · Technology-enhanced learning environment · Student model · Hidden-Markov model · Software-design problem-solving skills

1 Introduction

Ill-structured problems are complex because they have vaguely defined or unclear goals and unstated constraints; they possess multiple solutions and solution paths and involve multiple criteria for evaluating solutions [9]. Software design is a complex and ill-structured activity in which a software designer has to deal with issues such as understanding the unknown problem domain, eliciting requirements from multiple stakeholders' viewpoints, identifying alternative solutions, and making decisions based on selection criteria [1].

Novices find design daunting and face some difficulties like – the inability to structure a problem, fixation while creating a solution, and evaluation of the solution. Research [1, 11, 12] shows that experts are able to deal with these issues by implicitly applying cognitive skills such as drawing diagrams to simulate scenarios that aid in eliciting requirements and constraints which may not be directly stated initially. Hence, in addition to content knowledge, students need to be explicitly trained to effectively use these practices while solving software design problems.

M. M. Rodrigo et al. (Eds.): AIED 2022, LNCS 13355, pp. 540–545, 2022.
https://doi.org/10.1007/978-3-031-11644-5_46

We have designed and developed a technology-enhanced learning environment named Fathom [3], for the teaching-learning of ill-structured problem-solving skills in the context of solving software design problems. The targeted software design skills were: the ability to visualize the problem as a whole before formulating sub-problems and the ability to generate alternative design options before selecting one solution based on evaluation criteria. The learning activities are designed with both cognitive and metacognitive scaffolds to aid learners in not just solving the problem but to monitor and improve their skills. The cognitive scaffolds include: prompts, hints, case-study, study material, drawing tools to aid visualization, workspace to record learners' responses, and metacognitive scaffolds include: system-evaluated feedback.

We conducted research studies with undergraduate engineering students ($N = 50$) to evaluate the effectiveness of the Fathom in learning these skills. The methodology used is a pretest-intervention-posttest research design. The scores show significant gain from pretest to posttest in quality of problem formulation ($p = 0.05$, effect size = 0.66), solution quality ($p = 0.00$, effect size = 1.23) and justification ($p = 0.01$, effect size = 1.24).

However, the scores were not helping in providing insights into the interaction behavior of the high and low-performing students. To investigate the relationship between learning performance and the use of strategies by low and high performers, it became important to examine how these activities came together as larger behavior patterns and strategies. The aim of this paper is to discuss the process of building a student model for high and low-performing groups of learners using HMM and analyze their interaction behavior.

2 Related Work in Learner Modeling

A major area of educational data mining research [4, 5, 7, 8] is done to analyze MOOC and learning management system log data to identify patterns of learning behavior that can provide insights into educational practice. Research in educational data mining [2, 6] is done towards building student models and analyzing student behaviors in various interactive learning environments to predict student learning behaviors. For instance, hidden Markov models (HMM) were used to model school students' behavior based on the trace data generated from Betty's brain system which used the pedagogy of learning by teaching [2]. In a later study, Jeong et al. (2010) applied the same HMM approach to study the learning behavior of adult professionals in an asynchronous online learning environment. In particular, their exploratory study was aimed at identifying the main phases of the students' learning process in the examined course, and investigating the differences between high and low-performing students in terms of their transitions through the identified phases of the course.

We propose to use HMM similar to the work proposed by Jeong (2008) to investigate how engineering students interact with learning environments designed for complex problem solving and analyze student behaviors to get insights into how the learning environment facilitates learning of complex problem solving among high and low performers.

3 Methodology for Obtaining Behavior Model

To achieve our objective of modeling student behavior from log data, we chose a probabilistic model that could mathematically and diagrammatically describe the chain of activities followed by the students. Our approach involves four steps which are explained in greater detail in this section.

3.1 Log Data Collection and Processing

The log data was collected from Fathom in the form of triplet <learner_id, timestamp clicked_button>. The sample log data collected is as shown in Fig. 1.

```
115a1086 Time August 29, 2018, 10:32 am Id understandbutton
115a1086 Time August 29, 2018, 10:33 am Id showproblem
115a1086 Time August 29, 2018, 10:33 am Id understandhintButton
115A1085 Time August 29, 2018, 10:33 am Id SaveButton
115A1085 Time August 29, 2018, 10:33 am Id showproblem
115A1083 Time August 29, 2018, 10:33 am Id showproblem
115a1089 Time August 29, 2018, 10:34 am Id SaveButton_goal
115a1089 Time August 29, 2018, 10:34 am Id next
```

Fig. 1. Raw log data collected in Fathom

The log data in its raw form is very difficult to comprehend and needs to be processed before we can perform any operations on it. The log files consist of all the activities carried out by the students in the form of button clicks, edits made in the drawing tools and the text fields, access to hints, examples, etc.

We have removed the repetitive sequences that occur as a result of clicking the same button multiple times. This helps to reduce the length of the sequences and at the same time focus more on the transitions between states. The other dataset we worked on was the score sheet of post-test to identify low and high scorers. The students scoring low (score < 2) in quality of problem formulation and solution were categorized as low scorers and others were categorized as high scorers. Out of 50 students, 5 students did not complete the posttest, hence we considered only 45 students, out of which 13 students were categorized as low scorers and 32 as high scorers. The log sequences were then assigned to each student and two separate input dataset was created as input for the hidden Markov model.

3.2 Parsing the Log Files

In this study, we derive learners' behavior patterns by analyzing the sequence of their interactions with the system. To simplify the interpretation task we mapped learners' actions in each activity into one aggregate activity. For example, all the edits made in the understand_problem activity, like drawing the diagrams and saving in the first attempt as UP, accessing resources (hints, notes, examples, etc.) as RA, and then redoing after saving the responses as REDO, etc. All student activities were expressed as the six activities summarized in Table 1.

Table 1. Student activities and related actions

Activity	Student actions
UP	Saving the diagram drawn in the understand_problem activity
FG	Saving the formulated goals in the formulate_problem activity
GS	Saving the solutions generated in the generate_solutions activity
EV	Saving the evaluation of solutions
RA	Accessing resources in the form of hints, examples, notes, etc.
REDO	Modifying the responses in multiple attempts

Examples of the resultant sequences of a student is shown in Fig. 2.

"115A1086": ["UP", "FG", "UP", "FG", "UP", "FG", "FG", "GS", "REDO", "EV", "GS", "FG", "RA", "GS", "REDO", "EV", "REDO", "EV", "UP", "FG", "GS"],

Fig. 2. Parsed data of a student

3.3 Constructing the HMMs

The first step in interpreting this behavior data was to build hidden Markov models from the sequence of observable events. A hidden Markov model is characterized by three sets of parameters: initial probability vector π, state transition probability matrix, A, and output probability matrix, B [10].

The difficult part of the modeling process is to determine the optimal set of parameters and the size of the model (number of states) that maximizes the likelihood of the input sequences. Jeong (2010) compared two common iterative convergence optimization schemes, the Baum-Welch and the segmental K-Means algorithms to achieve the optimal model parameters, which include (π, A, B) and the number of states in the model. The results showed that the optimal number of states is six using both Baum-Welch and the segmental K-Means algorithms. We used the Viterbi algorithm for sequential decoding and calculating transition probabilities between states.

The parsed activity sequences of two groups-low and high performers were used to derive two sets of hidden Markov models as shown in Fig. 3 and Fig. 4.

Each model is made up of a set of states, the activity patterns (the output probability) associated with each state, and the transition probabilities between states. The transition probability associated with a link between two states indicates the likelihood of the student transitioning from the current state to the indicated state. We investigate further by interpreting these models in terms of the cognitive and metacognitive learning behaviors of the students.

4 Analysis of HMM Patterns

The analysis of transition shows certain patterns in both low and high performers. The likelihood percentage of high scorers transitioning to REDO state in each activity is

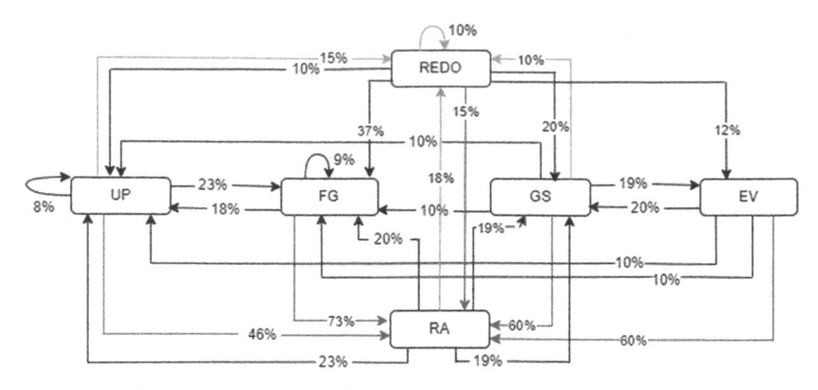

Fig. 3. HMM model of low performers (Color figure online)

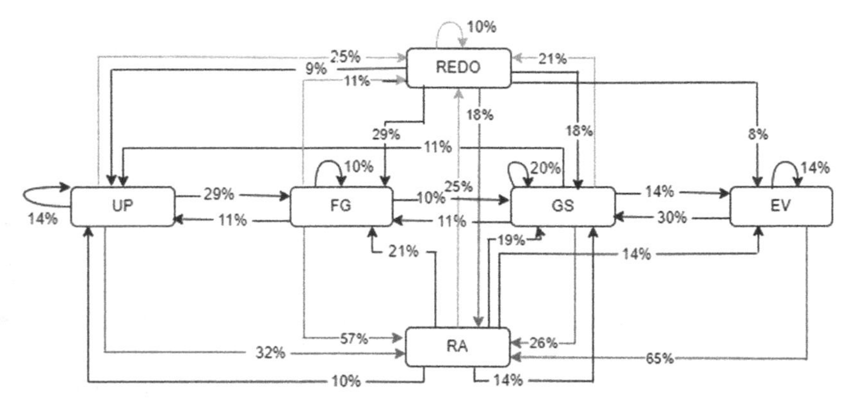

Fig. 4. HMM model of high performers (Color figure online)

more compared to low scorers, as seen in green color lines in Fig. 3 and 4. The high scorers' transitions from UP to REDO state with 25% likelihood compared to 15% in low scorers. This shows that high scorers were more responsive to the feedback given by the system and went back to the same activity to improve their responses.

The likelihood percentage of low scorers transitioning to RA state is higher than high scorers, shown in red color lines in Figs. 3 and 4. For example, the low scorers transitioned from UP to RA state with 46% likelihood compared to 32% in high scorers. The low scorers were accessing resources like hints, videos, and learning material more often which indicates that they had difficulty in comprehending the activity.

We find that the students in the low scorer group tend to stay mainly in the cognitive task of doing activities in UP, FG, GS, CR, and EV states, while the high scorer students tend to transition to the higher-level states such as REDO state. High scorers tend to transition between doing and redoing the task, and occasionally referring to the help provided by the system, and thus exhibit metacognitive behavior. While low scorers tend to be in a cognitive state of doing the activity and are less likely to monitor and reflect on their skills. The resource usage rate is high and REDO is low in low scores which indicate that low scorers have difficulty doing or comprehending the activities compared to high scorers.

Overall analysis shows that high scorers show more metacognitive behaviors while low scorers exhibit more help-seeking behaviors. This analysis is useful to predict student behaviors based on their interaction patterns in the learning environment and provide timely help to low performers.

5 Conclusion

In this paper, we discussed the process of creating student models representing learning patterns of high and low performers in learning ill-structured problem-solving skills in the technology-enhanced learning environment, named Fathom. The model was built using the hidden Markov model (HMM) using the log data generated in Fathom. The analysis shows that high scorers exhibit metacognitive behaviors in terms of the ability to do the activity, and monitor and reflect on their skills. While, low scorers tend to rely more on the resources given in the system and exhibit more help-seeking behavior, which implies that they have difficulty comprehending and doing the activity.

References

1. Tang, A., Aleti, A., Burge, J., van Vliet, H.: What makes software design effective? Des. Stud. **31**(6), 614–640 (2010)
2. Jeong, H., Biswas, G.: Mining student behavior models in learning-by-teaching environments. In: Educational Data Mining (2008)
3. Reddy, D., Iyer, S., Sasikumar, M.: Technology enhanced learning (TEL) environment to develop expansionist-reductionist (ER) thinking skills through software design problem solving. In: 2018 IEEE 10th International Conference on Technology for Education (T4E), pp. 166–173 (2018)
4. Mavrikis, M.: Data-driven modelling of students' interactions in an ILE. In: Educational Data Mining (2008)
5. Kovanović, V., Gašević, D., Joksimović, S., Hatala, M., Adesope, O.: Analytics of communities of inquiry: effects of learning technology use on cognitive presence in asynchronous online discussions. Internet High. Educ. **27**, 74–89 (2015)
6. Jeong, H., Biswas, G., Johnson, J., Howard, L.: Analysis of productive learning behaviors in a structured inquiry cycle using hidden Markov models. In: Educational Data Mining (2010)
7. Arpasat, P., Premchaiswadi, N., Porouhan, P., Premchaiswadi, W.: Applying process mining to analyze the behavior of learners in online courses. Int. J. Inf. Educ. Technol. **11**(10), 436–443 (2021)
8. Baker, R.S., Yacef, K.: The state of educational data mining in 2009: a review and future visions. J. Educ. Data Min. **1**(1), 3–17 (2009)
9. Jonassen, D., Strobel, J., Lee, C.B.: Everyday problem solving in engineering: lessons for engineering educators. J. Eng. Educ. **95**(2), 139–151 (2006)
10. Rabiner, L., Juang, B.: An introduction to hidden Markov models. In: IEEE ASSP Mag. **3**(1), 4–16 (1986)
11. Adelson, B., Soloway, E.: The role of domain experience in software design. IEEE Trans. Softw. Eng. **11**, 1351–1360 (1985)
12. Guindon, R.: Knowledge exploited by experts during software system design. Int. J. Man Mach. Stud. **33**(3), 279–304 (1990)

Mixing Backward- with Forward-Chaining for Metacognitive Skill Acquisition and Transfer

Mark Abdelshiheed(✉), John Wesley Hostetter, Xi Yang, Tiffany Barnes, and Min Chi

North Carolina State University, Raleigh, NC 27695, USA
{mnabdels,jwhostet,yxi2,tmbarnes,mchi}@ncsu.edu

Abstract. Metacognitive skills have been commonly associated with preparation for future learning in deductive domains. Many researchers have regarded *strategy-* and *time-awareness* as two metacognitive skills that address *how* and *when* to use a problem-solving strategy, respectively. It was shown that students who are both strategy- and time-aware (*StrTime*) outperformed their *nonStrTime* peers across deductive domains. In this work, students were trained on a logic tutor that supports a default forward-chaining (FC) and a backward-chaining (BC) strategy. We investigated the impact of mixing BC with FC on teaching strategy- and time-awareness for *nonStrTime* students. During the logic instruction, the experimental students (*Exp*) were provided with two BC worked examples and some problems in BC to practice *how* and *when* to use BC. Meanwhile, their control (*Ctrl*) and *StrTime* peers received no such intervention. Six weeks later, all students went through a probability tutor that only supports BC to evaluate whether the acquired metacognitive skills are transferred from logic. Our results show that on both tutors, *Exp* outperformed *Ctrl* and caught up with *StrTime*.

Keywords: Strategy awareness · Time awareness · Metacognitive skill instruction · Preparation for future learning · Backward chaining

1 Introduction

One fundamental goal of education is being prepared for future learning [6] by transferring acquired skills and problem-solving strategies across different domains. Despite the difficulty of achieving such transfer [6], prior research has shown it can be facilitated by obtaining metacognitive skills [1–3,8]. It has been believed that metacognitive skills are essential for academic achievements [5], and teaching such skills impacts learning outcomes [8] and strategy use [13]. Much prior research has categorized knowing *how* and *when* to use a problem-solving strategy as two metacognitive skills [15], referred to as strategy- and time-awareness, respectively. Our prior work found that students who were both strategy- and time-aware—referred to as *StrTime*—outperformed their

M. M. Rodrigo et al. (Eds.): AIED 2022, LNCS 13355, pp. 546–552, 2022.
https://doi.org/10.1007/978-3-031-11644-5_47

nonStrTime peers across deductive domains [1,2]. In the current work, we provide interventions for the latter students to catch up with their *StrTime* peers. Deductive domains such as logic, physics and probability usually require multiple problem-solving strategies. Two common strategies in these domains are forward-chaining (FC) and backward-chaining (BC). Early studies showed that experts often use a mixture of FC and BC to execute their strategies [12]. This work investigates the impact of mixing FC and BC on teaching strategy- and time-awareness for *nonStrTime* students.

Our study involved two intelligent tutoring systems (ITSs): logic and probability. Students were first assigned to a logic tutor that supports FC and BC, with FC being the default, then to a probability tutor six weeks later that only supports BC. During the logic instruction, *nonStrTime* students were split into experimental (*Exp*) and control (*Ctrl*) conditions. For *Exp*, the tutor provided two worked examples solved in BC and presented some problems in BC to practice *how* and *when* to use BC. *Ctrl* received no such intervention as each problem was presented in FC by default with the ability to switch to BC. Our goal is to inspect whether our intervention would make *Exp* catch up with the golden standard —*StrTime* students— who already have the two metacognitive skills and thus need no intervention. All students went through the same probability tutor to evaluate whether the acquired metacognitive skills are transferred from logic. Our results show that *Exp* outperformed *Ctrl* and caught up with *StrTime* on both tutors.

1.1 Metacognitive Skill Instruction

Metacognitive skills regulate one's awareness and control of their cognition [7]. Many studies have demonstrated the significance of metacognitive skills instruction on academic performance [5], learning outcomes [2,3,8] and regulating strategy use [13]. Schraw and Gutierrez [13] argued that metacognitive skill instruction involves feeling what is known and not known about a task. They stated that such instruction should further compare strategies according to their feasibility and familiarity from the learner's perspective. Chi and VanLehn [8] found that teaching students principle-emphasis skills closed the gap between high and low learners, not only in the domain where they were taught (probability) but also in a second domain where they were not taught (physics).

Strategy- and time-awareness have been considered metacognitive skills as they respectively address *how* and *when* to use a problem-solving strategy [5,15]. Researchers have emphasized the role of strategy awareness in preparation for future learning [2,4] and the impact of time awareness on planning skills and academic performance [5,9]. Belenky and Nokes [4] showed that students who had a higher aim to master presented materials and strategies outperformed their peers on a transfer task. Fazio et al. [9] revealed that students who knew when to use each strategy to pick the largest fraction magnitude had higher mathematical proficiency than their peers. de Boer et al. [5] showed that students who knew when and why to use a given strategy exhibit long-term metacognitive knowledge that improves their academic performance. de Boer et al. emphasized that knowing *when* and *why* has the same importance as knowing *how* when it comes to strategy choice in multi-strategy domains.

1.2 Forward- and Backward-Chaining

FC and BC are two standard problem-solving strategies in deductive domains. In FC, the reasoning proceeds from the given propositions toward the target goal, whereas BC is goal-driven in that it works backward from a goal state to a given state. Substantial work has investigated the impact of FC and BC strategies in two research categories: empirical studies and post-hoc observations.

Prior empirical studies have shown the significance of FC over BC in learning physics [10] and weightlifting movements [11]. Moore and Quintero [11] compared FC and BC in teaching the clean and snatch movements to novice weight lifters. The participants showed mastery performance with the FC training but showed substantially fewer improvements in performance accuracy via the BC training. All participants mastered the movements when some BC lifts were changed to FC. Conversely, some studies reported no significant difference between the two strategies [14]. Slocum and Tiger [14] assessed the children's FC and BC strategy preferences on various learning tasks. They found that children were equally efficient on both strategies and had similar mixed strategy preferences.

Early research has observed the impact of mixing FC and BC strategies [12]. Priest and Lindsay [12] compared how experts and novices solve physics problems. Although both groups used a mixture of FC and BC, *only* the experts knew how and when to use each strategy and significantly produced more complete plans and stages than their novice peers. In brief, while no consensus has been reached on whether FC or BC is most effective in problem-solving, prior work has observed that the mixture of FC and BC yields the highest performance accuracy as learners know how and when to use each strategy.

2 Methods

Participants. They are Computer Science undergraduates at North Carolina State University. Students were assigned each tutor as a class assignment and told that completion is required for full credit. Similar to our prior work, we utilize the random forest classifier (RFC) that, based on pre-test performance, predicts the metacognitive label ($StrTime$ or otherwise) before training on logic and was previously shown to be 96% accurate [2]. Specifically, $StrTime$ students frequently follow the desired behavior of switching *early* (within the first 30 actions) to *BC*, while their peers either frequently switch late (after the first 30 actions) or stick to the default *FC* [1–3]. A total of 121 students finished both tutors and were classified by the RFC into 26 $StrTime$ and 95 otherwise. The latter students were randomly assigned to *Experimental* (*Exp*: N = 49) and *Control* (*Ctrl*: N = 46) conditions. The RFC was 97% accurate in classifying students who received no intervention—*Ctrl* and $StrTime$.

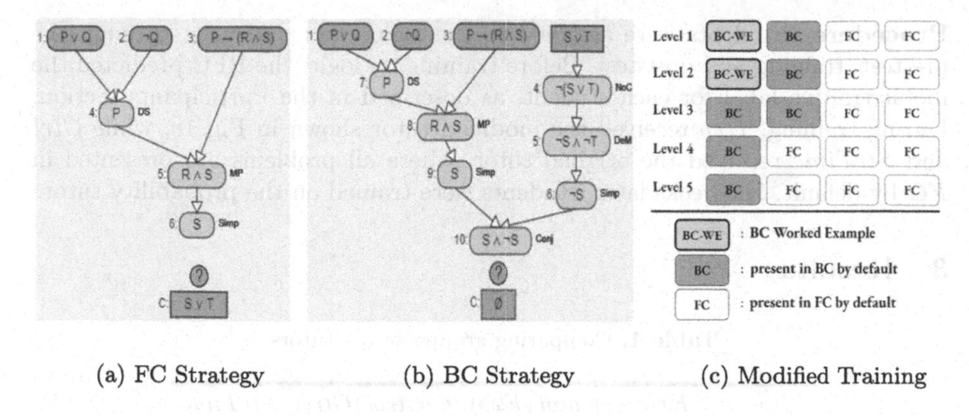

(a) FC Strategy (b) BC Strategy (c) Modified Training

Fig. 1. Logic tutor

Logic Tutor and Our Intervention. The logic tutor teaches propositional logic proofs by applying inference rules such as Modus Ponens. A student can solve any problem by either a **FC** or **BC** strategy. Students derive a conclusion at the bottom from givens at the top in *FC* (Fig. 1a), while they derive a contradiction from givens and the *negation* of the conclusion in BC (Fig. 1b). A problem is presented by *default* in FC with the ability to switch to BC by clicking a button. The tutor consists of two pre-test, 20 training and six post-test problems. The post-test is *much harder* than the pre-test, and the first two post-test problems are isomorphic to the two pre-test problems. The *pre-* and *post-test* scores are calculated by averaging the pre- and post-test problem scores, where a problem score is a function of time, accuracy, and solution length. The training consists of five ordered levels in an *incremental degree of difficulty*, and each level consists of four problems. We modified the training section to mix BC with FC (Fig. 1c). Specifically, two worked examples (WE) on BC were implemented, where the tutor provided a step-by-step solution, and six problems were presented in BC by default. The two WEs and the six problems are expected to teach students *how* and *when* to use BC. Note that the colored problems in Fig. 1c were selected based on the historical strategy switches in our data [1].

Probability Tutor. It teaches how to solve probability problems using ten principles, such as the Complement Theorem. The tutor consists of a textbook, pre-test, training, and post-test. The textbook introduces the domain principles, while training consists of 12 problems, each of which can *only* be solved by *BC* as it requires deriving an answer by *writing and solving equations* until the target is ultimately reduced to the givens. In pre- and post-test, students solve 14 and 20 open-ended problems graded by experienced graders in a double-blind manner using a partial-credit rubric. The *pre-* and *post-test* scores are the average grades in their respective sections, where grades are based *only* on accuracy. Like the logic tutor, the post-test is much harder than the pre-test, and each pre-test problem has a corresponding isomorphic post-test problem.

Procedure. Students were assigned to the logic tutor and went through the pre-test, training and post-test. Before training on logic, the RFC predicted the metacognitive label for each student, as described in the Participants section. During training, *Exp* received the modified tutor shown in Fig. 1c, while *Ctrl* and *StrTime* received the original tutor, where all problems are presented in *FC* by default. Six weeks later, students were trained on the probability tutor.

3 Results

Table 1. Comparing groups across tutors

	Experimental (*Exp*) ($N = 49$)	*Control* (*Ctrl*) ($N = 46$)	*StrTime* ($N = 26$)
Logic tutor			
Pre	61.7 (18)	58.7 (20)	62.1 (20)
Iso-Post	81 (11)	70.4 (14)	81.3 (10)
Iso-NLG	0.27 (.12)	0.09 (.31)	0.29 (.16)
Post	77.4 (11)	66.7 (14)	79 (9)
NLG	0.24 (.15)	0.06 (.37)	0.25 (.18)
Probability tutor			
Pre	74.8 (14)	74.2 (16)	75.8 (15)
Iso-Post	90.4 (10)	65.3 (16)	90.6 (8)
Iso-NLG	0.29 (.19)	−0.02 (.27)	0.26 (.17)
Post	89.5 (15)	62.5 (18)	88.8 (7)
NLG	0.26 (.21)	−0.08 (.3)	0.24 (.15)

Table 1 compares the groups' performance across the two tutors showing the mean and standard deviation of pre- and post-test scores, isomorphic scores, and the learning outcome in terms of the normalized learning gain (*NLG*) defined as ($NLG = \frac{Post-Pre}{\sqrt{100-Pre}}$), where 100 is the maximum test score. We refer to pre-test, post-test and NLG scores as *Pre*, *Post* and *NLG*, respectively. On both tutors, a one-way ANOVA found no significant difference on *Pre* between the groups.

To measure the improvement on isomorphic problems, repeated measures ANOVA tests were conducted using {*Pre*, *Iso-Post*} as factor. Results showed that *Exp* and *StrTime* learned significantly with $p < 0.0001$ on both tutors, while *Ctrl* did not perform significantly higher on *Iso-Post* than *Pre* on both tutors. These findings verify the RFC's accuracy, as *StrTime* learned significantly on both tutors, while *Ctrl* did not, despite both receiving no intervention.

A comprehensive comparison between the three groups was essential to evaluate our intervention. On the logic tutor, A one-way ANCOVA using *Pre* as covariate and group as factor found a significant effect on *Post*: $F(2, 117) = 14.5, p < .0001, \eta^2 = .18$. Subsequent post-hoc analyses with Bonferroni correction ($\alpha = .05/3$) revealed that *Exp* and *StrTime* significantly outperformed

Ctrl: $t(93) = 3.8, p < .001$ and $t(70) = 3.9, p < .001$, respectively. Similar patterns were observed on *NLG* using ANOVA and the post-hoc comparisons.

On the probability tutor, a one-way ANCOVA using *Pre* as covariate and group as factor showed a significant effect on *Post*: $F(2, 117) = 48.1, p < .0001$, $\eta^2 = .35$. Follow-up pairwise comparisons with Bonferroni adjustment showed that *Exp* and *StrTime* significantly surpassed *Ctrl*: $t(93) = 6.1, p < .0001$ and $t(70) = 5.9, p < .0001$, respectively. Similar results were found on *NLG* using ANOVA and the post-hoc comparisons.

4 Conclusion

We showed that mixing BC with FC on the logic tutor improved the experimental students' learning outcomes, as *Exp* significantly outperformed *Ctrl* on logic and on a probability tutor that only supports BC. Additionally, *Exp* caught up with *StrTime* on both tutors suggesting that *Exp* students are prepared for future learning [6] as they acquired BC mastery skills on logic and transferred them to probability, where they received no intervention. There is at least one caveat in our study. The probability tutor supported only one strategy. A more convincing testbed would be having the tutors support both strategies. The future work involves implementing FC on the probability tutor.

Acknowledgments. This research was supported by the NSF Grants: 1660878, 1651909, 1726550 and 2013502.

References

1. Abdelshiheed, M., et al.: Metacognition and motivation: the role of time-awareness in preparation for future learning. In: CogSci, vol. 42 (2020)
2. Abdelshiheed, M., et al.: Preparing unprepared students for future learning. In: CogSci, vol. 43 (2021)
3. Abdelshiheed, M., et al.: The power of nudging: exploring three interventions for metacognitive skills instruction across intelligent tutoring systems. In: CogSci, vol. 44 (2022)
4. Belenky, D.M., Nokes, T.J.: Motivation and transfer: the role of mastery-approach goals in preparation for future learning. J. Learn. Sci. **21**(3), 399–432 (2012)
5. de Boer, H., et al.: Long-term effects of metacognitive strategy instruction on student academic performance: a meta-analysis. Educ. Res. Rev. **24**, 98–115 (2018)
6. Bransford, J.D., Schwartz, D.L.: Rethinking transfer: a simple proposal with multiple implications. Rev. Res. Educ. **24**(1), 61–100 (1999)
7. Chambres, P., et al.: Metacognition: Process, Function, and Use. Kluwer Academic Publishers, Amsterdam (2002)
8. Chi, M., VanLehn, K.: Meta-cognitive strategy instruction in intelligent tutoring systems: how, when, and why. J. Educ. Technol. Soc. **13**(1), 25–39 (2010)
9. Fazio, L.K., et al.: Strategy use and strategy choice in fraction magnitude comparison. J. Exp. Psychol. Learn. Mem. Cogn. **42**(1), 1 (2016)
10. Larkin, J., et al.: Expert and novice performance in solving physics problems. Science **208**, 1335–1342 (1980)

11. Moore, J.W., Quintero, L.M.: Comparing forward and backward chaining in teaching Olympic weightlifting. J. Appl. Behav. Anal. **52**(1), 50–59 (2019)
12. Priest, A., Lindsay, R.: New light on novice-expert differences in physics problem solving. Br. J. Psychol. **83**(3), 389–405 (1992)
13. Schraw, G., Gutierrez, A.P.: Metacognitive strategy instruction that highlights the role of monitoring and control processes. In: Peña-Ayala, A. (ed.) Metacognition: Fundaments, Applications, and Trends. ISRL, vol. 76, pp. 3–16. Springer, Cham (2015). https://doi.org/10.1007/978-3-319-11062-2_1
14. Slocum, S.K., Tiger, J.H.: An assessment of the efficiency of and child preference for forward and backward chaining. J. Appl. Behav. Anal. **44**(4), 793–805 (2011)
15. Winne, P.H., Azevedo, R.: Metacognition. In: Sawyer, R.K. (ed.) The Cambridge Handbook of the Learning Sciences. Cambridge Handbooks in Psychology, 2 edn. Cambridge University Press (2014)

The Impact of Conversational Agents' Language on Self-efficacy and Summary Writing

Haiying Li[1](✉), Fanshuo Cheng[2], Grace Wang[3], and Art Graeser[4]

[1] Iowa College Aid, Des Moines, IA 50309, USA
haiying.li@iowa.gov
[2] Iowa City West High School, Iowa, IA 52246, USA
fache23@icstudents.org
[3] Liberty High School, North Liberty, IA 52317, USA
grwan23@icstudents.org
[4] University of Memphis, Memphis, TN 38152, USA
graesser@memphis.edu

Abstract. This study investigated the impact of conversational agent formality on summary writing and self-efficacy in a conversation-based intelligent tutoring system. Conversational agents guided learners to learn summarization strategies in one of three conditions: a formal language, an informal language, and a mixed language condition. Results showed no significant difference in summary writing gains between groups, but learners in the informal language group achieved higher self-efficacy gains than learners in the formal language group when controlling for demographic attributes, years of English learning, prior perception of summary writing, and prior reading and summary writing proficiency. Results also indicated a negative association between self-efficacy gains and summary writing gains with a marginal significance. Implications are discussed for the design of conversational agents in the ITS.

Keywords: Conversational agent · Summary writing · Self-efficacy

1 Introduction

Research on agent language in the intelligent tutoring system (ITS) investigates which language style elicits learning more than another, conversational language or formal language. Two primary designs of agent language are prevalent: personalized [11] and multi-level language principles [5]. The personalization principle adopts personal pronouns to differentiate conversational language from formal language. The former is represented by first- and second-person pronouns (e.g., *I*, *your*) whereas the latter is represented by third-person pronouns and impersonal articles (e.g., *he*, *their*). Personalized language directly addresses learners and creates a social partnership between the learner and instructor to motivate learners to learn [11]. The positive effect of personalized language on learning was consistently found among 14 out of 17 experimental tests [11] and from a meta-study across varied science topics [3], particularly for students with low

M. M. Rodrigo et al. (Eds.): AIED 2022, LNCS 13355, pp. 553–559, 2022.
https://doi.org/10.1007/978-3-031-11644-5_48

prior knowledge or low achievement and with short lessons up to 35 min. Other studies found positive personalization effects either on retention [10, 12] or transfer tests [13], but not on both. These inconsistent findings are likely due to different learning environments (research lab vs. MOOC), learners (college students vs. high school students), instructional languages (German vs. Chinese), etc.

The multi-level principle utilized multiple levels of language and discourse features such as word, syntax, referential cohesion, deep cohesion, and genre to distinguish conversational language from formal language [5, 7–9]. This principle defines conversational language as spontaneous, less organized, and more disjointed language, dependent on the contexts and common ground shared by the speaker and the listener, either in a spoken or written format. Formal language is defined as pre-planned, well-organized, and coherent language, used for academic communication with comparatively low reliance. Conversational language is more informal and easier to process with more concrete words and simple sentence structures, but fewer connectives and overlapping words and ideas in a narrative style. Empirical evidence in AutoTutor ARC (Adult Reading Comprehension) found no significant effect of agent language on summary writing, learners' use of language, or engagement [7–9]. Specifically, learners improved performance on summary writing, but this difference was not affected by agent language style [9].

Previous studies found a positive correlation between self-efficacy and academic success in reading and writing [14]. In ITS, social-oriented agents that spoke in encouraging language improved learners' self-efficacy [1]. Formal agents (task-oriented style), however, elicited greater self-efficacy among low-competency older users [2]. AutoTutor ARC models learners' cognitive states, designs a low-competence peer agent, and provides negative feedback to the peer agent if the human learner gives an incorrect answer [6–9]. The present study proposes that informal agents could build learners' self-confidence and are expected to yield greater improvements in self-efficacy than formal agents. This study aims to answer two research questions: (1) Does agent language affect learners' self-efficacy and summary writing performance? and (2) Is learners' self-efficacy associated with their summary writing performance?

2 Method

Conversations of the tutor agent, Cristina, and the peer agent, Jordan, were designed following the expectation and misconception-tailored (EMT) dialogue mechanism and a five-step tutoring frame: main question → answers → short feedback → multiple dialogue moves to reach the expectation → wrap-up [4]. To boost the learner's self-confidence, Jordan was designed with lower performance than the learner and received negative feedback if both learner and Jordan gave incorrect answers [6]. Human discourse experts generated agents' formal and informal conversations at the levels of word (e.g., *clarification* vs. *show*), syntax (e.g., subordinate clauses vs. simple subject-verb), referential cohesion (e.g., repeating content words vs. pronouns), deep cohesion (e.g., *additionally* vs. non-connectives), and genre (e.g., third vs. first- and second-person pronouns) [9] (see Table 1). Three conditions were formed: (1) a formal condition where both Cristina and Jordan spoke formally, (2) an informal condition where both agents spoke informally, and (3) a mixed language condition where Cristina spoke formally and Jordan spoke informally.

Table 1. Excerpts of formal and informal discourse.

Formal	Informal
Summarizing the consequences requires more general and important information rather than specific information. Muscle and energy can be categorized into consequences related to physical health	We should find more general and important information. This answer talks about physical health

The participants were 177 adult learners recruited from Amazon Mechanical Turk (AMT) with $30 compensation for a three-hour intervention (51.1% male; Age: $M = 33.55$, $SD = 8.69$; 82.6% English language learners) [7–9]. Participants were randomly assigned into one of three conditions and took a pre-survey/pretest, an intervention, and a post-survey/posttest. Pre-survey includes personal information (age, gender, education, country of birth, English learner, years of English learning, and years in a foreign country), perception of summary writing, perception of self-efficacy, and a pretest for reading comprehension and summary writing.

Prior reading comprehension on pretest was measured by ten-item three passages from a Test of Adult Basic Education (words: $M = 241.67$, $SD = 102.53$; FKGL: $M = 7.79$, $SD = 2.22$). Perception of summary writing in pre-survey was measured by six questions with 1–6 Likert scale from Never to Always. We measured summarization self-efficacy with the same 11 items [Bandura, 1997] on both pretest and posttest, with 1–6 Likert scale from *Strongly Disagree* to *Strongly Agree.* The questions involved the perception of capabilities for signal words, comprehension, and summary writing. Cronbach's alpha tests showed high reliability of questionnaire items, $\alpha = .935$ for pretest self-efficacy items, $\alpha = .926$ for posttest self-efficacy items, $\alpha = .832$ for reading. All items were reliable and kept in the analyses.

During a one-hour summarization intervention, agents interactively presented a mini-lecture on the function of signal words in comparison (e.g., *similarly*) and causation (e.g., *because*) texts with a text map to facilitate a better understanding of information in texts. Agents used four informational texts to evaluate whether participants could identify topic sentences, main ideas, and important and minor information through five MC questions. Participants wrote a summary for the text, evaluated their own summary, and then evaluated a peer's summary. Participants received personalized feedback and scaffolding only on MC and peer-rating questions.

On pretest and posttest, participants wrote a summary for each text they read with the requirement to state the main ideas with a topic sentence and specify important supporting information. They were required to use their own words and apply the appropriate signal words to explicitly express their ideas. Four English native speakers were trained to grade summaries and demonstrated satisfactory interrater reliabilities (Cronbach $\alpha = .82$) according to four-element criteria, each with 0–2 points: the presence of the topic sentence, inclusion of the important supporting information and exclusion of unimportant information, the presence of signal words of the text structures, and grammar and mechanics.

3 Results and Discussion

To answer the first part of the first question, a One-way ANCOVA was conducted to determine a statistically significant difference in self-efficacy gains (posttest self-efficacy – pretest self-efficacy) between formal, mixed, and informal language that agents spoke when controlling for gender, country of birth, years of English learning, prior reading proficiency, prior summarization perception, and prior summary writing proficiency. Results revealed a significant model, $F(8, 168) = 8.42, p < .001, R^2 = .286$ (see Table 2). Participants in the informal condition achieved significantly higher self-efficacy gains in the informal condition than those in the formal condition with a small effect size: $t_{(168)} = 2.44, p = .047$, Cohen's $d = 0.37$. This indicated that conversational language elicited higher self-efficacy than formal language, which is consistent with the previous findings that social-oriented agents improved learners' self-efficacy [1]. Conversational language is considered a social cue that primes a sense of social partnership between the computer agent and human learner and accordingly motivates the learner to make greater endeavors to understand the instruction and improve learning performance. Moreover, agents' conversational language is affiliated with an everyday oral conversation with the familiar words, sentences, and texts that learners easily understand [7–9]. This might cause participants to perceive that they are able to successfully complete summarization tasks. A sense of socialization and ease of information processing likely make learners

Table 2. ANCOVA for self-efficacy gains (*** p < .001, ** p < .01, * p < .05, † p < .10).

Source	*df*	*SS*	*MS*	*F*	*β*	*SE*	*t*	*M (SD)*
Intercept					1.95	0.50	3.92***	
Condition (base = Formal)	2	3.86	1.93	3.74*				0.25 (0.74)
Informal					0.33	0.14	2.44*	0.57 (0.94)
Mixed					0.09	0.14	0.66	0.26 (0.77)
Female (base = male)	1	4.84	4.84	9.38**	–0.21	0.11	–1.87†	
India (base = other)	1	1.62	1.62	3.15†	0.21	0.13	1.59	
Years of English learning	1	4.37	4.37	8.47**	–0.01	0.01	–2.20*	17.79
Prior reading comprehension	1	3.47	3.47	6.72*	–1.15	0.31	–3.71***	0.75
Prior summarization perception	1	14.71	14.71	28.54***	–0.27	0.05	–5.16***	4.22
Prior summary writing	1	1.88	1.89	3.66†	0.16	0.08	1.91†	4.11
Residuals	168	86.61	0.52					

perceive that they have learned about the summarization strategy from the agents and would write a better summary than before the intervention.

To answer the second part of the first question, a One-way ANCOVA test was performed to determine the difference in summary writing gains (posttest summary writing – pretest summary writing) between the formal, mixed, and informal conditions. Results did not show a significant difference. A trend, however, was found that informal agents elicited more learning gains than formal agents, which elicited more gains than mixed agents. These differences, however, were not statistically significant: $F(2, 164) = 0.175$, $p = .840$, Informal: $M = 0.24$, $SD = 0.67$; Formal: $M = 0.22$, $SD = 0.69$; Mixed: $M = 0.17$, $SD = 0.67$. These findings are inconsistent with previous findings that conversational language promotes learning outcomes [3, 11]. The inconsistency is likely due to the long duration of the intervention, the more challenging subject matter, and the different measures of agent language.

To answer the second question, a simple regression was performed with summary writing gains as a dependent variable and self-efficacy gains as an independent variable. Results showed a marginal significant relationship: $F\ (1, 175) = 3.01$, $p = .085$, $R^2 = .017$, $\beta = -0.11$, $SE = 0.06$. Adding the agent language condition and covariates did not improve the model performance. This means that the more self-efficacy gains learners had, the fewer summary writing gains they achieved. This finding is inconsistent with previous findings that self-efficacy had a positive correlation with reading and writing learning outcomes [14]. To identify what caused inconsistent findings, we examined the prior self-efficacy perception and found that 29% ($N = 51$) of learners had self-efficacy perception scores with more than 5, 50% ($N = 88$) 4, 20% ($N = 36$) 3, and only 1% ($N = 2$) less than 3. These findings imply that for learners with a high prior self-efficacy ($M = 4.70$, $SD = 0.82$), it is not easy to improve their self-efficacy for challenging learning tasks. Another explanation is that participants might conceive that they understand the agents' lecture, and therefore they believe they would be capable of writing a better-quality summary after the intervention, so their self-efficacy gains increased. However, their perception of self-efficacy might go beyond their actual capability when the tasks are extremely challenging.

4 Implications and Future Work

These findings provide implications for building conversation-based ITS by designing a conversational language for computer agents to increase learners' self-efficacy. Multiple levels of language and discourse components could be considered. For instance, using the first- and second-person pronouns increases social relationships. Adopting the narrative, everyday oral language creates an easy, familiar conversational atmosphere. Explicitly delivering ideas and concepts enhances information processing. Another implication is that agents need to provide immediate, real-time feedback and scaffolding on learning performance so that learners could have a more accurate perception of their capabilities for the learning tasks, especially for more challenging tasks. The lack of instant feedback on learners' performance of the task is likely to have learners perceive that they have mastered the skills and knowledge and they are capable of successfully completing the task.

Limitations of this study include the short-term intervention of one-hour training and learners recruited from AMT. Future studies will design multiple sessions through an instructional duration of around 20–35 min and provide a long-term intervention in traditional classroom settings. It would be interesting to conduct posttests and delayed effects tests and investigate whether a long-term intervention yields a conversational effect for summary writing in real classes. Another limitation is the self-reported efficacy. Future studies could add other self-efficacy instruments besides self-reported scales. The present study used the pre-designed feedback for the MC questions and peer rating questions but did not provide feedback on written summaries and self-rating summary. In the future, semi-automatically generated feedback would be adopted along with automated feedback on the quality of written summaries and self-rating summaries.

Acknowledgments. This work was funded by the Institute of Education Sciences (Grant No. R305C120001).

References

1. Kim, Y., Baylor, A.L.: Research-based design of pedagogical agent roles: a review, progress, and recommendations. Int. J. Artif. Intell. Educ. **26**(1), 160–169 (2015). https://doi.org/10.1007/s40593-015-0055-y
2. Chattaraman, V., Kwon, W.-S., Gilbert, J.E., Ross, K.: Should AI-based, conversational digital assistants employ social- or task-oriented interaction style? A task-competency and reciprocity perspective for older adults. Compt. Hum. Behav. **90**, 315–330 (2019). https://doi.org/10.1016/j.chb.2018.08.048
3. Ginns, P., Martin, A.J., Marsh, H.W.: Designing instructional text in a conversational style: a meta-analysis. Educ. Psychol. Rev. **25**(4), 445–472 (2013). https://doi.org/10.1007/s10648-013-9228-0
4. Graesser, A.C., Li, H., Forsyth, C.: Learning by communicating in natural language with conversational agents. Curr. Dir. Psychol. Sci. **23**, 374–380 (2014). https://doi.org/10.1177/0963721414540680
5. Graesser, A.C., McNamara, D.S., Cai, Z., Conley, M., Li, H., Pennebaker, J.: Coh-metrix measures text characteristics at multiple levels of language and discourse. Elem. Sch. J. **115**, 210–229 (2014). https://doi.org/10.1086/678293
6. Li, H., Cheng, Q., Yu, Q., Graesser, A.C.: The role of peer agent's learning competency in trialogue-based reading intelligent systems. In: Conati, C., Heffernan, N., Mitrovic, A., Verdejo, MFelisa (eds.) AIED 2015. LNCS (LNAI), vol. 9112, pp. 694–697. Springer, Cham (2015). https://doi.org/10.1007/978-3-319-19773-9_94
7. Li, H., Graesser, A.: Impact of pedagogical agents' conversational formality on learning and engagement. In: André, E., Baker, R., Hu, X., Rodrigo, M.M.T., du Boulay, B. (eds.) AIED 2017. LNCS (LNAI), vol. 10331, pp. 188–200. Springer, Cham (2017). https://doi.org/10.1007/978-3-319-61425-0_16
8. Li, H., Graesser, A.C.: Impact of conversational formality on the quality and formality of written summaries. In: Bittencourt, I.I., Cukurova, M., Muldner, K., Luckin, R., Millán, E. (eds.) AIED 2020. LNCS (LNAI), vol. 12163, pp. 321–332. Springer, Cham (2020). https://doi.org/10.1007/978-3-030-52237-7_26
9. Li, H., Graesser, A.C.: The impact of conversational agents' language on summary writing. J. Res. Technol. Educ. (2021). https://doi.org/10.1080/15391523.2020.1826022

10. Lin, L., Ginns, P., Wang, T., Zhang, P.: Using a pedagogical agent to deliver conversational style instruction: what benefits can you obtain? Comput. Educ. **143** (2020)
11. Mayer, R.E.: Designing multimedia instruction in anatomy: an evidence-based approach. Clin. Anat. **33**, 2–11 (2020). https://doi.org/10.1002/ca.23265
12. Reichelt, M., Kämmerer, F., Niegemann, H.M., Zander, S.: Talk to me personally: personalization of language style in computer-based learning. Comput. Hum. Behav. **35**, 199–210 (2014). https://doi.org/10.1016/j.chb.2014.03.005
13. Riehemann, J., Jucks, R.: Address me personally!: on the role of language styles in a MOOC. J. Comput. Assist. Learn. **34**, 713–719 (2018). https://doi.org/10.1111/jcal.12278
14. Shell, D.F., Colvin, C., Bruning, R.H.: Self-efficacy, attribution, and outcome expectancy mechanisms in reading and writing achievement: grade-level and achievement-level differences. J. Educ. Psychol. **87**(3), 386–398 (1995). https://doi.org/10.1037/0022-0663.87.3.386

Measuring Inconsistency in Written Feedback: A Case Study in Politeness

Wei Dai[1], Yi-Shan Tsai[1], Yizhou Fan[2], Dragan Gašević[1], and Guanliang Chen[1](✉)

[1] Monash University, Melbourne, Australia
{wei.dai1,yi-shan.tsai,dragan.gasevic,guanliang.chen}@monash.edu
[2] University of Edinburgh, Edinburgh, UK
yizhou.fan@ed.ac.uk

Abstract. Feedback, indisputably, has been widely recognized as one of the most important forms of communication between teachers and students and a significant lever to enhance learning experience and success. However, there is consistent evidence showing that higher education institutions struggle to deliver consistent, timely, and constructive feedback to students. This study aimed to investigate whether, and to what extent, feedback inconsistency manifested itself in terms of politeness displayed to students of different demographic attributes (i.e., gender and first-language background). To this end, a large-scale dataset consisting of longitudinal feedback given to 3,249 higher-education students in 35 courses were collected and analyzed by applying multi-level regression modeling. We demonstrated that there were significant differences between low-performing and high-performing students as well as between English-as-second-language and English-as-first-language students. However, the majority of variance measured in the politeness of feedback was explained by course-level and assessment-level characteristics, while student-level characteristics accounted for less than 1% variance.

Keywords: Automatic feedback analysis · Feedback inconsistency · Politeness · Hierarchical regression modeling

1 Introduction

It is widely acknowledged that quality feedback can be a significant lever to enhance learning experience and success [3,4,11]. Given the important role played by feedback, an increasing application of automatic approaches for assessing feedback quality has been developed, e.g., contrasting the effectiveness of immediate and delayed feedback [12], characterizing factors that are important to students' perception of feedback quality and effectiveness [8], and classifying feedback texts according to certain feedback-provision principles [5,6]. Noticeably, these automatic approaches have seldom been used to explore whether there exists any inconsistency in the feedback given to students of different demographic attributes, though researchers have pointed out that higher education

M. M. Rodrigo et al. (Eds.): AIED 2022, LNCS 13355, pp. 560–566, 2022.
https://doi.org/10.1007/978-3-031-11644-5_49

institutions often struggle to deliver consistent and constructive feedback that speaks to the needs of students in a large cohort [2,13].

Therefore, we argued that it is necessary to investigate feedback consistency by comparing feedback given to students of different protective attributes in higher education institutions. In particular, we were interested in measuring feedback consistency from the perspective of politeness, which has been documented to be of particular importance in various teaching and learning practices in the literature. In addition, we considered two types of protective attributes in this study, i.e., *gender* (female vs. male) and *first-language backgrounds* (English-as-first-language vs. English-as-second-language), both of which have been demonstrated to be related to students' attainment gap in various educational settings. Formally, this study was guided by the following **R**esearch **Q**uestion: *Does and if so, to what extent teacher feedback for students differ by students' gender and first-language backgrounds?*

Through extensive analyses, we contributed to the research on automatic feedback analysis with the following main findings: (i) multi-level regression modeling is effective in examining feedback inconsistency existing between student groups of different demographic attributes; (ii) the politeness of feedback was largely dependent on a student's performance in an assessment task; and (iii) there was no significant difference observed between female and male students, while English-as-first-language students tended to receive less polite feedback from instructors than their English-as-second-language counterparts.

2 Method

2.1 Dataset

The feedback data used in this study were retrieved from courses in the subject of Information Technology in a semester from July to October 2020 at a university in Australia. We cleaned the dataset by filtering out courses with less than 20 students, and removing students with missing data and those preferring not to reveal their gender or first-language background. As a result, 35 courses were kept, which including a total of 3,249 students, and the number of students enrolled in a course ranged from 21 to 455. These students made a total of 9,526 submissions to the assignments, i.e., the data used in this work contained both marks and feedback given to 9,526 assignment submissions.

2.2 Measuring Feedback Politeness

This study adopted a state-of-the-art tool [9] to measure feedback politeness, which was proposed to couple a bi-directional LSTM with a convolutional layer to capture not only the long-distance relationship in input text but also the linguistic features that are important for describing the text politeness.

With the tool, we calculated all the politeness scores of the feedback used in our study, whose distribution is depicted in Fig. 1 (a). Most of the feedback was rather direct, i.e., over 88% of them were with politeness score less than 0.08. To better work with skewed data for predictive modeling, as suggested in [1], we applied logarithm transformation to the politeness scores before serving them as inputs to the multi-level regression models. The distribution of the politeness scores after logarithm transformation is given in Fig. 1 (b).

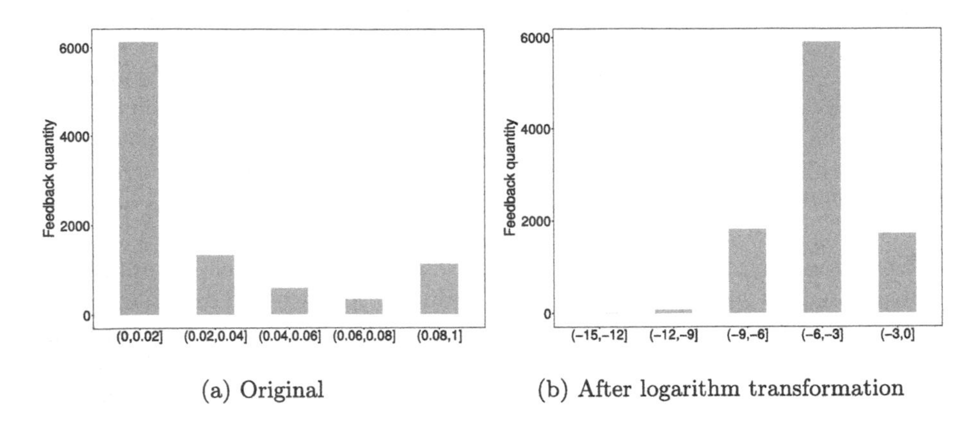

(a) Original (b) After logarithm transformation

Fig. 1. Distribution of politeness scores.

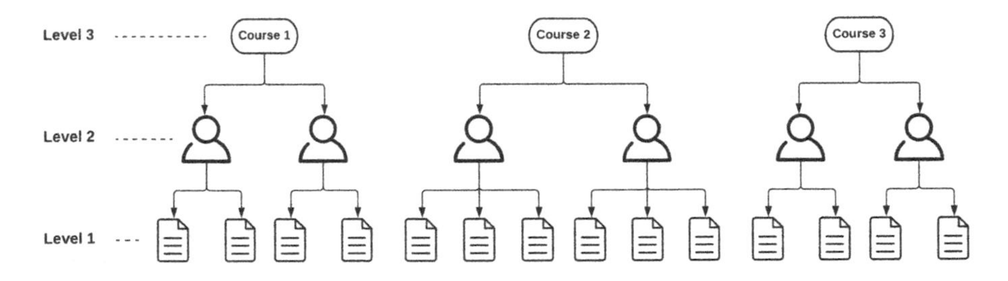

Fig. 2. A three-level data structure with assignments at level-1, students at level-2 and courses at level-3.

2.3 Measuring Feedback Inconsistency

We adopted multi-level regression models to investigate whether there existed any feedback inconsistency. All data analyses were performed with the aid of the MLWiN 3.05 software [10], as detailed below.

Firstly, before measuring feedback inconsistency, we verified the hierarchical structure in our dataset. We first constructed a linear regression model by fitting it on the longitudinal assessment data, which was then compared to the 2-level and 3-level models where the second-level variable was the students and the third-level was the courses in which the students enrolled. The nested 3-level model is depicted in Fig. 2. Secondly, we included the mark of an assignment into the 3-level model described above. Finally, we incorporated students' demographic attributes (i.e., gender and first-language background) to explore whether, and to what extent, instructors' politeness displayed in their written feedback were dependent on students' demographic attributes. To summarize, we constructed the following models for analyses and comparison:

- **Model-A** was solely built on students' longitudinal assessment order without considering their assessment performance or demographic attributes;
- **Model-B** was built by using the assignment marks as first-level predictors;
- **Model-C** was built by using students' gender as second-level predictors;
- **Model-D** was built by using students' first-language backgrounds as second-level predictors;

In particular, to examine whether the assessment performance of students of different demographic attributes had an impact on feedback politeness, interactions between students' assessment performance, and their demographic attributes were tested in Model-C and Model-D. Similarly, interactions between students' demographic attributes and time (i.e., the longitudinal order of the assignments) were also tested in these two models to investigate whether students' demographic attributes had any time-dependent impact on feedback politeness (Table 1).

Table 1. Examination of hierarchical structures.

	Regression	S.E.	2-level	S.E.	3-level	S.E.
VPC						
Level 3					28.31%	
Level 2			20.33%		0.93%	
Level 1			79.67%		70.76%	
Intercept	*** −4.97	0.05	*** −4.98	0.04	*** −4.73	0.17
Slope	*** 0.20	0.02	*** 0.21	0.02	*** 0.21	0.02
Deviance	37610.36		37313.73		35481.09	
X^2 change			*** 296.63		*** 1832.64	

*** $p < 0.001$

3 Results

Based on the variance change (X^2 *change*), we can easily conclude that the 3-level model fitted the data the best, which provides strong evidence for the hierarchical nature of the data and, more importantly, the examination of the effect of students' demographic attributes on feedback politeness should be based on this 3-level regression model.

Based on Table 2, we can make several interesting observations. Firstly, by scrutinizing Model-A, which only considered the longitudinal assessment data as input for regression modeling, we found that feedback politeness tended to increase throughout the running of a course (*Slope* = 0.22, $p < 0.001$). Furthermore, the results of Model-B demonstrated that the politeness level of feedback was positively correlated with a student's assessment performance (*Mark* = 1.98, $p < 0.001$). That is, the higher mark an assignment submission earned, the more polite feedback it received. This implies that there existed a significant gap between high-performing and low-performing students with respect to the politeness of feedback they received. When scrutinizing the results of Model-C, we did not observe any significant results. Also, there were no significant interactions between students' genders and their assessment performance throughout a course. Given the nature of the courses in our dataset, i.e., which were all related to the studies of Information Technology and often with an over-representation of male students, this showed some positive evidence that instructors' feedback politeness favored neither male nor female students. However, in Model-D, there were significant differences between students of different first-language backgrounds. Generally, English-as-second-language students were more likely to receive polite feedback than their English-as-first-language counterparts (*English – second* = 0.86, $p < 0.001$). This was probably because that instructors might have considered the potential impact of language barriers on performance and thus tried to support English-as-second-language students with more polite and encouraging feedback.

Table 2. Coefficients for the main effects of assignment marks (**Model-B**) and students' demographic attributes and their interaction with assignment marks over time (**Model-C** and **Model-D**).

	Model-A	S.E.	Model-B	S.E.	Model-C	S.E.	Model-D	S.E.
Intercept	*** −4.74	0.17	*** −6.33	0.19	−2.54	7.39	*** −7.02	0.26
Slope	*** 0.22	0.02	*** 0.26	0.02	−0.19	0.80	*** 0.37	0.04
Mark			*** 1.98	0.10	−2.03	9.52	*** 2.53	0.20
Male					−3.79	7.39		
Female					−3.81	7.39		
Mark * Male					4.05	9.52		
Mark * Female					3.85	9.52		
Time * Male					0.43	0.80		
Time * Female					0.51	0.80		
English-second							*** 0.86	0.22
Mark * English-second							** −0.71	0.23
Time * English-second							** −0.14	0.05
Deviance	35465.91		35074.40		35045.55		35016.24	
X^2 change			*** 391.51		*** 28.85		*** 29.32	

*** $p < 0.001$; ** $p < 0.01$

However, to our surprise, English-as-second-language students with good assessment performance tended to receive less polite feedback from instructors ($Mark{*}English{-}second = -0.71, p < 0.01$). On the other hand, throughout the running of a course, English-as-second-language students were likely to receive less and less polite feedback ($Time * English - second = -0.14, p < 0.01$). This showed that feedback inconsistency manifested itself not only among students of different first-language backgrounds but also among students of the same first-language background but with different assessment performances or at different times.

4 Conclusion

Overall, our observations suggest that inconsistency in feedback politeness occurs across student groups of different assessment performance, different first-language backgrounds, and different periods in the running of a course. More efforts are still required to understand the effectiveness of feedback in different politeness level and consistency issues on learning for different groups of students. For more in-depth insights, future research endeavors may be invested to measure feedback inconsistency from other perspectives, e.g., the presence of the four aspects of feedback proposed in [7], include different subjects, and explore how NLP techniques can be applied to assist instructors in avoiding feedback inconsistency in their feedback-writing practices.

References

1. Bandari, R., Asur, S., Huberman, B.: The pulse of news in social media: forecasting popularity. In: Proceedings of the International AAAI Conference on Web and Social Media, vol. 6 (2012)
2. Boud, D., Molloy, E.: Rethinking models of feedback for learning: the challenge of design. Assess. Eval. High. Educ. **38**(6), 698–712 (2013)
3. Butler, D.L., Winne, P.H.: Feedback and self-regulated learning: a theoretical synthesis. Rev. Educ. Res. **65**(3), 245–281 (1995)
4. Canavan, C., Holtman, M.C., Richmond, M., Katsufrakis, P.J.: The quality of written comments on professional behaviors in a developmental multisource feedback program. Acad. Med. **85**(10), S106–S109 (2010)
5. Cavalcanti, A.P., et al.: How good is my feedback? A content analysis of written feedback. In: LAK, pp. 428–437 (2020)
6. Cavalcanti, A.P., de Mello, R.F.L., Rolim, V., André, M., Freitas, F., Gaševic, D.: An analysis of the use of good feedback practices in online learning courses. In: ICALT, vol. 2161, pp. 153–157. IEEE (2019)
7. Hattie, J., Timperley, H.: The power of feedback. Rev. Educ. Res. **77**(1), 81–112 (2007)
8. Lizzio, A., Wilson, K.: Feedback on assessment: students' perceptions of quality and effectiveness. Assess. Eval. High. Educ. **33**, 263–275 (2008)
9. Niu, T., Bansal, M.: Polite dialogue generation without parallel data. Trans. Assoc. Comput. Linguist. **6**, 373–389 (2018)

10. Rasbash, J., et al.: A User's Guide to MLwiN, vol. 286. Institute of Education, London (2000)
11. Ryan, T., Henderson, M., Ryan, K., Kennedy, G.: Identifying the components of effective learner-centred feedback information. Teach. High. Educ., 1–18 (2021)
12. Smits, M.H., Boon, J., Sluijsmans, D.M., Van Gog, T.: Content and timing of feedback in a web-based learning environment: effects on learning as a function of prior knowledge. Interact. Learn. Environ. **16**(2), 183–193 (2008)
13. Yang, M., Carless, D., Salter, D., Lam, J.: Giving and receiving feedback: a Hong Kong perspective (2010)

Reducing Bias in a Misinformation Classification Task with Value-Adaptive Instruction

Nicholas Diana[1(✉)] and John Stamper[2]

[1] Colgate University, Hamilton, USA
ndiana@colgate.edu
[2] Carnegie Mellon University, Pittsburgh, USA
jstamper@cmu.edu
http://www.nickdiana.com, http://www.dev.stamper.org

Abstract. Instructional technology that supports the development of media literacy skills has garnered increased attention in the wake of recent misinformation campaigns. While critical, this work often ignores the role of myside bias in the acceptance and propagation of misinformation. Here we present results from an alternative approach that uses natural language processing to model the dynamic relationship between the user and the content they are consuming. This model powers a debiasing intervention in the context of a "fake news detection" task. Information about the user- and content-values was used to predict when the user may be prone to myside bias. The intervention resulted in significantly better performance on the misinformation classification task. These results support the development of content-general and embedded debiasing systems that could encourage informal learning and bias reduction in real-world contexts.

Keywords: Misinformation · Myside bias · Confirmation bias · Personalization · Media literacy · Civic technology · Civics education

1 Introduction

Modern digital media has novel features that set it apart from traditional media, including a lack of editorial oversight, the democratization of media sources, and the rapid propagation of stories (particularly stories that are emotionally charged). Many of these features provide an infrastructure that allows misinformation to flourish in ways that would be difficult in a pre-digital age [13]. In response to these new and pressing challenges, media literacy education has increasingly emphasized misinformation classification (i.e., the ability to accurately identify misinformation) as a critical civic skill. Recent successful misinformation campaigns illustrate both the public's susceptibility to believing so-called "fake news" [13] as well as the dire consequences that result from a failure to teach and exercise this fundamental media literacy skill.

M. M. Rodrigo et al. (Eds.): AIED 2022, LNCS 13355, pp. 567–572, 2022.
https://doi.org/10.1007/978-3-031-11644-5_50

While there has been an increased interest in the development of instructional tools designed to improve misinformation classification [9,11], few solutions account for the impact of bias. In this paper we present a contrasting approach that aims to model the dynamic relationship between the user and the content they are consuming. We used this model to predict when the user may be most susceptible to bias, and to provide adaptive recommendations in those moments.

We hypothesized that an intervention that leverages the *Alignment* between user and content values will reduce the impact of myside bias on ratings of plausibility. We expect that user ratings after seeing the value-adaptive intervention will be more accurate, and that, generally, the intervention will encourage any change in ratings to be a change in the right direction (i.e., towards the correct answer). This work may inform the development of tools for reducing bias when evaluating the veracity of information in real-world contexts.

Background. The specific skill isolated in the current experiment is one's ability to accurately estimate the plausibility of events, specifically in the realm of United States politics. These estimations are based on what we know about various political actors and how we believe they might behave. As such, these estimations can be honed with experience, by comparing what we believed was plausible with what was actually true. In the real world, a number of potential factors may play a role in determining plausibility. We attempt to control for these factors to isolate impact of bias, specifically *Myside Bias*, or one's tendency to evaluate claims or evidence more favorably if the claim or evidence supports one's own beliefs or worldview [15]. We expect myside bias to cause users to overestimate the plausibility of headlines that support their own beliefs, ultimately impacting the user's accuracy on the misinformation classification task.

We estimated user values using Moral Foundations Theory [7], which argues that moral judgements are driven by the importance we ascribe to a small set of *moral foundations*. These moral foundations have been empirically shown to be highly predictive of both general voting behavior [4] as well as specific political beliefs [10]. The output of the Moral Foundations Questionnaire (MFQ) is a vector of five scores, representing the degree to which the student values each of the five foundations when making moral judgments.

2 Method

Based on a power analysis, eighty-three (83) participants were recruited using the participant recruitment platform Prolific. Participants were required to be 18 years of age or older, U.S. citizens, and not have participated in any of our research group's prior studies. The estimated completion time was 28 min, and participants were paid \$3.15 (\$6.75/hour) for participating. Participants who failed reading-checks ($n = 2$) were excluded from analyses. The remaining 81 participants (36 female, 42 male, and 3 "Other/Prefer not to say") ranged in age from 18–68 years old ($M = 34.44$). These participants were drawn from a politically diverse population as evidenced by their scores on the MFQ.

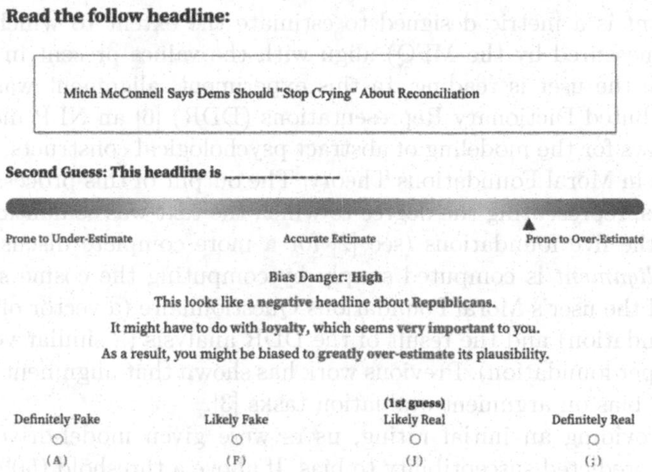

Read the follow headline:

Mitch McConnell Says Dems Should "Stop Crying" About Reconciliation

Second Guess: This headline is ________

Prone to Under-Estimate | Accurate Estimate | Prone to Over-Estimate

Bias Danger: High

This looks like a negative headline about Republicans.

It might have to do with loyalty, which seems very important to you.

As a result, you might be biased to greatly over-estimate its plausibility.

Definitely Fake (A) | Likely Fake (F) | (1st guess) Likely Real (J) | Definitely Real (;)

Fig. 1. A screenshot of the online interface including the value-adaptive intervention. The model-driven components of the adaptive intervention are displayed in bolded blue text. User performance improved significantly after seeing the intervention ($p < .05$). (Color figure online)

The entirety of the experiment was conducted through an online web interface. After viewing the consent form, participants were directed to a set of instructions that described the main task of the experiment (misinformation classification). Following the instructions, participants were given a pre-study survey that included the MFQ [8]. Participants were then directed to the misinformation classification task: a series of 52 news headlines taken from or based on Politifact headlines in the FakeNewsNet news misinformation dataset [12]. Headlines had two relevant features: authenticity (authentic or fabricated) and veracity (real or fake). *Authentic Real* and *Authentic Fake* headlines were actual news headlines classified as either "real" or "fake" (respectively) by Politifact. *Fabricated Real* and *Fabricated Fake* headlines were exact copies of authentic headlines, except that the subject of the headline was changed to a subject from the opposing side of the political spectrum. For example, the *authentic fake* headline *"BREAKING: Federal Judge Grants Permission To Subpoena Trump"* would be changed to the *fabricated fake* headline *"BREAKING: Federal Judge Grants Permission To Subpoena Obama"*.

For each item, users provided an initial rating of plausibility, then were shown additional (non-correctness) feedback (i.e., the debiasing intervention), and finally were asked to provide a second rating of plausibility in light of this information. After providing a second rating, users were given correctness feedback. In this way, the task closely resembles the Judge Advisor System [14] often employed in decision-making research. Following the classification task, participants were asked to complete a short post-study questionnaire that included questions about demographic information.

Alignment is a metric designed to estimate the extent to which the user's values (as measured by the MFQ) align with the values present in the text of the headline the user is reading. In this experiment, alignment was computed using Distributed Dictionary Representations (DDR) [6] an NLP method. This method allows for the modeling of abstract psychological constructs, such as the foundations in Moral Foundations Theory. The output of this process is a vector of five scores, representing the degree to which the text was semantically similar to each of the five foundations (see [2] for a more complete discussion of this process). *Alignment* is computed simply by computing the cosine similarity of the result of the user's Moral Foundations Questionnaire (a vector of five values, one per foundation) and the result of the DDR analysis (a similar vector of five values, one per foundation). Previous work has shown that alignment is a reliable predictor of bias on argument evaluation tasks [3].

After providing an initial rating, users were given model-driven feedback about their predicted susceptibility to bias. If above a threshold (50%), then the user was shown additional model-driven information including estimations of the headline's predominate value, political affiliation, valence, and the relationship to a user's values. Figure 1 shows a screenshot of the intervention. The blue text in the figure indicate the model-generated and/or user-adaptive feedback components. The generation of each of the elements of the intervention is detailed below.

The intervention consisted of two stages. First, a logistic regression model was used to predict the likelihood of a correct answer given alignment, the number of prior opportunities, and item-level effects (i.e., average difficulty). The beta values used in this model were derived from the results of a pilot study. This information is conveyed to users in the form of text reading "Bias Danger: [Level], where [Level] is low (>75% likelihood), moderate (50–75%) or high (<50%).

If the likelihood of a correct response was less than 75%, the user was shown an elaborated intervention that included information from two additional predictive models. Predictions about the text's valence, subject, and most relevant foundation were generated through the use of a SimCSE (Simple Contrastive Learning of Sentence Embeddings) model trained on the Stanford Natural Language Inference Corpus [5]. This model assessed the similarity of headline text to a set of archetypal sentences based on the topics explored in the Moral Foundations Vignettes [1]. The SimCSE model chose the archetypal sentence that most closely matched each news headline (e.g., the (fake) news headline *"BREAKING: Federal Judge Grants Permission To Subpoena Obama"* was most similar to the negative archetypal sentence *"A Democrat breaks the law."*).

Users were also given a prediction about their likelihood to over- or underestimate plausibility due to bias. This second likelihood prediction was derived from a second logistic regression model that predicts the likelihood of overestimation (coded as 1) or underestimation (coded as 0) given the user's alignment score and the number of previous opportunities. Again, the beta values used in this model were derived from the results of a pilot study. Additionally, a qualifier of either "slightly" or "greatly" was given to the prediction based on the likelihood score (either between 25% and 75% or outside of that range, respectively).

3 Results and Discussion

We hypothesized that user ratings after seeing the debiasing intervention would be more accurate than their initial ratings. A paired samples t-test was used to compare the outcomes of first and second attempts. There was a slight but significant improvement in outcomes between first ($M = .69, SD = .45$) and second attempts ($M = .71, SD = .45$) $t(80) = -2.56, p = .01$). Changes in ratings from one class to another were relatively rare. To get a more nuanced measure of the impact of the intervention, the direction of movement between the initial and second rating was also analyzed. That is, we assessed whether or not the intervention encouraged movement toward the correct answer – even if the user ultimately provided an incorrect class. We found that, of the instances in which a user changed their score between the first and second ratings, users were significantly more likely to "move" their ratings in the correct direction ($X^2(2, 2616) = 119.87, p < .001$) after seeing the intervention. Users were also asked to provide feedback about the quality and effectiveness of the AI assistance. Most users found the AI assistant's feedback to be helpful and mostly accurate.

These results provide additional evidence for the importance of *user-content alignment* in misinformation classification. Users more accurately classified misinformation after seeing the value-adaptive feedback, and the intervention encouraged movement towards the correct response. Taken together, these results suggest that the intervention resulted in both incremental and meaningful changes in user responses, a finding mirrored in the qualitative user feedback. This work has implications for the development of future media literacy instructional technologies, suggesting that accurate models of user learning in this require the consideration of bias. Future work will aim to provide similar value-adaptive debiasing interventions in real-world contexts. Integrating this *just-in-time* intervention into real-world settings where users encounter misinformation will shed light on the impact of this intervention in the presence of the numerous other factors that may play a role in a user's determination of plausibility.

The nature of the intervention's presentation in this study was limited by the fact that it leveraged item-level information (based off of previous experiments) in the initial outcome-based prediction stage. Including this item-level information provides greater accuracy, as it likely captures important baseline plausibility information. While alignment is included in this outcome prediction model, the practical result of this prediction is that users are seeing the elaborated intervention on items that are, on average, more difficult to classify. This is perhaps unavoidable as the specific context of an individual headline (i.e., the actors and their behavior) may always be the primary factor in determining plausibility. Nevertheless, to isolate the impact of bias, the second, bias-based prediction stage did not leverage any item-level information.

4 Conclusion

Identifying misinformation is a key media literacy skill, and one that may depend on the interaction between the user and the content they are consuming. We found that a value-adaptive debiasing intervention improved performance on a misinformation classification task. These results provide evidence for the importance of the dynamic relationship between user- and content-values, particularly in the media literacy domain.

References

1. Clifford, S., Iyengar, V., Cabeza, R., Sinnott-Armstrong, W.: Moral foundations vignettes: a standardized stimulus database of scenarios based on moral foundations theory. Behav. Res. Methods **47**(4), 1178–1198 (2015)
2. Diana, N., Stamper, J., Koedinger, K.: Towards value-adaptive instruction: a data-driven method for addressing bias in argument evaluation tasks. In: Proceedings of the 2020 CHI Conference on Human Factors in Computing Systems, pp. 1–11 (2020)
3. Diana, N., Stamper, J.C., Koedinger, K.: Predicting bias in the evaluation of unlabeled political arguments. In: CogSci, pp. 1640–1646 (2019)
4. Franks, A.S., Scherr, K.C.: Using moral foundations to predict voting behavior: regression models from the 2012 us presidential election. Anal. Soc. Issues Public Policy **15**(1), 213–232 (2015)
5. Gao, T., Yao, X., Chen, D.: SimCSE: simple contrastive learning of sentence embeddings. arXiv preprint arXiv:2104.08821 (2021)
6. Garten, J., Hoover, J., Johnson, K.M., Boghrati, R., Iskiwitch, C., Dehghani, M.: Dictionaries and distributions: combining expert knowledge and large scale textual data content analysis. Behav. Res. Methods **50**(1), 344–361 (2017). https://doi.org/10.3758/s13428-017-0875-9
7. Graham, J., et al.: Moral foundations theory: the pragmatic validity of moral pluralism. In: Advances in Experimental Social Psychology, vol. 47, pp. 55–130. Elsevier (2013)
8. Graham, J., Nosek, B.A., Haidt, J., Iyer, R., Spassena, K., Ditto, P.H.: Moral foundations questionnaire. J. Pers. Soc. Psychol. (2008)
9. Hone, B., Rice, J., Brown, C., Farley, M.: Factitious (2018). factitious.augamestudio.com
10. Koleva, S.P., Graham, J., Iyer, R., Ditto, P.H., Haidt, J.: Tracing the threads: how five moral concerns (especially purity) help explain culture war attitudes. J. Res. Pers. **46**(2), 184–194 (2012)
11. Literat, I., Chang, Y.K., Eisman, J., Gardner, J.: LAMBOOZLED!: the design and development of a game-based approach to news literacy education. J. Media Literacy Educ. **13**(1), 56–66 (2021)
12. Shu, K., Mahudeswaran, D., Wang, S., Lee, D., Liu, H.: FakeNewsNet: a data repository with news content, social context, and spatiotemporal information for studying fake news on social media. Big Data **8**(3), 171–188 (2020)
13. Silverman, C.: This analysis shows how viral fake election news stories outperformed real news on Facebook, November 2016. https://www.buzzfeednews.com/craigsilverman/viral-fake-election-news-outperformed-real-news-on-facebook
14. Sniezek, J.A., Buckley, T.: Cueing and cognitive conflict in judge-advisor decision making. Organ. Behav. Hum. Decis. Process. **62**(2), 159–174 (1995)
15. Stanovich, K.E., West, R.F., Toplak, M.E.: Myside bias, rational thinking, and intelligence. Curr. Dir. Psychol. Sci. **22**(4), 259–264 (2013)

A Generic Interpreting Method for Knowledge Tracing Models

Deliang Wang[1], Yu Lu[1,2](✉), Zhi Zhang[1], and Penghe Chen[1,2]

[1] School of Educational Technology, Faculty of Education, Beijing Normal University, Beijing 100875, China
luyu@bnu.edu.cn

[2] Advanced Innovation Center for Future Education, Beijing Normal University, Beijing 100875, China

Abstract. To interpret the deep learning based knowledge tracing models (DLKT), we introduce a generic method with four-step procedure. The proposed method and procedure are generally applicable to the DLKT models with diverse inner structures. The experiment results validate them on three existing knowledge tracing models, where the individual contributions of the input question-answer pairs to the models' decision are properly calculated. By leverage the calculated interpreting results, we explore the key information hidden in the DLKT models.

Keywords: Knowledge tracing · Explainable artificial intelligence · Deep learning

1 Introduction

Knowledge tracing (KT) aims to estimate learners' dynamic knowledge state and predict their future performance. Researchers have successfully adopted the Markov process, logistic regression, and deep learning techniques to build KT models. Deep learning based knowledge tracing (DLKT) models are the most recent ones and have been implemented in intelligent tutoring systems [4]. However, the existing DLKT models operate as a black box and lack transparency, which painfully hinders their practical usage. Recently, the post-hoc technique in explainable artificial intelligence (xAI) has been used to interpret DLKT models [3]. However, the previous work is primarily the model-specific method, and it is still lack of the interpreting method that can be applied to different DLKT models with diverse inner structure.

In this work, we introduce a widely applicable method to interpret DLKT models. Specifically, we employ a generic method and propose a four-step procedure to interpret three existing DLKT models, including DKT [7], DKVMN [9] and SAKT [6]. The experiment results show that the method is generally applicable to all the models, as it could properly compute the input contributions to the model's final prediction regardless of its inner structure. By leverage the

M. M. Rodrigo et al. (Eds.): AIED 2022, LNCS 13355, pp. 573–580, 2022.
https://doi.org/10.1007/978-3-031-11644-5_51

interpreting results, we find the skill effect and recency effect in these DLKT models, i.e., the DLKT models prefer using the recent input records on the same skill to make their predictions.

2 Related Work

Recently, deep learning techniques have been introduced into the KT domain. As the first DLKT model, deep knowledge tracing (DKT) model [7] adopts an recurrent neural networks (RNN) to model learners' knowledge state on multiple skills. Subsequently, the Dynamic key-value memory network (DKVMN) [9] and self-attentive model for knowledge tracing (SAKT) [6] has been designed, which are based on memory-augmented neural networks (MANN) and attention network, respectively.

Similar to other deep learning models, it is also difficult for human to understand DLKT model's untransparent prediction process. The interpretability issue in xAI domain can be generally classified into ante-hoc interpretability and post-hoc interpretability. In the KT domains, researchers start to investigate using the post-hoc method, such as layer-wise relevance propagation (LRP) [1] to interpret the DLKT model, but the solution is model-specific and hard to be generalized to other DLKT models. We thus introduce a new post-hoc method, called Deep SHAP [5], into KT domain and explore an interpreting solution that is applicable to different DLKT models.

3 Interpreting Method

3.1 Deep SHAP

Deep SHAP decomposes the prediction of a deep learning model to the sum of feature contributions through backpropagation. Specifically, Deep SHAP compares a sample to be interpreted with another sample, namely reference sample, and assumes that their output difference is caused by their difference in the input features. By backpropagating the prediction difference, Deep SHAP obtains the reference-specific contribution of each feature to the prediction. With multiple reference samples, Deep SHAP computes the mean value of these reference-specific feature contributions and obtains final feature contributions.

Specifically, given a deep learning model f, a sample x to be interpreted, and a reference sample r used to compare. We define Δx_i as the difference in feature i, namely $\Delta x_i = x_i - r_i$. The quantity Δy is defined as the difference in the output value, that is $\Delta y = f(x) - f(r)$. As Eq. 1 shows, Deep SHAP attributes Δy to feature contributions, where $C_{\Delta x_i \Delta y}$ represents an amount in Δy caused by Δx_i:

$$\Delta y = \sum_{i=1}^{n} C_{\Delta x_i \Delta y}. \tag{1}$$

To calculate $C_{\Delta x_i \Delta y}$, Deep SHAP defines multipliers. Given a network, the value difference in the input neuron x is Δx_i, and the value difference in the target neuron t is Δt. Multiplier $m_{\Delta x_i \Delta t}$ is defined as dividing the finite contributions of Δx_i to Δt by the finite changes Δx_i, namely $m_{\Delta x_i \Delta t} = \frac{C_{\Delta x_i \Delta t}}{\Delta x_i}$. In a complex multi-layer network, computing *multipliers* of an input neuron adopts the chain rule that is equivalent to the chain rule of partial derivative. Suppose a neural network has three layers, where the input layer has n neurons symbolized as x_i, the hidden layer also has n neurons symbolized as t_j, and the output layer only has a neuron symbolized as y. For every two neighbouring layers, we can easily obtain their multipliers $m_{\Delta x_i \Delta t_j}$ and $m_{\Delta t_j \Delta y}$ as shown in Eqs. 2 and 3:

$$m_{\Delta t_j \Delta y} = \frac{C_{\Delta t_j \Delta y}}{\Delta t_j} \tag{2}$$

$$m_{\Delta x_i \Delta t_j} = \frac{C_{\Delta x_i \Delta t_j}}{\Delta x_i}. \tag{3}$$

As Eq. 4 shows, the chain rule is adopted to calculate $m_{\Delta x_i \Delta y}$:

$$m_{\Delta x_i \Delta y} = \sum_{j=1}^{n} m_{\Delta x_i \Delta t_j} m_{\Delta t_j \Delta y}. \tag{4}$$

The details can be found in [8]. The feature contribution of Δx_i to the output difference Δy can be approximately obtained by using $m_{\Delta x_i \Delta y}$ multiplying Δx_i:

$$C_{\Delta x_i \Delta y} \approx m_{\Delta x_i \Delta y} \Delta x_i. \tag{5}$$

To obtain reasonable interpreting results, Deep SHAP selects multiple reference samples and the mean of these reference-specific feature contributions is the final feature contributions:

$$C_{\Delta x_i \Delta y} \approx \frac{1}{m} \sum_{k=1}^{m} m_{\Delta x_i \Delta y} \left(x_i - r_i^k\right), \tag{6}$$

where m is the number of reference samples and r^k is the k-th reference sample.

3.2 Interpreting DLKT Models

We adopt Deep SHAP to interpret the predictions of DLKT models. Figure 1 gives a toy example to illustrate the four-step interpreting procedure. Firstly, as Fig. 1(a) shows, given a sample x to be interpreted, we select reference samples (i.e., sequences in the training dataset whose last questions are on the same skill as sample x) and make predictions on the last questions. Secondly, as Fig. 1(b) shows, we compute the prediction difference between each reference sample and sample x. Thirdly, We backpropagate the prediction difference from the output

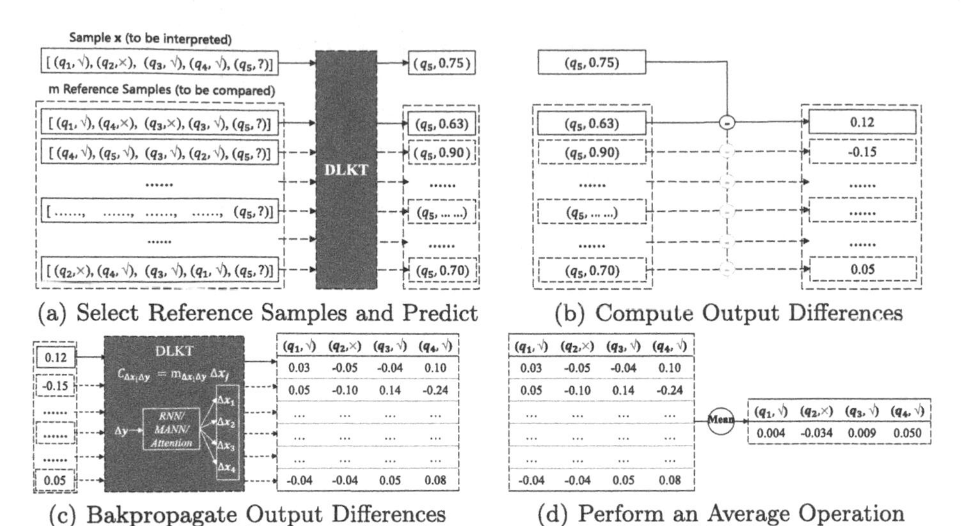

(a) Select Reference Samples and Predict (b) Compute Output Differences

(c) Bakpropagate Output Differences (d) Perform an Average Operation

Fig. 1. The four-step interpreting procedure on a DLKT model

layer to the input layer and compute the reference-specific feature contributions between each reference sample and the sample being interpreted, as shown in Fig. 1(c). Finally, to obtain the contribution of each question-answer pair in sample x to the DLKT model's prediction (i.e., the QA relevance), we perform an average operation on its reference-specific contributions, as Fig. 1(d) shows. The four-step interpreting procedure above is applicable to different DLKT models, regardless of their inner structure.

4 Multi-model Validation

4.1 DLKT Model Building

We build DKT, DKVMN and SAKT models on ASSISTment2009 [2] dataset. Specifically, we choose the *skill builder* dataset, eliminate repeated sequences and questions without skill labels. We set the maximum length between 10 and 200. The preprocessed dataset contains 320,488 interactions from 3,091 learners on 110 skills. 80% data is randomly selected for training and the remaining ones are used for testing. The training setting is similar to [3]. The built LSTM-based DKT model, whose hidden dimensionality is set to 64, achieves 0.74 in AUC and 0.72 in ACC. The built MANN-based DKVMN, whose state dimensionality and memory size are set to 64 and 110, achieves 0.76 in AUC and 0.74 in ACC. The built attention-based SAKT, whose hidden dimensionality is set to 64 with 4 heads, achieves 0.75 in AUC and 0.72 in ACC.

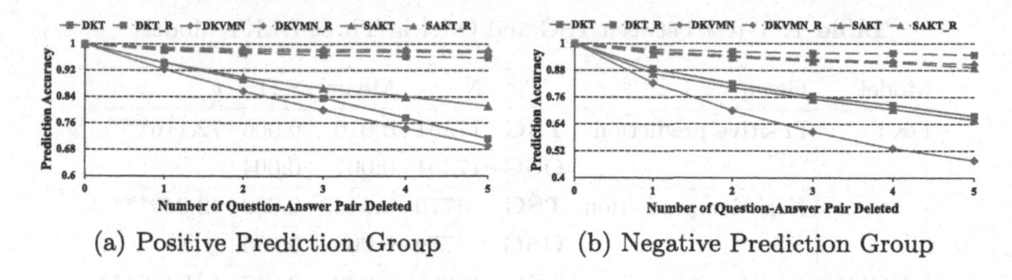

(a) Positive Prediction Group (b) Negative Prediction Group

Fig. 2. Pair deletion results for the correctly predicted sequences

4.2 Interpreting Result Validation

We split the test data into 48,670 sequences with a length of 15. The first 14 question-answer pairs serve as inputs to predict learners' performance on the 15th question. We thus obtain correctly and falsely predicted sequences. Then we select sequences with a length of 15 from the training dataset as reference samples, whose last question predicts the same skill as the test sequence. By Deep SHAP, we obtain the QA relevance of question-answer pairs in each test sequence. To examine the calculated QA relevance, we perform the deletion experiments and see the accuracy changes as the QA relevance value should reflect the amount of contributions to the DLKT model's prediction. For positive and negative predictions, we delete question-answer pairs in the descending and ascending order of their QA relevance, respectively. Figure 2 shows that among the correctly-predicted sequences of three DLKT models, deleting question-answer pairs based on QA relevance causes a significant drop in model accuracy compared to random deletions (represented by "_R"). It partially validates the interpreting method and procedure, as the calculated QA relevances reflect contributions of the question-answers pairs to the predictions of the DLKT models.

5 Hidden Information Explore

5.1 Skill Effect

With the interpreting results, we explore what is the role of skill in DLKT models. Specifically, we define the skill of the 15th question in each correctly predicted test sequence as the *Target Skill*. The question-answer pairs on the target skill are defined as *Target Skill Group* (TSG) and those on other skills are defined as *Other Skill Group* (OSG). We select the sequences with at least one question-answer pair on the target skill and other skills. We then compute and compare the mean of absolute QA relevance of *TSG* and *OSG* in three DLKT models. Table 1 shows that for all the DLKT models, the mean absolute QA relevance in TSG is significantly larger than that in OSG in both positive and negative prediction groups. The question-answer pairs on the target skill

Table 1. T-test between TSG and OSG in Three DLKT Models

Model	Group		N	Mean	S.D	t
DKT	Positive prediction	**TSG**	17351	0.010	0.006	72.116***
		OSG	17351	0.007	0.004	
	Negative prediction	**TSG**	4779	0.014	0.009	43.949***
		OSG	4779	0.008	0.004	
DKVMN	Positive prediction	**TSG**	18065	0.022	0.017	125.400***
		OSG	18065	0.007	0.004	
	Negative prediction	**TSG**	4260	0.026	0.020	64.468***
		OSG	4260	0.007	0.004	
SAKT	Positive prediction	**TSG**	18538	0.011	0.010	79.809***
		OSG	18538	0.006	0.004	
	Negative prediction	**TSG**	3606	0.0170	0.016	36.665***
		OSG	3606	0.007	0.004	

***$p < 0.001$

contribute more to the DLKT model's final prediction than the pairs on other skills. In other words, the DLKT models rely more on the learner's previous performance on the target skill to make decisions. We name it as *skill effect*.

5.2 Recency Effect

We further explore whether the position in the sequence of question-answer pairs affects the DLKT models' prediction. We split all the correctly predicted sequences into two parts, whose first half and second half both contain questions on the target skill. The first half of each sequence is simply named as *Far Group* (FG) and the second half is named as *Close Group* (CG). For pairs on the target skill, we compute and compare their mean absolute QA relevance in *FG* and *CG* in three DLKT models. Table 2 shows that in the three DLKT models, the mean absolute QA relevance in *CG* is significantly larger than that in *FG* in both positive prediction and negative prediction groups. The question-answer pairs closer to the prediction contribute more to the DLKT model's final prediction. In other words, the DLKT models tend to rely on the learner's recent performance to make decisions. We name it as *recency effect*.

Table 2. T-test between CG and FG in Three DLKT Models

Model	Group		N	Mean	S.D	t
DKT	Positive prediction	**FG**	11565	0.007	0.003	−107.824***
		CG	11565	0.010	0.004	
	Negative prediction	**FG**	4096	0.007	0.004	−79.147***
		CG	4096	0.012	0.006	
DKVMN	Positive prediction	**FG**	11777	0.006	0.003	−194.531***
		CG	11777	0.016	0.007	
	Negative prediction	**FG**	3906	0.007	0.003	−124.906***
		CG	3906	0.019	0.007	
SAKT	Positive prediction	**FG**	12380	0.005	0.003	−106.52***
		CG	12380	0.008	0.004	
	Negative prediction	**FG**	3136	0.007	0.004	−42.089***
		CG	3136	0.011	0.006	

***$p < 0.001$

6 Conclusion

In this work, we employ a generic interpreting method and propose a four-step procedure to interpret the DLKT models. The experiment results on all the three DLKT models validate the interpreting method. With the interpreting results, we discover the skill effect and recency effect from DLKT models. This study could serve as a solid basis to systematically interpret all the DLKT models.

Acknowledgements. This research is supported by the National Natural Science Foundation of China (No. 62077006, 62177009), Open Project of the State Key Laboratory of Cognitive Intelligence (No. iED2021-M007) and the Fundamental Research Funds for the Central Universities.

References

1. Bach, S., et al.: On pixel-wise explanations for non-linear classifier decisions by layer-wise relevance propagation. PLoS ONE **10**(7), e0130140 (2015)
2. Feng, M., et al.: Addressing the assessment challenge with an online system that tutors as it assesses. User Model. User-Adap. Inter. **19**(3), 243–266 (2009)
3. Lu, Yu., Wang, D., Meng, Q., Chen, P.: Towards interpretable deep learning models for knowledge tracing. In: Bittencourt, I.I., Cukurova, M., Muldner, K., Luckin, R., Millán, E. (eds.) AIED 2020. LNCS (LNAI), vol. 12164, pp. 185–190. Springer, Cham (2020). https://doi.org/10.1007/978-3-030-52240-7_34
4. Lu, Y., et al.: Radarmath: an intelligent tutoring system for math education. In: Thirty-Fifth AAAI Conference on Artificial Intelligence, pp. 16087–16090 (2021)
5. Lundberg, S.M., Lee, S.I.: A unified approach to interpreting model predictions. In: Advances in Neural Information Processing Systems 30 (2017)

6. Pandey, S., Karypis, G.: A self attentive model for knowledge tracing. In: 12th International Conference on Educational Data Mining, EDM (2019)
7. Piech, C., et al.: Deep knowledge tracing. In: Advances in Neural Information Processing Systems, pp. 505–513 (2015)
8. Shrikumar, A., et al.: Learning important features through propagating activation differences. In: International Conference on Machine Learning (ICML) (2017)
9. Zhang, J., et al.: Dynamic key-value memory networks for knowledge tracing. In: International Conference on World Wide Web, pp. 765–774 (2017)

An Automated Writing Evaluation System for Supporting Self-monitored Revising

Diane Litman(✉), Tazin Afrin, Omid Kashefi, Christopher Olshefski, Amanda Godley, and Rebecca Hwa

University of Pittsburgh, Pittsburgh, PA 15260, USA
{dlitman,taa74,kashefi,ca048,agodley,hwa}@pitt.edu

Abstract. This paper presents the design and evaluation of an automated writing evaluation system that integrates natural language processing (NLP) and user interface design to support students in an important writing skill, namely, self-monitored revising. Results from a classroom deployment suggest that NLP can accurately analyze where and what kind of revisions students make across paper drafts, that students engage in self-monitored revising, and that the interfaces for visualizing the NLP results are perceived by students to be useful.

Keywords: Writing · Revision · Natural language processing

1 Motivation

Automated writing evaluation (AWE) systems driven by natural language processing (NLP) are designed to provide formative feedback for students to revise, and ideally improve, their essays. However, although students do attempt to revise their essays in response to AWE feedback, student revisions often do not yield substantive essay improvements [12,15]. We envision that an enhanced automated writing evaluation system that analyzes and provides feedback on students' revision attempts can support the development of this critical skill. This paper presents the design and classroom evaluation of ArgRewrite, an AWE system that integrates NLP and user interface design to support students in *self-monitored revising*. The NLP backend automatically extracts all revised sentences between two paper drafts, then classifies whether the purpose of each revision was to make a surface (meaning-preserving) versus content (meaning-altering) change. The frontend uses visual interface components to convey the backend's revision analysis. A classroom deployment suggests that NLP provides accurate feedback and that students meaningfully revise.

Compared to prior research, while some AWE systems may detect revisions, they tend to provide feedback on a single essay draft [7,11,13] rather than on

Supported by the National Science Foundation under Grant #173572.

M. M. Rodrigo et al. (Eds.): AIED 2022, LNCS 13355, pp. 581–587, 2022.
https://doi.org/10.1007/978-3-031-11644-5_52

revisions between drafts. *Revision as a skill* is the target of feedback in our work. and more generally has been identified as an area for AWE research [3,13,14].

Most AWE systems provide feedback on task and performance rather than on process [13,14,16]. While recent work has analyzed keystroke logs [5], we analyze process at the level of sentences. Also, while most AWE systems provide actionable feedback messages [17], ArgRewrite instead visualizes NLP results to help students *self-monitor* their revisions, motivated by research on strategy instruction [4], self-regulation [10], and self-monitoring [16].

NLP-based writing revision analysis has focused on classifying a revision's purpose [9,19], assessing its quality [2], or understanding temporal patterns [14]. Our *revision extractor and purpose classifier* integrates binary purpose schemas, sentence alignment algorithms, and predictive NLP features from this literature [19]. An alternative approach not requiring revision extraction computes linguistic properties for essay drafts separately, then identifies changes [13].

Prior versions of ArgRewrite were either fully automated demonstration systems not evaluated with users [18] or Wizard of Oz semi-automated prototypes evaluated in lab contexts [1]. Generally, automated NLP revision analysis has not been used to trigger feedback in a target AWE system [13,14]. The ArgRewrite version described below is *fully-automated* and *deployed in a college class.*

2 ArgRewrite: System and Classroom Deployment

The ArgRewrite **backend** uses NLP algorithms developed in our prior work [19] to perform revision extraction and purpose classification. Given the raw text of two essay drafts, an algorithm aligns sentences across drafts based on similarity and global context; remaining added or deleted sentences are aligned to null. The pairs of non-identical aligned sentences are extracted as the essay *revisions.* Finally, a classifier predicts whether the purpose of each revision is to change the meaning of the essay or not [6], using the labels *content* or *surface*, respectively. The classifier was learned using a random forest algorithm due to the small size of the training corpus; four linguistic feature groups encoded each revision [19].

The **frontend** ArgRewrite interfaces were previously evaluated with positive outcomes in a wizarded lab study [1]. The present interfaces were slightly redesigned to better guide students in autonomously using ArgRewrite over the web in a class deployment. Students were first taken to the *Overview Interface* (Fig. 1a), which visualized the NLP results using a revision distribution pie chart (left) and a revision map (right). Next, students were taken to the *Review Interface* (Fig. 1b), which used color coding to convey the purpose labeling for each revised sentence (left). When a student clicked on a sentence revision, a details window showed the character-level differences between the original and revised sentence (top). By clicking the next button, students were taken to the *Revision Interface* (similar to the *Review Interface* but with an additional essay tab with no highlights) to further revise their essay. They were then returned to

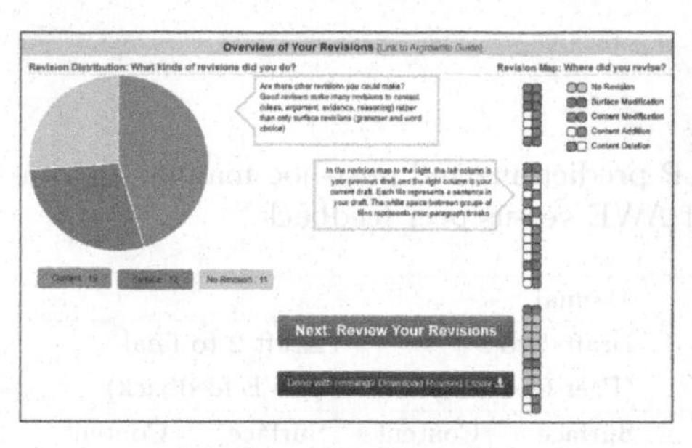

(a) Overview Interface

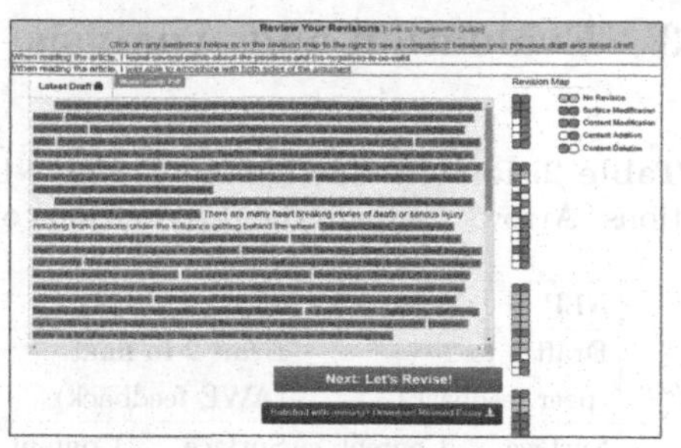

(b) Review Interface

Fig. 1. ArgRewrite interfaces

Table 1. ArgRewrite users providing IRB consent (column 1) and their revising summary (column 2). Revisions by students who created draft3 and beyond (column 3). NLP revision purpose classifier performance (column 4).

	Users who upload 2 drafts	User subset who revise (# additional drafts)	Total revisions (avg. per student)	Classifier (F1)
Assignment 1	7	4 (2.3)	26 (8.6)	91.1%
Assignment 2	16	7 (2.7)	291 (41.6)	94.7%
Assignment 3	14	5 (1)	38 (7.6)	90.4%

the *Overview Interface* to start a new cycle of revision (with the re-revised draft automatically uploaded as the latest draft), or to download their final essay.

We **deployed** ArgRewrite in a fall 2019 undergraduate cognitive psychology class that required three writing assignments involving two paper drafts. For each assignment, students 1) wrote draft1 of a paper in response to a prompt, 2) used a peer-review system to provide rubric-guided feedback on the papers of three other students, 3) wrote draft2 of their own paper after receiving peer feedback, and 4) engaged in a final round of peer review. Students were given the option to use ArgRewrite between steps 3 and 4, by submitting draft1 and draft2 of their papers to ArgRewrite, potentially creating further drafts based on the system's feedback, then downloading the final revised draft from ArgRewrite and using it (rather than draft2) for the second phase of peer review.

Although the use of ArgRewrite was completely voluntary, the instructor encouraged it in different ways across assignments: providing a demo in class for Assignment 1, then emailing low-performing students and offering extra credit for Assignment 2. Of the 157 students in the class, 31 used the system at least once. However, only 24 of these students gave IRB consent to use their data for the evaluation below. Of these, 2 students used the system to revise 1 assignment, 19 revised two assignments and only 3 used the system to revise all three assignments. Columns 1 and 2 of Table 1 show the user distribution per assignment.

3 Evaluation and Analysis

Table 2. Revision distributions, using NLP predictions and post-hoc manual annotations. Arrows compare the distributions of AWE versus peer feedback.[a]

	NLP				Manual			
	Draft 1 to 2 (peer feedback)		Draft 2 to final (AWE feedback)		Draft 1 to 2 (Peer feedback)		Draft 2 to final (AWE feedback)	
	Surface	Content	Surface	Content	Surface	Content	Surface	Content
A1	53 (32%)	113 (68%)	14 (54%↑)	12 (46%↓)	43 (27%)	119 (73%)	14 (56%↑)	11 (44%↓)
A2	166 (42%)	231 (58%)	89 (31%↓)	202 (70%↑)	148 (38%)	244 (62%)	94 (33%↓)	191 (67%↑)
A3	139 (51%)	134 (49%)	30 (79%↑)	8 (21%↓)	124 (47%)	138 (53%)	29 (76%↑)	9 (24%↓)

[a] Alignment errors in AWE cause the number of revisions to differ slightly.

NLP Revision Purpose Classifier. NLP performance was analyzed by comparing the classifier's purpose predictions to gold-standard labels that were manually annotated after the assignments were submitted. Each annotation was done by one of three experts familiar with the coding scheme (kappa > .7 in the lab study [1]). The last column of Table 1 shows that for all assignments, macro F1 in binary revision purpose prediction was greater than 90%. This was impressive as the classifier was developed by training on essays responding to two topics from high school English class assignments, but tested on essays responding to three topics from college psychology class assignments.

Student Revision Behavior. Table 1 shows that some students using ArgRwrite engaged in further revision (column 2) beyond peer review (column 1), and often engaged in multiple cycles of revision (column 2 in the parenthesis), e.g., writing 4th and even 5th drafts. Table 1 (column 3) also shows the total number of revisions made by the students who performed self-monitored revising after draft 2. In Table 2, the arrows show that for Assignment 2, the AWE system prompted more content revisions than peer feedback and a high percent of content revisions as compared to surface revisions. Content revisions are generally considered more important in revising and more difficult for students [6]. Possible reasons for the lower percentage of ArgRewrite content revisions in Assignments 1 and 3 could be the students' acclimation to the system during Assignment 1, the direct recruitment with extra credit for Assignment 2, and the low number of students for Assignments 1 and 3. Due to the high accuracy of the NLP classifier, the same inferences (represented by the arrows) can be drawn whether revisions purposes are predicted by NLP or are manually annotated.

Perceived Usability of ArgRewrite. 14 students who used the system for at least assignments 2 and 3 completed a survey at the end of the course. The survey (shown in Table 3) included educational technology usability items (1–7) [8] and items customized for ArgRewrite (8–14). The 'Class' column shows that students responded positively to 13/14 items (i.e., mean Likert values > 3, on a scale from 1–5). The highest score (item 13) indicates that students' perception of classifier

Table 3. Mean scores (1 = strongly disagree; 5 = strongly agree), comparing class deployment (n=14) to lab study (n=22). ** p < .01; * p <.05, + p < .1

	Survey Item	Class	Lab
1	System allows me to have a better understanding of my revision efforts	3.29	3.95*
2	I find the system easy to use	3.43	4.18*
3	My interaction with the system is clear and understandable	3.29	4.14**
4	The system helps me to recognize the weakness of my essay	2.71	3.32
5	System encourages me to make more revisions (quantity)	3.36	3.86
6	System encourages me to make more meaningful revisions (quality)	3.29	3.86
7	Overall the system is helpful to my writing	3.29	3.73
8	I found the "Overview of Your Revisions" page to be useful	3.43	4.14*
9	I found it useful to highlight my revision purposes in different colors	3.71	4.27+
10	I found the revision map visualization useful	3.21	4.09*
11	I found the small window of revision details to be useful	3.43	4.64**
12	I found it helpful to know whether my revision was a "surface" or "content" level change	3.57	4.05
13	My revisions were usually labeled correctly by the system	3.86	4.00
14	I trust the feedback that the system gave me	3.64	3.59

performance reflected the objective results in Table 1. The fact that items 9 and 12 had higher scores than item 10 suggests that feedback on revision purposes was more useful than feedback on revision location. The lowest score (item 4) focused on the essay rather than on the revisions. When focusing specifically on the revisions (e.g., items 1, 5, 6) and the writing process (item 7), the item responses were all positive. We also compared the students in the current study to 22 participants who responded to the same items in our wizarded lab study [1]. The mean 'Class' versus 'Lab' scores were compared using non-paired t-tests, with the results shown in the last two columns of Table 3. This analysis is quasi-experimental since there was no random assignment of survey respondents to the class versus lab conditions. Table 3 shows that for all but the last survey item, the average score in the classroom study was lower than in lab study. Perhaps the class participants are a more critical audience because their actual assignment grade was at stake. Finally, the relative pattern of response values across survey items demonstrated a moderate positive relationship across the class and lab responses (Pearson correlation R=.46, p <.1).

4 Conclusion and Future Directions

This paper described the ArgRewrite system for supporting self-monitored revising. NLP extracts revised sentences between paper drafts and classifies revision purposes, while visualizations convey the NLP results.

A classroom deployment suggests that NLP accurately analyzes revisions and that students engage in self-monitored revising and find the visualizations useful. Future plans include predicting fine-grained purposes using transformers, assessing a revision's quality and alignment with feedback, incorporating system guidance and tutoring, and evaluating via a controlled experiment rather than an 'in the wild' study.

References

1. Afrin, T., Kashefi, O., Olshefski, C., Litman, D., Hwa, R., Godley, A.: Effective interfaces for student-driven revision sessions for argumentative writing. In: Proceedings CHI Conference on Human Factors in Computing Systems, pp. 1–13 (2021)
2. Afrin, T., Litman, D.: Annotation and classification of sentence-level revision improvement. In: Proceedings 13th Workshop on Innovative Use of NLP for Building Educational Applications, pp. 240–246, New Orleans, Louisiana, June 2018
3. Burstein, J., Riordan, B., McCaffrey, D.: Expanding automated writing evaluation. In: Handbook of Automated Scoring, pp. 329–346. Chapman and Hall/CRC (2020)
4. Crossley, S.A., Allen, L.K., McNamara, D.S.: The writing pal: a writing strategy tutor. In: Adaptive Educational Technologies for Literacy Instruction, pp. 204–224, Routledge (2016)
5. Deane, P., Wilson, J., Zhang, M., Li, C., van Rijn, P., Guo, H., Roth, A., Winchester, E., Richter, T.: The sensitivity of a scenario-based assessment of written argumentation to school differences in curriculum and instruction. Int. J. Artif. Intell. Educ. **31**(1), 57–98 (2021)
6. Faigley, L., Witte, S.: Analyzing revision. Coll. Compos. Commun. **32**(4), 400–414 (1981)
7. Foltz, P.W., Rosenstein, M.: Data mining large-scale formative writing. In: Handbook of Learning Analytics, p. 199 (2017)
8. Holden, H., Rada, R.: Understanding the influence of perceived usability and technology self-efficacy on teachers' technology acceptance. J. Res. Technol. Educ. **43**(4), 343–367 (2011)
9. Kashefi, O., et al.: Argrewrite v. 2: an annotated argumentative revisions corpus. Language Resources and Evaluation, pp. 1–35 (2022)
10. MacArthur, C., Philippakos, Z., Ianetta, M.: Self-regulated strategy instruction in college developmental writing. J. Educ. Psychol. **107**(3), 855 (2015)
11. Mayfield, E., Butler, S.: Districtwide implementations outperform isolated use of automated feedback in high school writing. In: International Conference of the Learning Sciences, vol. 2128, London, UK (2019)
12. Roscoe, R.D., McNamara, D.S.: Writing pal: feasibility of an intelligent writing strategy tutor in the high school classroom. J. Educ. Psychol. **105**(4), 1010–1025 (2013)
13. Roscoe, R.D., Snow, E.L., Allen, L.K., McNamara, D.S.: Automated detection of essay revising patterns: applications for intelligent feedback in a writing tutor. Technol. Instr. Cogn. Learn. **10**(1), 59–79 (2015)
14. Shibani, A.: Constructing automated revision graphs: a novel visualization technique to study student writing. In: Bittencourt, I.I., Cukurova, M., Muldner, K., Luckin, R., Millán, E. (eds.) AIED 2020. LNCS (LNAI), vol. 12164, pp. 285–290. Springer, Cham (2020). https://doi.org/10.1007/978-3-030-52240-7_52
15. Wang, E.L., et al.: erevis(ing): students' revision of text evidence use in an automated writing evaluation system. Assessing Writing **44**, 100449 (2020)
16. Wilson, J., Huang, Y., Palermo, C., Beard, G., MacArthur, C.A.: Automated feedback and automated scoring in the elementary grades: Usage, attitudes, and associations with writing outcomes in a districtwide implementation of mi write. Int. J. Artif. Intell. Educ., 1–43 (2021)
17. Wingate, U.: The impact of formative feedback on the development of academic writing. Assess. Eval. High. Educ. **35**(5), 519–533 (2010)

18. Zhang, F., Hwa, R., Litman, D., B. Hashemi, H.: Argrewrite: a web-based revision assistant for argumentative writings. In: Proceedings f NAACL Conference: Demonstrations, San Diego, California, pp. 37–41 (2016)
19. Zhang, F., Litman, D.: Annotation and classification of argumentative writing revisions. In: Proceedings of the 10th Workshop on Innovative Use of NLP for Building Educational Applications, pp. 133–143. Denver, Colorado, June 2015

What Does Shared Understanding in Students' Face-to-Face Collaborative Learning Gaze Behaviours "Look Like"?

Qi Zhou[1(✉)], Wannapon Suraworachet[1], Oya Celiktutan[2], and Mutlu Cukurova[1]

[1] University College London, London, UK
qtnvqz3@ucl.ac.uk
[2] King's College London, London, UK
oya.celiktutan@kcl.ac.uk

Abstract. Several studies have shown a positive relationship between measures of gaze behaviours and the quality of student group collaboration over the past decade. Gaze behaviours, however, are frequently employed to investigate i) students' online interactions and ii) calculated as cumulative measures of collaboration, rarely providing insights into the actual *process* of collaborative learning in *real-world settings*. To address these two limitations, we explored *the sequences of students' gaze behaviours* as a process and its relationship to *collaborative learning in a face-to-face environment*. Twenty-five collaborative learning session videos were included from five groups in a 10-week post-graduate module. Four types of gaze behaviours (i.e., gazing at peers, their laptops, tutors, and undefined objects) were used to label student gaze behaviours and the resulting sequences were analyzed using the Optimal Matching (OM) algorithm and Ward's Clustering. Two distinct types of gaze patterns with different levels of shared understanding and collaboration satisfaction were identified, i) peer-interaction focused (PIF), which prioritise social interaction dimensions of collaboration and ii) resource-interaction focused (RIF) which prioritise resource management and task execution. The implications of the findings for automated detection of students' gaze behaviours with computer vision and adaptive support are discussed.

Keywords: Learning analytics · Face-to-face collaborative learning · Gaze behaviours · Process mining · Computer vision

1 Introduction

In recent years, multiple data sources and analytics techniques have been applied to extract insights from collaborative learning settings. However, the majority of existing research focuses on log data of student interactions in digital settings, followed by questionnaires and verbal documentation which are then analysed with descriptive and inferential statistics [1]. As presented in a recent systematic review on social learning, the dominant analytical approach researchers use is social network analysis, followed by inferential statistics and the dominant data source used is students' online traces [2]

M. M. Rodrigo et al. (Eds.): AIED 2022, LNCS 13355, pp. 588–593, 2022.
https://doi.org/10.1007/978-3-031-11644-5_53

while almost completely ignoring what is happening outside of the digital space. Nevertheless, the overreliance on digital traces from a single platform arguably provides insufficient information, overlooks learning as an ecosystem [1], and undervalues many real-world social context complexities that are crucial for social learning [2]. Investigation of students' real-world nonverbal behaviours from video data and computer vision techniques is an understudied area for AIED. Here, we investigated the sequences of students' gaze behaviours from a real-world face-to-face collaborative learning activity from videos and analysed their relationship to perceived shared understanding.

2 Background Research on Gaze Behaviours in Collaboration

Gaze behaviours are considered to be a crucial element for the building of shared understanding in collaborative learning. Learners use gaze to streamline speech, co-present, and disambiguate and direct others' attention. In eye-tracking research, gaze behaviours have been shown to have good potential for understanding and predicting the quality of collaboration through different measures such as joint visual attention (JVA) [3], gaze overlap [4], and attention similarity [5]. However, these features, which measure the cumulative frequency of whether learners are looking at the same object, can hardly be used to represent the complex process of disambiguating and directing attention [6]. As Fan and colleagues [7] argued, the establishment of "shared attention" in social contexts through gaze, consists of a sequence of gaze behaviours from involved agents rather than being a single act. It usually requires initial mutual attention in time, referring to the point of attention, following the reference, and shared attention. Considerations of gaze behaviours as a process might provide better insights into students' collaborative learning but are rarely considered in educational research studies.

Most existing gaze behaviour investigations in collaborative learning research come from eye-tracking studies. However, existing studies on gaze behaviours in collaborative learning are limited due to various inherent challenges. Firstly, limited by equipment and technology, most of the studies looked at collaboration in digital learning environments [8]. Previous work has used eye trackers [9] or markers in the real world [10] to capture learners' attentive region. These studies illustrated the close relationship between learners' visual attention and their collaborative learning outcomes. However, they focused more on the visual attention in the collaborative working space rather than the attention among peers, which also has been considered an important gaze behaviour during collaborative learning [11]. Secondly, nearly all published studies were conducted in a laboratory context rather than investigating natural real-world learning environments [10]. The effectiveness of the identified proxies has not been studied in an ecological setting which may have more interference and may be longer in duration than in studied experimental conditions.

Yet, understanding gaze communication dynamics *in face-to-face collaborative learning settings and interpreting students' gaze behaviours from video data with computer vision are understudied*. In this paper, we present a novel representation of gaze communication dynamics specific to real-world collaborative learning environments, which has significant implications for developing novel computer vision algorithms and AIED tools to provide timely and useful interventions and feedback.

3 Methodology

3.1 Context of Study, Data Collection, and Pre-processing

The data was collected from a 10-week postgraduate module. Students were assigned into groups of 4 or 5 students with interdisciplinary backgrounds, mixed-gender and varied first languages. Within each week, students were requested to attend a 1-h face-to-face session to discuss and complete a weekly task collaboratively on Miro (miro.com) which they accessed through their laptops/tablets.

During the sessions, students were seated as a group around a T-shaped table, facing a camera. Twenty-three sessions, lasting from about 33 min to about 67 min, have been used as the final dataset in this study. The first frames of each second from a particular session were extracted to generate a new video for the labelling of gaze behaviours in the analysis. After each session, students were asked to fill in a post-survey with 5-points Likert scale questions about their shared understanding. Ethics approval was received from the institution and individual consents were given by students before the start of the study.

3.2 Coding the Gaze Behaviours

We categorized the learners' gaze behaviours into four main categories: looking at a student (S), looking at a laptop (L), looking at a tutor (T), and looking at other objects (O). To be more specific, code S refers to the gaze behaviours of a student looking at another student in the same group. The learners in the group were labelled from 1 to n, where n is the number of students within the group. By using the code S1 to Sn, the actual learner who has been gazing at can be identified. Code L represents situations when the learner was looking at the laptop on the desk. L1 is used when the learner was looking at his/her own laptop while L2 is used when looking at another member's laptop. Code T refers to a situation when the learner was looking at the tutor who appeared in the video. Code O is used when the learner was looking at other objects which have not been defined above. For example, learners who were looking at their own gestures while speaking, or looking at food/cups on the table would be coded as O.

Computer Vision Annotation Tool (CVAT) (cvat.org) tool was used for video annotation. The coding scheme was implemented by two researchers. A sample video of 1000 frames was coded by both to achieve the consensus of coding with high reliability (Cohen's Kappa = 0.98).

3.3 Feature Engineering from Labelled Gaze Behaviours and Analysis

The shaping of shared understanding does not happen in a single gaze moment and requires to be analysed as a process. Here, we engineered a process feature named Shared Attention (SA) as a proxy to measure whether learners shared gaze attention in a specific time period. Ten-frame windows (representing ten seconds in original videos) were used to generate the process-based feature. In a specific window, the students who have been gazed at by over half of the students and the students who gazed at them were marked as "1", which means they might participate in building shared attention. To increase the accuracy of the processing, the overlapping window method was used. The window size was chosen as ten frames and the window was moved two frames further for each time. The output SA sequence for the whole group is consist of the ratio of shaping shared gaze attention for each frame in this session.

The Optimal Matching Algorithm (OMA) was applied to explore the gaze behaviour sequences. Based on the distance matrix obtained from OMA, further cluster analysis was applied. Before implementing OMA, numerical values in SA sequences were converted into codes. The numerical value "0" in the original SA sequences was labelled as "passive (P)" since learners showed no shaping shared gaze attention when this ratio is 0. On the contrary, the value "1" was coded as "active (A)". The values between 0 and 1 were coded as "Semi-active (S)". Since students had to follow the same set of activities regardless of their sessions and the length of activities varied, to avoid value loss, the first thirty minutes of each sequence were used. In total, 23 sequences with 1800 frames were included in the analysis.

A 23×23 matrix was the output of OMA at the session-level. Each cell in the matrix represented the "distance" between the following sequences. Then, Ward's Clustering was applied to hierarchically cluster the sequences with similar patterns across sessions. The agglomerative coefficient, which reflects the tightness of clustering, was 0.59.

4 Results and Discussion

Figure 1(a) shows the clusters of gaze behaviour sequences in sessions. According to this tree graph, we divided 23 input sessions into two types.

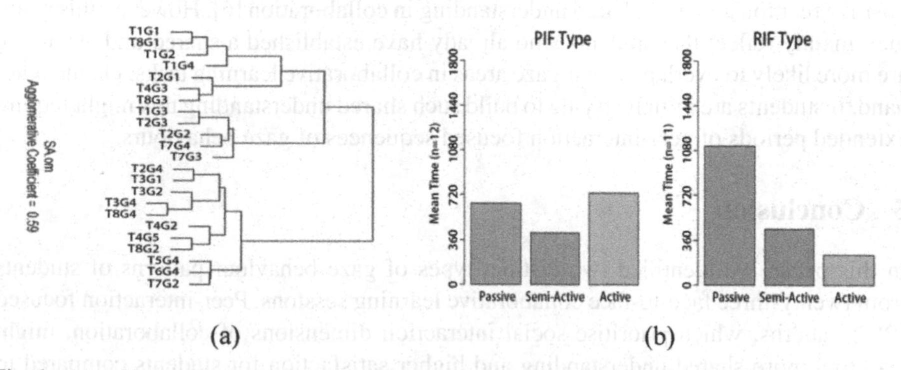

Fig. 1. (a) A hierarchical tree represents the result from Ward's clustering in which T represents a task number and G represents a group number. (b) A relative frequency of codes (1 = Passive, 2 = Semi-active, 3 = Active) in each cluster (Type 1 and Type 2).

The first type contains the top 12 sessions and the second type contains the bottom 11 sessions on the tree. Figure 1(b) shows the frequency of 3 codes in these two types. The green, purple and orange bars represent "Passive", "Semi-active", and "Active" shared gaze periods respectively. According to Fig. 1(b), these two types have a similar frequency of "semi-active" states. Meanwhile, type 1 presents more frequency of being "active" than type 2. The "active" status in type 1 appeared more frequently which means a longer period of active state was achieved compared to type 2. It can also be inferred that the shared gaze attention lasted longer in type 1 sessions. In other words, students from type 1 exhibited patterns of longer shared gaze periods. On the contrary, students from type 2 tended to focus more on completing the task on their Miro boards and gazing at their laptops. Type 1 sequences of gaze behaviours might be better associated with interactive and socio-emotional dimensions [12] of collaborative learning. These gaze sequences are more likely to occur when students are interacting with peers, actively listening to others, encouraging participation and inclusion of peers etc. On the other hand, Type 2 gaze behaviours may be better associated with the behavioural and regulative dimensions [13]. These gaze sequences are more likely to occur while students are doing resource management, taking actions on their laptops and during task execution phases. Therefore, we named type 1 sequences as the peer interaction focused (PIF) type and type 2 as the resource interaction focused (RIF) type. It is worth noting that, the types of tasks and groups did not show significantly different distribution between PIF and RIF patterns (Fig. 1(a)). This illustrates the potential of these sequences to be task and group size-independent features.

The PIF type ($m = 3.54$, $SD = 0.41$) and the RIF type ($m = 3.54$, $SD = 0.31$) did not show statistically significant difference in terms of their perceived shared understanding (SU). It means that a higher frequency of shared gaze attention with peers may not always lead students to perceive a better shared understanding in collaboration. Rather, groups that lack shared understanding might spend long periods of PIF sequences of gaze behaviours, trying to establish a shared understanding. Meanwhile, the shared understanding values of the PIF type are distributed wider than in the RIF type. Previous research illustrated that JVA (measured as overlapping gaze areas) had a significantly positive relationship with shared understanding in collaboration [6]. However, this result may mainly reflect that students who already have established a shared understanding are more likely to overlap in their gaze areas in collaborative learning tasks. On the other hand, if students are initially trying to build such shared understanding this might require extended periods of peer-interaction focused sequences of gaze behaviours.

5 Conclusion

In this paper, we identified two distinct types of gaze behaviour patterns of students from twenty-three face-to-face collaborative learning sessions. Peer-interaction focused (PIF) patterns, which prioritise social interaction dimensions of collaboration, might lead to a more shared understanding and higher satisfaction for students compared to resource-interaction focused (RIF) patterns, which prioritise resource management and task execution. This work has significant implications for developing novel computer vision algorithms and hence designing fully automatic behavioural analytics tools to provide intervention and feedback in real-world learning environments.

References

1. Mangaroska, K., Giannakos, M.: Learning analytics for learning design: a systematic literature review of analytics-driven design to enhance learning. IEEE Trans. Learn. Technol. **12**, 516–534 (2019). https://doi.org/10.1109/TLT.2018.2868673

2. Kaliisa, R., Rienties, B., Mørch, A.I., Kluge, A.: Social learning analytics in computer-supported collaborative learning environments: a systematic review of empirical studies. Comput. Educ. Open. **3**, 100073 (2022). https://doi.org/10.1016/j.caeo.2022.100073
3. Schlösser, C., Schlieker-Steens, P., Kienle, A., Harrer, A.: Using real-time gaze based awareness methods to enhance collaboration. In: Baloian, N., Zorian, Y., Taslakian, P., Shoukouryan, S. (eds.) CRIWG 2015. LNCS, vol. 9334, pp. 19–27. Springer, Cham (2015). https://doi.org/10.1007/978-3-319-22747-4_2
4. D'Angelo, S., Gergle, D.: An eye for design: gaze visualizations for remote collaborative work. In: Proceedings of the 2018 CHI Conference on Human Factors in Computing Systems, Montreal, QC, Canada, pp. 1–12. ACM (2018). https://doi.org/10.1145/3173574.3173923
5. Papavlasopoulou, S., Sharma, K., Giannakos, M., Jaccheri, L.: Using eye-tracking to unveil differences between kids and teens in coding activities. In: Proceedings of the 2017 Conference on Interaction Design and Children, Stanford, California, USA, pp. 171–181. ACM (2017). https://doi.org/10.1145/3078072.3079740
6. Sharma, K., Olsen, J.K., Verma, H., Caballero, D., Jermann, P.: Challenging joint visual attention as a proxy for collaborative performance, pp. 91–98 (2021)
7. Fan, L., Wang, W., Zhu, S.-C., Tang, X., Huang, S.: Understanding human gaze communication by spatio-temporal graph reasoning. In: 2019 IEEE/CVF International Conference on Computer Vision (ICCV), Seoul, Korea (South), pp. 5723–5732. IEEE (2019). https://doi.org/10.1109/ICCV.2019.00582
8. D'Angelo, S., Schneider, B.: Shared gaze visualizations in collaborative interactions: past present and future. Interact. Comput. **33**, 115–133 (2021). https://doi.org/10.1093/iwcomp/iwab015
9. Yang, C.-W., Cukurova, M., Porayska-Pomsta, K.: Dyadic joint visual attention interaction in face-to-face collaborative problem-solving at K-12 maths education: a multimodal approach, pp. 61–70 (2021)
10. Sung, G., Feng, T., Schneider, B.: Learners learn more and instructors track better with real-time gaze sharing. Proc. ACM Hum. Comput. Interact. **5**, 1–23 (2021). https://doi.org/10.1145/3449208
11. Joiner, R., Scanlon, E., O'Shea, T., Smith, R.B., Blake, C.: Evidence from a series of experiments on video-mediated collaboration: does eye contact matter? In: Proceedings of CSCL, p. 371 (2002). https://doi.org/10.3115/1658616.1658669
12. Rogat, T.K., Adams-Wiggins, K.R.: Interrelation between regulatory and socioemotional processes within collaborative groups characterized by facilitative and directive other-regulation. Comput. Hum. Behav. **52**, 589–600 (2015). https://doi.org/10.1016/j.chb.2015.01.026
13. Malmberg, J., Järvelä, S., Järvenoja, H.: Capturing temporal and sequential patterns of self-, co-, and socially shared regulation in the context of collaborative learning. Contemp. Educ. Psychol. **49**, 160–174 (2017). https://doi.org/10.1016/j.cedpsych.2017.01.009

Improving Prediction of Student Performance in a Blended Course

Sergey Sosnovsky[1(✉)] and Almed Hamzah[1,2]

[1] Utrecht University, Utrecht, The Netherlands
s.a.sosnovsky@uu.nl
[2] Universitas Islam Indonesia, Yogyakarta, Indonesia

Abstract. Traditionally, systems supporting blended learning focus only on one portion of the course by tracing students' interaction with learning content at home. In this paper, we argue that in-class activity can be also instrumental in eliciting the true state of students' knowledge and can lead to more accurate models of their performance. Quizitor is an online platform that delivers both the at-home and the in-class assessment. We show that a combination of the two streams of data that Quizitor collects from students can help build more accurate models of students' mastery that help predict their course performance better than models separately trained on either of these two types of activity.

Keywords: Self-assessment · Blended learning · Student modelling · Adaptive learning support · Voting tool

1 Introduction

Effective learning support in a large blended course can be challenging, especially, when a course population is diverse. An adaptive system aiming to facilitate such support should be able to accurately predict student performance in the course as the first step in administering effective adaptive interventions [1]. Intelligent tutoring systems [2] and adaptive educational hypermedia systems [3] have proven their effectiveness in various subjects and learning contexts. Unfortunately, such systems primarily focus only on one portion of the course by tracing students' interaction with learning content at home. Such a focus on the at-home part of the blended learning is understandable, as in most models of blended learning, the online component assumes individual, self-regulated work; which means, students may struggle with planning their learning, engaging in learning activities, reflecting on potential mistakes, etc. In fact, effective regulation of independent studying becomes the biggest challenge for students in blended learning scenarios [4].

Somewhat counter-intuitively, there have not been many effective attempts to propose working solutions for a unified support of the both components of

The research presented in this paper is partially supported by Universitas Islam Indonesia under Doctoral Grant for Lecturer 2019 (grant no 1296).

M. M. Rodrigo et al. (Eds.): AIED 2022, LNCS 13355, pp. 594–599, 2022.
https://doi.org/10.1007/978-3-031-11644-5_54

blended learning: in-class and at-home. Most of the existing literature focused on theoretical frameworks and architectures [5,6]. This paper is trying to make a more practical step in this direction by describing and evaluating an assessment tool that can be used both in class and at home and demonstrating the potential value of blending the two respective data streams.

A combination of in-class and at-home assessment coupled with adaptive support has a potential to significantly improve learning experiences in a blended course. In-class assessment and at-home self-assessment have different purposes, but they both can provide valuable information about student progress and opportunities for targeted interactions. The in-class assessment keeps students engaged and can serve as initial input on their conceptual understanding. The at-home self-assessment helps students practice acquired skills at individual pace and receive adaptive guidance. Combining these two streams of data in a single system could directly benefit students by enabling their reflection on the current progress and building a stronger link between knowledge and skills thus facilitating deeper understanding of the subject.

This paper presents Quizitor - a system that supports two modes of assessment in a blended course. It can be used by a teacher during a lecture for a pop-up synchronised assessment of the entire class, and by a student at home for individual self-paced assessment. We have evaluated Quizitor in an undergraduate programming course. An analysis of the collected data shows that a model integrating student activity from both at-home and in-class assessment can predict students' performance better than models trained on individual streams of activity. This effect persists when the data are aggregated on the level of the course as well as when we narrow down to topic-based models of student performance. Hence, by tracking students' attempts across the both modes and integrating the both streams of data, Quizitor has a potential to maintain a more accurate model of student performance in a blended course and a more holistic adaptive support of blended learning.

2 Quizitor

Quizitor is a hybrid quiz platform that can be used for both in-class and at-home assessments. The main components of its interface are depicted in Fig. 1.

The in-class assessment mode facilitates synchronous assessment where students take a quiz in a class with their teacher. The aims of such assessment can include: taking a short break from a lecture routine, asking students to recall the learning material that has been recently taught, helping students reflect on their understanding of the material, and giving the teacher information on how well students understand the material. A teacher controls when an in-class quiz (and every question within it) starts and finishes. The top-left screen on Fig. 1 shows the teacher interface of an in-class quiz. On this screen, the teacher can see the current question, monitor the time spent on it, the number of submitted answers, and the number of students currently participating in the quiz. Students can see the current question on their devices as well (top-left screen on

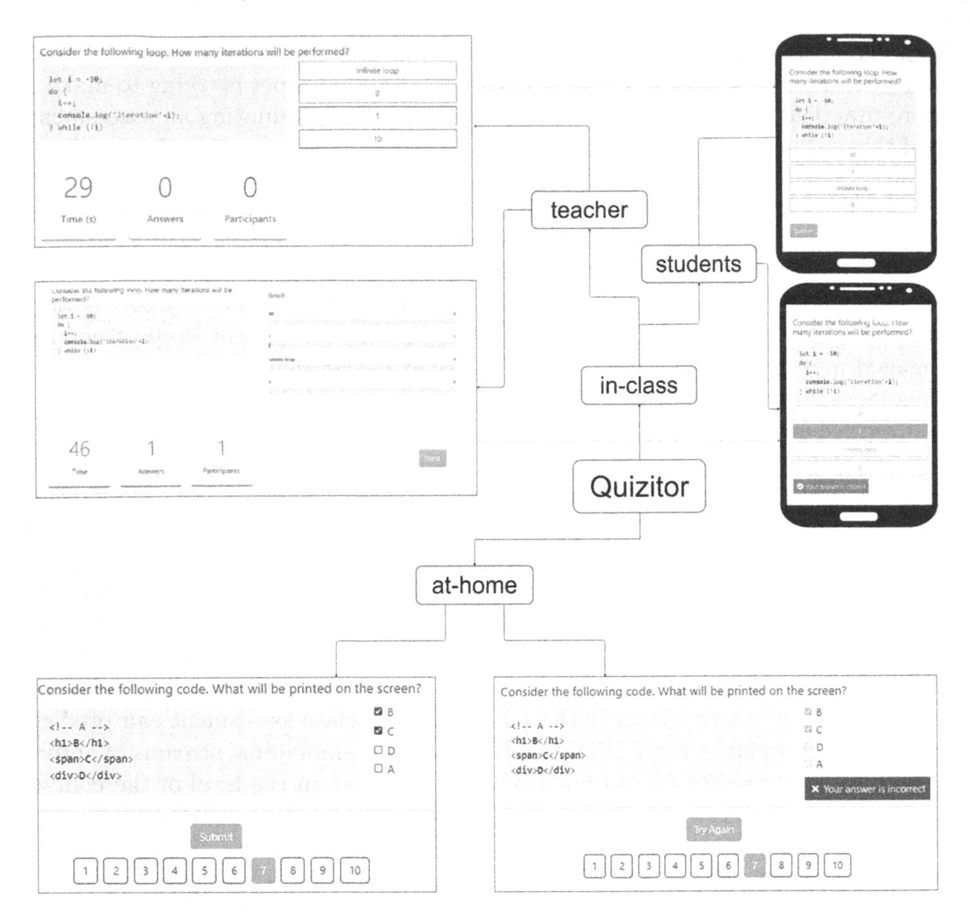

Fig. 1. User interfaces of Quizitor.

Fig. 1). They can submit an answer to the question once it is started, but will not move to the next question until the teacher decides to start it for the entire class. Once a teacher stops a question, the summary of its results is presented to the class on both the teacher's screen (middle-left), and individual students' screens (middle-right). The summary shows a distribution of different answers and indicates which answer is correct. After a brief discussion, the teacher can move to the next question.

The at-home mode is designed as a typical tool for individual self-assessment. The primary aims of Quizitor in this mode are to help students practice, reflect, identify knowledge gaps and prepare for exams. In contrast with the in-class questions, the at-home questions can be more complex, as students are not under time pressure when answering them. They can choose the day, time, and location where they want to take the quiz. For each question, students can submit as many attempts as they want. The feedback indicates only the correctness of the attempt and invites a student to repeat the question if the attempt was not correct.

3 Evaluation

The main hypothesis of this study is that models of student mastery taking into account the two streams of data coming from students' in-class and at-home assessment activity would be able to predict student course performance better than the models taking into account only individual streams of data.

3.1 Data Collection

The data were collected in the undergraduate course on Web technology taught in Utrecht University from February until March 2021. The overall number of students was 198. To participate in the study, students had to sign a consent form. We excluded from the analysis students who did not use the tool actively enough (attempted 75% of at-home questions) and those who did not pass the midterm exam. The resulting number of subjects in this study was 61. The use of Quizitor started during the third lecture and continued for six lectures until the midterm. The topics included basics of HTML, CSS, DOM, and Javascript.

3.2 Models of Students' Mastery

To estimate students' mastery based on their activity with Quizitor, we applied Elo Rating System (ERS), which is a relatively easy yet accurate method for modelling an ability. It has been recently gaining popularity in the educational data mining and student modelling community [7]. It can dynamically assess students' ability in a certain field based on the results of their continuous assessment. While assessing student ability, ERS also keeps adjusting the difficulty of questions that students answer. Essentially, ERS constantly balances the "strength" (=ability) of a student vs. the "strength" (=difficulty) of a question.

Two sets of student models have been built: the in-class (IC) models and the at-home (AH) models. The combined IC model is trained based on all students' in-class attempts. The combined AH model represents students' mastery as a result of their overall at-home self-assessment. Individual topical AH and IC models have been trained only based on the data from AH and IC quizzes pertaining to corresponding topics. In order to compute more accurate students' Elo scores, first we have estimated the Elo scores of all questions, i.e., their levels of difficulty. First, we split all students into two groups of 80% and 20%. The question difficulty is estimated by calculating their Elo ratings based on the answers from 80% of students. Then, the obtained question model is used to estimate the Elo scores of the remaining 20% of students. Then, another group of 20% of students is selected and the processes restarts. After five iterations, mastery of all students have been modeled. We have repeated this process separately to compute the IC and AH models.

3.3 Results

Simple linear regression models have been used to predict students' midterm performance based on their mastery estimates. There are four pairs of models (AH and IC) for course topics and one more pair of combined models, hence the simple regression has been computed ten times. After that five multiple regression models have been computed to verify the main hypothesis. Significant positive regression coefficients have been found for almost all models (except for IC model for the topic DOM). Table 1 provides the summary of all fifteen regression models. It is easy to see, that the main hypothesis is confirmed. Bigger portions of the variability in the predicted variables are explained by the joint models. Both for the overall case and for each individual topic. The results are consistent across all four target topics and the overall case. This means that both modes of students' work with Quizitor can provide mutually enriching sources of data. An effective "blend" of these data can inform an adaptive tool truly supporting blended learning. The adjusted R^2 of the combined models are also much higher compared to individual models indicating absence of overfitting.

Table 1. Result from regression model

Source	Model	R^2	R^2-adj	p-value
Overall	IC	0.117	0.102	0.007
	AH	0.114	0.099	0.008
	IC-AH	0.21	0.182	0.001
HTML	IC	0.1	0.089	0.003
	AH	0.047	0.036	0.042
	IC-AH	0.136	0.115	0.002
CSS	IC	0.152	0.13	<0.001
	AH	0.113	0.09	0.03
	IC-AH	0.237	0.198	0.005
DOM	IC	0.073	0.051	0.079
	AH	0.142	0.11	0.044
	IC-AH	0.218	0.157	0.041
JS	IC	0.257	0.231	0.004
	AH	0.154	0.123	<0.001
	IC-AH	0.357	0.309	0.003

4 Discussion and Conclusion

In this paper, we have presented Quizitor - an assessment tool that can deliver both in-class and at-home quizzes. Quizitor has been built as the first step in an attempt to organise truly blended adaptive support in a blended course.

While Quizitor at the moment does not have any adaptive capabilities, its initial evaluation has demonstrated that a combination of data coming from the both face-to-face and online components of a blended course can help achieve a more accurate estimation of student ability than models limited to only one of these components.

There are several directions for future research. First, based on the result, there is an evidence that the two streams of data coming from the in-class and at-home activities have an effect on students' grade. We plan to conduct another experiment with different approaches of student modelling where the in-class and at-home activities are merged as an integrated representation of student ability. Second, we plan to add into Quizitor an adaptive functionality that will support students in working with the question material based on their current levels of knowledge. Such an adaptive support can happen not only during students' at-home activity, but also during their in-class question answering in the form of personalised feedback. This can be done at first on the level of coarse-grained topics. The topic-based analysis described in this paper can be viewed as the first step in this direction.

References

1. Herder, E., Sosnovsky, S., Dimitrova, V.: Adaptive intelligent learning environments. In: Duval, E., Sharples, M., Sutherland, R. (eds.) Technology Enhanced Learning, pp. 109–114. Springer, Cham (2017). https://doi.org/10.1007/978-3-319-02600-8_10
2. Koedinger, K.R., Anderson, J.R., Hadley, W.H., Mark, M.A., et al.: Intelligent tutoring goes to school in the big city. Int. J. Artif. Intell. Educ. **8**(1), 30–43 (1997)
3. Brusilovsky, P., Sosnovsky, S., Yudelson, M.: Addictive links: the motivational value of adaptive link annotation. New Rev. Hypermedia Multimed. **15**(1), 97–118 (2009)
4. Rasheed, R.A., Kamsin, A., Abdullah, N.A.: Challenges in the online component of blended learning: a systematic review. Comput. Educ. **144**, 103701 (2020)
5. Howard, L., Remenyi, Z., Pap, G.: Adaptive blended learning environments. In: International Conference on Engineering Education, pp. 23–28 (2006)
6. Gynther, K.: Design framework for an adaptive MOOC enhanced by blended learning: supplementary training and personalized learning for teacher professional development. Electron. J. e-Learning **14**(1), 15–30 (2016)
7. Yudelson, M.: Elo, i love you won't you tell me your K. In: Scheffel, M., Broisin, J., Pammer-Schindler, V., Ioannou, A., Schneider, J. (eds.) EC-TEL 2019. LNCS, vol. 11722, pp. 213–223. Springer, Cham (2019). https://doi.org/10.1007/978-3-030-29736-7_16

Examining Gender Differences in Game-Based Learning Through BKT Parameter Estimation

Saman Rizvi(✉), Andrea Gauthier, Mutlu Cukurova, and Manolis Mavrikis

UCL Knowledge Lab, University College London, London, UK
{saman.rizvi,m.mavrikis}@ucl.ac.uk

Abstract. The increased adoption of digital game-based learning (DGBL) requires having a deeper understanding of learners' interaction within the games. Although games log data analysis can generate meaningful insights, there is a lack of efficient methods for looking both into learning as a dynamic process and how the game- and domain-specific aspects relate to contextual or demographic differences. In this paper, employing student modelling methods associated with Bayesian Knowledge Tracing (BKT), we analysed data logs from Navigo, a collection of language games designed to support primary school children in developing their reading skills. Our results offer empirical evidence on how contextual differences can be evaluated from game log data. We conclude the paper with a discussion of design and pedagogical implications of the results presented.

Keywords: Game-based learning · Gender differences · BKT

1 Introduction

There is a growing interest in digital educational games as an emerging pedagogy often referred to as digital game-based learning (DGBL). While data from DGBL environments have been used for automating sequencing and feedback (e.g., in adaptive systems) [1], to address research questions around how well games were aligned to educational objectives [2], research has also demonstrated the potential of using such data to derive design implications [3]. For example, [4] presents an approach relying on learning curves analysis to evaluate how the skills targeted within the game fit student performance data. Similarly, [5] explore different design variations to optimize challenge.

Concerning methodology, previous research on AIED (Artificial Intelligence in Education) algorithms showed the significant value of Bayesian methods (such as Bayesian Knowledge Tracing and variations) in terms of their effectiveness in predicting if a student has learned a specific skill by using students' logged data [6]. However, the models built for tracing students' performance might show different results for students with different demographics and individual characteristics. For instance, recently, [7] explored the equitability of knowledge tracing in relation to 'slow' or 'fast' learners. Here, we are particularly interested in applying BKT to understand the variations in student performance and engagement within different game mechanics and between different gender groups of students. The identification of potential gender differences

M. M. Rodrigo et al. (Eds.): AIED 2022, LNCS 13355, pp. 600–606, 2022.
https://doi.org/10.1007/978-3-031-11644-5_55

in student behaviours in DGBL environments requires a robust, practical, and reliable methodology for looking into the interaction of gender with DGBL performance and in-game behaviours for specific games. In this paper, we present an approach to analysing such differences in the context of a language learning DGBL environment with the help of pyBKT, an accessible implementation of BKT. Identification of gender bias is the first step towards its potential mitigation, and it may have important design implications in relation to specific games that tend to appeal to one gender.

2 Methodology

2.1 The Context, Data and Participants

To examine gender disparities in game-based reading development, we leveraged player demographics and data logs from *Navigo*, a collection of adaptive educational games, as part of the iRead project [8]. The iRead database comprised data collected from digital learning products in four European languages (English, Greek, Spanish and German). This study used a subset of data from the English language domain model, generated by 127 students playing *Navigo* games in ten UK primary schools. The students in our data subset belonged to three different year groups as follows: 26 students from year 1 (13 female, 13 male, age range: five to six years old), 28 students from year 2 (15 female, 13 male, age range: six to seven years old) and 73 students from year 3 (37 female, 36 male, age range: seven to eight years old). As a selection criterion, the records were selected where students' contextual information (e.g., school, class, year group, and gender) was complete. There were around equal proportion of female (n = 65, ~ 51%) and male students (n = 62, ~ 49%).

The included data were collected across five *Navigo* games, targeting two linguistic *levels:* Phonology and Word Recognition. Each linguistic level was further categorized into language *categories*: (1) Consonant Clusters, (2) Grapheme Phoneme Correspondence (GPC), (3) Syllabification from the Phonology linguistic level, and (4) Frequency and (5) Irregular GPCs from Word Recognition. Within each category, were different language *features*, the most granular unit of language. Each language feature could be exercised through different types of games, designed to promote one of three increasingly challenging linguistic *skills*: accuracy, blending, and automaticity [8]. Each of these linguistic skills required a unique approach to learning (e.g., multiple-choice questions, mix-and-match questions), consequently leading to distinct game mechanics (e.g., puzzle game, hit-the-target game) seen in the selected five games. Playing one round of any given *Navigo* game might involve several *questions*, generating multiple game *events* (e.g., start/end, correct/incorrect) (for more details see [8]).

While playing, the students covered a total of 5,027 questions across the five *Navigo* games. These students generated 3,760 unique game log entries that were further decomposable to questions, content, and game-event tables. Moreover, each game *log* recorded general information about the game *activity*, which was defined by the specific game *identifier* (game name, id), the targeted language *feature* (feature id), and the type of linguistic *skill* trained by the game mechanic, including information such as the specific question text and answer options. It also included information about the student's performance, e.g., duration, correct/incorrect responses, and whether any in-game feedback was received. Male students were relatively more active, generating more game log entries (2,159 games covering 2,885 questions, 57.4% game log entries) vs (1,601 games covering 2,142 questions, 42.6% game log entries) for female students.

2.2 Data Analysis

The Bayesian Knowledge Tracing (BKT) algorithm functions as a Hidden Markov Model (HMM) in its traditional form and assumes a student's knowledge (often referred to as the "mastery level" in prior literature) as a binary variable showing whether or not a student has mastered a skill. The knowledge here implies a latent variable that is updated every time a student answers a problem in a learning environment, questioning their understanding of a specific skill. BKT uses four key skill-specific parameters: $\mathbf{p(L_0)}$ or p-init, also known as p(know), is the probability that the student understands the skill beforehand. **p(T)** or p-transit, also known as p(will learn), learning probability or learn rate, is the probability that the student will demonstrate skill mastery on the next opportunity. **p(S)** or p-slip, is the probability that the student will make will answer incorrectly despite having mastered the skill. **p(G)** or p-guess, is the probability that the student correctly applies an unknown skill (a lucky guess aka guessing probability).

A recent BKT variant, pyBKT [6] is a probabilistic framework where the parameters are trained (learned or estimated) using the data from students' interactions with the learning system. The framework uses the Expectation-Maximization (EM) algorithm to achieve convergence in parameter estimation. While several useful class abstractions are possible in pyBKT (for example, creating, fitting, predicting, cross-validating, and evaluating BKT models), this study used *parameter fitting* class abstraction. In the preprocessing stage, we converted the input columns from games data using the default column mapping setting. For each linguistic category, the learners started with lowest level for a particular feature (e.g., competence level = 0) and their learn rate (i.e., p-transit) was operationalized as a proxy of students' performance. We estimated guessing and slipping probabilities in five categories (referred to as skills in BKT literature) played across three game mechanics (multiple-choice, target, and puzzle). For implementation details see https://github.com/Samanzehra/iRead.

3 Results and Discussion

3.1 Learn Rate

An estimated learn rate was derived for each student. We found that, overall, female learn rate (Mean = 0.26, SD = 0.30) was higher than male students (Mean = 0.20, SD = 0.28). Figure 1(a–c) illustrates the differences in mean learn rate in the linguistic level Phonology when three categories from Phonology were learned across different games (GPC, Clusters, and Syllabification). Figure 1(d–e) shows the differences in learn rate between both genders in the linguistic level Word Recognition.

Regardless of the categories, students' learn rate remained higher in the games designed to exercise reading *accuracy,* such as those employing puzzle (e.g., Hearoglyphs) or multiple-choice (e.g., Perilous Paths) learning strategies. The learn rate was relatively lower in the dynamic games where a student was supposed to hit a target while moving (e.g., Raft Rapid Fire), which were designed to exercise automaticity in language skills. One small exception was the seldomly played category of Syllabification, a particularly challenging language category usually taught in class in advanced years groups, where the learn rate remained consistently low.

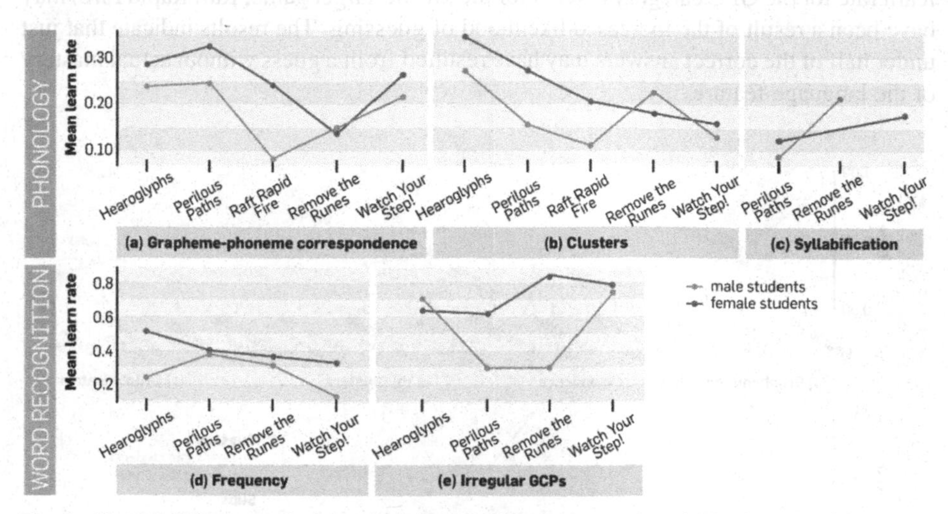

Fig. 1. Gender differences in learn rate in the Phonology (a–c) and Word Recognition (d–e) linguistic levels.

3.2 Guess and Slip Rate

While learn rates were calculated for individual students, a slightly different approach was required for guessing and slipping probabilities because these probabilities may be influenced more by the type of game and its mechanics. Therefore, we evaluated guess rates and slip rates for each of the five games individually and then compared them across various categories. Figure 2(a) to (e) illustrate the guess and slip rate in different games; each figure reports the result for one language category.

GPC was the most played language category (Fig. 2(a)), and no statistically significant gender gap was identified in the guess and slip rates. Overall, in this category, guessing probability remained higher than 0.5 except for the multiple-choice game *Perilous Paths* (where we noted p(G) = 0.43 for females, 0.48 for males). The slipping probability was also the highest for GPC in this game (p(S) = 0.33 for females, 0.38 for males). One potential reason could be that this game was the most played game by students, as it was designed for practicing skills, and in each round of *Perilous Paths*, students were supposed to cross three rope bridges (each bridge representing one question) to complete the game but perhaps due to rushing to cross the path the students made incorrect choices despite knowing the corresponding skill.

The high slipping rate remained consistent in *Perilous Paths*. However, the exact opposite trend was noticed in the puzzle game *Hearoglygh*, where, regardless of gender, students made the most guesses (p(G) = 0.77), and slipping probability remained the lowest (p(S) = 0.17). To aid focus in this puzzle game, the students were supposed to click the on-screen button to 'hear' the hidden word. Therefore, one potential reason for this contrast could be the unavailability or decreased volume of the game sounds that, particularly in classroom environments, may have been problematic. Overall, the high learn rate for the GPC category (except for the hit-the-target game; Raft Rapid Fire) may have been a result of the increased likelihood of guessing. The results indicate that just under half of the correct answers may have resulted from a guess without actual mastery of the language feature.

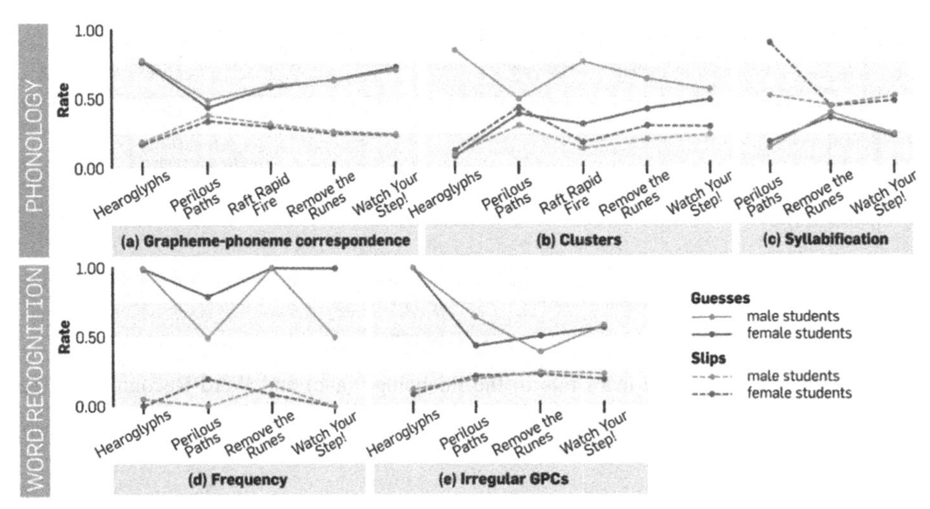

Fig. 2. Gender differences in guess and slip rate in the Phonology (a–c) and Word Recognition (d–e) linguistic levels.

Comparing the learning behaviour when students were learning *Clusters* (Fig. 2(b)), we found identical patterns, i.e., the lowest slipping probability in the puzzle game (*Hearoglygh*) and the highest slipping probability in the multiple-choice game of *Perilous Paths* which was played most for learning new skills and practice previously learnt skills (p(S) = 0.12 for females and 0.09 for males). Also identical to the GPC category, male students made most guesses in the *Hearoglyphs* puzzle game (p(G) = 0.85), closely followed by the target game *Raft Rapid Fire* (p(G) = 0.76) when practicing Cluster-related language features. Next, the Syllabification category was played across three multiple-choice games. Unlike the rest of the two skills from the Phonology linguistic level (GPC and Clusters), students made more slips than guesses, eventually resulting in a relatively low learn rate (see Fig. 1(a–c)). One potential reason could be that this category was designed to be played by relatively older students from year 3 onwards. Those students may already have some knowledge of this skill and therefore made relatively fewer guesses especially while learning in the practice game *Perilous Paths*. However, they still slipped more (for example, p(s) = 0.89 for female students, the highest slipping rate in the Phonology linguistic level).

From the Word Recognition level, the category *Frequency* was played most and across four distinct games. Consistent with the above discussed findings, students guessing probability remained highest in the puzzle game (*Hearoglyphs*). While the overall learn rate remained high for female students (Figs. 1 and 2), male students made relatively more attempts to guess the correct answers in most games. Yet, the only statistically significant difference between male and female students was in the *learn rates* in the games under the *Clusters* category ($H(1) = 3.844$, $p < 0.05$). Further research is required to investigate why; it could be that, compared to other categories, "Clusters" is the least frequently explicitly taught category in English schools. This result further provides leverage for the hypothesis that the implicit learning opportunities afforded by different game mechanics may be benefitting male students more than female students (c.f. [7]).

4 Conclusion and Future Work

Like most other educational games, *Navigo* was designed with difficulty levels and game mechanics that do not necessarily favour a specific player gender on purpose. However, such differences might indeed emerge in practice [e.g., 3]. The parameter estimation methodology used in this study is a promising start in answering these and similar empirical questions in dynamic learning environments in general and DGBL in particular. The findings from this study generate data-driven insights and raise further research questions that could have been difficult to derive otherwise (e.g., through classroom observations, painstaking video analyses or other qualitative methods).

Funding. Funded from the Strategic Investment Board of the IOE, UCL's Faculty of Education and Society. For data collection and game development funding see http://iread-project.eu.

References

1. Lester, J.C., Spain, R.D., Rowe, J.P., Mott, B.W.: Instructional support, feedback, and coaching in game-based learning. In: Plass, J.L., Mayer, R.E., Homer, B.D. (eds.) Handbook of Game-Based Learning, pp. 209–237 (2020)
2. Harpstead, E., Myers, B.A., Aleven, V.: In search of learning: facilitating data analysis in educational games. In: Proceedings of the SIGCHI Conference on Human Factors in Computing Systems, pp. 79–88 (2013)
3. Hou, X., Nguyen, H.A., Richey, J.E., McLaren, B.M.: Exploring how gender and enjoyment impact learning in a digital learning game. In: Bittencourt, I.I., Cukurova, M., Muldner, K., Luckin, R., Millán, E. (eds.) AIED 2020. LNCS (LNAI), vol. 12163, pp. 255–268. Springer, Cham (2020). https://doi.org/10.1007/978-3-030-52237-7_21
4. Harpstead, E., Aleven, V.: Using empirical learning curve analysis to inform design in an educational game. In: Proceedings of the 2015 Annual Symposium on Computer-Human Interaction in Play, pp. 197–207 (2015)
5. Lomas, D., Patel, K., Forlizzi, J.L., Koedinger, K.R.: Optimizing challenge in an educational game using large-scale design experiments. In: Proceedings of the SIGCHI Conference on Human Factors in Computing Systems, pp. 89–98 (2013)
6. Badrinath, A., Wang, F., Pardos, Z.: pyBKT: an accessible python library of Bayesian knowledge tracing models. arXiv preprint arXiv:2105.00385 (2021)

7. Doroudi, S., Brunskill, E.: Fairer but not fair enough on the equitability of knowledge tracing. In: Proceedings of the 9th International Conference on Learning Analytics & Knowledge, pp. 335–339 (2019)
8. Benton, L., et al.: Designing for 'challenge' in a large-scale adaptive literacy game for primary school children. Br. J. Edu. Technol. **52**(5), 1862–1880 (2021)

Popularity Prediction in MOOCs: A Case Study on Udemy

Lin Li, Zachari Swiecki, Dragan Gašević, and Guanliang Chen(✉)

Centre for Learning Analytics, Monash University, Melbourne, Australia
{lin.li,Zachari.Swiecki,Dragan.Gasevic,Guanliang.Chen}@monash.edu

Abstract. Massive Open Online Courses (MOOCs) have dramatically changed how people access education. Though substantial research works have been carried out to improve students' learning experiences, very little attention was directed to the characterization and identification of quality MOOCs for students to undertake (e.g., those with a large enrolment of students), which, we argue, is vital to empower students to make use of MOOCs to reskill and upskill. To fill the gap, this study aimed to investigate the extent to which ML models can be used to automatically identify the popularity of a MOOC before or upon its publication. Specifically, we collected data about more than $50K$ courses from Udemy, based on which we engineered a total of 21 features as input to four widely-used ML models for MOOC popularity prediction, namely Linear Regression, Random Forests, XGBoost, and Multi-Layer Perceptron Neural Network. Through extensive evaluations, we demonstrated that (i) XGBoost gave the best performance in predicting MOOC popularity; (ii) features like *the number of captions* and *enrolment fee* were strongly correlated with MOOC popularity; (iii) the prediction results were mostly inferior to those reported on predicting the popularity of social media posts and news articles, and thus more research effort is needed to boost the prediction performance.

Keywords: MOOCs · Course popularity · Gradient tree boosting

1 Introduction

Since their inception, Massive Open Online Courses (MOOCs) have been designed to educate the world. In MOOCs, an unlimited number of participants can simultaneously access learning materials anytime and anywhere as long as they can connect to the Web [3]. To improve learners' learning experiences in MOOCs, a substantial amount of research has been conducted on modeling students' learning behaviors [6,7]. However, relatively few have attempted to investigate how to proactively direct students to quality courses that align with their interests *before* they begin their learning. To bridge this gap, researchers have started constructing personalized recommender systems to match students to courses based on their learning interests. Though certain advancements have

M. M. Rodrigo et al. (Eds.): AIED 2022, LNCS 13355, pp. 607–613, 2022.
https://doi.org/10.1007/978-3-031-11644-5_56

been made, these course recommender systems are often require the input of rich historical data from students. However, in reality, a significant fraction of students might not have sufficient historical data to empower course recommender systems. As shown in later analyses based on Udemy (Fig. 1(b) in Sect. 3), over 97% students wrote less than five reviews to share their learning experiences in MOOC learning.

Inspired by the research on web content popularity prediction [9,10], we proposed that an alternative approach to matching students to quality courses is to first identify popular MOOCs and then recommend these MOOCs to students. As a pilot study, we investigated the predictability of course popularity before publication by applying machine learning methods on its inherent characteristics, where we defined two types of popularity measures, i.e., the average monthly number of students enrolled in a MOOC and the average monthly number of student-authored reviews received by a MOOC. Specifically, we collected a large-scale dataset consisting of over 51,826 MOOCs from Udemy, and engineered a total of 21 features to empower four ML models to predict MOOC popularity, i.e., Linear Regression, Random Forests, XGBoost, and Multi-Layer Perceptron Neural Network. To our knowledge, our study is the first research which focused on predicting MOOC popularity. Experimental results showed that (i) among the selected ML models, XGBoost performed the best in predicting MOOC popularity; (ii) the popularity of a MOOC was associated with factors such as the number of captions offered in the MOOC, its enrolment fee, and its content description; (iii) the prediction of enrolment-based popularity was more challenging than that of review-based popularity and further research efforts are needed to boost the prediction performance.

2 Method

2.1 Dataset

The dataset was collected in late September 2021 via the APIs provided by Udemy[1], which enabled us to retrieve information about MOOCs, the number of students enrolled in MOOCs and the ratings/reviews authored by the students. As we aimed to predict the popularity of MOOC based on its inherent characteristics (e.g., those with meaningful content description and carefully-designed learning materials and activities), we excluded courses that: (i) were taught in languages other than English; (ii) were particularly designed for standardized tests (e.g., GRE and GMAT) and with only questions or simulation tests as the main learning materials; and (iii) lacked important textual information to characterize the learning content (e.g., learning objectives and course description). Also, considering the natural time effect on a MOOC's popularity, we decided to only include the courses published between 2017/01/01 and 2021/03/01 for evaluation. After removal, there were 51,826 MOOCs left.

[1] https://www.udemy.com/developers/affiliate/models/course/.

In line with [2,5,8], we tackled MOOC popularity prediction as a regression problem. In particular, we defined two types of popularity measures, i.e., the average monthly number of students enrolled in a MOOC and the average monthly number of student-authored reviews received by a MOOC, both of which were calculated by averaging the total number of students/reviews over the number of months between the time when a MOOC was published and the time the dataset was collected. We chose to average the number of students/reviews to account for the natural time effect on accumulating a MOOC's student enrolment and course reviews.

2.2 MOOC Popularity Prediction

Feature Engineering. We engineered three categories of features, i.e., (i) *content-related features*, which include information about course title & headline, learning objectives, prerequisites, description, and target audience of a MOOC; (ii) *structure-related features*, which include information about the composition of a MOOC, namely the duration of all lectures, and the number of video lecture, articles, assessment tests, practice tests, questions, coding exercises and additional resources contained in a MOOC; and (iii) *metadata-related features* including the subject category of a MOOC, enrolment fee, number of captions and instructors, instructional level, published year/month and published date. There are three data types: *numerical*, *categorical*, and *textual*. For categorical features, we used one hot encoding to represent them. For textual features, we applied BERT [4] on them to generate their vector-based embeddings. In particular, we concatenated *Title* and *Headline* as input to BERT as they were closely related to each other. After pre-processing, all features were concatenated and used as input for the four selected ML models, before which we applied Principle Component Analysis [11] on the concatenated text embeddings with 0.95% variance explained to reduce the dimensions.

Machine Learning Models. In line with previous studies on content popularity prediction [2,5,8], we selected four well-established ML models for the prediction of MOOC popularity prediction: **Linear Regression**, **Random Forests**, **XGBoost** and **Multi-Layer Perceptron Neural Network**.

2.3 Study Setup

As suggested in [1,5], we applied logarithm transformation to the two popularity measures to deal with their highly skewed distributions in the data. All selected techniques were implemented using the *scikit-learn* package in Python. We used *Bert-as-a-service*[2] to generate the embedding representations for all textual features. We randomly split the data in the ratio of 8:2 as the training and testing set for model construction. When training a model, we used 5-fold cross validation and grid search to perform parameter tuning. The performance of each model was measured by two widely-used metrics for web content popularity prediction, i.e., Root Mean Squared Error (*RMSE*), and Coefficient of

[2] https://github.com/hanxiao/bert-as-service.

Determination (R^2). We report model performance on the testing set. All of the data, code, and parameter settings used in this study can be accessed via https://github.com/dscodepad/mooc-popularity for replication.

3 Results

The Necessity of Identifying Popular MOOCs. We plot (i) the total number of courses available on `Udemy` and (ii) the number of students who wrote at least one course review at different timestamps between 2017/01/01–2021/09/30 (i.e., the date we finished data collection) in Fig. 1, from which we can observe that though the number of MOOCs and students have greatly increased over the past 5 years, the majority of students wrote less than 5 reviews (97% until 2021/09/30). Such infrequent interaction data might imply that a substantial amount of students did not generate enough data to empower recommender systems, thus called for the need of identifying popular courses.

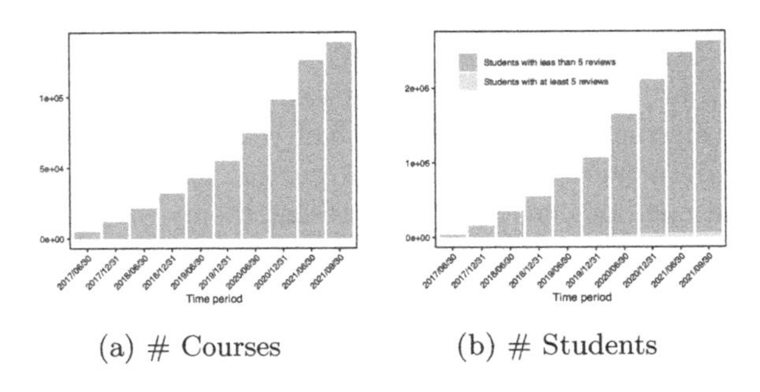

(a) # Courses (b) # Students

Fig. 1. The number of courses (a) and students (b) at different timestamps.

Results on MOOC Popularity Prediction. Table 1 presents the results of the four selected ML models. For both popularity measures, XGBoost gave the best performance while Linear Regression was the worst. When delving into the results reported in R^2 score for predicting *# Avg. Monthly Reviews*, XGBoost demonstrated 28.02%, 17.47% and 4.96% of improvement over Linear Regression, Random Forest, and MLP, respectively. We can make similar observations on the results reported on *# Avg. Monthly Enrolments*, though the performance of MLP was very close to that of XGBoost. The results reported on review-based popularity measure are much better than those reported on enrolment-based popularity measure, suggesting that it may be more challenging to identify popular courses measured in terms of *# enrolments* versus *# reviews*. Noteworthy, the results reported on review-based popularity measure in our study was comparable to those reported in some relevant studies. For instance, [1] reported a best R^2 score of 0.43 on predicting new articles' popularity by using a Linear Regression model, while the R^2 score achieved in our study was in the range of $[0.34, 0.43]$. However, our results were relatively inferior comparing to those reported in predicting popularity of posts and videos in social media platforms, e.g., [5] reported a best R^2 score of 0.8678 and a best RMSE value of 0.5978 obtained by Gradient Boost Regressor. This might be explained by the use of different features to empower ML models as other studies often used various kind of social interaction information.

Table 1. Results on MOOC popularity prediction. The best results are in bold.

Models	# Avg. monthly reviews		# Avg. monthly enrolments	
	RMSE	R^2	RMSE	R^2
Linear Regression	2.0617	0.3419	2.8731	0.2851
Random Forests	2.0129	0.3726	2.8313	0.3058
XGBoost	**1.9057**	**0.4377**	**2.7235**	**0.3576**
MLP	1.9404	0.4170	2.7258	0.3566

We also conducted an ablation test based on XGBoost to examine the importance of each feature in predicting MOOC popularity, which was measured by removing the feature from the input and then calculated the performance change caused due to the removal of the feature. The results are given in Table 2. Firstly,

Table 2. Feature importance analysis on predicting MOOC popularity. Features positively contributed to the prediction performance are marked with *. The top-3 decreased performance results are in bold.

Feature category	Feature input	# Avg. monthly reviews		# Avg. monthly enrolments	
		RMSE	R^2	RMSE	R^2
	All feature	1.9057	0.4377	2.7235	0.3576
Content	w/o Title & Headline	* 1.9114 (−0.18%)	* 0.4343 (−0.78%)	* 2.7397 (−0.59%)	∗**0.35(−2.13%)**
	w/o Content description	∗**1.9266(−0.65%)**	∗**0.4253(−2.83%)**	* 2.7352 (−0.43%)	* 0.3521 (−1.54%)
	w/o Target audience	* 1.9080 (−0.07%)	* 0.4363 (−0.32%)	* 2.725 (−0.06%)	* 0.3569 (−0.2%)
	w/o Prerequisites	1.9018 (0.12%)	0.4400 (0.53%)	* 2.7306 (−0.26%)	* 0.3543 (−0.92%)
	w/o Objectives	* 1.9060 (−0.01%)	* 0.4375 (−0.05%)	2.7201 (0.12%)	0.3592 (0.45%)
Structure	w/o # Lectures	* 1.9131 (−0.23%)	* 0.4333 (−1.01%)	* 2.7291 (−0.21%)	* 0.355 (−0.73%)
	w/o Lecture duration	* 1.9074 (−0.05%)	* 0.4367 (−0.23%)	* 2.7266 (−0.11%)	* 0.3562 (−0.39%)
	w/o # Articles	* 1.9143 (−0.27%)	* 0.4326 (−1.17%)	* 2.733 (−0.35%)	* 0.3532 (−1.23%)
	w/o # Assessment tests	1.9044 (0.04%)	0.4385 (0.18%)	2.7192 (0.16%)	0.3597 (0.59%)
	w/o # Practice tests	1.9036 (0.11%)	0.4398 (0.48%)	2.7194 (0.15%)	0.3596 (0.56%)
	w/o # Coding exercises	1.9047 (0.03%)	0.4383 (0.14%)	2.7232 (0.01%)	0.3578 (0.06%)
	w/o # Questions	* 1.9080 (−0.07%)	* 0.4363 (−0.32%)	2.7207 (−0.1%)	0.3592 (0.45%)
	w/o # Additional resources	* 1.9084 (−0.08%)	* 0.4361 (−0.37%)	2.7228 (0.03%)	0.3580 (0.11%)
Metadata	w/o Subject category	1.9029 (0.08%)	0.4393 (0.37%)	* 2.7269 (−0.12%)	* 0.356 (−0.45%)
	w/o # Instructors	1.9120 (−0.19%)	0.4340 (−0.85%)	∗**2.7442(−0.76%)**	∗**0.3479(−2.71%)**
	w/o Enrolment fee	∗**1.9408(−1.10%)**	∗**0.4168(−4.77%)**	* 2.7349 (−0.42%)	* 0.3523 (−1.48%)
	w/o # Captions	∗**1.9445(−1.22%)**	∗**0.4145(−5.30%)**	∗**2.742(−0.68%)**	∗**0.3489(−2.43%)**
	w/o Published date	* 1.9066 (−0.03%)	* 0.4372 (−0.11%)	* 2.7316 (−0.3%)	* 0.3538 (−1.06%)
	w/o Published year	1.9028 (0.09%)	0.4394 (0.39%)	2.7235 (0.0%)	0.3576 (0.0%)
	w/o Published_month	1.9047 (0.03%)	0.4383 (0.14%)	2.7233 (0.01%)	0.3577 (0.03%)
	w/o Instructional level	1.9027 (0.09%)	0.4394 (0.39%)	2.7226 (0.03%)	0.3581 (0.14%)
	All features marked *	1.9057	0.4377	2.7185	0.3600

many features did not contribute to the popularity prediction of MOOCs, such as *# Assessment tests*, *Published year* and *Instructional level*. Secondly, there were common features that were useful for predicting both types of popularity measures, such as *Content description*, *Target Audience*, *# Lectures*, *Lecture duration*, *# Captions*, and *Enrolment fee*. In particular, the performance drop occurred after removing *# Captions* are significant for both popularity measures. *Enrolment fee* also played an important role here. For enrolment-based popularity, the largest performance drop was found on the removal of *# Instructors* with the R^2 decreasing 2.71%. Overall, among the three categories of features, content-related features generally tended to be effective. By only taking features that contributed to the prediction performance into account, XGBoost achieved a similar or slightly better performance compared to that of using all available features.

4 Discussion and Conclusion

In this study, we investigated the feasibility of applying ML models to predict the popularity of a MOOC based on its inherent characteristics. We can see that *# Captions* and *Enrolment fee* were essential in predicting MOOC popularity based on reviews. This implied that it is beneficial to offer lecture captions in various languages. MOOC platforms may consider assist course instructors to produce captions of different languages. Besides, for *Enrolment fee* the Pearson correlation coefficients with the number of reviews and the number of enrollments in a MOOC were 0.1549 and 0.2125, showing that a higher enrolment fee does not necessarily degrade the popularity of a course. Lastly, it should be noted that MOOCs essentially can be regarded as a special type of web content. Given the much longer longevity of MOOCs compared to tweets and news articles (e.g., at least a few years), it would be worth incorporating additional information and engineer more complex features in the early stage after publication for popularity prediction in the future.

References

1. Bandari, R., Asur, S., Huberman, B.: The pulse of news in social media: forecasting popularity. In: ICWSM, vol. 6 (2012)
2. Borges, H., Hora, A., Valente, M.T.: Predicting the popularity of github repositories. In: PROMISE, pp. 1–10 (2016)
3. Christensen, G., Steinmetz, A., Alcorn, B., Bennett, A., Woods, D., Emanuel, E.: The MOOC phenomenon: who takes massive open online courses and why? Available at SSRN 2350964 (2013)
4. Devlin, J., Chang, M.W., Lee, K., Toutanova, K.: BERT: pre-training of deep bidirectional transformers for language understanding. arXiv preprint arXiv:1810.04805 (2018)
5. Gayberi, M., Oguducu, S.G.: Popularity prediction of posts in social networks based on user, post and image features. In: Proceedings of the 11th International Conference on Management of Digital EcoSystems, pp. 9–15 (2019)

6. Guo, P.J., Reinecke, K.: Demographic differences in how students navigate through MOOCs. In: L@S, pp. 21–30 (2014)
7. Kizilcec, R.F., Piech, C., Schneider, E.: Deconstructing disengagement: analyzing learner subpopulations in massive open online courses. In: LAK, pp. 170–179 (2013)
8. Mazloom, M., Pappi, I., Worring, M.: Category specific post popularity prediction. In: Schoeffmann, K., et al. (eds.) MMM 2018. LNCS, vol. 10704, pp. 594–607. Springer, Cham (2018). https://doi.org/10.1007/978-3-319-73603-7_48
9. Moniz, N., Torgo, L.: A review on web content popularity prediction: issues and open challenges. Online Soc. Netw. Media **12**, 1–20 (2019)
10. Tatar, A., de Amorim, M.D., Fdida, S., Antoniadis, P.: A survey on predicting the popularity of web content. J. Internet Serv. Appl. **5**(1), 1–20 (2014). https://doi.org/10.1186/s13174-014-0008-y
11. Wold, S., Esbensen, K., Geladi, P.: Principal component analysis. Chemom. Intell. Lab. Syst. **2**(1–3), 37–52 (1987)

Extensive Reading at Home: Extracting Self-directed Reading Habits from Learning Logs

Chia-Yu Hsu[1], Rwitajit Majumdar[2](✉), Huiyong Li[2], Yuanyuan Yang[1], and Hiroaki Ogata[2]

[1] Graduate School of Informatics, Kyoto University, Kyoto, Japan
[2] Academic Center for Computing and Media Studies, Kyoto University, Kyoto, Japan
majumdar.rwitajit.4a@kyoto-u.ac.jp

Abstract. Habit is an important concept attracting researchers' attention in many fields. In the past, habit was commonly measured by tools such as questionnaires and diaries which involve subjects' self-reported behaviors. However, attempts to extract and support good learning habits with actual learners' data are still limited. Therefore, in this study, we aim to identify students' reading habits using learning analytics, which enables fine description of learners' behaviors with accumulated amounts of log data. We define students' reading logs based on a psychological framework in which habit is defined as a repetitive behavior responding to a certain context. The learning context in this study is the promotion of extensive reading in English implemented in a Japanese public junior high school and supported by a digital learning environment. We extract 175 students' daily reading logs recorded from May 31, 2020 to April 9, 2021 (314 days) to examine whether periodicity occurs in the data so as to identify the students who newly form an English reading habit as a result of the promotion as well as their behavior patterns in the process.

Keywords: Learning analytics · Reading habits · Habit measurement · Extensive reading · Self-directed learning

1 Background and Motivation

Habit is defined in psychology as a repetitive behavior which responds to a certain context [1]. In the past, habit was commonly measured by tools such as questionnaires and diaries. That is, researchers examined whether and how subjects formed a habit via their self-reported behaviors [2, 3]. The technique of learning analytics, on the other hand, enables fine description of learners' behaviors with accumulated amounts of log data [4]. Thus, we argue that this technique can inform us of learners' learning habits with more fidelity. Even though issues on learning habits are not new in the field of learning analytics, it is innovative to conceptualize this idea in the context and examine learners' possession and formation of learning habits with this technique. Many past studies focused on the activation and facilitation of learning habits. Tools addressing these objectives were designed accordingly [5, 6]. However, the claims on the effects

M. M. Rodrigo et al. (Eds.): AIED 2022, LNCS 13355, pp. 614–619, 2022.
https://doi.org/10.1007/978-3-031-11644-5_57

need more verification, in which the measurement of habits play an important role. That is, to evaluate these various designs, it is essential to understand what is attributed to existing habits and what are newly-developed habits. The identification can also imply clusters with different behavior patterns, to which different designs can be adapted. Thus, we explore the applicability of learning analytics to identify learners' habits in an extensive reading (ER) context, where the learning habits are interpreted as reading habits in this study.

To conceptualize habits in the context of learning analytics, we refer to the framework proposed by [1]. In the framework, habits connect with contexts. A behavior is habitual when it occurs repetitively in a specific context. The initiation of the behavior may serve as a result of the attempt on a certain goal. However, as time goes on, the behavior can still occur automatically responding to the context even if the goal is no longer pursued. Thus, [1] argued that habits and goals affect each other on this premise. In this study, we fit home ER activities in the framework considering practical issues which ER educators in many schools are constrained by. In an ER program, students are expected to self-select reading materials, read at their own pace, and have sufficient time to read, which requires students to have their own goals in case of losing their way before they form good reading habits. However, it's difficult for teachers to know about students' habits in such activities as are often implemented out of class [7]. Therefore, to support both teachers and students to achieve the expected outcomes in the ER program, we aim to answer the following research questions: (1) How can reading habits of extensive reading be measured using the technique of learning analytics? (2) What are the observed habits of extensive reading at home for Japanese junior high school students?

2 Research Study: Extracting Self-directed Reading Habits

2.1 Self-directed Extensive Reading Activity with BookRoll and GOAL

The learning context in this study is the promotion of extensive reading in English implemented in a Japanese public junior high school. Figure 1 describes the reading activity with BookRoll and GOAL. Students are provided with more than 500 digital picture books to read in BookRoll, an ebook-based learning platform [8] and supported by GOAL system, a self-directed activity support environment [9]. The system aims to facilitate students' self-directed learning involving a cycle of setting goals, making plans, analyzing and monitoring behaviors and reflecting on processes. Through practicing this

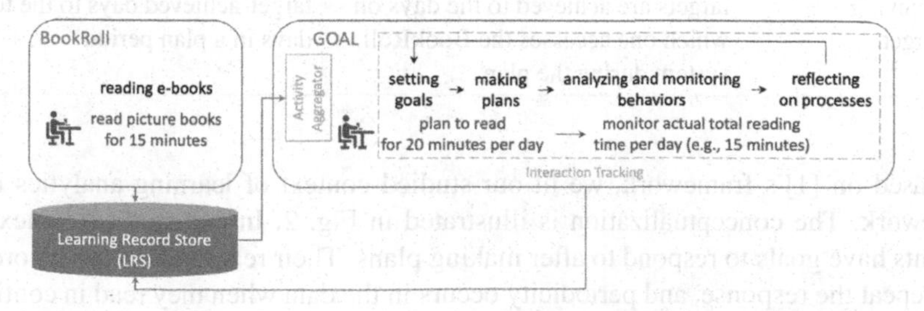

Fig. 1. Self-directed extensive reading activity with BookRoll and GOAL

cycle, students are expected to develop habits of English reading. Thus, the students can plan their reading schedule along with the target set by themselves in the GOAL system (e.g., reading for 20 min per day) and carry out their plan by reading the picture books in the BookRoll system, which can be accessed by digital devices such as tablets, computers, etc.

The GOAL system automatically synchronizes learners' BookRoll reading activity logs that are stored in the Learning Record Store (LRS). GOAL's activity aggregator calculates various indicators at different time scales: for instance, daily reading time spent in extensive reading, total e-books read in a period [7]. Thus, the total minutes for which a student reads per day are recorded by aggregating the time from when he or she opened and then closed a book. Details of plans made in GOAL, such as numbers, frequency, and period, are also tracked. With such log data, we aim to identify the students who form an English reading habit and the behavior patterns of that habit.

2.2 Operationalizing Model for Self-directed Reading Habits

Table 1 summarizes the concepts and the measurement of how we consider habits in the field of learning analytics.

Table 1. Concepts and measurement of reading habits in this study

Concepts	Description	Measurement
Making plans	Makes plans via the GOAL system	The numbers of the plans
Reading repeatedly	Accesses the BookRoll system more than 1 day during the period between the ongoing plan and its subsequent plan	The sum of the reading time in each plan period
		The length of the plan period
Reading in contiguity with context	Accesses the BookRoll system periodically during the period between the ongoing plan and its subsequent plan	The frequency of significant sine or cosine waves in an ongoing plan period
Probability to read periodically in plan period	The ratio of the plan periods in which one reads periodically to the total reading plans he or she makes	The ratio of significant sine or cosine waves to the total numbers of plan periods
Efficiency in achieving the target	The ratio of the days on which the targets are achieved to the days on which one accesses the BookRoll system during the plan	The ratio of the numbers of target-achieved days to the total days in a plan period

Based on [1]'s framework, we fit our studied context of learning analytics in the framework. The conceptualization is illustrated in Fig. 2. In our studied context, the students have goals to respond to after making plans. Their reading logs are recorded if they repeat the response, and periodicity occurs in the data when they read in contiguity with the BookRoll system. Then, they are identified to have the potential for acquiring reading habits. The formation of the habits can be verified if periodicity can still be identified in the subsequent logs of the students' reading data. This means that their response to the reading behaviors is cued in the e-book environment. During the period in which different plans are made, the habits extracted from the students' behaviors may be inconsistent since the students may be exploring their own ways to develop a habit of reading which are different in each plan period. Therefore, we verify these students' habit formation in terms of the probability that they read periodically in the plan periods. The students' efficiency in achieving their targets is also considered important and informative. The above approach considers a chance that a habit might form as long as periodicity is identified, though the outcome efficiency may be low. For example, if one sets 20 min reading per day in the plan period and ends up reading only 15 min on alternate days, such a case can be considered as having a habit but not efficient to achieve the target. In this study we identify the students' efficiency in achieving their target as the factor which profiles the students.

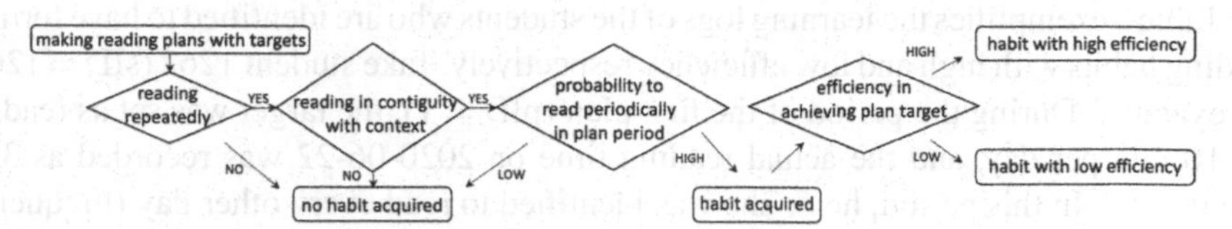

Fig. 2. Conceptualizing reading habit extraction based on [1]

To measure the concepts illustrated in Fig. 2, we define them based on the attributes of the log data (see Measurement in Table 1).

2.3 Findings from Japanese Junior High School Dataset

We extract 175 Japanese junior high school students' daily reading logs recorded from May 31, 2020 to Apr. 9, 2021 (314 days), in which the students made approximately 5 plans on average (M = 5.33, SD = 2.86) in the GOAL system. In terms of data processing, the students' reading logs on weekdays are limited to those from 6 p.m. to 8 a.m. the next day, while those on weekends are not limited to a specific time slot since the aim of this study is to extract the students' reading habits outside their school hours.

Figure 3 illustrates the procedure of the data analysis in this study. In the first round of omission, 167 students are identified to further examine whether periodicity occurs in their reading logs. We use the frequency with the maximum of the spectrum in the power spectrum, which indicates the power distribution of a time-series over frequency, to identify the potential periodicity of the students. In terms of these frequencies, we calculate the sine and cosine waves of the log data in the respective plan periods of each student so that the frequencies can be tested statistically with harmonic regression, where the time series are regressed on waves with a certain frequency. 104 students are identified as potential habit holders in this step with the statistically significant frequencies. In terms of these frequencies, we calculate each students' probability to read periodically in the plan period and identify 48 habit holders based on the median split of the data set (median = 0.32). Then, we consider the average ratio of the numbers of target-achieved days to the total days in a plan period as a student's overall efficiency in achieving targets. Based on the median split of efficiency (median = 0.15), participants were divided into high-efficiency habit holders (n = 24, M = 0.31) and low-efficiency habit holders (n = 24, M = 0.07).

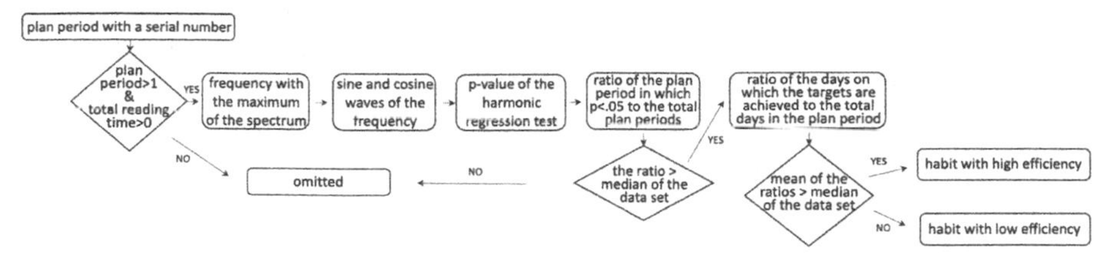

Fig. 3. Procedure of reading habits extraction using learning analytics

Table 2 exemplifies the learning logs of the students who are identified to have formed reading habits with high and low efficiency respectively. Take student 1262 (sID = 1262) for example. During the period of the first plan (pID = 1), the target was set as reading for 10 min per day, and the actual reading time on 2020-06-22 was recorded as 32.4 min in total. In this period, he or she was identified to read every other day (Frequency = 2.0). Considering all the plans made, the probability of his or her periodic reading behaviors is 0.63, while the average rate to achieve the set target is 0.19. Since both of the values are higher than the median splits of the respective data sets, student 1262 was grouped as a high-efficiency habit holder.

Table 2. Example learning logs of students with different efficiency

sID	pID	Date	Reading time	Target	Frequency	Probability	Efficiency	Group
1262	1	2020-06-22	32.4	10	2.0	0.63	0.19	High
351	1	2020-06-22	0.0	5	15.0	0.33	0.05	Low

3 Discussion

In this study, we aim to identify students' reading habits using learning analytics. We refer to a psychological framework as the theoretical basis and conceptualize students' habits in the context of promoting extensive reading. Supported by a digital learning environment, the learning logs capture the student's actions which can be interpreted as behaviors responding to the context. We define the concepts based on the logs and measure them using the logs to derive criteria which filter the students who are verified to form reading habits in the promotion (see Table 1). In the past, habit was commonly measured by tools such as questionnaires and diaries which involve subjects' self-reported behaviors. However, attempts to extract and support good learning habits with actual learners' data are still limited. In this study, we demonstrate the operationalized definition of habits in the learning context and use previous log data to examine the conceptualization. With 175 students' reading logs in 314 days, we extracted 48 students' reading habits during extensive reading in English. We consider the students' efficiency in achieving learning goals as the factor which profiles different groups of students. With these profiles, future works can further explore different behavior patterns and design adaptive support for different students. In addition, the behavior patterns of those who are considered as having no habit learning are not discussed in this study. However, the students with no reading habits can be identified by the proposed technique in this study, and better instructional and technological support can be provided based on the behavior patterns. Finally, the proposed technique of extracting students' reading habits needs further validation. Therefore, comparison between other parallel methods should be done in the future.

Acknowledgement. This study is supported by JSPS16H06304, 22H03902, NEDO JPNP18013, JPNP20006, SPIRITS 2020 of Kyoto University.

References

1. Wood, W., Neal, D.T.: A new look at habits and the habit-goal interface. Psychol. Rev. **114**(4), 843–863 (2007)
2. Okado, K., Yoshida, H., Kida, N.: Correlations between changes in study habits and academic results in junior high school students—a longitudinal survey at a private junior high school. Psychology **8**(13), 2102–2113 (2017)
3. Verplanken, B.: Beyond frequency: habit as mental construct. Br. J. Soc. Psychol. **45**(3), 639–656 (2006)
4. Li, J., Li, H., Majumdar, R., Yang, Y., Ogata, H.: Self-directed extensive reading supported with GOAL system: mining sequential patterns of learning behavior and predicting academic performance. In: LAK22: 12th International Learning Analytics and Knowledge Conference, pp. 472–477 (2022)
5. Vinay, M., Rassak, S.: A technological framework for teaching-learning process of computer networks to increase the learning habit. Int. J. Comput. Appl. **117**(4), 1–5 (2015)
6. Wang, R.: Design of web-based English learning support system. In: 2014 IEEE Workshop on Advanced Research and Technology in Industry Applications (WARTIA), pp. 771–773 (2014)
7. Li, H., Majumdar, R., Chen, M.R.A., Yang, Y., Ogata, H.: Analysis of self-directed learning ability, reading outcomes, and personalized planning behavior for self-directed extensive reading. In: Interaction Learning Environment, pp. 1–20 (2021)
8. Ogata, H., et al.: E-Book-based learning analytics in university education. In: International Conference on Computer in Education (ICCE 2015), pp. 401–406 (2015)
9. Li, H., Majumdar, R., Chen, M.R.A., Ogata, H.: Goal-oriented active learning (GOAL) system to promote reading engagement, self-directed learning behavior, and motivation in extensive reading. Comput. Educ. **171**, 104239 (2021)

Extraction of Useful Observational Features from Teacher Reports for Student Performance Prediction

Menna Fateen(✉) and Tsunenori Mine

Kyushu University, Fukuoka, Japan
menna.fateen@m.ait.kyushu-u.ac.jp

Abstract. Performance prediction models have been proposed countless times due to the benefits that they can provide to educational stakeholders. While many factors have been taken into account when predicting student performance, teachers' assessment or observation reports have not been commonly used. A teacher's assessment is a fundamental part of the educational process and has a direct impact on students' success. In this study, we analyze the topics, and psychological features in teachers' daily written reports and apply them to the student performance prediction model. Experimental results show the capability of this approach in contributing to the accuracy of performance prediction models.

Keywords: Teacher reports · Topic modelling · Sentiment analysis · Performance prediction · Text mining · Machine learning

1 Introduction

"What if educators had the ability to predict their students' performance ahead of time" This question has been proposed countless times which is unsurprising given the numerous benefits that performance prediction models can deliver. With accurate predictions, educators would be able to easily monitor their students' progress and verify whether they are on track to meet their learning objectives. Having the ability to detect students that may need further help enables teachers to take preventive measures. Resources and instruction can then be allocated more efficiently. Not only would prediction models benefit both teachers and students, but they would also have a positive impact on the educational institution as a whole, since students' success is often incorporated in metrics for evaluating effectiveness. An accurate prediction model can therefore benefit all educational stakeholders.

In prior work, many different types of factors have been taken into account. A factor that has not been commonly employed in performance prediction models, however, is teachers' observation reports. Teachers' assessment of their students is an integral part of the educational process and can be defined as "an essentially interactive process, in which the teacher can find out if what has been taught

M. M. Rodrigo et al. (Eds.): AIED 2022, LNCS 13355, pp. 620–625, 2022.
https://doi.org/10.1007/978-3-031-11644-5_58

has been learned, and if not, to do something about it" [7]. However, assessment routines can be highly affected by teachers' own mindsets along with their beliefs about the malleability of their students' abilities. A teacher that has a *fixed mindset*, views intelligence as static and believes that there is very little that can be done to change it [5]. Self-theory research has shown that such belief could have a negative effect on students' success [8]. On the other hand, teachers with a *growth mindset* could help their students understand that with hard work and persistence, their own intelligence can increase. With such a correlation between teachers' mindsets and students' success in mind, we take advantage of daily teacher observation reports and investigate whether they can contribute to performance prediction models.

Using a subset of our data, Fateen and Mine [3] directly estimated performance based on individual reports. Fateen et al. [4] further presented an improved student-based model where each instance represents a student instead. Inspired by Fateen et al.'s improved model, we propose an approach where we construct an interpretable student-based vector that summarizes and highlights the sentiments in different extracted topics to predict student performance.

We aim to address two main research questions: **RQ1:** Can we extract an interpretable input vector that summarizes teachers' reports and use it for student performance prediction? **RQ2**: How early can we predict students' performance using our model? To address these questions, we conduct three experiments on different subsets of the data where we investigate how early we can predict performance, and verify the effectiveness of the model across two different periods.

To conclude, (1) we present a topic modeling approach with sentiment analysis to represent teachers' reports, (2) we show that by using our approach we can outperform the accuracy of a prediction model that only relies on previous grades. Finally, we compare our results with previous study results [4] that used the teachers' comments as inputs to identify students that have a tendency to receive lower or higher grades.

2 Data Description

The teacher reports used in the experiments were provided by a cram school for middle school students in Japan. The teachers closely observe their students and provide feedback written for their parents about the students' progress from the teacher's viewpoint. The reports include information about what was studied during class, evaluation scores, and the teachers' written comments. Overall, the nature of the comments is encouraging and supportive while occasionally highlighting any potential concerns. We conduct three experiments, EXP1 to EXP3 with different subsets of the teachers' reports. The number of reports and students attending in each subset is shown in Table 1.

Students that attend the cram school are enrolled in different regular schools. The results of their regularly taken exams at school were also provided. We considered these regular test scores the features that would normally be used for prediction. The students that had their regular score provided were only a subset. In all experiments, we compared the performance of using the regular score as the only feature and adding the regular score to the vector extracted from the teachers' comments. The number of students with regular scores provided along with their corresponding number of reports is displayed in Table 1.

Table 1. Dataset information in each experiment

	Date range of reports	Simulation exam	Complete dataset		Regular score		Regular score provided	
			Students	Reports	Grade	Year	Students	Reports
EXP1	5/2020–5/2021	9/2021	321	24,477	8	2020	175	18,566
EXP2	5/2020–9/2020	9/2020	441	19,653	9	2020	377	17,545
EXP3	5/2021–9/2021	9/2021	433	19,866	9	2021	279	13,988

Since the actual performance in the entrance exam was unattainable. The provided results of the simulation exams that the students had taken before the actual entrance exam were utilized. These exams were taken twice, in September 2020 and September 2021. The simulation scores for the students were recorded for each subject and provided as a total score.

2.1 Feature Selection

In the experimental settings, we adopt three feature sets for comparison. The first set, FS_1 consists of the vector extracted from the teachers' comments. The method is explained in Sect. 3. FS_2 contains the students' regular scores. We investigate using the vector from the teachers' comments with the regular scores to verify whether adding teachers' reports contributes to the accuracy of the prediction model. FS_3 therefore is a concatenation of FS_1 and FS_2.

3 Methodology

3.1 Seed-Guided Topic Extraction

In a report, a teacher highlights different aspects to summarize the students' development such as their in-class progress or homework efforts. Five topics of discussion were pre-defined: (1) Understanding, (2) Classwork, (3) Tests, (4) Concentration, and (5) Homework. Using TopicRank, the most used key-phrases by the teachers were identified. To finalize the topic representations, the most representative key-phrases for the topics were chosen based on the cosine similarity of their BERT embeddings [2] to the topic name embeddings. Since a report may discuss several topics, the reports were divided into sentences as a preprocessing step. After transforming the sentences and topic phrases into BERT embeddings, each sentence was assigned to the topic with the highest cosine similarity. The most prevalent topics assigned were 'Understanding' and 'Classwork,' with 47.3% and 47% allocated to each, respectively with the remaining 6% split between the rest of the topics. We may deduce from these figures that the most often addressed themes in the instructors' comments are the teachers' perspectives on their students' understanding and summarizing daily progress.

3.2 BERT Sentiment Analysis

To obtain a sentiment score for the comments sentences, we deployed a pretrained BERT model on sentiment analysis [1]. The model generates a polarity label and a confidence score. The distribution of the scores exhibited a high negative skew where positive scores accounted for an average of 96%. Since the input vector is student-based [4], the topic sentiment for each student was estimated by taking the average of the scores corresponding to each topic. Taking the median, we found that 'Test' and 'Classwork' topics had the highest scores, 0.95 and 0.93, while the 'Understanding' and 'Homework' topics had lower average scores, 0.87 and 0.82, respectively. Therefore, we assume that the teachers tend to be more supportive when discussing tests and classwork while occasionally expressing negative feelings about the students' understanding and homework performance. Lastly, we conducted a bivariate correlation analysis of the average sentiment score in each topic with the simulation score and found that the 'Understanding' topic held the greatest correlation value with a maximum of 0.35.

3.3 Linguistic and Psychological Analysis

The nature of the comments is generally encouraging and supportive, resulting in a significant skew in sentiment score distribution. Hence, we extracted linguistic and psychological features using the LIWC software [6]. LIWC calculates the percentage of usage of words that belong to more than 70 categories and generates an output measure for each. For each student, an average of each output category was then computed. Next, a bivariate correlation analysis was conducted on each of the LIWC categories with the students' final scores. From 89 variables, we selected 21 which had a significant correlation in the 3 experiments. The final vector for FS_1, includes the average count in each of the 21 variables along with the average sentiment in the five topics. After constructing the vector for the feature sets, we employ a Gradient Boosting machine to predict the students' scores. Figure 1 shows a flowchart of the methodology described.

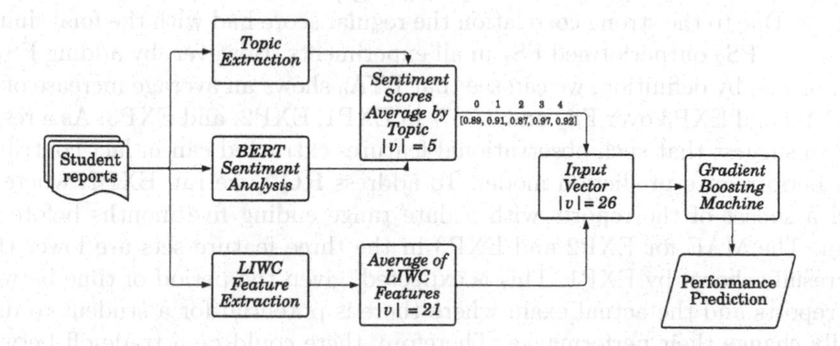

Fig. 1. Flowchart of the described methodology

3.4 Evaluation Metrics

We evaluate our experiments via two metrics: Mean Absolute Error (MAE) and Percentage by Tick Accuracy (PTA). A tick is the difference between two successive grades. We employ PTA_0 and PTA_1, which mean the model successfully predicted the grade or predicted it 1 tick away from the true grade, respectively.

4 Experiments

All experiments were evaluated using 10-fold cross-validation. The average MAE and PTA of the ten folds were computed and results of the three experiments are shown in Table 2. To compare our approach with the Students Model [4], we constructed input vectors with a similar method. For FS_1, a bag-of-words (BoW) vector was constructed to represent all teachers' comments for each student. FS_2 consisted of the regular score and FS_3 as defined.

Table 2. Average MAE and PTA in EXP1–EXP3 using the extracted feature vector and BoW vector. Bold values indicate best MAE and PTA_0 results using either approach, underlined values indicate best overall results

		EXP1			EXP2			EXP3		
		MAE	PTA_0	PTA_1	MAE	PTA_0	PTA_1	MAE	PTA_0	PTA_1
Feature vector	**FS_1**	66.29	0.36	0.45	56.65	0.56	0.30	65.82	0.40	0.40
	FS_2	42.39	0.55	0.37	38.61	0.63	0.29	**36.22**	**0.60**	0.35
	FS_3	**40.07**	**0.61**	0.33	36.03	0.68	0.26	37.97	0.59	0.35
BoW	**FS_1**	39.68	0.71	0.29	65.5	0.45	0.36	70.57	0.27	0.50
	FS_2	42.39	0.55	0.37	38.61	0.63	0.29	36.22	**0.60**	0.35
	FS_3	**39.07**	0.61	0.31	**36.14**	**0.67**	0.27	35.28	0.59	0.35

To answer **RQ1**, FS_1, EXP2 and EXP3 were constructed to evaluate the effectiveness of the input vector in predicting performance across two different periods. Due to the strong correlation the regular score had with the final simulation score, FS_2 outperformed FS_1 in all experiments. However, by adding FS_1 to FS_2, or FS_3 by definition, we can see that PTA_0 shows an average increase of 2% in EXP2 and EXP3 over FS_2 and 3.33% in EXP1, EXP2, and EXP3. As a result, we can suggest that such observational features extracted can in fact contribute to a performance prediction model. To address **RQ2**, we ran EXP1, where we used a subset of the reports with a date range ending five months before the exam. The MAE for EXP2 and EXP3 in the three feature sets are lower than the results shown by EXP1. This is expected, given the period of time between the reports and the actual exam where there is potential for a student to drastically change their performance. Therefore, there could be a trade-off between low MAE and early performance prediction. To compare with the baseline, we can see in Table 2 that in EXP1, BoW gained a higher performance in MAE and PTA_0, outperforming FS_2. On the contrary, the features FS_1 extracted were able to compete with BoW in both EXP2 and EXP3. We conclude that FS_1 provides in-depth comprehension of the reports while giving a competitive capacity for performance prediction. The features extracted reduce time and size complexity compared to other embeddings.

5 Conclusion

In this study, we attempted to extract sentiments and psychological features from teacher observation reports and use those as features for a performance prediction model. We held three subsets of the reports and ran experiments to verify the effectiveness of the model. The features extracted gave an in-depth understanding of the teacher comments while showing a correlation with the final student performance. Results showed that this approach can compete with approaches previously used. However, since teachers' ideas about their students' performance are limited, we suggest combining such features with student-related data such as previous scores to gain a prediction model with exceeding accuracy.

Acknowledgements. This work was supported by JST, the establishment of university fellowships towards the creation of science technology innovation, Grant Number JPMJFS2132, in part by e-sia Corporation and by Grant-in-Aid for Scientific Research proposal numbers (JP21H00907, JP20H01728, JP20H04300, JP19KK0257).

References

1. Daigo: BERT-base Japanese sentiment. https://huggingface.co/daigo/bert-base-japanese-sentiment. Accessed 29 Jan 2022
2. Devlin, J., Chang, M.W., Lee, K., Toutanova, K.: BERT: pre-training of deep bidirectional transformers for language understanding. arXiv:1810.04805 (2018)
3. Fateen, M., Mine, T.: Predicting student performance using teacher observation reports. In: The Fourteenth International Conference on Educational Data Mining, pp. 481–486 (2021)
4. Fateen, M., Ueno, K., Mine, T.: An improved model to predict student performance using teacher observation reports. In: ICCE 2021, pp. 31–40 (2021)
5. Murphy, L., Thomas, L.: Dangers of a fixed mindset: implications of self-theories research for computer science education. In: Proceedings of the 13th Annual Conference on Innovation and Technology in Computer Science Education, pp. 271–275 (2008)
6. Pennebaker, J.W., Francis, M.E., Booth, R.J.: Linguistic inquiry and word count: LIWC 2001. Mahway Lawrence Erlbaum Assoc. **71**(2001), 2001 (2001)
7. Wiliam, D., Lester, F.: Keep learning on track: classrooms assessment and the regulation of learning. In: Lester, F.K. (ed.) Second Handbook of Research on Mathematics Teaching and Learning, pp. 1051–1098 (2007)
8. Yorke*, M., Knight, P.: Self-theories: some implications for teaching and learning in higher education. Stud. High. Educ. **29**(1), 25–37 (2004)

Two-Stage Uniform Adaptive Testing to Balance Measurement Accuracy and Item Exposure

Maomi Ueno[1(✉)] and Yoshimitsu Miyazawa[2]

[1] The University of Electro-Communications, Tokyo, Japan
ueno@ai.is.uec.ac.jp
[2] The National Center for University Entrance Examinations, Tokyo, Japan
miyazawa@rd.dnc.ac.jp

Abstract. Computerized adaptive testing (CAT) presents a tradeoff problem involving increasing measurement accuracy vs. decreasing item exposure in an item pool. To address this difficulty, we propose two-stage uniform adaptive testing. In the first stage, the proposed method partitions an item pool into numerous uniform item groups using a state-of-the-art uniform test assembly technique based on the Random Integer Programming Maximum Clique Problem. Then the method selects the optimum item from a uniform item group. In the second stage, when the standard error of an examinee's ability estimate becomes less than a certain value, it switches to selecting and to presenting an optimum item from the whole item pool. Results of numerical experiments underscore the effectiveness of the proposed method.

Keywords: Computerized adaptive testing · Integer programming · Item response theory · Maximum clique algorithm · Uniform adaptive testing

1 Introduction

Computerized adaptive testing (CAT) selects and presents the optimal item which maximizes the test information (Fisher information measure) at the current estimated ability based on item response theory (IRT) from an item pool. However, in conventional CATs, the same items tend to be presented to examinees who have similar abilities. This tendency leads to bias of the item exposure frequency in an item pool.

To resolve this difficulty, various methods have been proposed (e.g. [1–3]). Recent studies by Songmuang and Ueno [4] and by Ishii and several collaborators [5–7] have explored several techniques using AI technologies to generate numerous uniform test forms from an item pool. Regarding the uniform test forms, each form consists of a different set of items, but the forms have equivalent measurement accuracy (i.e. equivalent test information based on item response

M. M. Rodrigo et al. (Eds.): AIED 2022, LNCS 13355, pp. 626–632, 2022.
https://doi.org/10.1007/978-3-031-11644-5_59

theory). Ueno and Miyazawa [8] proposed uniform adaptive testing (UAT) using the Maximum Clique Problem (MCP) described by Ishii et al. [6] to divide an item pool into several equivalent groups of items (uniform item groups) and then select the optimum item from a uniform item group. They demonstrated that the UAT reduced test length, item exposure, and bias of measurement accuracies among examinees although they did not evaluate the measurement accuracy directly. However, the UAT must degrade the measurement accuracy of examinees' abilities because decreasing the test length necessarily increases the measurement error.

To resolve that shortcoming, we propose two-stage uniform adaptive testing. In the first stage, the proposed method partitions an item pool into numerous uniform item groups using the Random Integer Programming Maximum Clique Problem (RIPMCP) presented by Ishii and Ueno [9], which is known to generate the greatest number of uniform tests. Then the method selects the optimum item from a uniform item group. In the second stage, when the examinee's ability estimate error becomes less than a certain value, designated as Switching Stage Criterion (SSC), in the uniform item group, the proposed method switches to the selection and presentation of the optimum item from the whole item pool until the update difference of the examinee's ability estimate becomes less than a constant value. Numerical experiments demonstrate that the proposed method reduces item exposure without increasing the measurement error.

2 Computerized Adaptive Testing Based on Item Response Theory

In CAT, an examinee's ability parameter is estimated based on Item Response Theory (IRT) ([10]) to select the optimum item with the highest information. In the two-parameter logistic model (2PLM), the most popular IRT model, the probability of a correct answer to item i by examinee j with ability $\theta \in (-\infty, \infty)$ is assumed as

$$p(u_i = 1|\theta) = \frac{1}{1 + exp[-1.7a_i(\theta - b_i)]}. \quad (1)$$

Therein, u_i is 1 when an examinee answers item i correctly; it is 0 otherwise. Furthermore, $a_i \in [0, \infty)$ and $b_i \in (\infty, \infty)$ respectively denote the discrimination parameter of item i and the difficulty parameter of item i. The asymptotic variance of estimated ability based on the item response theory was shown by [10] to approach the inverse of Fisher information. Accordingly, item response theory usually employs Fisher information as an index representing the accuracy. In 2PLM, the Fisher information is defined when item i provides an examinee's ability θ using the following equations.

$$I_i(\theta) = \frac{[\frac{\partial}{\partial \theta} p(u_i = 1|\theta)]^2}{p(u_i = 1|\theta)[1 - p(u_i = 1|\theta)]} \quad (2)$$

The results imply that the examinee's ability can be discriminated using an item with high Fisher information $I_i(\theta)$. Accordingly, that ability estimation can be expected to be implemented by selecting items with the highest amount of Fisher information given an examinee's ability estimate $\hat{\theta}$. The test information function $\mathrm{I}_T(\theta)$ of a test form T is defined as $\mathrm{I}_T(\theta) = \sum_{i \in T} \mathrm{I}_i(\theta)$. The asymptotic error of ability estimate $\hat{\theta}$, $\mathrm{SE}(\hat{\theta})$, can be obtained as the inverse of square root of the test information function at a given ability estimate $\hat{\theta}$ as $\mathrm{SE}(\theta) = \frac{1}{\sqrt{\mathrm{I}_T(\theta)}}$.

In conventional CAT, adaptive items are selected from an item pool using the following procedures.

1. An examinee's ability is initialized to $\hat{\theta} = 0$.
2. An item maximizing Fisher information for a given ability is selected from the item pool. It is then presented to the examinee.
3. The examinee's ability estimate is updated from the correct and incorrect response data to the item.
4. Procedures 2 and 3 are subsequently repeated until the update difference of the examinee's ability estimate decreases to a constant value of ϵ or less.

Consequently, CAT can reduce the number of items examined, but it does not reduce the test accuracy in comparison to that of the same fixed test.

3 Two-Stage Uniform Adaptive Testing

In a conventional CAT, it is highly likely that the same set of items will be presented to examinees exhibiting similar abilities. Therefore, conventional CAT cannot be used practically in situations where the same examinee can take a test multiple times. Furthermore, because the ability variable follows a standard normal distribution, items with higher information around $\theta = 0$ tend to be exposed frequently. Therefore, bias of the item exposure frequency occurs in an item pool. To resolve the shortcoming, various constrained CATs with item exposure control have been proposed (e.g. [1–3]). Earlier methods have mitigated the bias of item exposure frequency in an item pool. Unfortunately, they also entailed the important difficulty of increased measurement error for examinees. In fact, a tradeoff exists between minimizing item exposure and maximizing the measurement accuracy. Nevertheless, earlier methods did not resolve the tradeoff. For that reason, we propose a new CAT framework that can resolve the tradeoff: two-stage uniform adaptive testing.

3.1 First Stage Procedure

In the first stage, the proposed method partitions an item pool into numerous uniform item groups similarly to UAT, a method presented by Ueno and Miyazawa [8]. Although UAT employs MCP, which was introduced by Ishii et al. [6], the number of generated uniform item groups remains limited because of its heavy space complexity. In addition, MCP tends to engender a bias of item exposure frequency because it does not consider the bias.

A state-of-the-art uniform test assembly method, Random Integer Programming Maximum Clique Problem (RIPMCP), has been demonstrated by Ishii and Ueno [9] to generate the greatest number of uniform tests. Although the shadow-test method [3] maximizes the test information using integer programing, it increases the difference of measurement accuracies between the first assembled shadow test and the last one. In contrast, the proposed method maximizes the number of uniform item groups with the test constraints, so as not to increase the bias of measurement accuracy for the groups. In the first stage, the proposed method partitions an item pool into numerous uniform item groups using the RIPMCP. The method then selects the optimum item from a uniform item group as described below.

1. An arbitrary uniform item group is selected from a set of unused groups.
2. The optimal item maximizing Fischer information is selected from the group and is presented to an examinee in Procedure 1.
3. The examinee's ability estimate is updated from the examinee's response.
4. Procedures 2 and 3 are repeated until the asymptotic error of ability estimate $\mathrm{SE}(\hat{\theta})$ reaches a constant value of ε or less.

If a set of unused groups is empty in Procedure 1, then the algorithm resets it as a universal set of uniform item groups. The number of groups is optimized by comparing the respective performances of several numbers of groups. Item selection from a uniform item group accelerates convergence of the ability estimate to the neighborhood of the true ability value because the item difficulties in each group are distributed sparsely and uniformly over all the examinees' abilities.

3.2 Second Stage Procedure

The first stage rapidly provided a roughly approximated ability estimate of an examinee. The second stage reaches a more accurate ability estimate of the examinee. More specifically, when the examinee's ability estimate error becomes less than the determined value, designated as Switching Stage Criterion (SSC), in the first stage, it switches to the second stage, which selects and presents the optimum item from the whole item pool. The second stage is conducted until the update difference of the examinee's ability estimate becomes less than a constant value or less, just as traditional CATs do. The SSC is optimized by changing the value to compare performance. For this study, we use the Fischer information measure as an item selection criterion that becomes accurate for the second stage because it is an asymptotic approximation. Therefore, the second stage is expected to approach the true ability value efficiently and rapidly without greatly increasing the item exposure.

4 Numerical Evaluation

This section presents a comparison of the performances of the proposed method (designated as Proposal) to those of other computerized adaptive testing methods (conventional adaptive testing in 2 (designated as CAT), Kingsbury and

Zara [1] CAT (designated as KZ), van der Linden's IP-based CAT [3] (designated as IP), Linden and Choi's item-eligibility probability method [2] (designated as Prob) and the method described by Ueno and Miyazawa [8] (designated as UAT). Additionally, we evaluate the performances of the UAT employing RIPMCP to generate uniform item groups designated as UAT-RIPMCP. Furthermore, we employ OC = 5 for proposal, UAT, and UAT-RIPMCP.

Table 1. Experiment results obtained using an actual item pool

Test length	Method	No. item-groups	Avg. exposure item	Measurement error (RMSE)	No. non-presented items
30	CAT	–	227.27 (227.99)	0.24	846
	KZ(20)	48	131.58 (140.35)	0.29	750
	IP	–	80.86 (33.28)	0.33	607
	Prob.	–	95.85 (40.83)	0.34	665
	UAT(20)	215	20.94 (12.05)	0.50	23
	UAT-RIPMCP(20)	342	20.47 (8.91)	0.54	1
	Proposal(20, 0.225)	342	80.21 (163.75)	0.24 (0.69)	604
50	CAT	–	243.90 (233.59)	0.20	773
	KZ(25)	39	165.56 (198.94)	0.23	676
	IP	–	83.61 (31.66)	0.29	380
	Prob.	–	104.60 (39.98)	0.27	500
	UAT(20)	215	20.94 (12.06)	0.48	23
	UAT-RIPMCP(20)	342	20.47 (8.91)	0.52	1
	Proposal(20, 0.075)	342	69.83 (151.16)	0.20 (0.57)	284

An experiment was conducted using the item pool of real data, with 978 items, and a test constraint. Table 1 presents the results. In Table 1, the values in parentheses for KZ, UAT, and UAT-RIPMCP denote the group sizes. Those for Proposal represent the uniform item group sizes and SSC values. "Avg. exposure item" expresses the average exposure count of an item (the standard error of numbers of exposure items in parentheses), and "No. non-presented items" represents the number of items that have not been presented. The average test lengths (the standard error in parentheses) in the first stage for the total test lengths 30 and 50 are, respectively, 3.83 (1.11) and 9.65 (2.30). Those in the second stage for the total test lengths 30 and 50 are the remaining test lengths, respectively, 26.17 and 40.35. The average test lengths for the total test lengths 30 and 50 show large differences when compared to those in the simulation experiments because of their large difference of the optimum SSC values. Otherwise, the table lays out results that are almost identical to those obtained from the simulation experiment. The RMSEs in the first stage for the total test lengths 30 and 50 are, respectively, 0.69 and 0.57 and those in the second stage for the total test lengths 30 and 50 are, respectively, 0.24 and 0.20. In fact, results indicate that the proposed method reduces item exposure without increasing the measurement error. The results demonstrate that only the proposed method resolves the tradeoff problem between increasing measurement accuracy and decreasing item exposure.

5 Conclusion

The discussion and results presented herein have demonstrated that CAT entails tradeoff difficulties between increasing measurement accuracy and decreasing item exposure in an item pool. To address this difficulty, we proposed two-stage uniform adaptive testing. Experiments were conducted to compare the performance of the proposed method with that demonstrated by conventional methods. Results of those experiments demonstrated that, among all methods, only the proposed method resolved the tradeoff. We expect to apply the proposed uniform adaptive testing method to adaptive learning systems [11,12] and Deep IRT [13,14] in future studies.

References

1. Kingsbury, G.G., Zara, A.R.: Procedures for selecting items for computerized adaptive tests. Appl. Measur. Educ. **2**(4), 359–375 (1989)
2. van der Linden, W.J., Choi, S.W.: Improving item-exposure control in adaptive testing. J. Educ. Meas. **57**(3), 405–422 (2020)
3. van der Linden, W.J.: Review of the shadow-test approach to adaptive testing. Behaviormetrika (2021). https://doi.org/10.1007/s41237-021-00150-y
4. Songmuang, P., Ueno, M.: Bees algorithm for construction of multiple test forms in e-testing. IEEE Trans. Learn. Technol. **4**(3), 209–221 (2011)
5. Ishii, T., Songmuang, P., Ueno, M.: Maximum clique algorithm for uniform test forms assembly. In: International Conference on Artificial Intelligence in Education (AIED), LNAI 7926, pp. 451–462 (2013)
6. Ishii, T., Songmuang, P., Ueno, M.: Maximum clique algorithm and its approximation for uniform test form assembly. IEEE Trans. Learn. Technol. **7**(1), 83–95 (2014)
7. Ishii, T., Ueno, M.: Clique algorithm to minimize item exposure for uniform test forms assembly. In: International Conference on Artificial Intelligence in Education (AIED). LNCS 9112, pp. 638–641 (2015)
8. Ueno, M., Miyazawa, M.: Uniform adaptive testing using maximum clique algorithm. In: International Conference on Artificial Intelligence in Education (AIED). LNAI 11625, pp. 482–493 (2019)
9. Ishii, T., Ueno, M.: Algorithm for uniform test assembly using a maximum clique problem and integer programming. In: International Conference on Artificial Intelligence in Education (AIED). LNAI 10331, pp. 102–112 (2017)
10. Lord, F., Novick: Statistical Theories of Mental Test Scores. Addison-Wesley, M.R. (1968)
11. Ueno, M., Miyazawa, Y.: Pobability based scaffolding system with fading. In: Artificial Intelligence in Education (AIED). LNAI 9112, pp. 492–503 (2015)

12. Ueno, M., Miyazawa, Y.: IRT-based adaptive hints to scaffold learning in programming. IEEE Trans. Learn. Technol. **11**(4), 415–428 (2018)
13. Tsutsumi, E., Kinoshita, R., Ueno, M.: Deep item response theory as a novel test theory based on deep learning. Electronics **10**(9), 1020 (2021)
14. Tsutsumi, E., Kinoshita, R., Ueno, M.: Deep-IRT with independent student and item networks. In: Proceedings of the 14th International Conference on Educational Data Mining (EDM) (2021)

Multi-label Disengagement and Behavior Prediction in Online Learning

Manisha Verma(✉), Yuta Nakashima, Noriko Takemura, and Hajime Nagahara

Osaka University, Osaka, Japan
{mverma,n-yuta,takemura,nagahara}@ids.osaka-u.ac.jp

Abstract. Student disengagement prediction in online learning environments is beneficial in various ways, especially to help provide timely cues to make some feedback or stimuli to the students. In this work, we propose a neural network-based model to predict students' disengagement, as well as other behavioral cues, which might be relevant to students' performance, using facial image sequences. For training and evaluating our model, we collected samples from multiple participants and annotated them with temporal segments of disengagement and other relevant behavioral cues with our multiple in-house annotators. We present prediction results of all behavior cues along with baseline comparison.

Keywords: E-learning · Facial behavior analysis · Student disengagement

1 Introduction

Under the current COVID-19 situation, online learning draws great attention. A matter of concern is its feasibility to provide the same quality of care as attending a physical classroom. It may be more difficult to watch video feeds from all students to check their status than to cast a glance to each student in a physical classroom due to missing eye contact, limited screen sizes, *etc.*

This difficulty has escalated the necessity of automatically analyzing students' (dis)engagement. Vision-based approaches can be promising as cameras are almost ubiquitous, as well as they are not invasive and less intrusive. Various vision-based modalities have been considered for engagement analysis, such as facial, body gesture, and motion features [1,7,11,12].

The first in-depth work in vision-based engagement prediction utilized facial features based on box filters to assess engagement intensity [16]. Action units (AUs), which encodes the movement of facial muscles and provides rich information on facial expressions, can be extracted from facial images [14] and have been used for engagement prediction [2,3]. Another work additionally takes facial expressions into account for modeling engagement via students' learning gain [13]. Facial features can also be combined with biometrics, such as heart rate [10].

M. M. Rodrigo et al. (Eds.): AIED 2022, LNCS 13355, pp. 633–639, 2022.
https://doi.org/10.1007/978-3-031-11644-5_60

With the proliferation of deep neural networks (DNNs), the importance of datasets has been ever increasing. DAiSEE [6] and EmotiW [7] are such datasets that have been used for engagement analysis. Deep convolutional neural networks along with hand-crafted features and recurrent neural networks have been utilized for spatio-temporal feature learning for engagement analysis [4,8,17].

Most of these methods directly model students' engagement status; however, the engagement status of a student may depend on various factors (*e.g.*, tiredness, course content, external stimuli) and the cue from the student may vary, which makes the engagement analysis extremely challenging. We hypothesize that experienced teachers (unconsciously) use multiple signals from a student, such as eye movement, facial expression, body motion, *etc.* and make their guesses as a whole. This assumption may decompose the problem of disengagement detection into a series of relevant lower-level detection tasks, which may be easier than directly modeling disengagement itself.

In this paper, we propose to detect students' disengagement together with facial and body behaviors cues, applying a multi-task learning approach to train a model to simultaneously detect disengagement and relevant behavioral cues. For this, we build a video dataset and frame-level labels on disengagement and relevant cues, including strange eye movement, presence of facial expression, *etc.* Our model is a modified version of the SlowFast network [5] for its capability of learning local and global temporal features. We train our model with our dataset in an end-to-end supervised learning. The detection task is cast to frame-level classification, which can be helpful for real-time applications.

2 Dataset

We created a dedicated dataset for our problem setup using a pseudo learning task. Our learning task consists of reading and recall test sessions. During the reading session, a participant is required to read text in Japanese (1,600–1,800 words), while in the recall test, they are asked if certain words are present in the text or not. For doing this, we obtained approval from our IRB and 26 participants with informed consent are recruited.

Table 1. Statistics of annotated segments after cleaning

Cue	# seg.	Duration (s)		
		Min	Max	Avg
DE	2,497	0.53	684.63	15.26
SE	5,168	0.20	558.37	3.96
FE	2,415	0.20	88.97	4.47
YA	387	0.50	15.13	5.41
OC	1,583	0.20	299.27	15.37
BM	5,669	0.20	81.20	4.68
SC	3,247	0.20	1.45	0.75

We use the videos captured in the reading session to predict disengagement. The number of recorded videos is 1,560 in total. Due to some technical difficulties, a few of them should be discarded, which ends up in 1,264 videos with 5.7M frames in total. More information about the task to create videos for this purpose can be found in [15].

2.1 Labels

To help prediction of disengagement, we designed a set of seven behavioral cues such as disengagement (DE), strange eye movements (SE), presence of some kind of facial expression (FE), yawning (YA), face occlusion (OC), body movements (BM), and when participant needs to press keyboard key to change screen (SC).

All these behavioral cues can be relevant to disengagement. For example, frequent yawning can be a sign of boredom; frequently showing some facial expressions and strange eye movements may be a sign of discomfort.

Each video is annotated by at least three of our in-house annotators (A1, A2, A3, and A4). Annotators were provided with same set of instructions. They were asked to make temporal segments that contain one of seven behavioral cues by identifying start and end times of each occurrence of the cue. A video can have multiple occurrences of a certain behavioral cue.

We explored the annotator agreement for each cue and found that different cues have different agreement levels, falling into one of the following categories:

- **Subjective cues** include DE, FE, and SE which are highly ambiguous and subjective where disagreement may be induced by different opinions of annotators and may not necessarily imply just annotation noises.
- **Objective cues** include YA, BM, and OC of which definition is rather obvious and can be consistently recognized by all annotators.
- **Impulsive cue** includes only SC. Its definition is clear but the annotators can easily miss some of them because it lasts only for short time.

We investigated individual annotators and aggregated A1, A2, and A4's (left A3 because of low annotators agreement) annotations for subjective cues where overlapping labels for same cue merged together. For objective cues, we kept labels by all four annotators and assigned ground-truth label based on majority of agreement (when 2/4 or 2/3 agreed). For SC, we chose majority agreement, however, we kept all A4's labels as A4 was extremely exhaustive to spot this cue. The total number of temporal segments in each label along with duration statistics are presented in Table 1.

3 Methodology

Our task is to detect behavioral cues. We reformulate this detection task into frame-level multi-label classification, where a sliding window-based approach is employed to model the temporal dependency. Formally, for the k-th frame v_k, we aggregate surrounding n frames centered at k, forming a set $V_k = \{v_j \mid j = k - n/2, \dots, k + n/2\}$ of frames. Our model g takes V_k as input and make predictions $p_k = g(V_k) \in \mathbb{R}^{|\Omega|}$, where $\Omega = \{\text{DE}, \text{SE}, \text{FE}, \text{YA}, \text{OC}, \text{BM}, \text{SC}\}$ is the set of our behavioral cues. Our ground-truth label $t_k \in \mathbb{R}^{|\Omega|}$, associated with the k-th frame, is based on our temporal segment labels, *i.e.*, if the k-th frame is included in one of temporal segments of behavioral cue $l \in \Omega$, t_{kl} is set to 1, and 0 otherwise.

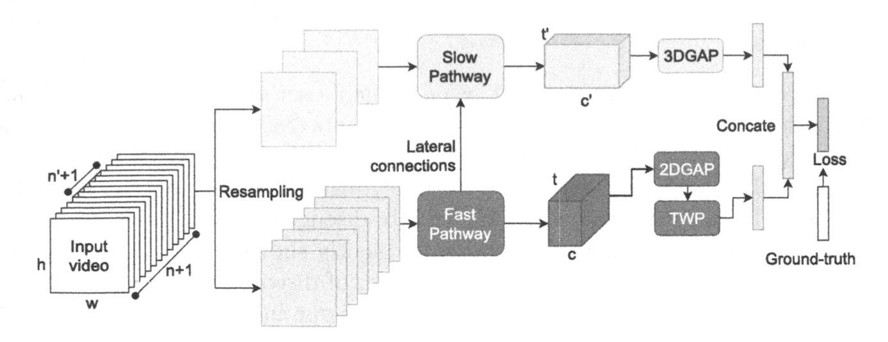

Fig. 1. Our network architecture based on SlowFast [5]. Here $h \times w \times n+1 \times 3$ is dimension (height, width, temporal, channel) of input, (t', c') and (t, c) are (temporal, channel) dimension of slow and fast pathway feature output.

We use a modified version of SlowFast network [5] as its architecture suits our task because some cues (*e.g.*, screen change) consist in rapid motion, while some others (*e.g.*, strange eye movement, like frequent blinks) involve both rapid motions with longer temporal dependency. We make two following changes to adapt it for our task.

Input Strategy. We reduce the number of input frames for the slow pathway and resample them as complete set of frames may bring noise when treated equally. For the fast pathway, original input is kept with a different resampling rate. For our 30 FPS videos, the resampling rate for the slow and fast pathways are 7.5 FPS and 15 FPS. Value of n' and n in Fig. 1 are 16 and 64 respectively.

Temporally-Weighted Average Pooling. Contribution of surrounding frames can be different for different cues. For example, screen change comes with a quick motion while body movement may last longer. Such long-range temporal dependency is modeled in the fast pathway; however, the required length of temporal dependency and contributions of different frames are not obvious, especially for the subjective cues. We, therefore, replace 3D global average pooling (GAP) in the fast pathway with 2D global average pooling (for spatial dimensions) and a trainable temporally-weighted pooling (TWP) layer. We initialize weights as Gaussian window and make it trainable to adapt to the nature of specific cue.

We also change hyper-parameter β of SlowFast network to 1/2 to handle the ratio of number of channels in slow and fast pathways. Further, both features from slow and fast pathways are concatenated and seven binary classifiers are used to predict previously described cues. To handle the data imbalance problem, we use focal loss [9]. The final loss is computed by combining all seven losses.

4 Experiments

Pre-processing. We cropped the face region in every frame of videos using MTCNN [18] or manually and resized to 256×256.

Training. We employed the 8-fold cross-validation scheme for training and evaluation. Each fold contains videos of 3–4 participants. We randomly sampled one frame from each ground-truth temporal segment to make V_k's (if a segment is longer than 7 s, we divided it into equal-length segments until each segment becomes shorter than 7 s). This frame sampling process was done for every epoch, which can work as data augmentation. The total number of samples extracted in this way was roughly 55K, which were split into training and validation sets, according to the participant-based folds. As a result, roughly 90% of samples were used for training and the rest for validation in each fold. We also used a random spatial crop from face regions for training as augmentation.

Evaluation. For evaluation, every fifth frame of a video in the test set was used as a center frame of V_k. With this, roughly 70k to 150k samples were in the test set of each fold.

Table 2. Average results of precision, recall, f1-score, AUC-PR, and AUC-ROC over the 8-folds.

	Pre.	Rec.	F1.	PR.	ROC.
DE	0.4521	0.6121	0.4955	0.5463	0.8062
SE	0.4914	0.6714	0.5220	0.6056	0.8847
FE	0.4849	0.3895	0.4093	0.3972	0.8791
YA	0.7221	0.6136	0.6274	0.7081	0.9748
OC	0.6557	0.8804	0.7027	0.8202	0.9712
BM	0.7435	0.6177	0.6667	0.7474	0.9480
SC	0.7295	0.5726	0.6317	0.7013	0.9832

Results and Discussion. The average performance over 8-folds in terms of precision, recall, f1-score, AUC-PR, and AUC-ROC is presented in Table 2. The difficulty of subjective cues is evident through the results which is compatible with their definition. We observe that a very small number of samples of a particular cue in a specific test set is the reason for destabilized performance sometimes for precision, recall, f1-score, and AUC-PR as they focus on positive class. Another possible reason for the performance degradation is the ambiguity in temporal segment boundaries; for our frame-level evaluation, frames on the boundary of a temporal segment are inherently error-prone.

We also compare our model with some baselines over a certain fold that shows an average performance on our model. Our modified SlowFast (Ours) is compared with SlowFast (SF), 3D-ResNet, and 2D-ResNet. Figure 2 summarizes the AUC-PR scores. Comparison between ours and SF vs. others shows the significance of two-stream structures. Also, the comparison between ours and SF highlights the advantages of our modifications. 2D-ResNet shows great performance drops for behavioral cues for which temporal information can be desirable, such as disengagement, change of screen, and body move. In con-

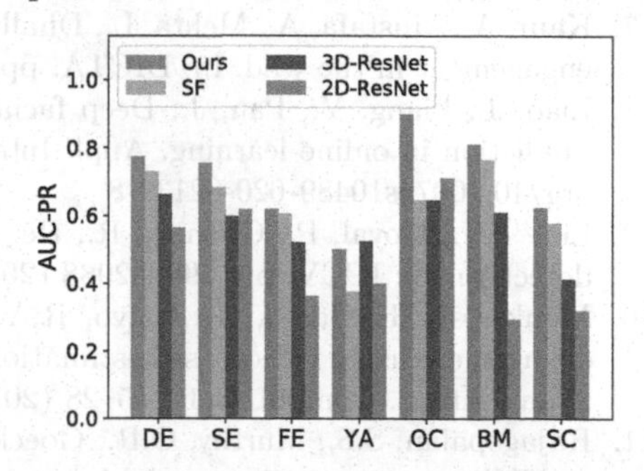

Fig. 2. AUC-PR comparison with baseline.

trast, temporal information may not be necessary for some cues like yawning and occlusion, which is consistent with our intuition.

5 Conclusion and Future Scope

In this work, we propose to predict disengagement along with multiple facial and body behavior cues. We collected participants' videos and assigned frame-level labels with in-house annotators. We employ the SlowFast network and make some modifications to learn spatio-temporal cues from videos. We show the importance of our modifications by baseline comparison. In future, we will focus more on the mutual relationships among behavioral cues and disengagement prediction in detail.

References

1. Alyuz, N., Aslan, S., D'Mello, S.K., Nachman, L., Esme, A.A.: Annotating student engagement across grades 1–12: associations with demographics and expressivity. In: AIED, pp. 42–51 (2021)
2. Bosch, N., et al.: Automatic detection of learning-centered affective states in the wild. In: IUI, pp. 379–388 (2015)
3. Bosch, N., D'mello, S.K., Ocumpaugh, J., Baker, R.S., Shute, V.: Using video to automatically detect learner affect in computer-enabled classrooms. ACM Trans. Interact. Intell. Syst. **6**(2), 1–26 (2016)
4. Dresvyanskiy, D., Minker, W., Karpov, A.: Deep learning based engagement recognition in highly imbalanced data. In: Karpov, A., Potapova, R. (eds.) SPECOM 2021. LNCS (LNAI), vol. 12997, pp. 166–178. Springer, Cham (2021). https://doi.org/10.1007/978-3-030-87802-3_16
5. Feichtenhofer, C., Fan, H., Malik, J., He, K.: SlowFast networks for video recognition. In: ICCV, pp. 6202–6211 (2019)
6. Gupta, A., D'Cunha, A., Awasthi, K., Balasubramanian, V.: DAiSEE: towards user engagement recognition in the wild. In: CVPR Workshops (2018)
7. Kaur, A., Mustafa, A., Mehta, L., Dhall, A.: Prediction and localization of student engagement in the wild. In: DICTA, pp. 1–8 (2018)
8. Liao, J., Liang, Y., Pan, J.: Deep facial spatiotemporal network for engagement prediction in online learning. Appl. Intell. **51**(10), 6609–6621 (2021). https://doi.org/10.1007/s10489-020-02139-8
9. Lin, T.Y., Goyal, P., Girshick, R., He, K., Dollár, P.: Focal loss for dense object detection. In: ICCV, pp. 2980–2988 (2017)
10. Monkaresi, H., Bosch, N., Calvo, R.A., D'Mello, S.K.: Automated detection of engagement using video-based estimation of facial expressions and heart rate. IEEE Trans. Affect. Comput. **8**(1), 15–28 (2016)
11. Rajagopalan, S.S., Murthy, O.R., Goecke, R., Rozga, A.: Play with me-measuring a child's engagement in a social interaction. In: FG, vol. 1, pp. 1–8 (2015)
12. Sanghvi, J., Castellano, G., Leite, I., Pereira, A., McOwan, P.W., Paiva, A.: Automatic analysis of affective postures and body motion to detect engagement with a game companion. In: HRI, pp. 305–312 (2011)

13. Sawyer, R., Smith, A., Rowe, J., Azevedo, R., Lester, J.: Enhancing student models in game-based learning with facial expression recognition. In: UMAP, pp. 192–201 (2017)
14. Tian, Y.I., Kanade, T., Cohn, J.F.: Recognizing action units for facial expression analysis. IEEE Trans. Pattern Anal. Mach. Intell. **23**(2), 97–115 (2001)
15. Verma, M., et al.: Learners' efficiency prediction using facial behavior analysis. In: ICIP, pp. 1084–1088 (2021)
16. Whitehill, J., Serpell, Z., Lin, Y.C., Foster, A., Movellan, J.R.: The faces of engagement: Automatic recognition of student engagement from facial expressions. IEEE Trans. Affect. Comput. **5**(1), 86–98 (2014)
17. Yang, J., Wang, K., Peng, X., Qiao, Y.: Deep recurrent multi-instance learning with spatio-temporal features for engagement intensity prediction. In: ICMI, pp. 594–598 (2018)
18. Zhang, K., Zhang, Z., Li, Z., Qiao, Y.: Joint face detection and alignment using multitask cascaded convolutional networks. IEEE Signal Process. Lett. **23**(10), 1499–1503 (2016)

Real-Time Spoken Language Understanding for Orthopedic Clinical Training in Virtual Reality

Han Wei Ng[1(✉)], Aiden Koh[1,2,3], Anthea Foong[1,2,3], Jeremy Ong[1,2,3], Jun Hao Tan[2,3], Eng Tat Khoo[2], and Gabriel Liu[2,3]

[1] Nanyang Technological University, 50 Nanyang Avenue, Singapore 639798, Singapore
hanwei001@e.ntu.edu.sg
[2] National University of Singapore, 21 Lower Kent Ridge Road, Singapore 119077, Singapore
[3] National University Health System, 1E Kent Ridge Road, Singapore 119228, Singapore

Abstract. With the increasing limitation on healthcare resources, lack of real or standardized patients' willingness to participate in medical education and a high medical litigation environment, the traditional learning of "see one" and "do one" is no longer acceptable. Virtual Reality training is becoming commonly used to address this issue. However, some main challenges such as the lack of a dynamic comprehensive conversation and poor language understanding prevents virtual reality from being adopted into mainstream education. In this study, a medical Natural Language Processing (NLP) pipeline is developed. The proposed pipeline displays a total accuracy of 95.6%, with significant improvements compared to the original baseline by an increase of 15% accuracy. The resultant NLP model is subsequently implemented into a medical clinical training virtual simulation.

Keywords: Natural Language Processing · Spoken language understanding · Medical education · Virtual reality · Virtual patient

1 Introduction

Previous research describing the use of technology aims to compensate for students' lack of experience with patients have been reported [1]. The use of chatbots, a text-based intent recognition program, can "impart" knowledge, but does not provide the skills required for a realistic patient-doctor conversation necessary for qualitative medical training. Other researchers have reported preliminary success in non-medical programs using Bidirectional Encoder Representations from Transformers (BERT) [2] mediated natural language processing (NLP) platforms to create conversations. However, this has not been tested in clinical conversations heavily depending on specific medical terminologies. Furthermore, large datasets tend to be biased towards intentions which have more training examples. Augmentation of the dataset can be resource intensive.

M. M. Rodrigo et al. (Eds.): AIED 2022, LNCS 13355, pp. 640–646, 2022.
https://doi.org/10.1007/978-3-031-11644-5_61

Therefore, in this research study, we propose and show that the use of a Dual Intent and Entity Transformer (DIET) NLP model [3] alongside a deep learning sentence augmentation model is able to achieve higher accuracy intention classification in the medical context. The use of synthetic augmentation of sentences serves to increase the amount of training data as well as to ensure a more homogenous dataset, achieving better overall accuracy for a high number of intents.

2 Literature Review

Assistive chatbot technology has seen an increase in uptake over recent years, accelerated by the COVID-19 pandemic. One such use of the assistive chatbot is to deploy chatbots in medical professional training [1]. Medical institutions globally have begun exploring ways to utilize chatbot technology to enhance their current curriculum for training doctors by simulating patients, revising and standardizing examinations. Chatbot technology has also seen much use in training other forms of healthcare professionals such as nurses in terms of patient management [4].

However, chatbot technology remains limited in terms of integration into medical curriculums due to the lack of dynamic conversational capabilities and realism to depict the environment and patient. Since most chatbot systems are founded on rule-based programming, the flow of the conversation and the responses given by the chatbots are linear in nature. This prevents users from being able to feel fully engaged.

Previous work done by Rojowiec et al. showed that the use of Bidirectional Encoder Representations from Transformers (BERT) based models achieved a maximum accuracy of 71.35% on their dataset when tested for intention recognition of the doctor during doctor-patient clinical interviews [5]. It is noted that their collected data shows a high imbalance in the dataset, with a majority of the intent classes having fewer than 20 sample sentences per intent while some intent classes reach above 30.

3 Experimental Methodology

3.1 Dataset Collection

In this study, a medical language corpus is collected from 35 clinical students in their fifth year at the NUS Yong Loo Lin School of Medicine. The medical language corpus consists of questions datasets belonging to questions asked by the doctor and answers given by the patients. In this case, the students are asked to roleplay as doctor and patient pairs. For this study, the disease focused on is an orthopedic disease, spondylolisthesis. To ensure consistency in the medical language corpus, the students are tasked to roleplay as an elderly Chinese female suffering from the disease.

To perform sentence-level intention classification, a total of 212 intention labels were collected and sorted with varying number of sentences per label from the roleplay via voice recording. Transcription was used to convert the audio to text data (Table 1).

Table 1. Representation of medical auditory data used to train and test the proposed Natural Language Processing model.

Intention label	Example sentence 1	Example sentence 2
Back Pain Location	Could you show me exactly where your back pain is?	Can you point to me where the backache is?
...	...	...
Drug Allergy	Do you have an allergy to any drugs?	Are you allergic to any medication?

4 Sentence-Level Speech Intention Classifier for Orthopedic Clinical Training

4.1 Sentence-to-Sentence Paraphrasing Generator

During the training stage of the NLP model, additional sentences are generated using a pre-trained text-to-text model with Pre-training with Extracted Gap-sentences for Abstractive SUmmarization Sequence-to-sequence models (PEGASUS) [6]. A transformer encoder-decoder mode is trained using self-supervised objective Gap Sentences Generation (GSG) for the purpose of generating unique sentences given a single sentence input. The original intention-to-sample ratio spread is shown in Fig. 1 (A).

By performing sentence augmentation on the original dataset, we are able to obtain a homogenous spread of sentences across all intention classes as shown in Fig. 1 (B). Thus, this reduces the amount of overfitting that the NLP model will have on certain classes. This enables the model to generalize better across all of the identified intent labels, leading to better overall performance of the model.

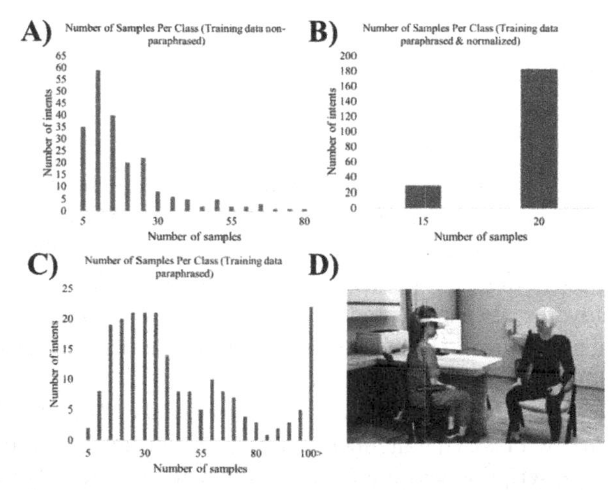

Fig. 1. (A) The original samples per intent distribution. (B) The distribution after performing paraphrasing augmentation. (C) The distribution after normalizing intents to a fixed number of samples. (D) Medical student interaction with virtual patient.

Additionally, it is a well-known issue that audio speech to text services may produce a non-trivial number of incoherent sentences. Thus, the pre-trained paraphraser model can also be used on the input sentences to correct for these mistakes.

The use of the paraphraser model can be seen to be useful in generating new example data using the original sentences as seen in Fig. 1 (B) and (C). However, a key limitation behind the paraphraser is its inability to generate a high number of unique examples given a single sentence example. Therefore, the paraphraser has an arbitrary upper limit in terms of the amount of data it can generate depending on the original input sentence. Figure 1 (C) shows the new data spread without homogenizing. Therefore, the paraphraser is unable to produce a perfectly evenly distributed augmented dataset since this would require producing a high number of augmented sentences to match the intent with the highest number of sample sentences, which is not possible.

4.2 Real-Time Speech Intention Classification

To perform sentence-level intention classification, the open-source DIET NLP model by RASA is implemented. The paraphraser model is used to generate synthetic training data in the medical training context for the DIET NLP model. Subsequently, the DIET NLP is optimized and trained using a combination of permutations involving both the original and generated datasets.

To perform real-time speech intention classification, the audio collected during the testing phase is converted to text using the Azure Speech to Text (Microsoft) service. Transcription error can still occur when converting audio to text. Thus, to reduce the inaccuracy of the trained NLP model, the paraphraser model as described in Sect. 4.1 is applied to the transcribed text which resolves the grammatical and semantic errors in the sentence while retaining as much of the original meaning as possible.

5 Virtual Reality Clinical Simulation

A virtual reality program for purpose of training a doctor in the clinical setting was created using Unity software as shown in Fig. 1 (D). The training simulation aims to mimic a real-life clinical scenario based on a patient who is suffering from an orthopaedic disease, whereby the trainee doctor is tasked to perform differential diagnosis and give a management plan.

The previously trained DIET NLP model gives the conversational capability to the patient avatar. As the DIET NLP model is trained with conversational stories, the appropriate response by the virtual patient is given as output after receiving the speech to text input from the user. If the confidence level of the model given is below that of 50%, the model is tasked to give a null fallback as the output indicating that the model is not certain of the answer. Otherwise, the NLP result which consists of the recognized intent is then collected and checked against a predefined list of actionable intents. The correct patient utterance and animation clip is then selected from a library of recorded audio and animation clips to be played.

6 Results and Discussion

6.1 Overall Accuracy and Ablation Studies

To study the effects of the augmentation as well as the effects of having a more evenly distributed training dataset, an ablation study was carried out.

Including the augmented sentences into the training data yielded a slight increase in the overall F1-score by 0.012. This strongly suggests that the resulting model did not improve significantly due to the inclusion of the paraphrased sentences into the training dataset. The slight improvement can be attributed to the model having a more evenly distributed dataset and a larger training dataset, encouraging the trained model towards a more generalized optimization.

By applying the paraphraser to only the test sentences, the improvement from the original baseline can be seen to be a significant increase of 0.119 in the F1-score. This is likely due to the paraphraser's model ability to convert the test sentence into a more common form that has a higher probability of having been seen by the model during the training phase, and thus result in higher accuracy (Table 2).

Table 2. Ablation studies result of the proposed spoken language understanding framework.

Model	Precision	Recall	F1-score
Rasa original	0.789	0.844	0.806
+ Paraphrased training	0.801	0.854	0.818
+ Paraphrased test	0.917	0.943	0.925
+ Both	**0.950**	**0.967**	**0.956**
+ Both (Normalised)	0.917	0.943	0.925

However, the usefulness of the augmented training data becomes apparent when combining both the training with paraphrased sentences alongside converting the test text input into a paraphrased sentence. This resulted in the highest increase in the model performance with an overall increase in 0.15 in the F1-score to achieve a total score of 0.956. This highly suggests that performing augmentation on both the training and test datasets allow for the mapping of the datasets towards a common domain.

Normalisation of the model did not result in any more increase in performance of the model. However, this is likely to be due to the size of the training dataset having been more than halved due to the forceful removal of sentences to ensure an evenly distributed data. Thus, the difference in the F1-scores suggest that the effect of having a sufficiently large training dataset is more significant as compared to having a more evenly distributed training data.

Therefore, to create the most optimal NLP model for understanding medical-focused intentions we propose the use of a customized paraphraser model that serves to both augment the dataset and to convert the incoming input into a more generalized domain format that is better understanding by the trained model and to reduce sentence errors.

6.2 Virtual Reality User Study

To assess the effectiveness of the NLP conversational model for orthopedic clinical training, a total of 23 undergraduate medical students were tasked to interact with the virtual patient. Minimal prompts were given to the students besides an introduction of how to operate the headset and interact with the virtual patient. After going through the entire clinical scenario, the students are then tasked to complete a survey to give their thoughts and opinions on the virtual patient simulator. A copy of the survey and the questions can be found below. The students were asked to rate their experience on a Likert scale from 1 to 5.

19 out of 23 (82.7%) of the students reported a good to excellent overall experience in the clinical training program to fulfil their learning objectives. Sub-group analysis showed that 69.6% of the students found the clinical interaction with the virtual patient to be similar to their experiences with real-life patient interactions. Furthermore, 86.9% of the student users found the graphics of the program to be realistic. 78% of the student users expressed that the clinical simulation would aid them in the clinical studies.

Overall, the preliminary studies with the undergraduate students strongly indicate a favorable response towards the use of the virtual reality program to enhance their medical clinical training. In addition, the use of the virtual patient simulation can be done at their own leisure, thus aiding in reinforcing the topics learnt and improving memory retention in the students.

7 Conclusion

In conclusion, the study has demonstrated the practical usefulness is the use of a paraphraser sentence augmentation model for the purpose of homogenizing the original training dataset and to convert transcribed audio into a more recognizable form by the model. This offers a significant improvement compared to the baseline, allowing for smoother conversation in a virtual clinical training scenario.

In this study, the trained model captured a total of 212 intents. The importance of training NLP models to recognize a large number of intentions is vital for progress towards building comprehensive interactable virtual agents. Through this study, we have highlighted the feasibility in using the DIET NLP model to perform highly accurate intent recognition on many unique intents. Additionally, the training and testing pipeline proposed by the study has shown that the use of sentence augmentation techniques can greatly improve the model's ability to generalize across testing labels.

References

1. Kaur, A., Singh, S., Chandan, J., Robbins, T., Patel, V.: Qualitative exploration of digital chatbot use in medical education: a pilot study. Digital Health **7**, 205520762110381 (2021)
2. Devlin, J., Chang, M., Lee, K., Toutanova, K.: BERT: pre-training of deep bidirectional transformers for language understanding. In: Proceedings of the 2019 Conference of the North (2019)
3. Bunk, T., Varshneya, D., Vlasov, V., Nichol, A.: DIET: Lightweight Language Understanding for Dialogue Systems (2020)

4. Chang, C., Hwang, G., Gau, M.: Promoting students' learning achievement and self-efficacy: a mobile chatbot approach for nursing training. Br. J. Edu. Technol. **53**(1), 171–188 (2021)
5. Rojowiec, R., Roth, B., Fink, M.C.: Intent recognition in doctor-patient interviews. In: LREC (2020)
6. Zhang, J., Zhao, Y., Saleh, M., Liu, P.: Pegasus: pre-training with extracted gap-sentences for abstractive summarization. In: International Conference on Machine Learning, pp. 11328–11339. PMLR, November 2020

Modeling Perspective Taking and Knowledge Use in Collaborative Explanation: Investigation by Laboratory Experiment and Computer Simulation Using ACT-R

Yugo Hayashi(✉) and Shigen Shimojo

Ritsumeikan University, 2-150 Iwakura-cho, Ibaraki 567-8570, Osaka, Japan
y-hayashi@acm.org

Abstract. This study focuses on modeling perspective taking behavior during collaborative explanation activities and the process of searching for relevant knowledge for constructive interaction. A laboratory experiment was conducted where dyads learned about concepts and engaged in a reasoning task using concept maps through discussions. We collected leaners conversational data and gaze using two eye trackers to examine perspective taking behavior. The results showed that leaners who considered others' diverse perspective expeditiously reflected upon their own knowledge and used the relevant knowledge for explaining the phenomenon to their partners. To further model the process of the perspective taking behavior and knowledge retrieval that occurred in the experiment, we developed a model using the cognitive architecture, Adaptive Control of Thought Rational (ACT-R). We further explained the mechanisms by implementing a model where perspective taking increases the opportunity to search and reconsidering the use common knowledge owing to the discrepancy triggered by diverse perspectives. This paper contributes to model-based research and provides knowledge on capturing the relationships of perspective taking and relevant knowledge retrieval in collaborative learning.

Keywords: Collaborative learning · Concept map · Computer model · ACT-R

1 Introduction

Studies on collaborative learning in the field of cognitive and learning science have shown that self-explanation activities are effective in triggering metacognition [2], and facilitate a deeper understanding of the learning materials [5], and facilitate abstract understanding of concepts. Past studies have shown that perspective-taking plays an important role in collaborative learning and problem-solving [6]. However, different types of cognitive bias have also been found; an example type is egocentric bias, which primarily entails the application of only accessible knowledge. Perspective-taking failure during a collaborative explanation activity leads to the formation of misunderstandings

M. M. Rodrigo et al. (Eds.): AIED 2022, LNCS 13355, pp. 647–652, 2022.
https://doi.org/10.1007/978-3-031-11644-5_62

and misconceptions about the knowledge. Thus, it is important to focus on how learners use relevant knowledge to develop a mutual understanding that will allow them to collaboratively develop knowledge.

How relevant knowledge is retrieved by dyads during explanation is another important topic of investigation. Explanation activities can be described as activities wherein one will perceive the knowledge of others and interpret one's own knowledge to make sense of the explained content [2]. From a cognitive processing perspective, memory retrieval-based knowledge searches may not always be successful, and may result in the retrieval of inadequate knowledge that leads to knowledge-sharing inconsistencies. Moreover, misconceptions may lead to more confusion and irrelevant memory and knowledge retrieval [3]. The failure to understand the perspective of their peer creates a discrepancy that may lead to a misunderstanding. Additionally, the memory retrieval process may be contaminated by noise, which can also be considered to be a source of discrepancy [1]. These points have not yet been comprehensively investigated in previous studies on collaborative learning.

This study was designed to investigate and model the process of relevant knowledge retrieval and perspective-taking behavior during constructive interactions in collaborative learning settings. To achieve these goals, we have focused on investigating dyad explanation activities wherein learners make inferences about a phenomenon by applying scientific concepts and concept maps. We collected conversational data from the learners and tracked their gazes using two eye trackers in order to investigate perspective-taking behavior. We apply results obtained from eye-tracking and protocol analysis to construct an ACT-R-based model of knowledge processing [1]. Our hypotheses were as follows:

H1: Learners who can take a different perspective from another person will reflect upon their own knowledge and apply relevant knowledge in a discussion with their partner.
H2: Learners who can take the perspectives of others and apply relevant knowledge during the explanation activity will collaboratively generate relevant knowledge in the concept map.

2 Experimental Method

2.1 Participants

Sixty university students majoring in psychology participated in a laboratory-based experiment in exchange for course credit. Hereafter, we will refer to these participants as learners.

2.2 Task and Procedure

The experiment was conducted in a remote environment that allowed them to communicate with each other using the Cmap software [4], a concept-mapping tool. The learners' goal in this experiment was to theoretically explain a particular case event using psychological theories. To enable data aggregation, the experiment included the following three phases: (1) the individual text learning phase, (2) the individual concept map generating phase, and (3) the collaborative reasoning phase wherein they collaborated to give a shared interpretation of their concept maps. The individual learning phase consisted of the following two sub-phases: (a) memorization of the theoretical concept of the attribution theory, and (b) memorization of the case story. In the individual learning phase, they were required to apply the attribution theory to a problem case story involving a student who participated in a school counseling program and to describe why the student has anxiety about the new academic year.

In the individual concept map generating phase, they were required to apply the theoretical knowledge to generate a concept map that explains the episode. Regarding the concept map, they were instructed to apply several different types of links to connect the nodes and describe the attribution process. A way to effectively apply the theory to explain the case is to input the text label for the nodes in the map using the words from the case study, and to use the text label for the links to connect the nodes. In the collaborative reasoning phase, learners worked in pairs by discussing the same task that they worked on in the individual concept map generating phase.

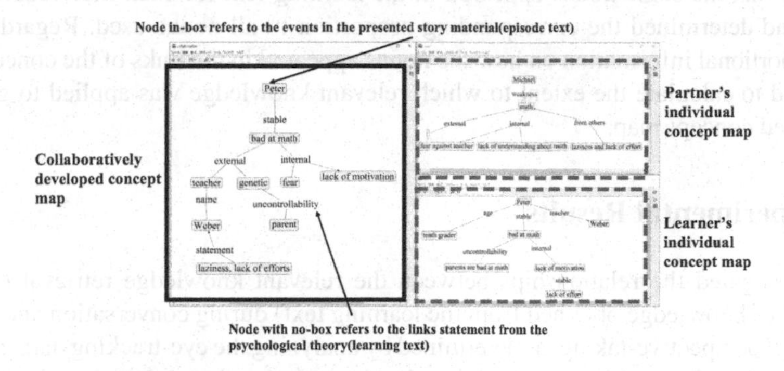

Fig. 1. Example of the screen viewed by a learner in the collaborative learning phase.

The right-hand side shows two windows of the previously generated individual concept maps. The left-hand side shows the shared concept map that was developed by collaborating during this phase. In the collaborative learning phase, the learners collaborated by providing oral explanations; particularly, they were instructed to explain the other's thoughts and develop another concept map. As can be seen in Fig. 1, the participants were able to see each other's concept maps (depicted on the right-hand side) that were developed during the individual learning phase as they worked on the common concept map (left-hand side).

2.3 Measures

We applied the following three dependent variables: (1) the gaze behavior, (2) the knowledge applied during explanations, and (3) the relevant knowledge used to generate the concept map. To obtain the data for (1), we employed two eye trackers (Tobii X2–30) to analyze where the learners were looking during the task. The screen was divided into three parts (i.e., Area 1: other, Area 2: self, and Area 3: shared area), and the number of fixations per area was counted. To quantify the extent to which each learner focused on their own knowledge to construct the shared concept map, we calculated the frequency of gaze fixations on Areas 1 and 2 during the explanation activities. This was achieved by using the following equation:

$$b = \frac{n_1 - n_2}{n_1 + n_2}, \tag{1}$$

where n_1 is the number of fixations on Area 2 (i.e., their partner's concept map), and n_2 is the number of fixations on Area 1 (i.e., their own concept map). A b value that is close to 0 implies that the learner spent roughly the same amount of time looking at the two areas.

For (2), we collected the learners' conversational data during the explanation activity. In accordance with the attribution theory, we conducted a morphological analysis and collected the nouns that were concise with the nouns that were used in the learning text, which was the learning material for the theoretical text. We calculated the number of times that the same nouns appeared in the learning text for each individuals' utterances and determined the corresponding proportion to all nouns used. Regarding (3), the proportional information on how the nouns appeared in the links of the concept map was used to calculate the extent to which relevant knowledge was applied to generate the shared concept map.

3 Experimental Results

We investigated the relationships between the relevant knowledge retrieval (i.e., the retrieval of knowledge obtained from the learning text) during conversation and the frequency of perspective-taking, as determined by analyzing the eye-tracking data. Figure 2 illustrates the relationship between the retrieval rate during the explanation (text) and the b index, which represents the degree of perspective-taking frequency, and thus indicates the extent to which relevant knowledge was applied to generate the shared concept map. The high/low data points in Fig. 2 were determined by dividing the values based on the output of the median of the relevant concept map (CM) links.

To investigate our two hypotheses, we conducted multiple regression analysis. The regression coefficient R^2 was 0.269, and the ANOVA F-value was 10.496, indicating statistical significance ($p = 0.000$). The experimental results support our two hypotheses and show that learners tend to apply relevant knowledge, which leads to the successful development of relevant concept maps. The data collected from the experiment indicate that learners may also fail to retrieve knowledge, but the reason for this cannot be inferred from the data. Thus, we developed a model that allowed us to fill in this gap of knowledge; we especially focused on the mechanisms of memory retrieval. We hypothesized that the frequency at which one considers the different perspective of the other, as reflected in the concept map (i.e., a low b value), is directly related to the frequency at which one tends to search for relevant knowledge that can be applied for explanation.

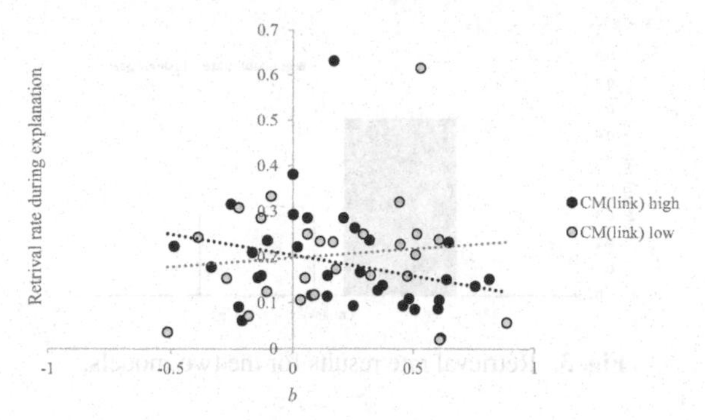

Fig. 2. Correlation between gaze behavior and knowledge retrieval (i.e., from the learning text) based on link generation.

4 ACT-R-Based Computer Simulation

In this model, we used two types of declarative knowledge, which are hereafter referred to as chunks. Specifically, these chunks are as follows: (1) the episode chunk, i.e., knowledge of the cause and result that was included in the episode text and have been described as nodes in the CM, and (2) the learning chunk, i.e., knowledge about the links that connect the two nodes that can be retrieved form the learning text. The model perceives the concept map of the self/other presented on the monitor. During each search loop, the model acquires the knowledge and adds a new chunk into their memory. Then, the model determines whether the new knowledge acquired from their partner matches any of their own. If the model detects a difference that indicates a discrepancy between the learners in the experiment, it begins to search for shared knowledge from the learning text. Then, the model searches for relevant knowledge chunks in the retrieval buffer by determining whether there are any matches with the knowledge chunks in the imaginal buffer. The model acquires new chunks by performing multiple observations of the self/other concept maps. This, in turn, provides more options for the initial knowledge search, and thus increases the likelihood of finding relevant knowledge. Note that noise and failure were taken into account in the ACT-R-based retrieval process, and that we implemented this by adjusting the parameters of the activation levels. We used the activation rate for memory retrieval to model the retrieval of learning text-related memories. Utility parameters in ACT-R were used as the reference to decide which knowledge chunks would be applied in the search.

We simulated two types of models based on the *b* index results obtained from the experimental data. One model (equitable gaze model), searched for both self/other concept maps (low *b*) and another model (single gaze model) observed only either self or other's concept map (high *b*). The retrieval rate results for the two models are shown in Fig. 3.

The results shown in the figure indicate that, when the model applies information from both the "self" and "other" concept maps, it will achieve a higher rate of relevant knowledge retrieval. Conversely, if the focus is maintained on either one of the concept maps, the retrieval rate will be lower. These results support our hypothesis.

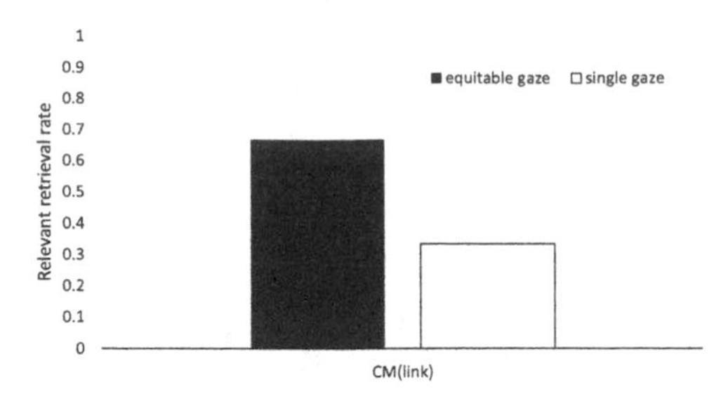

Fig. 3. Retrieval rate results for the two models.

5 Conclusion

We investigated and modeled the process of relevant knowledge retrieval and perspective-taking behavior during constructive interactions within the framework of a collaborative concept-mapping task. We collected the learners' conversational data, as well as used two eye trackers to collect their gaze data, to investigate their perspective-taking behaviors. The results revealed that learners who take the other's different perspective tend to reflect upon their own knowledge and more actively engage in discussions with their partner that entail the use of the relevant knowledge. Furthermore, such learners were also found to collaboratively generate a shared concept map that comprehensively reflects their relevant knowledge. To further investigate how knowledge retrieval rapidly occurred during perspective-taking, we developed an ACT-R-based model. This model was demonstrated to provide deeper insight into the mechanisms by which perspective-taking increases the opportunity to search for and reconsider the use of knowledge. Thus, the simulated results provide useful information on how reflective cognitive process may occur during collaborative learning activities.

Acknowledgements. This work was supported by Japan Society for the Promotion of Science (JSPS) KAKENHI Grant-in-Aid for Scientific Research (B), Grant Number JP20H04299.

References

1. Anderson, J.R.: How Can the Human Mind Occur in the Physical Universe. Oxford University Press, Oxford (2007)
2. Chi, M.T.: Self-explaining expository texts: the dual process of generating inferences and repairing mental models. In: Glaser, R. (ed.) Advances in Instructional Psychology. Educational Design and Cognitive Science, vol. 5, pp. 161–238. Lawrence Erlbaum, Mahwah (2000)
3. Clement, J.: Overcoming students' misconceptions in physics: fundamental change in children's physics knowledge. J. Res. Sci. Teach. **28**, 785–797 (1987)
4. Cmap Homepage. https://cmap.ihmc.us/, Accessed 4 Feb 2022
5. Greeno, J.G., van de Sande, C.: Perspectival understanding of conceptions and conceptual growth in interaction. Educ Psychol **42**, 9–23 (2007)
6. Hayashi, Y.: The power of a "maverick" in collaborative problem solving: an experimental investigation of individual perspective-taking within a group. Cogn. Sci. **42**(S1), 69–104 (2018)

Are Students Aware of How Much Help They Need? The Relationship Between Awareness, Help Seeking, and Self-assessment

Michael Smalenberger(✉)

Department of Mathematics and Statistics, University of North Carolina at Charlotte, Charlotte, USA
msmalenb@uncc.edu

Abstract. Facilitating students to become better self-regulated learners is a grand challenge in intelligent tutoring systems (ITS) research. Help-seeking (HS) and self-assessment (SA) are important in self-regulated learning. The literature has shown that students tend to have maladaptive HS behavior, e.g., overuse hints or not use hints when appropriate. Furthermore, student SAs of domain knowledge have been shown to be inaccurate a priori and persistently overconfident. Together these beckon the question of whether students are aware of how much help they need. To investigate this, 115 students in two introductory college statistics courses completed the Mindful Attention Awareness Scale (MAAS) questionnaire and completed a homework assignment in an ITS. Students with above-average MAAS scores used fewer hints, more accurately estimated the number of hints they would request on the subsequent similar question, and more accurately recalled the number of hints used on the previous question. However, differences in MAAS scores did not influence the number of errors made, nor the recall or prediction of errors. We contribute to the literature by showing how MAAS scores relate to HS and SA in ITS.

Keywords: Mindful Attention Awareness Scale (MAAS) · Help seeking · Self-assessment · Self-regulated learning · Intelligent tutoring system (ITS)

1 Introduction

In self-regulated learning (SRL), it is crucial for students to proficiently evaluate and improve their learning practices, including self-assessment (SA) and help-seeking (HS) [1]. [2] showed that increasing students' awareness of their mathematics learning behaviors positively affected their SRL behaviors. Hence, in this study, we establish a measurable relationship between awareness, self-assessment, and help-seeking.

SA refers to students' ability to accurately evaluate their knowledge while learning [3]. Accurate SA has been shown to correlate with productive help-seeking behaviors [4]. For students' SA to be accurate, students should be aware of the relative strengths and weaknesses of their knowledge, in relation to a target task [5]. However, students often overestimate their abilities [3]. Students who lack sufficient domain knowledge

M. M. Rodrigo et al. (Eds.): AIED 2022, LNCS 13355, pp. 653–659, 2022.
https://doi.org/10.1007/978-3-031-11644-5_63

are especially likely to make inaccurate SAs even when the solutions are presented to them [5]. However, the literature has shown that improving SA can be tutored using ITS [6]. Similarly, help-seeking may influence learning with ITS because many ITS provide on-demand, principle-based help messages [7]. The literature shows that students can have maladaptive help-seeking behavior. Improvements in help-seeking behavior can be tutored using an ITS [8, 9], but this does not always lead to better learning outcomes such as differences between pretest and post-test scores [10, 11].

If students are poor at self-assessing and can have maladaptive help-seeking behavior, the question beckons of whether students are aware of how much help they need. The Mindful Attention Awareness Scale (MAAS) is a widely used assessment for measuring self-regulation constructs [12] and has been repeatedly validated including on tasks requiring high working memory capacity [13]. The MAAS is a 15-item (1–6 Likert scale) questionnaire, where a higher average score represents greater mindfulness. Higher MAAS scores have been shown to be related to mathematics performance, including predicting higher exam scores in college samples [14]. To our knowledge, the MAAS has not been used to investigate student mathematics performance in ITS such as predicting the number of hints requested, errors made, learning gains, etc.

In this study, college students in two statistics courses complete the MAAS questionnaire and a homework assignment in an ITS which also routinely asked students to recall how many errors they made and hints used on the previous question, and to predict the number of errors they would make, and hints used on a subsequent similar question. We hypothesize that students with higher MAAS scores will better recall and predict the number of errors made and hints used, but not make fewer errors or use fewer hints. This is in line with past research which on one hand indicates that increased awareness may lead to changes in learning behaviors which in turn may lead to differences in learning outcomes, but on the other hand has not provided any evidence suggesting awareness is related to mathematics aptitude. Thus, our research questions are:

1. Do MAAS scores correlate with the average number of errors made or hints used?
2. Do higher MAAS scores lead to better recall of errors made or hints used on the previous question?
3. Do higher MAAS scores lead to better predictions of errors made or hints used on a subsequent similar question?

2 Methodology

2.1 Procedure

A set of 25 questions on normal random variables (NRV) was assigned as homework immediately after discussing that topic in class. The homework used a novel ITS created using Cognitive Tutor Authoring Tools (CTAT) integrated into the course through TutorShop [15]. Students were free to work at their own pace and were not required to complete the assignments in one sitting before the due date. Additionally, students completed the MAAS questionnaire using an online survey tool outside of class within a week of completing the ITS assignment.

2.2 Design

The NRV problem set used five question types, each having five isomorphic questions. An identical fixed sequence of questions was used that was grouped by isomorphism.

Solving a question requiréd a student to enter a numerical value into several "input" boxes. Each input box had multiple "levels" of on-demand, principle-based help available, and the last "bottom-out" hint provided the solution to that input box. Immediate correctness feedback was always given after entering a value, and an input box became "locked" after a student entered the correct solution for that box.

Immediately after completing a NRV question, students were asked four questions, one each to recall the number of hints and errors used on the previous question, and one each to predict the number of errors and hints on the subsequent question.

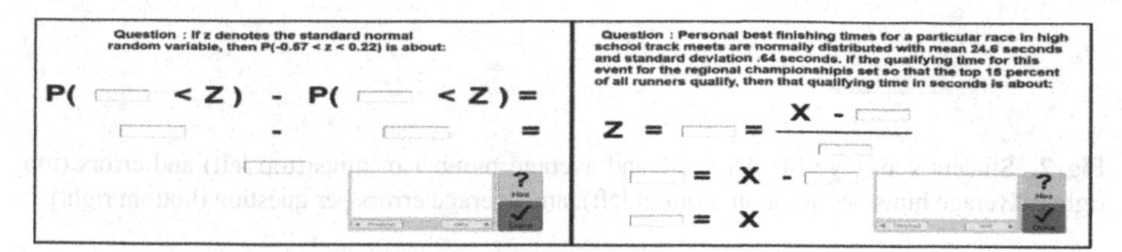

Fig. 1. Examples of 2 of the 5 NRV question types

3 Results

For analysis, we used the 'Stat 1222-004 S21 Chap 5 HW' and 'Stat 1220-010 S21 Chap 5 HW' datasets accessed via DataShop [16], and the MAAS questionnaire results. A total of 115 students enrolled in two courses participated in this study. Of those, 23 were removed for not completing the homework assignment, and 2 were removed for not completing the MAAS questionnaire. Prior research on the MAAS showed that in a large U.S. adult sample, the average MAAS score was 4.22 (S.D. = 0.63) [12]. In line with other studies involving MAAS [12], we restricted our analysis to students with a MAAS score within two standard deviations of this mean and removed 16 students who did not fall in this interval. After these exclusion criteria, 74 students remained for analysis. Of these 74 students, 20 had an above-average MAAS score, and 54 had a below-average MAAS score, with an overall sample mean of 3.90 (S.E. = 0.55).

3.1 MAAS and Proficiency

To address our initial research question, we calculate the average number of hints and the average number of errors by each student. The Pearson correlation coefficient between MAAS and average hints ($r = 0.149$, $p = 0.160$), and MAAS and average errors ($r = -0.004$, $p = 0.971$) were not statistically significant. After visual inspection of each scatterplot, we have no reason to believe a nonlinear correlation exists either.

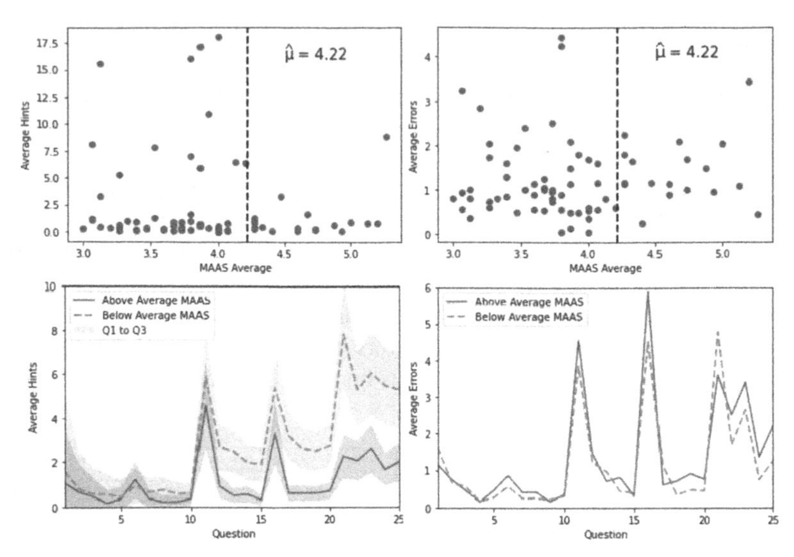

Fig. 2. Student's average MAAS score and average number of hints (top left) and errors (top right). Average hints per question (bottom left), and average errors per question (bottom right).

After bifurcating our sample into below average and above average MAAS scores ($\widehat{\mu}$ = 4.22) [12] (Fig. 2, top row, dotted line), several characteristics emerge. Students with below average MAAS scores on average use more hints and have more variation in the number of hints they use (M = 2.74, SD = 4.72) than their above-average MAAS score peers (M = 1.24, SD = 1.97) (Fig. 2, top left). The differences in means (Welch's t-test statistic = −2.08, Welch-Satterthwaite df. = 70.97, p = 0.041) and variances ($F_{53,19}$ = 5.82, $p < 0.001$) are both significant. Students with below average MAAS scores on average made a similar number of errors (M = 1.18, SD = 0.93) as students with above-average MAAS scores (M = 1.39, SD = 0.70) (Fig. 2, top right). The differences in means (t(72) = 0.91, p = 0.37) is not significant but the difference in variances ($F_{53,19}$ = 1.73, p = 0.095) approaches significance. While the number of hints used by students in both groups follow similar trends, their quantities diverge (Fig. 2, bottom left). There was not a similar divergence in the number of errors made (Fig. 2, bottom right).

3.2 MAAS and Recall

To address our second research question, we calculate the average absolute difference between the number of *hints* the student recalled requesting on the previous question and the actual number requested, and the average absolute difference between the number of *errors* the student recalled making on the previous question and the actual number made. After similarly bifurcating, students with above-average MAAS scores on average had more accurate recall of the number of hints used (M = 0.79, SD = 1.79) than students with below-average MAAS scores (M = 2.21, SD = 4.19). The differences in means (Welch's t-test statistic = −2.03, Welch-Satterthwaite df. = 70.40, p = 0.046) and variances ($F_{53,19} = 5.48$, $p < 0.001$) are statistically significant. However, students with above-average MAAS scores on average did not have more accurate recall on the number of errors made (M = 0.92, SD = 0.70) than students with below-average MAAS scores (M = 0.84, SD = 0.79). That is, the differences in means (t (72) = 0.44, p = 0.66) and variances ($F_{53,19} = 1.25$, p = 0.30) are not significant. Additionally, while the accuracy in recalling the number of hints used by students in both groups follows similar trends, their quantities diverge. However, there was no significant difference in the recall of the number of errors made between the two groups.

3.3 MAAS and Prediction

To address our third research question, we calculate the average absolute difference between the number of *hints* the student predicted they would request on the subsequent similar question and the actual number requested, and the average absolute difference between the number of *errors* the student predicted they would make on the subsequent similar question and the actual number made. After similarly bifurcating, students with above-average MAAS scores on average had more accurate predictions of the number of hints they would use (M = 1.10, SD = 1.86) than students with below-average MAAS scores (M = 2.56, SD = 4.39). The differences in means (Welch's t-test statistic = −2.03, Welch-Satterthwaite df. = 70.60, p = 0.048) and variances ($F_{53,19} = 5.58$, $p < 0.001$) are both significant. However, students with above-average MAAS scores on average did not have more accurate predictions of the number of errors they would make (M = 1.41, SD = 0.64) than their below-average MAAS score peers (M = 1.34, SD = 1.13). Specifically, the differences in means (Welch's t-test statistic = 0.34, Welch-Satterthwaite df. = 59.32, p = 0.74) and was not significant, but the difference in variances ($F_{53,19} = 3.07$, p = 0.004) was significant. Also, while the accuracy in predicting the number of hints used by students in both groups follows similar trends, their quantities diverge. There was no discernible difference in the predictions of the number of errors made.

4 Discussion

We take these results to support our hypothesis that students with higher MAAS scores on average better recall and predict the number of hints used, and not make fewer errors. However, these results do not support our hypothesis that students with higher MAAS scores on average better recall and predict the number of errors made, and do not request fewer hints. Succinctly, our results show that MAAS does better in predicting certain help-seeking behaviors than predicting accurate self-assessment.

5 Contributions, Limitations, and Future Research

We made important contributions, including exploring the relationship between MAAS scores and SA and HS in an ITS with detailed problem steps. Our results show that an ITS which supports awareness can contribute to better HS, and hence better SRL.

We also recognize that our study has limitations. For example, the quantitative metrics evaluated in this study minimally operationalized SA, HS, and awareness. However, these concepts have qualitative components which were not assessed herein.

Also, completing homework questions that are used to assess measures of SRL, and in close temporal proximity administer a questionnaire to assess awareness may lead to biased results. However, given the nature of the questions in the assignment and those on the MAAS questionnaire, we do not believe this significantly influenced our results.

Additionally, students with better mathematics abilities may make fewer errors and use fewer hints, particularly on the later longer, and potentially more challenging questions, which in turn may be easier to recall. Future research should account for such things by implementing an experimental design that includes a pretest and posttest.

Furthermore, the design of the tutor may have an impact on the results. For example, questions addressing SA and HS were given after the interface page was refreshed rendering the mathematics problem of inquiry invisible. Also, the fixed sequence of questions, coupled with the fact that a student's assignment completion time's proximity to the due date was not accounted for, may influence trends in the number of hints used on questions near the end of the assignment. While we do not believe this influenced our results, future research may be conducted to systematically investigate this.

Lastly, even though our sample mean was not significantly different than the estimated population mean in [12], our above-average MAAS score sample was larger than our below-average MAAS score sample. Furthermore, some students in our below-average MAAS sample had high numbers of average hints, and even though we are concerned about awareness, the impact of gaming should be taken into consideration. Future iterations of this study should also have more study participants, particularly from more than one instructor, to see if these sampling characteristics persist.

Acknowledgements. As always, K.H.S., J.M.S., E.M.S., and W.J.S. thank you and I l. y.

References

1. Zimmerman, B.: Motivational sources and outcomes of self-regulated learning and performance. In: Zimmerman, B.J., Shunk, D.H. (eds.) Handbook of Self-regulated Learning and Performance, pp. 49–64. Routledge, New York (2011)
2. Caswell, R., Steven, N.: Enhancing mathematical understanding through self-assessment and self-regulation of learning: the value of meta-awareness (2005)
3. Dunning, D., Heath, C., Suls, J.: Flawed self-assessment, implications for health, education, and the workplace. Psychol. Sci. Public Interest **5**(3), 69–106 (2004)
4. Nelson-Le Gall, S., Kratzer, L., Jones, E., DeCooke, P.: Children's self-assessment of performance and task-related help seeking. J. Exp. Child. Psychol. **49**, 245–263 (1990)
5. Kruger, J., Dunning, D.: Unskilled and unaware of It: how difficulties in recognizing one's own incompetence lead to inflated self-assessments. J. Pers. Soc. Psychol. **77**(6), 1121–1134 (1999)
6. Roll, I., Aleven, V., McLaren, B.M., Koedinger, K.R.: Metacognitive practice makes perfect: improving students' self-assessment skills with an intelligent tutoring system. In: Biswas, G., Bull, S., Kay, J., Mitrovic, A. (eds) Artificial Intelligence in Education. AIED 2011. LNCS, vol. 6738. Springer, Cham (2011). https://doi.org/10.1007/978-3-642-21869-9_38

7. Aleven, V.: Help seeking and intelligent tutoring systems: theoretical perspectives and a step towards theoretical integration. In: Azevedo, R., Aleven, V. (eds.) International Handbook of Metacognition and Learning Technologies. Springer International Handbooks of Education, vol. 28, pp. 311–335. Springer, New York (2013). https://doi.org/10.1007/978-1-4419-5546-3_21
8. Roll, I., Aleven, V., McLaren, B.M., Koedinger, K.R.: Can help seeking be tutored? Searching for the secret sauce of metacognitive tutoring. In: Luckin, R., Koedinger, K.R., Greer, J. (eds.) Proceedings of the 13th International Conference on Artificial Intelligence in Education (AIED 2007), pp. 203–210. IOS Press, Amsterdam (2007)
9. Roll, I., Aleven, V., McLaren, B.M., Koedinger, K.R.: Improving students' help-seeking skills using metacognitive feedback in an intelligent tutoring system. Learn. Instr. **21**, 267–280 (2011)
10. Aleven, V., Roll, I., McLaren, B.M., Koedinger, K.R.: Help helps, but only so much: research on help seeking with intelligent tutoring systems. Int. J. Artif. Intell. Educ. **26**, 205–223 (2016)
11. Aleven, V., Roll, I., McLaren, B., Ryu, E., Koedinger, K.: An architecture to combine metacognitive and cognitive tutoring: pilot testing the help tutor. In: Looi, C., McCalla, G., Bredeweg, B., Breuker, J. (eds.) ITS 2014, pp. 443–454. Springer, Cham (2004)
12. Brown, K., Ryan, R.: The benefits of being present: mindfulness and its role in psychological well-being. J. Pers. Soc. Psychol. **84**(4), 822–848 (2003)
13. Dubert, C.J., Schumacher, A.M., Locker, L., Gutierrez, A.P., Barnes, V.A.: Mindfulness and emotion regulation among nursing students: investigating the mediation effect of working memory capacity. Mindfulness **7**(5), 1061–1070 (2016)
14. Bellinger, D., DeCaro, M., Ralston, P.: Mindfulness, anxiety, and high-stakes mathematics performance in the laboratory and classroom. Conscious Cogn. **37**, 123–132 (2015)
15. Aleven, V., McLaren, B., Sewell, J., Koedinger, K.: A new paradigm for intelligent tutoring systems: example-tracing tutors. Int. J. Artif. Intell. Educ. **19**(2), 105–154 (2008)
16. Koedinger, K., Baker, S., Cunningham, K., Skogsholm, A., Leber, B., Stamper, J.: A data repository for the EDM community: the PSLC datashop. In Romero, C., Ventura, S., Pechenizkiy, M., Baker, S. (eds.) Handbook of Educational Data Mining. CRC Press, Boca Raton (2010)

Automated Support to Scaffold Students' Written Explanations in Science

Purushartha Singh[1(✉)], Rebecca J. Passonneau[1], Mohammad Wasih[1], Xuesong Cang[2], ChanMin Kim[1], and Sadhana Puntambekar[2]

[1] The Pennsylvania State University, State College 16801, USA
{pxs288,rjp49,mvw5820,cmk604}@psu.edu
[2] University of Wisconsin-Madison, Madison, WI, USA
xcang@uwisc.edu, puntambekar@education.wisc.edu

Abstract. In principle, educators can use writing to scaffold students' understanding of increasingly complex science ideas. In practice, formative assessment of students' science writing is very labor intensive. We present PyrEval+CR, an automated tool for formative assessment of middle school students' science essays. It identifies each idea in a student's science essay, and its importance in the curriculum.

Keywords: Science explanation · Natural language processing

1 Introduction

Secondary school science teachers face multiple demands in scaffolding students' learning of science ideas and science practices. Written explanation of science ideas is an important science practice, as well as a mechanism to assess students' understanding. However, formative assessment of writing is time consuming for teachers. This paper presents a natural language processing application for formative assessment of middle school students' physics essays on energy. It identifies the ideas they express, and the relative importance of these ideas.

Here we first briefly describe current automated support to scaffold science writing. Then we present PyrEval+CR, which extends PyrEval [3], an efficient tool originally developed to assess the content of summaries of the main ideas of source texts.[1] PyrEval+CR has a lightweight, modular design that can be easily adapted to new assignments or writing characteristics. It identifies propositions (statements) expressed in writing, and provides both quantitative and qualitative outputs to support feedback to students and teachers. Section 4 explains how we treat assessment as an optimization problem to match student propositions to propositions in a computable rubric (CR). To evaluate its performance before testing it in the classroom, we constructed a dataset mined from historical essays written by middle school students who used a similar curriculum. An experiment testing many configurations of PyrEval+CR on this data resulted in many settings that correlate well with a highly reliable manual assessment.

[1] PyrEval+CR is available at https://github.com/psunlpgroup/PyrEvalv2.

M. M. Rodrigo et al. (Eds.): AIED 2022, LNCS 13355, pp. 660–665, 2022.
https://doi.org/10.1007/978-3-031-11644-5_64

2 Automated Tools for Scaffolding Science Writing Skills

Previous work points to the potential for automated feedback on student science writing through identification of specific concepts and rubric components. Good agreement between human and automated output has been found in biology explanations from different institutions [6], on middle school explanations of why sugar dissolves in water [7], and on rubric elements for high school biology essays in Hebrew [1]. Below, we present high agreement of our tool with manual rubric scores for the modified middle school essays mentioned above.

Integration of formative assessment tools during science instruction is less well-studied. Teachers who used automated guidance in the WISE environment to help students revise science explanations found that teachers pursued a variety of guidance strategies [10]. A later case study of one teacher's use of automatically generated guidance found the teacher used multiple strategies, and students who revised made more substantial revisions [4]. We have begun a study to apply PyrEval+CR in nearly three dozen middle school classrooms to explore how teachers and students will utilize feedback in classroom settings.

3 PyrEval+CR Overview

PyrEval derives an assessment standard called a pyramid from several reference summaries written by experts. All propositions from the reference summaries are ranked for importance by the number of reference summaries each occurs in. PyrEval+CR relies on a computable rubric with the same form as a pyramid, but derived from a manual rubric. Here we describe the pre-processing that converts an essay to embeddings, and the computable rubric.

The first pre-processing step uses a special-purpose decomposition parser (DP) to decompose complex sentences. DP output consists of alternative ways to decompose the same sentence. For example, a complex sentence of two clauses will have at least two alternatives, one with two clauses, and the original (undecomposed) sentence. Decomposition supports more options for the optimization approach to align student propositions to the CR, as we later illustrate.

The DP uses context-free-grammar parses to extract all tensed verb phrases in a sentence, and dependency parses to identify the subjects of the main verbs. This ensures propositionally complete output clauses. A small set of rules handles traversal of the parses for different syntactic structures [3]. To adapt to middle school writing, we tested subsets of DP rules. We also added a parameter to constrain the minimum length in words of output clauses (MinSegLength).

The second preprocessing step converts DP output clauses to embeddings. Matching a student clause to a CU relies on the average pairwise cosine similarity (APCS) of the student's embedding to sets of CU embeddings. In our earlier work, we found WTMF [5] to give superior similarity results over other embedding methods, but we had not controlled for all factors [3]. Given the widespread use of GloVe [8], we decided to conduct a rigorous comparison between WTMF and GloVe on a standard benchmark, the SemEval semantic textual similarity

Table 1. Pearson correlations with human scores of WTMF and GloVe+SIF on three STS benchmarks. Vocabulary size (V) and total words (S) appear in parentheses.

Test data	WTMF (V = 81.8K; S = 4M)				Gigaword Sub. (V = 67.1K; S = 18.9M)			
	WTMF		GloVe + SIF		WTMF		GloVe + SIF	
	Sent	Win	Sent	Win	Sent	Win	Sent	Win
STS12	**0.7258**	0.6851	0.6859	0.6812	0.6400	**0.6482**	0.6256	0.6256
STS13	**0.7405**	0.6901	0.6426	0.6311	0.5909	**0.6224**	**0.6214**	**0.6214**
STS14	**0.7187**	0.7012	0.6299	0.6149	**0.6835**	**0.6835**	0.6223	0.6223

(STS) tasks (cf. [2]). For STS, humans rated pairs of sentences on a 6-point scale of semantic similarity. System predictions are compared to human ratings using Pearson correlation. We used three years of STS tasks.

WTMF applies weighted matrix factorization to a word-by-sentence matrix of tf.idf scores to compute word embeddings [5]. Using matrix reconstruction, phrase embeddings for unseen sentences can be constructed from the word embeddings. GloVe applies log-bilinear regression to co-occurrence data from a small moving context window over a training corpus [8]. We created GloVe phrase embeddings using a high-performing weighted average of a phrase's word embeddings (SIF) [2]. We trained WTMF and GloVe on a high-quality corpus created by the WTMF developers, and on an extract of the Gigaword news corpus, ensuring both methods used the same vocabulary list for a given corpus. The WTMF corpus combines a high proportion of definitional sentences with a small heterogeneous corpus (the Brown corpus). We sampled increasing amounts of Gigaword, but were unable to achieve matched vocabulary sizes. At nearly five times the size of the WTMF corpus, our Gigaword subset has only 82% of the WTMF corpus vocabulary (see table header in Table 1).

Table 1 shows that WTMF outperforms GloVe+SIF on the benchmark semantic similarity tasks, controlling for the same vocabulary list, corpus, context span, and vector dimensionality (100D). WTMF performs best with the WTMF corpus, using sentence contexts. GloVe results are more consistent across conditions. Due to these results, we use WTMF embeddings from the WTMF corpus.

The original PyrEval creates a set of content units (CUs), called a pyramid, extracted from four to five reference summaries written by experts. Each CU corresponds roughly to a set of paraphrases of the same idea. The number of reference summaries that express the same idea provides an importance weight on the idea. For PyrEval+CR we aimed for an assessment that would more closely resemble the application of an analytic rubric. As described elsewhere [9], we created a very reliable analytic rubric to assess essays that explain students' roller coaster designs with reference to energy concepts (e.g., potential vs. kinetic energy). Here, we describe how we created our computable rubric (CR).

For the CR, we mined phrases corresponding to rubric elements from middle school essays. Figure 1 illustrates a weight 4 CU for a rubric element that defines kinetic energy. In the CR, CU weights range from 5 for important ideas to

e_1	Kinetic energy is the energy of an object in motion
e_2	The energy of an object due to its motion is called kinetic energy
e_3	An object that is moving has kinetic energy
e_4	Kinetic energy is the energy of a moving object

Fig. 1. A weight 4 (w4) CU. The CR has 62 CUs: 3 w5, 4 w4, 13 w3, 16 w2, 26 w1.

1 for weak ideas. The weighted CUs in a PyrEval pyramid have a power law distribution, so we ensure that a CR also does (see Fig. 1 caption). All phrases i in a CU are converted to embeddings, represented schematically in column one of Fig. 1, as are all decomposed clauses from a student essay.

4 Assessment of Ideas as an Optimization Problem

An independent set for an undirected graph $G = (V, E)$ is defined as a subset $U \subset V$ such that no pair of vertices in U has an edge connecting them in E. A maximal independent set (MIS) is an independent set where no additional vertex from V can be added to U without violating the independent set constraint. The MIS problem is a well-documented NP-complete problem.

PyrEval+CR aims for the optimal way to match student sentences to the CR. Each essay sentence can have several decompositions, and each extracted clause can be more or less similar to each CU. Only one decomposition of a sentence can be used, and each CU can be matched at most once, to penalize repetition. Thus, our assessment task is equivalent to the MIS problem.

WMIN is a greedy weighted MIS algorithm that iteratively adds vertices with the next highest weights to the MIS. At each iteration, all neighbors of a recently added vertex are pruned from the graph. The process repeats until no remaining vertices can be selected. We extended WMIN to operate on a hypergraph (WMIN$_H$). Each hypernode corresponds to one way to decompose a sentence, and its internal nodes are the extracted clauses and their candidate CU matches. Pruning includes removing other occurrences of a matched CU from the rest of the graph, and recalculating node weights, which use CU weights.

Figure 2 illustrates two sentences in italics, S1 and S2, two of the several decompositions of S1, and the corresponding hypergraph. Hypergraph nodes are labeled by the sentence and decomposition (e.g., S1.2 vs. S1.3), and internal nodes by the clause indices (e.g., S1.2.1, S1.2.2). CU4 (weight 4) from Fig. 2 is shown inside the internal node for S1.2.2. Assume that CU5 (weight 2) is a vague statement about kinetic energy and also a potential match for two propositions: clauses S1.3.2 or S.2.1.1. The other internal nodes have anonymous weight 1 CUs (CUX, CUY).

The hypergraph has two edge types. The solid edge between S1.2 and S1.3 constrains selection of at most one decomposition of S1. The dashed edge between internal nodes S1.3.2 and S2.1.1 constrains selection of only one match to CU5. S1.2.2 is a good match to CU4: both state the relation between motion and

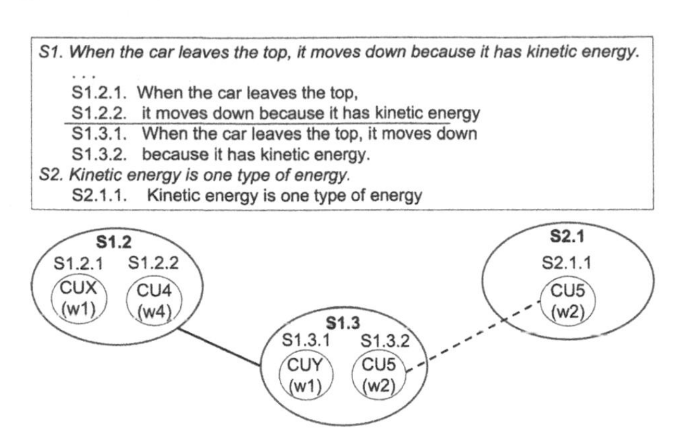

Fig. 2. Illustration of WMIN_H hypergraph nodes and edges.

kinetic energy. The weight of a hypernode is higher if the CU weights are higher, so WMIN_H selects S1.2 over S1.3, S1.3 is pruned, and CU5 only matches S2.1.1.

WMIN_H output for an essay is a log showing the decomposition that was selected for each sentence, and its matched CUs. The essay score is a normalized sum of the weights of the matched CUs.

WMIN_H has parameters to control the greediness of node selection: k for the length of the ranked list of CUs matching each internal node, a sorting metric (s) for ranking this list, and a weighting metric (w) for weighting each hypernode. For both s and w, we tested APCS (see above), the standard deviation of APCS, and the product of APCS and the CU weight (Product).

We tested PyrEval+CR on a curated set of historical essays from a similar curriculum, modified to eliminate sentences that mention ideas not in our current rubric. A set of 76 was subdivided into Set A (N = 10) for mining phrases for the CR, Set B (N = 46) as a validation set for parameter tuning, and Set C (N = 20) for testing. Application of a manual rubric to all 76 is discussed in [9]. We measured performance as the Pearson correlation of the PyrEval+CR score with the manually assigned score. We also reviewed the quality of matches between student's essays and CUs.

We performed grid search on Set B for different subsets of DP rules, different values of MinSegLen (see above), and the three WMIN_H parameters (k, s and w). The DP configurations were all rules (All), all but VP conjunction (-VP) and no decomposition (None). Table 2 reports results for three parameter settings on

Table 2. Example parameter configurations for WMIN_H.

PyrEval configuration	Set B	Set C
-VP, MinSegLen = 5, k = 4, s = StDev, w = APCS	0.69	0.84
All, MinSegLen = 3, k = 4, s = StDev, w = Product	0.70	0.83
-VP, MinSegLen = 5, k = 2, s = StDev, w = APCS	0.72	**0.85**

Sets B and C. DP-VP performed best, but DP-All often worked well. Smaller values of k ($k = 2$, 4) yielded best results, corresponding to a more greedy approach that considers fewer CUs per node. Both APCS and Product worked well for w. The standard deviation of APCS usually worked best for s.

The quality of the matches between a randomly selected subset of clauses from set C and CUs was rated by one of the co-authors as poor, moderate, or good. About 93% of the matches were split evenly between moderate and good.

5 Conclusion

PyrEval+CR is intended to support formative assessment for middle school science writing. On a semi-synthetic dataset, the scores correlate very well with a manual rubric. PyrEval+CR produces log output to show which clauses in a student essay match CUs from the computable rubric, along with the relative importance of the CU. Our next steps continue our collaboration with middle school teachers to study how to use PyrEval+CR in a classroom setting.

Acknowledgements. NSF DRK12 2010351 and 2010483 funded this work.

References

1. Ariely, M., Nazaretsky, T., Alexandron, G.: Machine learning and Hebrew NLP for automated assessment of open-ended questions in biology. Int. J. Artif. Intell. Educ. **22** (2022). https://doi.org/10.1007/s40593-021-00283-x
2. Arora, S., Liang, Y., Ma, T.: A simple but tough-to-beat baseline for sentence embeddings. In: International Conference on Learning Representations (2016)
3. Gao, Y., Sun, C., Passonneau, R.J.: Automated pyramid summarization evaluation. In: Proceedings of the 23rd CoNLL, pp. 404–418 (2019). https://doi.org/10.18653/v1/K19-1038
4. Gerard, L., Kidron, A., Linn, M.C.: Guiding collaborative revision of science explanations. Int. J. Comput. Support. Collab. Learn. **14**(3), 291–324 (2019). https://doi.org/10.1007/s11412-019-09298-y
5. Guo, W., Diab, M.: Modeling sentences in the latent space. In: Proceedings of the 50th ACL, pp. 864–872 (July 2012). https://aclanthology.org/P12-1091
6. Ha, M., Nehm, R.H., Urban-Lurain, M., Merrill, J.E.: Applying computerized-scoring models of written biological explanations across courses and colleges: prospects and limitations. CBE-Life Sci. Educ. **10**(4), 379–393 (2011)
7. Haudek, K.C., et al.: Using automated analysis to assess middle school students' competence with scientific argumentation. In: National Conference on Measurement in Education, Toronto, ON (2019)
8. Pennington, J., Socher, R., Manning, C.: GloVe: global vectors for word representation. In: Proceedings of the 2014 EMNLP, pp. 1532–1543 (2014). https://doi.org/10.3115/v1/D14-1162
9. Singh, P., et al.: Design of real-time scaffolding of middle school science writing using automated techniques. In: 2022 ISLS (November 2022)
10. Tansomboon, C., Gerard, L.F., Vitale, J.M., Linn, M.: Designing automated guidance to promote productive revision of science explanations. Int. J. Artif. Intell. Educ. **27**(4), 729–757 (2017). https://doi.org/10.1007/s40593-017-0145-0

GARFIELD: A Recommender System to Personalize Gamified Learning

Luiz Rodrigues[1(✉)], Armando Toda[1,2], Filipe Pereira[2,6], Paula T. Palomino[1], Ana C. T. Klock[3], Marcela Pessoa[4], David Oliveira[4], Isabela Gasparini[5], Elaine H. Teixeira[4], Alexandra I. Cristea[2], and Seiji Isotani[1]

[1] University of São Paulo, São Carlos, Brazil
lalrodrigues@usp.br
[2] Durham University, Durham, UK
[3] Tampere University, Tampere, Finland
[4] Federal University of Amazonas, Manaus, Brazil
[5] Santa Catarina State University, Joinville, Brazil
[6] Federal University of Roraima, Boa Vista, Brazil

Abstract. Students often lack intrinsic motivation to engage with educational activities. While gamification has the potential to mitigate that issue, it does not always work, possibly due to poor gamification design. Researchers have developed strategies to improve gamification designs through personalization. However, most of those are based on theoretical understanding of game elements and their impact on students, instead of considering real interaction data. Thus, we developed an approach to personalize gamification designs upon data from real students' experiences with a learning environment. We followed the CRISP-DM methodology to develop personalization strategies by analyzing self-reports from 221 Brazilian students who used one out of our five gamification designs. Then, we regressed from such data to obtain recommendations of which design is the most suitable to achieve a desired motivation level, leading to our interactive recommender system: GARFIELD. Its recommendations showed a moderate performance compared to the ground truth, demonstrating our approach's potential. To the best of our knowledge, GARFIELD is the first model to guide practitioners and instructors on how to personalize gamification based on empirical data.

Keywords: Tailored gamification · Data-driven · Education · e-Learning

1 Introduction

Intrinsic motivation (IM) is a strong predictor of learning gains [6]. In this regard, gamification is one method with strong potential to improve motivational learning outcomes [8]. However, gamification's effect might vary from person to person, leading to adverse effects (e.g., demotivation) for some people [6]. Research

M. M. Rodrigo et al. (Eds.): AIED 2022, LNCS 13355, pp. 666–672, 2022.
https://doi.org/10.1007/978-3-031-11644-5_65

shows that if gamified designs are not tailored to users and contexts, they are likely to not achieve their full potential, which encourages studies on how to tailor gamification [2].

Most often, gamification is tailored through personalization: designers or the system itself change the gamified design according to predefined information [11], such as changing the game elements according to the learning task. However, personalization demands a user/task model, such as those developed in [11] and [7]. In common, those and similar models are based on *potential* experiences: they were built from data captured through surveys or after seeing mock-ups [2]. Thereby, they are limited because *potential* experiences might not reflect *real* experiences [4]. For instance, [9] developed a model based on both learners' profiles and motivation before using the gamification, but with no information of learners' real experiences (e.g., after actually using gamification). Hence, to the best of our knowledge, there is no data-driven model, based on users' real (instead of potential) experiences, for personalizing gamification designs.

To address that gap, this paper presents GARFIELD - Gamification Automatic Recommender for Interactive Education and Learning Domains, a recommender system for personalizing gamification built upon data from real experiences. Our goal was to indicate the most suitable gamification design according to students' intrinsic motivation due to its positive relationship with learning [6]. For this, we followed a two-step reverse engineering approach: we collected self-reports of users' intrinsic motivations from *actually* using a gamification design, then, regressed from such data (N = 221) to obtain recommendations of which design is the most suitable to achieve a desired motivation level given the user's information. To the best of our knowledge, GARFIELD is the first model that guides practitioners and instructors on how to personalize gamification based on empirical data from *real* usage. Therefore, this paper contributes by creating and providing a motivation-based model for personalizing gamification, informing educators on how to personalize their gamified practices and researchers by performing a first step towards developing experience-driven models for designing gamification.

2 Method: CRISP-DM

Because we had an apriori goal, we followed the CRISP-DM reference model, which is suggested for goal-oriented projects [14][1].

CRISP-DM's first phase is **business understanding**. In this phase, we first defined the project's goal: creating a model based on students' intrinsic motivation captured after real system usage to allow the personalization of gamified educational systems. Additionally, we defined two requirements: i) the model must consider user characteristics and ii) the model must be interactive. The former is based on research showing users characteristics affect their experiences with gamified systems [6,11]. The latter aims to facilitate practical usage.

[1] For transparency, this link details our dataset and all analysis: osf.io/nt97s.

The second phase is **data understanding**. Openly sharing data extends a paper's contribution because it enables cheaper, optimized exploratory analyses [13] and is especially valuable for educational contexts wherein data collection is expensive. Accordingly, we opted to work with a dataset collected and made available by [5]. This dataset has data from students enrolled in STEM undergraduate courses of three Brazilian northwestern universities (ethical committee approval: 42598620. 0.0000.5464). Students self-reported their motivations to complete in-lecture assessments after using one of the following gamified designs: i) points, acknowledgments, and competition (PBL)[2], ii) acknowledgments, objectives, and progression (AOP), iii) acknowledgments, objectives and social pressure (AOS), iv) acknowledgments, competition, and time pressure (ACT), and v) competition, chance, and time pressure (CCT). We analyzed those designs by convenience because we used data shared by a previous study, which aimed to tailor gamification to user characteristics and learning activity type [5].

When available, each game element functions as follows. Students received *points* after completing a mission. After finishing each mission, they were *acknowledged* with a badge depending on their performances (e.g., getting all items right). Students could *compete* with each other based on a leaderboard that ranked them based on the points they made during the week. Within the leaderboard, a clock provoked *time pressure* by highlighting the time available to climb the leaderboard before the week's end. Additionally, a progress bar indicated student's *progression* within missions, a notification aimed to provoke *social pressure* by warning that peers just completed a mission, and a skill tree represented short-term *objectives* (i.e., completing 10 missions).

The third phase is **data preparation**. First, we ran *attribute selection*, choosing columns related to students' characteristics, intrinsic motivation, status, and the game elements they interacted with. Next, we proceeded to *data cleansing*, removing answers from students with less than 18 years ($N = 1$) due to ethical aspects and participants that provided their motivations without using the system ($N = 4$). Then, we conducted *data transformations* by: i) transforming the intrinsic motivation variable (captured through a seven-point Likert-scale using the respective subscale of the Situational Motivation Scale (SIMS) [1]) to range between zero and six to facilitate regression coefficients' interpretation; and ii) removing observations ($N = 8$) from levels representing less than 5% of the dataset, unless grouping them with another level was feasible, to avoid overfitting. Additionally, we *constructed new attributes* for highly skewed continuous variables by categorizing: i) weekly playing time into whether the student plays an average of at least one hour per day or more than that; and ii) age, into those below Brazilian undergraduate STEM students average (i.e., <21) and those at or above it. Lastly, we analyzed the *game elements* column, our dependent variable, and found a single observation of the ACT design; we removed it, lead-

[2] We consider Badges and Leaderboards implementations of Acknowledgments and Competition, respectively [10], but use PBL to maintain the standard nomenclature.

ing to the prepared dataset featuring 221 observations (see our supplementary materials for details: osf.io/nt97s).

Phase four is **modeling**. Here, we used Multinomial Logistic Regression [3] through the *nnet* R package with the maximum number of iterations set to 1000 to ensure the algorithm's convergence. This form of machine learning enables working with nominal dependent variables, such as gamification designs, based on the null hypothesis significance testing framework. Hence, allowing us to evaluate coefficients' contributions to the model based on their significance. This technique works similarly to standard Logistic Regression, but comparing the dependent variable's reference value to all others. In our analysis, we defined the PBL design as the reference value because PBL is the most used gamification design in educational contexts [8]. As independent variables, we started with all of those of the prepared dataset. Additionally, because recommendations should consider how students' intrinsic motivation from using a gamification design change depending on their characteristics, our model assumes intrinsic motivation interacts with all other variables.

Phase five **evaluates** modeling alternatives to determine the best option. Here, we used recursive feature elimination with p-values as the elimination criteria because we followed the standard of working within the null hypothesis significance testing framework. As this project has an exploratory nature, we considered a 90% confidence level, following similar research (e.g., [6]). After selecting the final model, we evaluated it based on its predictions according to Cohen's Kappa and F-measure, calculated using R packages *vcd* and *caret*, respectively, because those metrics are reliable for multi-class problems wherein data is unbalanced.

3 Evaluation Results and Deployment

After running the Multinomial Logistic Regression, we found significant interactions between all user's characteristics and intrinsic motivation. Hence, we removed no features and defined the initial model as the final one. In evaluating the model, we found the Cohen's Kappa for the agreement between its predictions and the ground truth is 0.43. This value is significantly different from zero ($p < 0.001$), with its 95% confidence interval ranging from 0.34 to 0.52, revealing a moderate agreement [12]. To further understand the model's predictions, Table 1 shows the confusion matrix along with the F-measure of each category, demonstrating the model performed the best for designs AOP and CCT. Differently, its performance for designs PBL and AOS were slightly worse. Additionally, the confusion matrix reveals the model's misclassifications (e.g., wrongly predicting AOS design should be PBL and AOP 13 and 18 times, respectively). Therefore, phase five shows the model recommends gamified designs with moderate performance, despite variations from one design to another. Thus, demonstrating its potential as well as room for improvement.

In terms of deployment, we developed GARFIELD, our interactive recommender system (access it here: osf.io/nt97s). Its interface receives user input and

Table 1. Confusion Matrix of the models predictions against the ground truth.

	PBL	AOP	AOS	CCT	Balanced Accuracy	F-measure
PBL	40	11	13	04	0.68	0.57
AOP	17	54	18	01	0.75	0.67
AOS	14	06	24	02	0.65	0.48
CCT	02	01	00	14	0.83	0.74

passes it to our model. Then, our model predicts the probability of recommending each possible design and presents it as a barplot. Accordingly, practitioners can use it to get recommendations for personalizing their gamified designs in a simple, interactive way. Thus, attending to our project's second requirement.

4 Discussion

Overall, our goal was to facilitate the personalization of gamification with a model that recommends a gamified design given an expected intrinsic motivation level. Additionally, we aimed that such recommendations considered user characteristics and could be used interactively. Ultimately, our recommender system - GARFIELD - achieves these goals, allowing educators to use it in an interactive, web-based way to receive design recommendations based on the aforementioned input. Thus, this research expands the literature by i) creating personalization guidelines from feedback collected after *real* experiences, in contrast to prior research that developed personalization guidelines based on potential experiences (e.g., [7,9] and ii) providing concrete, interactive recommender system unlike the conceptual tools related work has contributed (e.g., [2]).

As implications for future research, our contribution is twofold. First, the lack of data-driven strategies likely poses a challenge for researchers interested in developing similar approaches. In developing our approach, we demonstrate how one can create personalization strategies step-by-step through the CRISP-DM reference model, contributing with a concrete example that can be followed to implement data-driven personalization guidelines. Second, we understand that modeling users efficiently is challenging, especially for tasks that depend on people's subjective experiences (e.g., intrinsic motivation). In this paper, we created a model using 221 observations with inputs of self-reported intrinsic motivation and demographic characteristics (e.g., age, gender, and gaming preferences). Yet, our model yielded a moderate predictive power (Cohen's Kappa = 0.43). Thus, our results inform future research that while such information contributes to understanding which gamification design to use, we likely need additional information to personalize gamification more accurately.

In summary, with our results practitioners have technological support to help them personalize their gamified practices. This can be achieved using

GARFIELD, an interactive, ready-to-use recommender system to get design suggestions. Additionally, with this paper, researchers have a concrete guide on how to use CRISP-DM for creating data-driven personalization strategies based on real (instead of potential) experiences. Note, however, that our recommender's predictions are limited to moderate predictive power. We understand that limits its practical usage as it is. Nevertheless, to our best knowledge, GARFIELD is the first tool to provide gamification design recommendations based on real experiences. Thus, we believe it provides practitioners with a reliable starting point and paves the way for researchers to expand and improve it in future research.

Acknowledgments. This research received financial support from the following Brazilian institutions: CNPq (141859/2019-9, 163932/2020-4, 308458/2020-6, 308513/2020-7, and 308395 /2020-4); CAPES (Finance code - 001; PROAP/AUXPE); FAPESP (2018/ 15917-0, 2013/07375-0); Samsung-UFAM (agreements 001/2020 and 003/2019).

References

1. Guay, F., Vallerand, R.J., Blanchard, C.: On the assessment of situational intrinsic and extrinsic motivation: the situational motivation scale (sims). Motiv. Emot. **24**(3), 175–213 (2000)
2. Klock, A.C.T., Gasparini, I., Pimenta, M.S., Hamari, J.: Tailored gamification: a review of literature. Int. J. Hum.-Comput. Stud. (2020)
3. Kwak, C., Clayton, A.: Multinomial logistic regression. Nurs. Res. **51** (2002)
4. Palomino, P., Toda, A., Rodrigues, L., Oliveira, W., Isotani, S.: From the lack of engagement to motivation: gamification strategies to enhance users learning experiences. In: 19th Brazilian Symposium on Computer Games and Digital Entertainment (SBGames)-GranDGames BR Forum, pp. 1127–1130 (2020)
5. Rodrigues, L., et al.: How personalization affects motivation in gamified review assessments, May 2022. https://doi.org/10.17605/OSF.IO/EHM43
6. Rodrigues, L., Toda, A.M., Oliveira, W., Palomino, P.T., Avila-Santos, A.P., Isotani, S.: Gamification works, but how and to whom? an experimental study in the context of programming lessons. In: Proceedings of the 52nd ACM Technical Symposium on Computer Science Education, pp. 184–190 (2021)
7. Rodrigues, L., Toda, A.M., dos Santos, W.O., Palomino, P.T., Vassileva, J., Isotani, S.: Automating gamification personalization to the user and beyond. IEEE Trans. Learn. Technol. (2022)
8. Sailer, M., Homner, L.: The gamification of learning: a meta-analysis. Educ. Psychol. Rev. **32**(1), 77–112 (2019). https://doi.org/10.1007/s10648-019-09498-w
9. Stuart, H., Lavoué, E., Serna, A.: To tailor or not to tailor gamification? An analysis of the impact of tailored game elements on learners' behaviours and motivation. In: 21th International Conference on Artificial Intelligence in Education (2020)
10. Toda, A.M., et al.: Analysing gamification elements in educational environments using an existing gamification taxonomy. Smart Learn. Environ. **6**(1), 16 (2019). https://doi.org/10.1186/s40561-019-0106-1
11. Tondello, G.F.: Dynamic personalization of gameful interactive systems. Ph.D. thesis, University of Waterloo (2019)
12. Viera, A.J., Garrett, J.M., et al.: Understanding interobserver agreement: the kappa statistic. Fam. Med. **37**(5), 360–363 (2005)

13. Vornhagen, J.B., Tyack, A., Mekler, E.D.: Statistical significance testing at chi play: challenges and opportunities for more transparency. In: Proceedings of the Annual Symposium on Computer-Human Interaction in Play, pp. 4–18 (2020)
14. Wirth, R., Hipp, J.: CRISP-DM: towards a standard process model for data mining. In: Proceedings of the 4th International Conference on the Practical Applications of Knowledge Discovery and Data Mining, vol. 1. Springer, London, UK (2000). https://www.tib.eu/en/search/id/BLCP:CN039162600/CRISP-DM-Towards-a-Standard-Process-Model-for-Data?cHash=16bb58abe6455a858f809fd55fb0ca8f

Technology Ecosystem for Orchestrating Dynamic Transitions Between Individual and Collaborative AI-Tutored Problem Solving

Kexin Bella Yang[1](✉), Zijing Lu[1], Vanessa Echeverria[2], Jonathan Sewall[1], Luettamae Lawrence[3], Nikol Rummel[4](✉), and Vincent Aleven[1](✉)

[1] Carnegie Mellon University, Pittsburgh, PA, USA
{kexiny,zijinglu,sewall,va0e}@andrew.cmu.edu
[2] Escuela Superior Politécnica del Litoral (ESPOL), Guayaquil, Ecuador
vanechev@espol.edu.ec
[3] University of California, Irvine, CA, USA
[4] Ruhr-Universität Bochum, Bochum, Germany
nikol.rummel@rub.de

Abstract. It might be highly effective if students could transition *dynamically* between individual and collaborative learning activities, but how could teachers manage such complex classroom scenarios? Although recent work in AIED has focused on teacher tools, little is known about how to orchestrate dynamic transitions between individual and collaborative learning. We created a novel technology ecosystem that supports these dynamic transitions. The ecosystem integrates a novel teacher orchestration tool that provides monitoring support and pairing suggestions with two AI-based tutoring systems that support individual and collaborative learning, respectively. We tested the feasibility of this ecosystem in a classroom study with 5 teachers and 199 students over 22 class sessions. We found that the teachers were able to manage the dynamic transitions and valued them. The study contributes a new technology ecosystem for dynamically transitioning between individual and collaborative learning, plus insight into the orchestration functionality that makes these transitions feasible.

Keywords: Classroom orchestration · Dynamic transitions · Differentiated learning · Collaborative learning

1 Introduction

Combining individual and collaborative activities is very common in educational practice (e.g., Think-Pair-Share [4]). Such combinations can be more effective than learning solely in one mode [6]. An exciting vision for the smart classroom of the future is to *dynamically* combine collaborative and individual learning [1]. In dynamic transitions, students switch between collaborative and individual learning when the need arises (e.g., when a student is no longer progressing productively in one mode of learning). Such transitions are not pre-planned, but can happen opportunistically in order to address

M. M. Rodrigo et al. (Eds.): AIED 2022, LNCS 13355, pp. 673–678, 2022.
https://doi.org/10.1007/978-3-031-11644-5_66

students' in-the-moment needs. Dynamic transitions hold potential to be maximally responsive to the fact that students learn at their own pace and may achieve more personalized learning for students than pre-planned transitions [7]. For example, teachers may team up students to work together if one of them is struggling and can use a partner's help. However, orchestrating dynamic transitions in classrooms is a major challenge for teachers [7], as it involves not only understanding students' in-the-moment needs, but also managing the transitions in real time while attending to the ongoing class activities.

Prior research has produced many tools that support teachers in orchestrating complex learning scenarios (e.g., [2]). These tools have, however, typically been designed with the assumption that a class of students progresses through instructional activities in a relatively synchronized manner [5]. Furthermore, existing orchestration tools generally focus on enhancing teacher *awareness* by providing teachers with real-time analytics [5]. Few provide intelligent support for teachers' in-the-moment, dynamic *decision-making* [8], with some exceptions (e.g., [9]). Providing intelligent support to teachers when orchestrating highly-differentiated, self-paced classrooms remains a challenging research problem [7], with little prior work in this area. Our own prior study explored the potential of supporting dynamic transitions between individual and collaborative learning in the classroom [1]. We found a need for sharing control over these transitions between students, teachers and AI systems. The study was a technology probe "Wizard of Oz" study, where a researcher mimicked part of the orchestration functionality. In the current study, we test a fully functioning system, without a wizard.

Specifically, we created a technology ecosystem (Fig. 1) that supports teachers in orchestrating students' dynamic transitions between individual and collaborative learning, both supported by intelligent tutoring software (ITS). We conducted an exploratory classroom study with 5 teachers and 199 middle-school students to gain insight into the feasibility of dynamically transitioning between individual and collaborative learning. The work extends prior work in orchestration technologies with AI support by implementing an orchestration tool that allows teachers to manage dynamic transitions between individual and collaborative learning and demonstrating that the combination of awareness support and AI-based pairing suggestions can feasibly support these dynamic transitions.

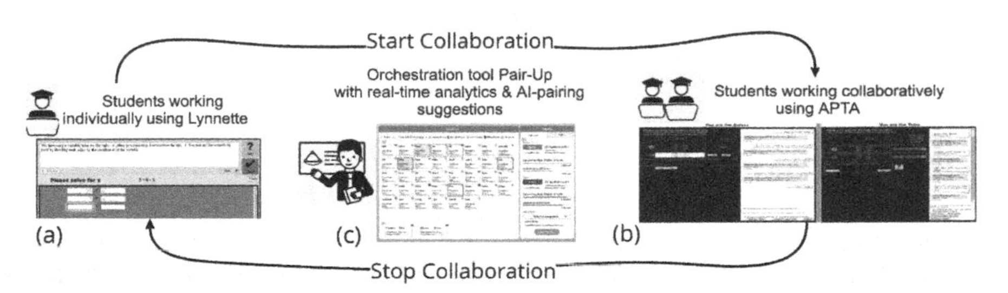

Fig. 1. Technology ecosystem for supporting dynamic transitions, including individual tutor (a), collaborative tutor (b) and the orchestration tool (c)

2 Technology Ecosystem for Dynamic Transitions

The technology ecosystem consists of two tutoring software which respectively support students' individual and collaborative learning, and a teacher-facing orchestration tool.

2.1 Support for Students' Individual and Collaborative Learning

A standard ITS, Lynnette (Fig. 1, a), offers support for *individual* learning of basic equation solving. Lynnette provides step-by-step guidance, in the form of adaptive hints, correctness feedback, and error specific messages, and has been proven to improve students' equation-solving skills in several classroom studies (e.g., [3]).

The Adaptive Peer Tutoring Assistant, APTA (Fig. 1, b), extends Lynnette's functionality to support *collaborative learning,* specifically, reciprocal tutoring. When using APTA, two students respectively take the role of "solver" and "tutor". The "solver" solves the math problem and can seek help from their partner. The "tutor" helps the "solver" through step by step evaluation and feedback via chat window. APTA supports the student in the "tutor" role with both math advice and advice on how to tutor. Classroom studies with an earlier version of APTA demonstrated that adaptive support (in the form of system-generated chat messages) can improve the quality of help peer tutors give and improve their domain learning, compared to the parallel non-adaptive condition [10]. APTA is a reimplementation of the earlier version and covers the same equation solving skills as in Lynnette.

2.2 Orchestration Tool for Dynamic Transitions (Pair-Up)

Orchestration of the dynamic transitions is through a tool (Pair-Up) that synergistically leverages strengths of teachers and AI (Fig. 2). The design of the tool is informed by previous user research on teacher preferences [12], log data simulation [11], and co-design sessions with teachers [13]. Pair-Up has two key features: real-time analytics of students' learning status, and the option of AI-suggested pairing partners for teachers to decide. It helps teachers make judgments about which students might benefit from transitioning from one mode of learning (individual or collaborative) to the other and (in the case of transitioning from individual to collaborative learning), who might be good partners to team up and what they should work on collaboratively. The teacher has the final say over all pairing decisions.

Real-Time Analytics of Students Learning Status. Our previous user research found that teachers would like to be able to view student progress in an easily glanceable way, when orchestrating the dynamic transitions [13]. Pair-Up indicates students' recent learning behaviors such as idling, misusing the software, making lots of errors, making many attempts, and doing well through icons attached to individual student cards [3]. Additionally, in both individual and collaborative modes, teachers can see the number of math problems that student(s) completed in a progress bar in the student(s) card. To further assist in monitoring, teachers can sort students alphabetically, based on the number of math problems solved (least to most or most to least), or based on the learning status indicators.

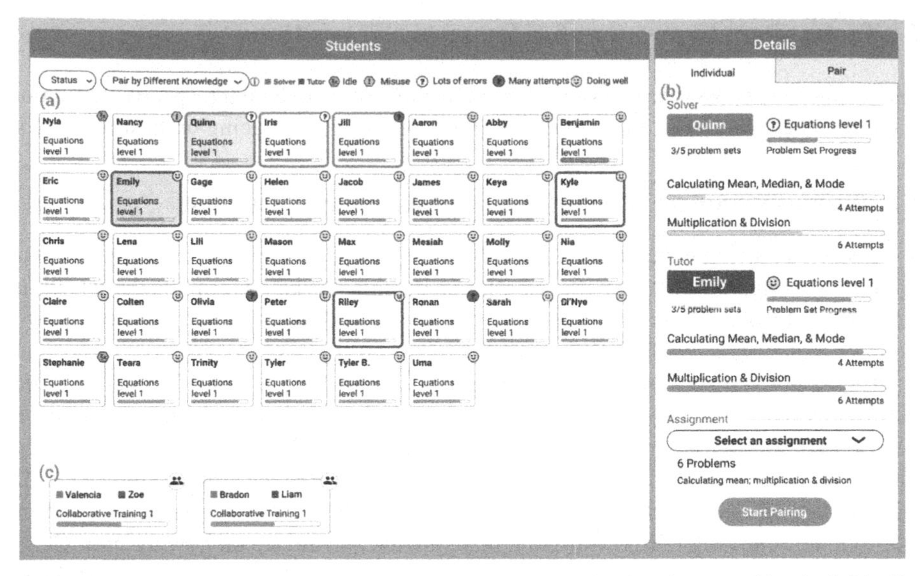

Fig. 2. The teacher-facing orchestration tool *Pair-Up*: (a) cards of students working individually, including system-suggested "solvers" (teal) and "tutor" (purple); (b) panel where the teacher can evaluate a potential match between two students by comparing their skill before deciding whether to team them up; in this panel the teacher also selects the math content for collaboration; (c) collaborating students pairs. (Color figure online)

AI-Suggested Pairing Partners for Teachers to Decide. Previous user research found that teachers prefer to have the AI system suggest potential candidates to pair up. However, they very strongly prefer to have the final decision over all dynamic transitions [12]. They also like to be able to select an appropriate pairing algorithm (for making suggestions) based on the learning goals [13]. Based on surveying of 54 math teachers [12], teachers most commonly used two pairing strategies in collaborative activities: random pairing (so students work with new partners) and pairing students with different knowledge levels, so that students who are wheel-spinning or making slow progress [11] can work with a partner who is further along learning the particular skills at issue. Accordingly, the tool has two pairing policies: *random pair*, and *pair by different knowledge*. In the *random pair* policy, Pair-Up suggests random students in the class as solvers and tutors. In the *pair by different knowledge* policy, Pair-Up suggests students who are making slow progress on some of the knowledge components to be solvers. Once the teacher selects a solver, Pair-Up then suggests three "tutors" who are ahead of the "solvers" in the knowledge components they are struggling with.

Teachers have full agency over choosing which policy to use, as well as whether to follow system pairing suggestions or override suggestions and pair students based on their judgment. If they activate the system suggestions function, Pair-Up will suggest students take on the role of "solver" or a "tutor", by highlighting them in teal and purple outline,. The teacher however will make the final pairing decision. In addition, teachers can pair students without tool suggestions. Teachers can pair students to work collaboratively, and select an assignment (which contains three equation solving problems) they see as fit for the pair (e.g., an assignment focused on skills that the "solver" is struggling with). Based on students' progress, teachers can also choose when to unpair to stop the collaboration.

3 Feasibility Testing in Classroom

We conducted an in-person classroom study in a suburban public school near Pittsburgh, with five middle school math teachers and 199 students participants from 11 classes. One teacher teaches special education with 7 students who have an Individual Education Program (IEP). Each class participated for 2 sessions, each lasting 33–37 min. After a short video tutorial, students started with individual equation solving. The teachers paired up students as they wished. When students were done with the collaborative assignment or when they were unpaired by the teacher, they switched back to individual work.

We analyzed log data to study students' dynamic transitions. During the 22 class sessions, 210 collaboration episodes (defined as two students teamed up to work collaboratively on one assignment) happened, with on average of 18 episodes in each class over the duration of the study. The teachers generally were able to use the orchestration tool autonomously. Similarly, the students were able to work with tutoring softwares. Two teachers in the study teamed up all students at the same time, and three paired students up at different times as they saw fit. All participating teachers stated that they see pedagogical value in dynamic transitions. However, the special education teacher expressed that transitioning between learning activities may be challenging for her students. Still, all five teachers reported being likely to use such a technology ecosystem in their regular classrooms.

4 Conclusion

In this study, we introduced a new technology ecosystem to support dynamic transitions between individual and collaborative learning, which has not been tried before in the AIED literature, to our knowledge. We tested the feasibility of the ecosystem in 11 classrooms. The substantial number of collaboration episodes (on average 18 per class) is one piece of evidence of feasibility, showing that all teachers were able to use the orchestration tool to initiate dynamic transitions between individual and collaborative learning. All participating teachers reported being likely to use the technology ecosystem in their daily practice. Thus, the study provides insight into what orchestration tool functionality makes it feasible for teachers to manage dynamic transitions between individual and collaborative learning: a combination of (1) support for monitoring students' real-time learning progress in both individual and collaborative learning modes; (2) AI-generated pairing suggestions regarding whom to team up, with (3) full control by the teacher over pairing and unpairing decisions, Future work will further analyze students' learning process in dynamic transitions, and further improve the tools as our understanding of how to support teachers continually evolve. The ecosystem will support further research into the value of dynamic transitions, including how they affect students' learning outcomes, compared to for example pre-planned transitions. This exploratory study brings us closer to the vision of the smart classroom of the future, where the students transition dynamically between different learning modes, at moments that such transitions may be most helpful.

References

1. Echeverria, V., Holstein, K., Huang, J., Sewall, J., Rummel, N., Aleven, V.: Exploring human–AI control over dynamic transitions between individual and collaborative learning. In: Alario-Hoyos, C., Rodríguez-Triana, M.J., Scheffel, M., Arnedillo-Sánchez, I., Dennerlein, S.M. (eds.) EC-TEL 2020. LNCS, vol. 12315, pp. 230–243. Springer, Cham (2020). https://doi.org/10.1007/978-3-030-57717-9_17
2. Holstein, K., Hong, G., Tegene, M., McLaren, B., Aleven, V.: The classroom as a dashboard: co-designing wearable cognitive augmentation for K-12 teachers. In: Proceedings of LAK, pp. 79–88 (2018)
3. Holstein, K., McLaren, B.M., Aleven, V.: Student learning benefits of a mixed-reality teacher awareness tool in AI-enhanced classrooms. In: Penstein Rosé, C., et al. (eds.) AIED 2018. LNCS (LNAI), vol. 10947, pp. 154–168. Springer, Cham (2018). https://doi.org/10.1007/978-3-319-93843-1_12
4. Kothiyal, A., Murthy, S., Iyer, S.: Think-pair-share in a large CS1 class: does learning really happen? In: Proceedings of ITiCSE, pp. 51–56 (2014)
5. Van Leeuwen, A., Janssen, J., Erkens, G., Brekelmans, M.: Teacher regulation of multiple computer-supported collaborating groups. Comput. Hum. Behav. **52**, 233–242 (2015)
6. Olsen, J.K., Rummel, N., Aleven, V.: It is not either or: an initial investigation into combining collaborative and individual learning using an ITS. Int. J. Comput. Support. Collab. Learn. **14**(3), 353–381 (2019). https://doi.org/10.1007/s11412-019-09307-0
7. Olsen, J., Rummel, N., Aleven, V.: Designing for the co-orchestration of social transitions between individual, small-group and whole-class learning in the classroom. Int. J. Artif. Intell. Educ. **31**, 24–56 (2021). https://doi.org/10.1007/s40593-020-00228-w
8. Sergis, S., Sampson, D.G.: Teaching and learning analytics to support teacher inquiry: a systematic literature review. In: Peña-Ayala, A. (ed.) Learning Analytics: Fundaments, Applications, and Trends. SSDC, vol. 94, pp. 25–63. Springer, Cham (2017). https://doi.org/10.1007/978-3-319-52977-6_2
9. Tissenbaum, M., Slotta, J.: Supporting classroom orchestration with real-time feedback: a role for teacher dashboards and real-time agents. Int. J. Comput.-Support. Collab. Learn. **14**(3), 325–351 (2019). https://doi.org/10.1007/s11412-019-09306-1
10. Walker, E., Rummel, N., Koedinger, K.: Adaptive intelligent support to improve peer tutoring in algebra. Int. J. Artif. Intell. Educ. **24**, 33–61 (2014). https://doi.org/10.1007/s40593-013-0001-9
11. Yang, K.B., et al.: Exploring policies for dynamically teaming up students through log data simulation. In: Proceedings of EDM (2021)
12. Yang, K.B., Lawrence, L., Echeverria, V., Guo, B., Rummel, N., Aleven, V.: Surveying teachers' preferences and boundaries regarding human-AI control in dynamic pairing of students for collaborative learning. In: De Laet, T., Klemke, R., Alario-Hoyos, C., Hilliger, I., Ortega-Arranz, A. (eds.) EC-TEL 2021. LNCS, vol. 12884, pp. 260–274. Springer, Cham (2021). https://doi.org/10.1007/978-3-030-86436-1_20
13. Lawrence, L., et al.: Co-designing AI-based orchestration tools to support dynamic transitions: design narratives through conjecture mapping. In: Proceedings of ISLS (2022)

Generating Personalized Behavioral Feedback for a Virtual Job Interview Training System Through Adversarial Learning

Alexander Heimerl[1(✉)], Silvan Mertes[1], Tanja Schneeberger[2], Tobias Baur[1], Ailin Liu[1], Linda Becker[3], Nicolas Rohleder[3], Patrick Gebhard[2], and Elisabeth André[1]

[1] Lab for Human-Centered AI, Augsburg University, 86159 Augsburg, Germany
{alexander.heimerl,silvan.mertes,tobias.baur,elisabeth.andre}@uni-a.de

[2] German Research Center for Artificial Intelligence (DFKI), Saarland Informatics Campus D3.2, Saarbrücken, Germany
{schneeberger,gebhard}@dfki.de

[3] Department of Psychology, Friedrich-Alexander University Erlangen-Nürnberg, 91054 Erlangen, Germany
{linda.becker,nicolas.rohleder}@fau.de

Abstract. Job interviews are usually high-stakes social situations where professional and behavioral skills are required for a satisfactory outcome. In order to increase the chances of recruitment technological approaches have emerged to generate meaningful feedback for job candidates. We extended an interactive virtual job interview training system with a Generative Adversarial Network (GAN)-based approach that first detects behavioral weaknesses and subsequently generates personalized feedback. To evaluate the usefulness of the generated feedback, we conducted a mixed-methods pilot study using mock-ups from the job interview training system. The overall study results indicate that the GAN-based generated behavioral feedback is helpful. Moreover, participants assessed that the feedback would improve their job interview performance.

Keywords: Job interview training · Generative adversarial networks · Counterfactual explanations · Engagement

1 Introduction

In stressful situations, such as job interviews, many people tend to show nervous and uncontrolled behaviours. This circumstance most often affects their performance in a negative way. Especially in job interviews, the goal is to convince a recruiter of ones fit in a company by actively engaging in the conversation. Recruiters hereby consciously or unconsciously evaluate the candidate's social cues. The amount of positive engagement a candidate shows towards the interviewer may play a central role in deciding whether the candidate is suitable.

M. M. Rodrigo et al. (Eds.): AIED 2022, LNCS 13355, pp. 679–684, 2022.
https://doi.org/10.1007/978-3-031-11644-5_67

Delroy et al. [7] found that active integration behaviors such as engagement, laughing, and humor led to better performance ratings and, therefore, to a higher chance of recruitment. In recent years, technology-based job interview training systems have been developed to improve the performance of candidates (e.g. [1,4,11]).

This paper presents a feedback extension to an existing job interview training environment that uses a socially interactive agent as a recruiter and an engagement recognition component to enable the virtual agent to react and adapt to the user's behavior, and emotions [2]. This training aims to help improve social skills that are pertinent to job interviews. The new feedback extension employs an eXplainable AI (XAI) method based on counterfactual reasoning for generating verbal feedback about observed social behavior.

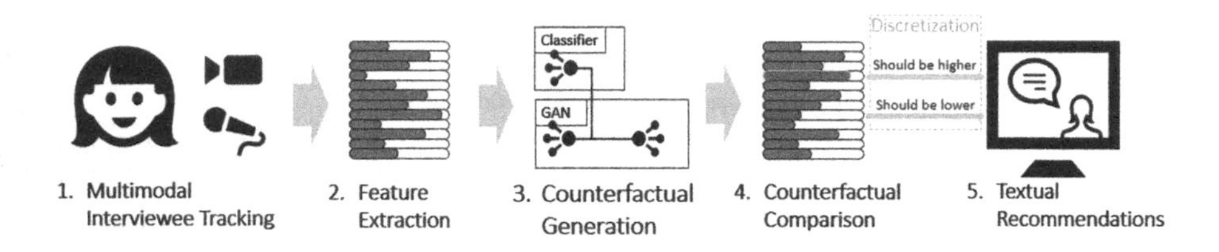

Fig. 1. Job interview training system with GAN-generated recommendations.

The introduced feedback extension is based on a deep learning classifier predicting the user engagement in job interview situations that uses multimodal feature (e.g., gaze, body posture, or gestures) representations of the trainee as input. We exploit the concept of counterfactual explanations to show what the user would need to change to appear more engaged. Therefore, a GAN-driven counterfactual explanation model is trained that transforms those feature representations to corresponding counterfactual explanations, i.e., the feature representations are changed so that the user would have appeared engaged. The explanation generation compares the counterfactual feature vectors with the original feature vectors to derive textual recommendations automatically. Finally, they are presented to the trainee by a socially interactive agent in the role of a job interview coach. Figure 1 shows a schematic overview of our approach.

2 Recommendation Generation

The next sections offer an overview of the different components we implemented to generate behavioral recommendations that point out how the user should have behaved to appear more engaged.

Feature Extraction. In order to train a model for engagement recognition and recommendation generation, we modeled a high-level engagement feature set. The feature set consists of 18 metrics mapping facial behavior, body language and conversation dynamics.

Engagement Model. Based on the introduced feature set we trained a simple feedforward neural network with two dense layers for the recognition of low and high engagement on the NoXi database [3]. We decided on the NoXi corpus since it contains multi-modal multi-person interaction data and its transferability to social coaching scenarios. Moreover, the setup of the corpus allowed for both engaging as well as non-engaging interactions. The 10.5 h of data has been randomly split into training and test sets, so that no sample of the same participant is present in the training and the test set. The corresponding classifier achieved an accuracy of 70.5%.

Counterfactual Features. In a next step, to be able to give recommendations on how the user should have behaved to appear more engaged, we apply a counterfactual explanation generation algorithm, i.e., we aim to modify the input feature vectors that were classified as *low engaged* in a way that the classifier would change it's decision to *high engaged.* As described in Sect. 1, the recommendations that we aim for can be seen as counterfactual explanations for the engagement model presented in Sect. 2. To generate these counterfactual feature vectors, we used an adversarial learning approach. In prior work, Mertes et al. [5] presented their *GANterfactual* architecture, which is an adversarial approach to transforming original samples to counterfactual samples that are classified in a different way by a specific decision system to be explained. For our system, we built a network architecture adapted from the GANterfactual framework, which was originally implemented for generating counterfactual explanations in the image domain. The use of the GANterfactual framework has multiple benefits for the recommendation quality: Firstly, the cycle-consistency loss that is an integral part of the underlying adversarial architecture forces that the learned transformation is minimal, i.e., only relevant features are changed. In the context of recommendation generation, this implies that the generated behavioral recommendations are highly personalized. Secondly, the adversarial loss component that is part of every GAN architecture leads to highly realistic results. Thus, recommendations are not drawn from highly exaggerated or oversimplified feature vectors. Thirdly, the counterfactual loss introduced by Mertes et al. enforces that the counterfactual explanations (in our case, the behavioral recommendations), are valid. For technical details of our modifications to the original GANterfactual framework, please refer to our implementation.[1] For the GAN-training, we relied on the NOXI dataset, which we also used for training the engagement classifier. Thus, the adversarial framework learns to convert feature vectors that show low engagement to feature vectors that show high engagement.

Textual Recommendations. After generating the counterfactual feature vectors we compare them to the original feature vectors that represent the shown nonverbal behavior. Depending on the demanded detail of feedback we return the features that had undergone the greatest value transformation and convert them into textual feedback. For this purpose, we discretize the features based

[1] Our implementation is available at https://github.com/hcmlab/FeatureFactual.

on a defined textual template. For example, the feature representing the overall activity of the head gets translated into "try to keep your attention on your interlocutor" or "try to use more nonverbal feedback" depending on the present feature value. The generated feedback is provided verbally to the user by the virtual coach inside the job interview training environment.

3 Evaluation

Pilot Study. The present pilot study's goal was to get preliminary insights about the assessment of a possible job interview training applying GAN driven recommendations. We gathered data from 12 volunteering student participants (7 female, 5 male). Participants' age was between 21 and 29 years ($M = 23.83$, $SD = 2.66$). The participants were presented with videos of our job interview training system applied to a multi-modal job interview role-play dataset [10]. Participants were asked to imagine that they were the trainees using the training to practice a job interview. Next, participants filled in questionnaires about Demographics (age, sex), Usefulness (MeCUE [6]), Transfer motivation (four items adapted from [9]) and Feedback Quality ("I felt the feedback was accurate.", "I would have given similar feedback.", "I feel like the feedback is helpful.", "I don't think the computer can give me accurate feedback."). Then, a semi-structured interview was held, which covered five areas: 1) general impression, 2) other possible use-cases, 3) suggestions for improvement, 4) intention for further use, and 5) added value.

In the three questionnaires, the following descriptive data was found: Usefulness ($M = 4.72$, $SD = 1.17$); Transfer motivation ($M = 4.92$, $SD = .94$); Feedback Quality ($M = 4.60$, $SD = 1.26$). The answers gathered in the semi-structured interview were analyzed and categorized for each of the five areas separately. Regarding the *General impression*, the majority of participants mentioned that the recommendations were useful (6)/feasible or comprehensible (2). As *other possible use-cases* participants named training to improve communication skills in general (8) and for more specific groups, like patients with anxiety disorders or people with social phobias. Participants mentioned seven times that they would like to have more specific recommendations, e.g. "The agent could say something like: Nonverbal feedback is nodding, for example.". *Intention for further use* was indicated by 9 participants. The *added value* of the training was for most of the participants that the recommendations are given directly on a specific behavior shown in a specific situation during the job interview.

Recommendation Generation. In order to verify the validity of our approach, we examined whether the counterfactuals generated by the GAN are modifying the features that the incorporated engagement classifier identified as important for the classification of low and high engagement. For this evaluation, we used five sessions of the multi-modal job interview role-play dataset [10] that have also been used in the pilot study and extracted the importance scores of every feature in regard to the model's classification with LIME [8]. We calculated the pearson correlation between the absolute value change of how much each feature has been modified by the counterfactual transformation and the importance score of every feature, see Table 1. High correlation scores indicate that the counterfactual feature transformation is in line with the corresponding importance of the feature. Seven features showed a strong positive correlation (GZ_DR, AM_CR, HD_TH, DIST_RW, YROT_LE, SDX_HD, SDXROT_HD), six features had a moderate positive correlation (HD_AC, YROT_RE, XROT_RE, TN_HD, CONT_MOV, EN_HA) and two features presented with a low positive correlation (DIST_LW, XROT_LE). Moreover, FO_RW had a strong negative correlation, VAL_F showed a moderate negative correlation and FO_LW had a weak negative correlation. Moreover, we conducted a computational evaluation to investigate how well the generated counterfactual features change the decision of the engagement classifier. We found that 96.49% of the generated counterfactual feature vectors led to a different decision of the engagement model as the original input features.

Table 1. Pearson correlation between the absolute change of the feature values and the LIME classification relevance scores for every feature. The features are from left to right: *Valence Face, Gaze behavior, Head activity, Arms crossed, Head touch, X distance of left/right wrist and hip, Y rotation left/right elbow, Y distance of left/right wrist and hip, X rotation left/right elbow, Standard deviation head movement in X axis, Standard deviation Head X rotation, Turn hold, Continuous movement, Gesticulation.*

Feature	VAL _F	GZ _DR	HD _AC	AM _CR	HD _TH	DST _LW	DST _RW	YR _LE	YR _RE	FO _LW	FO _LW	XR _LE	XR _RE	SDX _H	SDXR _H	TN _HO	CNT _MV	EN _HA
r	−0.59	0.98	0.63	0.80	0.93	0.22	0.74	0.77	0.62	−0.04	−0.77	0.29	0.64	0.74	0.87	0.43	0.48	0.35

4 Discussion and Conclusion

We introduce a novel approach for generating textual nonverbal behavior recommendations in job interview training environments. In a pilot study, we presented the approach to participants. The results indicate that such training could be helpful to prepare for job interviews successfully. The recommendations given by the system were found to be helpful and comprehensible, and transferable to other use cases. Moreover, most participants noted that the proposed approach adds additional value to the training by giving recommendations directly on a specific behavior in a specific situation. Part of the underlying training system automatically extracts situations that could be improved and displays them alongside the recommendation presented by the virtual coach. Moreover, we examined the validity of our GAN-driven recommendation generation approach by calculating the Pearson correlation coefficient between the absolute changes of the feature values after counterfactual transformation and the importance of the features the classifier attributed to them regarding the classification result. We showed that most of the features (15 out of 18 features) had a moderate to strong correlation, which emphasizes the validity of the proposed approach. Only the two features corresponding to the relative position and movement of the left wrist and the feature representing the flexion of the left elbow presented a weak correlation. Further, we also investigated how well the generated counterfactual features can change the decision of the engagement classifier. Overall, 96.49% of the counterfactual feature vectors led to a different decision of the engagement classifier as the original input features. This indicates that our GAN-driven approach enables to generate recommendations that, when being adopted, are consistently leading to a perception of high engagement. The computational evaluation, as well as the user study, indicate that the generated recommendations are valid and helpful in the context of job interview coaching scenarios.

Acknowledgments. This work presents and discusses results in the context of the research project ForDigitHealth. The project is part of the Bavarian Research Association on Healthy Use of Digital Technologies and Media (ForDigitHealth), funded by the Bavarian Ministry of Science and Arts. Further, the work described in this paper has been partially supported by the BMBF under 16SV8688 within the MITHOS project, and by the BMBF under 16SV8493 within the AVASAG project.

References

1. Baur, T., Damian, I., Gebhard, P., Porayska-Pomsta, K., Andre, E.: A job interview simulation: social cue-based interaction with a virtual character (2013)
2. Baur, T., et al.: Context-aware automated analysis and annotation of social human-agent interactions. ACM Trans. Interact. Intell. Syst. **5**(2) (2015)
3. Cafaro, A., et al.: The noxi database: multimodal recordings of mediated novice-expert interactions. In: ICMI 2017, November 2017
4. Hoque, E., Courgeon, M., Claude Martin, J., Mutlu, B., Picard, R.W.: Mach: my automated conversation coach. In: Proceedings of the 2013 ACM International Joint Conference on Pervasive and Ubiquitous Computing (2013)
5. Mertes, S., Huber, T., Weitz, K., Heimerl, A., André, E.: Ganterfactual-counterfactual explanations for medical non-experts using generative adversarial learning. Front. Artif. Intell. **5** (2022)
6. Minge, M., Riedel, L.: meCUE-Ein modularer Fragebogen zur Erfassung des Nutzungserlebens. In: Mensch and Computer 2013-Tagungsband, pp. 89–98. Oldenbourg Wissenschaftsverlag (2013)
7. Paulhus, D.L., Westlake, B.G., Calvez, S.S., Harms, P.D.: Self-presentation style in job interviews: the role of personality and culture. J. Appl. Soc. Psychol. **43**(10), 2042–2059 (2013)
8. Ribeiro, M.T., Singh, S., Guestrin, C.: "why should I trust you?": Explaining the predictions of any classifier. In: Proceedings of the 22nd ACM SIGKDD International Conference on Knowledge Discovery and Data Mining, San Francisco, CA, USA, 13–17 August 2016, pp. 1135–1144 (2016)
9. Rowold, J., Hochholdinger, S., Schaper, N.: Evaluation und Transfersicherung betrieblicher Trainings: Modelle. Methoden und Befunde, Hogrefe (2008)
10. Schneeberger, T., Scholtes, M., Hilpert, B., Langer, M., Gebhard, P.: Can social agents elicit shame as humans do? In: 2019 8th International Conference on Affective Computing and Intelligent Interaction (ACII), pp. 164–170. IEEE (2019)
11. Takeuchi, N., Koda, T.: Initial assessment of job interview training system using multimodal behavior analysis. In: Proceedings of the 9th International Conference on Human-Agent Interaction. HAI 2021, pp. 407–411. Association for Computing Machinery, New York (2021)

Enhancing Auto-scoring of Student Open Responses in the Presence of Mathematical Terms and Expressions

Sami Baral[1(✉)], Karthik Seetharaman[1], Anthony F. Botelho[2], Anzhuo Wang[1], George Heineman[1], and Neil T. Heffernan[1]

[1] Worcester Polytechnic Institute, Worcester, MA, USA
{sbaral,kvseetharaman,awang6,heineman,nth}@wpi.edu
[2] University of Florida, Gainesville, FL, USA
a.botelho@ufl.edu

Abstract. Prior works have led to the development and application of automated assessment methods that leverage machine learning and natural language processing. The performance of these methods have often been reported as being positive, but other prior works have identified aspects on which they may be improved. Particularly in the context of mathematics, the presence of non-linguistic characters and expressions have been identified to contribute to observed model error. In this paper, we build upon this prior work by observing a developed automated assessment model for open-response questions in mathematics. We develop a new approach which we call the "Math Term Frequency" (MTF) model to address this issue caused by the presence of non-linguistic terms and ensemble it with the previously-developed assessment model. We observe that the inclusion of this approach notably improves model performance, and present an example of practice of how error analyses can be leveraged to address model limitations.

Keywords: Math-terms · Open-ended responses · Automated assessment · Machine learning · Natural language processing · Mathematics

1 Introduction

Advancements in artificial intelligence and machine learning research have led to greater integration of prediction models into educational contexts through computer-based learning systems. These systems are being used in educational settings to support teachers and students in a variety of ways. Most prominent of the supports offered by most learning systems is that of automated assessment.

When assessing open-ended problems, however, the correctness of student responses can be subjective, where teachers commonly assess students based on

S. Baral and K. Seetharaman—Both authors contributed equally to this research.

M. M. Rodrigo et al. (Eds.): AIED 2022, LNCS 13355, pp. 685–690, 2022.
https://doi.org/10.1007/978-3-031-11644-5_68

an explicit or implicit rubric that identifies key points that must be included in a student response to sufficiently demonstrate comprehension. Current automatic assessment methods commonly apply natural language processing (NLP) to build a high-dimensional representation of student responses that is then combined with various machine learning approaches (e.g. [2,3,8,11]).

In consideration of the challenges in assessing open-ended problems, mathematics based domains make developing automated assessment models even more difficult, as most traditional NLP techniques were not designed for such a context, with a few recent exceptions [4,7,9]. Recent work has identified that the existence of non-linguistic terms is positively correlated with model prediction error in models that have outperformed existing benchmarks in this context [1].

This work presents a simple, targeted method to resolve this problem. We call this proposed method the "Math Term Frequency" (MTF) model and demonstrate how it can be combined with previously-developed assessment models to improve performance. Specifically, this work addresses the following research questions: 1) How does accounting for non-linguistic terms through our MTF model affect the performance of auto-assessment methods on existing benchmarks? and 2) Does our MTF method reduce the correlation between non-linguistics terms and model prediction error?

2 Dataset

To explore and examine the methods proposed in this work, we observe two datasets consisting of student answers to mathematics open-response questions. These datasets were collected from ASSISTments [6] and contains 150,477 student responses from 27,199 students for 2,076 open-ended math problems scored by 970 unique teachers (where each response was scored by a single teacher); this dataset is the same used to establish benchmark results [4] and is used to directly compare performance against models presented in prior work [1,4]. Teachers scored responses based on a 5-point integer scale ranging from 0 to 4, with a 4 indicating a very strong and a 0 indicating a very weak response. The second dataset used in this paper was similarly used in prior work to conduct an error analysis to identify factors that correlate with prediction error [1]. This dataset is comprised of student open responses collected in a pilot study of the QUICK-Comments tool and contains 30,371 scored student responses from 1,628 students for 915 unique open-response questions assessed by 12 teachers.

3 The SBERT-MTF Model

The methods presented in this work target the specific problem of non-linguistic terms contributing to prediction error. The previously-developed SBERT-Canberra model outperformed previous decision-tree- and deep-learning-based approaches [4] by leveraging pre-trained Sentence-BERT embeddings. The challenge, however, is that only a finite number of words (and sentences, by extension) can be recognized by these methods. When observing non-lingustic terms such as numbers and expressions, many such terms may not be represented within the embeddings (e.g. representing "the answer is 4.3333" with the same embedding as, for example, "the answer is 2.987" if neither of the numbers are recognized). Instead, we propose the "Math Term Frequency" (MTF) method which takes a much simpler approach, drawing inspiration from assessment methods applied for close-ended problems. The goal of this method is to supplement the previously-developed SBERT-Canberra model through ensembling, resulting in what we are calling the "SBERT-MTF" model.

The MTF method works by first parsing student answers to identify non-linguistic terms. The function[1] splits each student answer by spaces, removes alphabet-only terms (accounting for punctuation), removing spaces around math operators, and rounding off large decimals. Once the non-linguistic terms have been identified, the MTF method involves identifying the most frequently-occurring terms for each possible integer score as a means of learning a kind of rubric. There will likely be some terms that are common throughout all scored answers, but there are likely to be some terms that demonstrate comprehension; similarly, students exhibiting common misconceptions may arrive at a similar set of incorrect answers. With this in mind, we select the five most-frequent terms from the list of parsed non-linguistic terms for each problem. With these, for a new response for which we want to generate a score, we calculate a set of 5 indicator values representing whether the response contains each of the most-frequent terms. These features are used in a multinomial logistic regression (following previous works) that is trained separately for each problem.

The score predictions from the MTF model are then ensembled with the SBERT-Canberra predictions using another logistic regression model, referred to as the SBERT-MTF model; to clarify, this ensemble regression model observes ten features corresponding to the probability estimates produced for each of the five possible scores for each of the two observed models. The goal of this is to combine the semantic representation captured by the SBERT method, while taking advantage of the non-linguistic term matching from the MTF method.

3.1 SBERT-MTF Model Performance

As to directly compare the existing method to the prior works, we use similar evaluation method and dataset used in [1,4]. This evaluation method utilizes a 2-parameter IRT model to compare model estimates [10]. The model predictions are used as covariates within the IRT model allowing for the comparison of scoring methods that controls for variables of general student ability and problem difficulty; the number of words in the response is also added as a covariate in this evaluation model in an attempt to further compare models on their ability to interpret student answers rather than be based on other more superficial response features. This evaluation method allows for a fair comparison that accounts for factors that likely impact score that are external to the observed text of the

[1] All code used in this work is available at https://github.com/ASSISTments/SBERT-MTF.

student response. For comparison to previous works, we evaluate our method using three metrics: AUC (see [5]), Root Mean Squared Error (RMSE; calculated using model estimates as a continuous-valued integer scale), and Cohen's Kappa.

The IRT model performance of the Math terms frequency model as compared to the performance of the prior models for scoring open-ended responses is presented in Table 1. The results suggests that the proposed SBERT-MTF model outperforms the previous highest-performing model across evaluation metrics.

Table 1. IRT model performance compared to the models developed in prior works related to auto-scoring of student open responses in mathematics.

Model	AUC	RMSE	Kappa
Baseline IRT	0.827	0.709	0.370
IRT + SBERT-Canberra	0.856	0.577	0.476
IRT + SBERT-MTF	**0.871**	**0.524**	**0.508**

3.2 Error Analysis of SBERT-MTF

The proposed MTF method was designed to address a very targeted problem exhibited by the previously-developed SBERT-Canberra model. We therefore conduct a similar error analysis to observe whether this method impacts the observed positive correlation between the presence of non-linguistic terms and model error. For this analysis, we use the second dataset as described in Sect. 2 for a direct comparison with the previous work. While the modeling task treats scoring as a categorization task, we convert the model predictions to a ordinal-scale integer value (i.e. 0–4). We calculate model prediction error as the absolute value of the teacher-provided score minus the predicted score. In this way, positive values correspond with higher error and values close to 0 represent low error (high performance) and conduct a linear regression observing absolute error as the dependent and answer-level features as independent variables.

We compare three models within this analysis to identify how two modeling decisions presented in this work correspond with observed changes in feature coefficients. The first model observed is that of the SBERT-Canberra model reported in [1] as a baseline for comparison. The second model uses the same SBERT-Canberra method, but trains a logistic regression per problem with the model predictions as covariates (e.g. similar to the ensembled method described earlier, without MTF); the intuition here is that problem-specific adjustments may itself help to account for error in the model. Finally, we observe the ensembled SBERT-MTF model for impacts beyond these other two methods.

The results of the error analysis is presented in Table 2. The results indicate that the linear model for both Logistic SBERT and SBERT-MTF explains 34.8% of the variance of the outcome as given by r-squared; this alone suggests that there is a large portion of variance in the error unexplained by the observed features. Among the observed features, similar to the results from [1], nearly all were

Table 2. The resulting model coefficients for the uni-level linear regression model of absolute error for SBERT Canberra, Logistic SBERT and MTF model.

	SBERT-Canberra		Logistic SBERT		SBERT-MTF	
	B	Std. error	B	Std. error	B	Std. error
Intercept	0.581***	0.017	0.738***	0.017	0.776***	0.070
Answer length	−0.008***	0.001	−0.008***	0.001	−0.009***	0.001
Avg. word length	−0.014***	0.003	−0.013***	0.003	−0.014***	0.003
Numbers count	<0.001	<0.001	<0.001	<0.001	<0.001	<0.001
Operators count	−0.006***	0.001	0.001	0.001	0.004**	0.001
Equation percent	0.443***	0.018	−0.062***	0.019	−0.128***	0.019
Presence of images	2.248***	0.021	2.058***	0.022	2.018***	0.022

*p < 0.05 **p < 0.01 ***p < 0.001

statistically reliable in predicting the model error. However, it is arguable that from the relatively small scale of most coefficients, two of the features exhibit more meaningful impacts in comparison to the others: the presence of mathematical expression and presence of images in the student answers. However, with the introduction of a logistic regression model that follows the SBERT-Canberra method, the coefficient value of presence of mathematical terms has changed; it would appear that accounting for problem-level adjustments alone removes much of the impact of non-linguistic terms in the dataset. Most notably, however, is that the addition of our MTF method exhibits an even stronger negative correlation between the presence of non-linguistic terms and model error; what once was a weakness now appears to be a potential strength of the model.

4 Discussion and Future Work

The results of all of the presented analyses illustrate MTF (specifically, SBERT-MTF) as a promising method to mitigate model error attributed to the presence of non-linguistic terms. The MTF method represents an intentionally-simple approach to address a targeted weakness observed in previously-developed models and seemingly led to positive impacts.

With that, there are still several areas in which these models could be improved, in addition to improving the accuracy of the parsing function. Most notably, is the remaining correlation between the presence of images and model error. While this is not surprising, as the models do nothing to account for images, this remains an unhandled case that cannot be ignored. As it is also the case that some students include mixtures of natural language, non-linguistic terms, and images all in the same answer, developing methods to handle such cases fairly is important for future work.

Similarly, the error analysis suggests that there is a large amount of variance in model error left unexplained. Previous work [1] identified problem- and teacher-level factors that seemingly account for much of this unexplained error, but this does not provide clear guidance as to how to account for these external factors fairly within an automatic assessment model.

Acknowledgements. We thank multiple grants (e.g., 1917808, 1931523, 1940236, 1917713, 1903304, 1822830, 1759229, 1724889, 1636782, 1535428, 1440753, 1316736, 1252297, 1109483, & DRL-1031398); IES R305A170137, R305A170243, R305A180401, R305A120125, R305A180401, & R305C100024, P200A180088 & P200A150306, as well as N00014-18-1-2768, Schmidt Futures and a second anonymous philanthropy.

References

1. Baral, S., Botelho, A., Erickson, J., Benachamardi, P., Heffernan, N.: Improving automated scoring of student open responses in mathematics. In: Proceedings of the Fourteenth International Conference on Educational Data Mining, Paris, France (2021)
2. Chen, H., He, B.: Automated essay scoring by maximizing human-machine agreement. In: Proceedings of the 2013 Conference on Empirical Methods in Natural Language Processing, pp. 1741–1752 (2013)
3. Dikli, S.: An overview of automated scoring of essays. J. Technol. Learn. Assess. **5**(1) (2006)
4. Erickson, J.A., Botelho, A.F., McAteer, S., Varatharaj, A., Heffernan, N.T.: The automated grading of student open responses in mathematics. In: Proceedings of the Tenth International Conference on Learning Analytics and Knowledge, pp. 615–624 (2020)
5. Hand, D.J., Till, R.J.: A simple generalisation of the area under the roc curve for multiple class classification problems. Mach. Learn. **45**(2), 171–186 (2001)
6. Heffernan, N.T., Heffernan, C.L.: The assistments ecosystem: building a platform that brings scientists and teachers together for minimally invasive research on human learning and teaching. Int. J. Artif. Intell. Educa. **24**(4), 470–497 (2014)
7. Lan, A.S., Vats, D., Waters, A.E., Baraniuk, R.G.: Mathematical language processing: automatic grading and feedback for open response mathematical questions. In: Proceedings of the Second (2015) ACM Conference on Learning@ Scale, pp. 167–176 (2015)
8. Riordan, B., Horbach, A., Cahill, A., Zesch, T., Lee, C.: Investigating neural architectures for short answer scoring. In: Proceedings of the 12th Workshop on Innovative Use of NLP for Building Educational Applications, pp. 159–168 (2017)
9. Shen, J.T., et al.: MathBert: a pre-trained language model for general NLP tasks in mathematics education. arXiv preprint arXiv:2106.07340 (2021)
10. van Schuur, W.: Ordinal Item Response Theory: Mokken Scale Analysis, vol. 169. SAGE Publications, Thousand Oaks (2011)
11. Zhao, S., Zhang, Y., Xiong, X., Botelho, A., Heffernan, N.: A memory-augmented neural model for automated grading. In: Proceedings of the Fourth (2017) ACM Conference on Learning@ Scale, pp. 189–192 (2017)

Automated Scoring for Reading Comprehension via In-context BERT Tuning

Nigel Fernandez[1], Aritra Ghosh[1], Naiming Liu[2], Zichao Wang[2], Benoît Choffin[3], Richard Baraniuk[2], and Andrew Lan[1(✉)]

[1] University of Massachusetts Amherst, Amherst, MA 01003, USA
andrewlan@cs.umass.edu

[2] Rice University, Houston, TX 77005, USA

[3] Paris, France

Abstract. Automated scoring of open-ended student responses has the potential to significantly reduce human grader effort. Recent advances in automated scoring leverage textual representations from pre-trained language models like BERT. Existing approaches train a separate model for each item/question, suitable for scenarios like essay scoring where items can be different from one another. However, these approaches have two limitations: 1) they fail to leverage item linkage for scenarios such as reading comprehension where multiple items may share a reading passage; 2) they are not scalable since storing one model per item is difficult with large language models. We report our (grand prize-winning) solution to the National Assessment of Education Progress (NAEP) automated scoring challenge for reading comprehension. Our approach, in-context BERT fine-tuning, produces a single shared scoring model for all items with a carefully designed input structure to provide contextual information on each item. Our experiments demonstrate the effectiveness of our approach which outperforms existing methods. We also perform a qualitative analysis and discuss the limitations of our approach. (Full version of the paper can be found at: https://arxiv.org/abs/2205.09864 Our implementation can be found at: https://github.com/ni9elf/automated-scoring)

Keywords: Automated scoring · BERT · Reading comprehension

1 Introduction

Automated scoring (AS) refers to the problem of using algorithms to automatically score student responses to open-ended items. AS approaches have the potential to significantly reduce human grading effort and scale well to an increasing student population. AS has been studied in many different contexts, including automated essay scoring (AES) [1] for essays, and automatic short answer grading (ASAG) for humanities, social sciences [13], and math [3].

M. M. Rodrigo et al. (Eds.): AIED 2022, LNCS 13355, pp. 691–697, 2022.
https://doi.org/10.1007/978-3-031-11644-5_69

Existing approaches for AS use feature engineering, i.e., developing features that summarize the length, syntax [2], cohesion [6], relevance to the item [11], and semantics of student responses combined with machine learning-based classifiers [2,9] to predict the score. These approaches have excellent interpretability but require human expertise in feature development. Recent approaches leverage advances in deep learning to produce better textual representations of student responses, alleviating the need to rely on human-engineered features. Examples include combining neural networks with handcrafted features for score prediction [14] and especially approaches that fine-tune neural language models (LMs) such as BERT [5] and GPT-2 [12] on the downstream AS task [7,8]. These approaches have performed well on publicly available datasets like the automated student assessment prize (ASAP) for essay scoring [1].

However, existing AS approaches are limited in many ways, including a significant one which we address in this paper: They train a separate AS model for each item. This approach is acceptable in contexts such as AES since the items (essay prompts) are likely not highly related to each other. However, in other contexts such as reading comprehension, multiple items may share the same background passage; Training a separate model for each item fails to leverage this shared background information. More importantly, this approach results in a separate model for each item; For large LM-based models that have millions of parameters, this approach creates a significant model storage problem.

Contributions. First, we present our (grand prize-winning) solution to the NAEP AS challenge[1] for reading comprehension items. We develop a novel AS approach based on meta-learning ideas via in-context fine-tuning of LMs. A carefully designed input format using in-context examples provides context on each item to leverage linkage across items. Second, our experiments demonstrate the effectiveness of our AS approach which outperforms existing LM-based AS approaches and significantly outperforms other non-LM-based baselines. Third, we perform a qualitative analysis to identify common scoring error types and discuss limitations of our approach with insights for future work.

2 Methodology

2.1 Problem Formulation and Dataset

The NAEP AS challenge features 20 reading comprehension items from grades 4 and 8. Each item, indexed by i, has 1) a long reading passage P_i of around 657 words, 2) a short question Q_i of around 27 words, 3) a scoring rubric, and 4) a large training dataset of around 18,000 human scored student responses $D_i^{train} = \{(x_j, y_j)\}$. The average response length is 19 and 37 words for grades 4 and 8, respectively. The goal is to predict human scores for responses in the test dataset $D_i^{test} = \{(x_j^{target})\}$ for all tasks i. A key observation is that pairs of

[1] Ran by the US Dept. of Education: https://github.com/NAEP-AS-Challenge/info.

items (l, m), one from grade 4 (l) and one from grade 8 (m), often share passages $(P_l = P_m)$ and questions $(Q_l = Q_m)$. This linkage across items motivates our approach of leveraging shared semantics by not treating items independently.

2.2 Our Approach: Meta-trained BERT with In-context Tuning

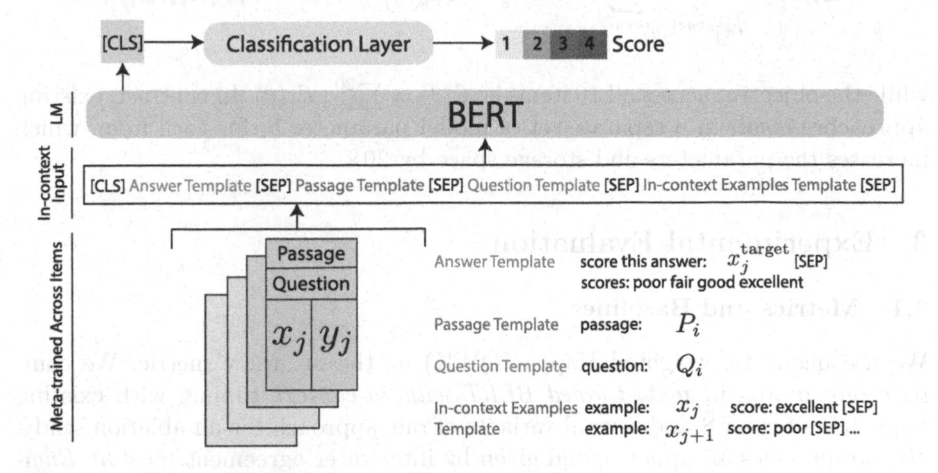

Fig. 1. Meta-trained BERT via in-context tuning. Best viewed in color. (Color figure online)

Meta-trained BERT with In-context Tuning. We use a pre-trained BERT model [5] as the base LM (we also tried GPT-2 [12] which gave similar performance). The key idea of our approach, in-context learning of a shared AS model across all items, uses a carefully-designed input structure to provide context to the shared model and associate it with each specific item. For this, following recent approaches for meta-training-based in-context learning [4,10], we add *in-context examples* which are (response, score) pairs from the training set, to the input. Intuitively, these examples provide further context about each item and enable the AS model to focus on learning to associate responses to scores given the passage and the question, thus enabling knowledge sharing across items.

We build the model input as shown in Fig. 1 by concatenating the target student response x_j^{target} to be scored, passage P_i, question Q_i, and K (response, score) examples $E_i \subseteq D_i^{train}$. We add separator tokens [SEP] to help the model differentiate between input segments. We convert the numeric scores y_j in the examples E_i to meaningful words as: $\{1 : \text{poor}, 2 : \text{fair}, 3 : \text{good}, 4 : \text{excellent}\}$. We add semantically meaningful task instructions to each input component.

Since each item i has a different score range $y_j \in [s_{min} = \{1\}, s_{max} = \{2, 3, 4\}]$, we add valid score classes for x_j^{target} as explicit options in the input to make the model aware of valid output scores. We mask out invalid scores for each response before computing the softmax to ensure that training loss is backpropagated over valid scores only, enabling us to meta-train a single shared model for all items.

We train our AS model on the union of training datasets for all items $\bigcup_{i=1}^{20} D_i^{train}$. Let θ represent the parameters to be learned, i.e., BERT parameters fine-tuned and classification layer parameters learned from scratch. The in-context learning objective $\mathcal{L}_i$ for an item is the negative log-likelihood loss:

$$\mathcal{L}_i(\theta) = \sum_{(x_j^{target}, y_j^{target}) \in D_i^{train}} [-\log p_\theta(y_j^{target} | x_j^{target}, P_i, Q_i, E_i)],$$

while the objective across all 20 items is: $\mathcal{L}(\theta) = \sum_{i=1}^{20} \mathcal{L}_i(\theta)$. In contrast, existing approaches result in a separate set of model parameter θ_i for each item, which increases the parameters and storage space by 20×.

3 Experimental Evaluation

3.1 Metrics and Baselines

We use quadratic weighted Kappa (QWK) as the accuracy metric. We compare our approach, *meta-trained BERT with in-context* tuning, with existing approaches for AES and several variants of our approach for an ablation study. *Human* provides an upper bound given by inter-rater agreement. *Feature Engineering* uses handcrafted features [14] with random forests. *Stacked LSTM* uses a stack of two LSTMs. *Clusering + Classification* uses cluster indicators of responses with random forests. *BERT (response)* fine-tunes 20 independent BERT models, one for each item, using only responses as input. *BERT (passage+question+response)* adds passage and question text. *BERT in-context* adds in-context examples. *BERT multi-task* uses multi-task learning to fine-tune a single shared model with 20 separate classification layers, one for each item.

3.2 Implementation Details

We spell check since 1) the rubric ignore spelling errors, and 2) to transform student responses to be similar to what BERT sees in pre-training. Spell check improves performance by 0.5%. We perform a five-fold cross-validation for each item. We use a pre-trained BERT [5] model with default parameters as our base LM. For our meta-trained BERT model with in-context tuning, for each training response, we randomly sample up to 25 in-context examples per score class from the training dataset of the corresponding item. At testing we use 8 different sets of randomly sampled examples and average the predicted scores.

3.3 Results and Analysis

Table 1. Results averaged across items. Model performance in Quadratic Weighted Kappa (higher is better). **Bold** indicates best result except human performance.

Approach	Avg. QWK	p-value
Human	0.878	–
Feature Engineering + Random Forest	0.443	–
Stacked LSTM	0.657	–
Clustering + Classification	0.709	–
BERT (response)	0.828	–
BERT (passage+question+response)	0.828	0.414
BERT in-context	0.833	0.001
BERT multi-task	0.833	1.6×10^{-4}
Meta-trained BERT in-context	**0.841**	9.6×10^{-5}

We report the average QWK for all approaches in Table 1. LM-based approaches outperform non-LM-based approaches by a significant margin. Our approach *meta-trained BERT in-context* performs best. We perform paired t-tests using *BERT (response)* as the baseline. We observe that learning a single shared model across items achieves statistically significant improvement over learning one model per item. This observation validates our intuition that leveraging shared information across items improves accuracy. Surprisingly, adding passage and question text as input does not lead to improved performance suggesting that our model has not mastered reading comprehension. Our approach performs better than *BERT (response)*, more on shared items across grades 4 and 8 and less on non shared items as seen in Table 2. This observation suggests that our approach may struggle to generalize to new items.

Table 2. Results averaged over shared vs. non-shared items from grade 4 and 8.

Approach	Non-shared G4	Shared G4	Non-shared G8	Shared G8
BERT (response)	0.825	0.842	0.763	0.840
Meta-trained BERT in-context	**0.826**	**0.856**	**0.771**	**0.853**

Qualitative Error Analysis. We randomly sample 100 responses that our approach scores incorrectly. We identify 5 error types shown in Table 3 with made-up examples that reflect our observation: 1) spelling and grammar errors (that remain after the spell check), 2) human error, 3) infrequent (our approach struggles on correct responses that occur rarely), 4) imitation (incorrect responses with content/structure that mimic correct responses), and 5) character coreference (referring to a different character).

Table 3. Illustration of responses that our approach tends to score incorrectly.

Error type	Example student response	Prediction	Actual
Spelling/grammar	"*mearchant are* a good man because ..."	2	3
Human error	"Long ago a poor country boy left ..."	1	3
Infrequent	"merchant is described as *brave* as ..."	2	4
Imitation	"The merchant is dishonest because ..."	3	1
Coreference	"*merchant*, not *innkeeper*, is greedy ..."	4	1

4 Conclusions and Future Work

In this paper, we presented our (grand prize-winning) solution to the NAEP AS challenge, an in-context BERT tuning model meta-trained on all items that leverages item linkage and shared semantics for reading comprehension items. There are several avenues for future work. First, adding passage and question text as input does not improve performance indicating that our model has not mastered reading comprehension, which suggests that future work should leverage passage information and score responses in context of the passage for better generalization to new items. Second, spell checking student responses improves accuracy, which suggests that future work should pre-train LMs on student-generated text. Third, our model does not offer any explanation on score predictions, which suggests that future work should improve the interpretability of automated scoring methods via e.g., incorporating scoring rubrics or visualizing the model.

References

1. The hewlett foundation: Automated essay scoring. Online: https://www.kaggle.com/c/asap-aes. Accessed May 2022
2. Attali, Y., Burstein, J.: Automated essay scoring with e-rater®. J. Technol. Learn. Assess. **4**(3) (2006)
3. Baral, S., Botelho, A.F., Erickson, J.A., Benachamardi, P., Heffernan, N.T.: Improving automated scoring of student open responses in mathematics. In: 14th International Conference on Educational Data Mining (EDM) (2021)
4. Chen, Y., Zhong, R., Zha, S., Karypis, G., He, H.: Meta-learning via language model in-context tuning. In: 60th Annual Meeting of the Association for Computational Linguistics (ACL) (2022)
5. Devlin, J., Chang, M.W., Lee, K., Toutanova, K.: BERT: pre-training of deep bidirectional transformers for language understanding. In: NAACL-HLT (2019)
6. Graesser, A.C., McNamara, D.S., Louwerse, M.M., Cai, Z.: Coh-Metrix: analysis of text on cohesion and language. Behav. Res. Meth. Instrum. Comput. **36**(2), 193–202 (2004)
7. Lottridge, S., Godek, B., Jafari, A., Patel, M.: Comparing the robustness of deep learning and classical automated scoring approaches to gaming strategies. Cambium Assessment Inc, Technical report (2021)

8. Mayfield, E., Black, A.W.: Should you fine-tune BERT for automated essay scoring? In: 15th Workshop on Innovative Use of NLP for Building Educational Applications, pp. 151–162 (2020)
9. McNamara, D.S., Crossley, S.A., Roscoe, R.D., Allen, L.K., Dai, J.: A hierarchical classification approach to automated essay scoring. Assess. Writ. **23**, 35–59 (2015)
10. Min, S., Lewis, M., Zettlemoyer, L., Hajishirzi, H.: MetaICL: learning to learn in context. In: NAACL-HLT (2022)
11. Persing, I., Ng, V.: Modeling prompt adherence in student essays. In: 52nd Conference of the Association for Computational Linguistics (ACL) (2014)
12. Radford, A., Wu, J., Child, R., Luan, D., Amodei, D., Sutskever, I.: Language models are unsupervised multitask learners. OpenAI blog (2019)
13. Sung, C., Dhamecha, T., Saha, S., Ma, T., Reddy, V., Arora, R.: Pre-training BERT on domain resources for short answer grading. In: 24th Conference on Empirical Methods in Natural Language Processing (EMNLP), pp. 6071–6075 (2019)
14. Uto, M., Xie, Y., Ueno, M.: Neural automated essay scoring incorporating handcrafted features. In: 28th Conference on Computational Linguistics (COLING), pp. 6077–6088 (2020)

Deep Learning for Automatic Detection of Qualitative Features of Lecturing

Anna Wróblewska[1(✉)], Józef Jasek[1], Bogdan Jastrzebski[1], Stanisław Pawlak[1], Anna Grzywacz[1], Siew Ann Cheong[2], Seng Chee Tan[2], Tomasz Trzciński[1,3,4], and Janusz Hołyst[1]

[1] Warsaw University of Technology, Warsaw, Poland
anna.wroblewska1@pw.edu.pl
[2] Nanyang Technological University, Singapore, Singapore
[3] Jagiellonian University of Cracow, Cracow, Poland
[4] Tooploox, Wrocław, Poland

Abstract. Artificial Intelligence in higher education opens new possibilities for improving the lecturing process, such as enriching didactic materials, helping in assessing students' works or even providing directions to the teachers on how to enhance the lectures.

This research explores how an academic lecture can be assessed automatically by quantitative features. First, we prepare a set of qualitative features based on teaching practices and then annotate the dataset of academic lecture videos. We then show how these features could be detected automatically using machine learning and computer vision techniques. Our results show the potential usefulness of our work.

Keywords: Didactic features · Qualitative analysis · Video recognition · Deep learning

1 Introduction

Digital lecturing proliferation during the pandemic drew public attention to the higher education situation and its possible improvements [3]. Numerous newly created programs aim to enhance education quality using the advantage of artificial intelligence and deep learning methods [2]. Adopting artificial intelligence in higher education is still a significant challenge. We follow this research path, and the principal aim of this research is to improve the academic lecturing process. When the majority of newly proposed solutions focus on the learners, their performance [1], or learning sources [5], this research concentrates on the teachers.

Our research explores how an academic lecture can be assessed automatically by quantitative features. We define the features following teaching practices and their crucial aspects. The features also regard such techniques as a slide presentation and other visualizations, organizational information, and teacher interactions with the public. The main goal of designing these features is to give objective feedback to the lecturers to improve their didactic behaviours or the

M. M. Rodrigo et al. (Eds.): AIED 2022, LNCS 13355, pp. 698–703, 2022.
https://doi.org/10.1007/978-3-031-11644-5_70

content of their lecture materials. On the other hand, we designed the features to be feasible to detect automatically using artificial intelligence methods, especially computer vision or machine learning. After designing features, we gather a dataset of video lecture recordings and annotate it to obtain these features. Then we tested how we need to preprocess and prepare deep learning models to detect these features, e.g. via computer vision, audio or text processing. Our general goal is to build a solution that can automatically help teachers get feedback about their lecture assessment and suggest ways to improve, offering particular features that can be easily indicated during the lectures.

Thus, the main contributions of this research are: (1) preparing a set of didactic features feasible to annotate and detect automatically, (2) collecting a dataset of lecture video recordings annotated with these features, and (3) designing a set of deep learning models that were trained to determine the presence of the selected didactic feature in our dataset.

2 Proposed Approach

For our experiments, we choose the most frequent features that can be detected across two modalities. Experiment I concerns features from an audio stream or its transcription, and Experiment II – features describing slide content or visual characteristics of the teachers' behaviours (see Table 1).

At first, we selected and analyzed teaching practices thoroughly, following the approach in [6,7]. Based on these sources, we defined a set of valuable didactic features from the teaching point of view and feasible for automatic detection in lecture videos. Table 1 presents the designed features that are related to the main categories of Singapore Teaching Practices taxonomy [7], e.g., "activating prior knowledge," "arousing interest," and "deciding on teaching aids and learning resources." Once the features were defined, we chose to annotate a dataset that contains 128 lectures recorded in English (on average, one lecture lasts 1.5 h).

2.1 Experiment I: Detection of Audio-Based Features

This experiment is dedicated to the selected audio behavioural features depending on uttered words or sounds in the lecture room. Most open datasets for video annotations, especially behavioural feature recognition, are fine-grained and contain short diversified behaviours. On the contrary, our end goal is to detect the features that last a considerable time slot during a lecture. Thus, we can use very coarse-grained portions of videos (within a range of a few seconds) to process, optimize hyperparameters and train models faster. We used the Azure automatic speech recognition (ASR) service to provide data for building text models. After linking our annotations with the transcription from Azure, we split the dataset into a train, dev, and test sets; these sets contain video samples in which samples from the same lecture and annotations made by the person are within only one of these sets.

Table 1. Features and number of their occurrences in our dataset, and experiments for their automatic detection

Feature	Occurrences	Experiment
Asking questions (AQ)	4926	I & I-Q
Giving questions to students (GQ)	3616	I & I-Q
Organization: giving class outline (O)	1211	I
Test session (S)	854	I
Active teacher stands by slides and explains them (AT)	835	I
Summing up (SU)	72	I
Students' laughter	315	I (AUDIO only)
Use of intonation to emphasise important issues	200	I (AUDIO only)
Teacher is moving across a podium	1379	II
Films or animations in slides	583	II
Images in slides	2793	II
Test session	854	II
Charts in slides	3356	II
Showing websites (on a slide presentation)	307	II
Writing on a whiteboard	3059	II
Writing on slides	6738	II
Eye contact with students	9943	II

We trained our models on two tasks: "Questions only" and "Full task." In the "Questions only" task, we tried to predict whether a given sample depicts a question (i.e. AQ or GQ). In the "Full task," the task is to predict all eight features for audio models and six features/classes for text models without those based on non-verbal audio aspects (see Experiment I in Table 1). We employed M5 and Wav2Letter audio networks to accomplish the feature detection on the audio stream [4]. For the text-based approach, we used BERT-based models.[1] Our feature detection tasks were defined as classification tasks for downstream tasks. Contextual bandits algorithms, which we also used to perform the classification task, was from the Vowpal Wabbit framework.[2] To tokenize the data for TF-IDF, each word and punctuation mark was treated as a token. We extracted the top 10,000 most common words and bigrams, as an input to a classification feed-forward network.

The text-based models significantly outperform audio-based networks, which struggled to get any predictions correctly. The poor performance of audio-based detection is mainly because of no open pre-trained networks and a very long training time (about 48 h compared to text-based networks trained within minutes). The text-based networks achieve much better results despite the propagation of automatic speech recognition (ASR) transcription errors (see Table 2). Vowpal Wabbit has the best results on the questions-only task from all text-based networks. recognition task.

[1] https://huggingface.co/bert-base-uncased,roberta-base.
[2] https://vowpalwabbit.org/.

Table 2. Results for the text-based features detection in Experiment I. Note: Acc – Accuracy, Prec – Precision, Rec - recall, the results for selected features is reported in F1 score. Note: for feature labels refer to Table 1

"Questions only" (AQ & GQ)					"Full task"					
Model	Acc	Prec	Rec	F1	AQ	GQ	O	S	AT	SU
BERT	0.831	0.332	0.453	0.383	0.29	**0.27**	0.03	0.08	0.31	0.02
VowpalWabbit	0.757	0.481	0.387	**0.429**	**0.32**	**0.27**	0.17	**0.37**	0.52	0.02
RoBERTa	0.116	0.116	**1.00**	0.207	0.28	0.25	**0.20**	0.20	0.45	**0.06**
TF-IDF	0.884	0.461	0.100	0.164	0.08	0.09	0.05	0.21	**0.63**	0.0

The difference between accuracy of feature detection roughly corresponds to the number of occurrences in our dataset (see Fig. 1). Figure 2 presents F1 scores

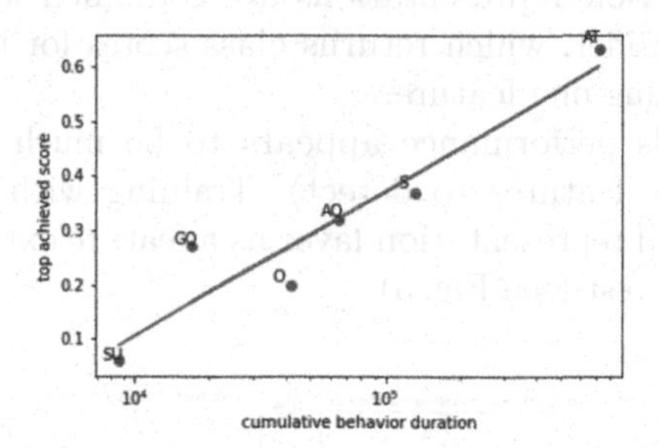

Fig. 1. Correlation between cumulative duration of a behaviour and top achieved F1 score. Note: for feature labels refer to Table 1

when training is performed on BERT using only a subset of available data; here, we compared the results of our dataset with a similarly prepared TED talks dataset (in which transcription is very precise made by humans). These figures suggest that results can be significantly improved by increasing the size of our datasets and the consistency of annotations.

2.2 Experiment II: Multi-task Learning for Visual Features

This experiment approaches differentiating visual features, which appear primarily in the view of the lecturer and presentation screen (see Experiment II in Table 1). This experiment explores the possibility of learning distinctions between visual features with one deep learning model which can share characteristics between different features and thus be more efficient.

Here, the prepared dataset consists of selected frames from videos. Every frame was selected from the middle of every occurrence of the feature (an event recorded in our dataset). It creates a dataset of less correlated events, capturing exactly one observation per recorded event. We have roughly 25,764 events relevant to visual data.

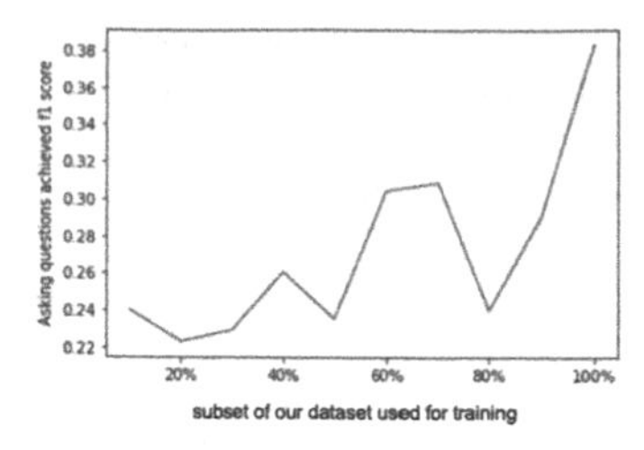

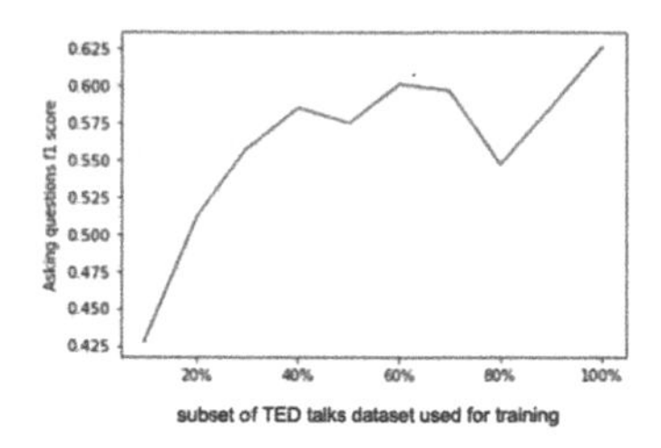

Fig. 2. F1 score achieved on "Questions only" task using BERT on a subset of available data in our datasets and similarly prepared dataset from TED talks

In our model architecture, each video frame is embedded by a feature extractor (e.g. AlexNet, ResNet), and then the deep view representations go into the siamese encoder – a neural network that extracts relevant features for our problem. In the end, the view representations are combined with max-pooling and are passed into a classifier, which returns class scores for the nine classes, each responsible for detecting one feature.

The tested models performance appears to be much better than random overall (we have nine features to detect). Training with imbalanced loss the AlexNet with the third representation layer as a feature extractor achieves above 70% accuracy for the test (see Fig. 3).

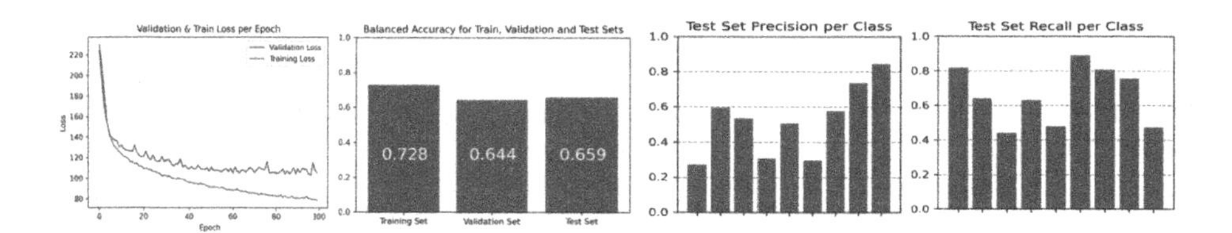

Fig. 3. Results of one of our best networks. From the left to the right: network learning curve, balanced accuracies, precision and recall on the test set for each class/feature given in the same order as in Table 1

Another vital thing our analysis shows is that classes/features might fight. First, the distribution of network prediction trained on balanced loss appears to be more uniform. Those networks do not favour the most frequent classes and guess much more often than a less frequent feature is visible in the video lecture. Generally, the networks commonly confuse features relying on similar data and often co-occurring, e.g. "images in slides" and "charts in slides", or "movement across a podium" and "teacher's eye contact".

3 Conclusions

Our contribution in this work was defining features that can give tips to assess qualitatively didactic aspects of lectures. Then we collected and annotated a dataset that allowed training and testing of deep learning networks for automatic detection and recognition of these features. Our results showed the usefulness of our work. We delivered an approach to preparing and preprocessing the data of video recording. Also, we showed the main directions for deep learning methods. Especially we bypassed audio processing and modelling because it needed extensive computing performance, much more than text processing.

Given our results that are not yet acceptable for professional use but as a proof-of-concept, it is still possible that this approach can work if supplied with better pre-trained feature extractors and trained on a more extensive dataset. The dataset should be more diversified to achieve stable results. Furthermore, our results showed an urge for more effort to get proper annotations, such as: defining extended annotation protocol, employing more annotators, and training them to be more consistent.

Acknowledgements. The research was funded by the Centre for Priority Research Area Artificial Intelligence and Robotics of Warsaw University of Technology within the Excellence Initiative: Research University (IDUB) programme (grant no 1820/27/Z01/POB2/2021) and by the RENOIR project under the EU program Horizon 2020 under project contract no 691152. We want to thank Sylwia Sysko-Romanczuk for her ideas and help in the structuralization of this research.

References

1. Alamri, R., Alharbi, B.: Explainable student performance prediction models: a systematic review. IEEE Access **9**, 33132–33143 (2021)
2. Cheng, X., Sun, J., Zarifis, A.: Artificial intelligence and deep learning in educational technology research and practice. Br. J. Edu. Technol. **51**(5), 1653–1656 (2020)
3. Crawford, J., et al.: COVID-19: 20 countries' higher education intra-period digital pedagogy responses. J. Appl. Learn. Teach. **3**(1), 9–28 (2020)
4. Dai, W., Dai, C., Qu, S., Li, J., Das, S.: Very deep convolutional neural networks for raw waveforms. In: IEEE International Conference on Acoustics, Speech and Signal Processing (ICASSP), pp. 421–425. IEEE (2017)
5. Gari, A.D., Mylonas, K., Nikolopoulou, V., Mrvoljak, I.: Educational and learning resources in a Greek student sample: QELC factor structure and methodological considerations. Psychol. Test Assess. Model. **63**(2), 205–226 (2021)
6. Piburn, M., Sawada, D.: Reformed teaching observation protocol (RTOP) reference manual. Technical report, Arizona Collaborative for Excellence in the Preparation of Teachers (ERIC ED447205) (2000)
7. Singapore Ministry of Education: The Singapore Teaching Practice (STP) (2018)

Predicting Second Language Learners' Actual Knowledge Using Self-perceived Knowledge

Yo Ehara(✉)

Tokyo Gakugei University, Koganei, Tokyo 1848501, Japan
ehara@u-gakugei.ac.jp

Abstract. Self-perceived knowledge refers to the knowledge that learners believe they possess. To measure actual knowledge, developing a valid test for subjects is typically necessary. Self-perceived knowledge is considerably easier to measure than actual knowledge because it can be obtained by simply asking the participants if they possess the knowledge. Hence, if the actual knowledge of a subject can be predicted with high accuracy from their self-perceived knowledge, the burden of formulating test questions and building a dataset can be reduced. In this study, we created a reliable dataset for predicting actual knowledge from self-perceived knowledge in the field of second-language vocabulary learning; this dataset, to the best of our knowledge, is the first of its type. Herein, we provide detailed item response theory (IRT)–based analyses for our datasets as well as simple IRT-based methods for predicting the responses of learners to actual knowledge. We also demonstrate a deep transfer learning–based approach that slightly outperforms the IRT-based approach in terms of predictive accuracy.

Keywords: Self-perceived knowledge · Item response theory · Transformer models · Vocabulary testing

1 Introduction

Learning a second language requires acquiring many words. To determine words that a learner knows, two primary approaches have been previously proposed. The first method is to simply present words to the learner and ask them directly, "Do you know these words?", which is called *self-report testing* [9,10]. If the learner answers that they know the word, the word is said to be a part of their "self-perceived knowledge" because the learner believes that they know the word.

The second is to accurately measure a learner's knowledge of a word by asking them to answer *multiple-choice questions*, as in Table 1 [2,8]. The words that learners correctly answer to vocabulary questions, such as those in Table 1, are the words that they likely know. However, multiple-choice questions are quite burdensome for testers to create and participants to answer, especially when

M. M. Rodrigo et al. (Eds.): AIED 2022, LNCS 13355, pp. 704–710, 2022.
https://doi.org/10.1007/978-3-031-11644-5_71

Table 1. Example of a multiple-choice test question. A test-taker is asked to select the option with the closest meaning to that of the underlined word in the sentence.

It was a difficult period.
a) question b) time c) thing to do d) book

many words need to be asked. For example, an excellent second-language learner knows more than 10,000 words [6,7], and asking learners to answer 10,000-word questions is excessively labor-intensive and unrealistic.

To reduce the required effort, we can consider another approach that involves self-report testing. The results of self-report testing can be used to predict the outcomes of multiple-choice questions with high accuracy using machine learning, which reduces the required effort. Is this approach more plausible? How accurate is the prediction? Can accuracy be improved using recent neural machine learning techniques? To measure prediction accuracy, we require a dataset that contains the results of both self-report and multiple-choice testing on the same word set for the same learner set. To the best of our knowledge, such a dataset does not currently exist, and this study provides the first dataset with desirable properties to answer these challenging research questions. Moreover, this study proposes detailed analysis and prediction methodologies for this dataset.

Dataset: We used the vocabulary size test (VST) [7] to build our dataset[1]. The VST comprises 100 questions, each asking about a word sampled from 20,000 words. As its name suggests, the VST was designed to measure the vocabulary size of language learners. To calculate the vocabulary size of a learner using the VST, the number of correctly answered questions is multiplied by 20. There are two versions of the VST: A and B. We used A throughout this paper.

In [4], a publicly available dataset for vocabulary testing was created and published, wherein learners were asked to take a vocabulary test. For comparison, we used a similar setting. To create our dataset, we used the Lancers (https://lancers.co.jp/) crowdsourcing service as [4] did. As Lancers is native to Japan, most of the participants were native Japanese speakers. However, not all of them are English learners. Hence, if workers are randomly recruited, those who do not learn English may take the test. To exclude such workers, we explicitly specified that only learners who had previously taken the Test of English for International Communication (TOEIC) (https://www.ets.org/toeic), which is conducted by the English Testing Service, could complete the test. A Japanese learner is typically required to pay approximately 70 USD to take the TOEIC test. Therefore, if the learner has previously taken the TOEIC, they are likely motivated to measure their English ability, even at the given cost. For version A of the VST, we obtained 191 participants (i.e., test-takers).

As the VST comprises 100 words sampled from 20,000 words, each question is ordered by the frequency of the word that the question asks. Hence, most

[1] The dataset will be released at http://yoehara.com/ or http://readability.jp/.

learners were unable to correctly answer the final questions of the test. These questions were not the target of this study because without testing, the learner was unlikely to know them. Moreover, if the learners become tired, they may start randomly answering the questions. Hence, we adjusted the number of questions to 65 by removing the 35 most difficult ones.

The test-takers were first presented with a list of 65 words, which was a self-report test, and then asked to answer whether they knew the words. In the next test, the test-takers were asked to answer 65 vocabulary questions for the 65 words in the multiple-choice format of the VST, as in Table 1. Note that the test-takers were not allowed to return to the first self-report testing part after moving to the multiple-choice question. Hence, each word had two questions, self-report and multiple choice, both of which are responded by each test-taker. Thus, each word has two difficulty parameters in item response theory (IRT).

2 Experiments

IRT Experiments: IRT is a widely used probabilistic model that analyzes vocbulary tests and estimates learner ability and item difficulty [1]. The 2PL model was used to obtain the parameters. We used Pyirt (https://github.com/17zuoye/pyirt) to obtain the difficulty and discrimination parameters. The difficulty parameters for the self-report and multiple-choice questions are indicated on the horizontal and vertical axes, respectively, and plotted at the same scale and range on the two axes, as shown in Fig. 1. Each point represents a single word.

The dotted diagonal line was constructed from the lower left to the upper right of Fig. 1. Both the horizontal and vertical axes of Fig. 1 show the values of the difficulty parameters, where a higher value indicates that the item is judged to be more difficult. Therefore, the point to the upper right of the diagonal line indicates that the difficulties of the multiple-choice questions was lower than those of the self-report questions. This result appears counterintuitive as multiple-choice questions are typically more difficult. However, multiple-choice questions can sometimes be correctly answered by randomly selecting the answer. Hence, difficult self-report questions can be sometimes correctly answered by chance. This presumably causes the lower estimated difficulties in multiple-choice questions in Fig. 1. The results of the Wilcoxon test demonstrated that the column of values on the vertical axis was statistically significantly smaller than that on the horizontal axis ($p < 0.01$).

To investigate whether the vertical axis questions were more difficult than the horizontal axis questions, we conducted the following experiment, which is illustrated in Fig. 2. First, the 191 test-takers for version A were divided into 100 and 91 test-takers. The parameters of the actual knowledge questions were estimated only from the responses of the former 100 subjects, while those of the perceived knowledge questions were estimated from the responses of all 91 subjects. Note that we did not use the latter data of 91 subjects $\times$ 65 questions, or 5,915 responses in total. The predictive accuracy of the responses of the dashed area in Fig. 2 was used for evaluation.

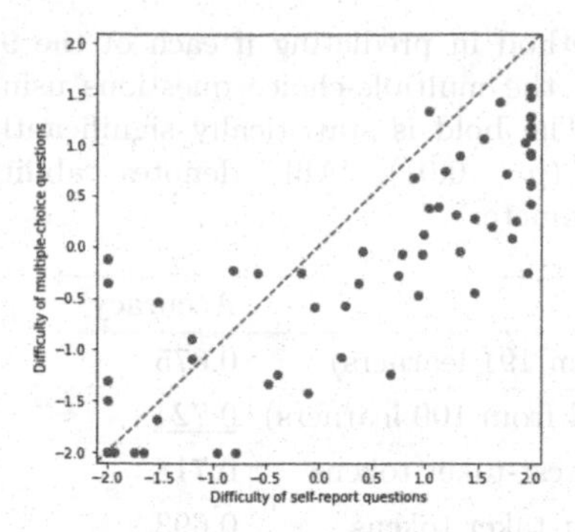

Fig. 1. Plot of the difficulties of the self-report questions vs. those of the multiple-choice questions (vertical axis). Each point represents a word.

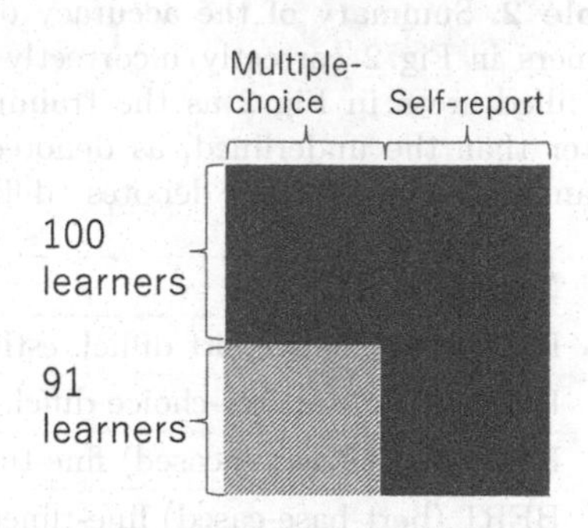

Fig. 2. Experiment setting. The filled area is the *training* data, used to estimate parameters. The accuracy of the dashed *test-data* area was used for evaluation.

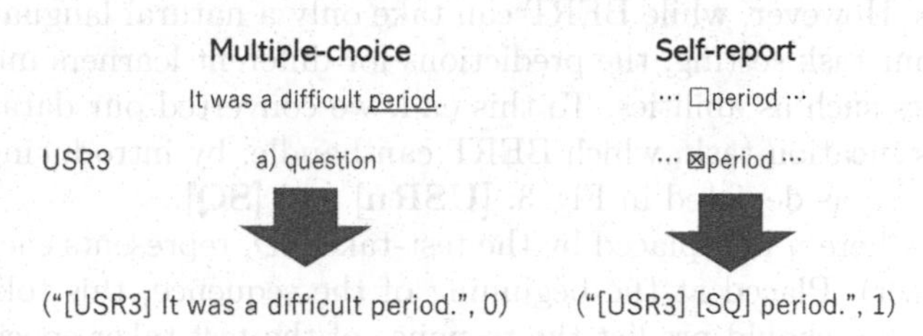

Fig. 3. Example of converting a vocabulary test result dataset into BERT inputs.

Two main methods were applied when using IRT to predict the left-bottom dashed area in Fig. 2. IRT can calculate the ability value θ_j for each learner and the difficulty parameter d_i for each question. *Note that each word has multiple-choice and self-report questions*, and hence, two difficulty parameters. By changing which difficulty parameter to use for predictions, we compared the following two settings. The first method uses the multiple-choice question difficulty estimated from the responses of only 100 learners in Fig. 2. The second method uses the self-report question difficulty estimated from all responses.

According to the experimental results, the first method obtained 0.724, whereas the second method obtained 0.697 for predicting the accuracy of the dashed area. The difference between these results was statistically significant according to the Wilcoxon test ($p < 0.01$). This result indicates that, to predict responses to multiple-choice questions, we should rely on the item parameters estimated from the responses of other test-takers rather than the self-report parameters estimated from the responses of all test-takers.

BERT-Based Methods: To predict the dashed areas in Fig. 2, we also want to use deep transfer learning methods such as bidirectional encoder representations from transformers (BERT) [3] because they reportedly achieved high predictive

Table 2. Summary of the accuracy of each method in predicting if each of the 91 learners in Fig. 2 correctly/incorrectly answered the multiple-choice questions using the filled areas in Fig. 2 as the training data. The bold is statistically significantly better than the underlined, as denoted by (**) ($p < 0.01$). "Abl." denotes "ability parameter", and "difficl." denotes "difficulty parameter".

Method	Accuracy
IRT (Abl. - Self-report difficl. estimated from 191 learners)	0.675
IRT (Abl. - Multiple-choice diffcl. estimated from 100 learners)	<u>0.724</u>
BERT (bert-base-uncased) fine-tuned with test-taker tokens	0.718
BERT (bert-base-cased) fine-tuned with test-taker tokens	0.693
BERT (bert-large-uncased) fine-tuned with test-taker tokens	0.722
BERT (bert-large-cased) fine-tuned with test-taker tokens	**0.729** (**)

performances. However, while BERT can take only a natural language sequence as input, in our task setting, the predictions for different learners must consider learners' traits such as abilities. To this end, we converted our dataset into the sequence classification task, which BERT can handle, by introducing two types of *special tokens* as depicted in Fig. 3: [**USRn**] and [**SQ**].

[**USRn**], where n in replaced by the test-taker ID, represents each test-taker or learner (user). Placed at the beginning of the sequence, this token specifies that the classifier should predict the response of the test-taker specified by this token. The example in Fig. 3 implies that we want to predict the response of **USR3** for the sequence "It is a difficult period." In this case, because **USR3** answered the multiple-choice question incorrectly, the label for the multiple-choice question was 0. The option that **USR3** selected incorrectly, namely "question" in the example of Fig. 3, was ignored in the sequence. This was intentionally done for a fairly accurate comparison, as the IRT-based methods also did not consider which *incorrect* option, or distractor, the test-taker selected. There are as many test-taker tokens as the number of test-takers involved.

The sequence classifiers also need to be able to handle a self-report testing. For this purpose, we used the special token [**SQ**], which denotes a self-report question. Placing this token immediately after the [**USRn**] token would let the classifier know that this was a self-report vocabulary question, such that only the token following [**SQ**] would be considered. As this test-taker answered "yes" for the word period in the self-report test, the label is 1 in Fig. 3.

We implemented our method using **BertForSequenceClassification** function in the **transfomers** library (https://github.com/huggingface/transformers), which is a standard library that implements BERT [3] and other transformer-based transfer learning models. We used **bert-base/large-uncased/cased**, downloadable from the **transformers** library, for our pre-trained models. The special tokens were added using the **add_tokens** function. Following conversion, we simply used **BertForSequenceClassification** for fine-tuning to build the sequence classifiers. The Adam optimizer [5] was used to fine-tuning. The learning rate was set to 10^{-5}. The number of epochs was set to nine for all models.

The accuracy of this BERT-based method was slightly but statistically significantly superior to that of the IRT-based method with an accuracy of 0.729 (Wilcoxon test, $p < 0.01$) at the 9th epoch. This result indicates that although the task appeared to not be considerably dependent on the context of the vocabulary questions, its accuracy can still be improved by considering the contexts. The final results are summarized in Table 2. We can observe that BERT (bert-large-cased) with test-taker tokens achieved the best statistically significant accuracy in our experiments. The lower accuracy values of **bert-base-cased/uncased** show that the size of the pre-trained models is important for achieving high prediction accuracy.

3 Conclusions

In this study, we developed a dataset to compare multiple-choice and self-report questions for predicting costly actual test results. Experimental results demonstrated that the best IRT-based method achieved an accuracy slightly but statistically significantly lower than that of the BERT-based method. This result suggests that, by considering the semantic contexts of the words in vocabulary questions, the accuracy can be improved although their lengths are typically short. Our future work includes additional analysis to investigate how the short contexts contribute to achieving high predictive performances.

Acknowledgements. This work was supported by JST ACT-X Grant Number JPMJAX2006, Japan. We used the AIST ABCI infrastructure and RIKEN miniRaiden system for computational resources.

References

1. Baker, F.B.: Item Response Theory: Parameter Estimation Techniques, Second Edition. CRC Press, July 2004
2. Beinborn, L., Zesch, T., Gurevych, I.: Predicting the difficulty of language proficiency tests. TACL **2**, 517–530 (2014)
3. Devlin, J., Chang, M.W., Lee, K., Toutanova, K.: BERT: pre-training of deep bidirectional transformers for language uderstanding. In: Proceedings of NAACL (2019)
4. Ehara, Y.: Building an English vocabulary knowledge dataset of Japanese English-as-a-second-language learners using crowdsourcing. In: Proceedings of LREC, May 2018
5. Kingma, D.P., Ba, J.: Adam: A method for stochastic optimization. In: Proceedings of ICLR (2015)
6. Laufer, B., Ravenhorst-Kalovski, G.C.: Lexical threshold revisited: lexical text coverage, learners' vocabulary size and reading comprehension. Reading Foreign Lang. **22**(1), 15–30 (2010). https://eric.ed.gov/?id=EJ887873
7. Nation, I.: How large a vocabulary is needed for reading and listening? Canad. Mod. Lang. Rev. **63**(1), 59–82 (2006)

8. Nation, I., Beglar, D.: A vocabulary size test. Lang. Teach. **31**(7), 9–13 (2007)
9. O'Dell, F., Read, J., McCarthy, M., et al.: Assessing vocabulary. Cambridge University Press (2000)
10. Wesche, M., Paribakht, T.S.: Assessing second language vocabulary knowledge: depth versus breadth. Can. Mod. Lang. Rev. **53**(1), 13–40 (1996)

Investigating Student Interest and Engagement in Game-Based Learning Environments

Jiayi Zhang[1](✉), Stephen Hutt[1], Jaclyn Ocumpaugh[1], Nathan Henderson[2], Alex Goslen[2], Jonathan P. Rowe[2], Kristy Elizabeth Boyer[3], Eric Wiebe[2], Bradford Mott[2], and James Lester[2]

[1] University of Pennsylvania, Philadelphia, USA
joycez@upenn.edu
[2] North Carolina State University, Raleigh, USA
[3] University of Florida, Gainesville, USA

Abstract. As a cognitive and affective state, interest promotes engagement, facilitates self-regulated learning, and is positively associated with learning outcomes. Research has shown that interest interacts with prior knowledge, but few studies have investigated these issues in the context of adaptive game-based learning environments. Using three subscales from the User Engagement Scale, we examine data from middle school students (N = 77) who interacted with Crystal Island in their regular science class to explore the relationship between interest, knowledge, and learning. We found that interest is significantly related to performance (both knowledge assessment and game completion), suggesting that students with high interest are likely to perform better academically, but also be more engaged in the in-game objectives. These findings have implications both for designers who seek to identify students with lower interest and for those who hope to create adaptive supports.

Keywords: Interest · Science learning · Learning technology

1 Introduction

Interest is a construct with both cognitive and affective components [1] that has been repeatedly found to influence learning [2]. It is known to affect student attention and self-regulation [3], and it has been found to motivate student engagement with science content and practices [4].

However, more careful attention to the types of interest is needed, as these may be critical for fosternig equitable learning outcomes [5]. For example, Hidi & Renninger [4] describe a four-phase model of interest development that distinguishes between triggered situational interest, maintained situational interest (sustained over time), emerging individual interest, and well-developed individual interest (which endures regardless of context). Students who have not developed individual interest likely need more extrinsic rewards and stimulation to trigger situational interest [6]. Thus, understanding how

M. M. Rodrigo et al. (Eds.): AIED 2022, LNCS 13355, pp. 711–716, 2022.
https://doi.org/10.1007/978-3-031-11644-5_72

interest emerges—and how it relates to student learning behaviors within an online system—could lead to improved learning designs and more effective adaptive systems.

This paper investigates students' situational interest and engagement in a game-based learning environment for middle school science. We combine survey measures of these constructs with student knowledge assessments and interaction logs to explore potential relationships, showing how interest is related to knowledge and engagement.

1.1 Related Literature

Interest and curiosity are both constructs that facilitate learning [7]. Both have factors related to cognition, affect, and the desire to close a knowledge gap [6]. Though not universally recognized as separate constructs (see [7]), interest measures tend to address content-related factors. Similarly, curiosity may reflect an immediate knowledge deficit (aligning with situational interest) as opposed to a long-term propensity to re-engage with the topic at hand (individual interest). Hidi & Reninger's four-phase model of interest development [7] describes two phases of situational interest and two phases of individual interest. As learners progress from Phase 1 to 4, they become increasingly motivated to re-engage with the topic without needing external support.

In science learning, interest is associated with intrinsically motivated engagement [4], and behavioral engagement with science in non-academic contexts [8]. Interested students are also more likely to engage in self-regulated learning, showing increased attention and better goal-setting abilities [3].

Dimensions of Interest. Studies grounded in different theoretical frameworks operationalize interest in different ways, leading to measures that do not always align. In general, however, researchers tend to agree that interest is driven both by cognitive and affective processes [1]. That is, even in the early stages of situational interest, students experience curiosity, or the desire to close an information gap. While this experience may sometimes be frustrating, by the time students have achieved a well-developed individual interest (i.e., [4]'s fourth phase), we might expect students to regulate their emotions well enough to maintain a flow-like state. Not surprisingly, this development coincides with increased knowledge. Zhang et al.'s study [9] of middle school science found prior knowledge slows the decline in interest and facilitates the growth of interest in more knowledgeable students. Additionally, prior knowledge interacts with interest predicting the level of conceptual change [10]. In other words, core components of individual interest are increasing curiosity (the desire to close knowledge gaps) and sustained affective engagement. Subject knowledge accumulates as students grow from situational interest into a more sustained form of interest. Yet, prior knowledge likely drives the kinds of questions a student is capable of asking and therefore is a necessary ingredient (and not just a biproduct) in the later stages of interest.

Interest in Game-Based Learning. Previous work has sought to make connections between the research on interest and the research on student engagement [11]. In particular, researchers have considered how game-like elements trigger students' situational interest [12]. Such investigations can lead to adapting learning technologies to promote and sustain student science interest [13], which can be accomplished by personalizing questions [14] and feedback [15].

1.2 Current Study

This study investigates interest using data from an inquiry-based learning game for middle school microbiology, Crystal Island [11]. The analyses use three scales of the User Engagement Survey [16] to operationalize the cognitive and affective engagement aspects of student interest. Specifically, we examine the relationship between these scales and student performance measures (both external knowledge assessments and game completion). The findings are relevant for the design of learning technologies that can adapt to student interest.

2 Methods

Research was conducted using Crystal Island, a game-based learning environment for middle school microbiology that supplements classroom instruction by combining inquiry-based learning and direct instruction. The first-person, single-player game places students in a research camp on a remote island where a mysterious infectious disease has caused widespread illness [11]. Students play the role of a medical detective tasked with identifying the disease and its transmission source. Students must navigate the island, gather information, form hypotheses, conduct tests, and synthesize their findings to solve the mystery. As they do, they interact with non-player characters and virtual objects, including posters, research articles, and books that impart knowledge about microbiology and specific information about the mysterious disease.

2.1 Data Collection

Gameplay took place in a middle school science class in the southeastern US, as previously reported in [17], who sought to detect and prevent dialogue breakdown with an non-player characher. Interaction data was collected as 92 students used the game over three days or until they completed the game. An identical pretest and posttest on microbiology were given at the start and end of the study. To account for prior knowledge we computed normalized learning gain using the method described in [18]. Students with incomplete surveys were excluded, resulting in 77 students analyzed.

2.2 Survey Measures of Interest and Engagement

Three survey scales (collected immediately after students use the program) were used as a proxy for the related constructs of interest and engagement. These were drawn from the original version of the User Engagement Scale (UES; [19]) and a revised version (UESz, re-validated specifically in a video-game environment; [16]). We focus on the three scales (Table 1). The Novelty (NO) scale measures students' interest and curiosity in the game, while the others measure students' engagement. In fact, Focused Attention (FAz) is strongly correlated with the Flow State Scale [20], modeled after Csikszentmihalyi's original conception of flow [16]. There is a one-item overlap between the FAz and the Felt Involvement (FI) scale, which had initially been characterized as capturing the enjoyment and interest of the gameplay experience. In summary, NO appears to capture a basic measure of situational interest, FAz captures flow-like engagement, and FI might be described as the enjoyment at the intersection of those two constructs.

Table 1. UES and UESz subscales used to operationalize student interest and engagement

Scale name/Construct	Items
Novelty (NO): used to operationalize situational interest/curiosity	I continued to play the game out of curiosity
	The content of the game incited my curiosity
	I felt interested in the game
Felt Involvement (FI): used as secondary measure of flow/engagement	I was really drawn into the game*
	I felt involved in the game
	The gaming experience was fun
Focused Attention (FAz): used to operationalize flow/engagement	I lost myself in this gaming experience
	I was so involved in the game that I lost track of time
	I blocked out things around me when I was playing the game
	When I was playing the game, I lost track of the world around me
	The time I spent playing the game just slipped away
	I was absorbed in the game
	During the gaming experience I let myself go
	I was really drawn into the game*

3 Results

We first examine the Spearman correlations between interest survey scales and external knowledge assessments (pretest, posttest and normalized learning gain) and then use t-tests to compare these scales to game completion. Table 2 shows 10 significant positive correlations among the knowledge assessments and the survey scales. Specifically, pretest is associated with all three survey scales, and posttest is associated with NO and FI. Learning gain was not significantly correlated with any of the survey scales and is only related to the posttest (but not the pretest).

We next considered a knowledge measure internal to the game, namely completing the game by solving the mystery. Game completion provides a holistic measure of students' in-game achievement while also demonstrating the extent to which their behavioral engagement aligned with the goals of the learning task. To solve the mystery, students must both acquire relevant science knowledge while also engaging in a series of experiments and scientific reasoning processes to derive a conclusion. As the measure was binary, Welch two sample t-test was conducted to examine any difference between each of the interest measures. Cohen's d was used to test the effect size [21]. Results show that students who solved the mystery ($N = 42$) reported higher values for all three interest scales (FAz: $t(74.72) = -2.47$, $p = .016$, $d = -0.55$; NO: $t(74.68) = -2.21$, $p = .030$, $d = -0.50$; and FI: $t(75.00) = -2.59$, $p = .012$, $d = -0.58$).

Table 2. Correlations between Knowledge Assessments and Survey Measures

Variable	*Mean*	*SD*	1	2	3	4	5
1. PreTest	6.55	2.74					
2. PostTest	6.79	3.12	.73**				
3. Learning Gain	0.03	0.37	.005	.64**			
4. Focused Attention (FAz)	24.35	7.36	.26*	.16	−.03		
5. Novelty (NO)	10.34	3.01	.23*	.23*	.11	.77**	
6. Felt Involvement (FI)	10.51	3.05	.28*	.29**	.12	.83**	.89**

Note. SD = standard deviation; * = $p < .05$; ** = $p < .01$

4 Discussion and Conclusions

Understanding the relationship between interest and behavior can help developers create additional game features to promote situational interest and tackle cognitive and behavioral disengagement. Our results suggest that future development of learning games should consider measuring interest explicitly and comparing that to real-time student patterns and feedback so that adaptive technologies can match game challenge to interest, scaffolding the latter.

Specifically, we find that interest and engagement measures are positively correlated with a student's science content knowledge. This result is in accordance with prior work showing a symbiotic relationship between interest and content knowledge. In this study, students with higher knowledge of microbiology showed higher interest in the game. While correlation cannot imply causality, this finding contributes to ongoing debates surrounding knowledge and interest's reciprocal development. Likewise, students with high interest were more likely to solve the mystery and thus complete the game. Game completion speaks both to student knowledge and also to their broader engagement, since post-test measures show improvement even among students who did not complete the game. This finding implies that those with higher interest, in addition to learning more, were more motivated/engaged to complete the objectives of the game.

Future work should consider ways in which surveys of constructs like interest align with student behaviors in other adaptive learning systems. For example, this study used retrospective UES scales to measure student interest and engagement, but measures designed to capture situational interest *in situ* or to capture interest more broadly (e.g., IMI, [22]) might produce different results. Future work should also explore new ways to connect to students' existing prior knowledge and interests. That is, this study asked specifically about the interest students have in the game they were presented with (in line with a substantial body of literature on interest and engagement), but did not ask about students' interests outside the game, a method supported by a growing body of research on the relationship between interest and prior knowledge.

As we continue to develop AI-based learning technologies, we should consider ways to adapt and respond to student assets (e.g., engagement or interest), rather than deficits (such as disengagement). Responding directly to student interest and prior knowledge appears to be a critical step in that process.

References

1. Hidi, S., et al.: Interest, a motivational variable that combines affective and cognitive functioning (2004)
2. Ainley, M., et al.: Interest, learning, and the psychological processes that mediate their relationship. J. Educ. Psychol. **94**, 545 (2002)
3. Hidi, S., Ainley, M.: Interest and self-regulation: relationships between two variables that influence learning (2008)
4. Hidi, S., Renninger, K.A.: The four-phase model of interest development. Educ. Psychol. **41**, 111–127 (2006)
5. Renninger, K.A., Hidi, S.: Student interest and achievement: developmental issues raised by a case study. In: Development of Achievement Motivation, pp. 173–195. Elsevier (2002)
6. Renninger, K.A., Hidi, S.E.: To level the playing field, develop interest. Policy Insights Behav. Brain Sci. **7**, 10–18 (2020)
7. Hidi, S.E., Renninger, K.: Interest development and its relation to curiosity: needed neuroscientific research. Educ. Psychol. Rev. **31**, 833–852 (2019)
8. Fortus, D., Vedder-Weiss, D.: Measuring students' continuing motivation for science learning. J. Res. Sci. Teach. **51**, 497–522 (2014)
9. Zhang, T., et al.: Prior knowledge determines interest in learning in physical education: a structural growth model perspective. Learn. Individ. Differ. **51**, 132–140 (2016)
10. Linnenbrink-Garcia, L., et al.: Measuring situational interest in academic domains. Educ. Psychol. Measur. **70**, 647–671 (2010). https://doi.org/10.1177/0013164409355699
11. Rowe, J.P., et al.: Integrating learning, problem solving, and engagement in narrative-centered learning environments. Int. J. Artif. Intell. Educ. **21**, 115–133 (2011)
12. Rowe, J.P., et al.: Off-task behavior in narrative-centered learning environments. In: AIED, pp. 99–106 (2009)
13. Darling-Hammond, L., et al.: Implications for educational practice of the science of learning and development. Appl. Dev. Sci. **24**, 97–140 (2020)
14. Bernacki, M.L., Walkington, C.: The role of situational interest in personalized learning. J. Educ. Psychol. **110**, 864 (2018)
15. Koenka, A.C., Anderman, E.M.: Personalized feedback as a strategy for improving motivation and performance among middle school students. Middle Sch. J. **50**, 15–22 (2019)
16. Wiebe, E.N., et al.: Measuring engagement in video game-based environments: investigation of the user engagement scale. Comput. Hum. Behav. **32**, 123–132 (2014)
17. Min, W., et al.: Multimodal goal recognition in open-world digital games. In: Thirteenth Artificial Intelligence and Interactive Digital Entertainment Conference (2017)
18. Vail, A.K., Grafsgaard, J.F., Boyer, K.E., Wiebe, E.N., Lester, J.C.: Predicting learning from student affective response to tutor questions. In: Micarelli, A., Stamper, J., Panourgia, K. (eds.) ITS 2016. LNCS, vol. 9684, pp. 154–164. Springer, Cham (2016). https://doi.org/10.1007/978-3-319-39583-8_15
19. O'Brien, H.L., Toms, E.G.: The development and evaluation of a survey to measure user engagement. J. Am. Soc. Inform. Sci. Technol. **61**, 50–69 (2010)
20. Jackson, S.A., Marsh, H.W.: Development and validation of a scale to measure optimal experience: the flow state scale. J. Sport Exerc. Psychol. **18**, 17–35 (1996)
21. Cohen, J.: Statistical Power Analysis for the Behavioral Sciences. Taylor & Francis (2013)
22. McAuley, E., et al.: Psychometric properties of the intrinsic motivation inventory in a competitive sport setting: a confirmatory factor analysis. Res. Q. Exerc. Sport **60**, 48–58 (1989)

Adopting Automatic Machine Learning for Temporal Prediction of Paid Certification in MOOCs

Mohammad Alshehri(✉), Ahmed Alamri, and Alexandra I. Cristea

Department of Computer Science,
Durham University, Lower Mountjoy, South Rd, Durham D1 3LE, UK
{mohammad.a.alshehri,ahmed.s.alamri,
alexandra.i.cristea}@durham.ac.uk

Abstract. Massive Open Online Course (MOOC) platforms have been growing exponentially, offering worldwide low-cost educational content. Recent literature on MOOC learner analytics has been carried out around predicting either students' dropout, academic performance or students' characteristics and demographics. However, predicting MOOCs certification is significantly underrepresented in literature, despite the very low level of course purchasing (less than 1% of the total number of enrolled students on a given online course opt to purchase its certificate) and its financial implications for providers. Additionally, the current predictive models choose conventional learning algorithms, randomly, failing to finetune them to enhance their accuracy. Thus, this paper proposes, for the first time, *deploying automated machine learning (AutoML) for predicting the paid certification in MOOCs.* Moreover, it uses a temporal approach, with prediction based on first-week data only, and, separately, on the first half of the course activities. Using 23 runs from 5 courses on FutureLearn, our results show that the AutoML technique achieves promising results. We conclude that the dynamicity of AutoML in terms of automatically finetuning the hyperparameters allows to identify the best classifiers and parameters for paid certification in MOOC prediction.

Keywords: MOOCs · Certification prediction · AutoML · Auto-sklearn

1 Introduction

Online courses have been revolutionising and reforming education for decades. More recently, massive open online courses (MOOCs) were explicitly introduced, to democratise access to education and reach a massively unlimited number of potential learners from around the world. The first official emergence of MOOCs was with the launch of Stanford's Coursera in 2011 [1, 2], although the following year was coined as "the year of the MOOCs" when many of today's successful platforms, such as *FutureLearn*, *edX*, *Udemy* and *Coursera* went live [3, 4], offering scalable world-wide online courses to the public [5, 6].

M. M. Rodrigo et al. (Eds.): AIED 2022, LNCS 13355, pp. 717–723, 2022.
https://doi.org/10.1007/978-3-031-11644-5_73

Although MOOCs have been successful, attracting many online learners, the staggeringly *low completion and certification rates* are still one of the more concerning aspects to date, a funnel with students "leaking out" at various points along the learning pathway [7, 8]. While the high dropout rate has been the focus of many studies, the race towards identifying precise predictors of completion as well as the *predictors of course purchasing* continues. Importantly, although MOOCs have started being analysed more thoroughly in the literature, few studies have investigated the characteristics and temporal activities for modelling learners' certification decision behaviours.

Another objective this study attempts to address is examining the extent to which AutoML can help achieve competitive performance in predicting certification in MOOCs. With machine learning becoming more mainstream in data science, there has been an increasing demand for automated tools that can automate the process of designing and optimising machine learning pipelines with less human intervention [9]. In response to this demand, many AutoML frameworks have been introduced [10–12].

Considering the recent MOOCs' transition towards paid macro-programmes and online degrees, with affiliate university partners, along with the advancements in the automation and explanation of learners' activities prediction, this paper presents an automated predictor of MOOC paid certification. Specifically, this paper attempts to answer the following research question:

- *To what extent can AutoML predict MOOC learners' purchase decisions (certification)?*

It is worth mentioning that the contribution of this study goes beyond randomly comparing different classifiers for predicting paid certifications in MOOCs to *proposing a stable, comprehensive automated model for dynamically optimising hyperparameters during the learning process*. Additionally, we are *investigating the classification performance temporally*, using different periods (early and middle) during the course. This is the *first study that employs AutoML to predict paid certification in MOOCs* to the best of our knowledge.

2 Related Works

While several studies have predicted learners' behaviours in MOOCs, the number of studies that use AutoML for this purpose remains relatively low. Concerning the previous studies that used AutoML to predict or classify learners, [13] investigates the potential of Auto-Weka (one of the standard AutoML systems) in early predicting learning outcomes (pass/fail) based on learners' participation in the Moodle e-learning platform. The study limited the experiment to tree-based and rule-based models for more transparent and interpretable results, using data from 591 students over 3 courses. For the purpose of initial comparison, one predictor of each main category of learners (Bayes classifiers, rule-based, tree-based, function-based, lazy and meta classifiers) has been randomly chosen to compare against Auto-Weka performance. The results show that the latter significantly achieved better results on the classification task.

[14] proposes a generic automated weak supervision framework (AutoWeakS), using reinforcement learning, to build a MOOC course recommender for job seekers. The framework allows training multiple supervised ranking models and automatically searching for the best combination of supervised and unsupervised models. With experiments on 1951 course descriptions of different disciplines obtained from Xuetang X[1], a Chinese MOOC platform, the model significantly outperforms the classical unsupervised, supervised and weak supervision baseline.

Recently, [15] assisted the impact of adopting an AutoML strategy on feature engineering, model selection, and hyperparameters tuning in predicting student success. The researchers replicated a previous experiment to involve hyperparameter tuning via an AutoML technique for hyperparameter tuning with the data cleaning, preprocessing, feature engineering and time segmentation approach from the previous experiment as-is. The study showed significant general improvement, with specific classifiers (Decision Tree, Extra Tree, Random Forest) performing the best. This is another indicator that AutoML can outperform even carefully planned educational prediction models. However, none of the previous works has addressed the issue of the low certification rate in MOOCs using AutoML. Unlike previous studies, our proposed model aims to predict the financial decisions of learners on whether to *purchase* the course certificate. Also, our work is applied to a less frequently studied platform, FutureLearn [16, 17]. Our study additionally identifies the most representative factors for certification purchase prediction. It also proposes an AutoML-based collection of tree-based and regression classifiers to predict MOOC purchasability using relatively few input features.

3 Methodology

3.1 Data Collection and Preprocessing

The current study is analysing data extracted from a total of 23 runs spread over 5 MOOC courses, on four distinct topic areas, all delivered through FutureLearn, by the University of Warwick. These topic areas are Literature, Psychology, Computer Science and Business [18].

These courses were delivered repeatedly in consecutive years (2013–2017); thus, we have data on several '*runs*' for each course.

The dataset obtained went through several processing steps to be prepared and fed into the learning model. Since some students were enrolled on more than one run of the same course, the run number was attached to the student's ID, to avoid any mismatch during joining student activities over "several runs" with their current activities. Additionally, we eliminated irrelevant data generated by organisational administrators (455 administrators across the 23 runs analysed) and applied other standard preprocessing.

[1] http://www.xuetangx.com.

3.2 AutoML Systems

The fundamental purpose of AutoML systems is reducing human intervention via automating feature preprocessing, hyperparameters finetuning and best-performing algorithm selection, with the ultimate goal of maximising classification accuracy on a supervised classification task [9]. Auto-sklearn is a scikit-learn-based framework that uses 15 classifiers, 14 feature preprocessing methods, and 4 data preprocessing methods, giving rise to a structured hypothesis space with 110 hyperparameters. It improves on other existing AutoML methods, by automatically considering the past performance on similar datasets and constructing ensembles from the models evaluated during the optimisation.

3.3 Setting the Auto-sklearn Hyper Parameters

Although AutoML systems automatically optimise pipelines with less human intervention, there are some Auto-sklearn-specific hyperparameters that master the overall learning process and already have default values for a higher level of automation. However, these parameters can be manually finetuned to further improve the pipeline's performance.

After training and testing the models, Auto-sklearn automatically nominated the best performing models beside a set of hyper-parameters for each one of the five courses. Our best performing classifiers include *Bernoulli_nb, Adaboost, Extra_trees, Decision_tree, Libsvm_svc (C-Support Vector Classification), Random_forest, Linear Discriminant Analysis (LDA), Gradient_boosting, Multinominal_nb, Passive_aggressive* and *Sgd (stochastic gradient descent) learning.*

4 Results and Discussion

We demonstrate that using the AutoML technique, each dataset has its own features, and thus, even the most common classifiers adopted among MOOC researchers may not be the best performing on each dataset. Our previous experiment [4], using the most common classifiers, has reached satisfactory results. However, the results below outperform the current MOOC paid certification state-of-art and introduce a promising approach to adopting AutoML in modelling learner behaviour prediction in MOOCs.

Table 1 shows the results of AutoML-based predicting certification using the first-week-only logged data, versus the first half of the course. It can be seen that, although some course results, such as Supply Chain (SC), were relatively high, the differences in recalls of class 0 and class 1 are high across the five courses. This means that the model is highly biased towards class 1; hencc the first-week data may not accurately predict certification.

Also, it can be seen that the performance improved between 1% to 9% across the five courses, when further data were added. The SC course has shown the lowest improvement. Nevertheless, both class recalls participated almost equally in the second experiment. It also can be seen that the gap between the recalls of the two classes has shrunk when further weekly activities have been included.

Table 1. Best optimised pipelines of Auto-sklearn, distributed by course, using the first-week-only activities, versus the first half of the course; class 0 = non-paying learners, class 1 = paid learners; metrics rounded to 2 decimal places.

C	Classifier	1st Week only			Classifier	1st Half of the Course		
		Rec_0	Rec_1	BA		Rec_0	Rec_1	BA
BIM	*Ber_NB*	0.63	0.95	0.78	*AdaBoost*	0.78	0.9	0.84
	AdaBoost	0.62	0.95	0.78	*RF*	0.78	0.9	0.84
	EXT	0.6	0.95	0.77	*DT*	0.8	0.89	0.84
	DT	0.6	0.95	0.77	*LIBSVM_SVC*	0.8	0.89	0.84
BD	*AdaBoost*	0.76	1.00	0.88	*RF*	0.87	0.98	0.92
	LIBSVM_SVC	0.77	0.98	0.88	*DT*	0.86	0.98	0.92
	RF	0.77	0.98	0.87	*EXT*	0.86	0.98	0.92
	DT	0.75	1.00	0.87	*GrBoost*	0.86	0.98	0.92
SC	*EXT*	0.84	1.00	0.92	*LIBSVM_SVC*	0.9	0.93	0.92
	RF	0.84	1.00	0.92	*PA*	0.9	0.93	0.92
	LDA	0.83	1.00	0.91	*GrBoost*	0.9	0.93	0.92
	GrBoost	0.82	1.00	0.91	*RF*	0.9	0.93	0.91
SP	*DT*	0.59	0.99	0.79	*RF*	0.79	0.97	0.88
	Mul_NB	0.57	1.00	0.78	*Ber_NB*	0.79	0.97	0.88
	PA	0.56	1.00	0.78	*PA*	0.79	0.97	0.88
	LIBSVM_SVC	0.56	1.00	0.78	*LIBSVM_SVC*	0.79	0.97	0.88
TMF	*EXT*	0.68	0.98	0.83	*EXT*	0.83	0.95	0.89
	PA	0.68	0.98	0.83	*LIBL_SVC*	0.82	0.96	0.89
	DT	0.68	0.98	0.83	*Ber_NB*	0.82	0.96	0.89
	SGD	0.68	0.98	0.83	*LIBSVM_SVC*	0.82	0.96	0.89

5 Conclusion and Future Work

There are few studies on using AutoML techniques to predict MOOC learners' activities. Thus, this paper *proposes, for the first time, automated machine learning (AutoML) for predicting paid certification in MOOCs*. Our results show that the AutoML technique achieved better results than the traditional approach of randomly selecting best-in-class predictive algorithms. In our subsequent work, we will further investigate the reason behind having different classifiers in each one of the temporal scenarios. It is known that each classifier initially has its own capability based on the data fed (here, the number of weekly features), but a deeper investigation of a range of parameters configuration is needed, in order to understand these varying results.

References

1. Ng, A., Widom, J.: Origins of the Modern MOOC (xMOOC). InL Hollands, H.F.M., Tirthali, D. (eds.) MOOCs: Expectations and Reality: Full Report, pp. 34–47 (2014)
2. Alshehri, M., Alamri, A., Cristea, A.I., Stewart, C.D.: Towards designing profitable courses: predicting student purchasing behaviour in MOOCs. Int. J. Artif. Intell. Educ. **31**(2), 215–233 (2021). https://doi.org/10.1007/s40593-021-00246-2
3. Gardner, J., Brooks, C.: Student success prediction in MOOCs. User Model. User-Adap. Inter. **28**(2), 127–203 (2018). https://doi.org/10.1007/s11257-018-9203-z
4. Alshehri, M., Alamri, A., Cristea, A.I.: Predicting certification in MOOCs based on students' weekly activities. In: Cristea, A.I., Troussas, C. (eds.) Intelligent Tutoring Systems: 17th International Conference, ITS 2021, Virtual Event, June 7–11, 2021, Proceedings, pp. 173–185. Springer International Publishing, Cham (2021). https://doi.org/10.1007/978-3-030-80421-3_20
5. Alamri, A., et al.: Predicting MOOCs dropout using only two easily obtainable features from the first week's activities. In: Coy, A., Hayashi, Y., Chang, M. (eds.) Intelligent Tutoring Systems: 15th International Conference, ITS 2019, Kingston, Jamaica, June 3–7, 2019, Proceedings, pp. 163–173. Springer International Publishing, Cham (2019). https://doi.org/10.1007/978-3-030-22244-4_20
6. Cristea, A.I., et al.: Earliest predictor of dropout in MOOCs: a longitudinal study of FutureLearn courses. In: Association for Information Systems (2018)
7. Clow, D.: MOOCs and the funnel of participation. In: Proceedings of the Third International Conference on Learning Analytics and Knowledge. ACM (2013)
8. Breslow, L., et al.: Studying learning in the worldwide classroom research into edX's first MOOC. Res. Pract. Assess. **8**, 13–25 (2013)
9. Olson, R.S., Moore, J.H.: TPOT: A tree-based pipeline optimization tool for automating machine learning. In: Workshop on Automatic Machine learning. PMLR (2016)
10. Bergstra, J., Yamins, D., Cox, D.D.: Hyperopt: a python library for optimizing the hyperparameters of machine learning algorithms. In: Proceedings of the 12th Python in Science Conference. Citeseer (2013)
11. Kotthoff, L., Thornton, C., Hoos, H.H., Hutter, F., Leyton-Brown, K.: Auto-WEKA: automatic model selection and hyperparameter optimization in WEKA. In: Hutter, F., Kotthoff, L., Vanschoren, J. (eds.) Automated Machine Learning. TSSCML, pp. 81–95. Springer, Cham (2019). https://doi.org/10.1007/978-3-030-05318-5_4
12. Feurer, M., Klein, A., Eggensperger, K., Springenberg, J.T., Blum, M., Hutter, F.: Auto-sklearn: efficient and robust automated machine learning. In: Hutter, F., Kotthoff, L., Vanschoren, J. (eds.) automated machine learning. TSSCML, pp. 113–134. Springer, Cham (2019). https://doi.org/10.1007/978-3-030-05318-5_6
13. Tsiakmaki, M., et al.: Implementing AutoML in educational data mining for prediction tasks. Appl. Sci. **10**(1), 90 (2020)
14. Hao, B., et al.: Recommending Courses in MOOCs for Jobs: An Auto Weak Supervision Approach. arXiv preprint arXiv:2012.14234 (2020)
15. Drăgulescu, B., Bucos, M.: Hyperparameter tuning using automated methods to improve models for predicting student success. In: Lopata, A., Butkienė, R., Gudonienė, D., Sukackė, V. (eds.) Information and Software Technologies: 26th International Conference, ICIST 2020, Kaunas, Lithuania, October 15–17, 2020, Proceedings, pp. 309–320. Springer International Publishing, Cham (2020). https://doi.org/10.1007/978-3-030-59506-7_25
16. Cristea, A.I., et al.: How is learning fluctuating? FutureLearn MOOCs fine-grained temporal analysis and feedback to teachers (2018)

17. Cristea, A.I., et al.: Can learner characteristics predict their behaviour on MOOCs? In: 10th international conference on education technology and computers (ICETC 2018), Tokyo Institution Technology, Tokyo, Japan. Association Computing Machinery (2018)
18. Alshehri, M., et al.: On the need for fine-grained analysis of Gender versus Commenting Behaviour in MOOCs. In: Proceedings of the 2018 the 3rd International Conference on Information and Education Innovations. ACM (2018)

Raising Student Completion Rates with Adaptive Curriculum and Contextual Bandits

Robert Belfer[1(✉)], Ekaterina Kochmar[1,2], and Iulian Vlad Serban[1]

[1] Korbit Technologies Inc., Montreal, Canada
robert@korbit.ai
[2] University of Bath, Bath, England

Abstract. We present an adaptive learning Intelligent Tutoring System, which uses model-based reinforcement learning in the form of contextual bandits to assign learning activities to students. The model is trained on the trajectories of thousands of students in order to maximize their exercise completion rates and continues to learn online, automatically adjusting itself to new activities. A randomized controlled trial with students shows that our model leads to superior completion rates and significantly improved student engagement when compared to other approaches. Our approach is fully-automated unlocking new opportunities for learning experience personalization.

Keywords: ITS · Contextual bandits · LinUCB · Personalized learning

1 Introduction

Intelligent Tutoring Systems (ITS) aim to provide personalized tutoring in a computer-based environment and are capable of selecting problems on an individual basis [14]. Many ITS consider the development of personalized curricula: a recommended sequence of learning activities adapted in real-time to the needs of each individual student [1,16]. Investigating novel methods for developing personalized curricula that can adapt to millions of students and thousands of courses or domains in real-time is key to further improvements in the learning experience of students interacting with ITS.

We present `Korbit`, an adaptive learning ITS leveraging reinforcement learning (RL) in order to automatically assign learning activities to students. `Korbit` is an online learning platform, where students follow a blended-learning framework combining problem-solving activities, lecture videos, Socratic tutoring and project-based learning [17]. We focus on ordering the text-based problem-solving activities, which students can answer in free-form text or as multiple-choice questions (MCQs). If the student answers correctly, they move on to a different exercise; otherwise, they are given feedback and may try again or skip the

M. M. Rodrigo et al. (Eds.): AIED 2022, LNCS 13355, pp. 724–730, 2022.
https://doi.org/10.1007/978-3-031-11644-5_74

activity and move on. The ordering of exercises and all other activities within the same continuous topic (called *learning unit*) is determined by a model-based RL system employing the LinUCB algorithm [10].

The main contributions of this paper are two-fold: (1) we present the design and implementation of a model-based RL system for ordering learning activities based on the LinUCB algorithm; (2) we evaluate this model in a randomized controlled trial and show that it attains superior completion rates and improved student engagement when compared to alternative approaches.

2 Background

Personalization is key to effective learning [2,4]. In computer-based learning environments (CBLEs), ITS have been shown to dramatically improve student learning outcomes and engagement [8,18] due to their ability to address individual needs and develop personalized feedback [1,5,13]. One of the most powerful families of algorithms deployed in CBLEs are RL algorithms,which have been successfully applied to personalize the curriculum and learning activities [7,9,12] and to assess different educational interventions through the use of multi-armed bandits [11,19,20].

In multi-armed bandit problems, an agent sequentially selects an action and observes a reward from it, with the ultimate goal of maximizing cumulative reward over the long term. Since actions taken by the agent at any particular time may be suboptimal, a mix between *exploration* (trying out new strategies) and *exploitation* (picking the action deemed optimal at the time) is required in practice to maximize observed long-term reward. Agents are evaluated using *regret*, which is defined as the cumulative expected difference between the rewards of the optimal action and the selected actions.

The *contextual bandit* model presents an agent with information about the current context that it can use to inform its decision. The `LinUCB` Algorithm [10], which we use in this work, achieves the theoretical regret bound of $O(\sqrt{T})$ (where T is the number of timesteps), while being relatively easy to implement and less prone to numerical instability issues throughout its runtime than alternatives [3]. At each timestep $t = 1, ..., T$, the `LinUCB` agent observes the current user u_t, a set A_t of actions, and a feature vector $\mathbf{x}_{t,a}$ for each $a \in A_t$. Each feature vector contains information about both the user u_t and its corresponding action a, and is referred to as a context. The algorithm then computes a score $p_{t,a}$ for each action, based on its expected reward and uncertainty determined by the context vectors and its internal parameters. It receives a reward r_t and uses it to update its internal parameters, thus improving its selection strategy.

The approach proposed here combines the `LinUCB` algorithm with model-based RL, where an internal model of the environment is learned by the RL agent. By learning an internal environment model the agent may be able to reduce the amount of trial-and-error learning and better generalize across states and actions. In particular, the internal environment model may be learned from historical data, if such is available. Several researchers have also investigated the application of model-based reinforcement learning for ITS, including learning effective pedagogical policies, selecting effective instructional sequences and personalizing curricula for students [6,7,15].

3 Methodology

We train a model that can predict a student's performance on an exercise and then use it to simulate student trajectories to pre-train the `LinUCB` exercise selection model.

Dataset: We first extract all previous solution attempts across all 1,977 students that created their accounts between November 2020 and July 2021 and that have attempted at least one exercise. We retrieve 129,000 exercise attempts across 971 unique exercises and 61 learning units. The majority of students on the platform at the time were free users, so we separate the free users and the customers in further experiments.

Exercise Affinity Model: The five possible outcomes for an interaction between a student and an exercise on our platform are defined as follows:

- **Instant success**: The student solved the exercise correctly on the first try.
- **Eventual success**: It took the student multiple attempts to get to a correct solution.
- **Eventual failure**: The exercise was attempted unsuccessfully until a solution was provided to the student.
- **Instant skip**: The exercise was skipped without any attempt.
- **Eventual skip**: The exercise was attempted but eventually skipped.

First, we build a logistic regression model that uses exercise features and students' performance on previous exercises to predict the outcome on future exercises. This model will act as the "world model" in the context of model-based RL, and will provide the agent with the outcomes when it offers an exercise to a student. We train this model by first extracting a student's exercise attempt history, which contains all of a student's attempts at solving the exercises they were presented with (both successful and unsuccessful). We then mask out an attempt on an exercise, and have the model predict the outcome of the student's attempt on this exercise. The model's input features relate to the *student behavior and skills* (including the student's performance on the previous exercise in the learning unit, the student's skip rate in the learning unit, and whether or not the student has watched the video that covers the learning unit), *exercise difficulty* (the historic success rate across all students on the exercise), and the *exercise type* (a one-hot encoding of the expected solution form and the context in which the exercise could be applied). We show an example of a free-form question in Fig. 1.

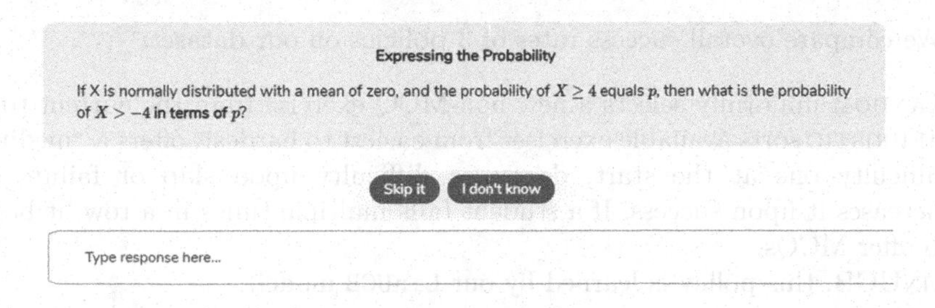

Fig. 1. Example of a free-form question on our platform.

We train the model on approximately 129,000 examples and evaluate it using 5-fold cross-validation, observing an accuracy of 66%. A baseline model that selects the majority class 100% of the time achieves an accuracy of 60%. Although the prediction model can be refined and improved, we believe that it is good enough to be used in the context of pre-training a bandit model. We then use this model to predict students' performance on all unattempted exercises on topics that they have started. This prediction takes the form of a probability vector across the 5 possible outcomes. A total of 165,000 exercise attempts are predicted. These predictions allow us to simulate what would happen if the student receives an exercise that we have no record of them attempting. To train the bandit model, we draw samples from the predictions.

Bandit Model: Using the dataset of student trajectories and attempt predictions, we train a bandit model with `LinUCB`. At each timestep, it selects which exercise to present to the student and whether it should be a MCQ or a free-form question. For each action, we compute a feature vector that encodes information about both the student and the exercise using the same features as the exercise prediction model. To simulate students' progressing through various learning units, for each unit a student has started we let the model sequentially select exercises to present to them. We define our reward function such that more desirable outcomes receive higher rewards, with the ultimate goal of maximizing average success rate for the students in the dataset, while restricting the frequency of MCQs. While doing a grid search, we observed that rewarding instant success higher than eventual success led to a higher average completion rate on the dataset. Our final reward function is as follows: 1.5 for instant success, 1 for eventual success, 0.5 for eventual failure, and 0 for instant and eventual skips. To discourage the model from always presenting an MCQ, we penalize the observed reward for MCQs by reducing it by 0.4. Since students are more likely to correctly answer such questions, the model is more likely to observe a positive outcome when presenting them. However, the free-form questions lead to higher learning outcomes and engagement in students. For a given (*student*, *exercise*) pair, we use the observed outcome if the student has attempted that exercise in our dataset. Otherwise, we sample an outcome from the probabilities computed by our prediction model.

We compare overall success rates of 3 policies on our dataset:

- RANDOM uniformly selects a new non-MCQ exercise from the current topic.
- HEURISTIC sorts available exercises from easiest to hardest, offers a "medium" difficulty one at the start, decreases difficulty upon skip or failure and increases it upon success. If a student fails multiple times in a row, it begins to offer MCQs.
- LINUCB: this policy is learned by our `LinUCB` model.

For each policy, we simulate every student attempting the exercise presented by the policy, and keep track of the average success rate. Due to randomness, we do this 20 times. We observe an average success rate of 58% for the RANDOM policy, 60% for the HEURISTIC policy, and 64% for LINUCB. These values are consistent throughout each run, deviating by no more than 0.5%. Both the HEURISTIC and LINUCB policies offered a MCQ 12% of the time. In conclusion, in our simulated environment, LINUCB noticeably improves student success rate compared to the other two policies.

4 Experiments

Following the successful experiments in a simulated environment, we perform a randomized controlled trial on students that have signed up on the platform between December 2021 and February 2022. On sign-up, each student is assigned exercises either by the adaptive HEURISTIC or by the LINUCB model. We study 2 cohorts of students: **free users**, a diverse set of 44 students (21 under LINUCB and 23 under the heuristic policy) from around the world who signed up on the learning platform for free, and users from a **customer** organization (15 assigned to the HEURISTIC and 11 to the LINUCB policy) using the platform to upskill in data science as part of a broader training program. Students from the second cohort have mandated modules they must finish and tend to be highly motivated regardless of selection policy. Within each cohort, we compare *completion rates*, *skip rates*, and *study time*. Completion and skip rates are local indicators that the exercises we give students are relevant, interesting, and achievable, while study time is a global indicator that the policy is effective at engaging students and motivating them to study on the platform for a longer duration.

Completion and Skip Rates: As Table 1 demonstrates, the students from both cohorts have a substantially higher success rate under LINUCB than the adaptive HEURISTIC model. We also observe that for both cohorts, students under the LINUCB policy have a substantially lower skip rate than under the adaptive heuristic baseline. These results demonstrate that the LINUCB model improves student outcomes and thus does a better job offering more relevant, interesting and achievable exercises to students than the HEURISTIC model.

Table 1. Exercise outcome rates for various groups of users.

Cohort	Policy	Skip	Fail	Success
Free	LinUCB	**7.8** ± 0.8%	4.8 ± 0.5%	**87.4** ± 1.3%
	Heuristic	12.5 ± 1.4%	5.2 ± 0.8%	82.4 ± 1.8%
Customer	LinUCB	**5.6** ± 0.4%	5.7 ± 0.4%	**88.6** ± 0.9%
	Heuristic	8.3 ± 0.4%	5.8 ± 0.3%	85.9 ± 0.7%

Study Time: Finally, we also observe that students under LinUCB across the free cohort spend noticeably more time on the learning platform: the average study time under the adaptive Heuristic model is 109 min, and the average study time under the LinUCB policy is 174 min. Similarly, for the students in the customer cohort, the average study time under the adaptive Heuristic is 265 min, and the average study time under the LinUCB policy is 258 min.

5 Conclusions

We have provided a framework for developing a model-based reinforcement learning agent based on the LinUCB algorithm, which is capable of both learning from historical student data and online. This approach outperformed competitive models by achieving significantly higher completion rates, while reducing the rate at which exercises are skipped in two diverse cohorts of students, while also leading to increased study time across cohorts. These findings demonstrate that the model leads to substantially higher engagement in students. In addition, we note that the reinforcement learning model learns autonomously and is expected to improve automatically as more and more students sign up, thus ensuring its scalability and continuous improvement.

In the future, we plan to validate our findings with a larger sample size. In addition, we will address one of the limitations of the this bandit model – the requirement that all available exercises pertain to the same topic and that sufficient data is available to reach a point where the bandit can mostly exploit rather than explore. Finally, we plan to explore application of more sophisticated bandit algorithms, such as those that incorporate collaborative filtering.

References

1. Albacete, P., Jordan, P., Katz, S., Chounta, I.-A., McLaren, B.M.: The impact of student model updates on contingent scaffolding in a natural-language tutoring system. In: Isotani, S., Millán, E., Ogan, A., Hastings, P., McLaren, B., Luckin, R. (eds.) AIED 2019. LNCS (LNAI), vol. 11625, pp. 37–47. Springer, Cham (2019). https://doi.org/10.1007/978-3-030-23204-7_4
2. Anania, J.: The influence of instructional conditions on student learning and achievement. Eval. Educ. Int. Rev. Ser. **7**(1), 3–76 (1983)

3. Auer, P.: Using confidence bounds for exploitation-exploration trade-offs. J. Mach. Learn. Res. **3**, 397–422 (2002)
4. Bloom, B.S.: The 2 sigma problem: the search for methods of group instruction as effective as one-to-one tutoring. Educ. Res. **13**(6), 4–16 (1984)
5. Brunskill, E., Mu, T., Goel, K., Bragg, J.: Automatic curriculum generation applied to teaching novices a short Bach Piano segment. In: NeurIPS Demonstrations (2018)
6. Chi, M., VanLehn, K., Litman, D., Jordan, P.: Empirically evaluating the application of reinforcement learning to the induction of effective and adaptive pedagogical strategies. User Model. User-Adap. Inter. **21**(1), 137–180 (2011)
7. Doroudi, S., Aleven, V., Brunskill, E.: Where's the reward? Int. J. Artifi. Intell. Educ. **29**(4), 568–620 (2019)
8. Kulik, J.A., Fletcher, J.: Effectiveness of intelligent tutoring systems: a meta-analytic review. Rev. Educ. Res. **86**(1), 42–78 (2016)
9. Lan, A.S., Baraniuk, R.G.: A contextual bandits framework for personalized learning action selection. In: EDM, pp. 424–429 (2016)
10. Li, L., Chu, W., Langford, J., Schapire, R.E.: A contextual-bandit approach to personalized news article recommendation. In: Proceedings of the 19th International Conference on World Wide Web - WWW 2010 (2010)
11. Liu, Y.E., Mandel, T., Brunskill, E., Popovic, Z.: Trading off scientific knowledge and user learning with multi-armed bandits. In: EDM (2014)
12. Lopes, M., Clement, B., Roy, D., Oudeyer, P.: Multi-armed bandits for intelligent tutoring systems (2013). http://arxiv.org/abs/1310.3174
13. Mu, T., Wang, S., Andersen, E., Brunskill, E.: Combining adaptivity with progression ordering for intelligent tutoring systems. In: Proceedings of the Fifth Annual ACM Conference on Learning at Scale, pp. 1–4 (2018)
14. Nye, B.D., Graesser, A.C., Hu, X.: AutoTutor and family: a review of 17 years of natural language tutoring. IJAIED **24**(4), 427–469 (2014)
15. Rowe, J.P., Lester, J.C.: Improving student problem solving in narrative-centered learning environments: a modular reinforcement learning Framework. In: Conati, C., Heffernan, N., Mitrovic, A., Verdejo, M.F. (eds.) AIED 2015. LNCS (LNAI), vol. 9112, pp. 419–428. Springer, Cham (2015). https://doi.org/10.1007/978-3-319-19773-9_42
16. Rus, V., Stefanescu, D., Niraula, N., Graesser, A.C.: DeepTutor: towards macro- and micro-adaptive conversational intelligent tutoring at scale. In: Proceedings of the First ACM Conference on Learning@ Scale Conference, pp. 209–210 (2014)
17. Serban, I.V., et al.: A large-Scale, open-domain, mixed-interface dialogue-based ITS for STEM. In: Bittencourt, I.I., Cukurova, M., Muldner, K., Luckin, R., Millán, E. (eds.) AIED 2020. LNCS (LNAI), vol. 12164, pp. 387–392. Springer, Cham (2020). https://doi.org/10.1007/978-3-030-52240-7_70
18. VanLehn, K.: The relative effectiveness of human tutoring, intelligent tutoring systems, and other tutoring systems. Educ. Psychol. **46**(4), 197–221 (2011)
19. Whitehill, J., Movellan, J.: Approximately optimal teaching of approximately optimal learners. IEEE Trans. Learn. Technol. **11**(2), 152–164 (2017)
20. Williams, J.J., Rafferty, A.N., Tingley, D., Ang, A., Lasecki, W.S., Kim, J.: Enhancing Online Problems Through Instructor-Centered Tools for Randomized Experiments, p. 1–12. Association for Computing Machinery, New York (2018)

Self-attention in Knowledge Tracing: Why It Works

Shi Pu[✉] and Lee Becker

Educational Testing Service, 660 Rosedale Road, Princeton, NJ 08540, USA
{spu,lbecker001}@ets.org

Abstract. Knowledge tracing refers to the dynamic assessment of a learner's mastery of skills. There has been widespread adoption of the *self-attention* mechanism in knowledge-tracing models in recent years. These models consistently report performance gains over baseline knowledge tracing models in public datasets. However, why the *self-attention* mechanism works in knowledge tracing is unknown.

This study argues that the ability to encode when a learner attempts to answer the same item multiple times in a row (henceforth referred to as *repeated attempts*) is a significant reason why *self-attention* models perform better than other deep knowledge tracing models. We present two experiments to support our argument. We use context-aware knowledge tracing (AKT) as our example *self-attention* model and dynamic key-value memory networks (DKVMN) and deep performance factors analysis (DPFA) as our baseline models. Firstly, we show that removing repeated attempts from datasets closes the performance gap between the AKT and the baseline models. Secondly, we present DPFA+, an extension of DPFA that is able to consume manually crafted repeated attempts features. We demonstrate that DPFA+ performs better than AKT across all datasets with manually crafted repeated attempts features.

Keywords: Deep knowledge tracing · Self-attention · Knowledge tracing

1 Introduction

Knowledge tracing refers to the dynamic assessment of a learner's mastery of skills. It is usually used with a recommendation policy to achieve personalized learning. For example, in mastery learning, a learner is only permitted to move on to the next skill when a knowledge-tracing model estimates the learner has achieved mastery of the prerequisite skills [10].

A knowledge-tracing task is structured as a sequential prediction problem in which a knowledge-tracing model tries to predict whether a learner will be able to correctly answer the next item given their past item responses. Throughout this paper, we use "item" to refer to a question or exercise that a learner completes and "item response" to refer to the correctness of the learner's answer to the item.

M. M. Rodrigo et al. (Eds.): AIED 2022, LNCS 13355, pp. 731–736, 2022.
https://doi.org/10.1007/978-3-031-11644-5_75

While early knowledge-tracing models are relatively straightforward [2,5,6], modern models are structured as complicated deep neural networks [1,4,9,11, 12]. A family of these deep neural networks has adopted the *self-attention* mechanism from the natural language processing (NLP) community and shown remarkable success in specific datasets.

What is *self-attention*? Deep neural networks represent a learner's item response (or an item) as a multidimensional vector. This vector is referred to as the embedding of an item response. This embedding is a blend of model-readable (not human-readable) features. Hypothetically, these features include the skill associated with the item, the difficulty of the item, and the mastery of skills from the item response, among other aspects. The *self-attention* mechanism models a learner item response embedding as a weighted sum of previous learner item response embeddings, followed by a non-linear transformation so that the item response embedding is a non-linear function of the previous item response embeddings.

While *self-attention* offers a remarkable performance boost to knowledge-tracing models, why the mechanism works is unknown. The *self-attention* mechanism originated from the NLP community, whose scientists seek to dynamically model a word's (or part of a word's) embedding (i.e., a multidimensional vector representing a word's semantic and syntactical features) based on its context. For example, "bank" may refer to a financial institution or a river bank depending on the context. Self-attention allows a word's embedding to change based on the other words in the sentence or paragraph.

Likewise, *self-attention* in deep knowledge tracing allows the embedding of a learner's response to change based on the learner's previous responses to items (because it is a non-linear function of previous item responses). However, this flexibility appears unnecessary for knowledge tracing, given that the skill associated with an item, the difficulty of an item, and the mastery of skills from a correct/incorrect item response all appear to be context-independent.

Therefore, we are interested in uncovering what context-dependent features *self-attention* extract from the data. One of the most promising candidates is the number of repeated attempts made on an item. For example, Ghosh et al. [4] reported AKT significantly outperformed deep knowledge tracing (DKT) [7], a recurrent neural network knowledge-tracing model, on ASSISTments 2017 where repeated attempts are widespread. Using the same dataset, Gervet et al. [3] documented the self-attentive knowledge-tracing (SAKT) model to perform similarly to DKT after repeated attempts were removed.

There is theoretical grounding to believe the number of repeated attempts is a valuable feature for knowledge tracing. First, how well a learner masters a skill is affected by how soon the learner answers an item correctly. In addition, items in public datasets are often multiple-choice questions, making the second or third attempt easier than the first attempt. Finally, learning systems often provide feedback after an unsuccessful trial (e.g., coding practices). Accordingly, later attempts at an item are significantly more likely to succeed than first attempts.

In the rest of the paper, we provide evidence that the ability to encode repeated attempts is a significant reason why *self-attention* models perform better than other deep knowledge tracing models. We demonstrate that AKT outperforms strong baseline models when there are considerable repeated attempts in datasets but that AKT performs on par with those same baseline models when the repetitions are removed. We then manually encode the number of repeated attempts as a feature into a strong baseline, showing that the baseline outperforms AKT in all datasets where repeated attempts are present.

Table 1. Data statistics

Datasets	Learners	Items	Responses	% of multiple attempts	% of repeated attempts
ASISSTments 2017	1709	4117	942K	58.36	52.12
CSEDM 2019	87	35	2.77K	68.20	66.37
Statics 2011	316	987	205K	34.23	31.62

2 Experiment Setup

We choose three public datasets for our experiments. These datasets provide item identifiers and timestamps for all item responses. More importantly, all the datasets have multiple attempts[1] and repeated attempts. Table 1 reports the descriptive statistics for each dataset. The details of the datasets are as follows:

ASSISTments 2017.[2] This dataset records secondary school students' answers to math problems using the ASSISTments tutoring system.

STATICS 2011.[3] This dataset is from an engineering course on statics in fall 2011. We follow [8] for data preprocessing. Unlike previous studies that keep only students' first responses to items, this study keeps all responses.

CSEDM 2019.[4] This dataset is from a study of novice programmers working with the ITAP intelligent tutoring system. We remove all hints in the data.

For each dataset, we evaluate a model with student-stratified five-fold cross-validation. For each fold, 60% of students are used as training data, 20% students are used as validation data, and 20% of students are used as test data. We use the average test area under the curve (AUC) as the evaluation metric.

We set the maximum length of a response sequence to be 200. Learner response sequences shorter than 200 are padded with 0 to the left, and sequences longer than 200 are folded. We used the hyperparameter reported by the original authors when possible. If the model has not been experimented with a dataset before, we use the validation data for hyperparameter tuning.

[1] Multiple attempts is different from repeated attempts that they may or may not happen consecutively.

[2] https://sites.google.com/view/assistmentsdatamining.

[3] https://pslcdatashop.web.cmu.edu/DatasetInfo?datasetId=507.

[4] https://pslcdatashop.web.cmu.edu/Files?datasetId=2865.

Table 2. Test AUC w/o repeated attempts

Datasets	ASSISTments 2017		CSEDM 2019		Statics 2011	
	With repeat	No repeat	With repeat	No repeat	With repeat	No repeat
AKT	**0.7620**	0.7862	**0.7602**	0.7720	**0.7958**	**0.8059**
DKVMN	0.7217	**0.7917**	0.7342	0.7843	0.7867	0.7951
DPFA	0.7167	0.7885	0.7412	**0.8031**	0.7795	0.7930

3 Experiment 1: Repeated Attempts and Self-attention

If having a representation of "repeated attempts" explains why self-attention models work better than other KT models, then we should observe *self-attention* models outperforming baseline models in a dataset with repeated attempts. More importantly, we should observe the closure of the performance gap if we remove repeated attempts from a dataset by keeping only learners' first item responses.

We choose AKT as our example self-attention model. We choose DKVMN and DPFA [8] (a modern version of performance factors analysis [5]) as our baselines. Neither baseline uses *self-attention*, and both baselines have outstanding performances on multiple public datasets.

We experiment with the models on two versions of the dataset, one that includes repetition and the other that excludes repetition by keeping only a learner's first response to an item. Table 2 shows the performance of the models on both versions of the dataset. Across all datasets, AKT outperforms the baseline models when we keep the repeated attempts in the data. The performance gap is most evident in ASSISTments 2017. However, if we remove the repeated attempts from the data by keeping only learners' first attempts, AKT's performances is very close to the baselines' performances. AKT even underperforms in a small dataset like CSEMD 2019. The result suggests that the presence of repeated attempts is crucial for AKT to achieve its performance potential.

4 Experiment 2: Repeated Attempts as Features

We present DPFA+, which is an extension of DPFA [8]. When predicting whether a learner can correctly answer the next question, DPFA considers the difficulty of the next item and the number of similar items the learner has answered correctly or incorrectly in the past. DPFA+ extends the model by incorporating model features for number of attempts made on previous items.

4.1 Model Specification

The original DPFA is a logistic regression:

$$p_{t+1} = \sigma(\beta_{t+1} + \sum_i w_i v_i) \quad \forall i \leq t \tag{1}$$

Table 3. Test AUC for DPFA+

Datasets	ASSISTments 2017	CSEDM 2019	Statics 2011
AKT	0.7620	0.7602	0.7958
DKVMN	0.7217	0.7342	0.7867
DPFA	0.7167	0.7412	0.7795
DPFA+	**0.7794**	**0.7932**	**0.8117**

where β_{t+1} represents the next item difficulty, v_i represents the estimated mastery of skill from past item response i, and w_i represents the relevance of item $t+1$ and the item i. In DPFA, β_{t+1} and v_i are functions of items and correctness of item response. DPFA+ argues that the number of proceeding success and failure on the item should also be part of the function parameters. Specifically, in DPFA+:

$$\beta_{t+1} = W_2(tanh(W_1[d_{t+1} \oplus s_t \oplus f_t] + b_1)) + b_2 \tag{2}$$

$$v_i = W_4(tanh(W_3[d_i \oplus s_i \oplus f_i] + b_3)) + b_4 \tag{3}$$

where d_i is the one hot encoding of item i, s_i is the number of successes in the repeated attempts, f_i is the number of failures in the repeated attempts. For example, if a student attempt an item k three times in a roll with $[success, failure, success]$, the values of s_k are $[1, 1, 2]$ and values of f_k are $[0, 1, 1]$. Note that when modeling the next item difficulty β_{t+1}, only *proceeding* success s_t and *proceeding* failures f_t are visible. $\oplus$ represents the concatenation operation. W_* and b_* are learned parameters. DPFA+ is identical to DPFA in other aspects. We therefore recommend readers to refer to the original paper [8] for model details due to page limit.

4.2 Results

Table 3 presents the average AUC of the DPFA+ model on the test data. DPFA+ performs better than the original DPFA model with the manually encoded repeated attempts features. More importantly, DPFA+ outperforms *self-attention* model AKT. This suggests that while a *self-attention* model like AKT can encode repeated attempts as part of its item embedding, the mechanism is not as effective as manually crafted features.

5 Discussion

In this study, we presented evidence that the ability to encode repeated attempts as features is the reason why *self-attention* models perform better than other deep knowledge models in datasets where such repetition is common.

Repeated attempts are common in systems where a learner can attempt an item multiple times (e.g., coding practice). The ability to model whether a learner can succeed on the nth attempt is useful for personalized learning systems. For example, a system may decide to provide scaffolding only if a learner has already failed a few times and the knowledge-tracing model believes the learner is not likely to succeed on the next attempt. In such situations, DPFA+ is a better choice than *self-attention* models due to its better performance and higher interpretability.

References

1. Choi, Y., et al.: Towards an appropriate query, key, and value computation for knowledge tracing. arXiv preprint arXiv:2002.07033 (2020)
2. Corbett, A.T., Anderson, J.R.: Knowledge tracing: modeling the acquisition of procedural knowledge. User Model. User-Adap. Inter. **4**(4), 253–278 (1994)
3. Gervet, T., Koedinger, K., Schneider, J., Mitchell, T., et al.: When is deep learning the best approach to knowledge tracing? J. Educ. Data Mining **12**(3), 31–54 (2020)
4. Ghosh, A., Heffernan, N., Lan, A.S.: Context-aware attentive knowledge tracing. In: Proceedings of the 26th ACM SIGKDD International Conference on Knowledge Discovery & Data Mining, pp. 2330–2339 (2020)
5. Pavlik Jr, P.I., Cen, H., Koedinger, K.R.: Performance factors analysis-a new alternative to knowledge tracing. Online Submission (2009)
6. Pelánek, R.: Applications of the ELO rating system in adaptive educational systems. Comput. Educ. **98**, 169–179 (2016)
7. Piech, C., et al.: Deep knowledge tracing. In: Advances in Neural Information Processing Systems, pp. 505–513 (2015)
8. Pu, S., Converse, G., Huang, Y.: Deep performance factors analysis for knowledge tracing. In: Roll, I., McNamara, D., Sosnovsky, S., Luckin, R., Dimitrova, V. (eds.) AIED 2021. LNCS (LNAI), vol. 12748, pp. 331–341. Springer, Cham (2021). https://doi.org/10.1007/978-3-030-78292-4_27
9. Pu, S., Yudelson, M., Ou, L., Huang, Y.: Deep knowledge tracing with transformers. In: Bittencourt, I.I., Cukurova, M., Muldner, K., Luckin, R., Millán, E. (eds.) AIED 2020. LNCS (LNAI), vol. 12164, pp. 252–256. Springer, Cham (2020). https://doi.org/10.1007/978-3-030-52240-7_46
10. Ritter, S., Yudelson, M., Fancsali, S.E., Berman, S.R.: How mastery learning works at scale. In: Proceedings of the Third (2016) ACM Conference on Learning@ Scale, pp. 71–79 (2016)
11. Shin, D., Shim, Y., Yu, H., Lee, S., Kim, B., Choi, Y.: Saint+: integrating temporal features for ednet correctness prediction. In: LAK21: 11th International Learning Analytics and Knowledge Conference, pp. 490–496 (2021)
12. Zhang, J., Shi, X., King, I., Yeung, D.Y.: Dynamic key-value memory networks for knowledge tracing. In: Proceedings of the 26th international conference on World Wide Web, pp. 765–774 (2017)

Student Low Achievement Prediction

Andrea Zanellati, Stefano Pio Zingaro(✉), and Maurizio Gabbrielli

University of Bologna, Bologna, Italy
{andrea.zanellati2,stefanopio.zingaro,maurizio.gabbrielli}@unibo.it

Abstract. In this paper, we propose a method for assessing the risk of low achievement in primary and secondary school. We train three machine learning models with data collected by the Italian Ministry of Education through the INVALSI large-scale assessment tests. We compare the results of the trained models and evaluate the effectiveness of the solutions in terms of performance and interpretability. We test our methods on data collected in end-of-primary school mathematics tests to predict the risk of low achievement at the end of compulsory schooling (5 years later). The promising results of our approach suggest that it is possible to generalise the methodology for other school systems and for different teaching subjects.

Keywords: Low achievement · Performance prediction · Assessment tests

1 Introduction

Low achievement at school is a widespread phenomenon which has long-term consequences, both for the individual and for society as a whole. In 2016, above 28% of students across OECD[1] countries underscored the minimum level of proficiency in at least one of the three core subjects according to the Programme for International Student Assessment (PISA), which are English reading and comprehension, mathematics, and science [14]. Low achievement is strongly related to school dropout, i.e., the discontinuation of education [7], and impact on the cultural and professional growth of the individual and citizen [3,12]. Indeed, school performance in first grade is already a significant indicator of future high dropout risk. In 2019, a study conducted by the National Institute for Assessment of the Education System (INVALSI) found that 20% percent of Italian students had a lower-than-expected achievement and, eventually, dropped out of school [17].

To counteract dropout as soon as possible and to detect low achievement, we address the following research questions:

RQ1 Is it possible to quantitatively represent a student's knowledge level and build a model of his or her skill attainment?
RQ2 Is it possible to develop a suitable AI-tool to predict, at an early stage, the risk of low achievement at secondary school for primary school students?

[1] OECD stands for Organization for Economic Co-operation and Development.

M. M. Rodrigo et al. (Eds.): AIED 2022, LNCS 13355, pp. 737–742, 2022.
https://doi.org/10.1007/978-3-031-11644-5_76

In the following, we present a case study, focusing on the Italian context and using data collected from the INVALSI national large-assessment tests in mathematics. In particular, from these tests, we aim to extract the relevant features related to students' learning in terms of their skill and competence level performance.

In **RQ2**, we refer to "early stage" meaning to detect risk as soon as possible, i.e., several years in advance, so that appropriate countermeasures can be taken, and to design an intervention aimed at reducing risk when it is detected. Concretely, we develop three models able to predict the risk of low achievement at K-10, using student data at K-5.

In selecting the predictive model, we strive for a balance between interpretability and performance. Hence, we consider state-of-the-art machine learning techniques that proved to be effective in preliminary experiments: random forests and neural networks [5,9]. On the one hand, we exploit random forests to extract rules that facilitate the process of interpreting the outcomes of the research and, on the other hand, we test neural networks for flexibility, e.g., exploring non-linear correlations, and performance gain.

2 Related Work

The topic of students' low achievement is a widely studied phenomenon in the social sciences and education [6,8]. The problem has also been addressed in terms of predictive models for low achievement or dropout risk for both high school and college students. These models exploit different machine learning techniques, including supervised learning, e.g., random forests, support vector machine and Bayesian network, unsupervised learning, e.g., k-means and hierarchical clustering, and reccommender systems, e.g., collaborative filtering [2,16]. Moreover, several kinds of data have been used to tackle the problem. In [13] the dataset for building the predictive model uses demographic data of the students and their grades. Other studies are based on students performance, i.e., grades, collected during first semester courses [1,11]. Some datasets include behavioural data supplemented with other features related to learning results [18], in a mix of cognitive and non-cognitive characteristics. In some studies data collected through large-scale assessment tests were used to design predictive models of student performance through several machine learning techniques. In [15], for example, the authors refer to data collected through the PISA international large-scale assessment tests.

We aim to contribute to the research field of Artificial Intelligence-based education solutions by presenting a case study for predicting the risk of low achievement of high school students using their performance data collected during primary school. In addition, we extract features directly related to students' learning in terms of knowledge and skills, privileged indicators for the study of learning [4], thus proposing a knowledge-based method for encoding students' learning. We believe that this element can improve the interpretability of the results and make this tool useful for students, teachers and instructional coordinators.

3 Methodology

The INVALSI dataset is the result of a large-scale assessment administered in Italy since the school year 2002/03 at the levels K-2, K-5, K-8, K-10, and K-13. In our case study, we considered data on maths test from two cohorts of students: K-5 of the 2012/13 school year and K-5 of the 2013/14 school year. For the same students, we collected data from five years later at grade K-10, to be used for the definition of the low achievement target, i.e., the students grade in the test is less than or equal to 2 on a scale from 1 to 5. After merging K-5 datasets with their correspondent K-10 targets, the K-5 2012/13 cohort is made up of 351746 students, while the K-5 2013/14 cohort of 354987 students.

There are several features in the dataset and we applied a feature selection process to determine a subset of relevant features. The datasets also contain a boolean feature for each test item, where the students' answers correctness are recorded. To enable the use of our predictive models on different cohorts of students it is necessary to release the dataset from the individual items that constitute a certain test. Therefore, we used a knowledge-based approach considering the items classification in terms of areas, processes and macro-processes according to the INVALSI framework for the design of math tests. In Table 1, we give for reference an overview of the areas, processes, and macro-processes that have been used in the encoding of the questions.

Table 1. Maths INVALSI framework for question encoding.

Areas
(NU) Numbers
(SF) Space and figures
(DF) Data and forecasts
(RF) Relations and functions
Process
(P1) Know and master the specific contents of mathematics
(P2) Know and use algorithms and procedures
(P3) Know different forms of representation and move from one to the other
(P4) Solve problems using strategies in different fields
(P5) Recognize the measurable nature of objects and phenomena in different contexts and measure quantities
(P6) Progressively acquire typical forms of mathematical thought
(P7) Use tools, models and representations in quantitative treatment information in the scientific, technological, economic and social fields
(P8) Recognize shapes in space and use them for problem solving
Macro-process
(MP1) Formulating
(MP2) Interpreting
(MP3) Employing

We define one new variable for each area, process, and macro-process. Each of these new features takes the value corresponding to the percentage of correct answers provided by the student for that specific group of items, namely, correctness rate. Last, we concatenate the computed values to obtain a new flattened representation of learning, where each item is a possible indicator and not its unique representative. Following our strategy, we represent each student's learning in the space of fifteen 15 dimensions, as shown in Table 2.

We use two techniques to develop our AI-tool. The first one is Random forest (RF) [5], which is widely used in Educational Data Mining for the high degree of explainability and effortless interpretation of the results. We trained our models through bootstrap aggregating (bagging) to reduce the overfitting of dataset and increase precision. To tune the model, we performed a grid search.

The second technique is based on neural networks, which has recently become widespread also in the field of Educational Data Mining and has also been applied in predictive models for student performance [10]. We firstly include a preprocessing step, aimed at encoding the values of categorical variables into numerical values with a "one-hot" encoding algorithm. After preprocessing, we implemented two neural networks based on different data transformation approaches. Categorical Embeddings (CE), is a neural network that treats the input depending on its type: categorical inputs are passed through an embedding layer, numerical ones are fed to a dense layer. Feature Tokenizer Transformer (FTT) [9] is able to identify the input or the group of inputs that most influence the output, thanks to attention maps.

Table 2. Example of the student's learning final encoding.

Id	NU	SF	DF	RF	P1	P2	P3	P4	P5	P6	P7	P8	MP1	MP2	MP3
1	0.86	0.75	0.90	0.80	0.71	0.80	1.00	0.89	1.00	0.67	0.91	0.75	0.81	0.73	0.94
2	0.50	0.25	0.50	0.53	0.29	0.60	0.50	0.22	1.00	0.33	0.73	0.25	0.50	0.47	0.44

4 Experimental Results

We carried out all the experiments using the Google Colaboratory Notebook environment, with the Python programming language and popular machine learning libraries, such as scikit-learn and pandas.

The dataset for all the experiments was preprocessed cleaning features with many missing values, highly correlation (computed by R^2 measure above 0.5) or specifically referred to a cohort of students, preventing the model to be transferred to new cohorts (e.g., identification code for a class). This features selection process, together with the engineering of the features related to the items in the tests, results in a set of 34 features, which refers both to socio-economic and cultural context, demographic data and learning dimension. For the definition of the training set we used the data from 2012/13 K-5 cohort. For the models based on neural networks we split this cohort to generate both training and validation sets (split in 80% and 20% respectively). Finally, we used the K-5 2013/14 cohort to test and measure the model performance. The dataset is unbalanced between underachievement/non-underachievement classes; therefore balancing techniques were applied. In the development of the RF models, a random undersampling technique was used, implemented in the imblearn library. We trained neural networks using a weighted random sampler, that samples the data to balance classes ratio in the training batches.

In Table 3, we present the overall results on the test dataset of the above mentioned models: RF, CE, and FTT. For RF, we considered the best hyper-parameters setting determined with the grid search technique: 50 estimators in the forest, trained with 30% of random samples, 60% of random features and max depth set to 11. The FTT outperforms the other predictive models with accuracy, precision and recall between 77% and 78%.

Table 3. Performance on test set

Models	Accuracy	Precision	Recall
Random Forest	0.77	0.62	0.67
CE neural network	0.76	0.76	0.76
FTT neural network	0.78	0.77	0.78

5 Conclusion and Future Work

Our results suggest that the challenge of predicting low achievement risk for primary and secondary school students can be effectively addressed through the use of well-curated datasets and the choice of reliable predictive models. Our abstract representation of (INVALSI) tests and the related encoding for the student achievement allowed us to transfer the trained models on different cohorts and therefore to obtain a accurate prediction. We believe that the ability to predict low school achievement with reasonable accuracy five years in advance offers a practical tool for policy makers, managers and educators.

We are interested in extending our work in several directions. First, we want to verify the transferability of the proposed methodology to other disciplines, using a representation for students' learning similar to the one proposed in this work. Second, we want to increase the quality of the information provided as input to the predictive models, e.g., by collecting more data and by integrating new data sources. We aim to improve the learning encoding—thus the feature extraction process—in a way that is not knowledge-based to limit the bias. Finally, we want to deepen the interpretability of the results of our models, by analysing the feature importance computed on RF model and comparing it with the interpretation of the weights that define the neural networks we have used.

References

1. Al-Barrak, M.A., Al-Razgan, M.: Predicting students final GPA using decision trees: a case study. Int. J. Inf. Educ. Technol. **6**(7), 528 (2016)
2. Albreiki, B., Zaki, N., Alashwal, H.: A systematic literature review of student performance prediction using machine learning techniques. Educ. Sci. **11**(9), 552 (2021)
3. Alexander, K.L., Entwisle, D.R., Olson, L.S.: Schools, achievement, and inequality: a seasonal perspective. Educ. Eval. Policy Anal. **23**(2), 171–191 (2001)
4. Baartman, L.K., De Bruijn, E.: Integrating knowledge, skills and attitudes: conceptualising learning processes towards vocational competence. Educ. Res. Rev. **6**(2), 125–134 (2011)
5. Breiman, L.: Random forests. Mach. Learn. **45**(1), 5–32 (2001)
6. Cassen, R., Kingdon, G.: Tackling Low Educational Achievement. Joseph Rowntree Foundation (2007)
7. Curtis, D.D., McMillan, J.: School non-completers: profiles and initial destinations (2008)
8. Geary, D.C.: Consequences, characteristics, and causes of mathematical learning disabilities and persistent low achievement in mathematics. J. Dev. Behav. Pediatr. JDBP **32**(3), 250 (2011)
9. Gorishniy, Y., Rubachev, I., Khrulkov, V., Babenko, A.: Revisiting deep learning models for tabular data (2021)
10. Hernández-Blanco, A., Herrera-Flores, B., Tomás, D., Navarro-Colorado, B.: A systematic review of deep learning approaches to educational data mining. Complexity **2019**, 1–22 (2019)
11. Ibrahim, Z., Rusli, D.: Predicting students' academic performance: comparing artificial neural network, decision tree and linear regression. In: 21st Annual SAS Malaysia Forum, 5th September (2007)
12. Ingels, S.J., Curtin, T.R., Kaufman, P., Alt, M.N., Chen, X., et al.: Coming of Age in the 1990s: The Eighth-Grade Class of 1988 12 Years Later. Eric (2002)
13. Kotsiantis, S., Pierrakeas, C., Pintelas, P.: Predicting students' performance in distance learning using machine learning techniques. Appl. Artif. Intell. **18**(5), 411–426 (2004)
14. OECD: Who and Where are the Low-Performing Students? OECD Publishing (2016)
15. Pejić, A., Molcer, P.S., Gulači, K.: Math proficiency prediction in computer-based international large-scale assessments using a multi-class machine learning model. In: 2021 IEEE 19th International Symposium on Intelligent Systems and Informatics (SISY), pp. 49–54. IEEE (2021)
16. Rastrollo-Guerrero, J.L., Gomez-Pulido, J.A., Durán-Domínguez, A.: Analyzing and predicting students' performance by means of machine learning: a review. Appl. Sci. **10**(3), 1042 (2020)
17. Ricci, R.: La dispersione scolastica implicita (2019)
18. Sultana, S., Khan, S., Abbas, M.A.: Predicting performance of electrical engineering students using cognitive and non-cognitive features for identification of potential dropouts. Int. J. Electr. Eng. Educ. **54**(2), 105–118 (2017)

Automatic Question Generation for Scaffolding Self-explanations for Code Comprehension

Lasang J. Tamang(✉), Rabin Banjade, Jeevan Chapagain, and Vasile Rus

University of Memphis, Memphis, TN, USA
{ljtamang,rbnjade1,jchpgain,vrus}@memphis.edu

Abstract. This work presents two systems, Machine Noun Question Generation (QG) and Machine Verb QG, developed to generate short questions and gap-fill questions, which Intelligent Tutoring Systems then use to guide students' self-explanations during code comprehension. We evaluate our system by comparing the quality of questions generated by the system against human expert-generated questions. Our result shows that these systems performed similarly to humans in most criteria. Among the machines, we find that Machine Noun QG performed better.

Keywords: Automatic question generation · Self-explanation · Program comprehension · Intelligent Tutoring System · Authoring

1 Introduction

This work is part of our larger effort to develop Intelligent Tutoring Systems (ITS) to help students learn computer programming. Such ITS uses questions to be provided as hints meant to scaffold students' self-explanation [9] during code comprehension. When done by human experts, which is currently the norm, authoring such questions is expensive and hard to scale, often taking 100–200 hours to prepare 1-h instructional content [1]. In this work, we develop two systems called Machine Noun Question Generator (QG) and Machine Verb QG for automatically generating short questions and gap-fill questions using expert generated code-block explanations that ITS employs to scaffold student self-explanation during code comprehension.

Some prior works [5,10,11] automatically create clones of programming exercises that provide opportunities to practice more as opposed to scaffolding students' self-explanation for the particular code, which is the focus of our work. Other works like [2,8] automatically generated short questions from static analysis of code, using the template-based QG approach, which requires significant

This work is supported by the National Science Foundation under grant number 1822816 and 1934745. All findings and opinions expressed are solely the authors'.

M. M. Rodrigo et al. (Eds.): AIED 2022, LNCS 13355, pp. 743–748, 2022.
https://doi.org/10.1007/978-3-031-11644-5_77

time to design the templates. Unlike past work, we do not use a template approach for question generation. Instead, we use the current state-of-art model ProphetNet [6] which inputs textual explanations of the code, leading to a more computer language-independent approach for question generation. Also, it can produce a more profound and broader variety of questions compared to the limited type of questions that the expert-provided templates can generate.

In sum, this paper answers the following research questions:

1. Is it possible to automatically generate short questions that are linguistically well-formed, pedagogically sound, and indistinguishable from human-generated questions?
2. Is it possible to automatically produce gap-fill questions useful for ITS?
3. How do questions generated by machines compare to expert questions?
4. How do Machine Noun QG and Machine Verb QG compare in teperformance?

2 Dataset

The dataset used for this work consists of 10 code examples with explanations followed by short and gap-fill questions for each code block, as shown in Fig. 1, prepared and refined by our group of subject experts in several iterations.

```
public class AverageOfNumbers {
 public static void main(String[] args) {

    /* Code-Block 1, Expert-Explanation, short and gap-fill questions*/
    double[] numArray = {8,6,11,7};
    double sum = 0.0; double average;

    /* Code-Block 2
    The sum of numbers is calculated using a for loop that iterates over
    each number in the numArray array and adds each number to the sum.
    When the for loop completes execution, the value of the sum is 32.
    1. How is a sum of numbers calculated?
    The sum of numbers is calculated using a _____ that iterates over
    each number of numArray and adds each number to the sum.
    2. What is the value of sum when the for loop completes execution?
    When the forloop completes execution, the value of the sum is __.
    */
    for (int i = 0; i < numArray.length; i++) {
        sum += numArray[i];
    }

    /*Code-Block 3, Expert-Explanation, short and gap-fill questions*/
    average = sum / numArray.length;
    System.out.format("The average is: %.2f", average);
}}
```

Fig. 1. Sample code example in our dataset

3 System Design

3.1 Machine Noun QG

First, Machine Noun QG segments the expert's explanation for each code block into individual sentences using a library called pySBD, a pipeline extension in spaCy v2.0. Then, we extract noun chunks for each sentence, also using spaCy. When a sentence has multiple noun chunks, the first step is to discard any noun chunk with more than four words; Chau and colleagues define "single words or short phrases of two to four words" as domain concepts [3,4] (i.e., ideally what we would like to target with our questions). Then, we select the longest noun chunk from the remaining noun chunks because longer inputs are beneficial for the question generator. If two noun chunks have the same length, we select the noun chunk that has appeared first in the sentence, assuming that an important keyphrase comes first.

Next, We pass a pair of <sentence, selected noun chunk for the sentence> to a pre-trained sequence-to-sequence model ProphetNet [6] fine-tuned for question generation tasks using the SQUAD [7] dataset. The model outputs the short question. The gap-fill question is created by masking the sentence's noun chunk.

3.2 Machine Verb QG

Machine Verb QG works the same way as Machine Noun QG except it targets verb phrase in the input sentences. We extract verb phrases in a sentence by matching the pattern = ['POS': 'VERB', 'OP': '?', 'POS': 'ADV', 'OP': '*', 'POS': 'AUX', 'OP': '*', 'POS': 'VERB', 'OP': '+'], using Matcher in the spacy library.

4 Evaluation

The two independent annotators (Ph.D. students in Computer Science) annotated a total of 450 questions, each 150 (75 short +75 gap-fill) questions generated by Machine Noun QG, Machine Verb QG, and experts (question in our dataset), using the evaluation criteria as described below. The inter-annotator agreement, measured by Cohen's Kappa, is 0.30, 0.39, 0.71, 0.93, 0.37, 0.37, and 0.91 for grammaticality, semantic correctness, domain relevancy, answerability, helpfulness, recognizability and gap-fill questions, respectively.

We evaluated short questions using the following criteria.

1. **Grammaticality**: Is the question grammatically correct?
2. **Semantic Correctness**: Is the question semantically correct?
3. **Domain Relevancy**: Is the question relevant to the target domain, i.e., does it target a programming concept?
4. **Answerability**: Does the question have a clear answer in the input text?
5. **Helpfulness**: Is the question likely to help the student think about the target concept and produce an answer close to the expert-provided explanation?

6. **Recognizability**: How likely is it that a human generated the question?

The scale for the first two, second two, and last two are 1 (Very Poor) to 5 (Very Good), Yes/No, and 1 (Not Likely) to 5 (Very Likely), respectively.

Each gap-fill question is labeled into one of the following categories.

1. **Good**: Asks about key concepts and would be reasonably difficult to answer.
2. **OK**: Asks about a) key concept but might be difficult to answer or b) likely key concept (weak concept).
3. **Bad**: Asks about 1) an unimportant aspect or 2) has an answer that can be figured out from the context of the sentence.
4. **Acceptable**: OK or Good questions are automatically labeled as acceptable.

5 Results

The overview of quality of short and gap-fill questions is shown in Table 1 and Table 2, respectively. To check whether the difference is significant, we use independent-samples t-tests for the mean score and the Chi-square test of independence for the proportion. We present below a detailed analysis and interpretation of these results in accordance with our research questions.

5.1 Short Questions

Table 1. Performance of Machine Noun QG, Machine Verb QG, and Human on Short-questions. SD = Standard Deviation.

	Machine Noun QG	Machine Verb QG	Human
Mean grammaticality	4.51 (SD = 0.62)	4.64 (SD = 0.48)	4.67 (SD = 0.48)
Mean semantic correctness	4.76 (SD = 0.46)	4.49 (SD = 0.80)	4.84 (SD = 0.57)
Mean helpfulness	4.27 (SD = 0.96)	3.44 (SD = 0.96)	4.31 (SD = 0.77)
Mean recognizability	3.49 (SD = 1.33)	2.76 (SD = 1.17)	4.49 (SD = 0.91)
Answerability (Yes)%	89.3	54.7	97.3
Domain relevancy (Yes)%	92	89.3	93.3

Both Machine Noun QG and Machine Verb QG generated linguistically well-formed, i.e., grammatically and semantically very good questions with mean scores for grammaticality of 4.51 and 4.64 and semantic correctness of 4.76 and 4.49, respectively.

Likewise, it is possible to automatically generate short questions which are pedagogically sound as measured by domain relevancy, answerability, and helpfulness criterion. The systems generated questions relevant to the domain in program comprehension in an impressive proportion: 92% by the Machine Noun QG and 89.3% by the Machine Verb QG. While the Machine Noun QG produced almost all, i.e., 93% answerable questions, the Machine Verb QG generated slightly more than half, i.e., 54.7% questions that are answerable. The average helpfulness score of Machine Noun QG questions is 4.27 and, therefore, is likely to help students articulate the expected answer. On the other hand, the Machine Verb QG's average helpfulness score is only 3.44, indicating it may or may not help students scaffold explanation for the code.

Also, it is possible for the system to automatically generate short questions that are indistinguishable from human-generated questions, measured by recognizability. The mean recognizability score for Machine Noun QG is 3.49, indicating that human annotators think humans likely generate these. On the other hand, the mean recognizability score for the Machine Verb QG system is 2.76, which signifies that it at least challenges or makes annotators hard to say who generated the questions, i.e., they think the question has equal chances of being created by human or machine.

Comparison: Compared to human, Machine Noun QG performed comparably; we did not find significant difference in mean or proportion in any criteria. However, Machine Verb QG significantly under-performed to human in helpfulness [$t(141.27) = -6.09, p = 0.00$] and answerablity [$\chi^2(1, n = 150) = 35.12, p = 0.00.$] criteria, but, no significant difference in rest of criteria. Between machines, Machine Noun QG significantly outperformed machine verb in semantic correctness [$t(118.62) = 2.51, p = 0.01$] and helpfulness [$t(148) = 5.26, p = 0.00$], and they pefromed similalry in rest of criterion.

5.2 Gap-Fill Questions

Table 2. Performance of Machine Noun QG, Machine Verb QG, and Human on Gap-Fill Questions.

	Bad %	Okay %	Good %	Acceptable %
Machine Noun QG	16	73.3	10.7	84
Machine Verb QG	20	38.7	41.3	80
Human	2.7	53.3	44	97.3

These systems can produce a majority of acceptable gap-fill-questions, i.e., 84% by Machine Noun QG and 80% by Machine Verb QG.

Comparison: Compared to human, both Machine Noun QG [$\chi^2(1, n = 150) = 6.38, p = 0.012$] and Machine Verb QG [$\chi^2(1, n = 150) = 9.55, p = 0.002$] significantly under-performed in gap-fill QG task. There is no significant difference in the proportions of acceptable gap-fill questions generated between Machine Noun QG and Machine Verb QG, $\chi^2(1, n = 150) = 0.19, p = 0.67$.

6 Conclusion

In this work, we developed Machine Noun QG and Machine Verb QG systems to automatically generate short and gap-fill questions that ITS can use to scaffold students by presenting them as a hint. Our evaluation shows that these systems can generate short questions which are linguistically well-formed, pedagogically sound, and likely indistinguishable from human-generated questions. We also found that most gap-fill questions generated by machines are of acceptable quality to be used by ITS. Compared to human experts, Machine Noun QG performed comparable for short questions but under-performed for gap-fill questions in almost all criteria. Between the systems, Machine Noun QG performed better.

In our future work, we plan to automate the generation of code explanations using code examples and the surrounding text in programming textbooks and then use the explanations to generate the questions automatically, thus making the process fully automated.

References

1. Aleven, V., Mclaren, B.M., Sewall, J., Koedinger, K.R.: A new paradigm for intelligent tutoring systems: example-tracing tutors. Int. J. Artif. Intell. Educ. **19**(2), 105–154 (2009)
2. Alshaikh, Z., Tamang, L.J., Rus, V.: Experiments with auto-generated socratic dialogue for source code understanding. In: CSEDU (2), pp. 35–44 (2021)
3. Banjade, R., Oli, P., Tamang, L.J., Chapagain, J., Rus, V.: Domain model discovery from textbooks for computer programming intelligent tutors. In: The International FLAIRS Conference Proceedings, vol. 34 (2021)
4. Chau, H., Labutov, I., Thaker, K., He, D., Brusilovsky, P.: Automatic concept extraction for domain and student modeling in adaptive textbooks. Int. J. Artif. Intell. Educ. **31**(4), 820–846 (2021)
5. Hsiao, I.H., Brusilovsky, P., Sosnovsky, S.: Web-based parameterized questions for object-oriented programming. In: E-Learn: World Conference on E-Learning in Corporate, Government, Healthcare, and Higher Education, pp. 3728–3735. Association for the Advancement of Computing in Education (AACE) (2008)
6. Qi, W., et al.: ProphetNet: predicting future n-gram for sequence-to-sequence pre-training. arXiv preprint arXiv:2001.04063 (2020)
7. Rajpurkar, P., Zhang, J., Lopyrev, K., Liang, P.: SQuAD: 100,000+ questions for machine comprehension of text. arXiv preprint arXiv:1606.05250 (2016)
8. Tamang, L.J., Alshaikh, Z., Khayi, N.A., Oli, P., Rus, V.: A comparative study of free self-explanations and socratic tutoring explanations for source code comprehension. In: Proceedings of the 52nd ACM Technical Symposium on Computer Science Education, pp. 219–225 (2021)
9. Tamang, L.J., Alshaikh, Z., Khayi, N.A., Rus, V.: The effects of open self-explanation prompting during source code comprehension. In: The Thirty-Third International Flairs Conference (2020)
10. Thomas, A., Stopera, T., Frank-Bolton, P., Simha, R.: Stochastic tree-based generation of program-tracing practice questions. In: Proceedings of the 50th ACM Technical Symposium on Computer Science Education, pp. 91–97 (2019)
11. Zavala, L., Mendoza, B.: On the use of semantic-based AIG to automatically generate programming exercises. In: Proceedings of the 49th ACM Technical Symposium on Computer Science Education, pp. 14–19 (2018)

Incorporating AI and Analytics to Derive Insights from E-exam Logs

Hatim Fareed Lahza(✉), Hassan Khosravi, and Gianluca Demartini

The University of Queensland, St Lucia, QLD 4072, Australia
h.lahza@uq.net.au

Abstract. In recent years, the use of electronic assessment platforms (EAPs), which allow exams to be administered on- or off-line, has become increasingly popular. A benefit of EAPs is that they capture a detailed log of an examinee's journey through their exams. However, methods of leveraging exam logs for developing analytical insights are still under-explored. In this paper, we employ AI and analytical techniques to investigate whether exam-takers exhibit distinct behaviours while taking e-exams. We evaluate our methods using an e-exam log of 90 multiple-choice questions administrated to 463 university-level medical students. Our findings indicate that the students exhibited distinctive test-taking tactics and strategies, and some of the tactics are associated with their performance. We discuss the implications for analytical techniques to support instructors' decisions in AI-supported EAPs and e-exams.

Keywords: Test-taking strategies · E-exam · Assessment analytics · Process mining

1 Introduction

Instructional interventions for test-taking enhancement have been found to reduce anxiety and help students perform optimally in exams [7]. However, for instructors to provide reliable interventions, they first need to understand examinees' behaviours [2] and monitor them to provide feedback. Traditionally, to track and understand those behaviours, researchers use self-report measures (e.g., interviews), which are known for their inclination to provide inaccurate data. Hence, as we assume in this study that *exam-takers exhibit distinctive behaviours while taking e-exams*, such measures cannot accurately validate this assumption nor the benefits of instructional interventions.

Electronic assessment platforms (EAPs) have taken the examination field to a new era and have addressed this gap by authentically documenting examinees' digital footprints as log files during exams. Recent studies show the potential of using exam log data as a new measurement to support instructors' oversight to better understand test-taking behaviours through data mining, machine learning and statistical methods. Examples include visualisation techniques for

M. M. Rodrigo et al. (Eds.): AIED 2022, LNCS 13355, pp. 749–755, 2022.
https://doi.org/10.1007/978-3-031-11644-5_78

understanding student-item interactions on different levels [5], statistical modelling for understanding examinees' engagement [13], supervised machine learning for understanding the power of test-taking behaviours to predict performance (e.g., [10]) and traditional statistical analysis for understanding the relationship between the behaviours and learning constructs or learner characteristics (e.g., [9]). For example, some of the studies under the categories above could identify examinees' disengagement by modelling response time [13], detect examinees' cheating incidents using text mining and temporal data analysis accompanied with visualisation techniques [3], predict examinees' performance using temporal response patterns [10] and establish the relationship between personality traits and response time [9]. Few studies contributed to the understanding of examinees' test-taking strategies. For example, [4] proposed a visualisation technique to depict examinees' progress and responses overtime to uncover test-taking strategies. However, this method involves human judgment to a considerable extent (i.e., instructors must examine individual data for all examinees). As most of the existing methods appear to focus on quantitative behaviours (e.g., response-time), more sophisticated analytical approaches are needed to look at examinees' behaviours from the *process perspective* [11] and support instructor oversight and decisions in e-assessment systems.

Here, we incorporate an emerging approach from the learning analytics field, which aims to understand learners' behaviours by representing their behaviours at two inspired theoretical levels of tactics and strategies. The approach has previously been used in various learning settings such as flipped classrooms to detect learning strategies and examine their association with performance [8] or in conjunction with process models to detect and interpret time management tactics and strategies [1]. Following their methodology, we employ sequence analysis and clustering techniques on examinees' trace data to identify distinctive behavioural patterns taken by students while taking e-exams. We evaluate our approach by analysing an e-exam log of 90 multiple choice questions administrated to 463 university-level medical students.

2 Proposed Approach

This section describes our approach to investigate whether examinees exhibit distinct behaviours in e-exams. For the sake of space limit, Algorithm 1 provides only high-level pseudocode for our approach, as discussed below.

Data Pre-processing and Transformation. [Steps 1 to 3]. The data pre-processing incorporates three main steps, data cleaning, data interpretation and labelling, and sequence generating. The data interpretation and labelling step encompasses two sub-steps: 1) finding a list of meaningful activities from the literature (e.g., [6,13]) and multiple brainstorming sessions and 2) creating a set of rules for labelling the log file as follows: (a) answering, if the examinee responded to a question for the first time; (b) rapid-answering, if the examinee responded to a question for the first time in less than certain time; (c) long-answering, if the examinee responded to a question for the first time in more than

Algorithm 1: Detecting test-taking tactics and strategies

Input : L, R, k, q [L: **Log**, R: **Rules**, k: **number of tactic clusters**, q: **number of strategy clusters**];
Output: TAC, $STRT$ [TAC: **A set of tactics**, $STRT$: **A set of strategies**];
1 $visitExamineeQuestion \leftarrow extractVisits(L)$ [**Extract visits, where a visit is an examinee-question pair consecutive events**];
2 $labeledLog \leftarrow labelVisits(visitExamineeQuestion)$ [**Label each visit using rules introduced under pre-processing in this section**];
3 $sequenceExamineeQuestion \leftarrow generateSequences(labeledLog)$ [**Generate sequences, where a sequence is an examinee-question pair labelled visits (i.e., activities) ordered in chronological order**];
4 $dissMatrix \leftarrow constructDissMatrix(sequenceExamineeQuestion)$ [**Compute the distance between each pair of sequences and construct a dissimilarity matrix**];
5 $tactiClusters \leftarrow clusterSequences(dissMatrix, k)$ [**Create k clusters and put similar tactics in one cluster**];
6 $examineeVectors \leftarrow createExamineeVectors(tactiClusters)$ [**Create vector with length k for each examinee that represents each tactic frequency**];
7 $strategyClusters \leftarrow clusterExamineeVectors(examineeVectors, q)$ [**Create k clusters and put similar examinees regarding their tactics use in one cluster**];
8 **return** $TAC, STRT$;

certain time; (d) answer-changing, if the examinee changed their first response; (e) skipping, if the examinee left a question blank; and (f) revisiting, if the examinee revisited a question with no changes.

Tactic Detection [Steps 4 and 5]. Winnie described a learning tactic as a format of the If-THEN rule such that a learner can use the proper technique at the right time to cope with a typical situation (e.g., if a question is easy then answer it without reviewing your answer but if a question is hard then answer it and then review it multiple times) [12]. We replicate the steps taken by [1] for detecting learning tactics TAC, which incorporates the optimal matching and agglomerative hierarchical clustering algorithms. The optimal number of clusters k is chosen based on several runs of the algorithm, where the highest average silhouette width is chosen.

Strategy Detection [Steps 6 and 7]. A learning strategy can be viewed as a set of learning tactics. We replicate the steps taken by [1] for detecting strategies $STRT$ and assigning them to learners, but with the K-means++ algorithm instead of hierarchical clustering. The number of clusters q is again specified based on several runs of the algorithm and examining the average silhouette width.

3 Evaluation

Data Set. We obtained L of 463 undergraduate medical students who took a 90 multiple-choice questions summative exam administrated on an EAP called Examplify in 2017 at The University of Queensland[1]. From L, we generated 173,761 events and generated and labelled 94,271 visits (see Sect. 2). We determined the cut-off value for rapid and long answering behaviour by analysing

[1] Approval from our Human Research Ethics Committee was received for conducting the study presented in this paper (2018000841).

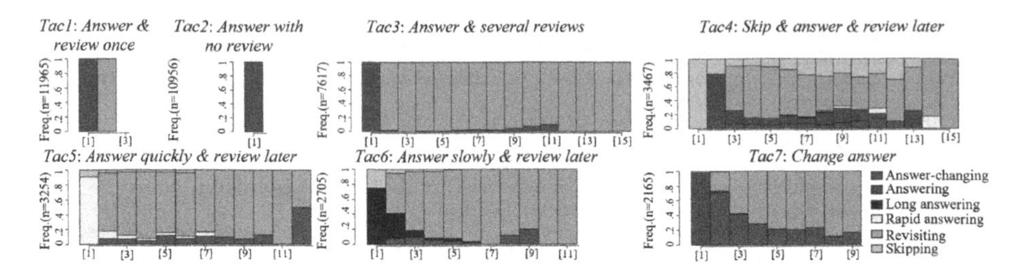

Fig. 1. An overview of the detected test-taking tactics.

time interaction data. Skipping and revisiting behaviours peak at three seconds which is an acceptable threshold [13]. Thus, three seconds were chosen for rapid-answering. Similarly, 51 s threshold for long-answering was chosen, representing the cut-off value of the intersection between the time of a normal visit for answering and a visit for answering with multiple answers. The set of visits was interpreted as (a) 35,773 answering; (b) 3,222 rapid answering; (c) 2,666 long answering; (d) 6,257 skipping; (e) 3,247 answer changing; and (f) 43,106 revisiting. The data was then transformed to 41,670 sequences with $min(sequence-length) = 1$ and $max(sequence - length) = 16$.

Results for Test-Taking Tactics. Figure 1 provides an overview of the seven identified tactics. The major difference between the tactics lies in the activity types comprising the tactics and the length of the tactics. However, most of the sequences belonging to each tactic varied in length ($1 \leq \mu \leq 3.7; 0 \leq \sigma \leq 1.5$). In what follows, Table 1 provides a brief description of each tactic and statistics, including whether or not a tactic was associated with examinees' grades using Spearman's rank correlation at .05 significance level.

Table 1. Tactics summary

ID	Name	Size %	Description	Association (grade)
Tac 1	Answer & review once	28.4	*Answering* is always followed by a *revisit*	($r_s = -.13, p < .05$)
Tac 2	Answer with no review	26	It consists of just *answering*	($r_s = -.20, p < .05$)
Tac 3	Answer & several reviews	18	It is the longest tactic chara-cterised by multiple *revisiting* after *answering*	($r_s = -.11, p < .05$)
Tac 4	Skip & answer & review later	8	It always starts with *skipping*	($r_s = -.07, p > .05$)
Tac 5	Answer quickly & review later	7.7	The dominant actions were *rapid-answering* and *revisiting*	($r_s - .03, p > .05$)
Tac 6	Answer slowly & review later	6.4	The use of *long-answering* and *revisits* was apparent	($r_s = .07, p > .05$)
Tac 7	Change answer	5	*Answering*, *answer-changing* and *revisiting* were prominent	($r_s = .19, p < .05$)

Results for Test-Taking Strategies. Following our approach, we chose $k = 6$ (silhouette score $= 30$). The generated strategies were significantly different. Figure 2 shows a general picture of the tactics used by each strategy. *Tac 5*, *tac 1* and *tac 3* were the most common tactics in the strategies. In addition, we investigate the differences between the strategies regarding exam grades where a Kruskal-Wallis test revealed a statistically significant association between the strategies and the grades at $p = .05(H(5) = 8.17, P = .02)$. Further post-hoc comparisons with Bonferroni correction showed no statistically significant association between each pair of the strategy groups. Table 2 briefly describes each strategy.

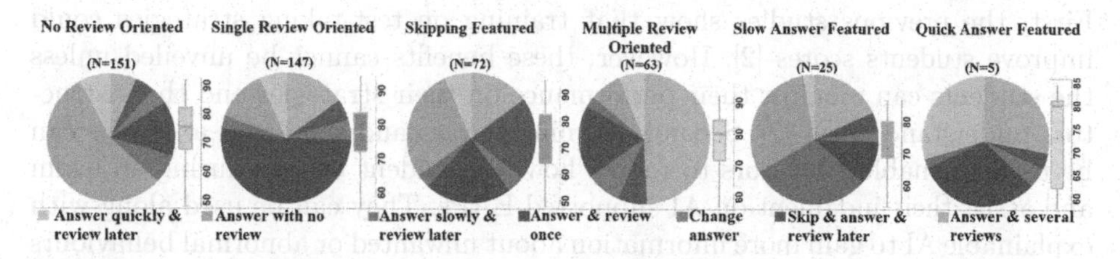

Fig. 2. An overview of the strategies (Pie charts: tactics use; box and whiskers: grades).

Table 2. Strategies' summary

ID	Name	Size %	Description
Strt 1	No review oriented	32	*Tac 2* was dominant
Strt 2	Single review oriented	32	*Tac 1* is the dominant tactic
Strt 3	Skipping featured	15	Unlike the other groups, *tac 4* was apparent
Strt 4	Multiple review oriented	13	*Tac 3* is the dominant tactic. *tac 7* was relatively higher than the other groups
Strt 5	Slow answer featured	5	Unlike the other groups, *tac 6* was frequently used
Strt 6	Quick answer featured	1	Unlike the other groups, *tac 5* was apparent. *Tac 2* was never used

4 Discussion and Conclusion

This paper used an approach that combines AI and LA to gain novel insight into test-taking behaviours. Our approach consists of analytics that could assist educators or examiners in gaining actionable insight into test-taking tactics and strategies. We evaluated our approach on real data of a high-stake exam. We uncovered seven test-taking tactics and six test-taking strategies that are distinctive. The tactics and strategies were easy to interpret and overlapped with previous research findings [4]. However, our approach has the additional benefit of being scalable as it does not require human intervention for visualising individual data for all examinees.

The small negative correlation between the tactics and grades should indicate that these tactics were common among students with different performance levels. However, we observed that *tac 2* and *tac 7* had a relatively larger correlation ($r_s = -.20$, $r_s = .19$, respectively) than other tactics. Hence, students should be recommended to review their answers and change them when in doubt. These results can also have implications for adaptive testing systems for permitting review and revision. Further, the absence of differences between the strategy groups' grades might indicate that there is no one way to approach the exam and score high.

Also, the findings have practical implications for improving instructional exam preparation and supporting instructors decisions in AI-supported EAPs. First, the previous studies show that training on test-taking strategies could improve students scores [2]. However, these benefits cannot be unveiled unless the students can monitor their performance on their strategies and the instructors understand them [2]. Second, learning tactics and strategies analytics can be used to enable educators to report how the student behave during an exam and assist their judgment on AI-supported EAPs. They can be used along-with explainable AI to gain more information about unwanted or abnormal behaviours (e.g., *strt 6*, where about one-third of the exam questions were answered in less than 4 s and a large discrepancy in exam scores was observed, see Fig. 2).

References

1. Ahmad Uzir, N., Gašević, D., Matcha, W., Jovanović, J., Pardo, A.: Analytics of time management strategies in a flipped classroom. J. Comput. Assist. Learn. **36**(1), 70–88 (2020)
2. Bumbálková, E.: Test-taking strategies in second language receptive skills tests: a literature review. Int. J. Instr. **14**(2), 647–664 (2021)
3. Cleophas, C., Hoennige, C., Meisel, F., Meyer, P.: Who's cheating? Mining patterns of collusion from text and events in online exams. Mining Patterns of Collusion from Text and Events in Online Exams, 12 April 2021
4. Costagliola, G., Fuccella, V., Giordano, M., Polese, G.: Monitoring online tests through data visualization. IEEE Trans. Knowl. Data Eng. **21**(6), 773–784 (2008)
5. Effenberger, T., Pelánek, R.: Visualization of student-item interaction matrix. In: Visualizations and Dashboards for Learning Analytics, pp. 439–456. Springer (2021). https://doi.org/10.1007/978-3-030-81222-5_20
6. George, T.P., Muller, M.A., Bartz, J.D.: A mixed-methods study of prelicensure nursing students changing answers on multiple choice examinations. J. Nurs. Educ. **55**(4), 220–223 (2016)
7. Hong, E., Sas, M., Sas, J.C.: Test-taking strategies of high and low mathematics achievers. J. Educ. Res. **99**(3), 144–155 (2006)
8. Jovanović, J., Dawson, S., Joksimović, S., Siemens, G.: Supporting actionable intelligence: reframing the analysis of observed study strategies. In: The 10th International Conference on Learning Analytics & Knowledge, pp. 161–170 (2020)
9. Papamitsiou, Z., Economides, A.A.: Exhibiting achievement behavior during computer-based testing: what temporal trace data and personality traits tell us? Comput. Hum. Behav. **75**, 423–438 (2017)

10. Papamitsiou, Z., Karapistoli, E., Economides, A.A.: Applying classification techniques on temporal trace data for shaping student behavior models. In: Proceedings of the Sixth International Conference on Learning Analytics & Knowledge, pp. 299–303. LAK 2016. ACM, New York, NY, USA (2016)
11. Pechenizkiy, M., Trcka, N., Vasilyeva, E., Van der Aalst, W., De Bra, P.: Process mining online assessment data. International Working Group on Educational Data Mining (2009)
12. Winne, P.H.: Learning strategies, study skills, and self-regulated learning in post-secondary education. In: Paulsen, M.B. (ed.) Higher Education: Handbook of Theory and Research, Higher Education: Handbook of Theory and Research, vol. 28, pp. 377–403. Springer, Dordrecht (2013). https://doi.org/10.1007/978-94-007-5836-0_8
13. Wise, S.L.: Rapid-guessing behavior: its identification, interpretation, and implications. Educ. Meas. Issues Pract. **36**(4), 52–61 (2017)

Multitask Summary Scoring with Longformers

Robert-Mihai Botarleanu[1], Mihai Dascalu[1,2](✉), Laura K. Allen[4], Scott Andrew Crossley[3], and Danielle S. McNamara[5]

[1] University Politehnica of Bucharest, 313 Splaiul Independentei, 060042 Bucharest, Romania
robert.botarleanu@stud.acs.upb.ro, mihai.dascalu@upb.ro
[2] Academy of Romanian Scientists, Str. Ilfov, Nr. 3, 050044 Bucharest, Romania
[3] Department of Applied Linguistics/ESL, Georgia State University, Atlanta, GA 30303, USA
scrossley@gsu.edu
[4] University of Minnesota, Department of Educational Psychology, 250 Education Sciences Bldg, 56 E River Rd, Minneapolis, MN 55455, USA
lallen@umn.edu
[5] Department of Psychology, Arizona State University, 871104, Tempe, AZ 85287, USA
dsmcnama@asu.edu

Abstract. Automated scoring of student language is a complex task that requires systems to emulate complex and multi-faceted human evaluation criteria. Summary scoring brings an additional layer of complexity to automated scoring because it involves two texts of differing lengths that must be compared. In this study, we present our approach to automate summary scoring by evaluating a corpus of approximately 5,000 summaries based on 103 source texts, each summary being scored on a 4-point Likert scale for seven different evaluation criteria. We train and evaluate a series of Machine Learning models that use a combination of independent textual complexity indices from the ReaderBench framework and Deep Learning models based on the Transformer architecture in a multitask setup to predict concurrently all criteria. Our models achieve significantly lower errors than previous work using a similar dataset, with MAE ranging from 0.10–0.16 and corresponding R^2 values of up to 0.64. Our findings indicate that Longformer-based [1] models are adequate for contextualizing longer text sequences and effectively scoring summaries according to a variety of human-defined evaluation criteria using a single Neural Network.

Keywords: Natural language processing · Text summarization · Automated summary scoring · Multitask learning

1 Introduction

Summary scoring is a common task in education that requires a significant amount of attention and time, but that represents a crucial skill for students since it evaluates their ability to discern the primary message of texts. Evaluating a summary requires the rater to read both the source text and the summary and then evaluate the extent to which the summary captures the essence of the source text in a concise manner. Summarization has been shown to be an important aspect of reading comprehension and learning to read [2–4], for both first and second language learners [5, 6].

M. M. Rodrigo et al. (Eds.): AIED 2022, LNCS 13355, pp. 756–761, 2022.
https://doi.org/10.1007/978-3-031-11644-5_79

Automated summary scoring can help reduce the load on teachers and represents a method to provide immediate feedback to students on the quality of their summaries. Semi-automatic methods, such as having humans select portions of texts to be evaluated [7, 8], have been proposed; however, they do not significantly lower the time impact for teachers. Automated methods involve the use of various NLP techniques for text representation, including similarities using Latent Semantic Analysis [9], word embeddings and linguistic indices generated by the ReaderBench framework [10], or divergences among probability distributions [11]. However, much work on automated summary evaluation systems has focused on evaluation methods for automatically generated summaries, and not for scoring summaries written by humans.

The current study expands on Botarleanu, Dascalu, Allen, Crossley and McNamara [10] where textual complexity indices were used to train a summary scoring system that measured how well a human-written summary covers the main idea of the original source text. We build on this work by using a larger corpus of summaries and build regressors to predict seven different summary evaluation criteria. The regressors implemented in the current study can handle source texts of relatively large lengths, which was the main limitation of Botarleanu et al. [10]. This study aims to answer the following research questions:

1. How well do Longformer-based architectures, as compared to the linguistic indices used in Botarleanu et al. [10], perform in automatically scoring summary elements?
2. What is the performance of a multi-task learning model that predicts all 7 scoring criteria simultaneously in contrast to 7 individual models?

2 Method

2.1 Corpora

Our corpus is an expanded version of the corpus considered by Botarleanu et al. [10] and includes 5,037 summaries (instead of the 2,976 previously used) corresponding to 103 source texts (instead of the 87 found in the aforementioned work). Our corpus was rated on a 1 to 4 Likert scale by expert raters for seven different scoring criteria: the cohesiveness of the summary text ("cohesion"), the appropriate use of objective language ("Objective Language"), the appropriate use of new paraphrasing ("Paraphrasing"), the use of language beyond that found in the source text ("Language Beyond Source Text"), how appropriate the length of the summary is in relation to the source text ("Summary Length"), the degree to which important details are captured from the source text ("Details") and whether the summary succeeds in capturing the main point of the reference text ("Main Point").

The corpus consists of summaries collected from a mix of unrelated studies: a) summaries collected in a study on Adult Literacy on general topics such as seat belt laws, disability services, and patients' rights, b) summaries collected using Amazon's Mechanical Turk service (MTurk) on science texts related to biology and climatology, c) summaries on heart disease and red blood cells collected in a study on Adult Literacy using MTurk, d) summaries on science texts collected using MTurk from primarily speakers of English as a second language, e) summaries on cellphone risk and climate change were collected as part of a study on multiple text comprehension and f) summaries on science and history written by undergraduate students.

In order to filter out summaries that were malformed, we searched for summaries that were either as long as the source text or significantly shorter than it. A plot showing the ratio between the source and summary text lengths is presented in Fig. 1.a. We elected to remove all summaries with lengths below 10% or above 80% of the source text length. These values were chosen to remove the tails of the distribution from Fig. 1.b which depicts the ratios between summary length and source text length. This pro-cess reduced the number of summaries from 5,037 to 4,233 without removing any of the 103 source texts.

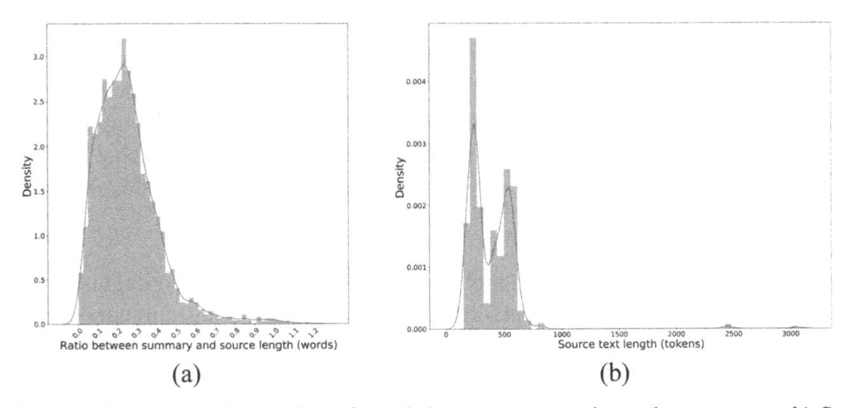

Fig. 1. a) Ratio between the number of words between summaries and source texts. b) Source text lengths in tokens.

Another consideration in corpus development is the length of the sequences that are used as inputs for transformer-based models. Due to internal constraints, models such as BERT (Devlin et al. 2018) are only suitable for sequences of up to 512 tokens, whereas models such as Longformer can work well with longer sequences, typically of up to 4,096 tokens. We utilized the pre-trained Longformer tokenizer provided by the "transformers" package [12] with the "allenai/longformer-base-4096" pretrained model to evaluate the lengths of the texts in our corpus (see Fig. 1.b). Indeed, a significant proportion of source texts in our corpus exceeded 512 tokens; however, the majority did not exceed 1000 tokens in length making the texts suitable for the Longformer model, but not for BERT.

2.2 Regression Models

We elected to construct our regression models around the Longformer model [1] to handle the source and summary texts that were often too long for BERT. The Longformer model employs an attention mechanism that combines local windowed attention with a global attention mechanism, that is designed to encode inductive bias about the task that the model is being trained to solve. Given the distribution in Fig. 1.b, we opted to use padded sequences of 2048 tokens formed by tokenizing and trimming both the source and summary lengths down to 1024 tokens. The overview of this architecture is illustrated in Fig. 2. The architecture is evaluated under two setups: one where the model predicts only one of the seven summary scoring criteria at a time, and a second where the model is tasked with predicting all seven objectives at once.

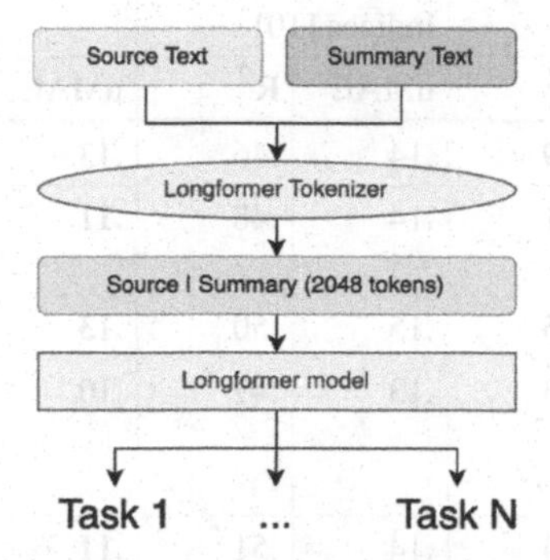

Fig. 2. The architecture of the Longformer-based model.

3 Results

We present the normalized Mean Absolute Error (nMAE) and the corresponding R^2 coefficients for the Longformer model used in both single-task and multi-task settings (see Table 1). We compare our results to those measured using the model presented by Botarleanu et al. [10], where a network with a single hidden layer with 256 units is applied to an input consisting of the textual complexity indices generated by the ReaderBench framework for both the summary and the source texts. This was trained using the same Once-Cycle Policy described in Botarleanu et al. [10], for 50 epochs with a batch size of 8.

The first observation is that the trained Longformer models outperform the models relying on the ReaderBench indices, with R^2 coefficients having values that are between .07 and .13 higher. Second, the multi-task model matches the performance of the individual models on average, and even exceeds the single-task models for the "Cohesion", "Language Beyond Source Text", and "Details" criteria. The most significant degradation in performance between the single-task and the multi-task setting is for the "Paraphrasing" score with the R^2 coefficient falling from .55 to .42, which is still higher than the performance of the model that uses ReaderBench indices.

Finally, the seven scoring criteria appear to have a relatively similar difficulty in terms of the models' capability to learn them. The lowest R^2 coefficient for the Longformer-based models is measured on the "Details" objective (.37 for the single task model), whereas the highest coefficient is observed for the "Summary Length" objective (.67 for the single task model). In contrast, the multi-task model appears to have a narrower variation in performance, with R^2 coefficients ranging from .42 for the "Paraphrasing" objective up to .64 for the "Summary Length" criterion, which may be explained by the fact that the multi-task model was trained with all 7 criteria being seen as equal, and no objective-specific weights were applied to the loss function.

Table 1. Normalized MAE and the R^2 score for the seven evaluation criteria.

Scoring criterion	Single task models using ReaderBench indices [10]		Multi-task model using ReaderBench indices [10]		Single task Longformer models		Multi-task Longformer model	
	nMAE	R^2	nMAE	R^2	nMAE	R^2	nMAE	R^2
Cohesion	.14	.49	.14	.46	.13	.52	**.12**	**.58**
Objective language	.13	.51	.14	.48	**.11**	**.59**	.13	.50
Paraphrasing	.15	.45	.15	.50	**.13**	**.55**	.16	.42
Language beyond ource text	.13	.43	.13	.47	.10	.59	**.10**	**.60**
Summary length	.13	.54	.14	.51	**.11**	**.67**	.12	.64
Details	.15	.46	.16	.39	.15	.37	**.13**	**.53**
Main point	.15	.53	.14	.52	.11	**.64**	**.12**	.59
Average	*.14*	*.49*	*.14*	*.48*	***.12***	***.58***	*.13*	*.55*

4 Conclusions and Future Work

In this paper, we analyzed the effectiveness of using Longformer-based regression models to perform automated summary scoring. Our models achieved significantly better results than the previous models in Botarleanu et al. [10]. Our results also indicate that a model trained in a multi-task setting achieved a performance that was on par with training seven different networks. With an average normalized mean absolute error of .13 and a corresponding R^2 of .55, our model predicts the human rating of a summary with an average deviation of 13%. The capability to perform automated summary scoring in a multi-task setting has several advantages. First, it reduces the computational load and supports the development of automated summary scoring systems that can analyze summaries more effectively. Second, it more closely matches the human expert scoring method because it forces the model to perform a holistic analysis of the text, instead of relying on patterns captured for each of the scoring criteria individually. One method of improving the performance of the model might be to combine the ReaderBench indices with the Longformer inferences into an ensemble model. Moreover, part of the ReaderBench indices might also benefit from the use of Longformer models to predict their values (e.g., intra- and inter- paragraph cohesion scores).

A potential avenue for future research lies in performing an interpretability analysis of the multi-task model. Through this, one might explore the degree to which the different summary scoring criteria presented in this work may complement each other. Additionally, studying the way in which the most relevant blocks of the summaries and source texts are selected by the model, and aligning these segments with human rater observations may provide valuable insight into what humans look for in summaries, which can help in providing targeted feedback to students.

Finally, the principal measure of the usefulness of such a system lies in the impact it has on the summarization skills of real students. An important future research direction would be the use of the model described in this work to help students improve their summary writing skills. Notably, the success of this model in predicting scores on seven different attributes, for such a wide range of source texts, bodes well for the eventual utility of this model within an automated tutoring system.

Acknowledgments. This research was supported by a grant from the Romanian National Authority for Scientific Research and Innovation, CNCS – UEFISCDI, project number TE 70 *PN-III-P1-1.1-TE-2019-2209,* ATES – "Automated Text Evaluation and Simplification", the Institute of Education Sciences (R305A180144 and R305A180261), and the Office of Naval Research (N00014-17-1-2300; N00014-20-1-2623; N00014-19-1-2424, N00014-20-1-2627). The opinions expressed are those of the authors and do not represent the views of the IES or ONR.

References

1. Beltagy, I., Peters, M.E., Cohan, A.: Longformer: the long-document transformer. arXiv preprint arXiv:2004.05150 (2020)
2. Bensoussan, M., Kreindler, I.: Improving advanced reading comprehension in a foreign language: summaries vs. short-answer questions. J. Res. Read. **13**(1), 55–68 (1990)
3. Brown, A.L., Campione, J.C., Day, J.D.: Learning to learn: on training students to learn from texts. Educ. Res. **10**(2), 14–21 (1981)
4. Bean, T.W., Steenwyk, F.L.: The effect of three forms of summarization instruction on sixth graders' summary writing and comprehension. J. Read. Behav. **16**(4), 297–306 (1984)
5. Karbalaei, A., Rajyashree, K.S.: The impact of summarization strategy training on university ESL learners' reading comprehension. Int. J. Lang. Soc. Cult. **30**, 41–53 (2010)
6. Pakzadian, M., Rasekh, A.E.: The effects of using summarization strategies on Iranian EFL learners' reading comprehension. Engl. Linguist. Res. **1**(1), 118–125 (2012)
7. Nenkova, A., Passonneau, R.J.: Evaluating content selection in summarization: the pyramid method. In: The Human Language Technology Conference of the North American Chapter of the Association for Computational Linguistics (HLT-NAACL 2004), Boston, Massachusetts, USA, pp. 145–152. ACL (2004)
8. Van Halteren, H., Teufel, S.: Examining the consensus between human summaries: initial experiments with factoid analysis. In: HLT-NAACL 03 Text Summarization Workshop, Edmonton, Canada, pp. 57–64. ACL (2003)
9. Steinberger, J., Jezek, K.: Using latent semantic analysis in text summarization and summary evaluation. In: Proceedings of the 7th International Conference ISIM, pp. 93–100 (2004)
10. Botarleanu, R.-M., Dascalu, M., Allen, L.K., Crossley, S.A., McNamara, D.S.: Automated summary scoring with ReaderBench. In: Cristea, A.I., Troussas, C. (eds.) Intelligent Tutoring Systems. Lecture Notes in Computer Science, vol. 12677, pp. 321–332. Springer, Cham (2021). https://doi.org/10.1007/978-3-030-80421-3_35
11. Torres-Moreno, J.-M., Saggion, H., Cunha, I.D, SanJuan, E., Velázquez-Morales, P.: Summary evaluation with and without references. Polibits **42**, 13–20 (2010)
12. Facebook Inc.: Transformers (n.d.). https://huggingface.co/docs/transformers/index. Accessed 20 Jan 2022

Investigating the Effects of Mindfulness Meditation on a Digital Learning Game for Mathematics

Huy A. Nguyen[1](✉), Zsofia K. Takacs[2], Enikő O. Bereczki[3], J. Elizabeth Richey[4], Michael Mogessie[1], and Bruce M. McLaren[1]

[1] Carnegie Mellon University, Pittsburgh, PA 15213, USA
hn1@cs.cmu.edu
[2] School of Health in Social Science, University of Edinburgh, Edinburgh E8 9AG, UK
[3] Institute of Education, ELTE Eötvös Loránd University, Budapest 1075, Hungary
[4] University of Pittsburgh, Pittsburgh, PA 15260, USA

Abstract. Mindfulness has been shown in prior studies to be an effective device to help students develop self-regulatory skills, including executive functions. However, these effects have been rarely tested at scale in technology-assisted learning systems such as digital learning games. In this work, we investigate the effects of mindfulness in the context of playing and learning with *Decimal Point*, a digital learning game for mathematics. We conducted a study with 5th and 6th grade students in which three conditions were compared - the game with short mindfulness meditations integrated, the game with similar-length, age-appropriate stories integrated, and the game in its original form. From the study results, we found no differences in time spent on the game, error rates while playing, or learning outcomes across the three conditions. Embedding mindfulness prompts within the game did not enhance learning or change students' gameplay behaviors, which suggests that we may not have successfully induced a state of mindfulness or that mindfulness is not beneficial for learning within digital learning games. We discuss the challenges of incorporating individual mindfulness meditations in elementary and middle school classrooms.

Keywords: Digital learning games · Mindfulness · Decimal numbers · Middle school math

1 Introduction

Mindfulness is the concept of attending to the present moment with focus and without judgment. Mindfulness meditation has been shown to support self-regulation, attention skills and executive function, especially working memory capacity and inhibitory control [5, 12], which could in turn contribute to math learning [4]. Additionally, mindfulness practice might reduce math anxiety, which could further enhance math performance [10].

M. M. Rodrigo et al. (Eds.): AIED 2022, LNCS 13355, pp. 762–767, 2022.
https://doi.org/10.1007/978-3-031-11644-5_80

Despite these strong theoretical reasons, the role of mindfulness for children's academic outcomes is less clear, especially due to the limited evidence so far. A meta-analysis by [7] found that five studies assessing the efficacy of mindfulness-based interventions for academic achievement showed a non-significant, small average effect. Although since then some promising preliminary results with older students [6] and students with ADHD [11] emerged, the evidence regarding the effects of mindfulness-based interventions on academic achievement and learning is inconclusive.

In the present study we aimed to test whether the addition of mindfulness practice within a math game *Decimal Point* would contribute to students' learning. Rather than employing longer mindfulness training *interventions* on academic achievement as in prior work [7], we applied short mindfulness *inductions* at the beginning of the learning sessions to induce a state of mindfulness. Additionally, instead of applying mindfulness exercises in groups [14], our students conducted those individually as part of the math game. Given these differences, we aimed to investigate the effectiveness of state mindfulness for math learning and test the feasibility of individual mindfulness practice with middle-school-aged children, built into a digital learning game.

2 A Mindfulness Study with the *Decimal Point* Learning Game

Decimal Point [8] is a digital learning game designed to help middle schoolers learn about decimals and decimal operations. The game is based on an amusement park metaphor in which students play a series of mini-games targeted at common decimal misconceptions. Each mini-game consists of a problem-solving activity (e.g., sorting a list of decimal numbers from smallest to largest), followed by a multiple-choice self-explanation question. Students get immediate feedback after each attempt and can continue making attempts if their current answer is incorrect; they need to finish all exercises in the current mini-game to move on to the next mini-game.

Extending on the initial study of the game [8], the current study examines whether mindfulness practices may have similar impacts in a learning game, given their benefits to executive control [5]. To this end, we compare a version of *Decimal Point* with embedded mindfulness inductions against two comparison conditions: the original game version and an active comparison condition that incorporates thematically-appropriate stories and jokes instead of mindfulness. The story version was created to control for the amount of time spent on mindfulness inductions and additional material that was not designed to induce mindfulness. Across the three versions of the game, we investigate the following research questions:

RQ1: *Do students who receive short mindfulness inductions during game play of a digital learning game demonstrate different behaviors than other game playing students?* We hypothesized that mindfulness would enhance students' executive control, leading to students in the mindfulness condition spending more time and exhibiting fewer errors on the mini-game problems and self-explanation questions than in the other two conditions.

RQ2: *Do the students who receive mindfulness inductions learn more than other game playing students?* We hypothesized that students who received the mindfulness treatment would show greater learning gains than the story and control treatments, as a result of enhanced executive functioning. Additionally, mindfulness may reduce anxiety and thus free up working memory to help students to focus more on constructive learning processes that enhance learning (e.g., repairing misconceptions, connecting new information to prior knowledge [3]).

3 Method and Materials

Our study was conducted in 5th and 6th grade classrooms across three public schools in a mid-sized U.S. city during the fall of 2021. A total of 243 students participated in the study; however, 77 were excluded from our analyses because they did not complete all materials. The final sample included 166 students (76 males, 90 females), with 57 students assigned to the control treatment, 56 to the story treatment and 53 to the mindfulness treatment. Students reported an average age of 10.84 ($SD = 0.65$). Students participated in the study for a total of six days as part of their regular class activities. On the first day, students completed a pretest. They then progressed through the materials at their own pace for up to four additional days. In the mindfulness and story conditions, students received their respective treatments (i.e., mindfulness induction or story) at the beginning of each class. Students then completed a posttest right after finishing the game and a delayed posttest one week later. The pretest, posttest, and delayed posttest consisted of three isomorphic versions of a decimal test that were counterbalanced across students and conditions. All tests included 42 items; as some items contained multiple components, students could earn a total of 52 points on each test. Test items targeted the same decimal misconceptions addressed in the mini-games.

Across the three treatment conditions, students played the same basic version of *Decimal Point* as described above, and the order and content of the mini-games was identical. The key differences between conditions are as follows. The *Mindfulness* and *Story* conditions both incorporated a brief, five-minute audio session that students listened to at the start of each day of the study, prior to playing the game. In the *Mindfulness* condition, the audio content entailed an alien friend sharing mindfulness advice that prompted students to close their eyes, focus on their breath and sounds in the environment, and let go of passing thoughts [13]. In the *Story* condition, the audio content was related to science fiction stories selected to be age appropriate, emotionally neutral (i.e., not emotionally arousing or upsetting) and unrelated to the learning content. In addition, both conditions featured an in-game minute-long reminder that shows up when the student has made three consecutive incorrect attempts while playing. In the *Mindfulness* condition, students would be reminded about mindfulness and encouraged to slow down, close their eyes, and focus on their breath for a moment. In the Story condition, students would instead listen to a series of jokes from the aliens. Each reminder would only show up at most once every 10 min to avoid overwhelming the students; the reminders and stories were also omitted when students were taking the tests. Finally, students in the *Control* condition did not receive any activity with aliens before beginning to play the game each day, nor did they receive any reminders based on errors. The content and structure of the game they completed was identical to the game in the other conditions, but without the story or mindfulness components.

4 Results

For RQ1, descriptive statistics of students' game play behaviors are included in Table 1. A series of one-way ANOVAs showed no significant condition effects on the number of errors made on problem-solving in the game, $F(2, 163) = 0.08, p = .93, \eta_p^2 = .001$, or the amount of time students spent completing the problem-solving portion of the mini-games, $F(2, 163) = 0.047, p = .95, \eta_p^2 = .001$. Similarly, there was no significant condition effect on the number of self-explanation errors made, $F(2,163) = 0.23, p = .79, \eta_p^2 = .003$, or the amount of time students spent on the self-explanation questions, $F(2, 163) = 0.31, p = .74, \eta_p^2 = .004$. Thus, our hypothesis that students in the mindfulness condition would take more time and make fewer errors was not confirmed.

For RQ2, descriptive statistics of students' test scores by condition are reported in Table 1. A repeated-measures ANOVA tested learning condition as a between-subjects factor and test time (pretest, posttest, and delayed posttest) as a within-subjects factor. Results indicated a significant effect of test time, with students' test scores improving significantly across tests, $F(2,162) = 27.39, p < .001, \eta_p^2 = .253$. There was no main effect of learning condition, $F(2, 163) = 0.009, p = .99, \eta_p^2 < .001$, and planned comparisons revealed no differences between the control and the other two conditions, $p = .93$, or between the mindfulness and story conditions, $p = .93$. There was also no interaction between test time and learning condition, $F(4, 324) = 1.52, p = .20, \eta_p^2 = .018$, indicating that students' test score improvements did not differ by condition. Follow-up repeated measures analyses indicated that students' test scores increased significantly from pretest to posttest, $F(1, 163) = 50.75, p < .001, \eta_p^2 = .24$, and from pretest to delayed posttest, $F(1,163) = 45.39, p < .001, \eta_p^2 = .22$, but not from posttest to delayed posttest, $F(1, 163) = 0.44, p = .51, \eta_p^2 = .003$. In summary, our prediction that students in the mindfulness condition would learn more than the other conditions was not confirmed.

Table 1. Test performance, game play measures, and enjoyment ratings by condition, reported in *M* (*SD*) format. The test scores are on a scale from 0–52 and the duration measures are in minutes.

Category\Condition	Control ($n = 57$)	Mindfulness ($n = 53$)	Story ($n = 56$)
Pretest scores	18.40 (9.40)	19.74 (9.33)	18.80 (8.08)
Posttest scores	22.51 (10.29)	21.68 (10.33)	22.48 (9.42)
Delayed posttest scores	23.02 (11.12)	22.36 (10.49)	21.98 (10.37)
Problem-solving duration	63.88 (21.39)	63.78 (26.05)	62.69 (22.58)
Problem-solving errors	113.32 (51.25)	115.83 (68.18)	117.91 (69.16)
Self-explanation duration	11.73 (3.45)	12.10 (3.29)	11.61 (3.46)
Self-explanation errors	16.40 (7.21)	15.81 (7.90)	15.45 (7.19)

5 Discussion

In contrast to prior work demonstrating the effectiveness of mindfulness inductions on middle schoolers [5], our study results show no evidence that the mindfulness or story treatments had an effect on students' game behaviors or on learning outcomes. A key element that could explain this difference is the study context. In previous studies, mindfulness inductions were conducted as a teacher-led synchronous group activity or when the student was alone. Our study instead examined mindfulness inductions as a self-guided and self-paced activity in a classroom context, where the lack of mindfulness feedback from teachers may undermine their effect. In addition, the presence of other classmates who may be engaging in different game activities likely introduced more distractions and may have made students more self-conscious about closing their eyes and following along with the mindfulness induction. While randomization of student learning conditions within each classroom provides greater statistical power by minimizing class-level effects, future research could test the same mindfulness intervention administered at a classroom level so that all students would begin each day completing the same intervention with their peers.

A different explanation is that the effect of mindfulness, if present, was only in the short term and therefore not reflected in aggregate game play measures or post-intervention assessments. Future work could validate this conjecture by administering a short survey which probes students to reflect on their current mindfulness state right after the daily induction [2]. Alternatively, it might be possible that mindfulness inductions are less powerful in digital learning games, because the mechanisms through which they enhance learning – i.e., improving executive functions, reducing students' rush and carelessness – are similar to those of learning games [1]. *Decimal Point*, in particular, was shown to improve learning via reducing students' cognitive disengagement [9]. In other words, mindfulness interventions might not be as useful in a digital learning game, where attention is already enhanced, compared to other learning contexts where student attention and engagement are lower. Future research could compare the same mindfulness induction within *Decimal Point* and a non-game digital control that covers the same content [8], to see whether mindfulness has a more pronounced effect in a non-game context.

At the same time, there are a number of limitations that may influence our interpretations of the results. First, due to COVID restrictions we could not be present in the classroom to ensure that students were following the meditation guidelines, rather than being idle. Likewise, we were not able to deploy sensing technologies (e.g., eye tracking) to capture more nuanced data about students' mindfulness practice. Finally, the effect of mindfulness inductions may require a larger sample size to detect.

In conclusion, the findings from our study provide an important first step toward identifying boundary conditions for when and how mindfulness meditation can be used to support learning in classroom contexts and in conjunction with digital learning games. As described above, future research on the classroom factors and types of digital learning environments that are best suited to mindfulness inductions will contribute important additional evidence for researchers and teachers seeking to understand the conditions under which mindfulness inductions can be a useful learning tool.

Acknowledgements. This project was partially funded by the Hungarian National Research, Development and Innovation Office (grant PD OTKA – PD – 138723) and by the National Science Foundation Award #DRL-1661121. The opinions expressed are those of the authors and do not represent the views of the funding agencies. Thanks also to Jodi Forlizzi, who contributed to the original *Decimal Point* work.

References

1. Bediou, B., Adams, D.M., Mayer, R.E., Tipton, E., Green, C.S., Bavelier, D.: Meta-analysis of action video game impact on perceptual, attentional, and cognitive skills. Psychol. Bull. **144**, 77 (2018)
2. Carsley, D., Heath, N.L.: Evaluating the effectiveness of a mindfulness coloring activity for test anxiety in children. J. Educ. Res. **112**, 143–151 (2019)
3. Chi, M.T., Wylie, R.: The ICAP framework: linking cognitive engagement to active learning outcomes. Educ. Psychol. **49**, 219–243 (2014)
4. Cragg, L., Gilmore, C.: Skills underlying mathematics: the role of executive function in the development of mathematics proficiency. Trends Neurosci. Educ. **3**, 63–68 (2014)
5. Dunning, D.L., et al.: Research review: the effects of mindfulness-based interventions on cognition and mental health in children and adolescents–a meta-analysis of randomized controlled trials. J. Child Psychol. Psychiatry **60**, 244–258 (2019)
6. Güldal, Ş., Satan, A.: The effect of mindfulness based psychoeducation program on adolescents' character strengths, mindfulness and academic achievement. Curr. Psychol. 1–12 (2020)
7. Maynard, B.R., Solis, M.R., Miller, V.L., Brendel, K.E.: Mindfulness-based interventions for improving cognition, academic achievement, behavior, and socioemotional functioning of primary and secondary school students. Campbell Syst. Rev. **13**, 1–144 (2017)
8. McLaren, B.M., Adams, D.M., Mayer, R.E., Forlizzi, J.: A computer-based game that promotes mathematics learning more than a conventional approach. Int. J. Game-Based Learn. (IJGBL) **7**, 36–56 (2017)
9. Richey, J.E., et al.: Gaming and confrustion explain learning advantages for a math digital learning game. In: Roll, I., McNamara, D., Sosnovsky, S., Luckin, R., Dimitrova, V. (eds.) Artificial Intelligence in Education. LNCS (LNAI), vol. 12748, pp. 342–355. Springer, Cham (2021). https://doi.org/10.1007/978-3-030-78292-4_28
10. Samuel, T.S., Warner, J.: "I can math!": reducing math anxiety and increasing math self-efficacy using a mindfulness and growth mindset-based intervention in first-year students. Commun. Coll. J. Res. Pract. **45**, 205–222 (2021)
11. Singh, N.N., Lancioni, G.E., Nabors, L., Myers, R.E., Felver, J.C., Manikam, R.: Samatha meditation training for students with attention deficit/hyperactivity disorder: effects on active academic engagement and math performance. Mindfulness **9**, 1867–1876 (2018)
12. Takacs, Z.K., Kassai, R.: The efficacy of different interventions to foster children's executive function skills: a series of meta-analyses. Psychol. Bull. **145**, 653 (2019)
13. Vekety, B., Kassai, R., Takacs, Z.K.: Mindfulness with children: a content analysis of evidence-based interventions from a developmental perspective. Educ. Dev. Psychol. (2022)
14. Vekety, B., Logemann, H.A., Takacs, Z.K.: The effect of mindfulness-based interventions on inattentive and hyperactive–impulsive behavior in childhood: a meta-analysis. Int. J. Behav. Dev. **45**, 133–145 (2021)

Author Index

Printed in the United States
by Baker & Taylor Publisher Services

Printed in the United States
by Baker & Taylor Publisher Services